Advances in Isotope Geochemistry

Series Editor

Jochen Hoefs, Geowissenschaftliches Zentrum, Universität Göttingen, Göttingen, Germany

In the last few decades, isotope geochemistry has become an essential part of geochemistry and has contributed significantly to the solution of a wide variety of geoscientific problems, which span the whole field of earth sciences. Continued improvements in mass spectrometry and the invention of new mass-spectrometer systems, such as multicollector-ICP mass spectrometers (MC-ICP-MS), has enabled investigations of isotope variations of a wide range of transition and heavy elements that could not previously be measured with adequate precision. This has allowed many of the stable and radioactive isotopic systems to be investigated and applied to a huge variety of inorganic and organic samples.
Advances in Isotope Geochemistry, seeks to provide in-depth reviews of isotopic systems, methods and applications to a degree which is not possible within journal articles. Methods are described in detail, from sample collection and preparation to the fine tuning and subtleties of the mass spectrometric methods, data reduction and interpretation. The AIG series is the first stop when establishing new methods and an excellent reference for every isotope laboratory, serving as textbooks in university courses as well as a source of information for professionals.

More information about this series at http://www.springer.com/series/8152

Clark Johnson · Brian Beard ·
Stefan Weyer

Iron Geochemistry: An Isotopic Perspective

Springer

Clark Johnson
Department of Geoscience
University of Wisconsin-Madison
Madison, WI, USA

Brian Beard
Department of Geoscience
University of Wisconsin-Madison
Madison, WI, USA

Stefan Weyer
Institute of Mineralogy
Leibniz Universität Hannover
Hannover, Germany

ISSN 2364-5105 ISSN 2364-5113 (electronic)
Advances in Isotope Geochemistry
ISBN 978-3-030-33827-5 ISBN 978-3-030-33828-2 (eBook)
https://doi.org/10.1007/978-3-030-33828-2

This Springer imprint is published by the registered company Springer Nature Switzerland AG
The registered company address is: Gewerbestrasse 11, 6330 Cham, Switzerland

Preface

Iron is a remarkable element in its relatively high abundance for its mass and its sensitivity to (and control of) redox on a planetary scale over geologic time. Numerous volumes have been written on iron geochemistry, and here we provide an in-depth review of iron geochemistry from the isotopic perspective. Stable iron isotope geochemistry is a relatively new field, belonging to what is sometimes called "non-traditional" stable isotopes. There have been a number of reviews of this field of isotope geochemistry, the most recent of which is found in Teng et al. (2017) in their Reviews in Mineralogy and Geochemistry volume 82, published by the Mineralogical Society of America and the Geochemical Society. In chapter 11 of this volume, Dauphas et al. (2017) review iron isotope systematics.

Although we attempt an in-depth review, we do not cite every study ever done, instead developing a theme-based approach to iron geochemistry as viewed through the lens of stable iron isotopes. We begin, in Chap. 1 (lead author: Johnson), with a brief overview of iron geochemistry, but leave details for later chapters that cover specific topics. In addition, we provide a brief review of the principles of isotope geochemistry, and then a brief overview of each of the subsequent chapters in the book; it is here that the reader can see what topics might interest them the most. In Chap. 2 (lead author: Beard), the wide range of analytical methods that are used for modern iron isotope analysis is reviewed. Experimental and theoretical approaches for determining iron isotope fractionation factors (key to understanding any isotope system) are covered in Chap. 3 (lead author: Beard). High-temperature applications of iron isotopes, from planetary formation to hydrothermal systems, are covered in Chap. 4 (lead author: Weyer). Chapter 5 (lead author: Johnson) discusses application of iron isotopes to the modern surficial world, from weathering to the oceans. Application of iron isotopes to understanding biogeochemical cycling of iron in Earth's ancient surface environments, from the Cenozoic to the Eoarchean, is covered in Chap. 6 (lead author: Johnson); the long time period covered in this chapter and the very large body of available data make this the longest chapter in the book.

Madison, USA — Clark Johnson
Madison, USA — Brian Beard
Hannover, Germany — Stefan Weyer

Acknowledgements

An in-depth review such as this requires assembly of thousands of data from hundreds of sources, and we are very grateful to those studies that have made their data, and its documentation, readily available. As has become clear to us, this is not always the case! A surprising number of studies do not document their data, or do not have data where they say it is, and journals do not enforce online appendices. The geochemical community, and science in general, is moving toward better documentation and availability of data, but assembly of the large datasets needed for this book has made it apparent that we are not there yet. To the authors of papers whose data are easily adapted for various uses and beautifully documented, we say Thank You!

We, of course, extend our appreciation to Prof. Jochen Hoefs for inviting us to write this review, and to the efforts and support of Springer and its staff. And we are thankful for their great patience in the (slow) speed with which we wrote the volume. Additional assistance at UW-Madison was provided by Lisa NurMarini Mohd Kamal and Victoria Khoo, and in Hanover by Lilian Meier and Chris Rosendahl.

A great debt is due to the reviewers who provided critical comments for various sections of the book. In several cases, we purposefully sought out reviewers with whom we have disagreed in print, in an attempt to provide as balanced a book as possible. That they are listed below, however, should not be taken as an endorsement of what we say in the book. We are extremely grateful for the comments provided by Annie Bauer, Gregory Brennecka, Piyali Chanda, Andy Czaja, FX D'Abzac, Ralf Dohmen, Ingo Horn, Seth John, Alan Mathews, Martin Oeser, Noah Planavsky, Jamie Robbins, Olivier Rouxel, Edwin Schauble, Ronny Schoenberg, Silke Severmann, Paolo Sossi, Fang-Zhen Teng, Helen Williams, and Xin-Yuan Zheng.

Finally, support from our family, respective home institutions, and funding agencies has been critical during the writing of this volume. For the latter, C.M.J. and B.L.B. acknowledge the NASA Astrobiology Institute and the National Science Foundation.

Contents

1 Introduction and Overview 1
1.1 Geochemistry of Fe 1
1.1.1 Fe Redox 2
1.2 Stable Isotope Geochemistry 4
1.2.1 Nomenclature 5
1.2.2 Isotopic Fractionation 6
1.2.3 Processes that Produce Isotopic Variations 8
1.3 Overview of the Chapters 10
References 13

2 Analytical Methods 17
2.1 Introduction 17
2.2 Iron Purification Methods 18
2.3 Mass Spectrometry 20
2.3.1 Beginnings of Fe Isotope Analysis 20
2.3.2 Multi Collector Inductively Coupled Plasma Mass Spectrometry (MC-ICP-MS) 21
2.3.3 Modern MC-ICP-MS Using Pseudo High Mass Resolution Methods 24
2.3.4 Matrix Effects and Instrumental Mass Fractionation Corrections 25
2.4 In Situ Techniques 28
2.5 Summary 32
References 33

3 Fe Isotope Fractionation Factors 39
3.1 Introduction 39
3.2 Deriving Fe Isotope Fractionation Factors from First Principles 40
3.3 Experimental Methods for Measuring Fe Isotope Fractionation Factors 41
3.3.1 The Role of Sorption in Isotope Exchange 46
3.4 Equilibrium Fractionation of Fe Isotopes: Working Toward a Unified Set of Fractionation Factors 47
3.4.1 Aqueous Fe Species 47
3.4.2 Aqueous Fe Mineral Fractionation 54

3.5 Biological Experiments . . . 64
3.5.1 Fe Oxidizing Experiments . . . 65
3.5.2 Magnetotactic Bacteria . . . 69
3.5.3 Fe Reducing Experiments . . . 71
3.6 Preferred Set of β-Values . . . 75
3.7 Summary . . . 77
References . . . 77

4 High-Temperature Fe Isotope Geochemistry . . . 85
4.1 Iron Isotope Variations in the Solar System . . . 85
4.1.1 Chondrites and Chondritic Components . . . 86
4.1.2 Differentiated Planetary Material . . . 89
4.2 The Silicate Earth . . . 91
4.2.1 The Mantle and Its Minerals . . . 91
4.2.2 Basalts and Komatiites . . . 93
4.2.3 Differentiated Crust . . . 96
4.2.4 Magmatic Minerals . . . 97
4.2.5 Hydrothermal Products and Ores . . . 98
4.2.6 Metamorphic Rocks . . . 101
4.3 Planetary Formation and Magmatic Processes . . . 102
4.3.1 Planetary Accretion . . . 102
4.3.2 Formation and Differentiation of Planetary Cores . . . 104
4.3.3 Partial Melting on Earth and Other Planets . . . 108
4.3.4 Mantle Metasomatism . . . 114
4.3.5 Differentiation of Melts . . . 119
4.3.6 The Mantle and Crust of the Earth as Compared to Other Planets . . . 133
4.4 Summary . . . 137
References . . . 137

5 The Modern Surficial World . . . 149
5.1 Weathering . . . 149
5.1.1 Mechanical Weathering . . . 149
5.1.2 Chemical Weathering . . . 149
5.1.3 Soils . . . 150
5.2 Rivers and Groundwater . . . 154
5.2.1 Rivers . . . 154
5.2.2 Groundwater and Terrestrial Hydrothermal Systems . . . 158
5.3 Redox-Stratified Water Bodies . . . 162
5.3.1 Lake Water . . . 163
5.3.2 Lake Sediments . . . 166
5.3.3 The Black Sea . . . 168
5.4 Marine Sediments . . . 171
5.4.1 Reactive Fe Inventories . . . 171
5.4.2 Pore Fluid-Sediment Interactions . . . 175
5.4.3 Solid-Phase Fe Components . . . 179
5.4.4 Benthic Fe Fluxes . . . 181

5.5 The Fe Budget of the Modern Oceans 186
5.5.1 Seawater Fe. 186
5.5.2 Riverine and Aeolian Sources 188
5.5.3 Benthic Sources. 192
5.5.4 Hydrothermal Sources 197
5.6 Summary 203
References. 204

6 The Ancient Earth. 215
6.1 The Cenozoic Marine System. 215
6.1.1 Global Changes in the Cenozoic 216
6.1.2 Fe–Mn Crusts as Archives of Paleo-Seawater Compositions. 220
6.1.3 Fe Isotope Variations in Cenozoic Seawater 221
6.2 Cretaceous Anoxic Events 228
6.2.1 Cenomanian-Turonian OAE-2. 230
6.3 Precambrian Earth: An Introduction 237
6.3.1 Broad Changes in the Surface Earth in the Precambrian. 238
6.3.2 Temporal Changes in Fe Abundance and Speciation. 242
6.3.3 Differences in Marine Fe Pathways Between Modern and Ancient Earth 246
6.3.4 Authigenic Fe Isotope and Reactive Fe Trends 252
6.4 Precambrian Earth: The Neoproterozoic 253
6.4.1 Neoproterozoic Clastic Marine Sedimentary Rocks 255
6.4.2 Revisiting Reactive Fe Speciation and $\delta^{56}Fe$. 258
6.4.3 Neoproterozoic Iron Formations (IFs) 263
6.5 Precambrian Earth: The Paleoproterozoic and Neoarchean Transition Through the GOE 267
6.5.1 The Post-GOE Sedimentary Record 268
6.5.2 Changes in Weathering Across the GOE. 273
6.5.3 Moving to a Low-Oxygen World: Key Issues of Fe Mass Balance, Fe Isotope Fractionation Factors, Fe^{2+}_{aq} Oxidation, and the Age of Redox Proxies 276
6.5.4 Early Paleoproterozoic Iron Formations (IFs) 283
6.5.5 Early Paleoproterozoic Rise of Mn Redox 293
6.5.6 Paleoproterozoic and Neoarchean Continental Margins: Relations Between Shales, Carbonate Platforms, and IFs 298

6.6 Precambrian Earth: The Early Archean Record 307
6.6.1 The Mesoarchean Witwatersrand and Pongola Basins 308
6.6.2 The Paleoarchean Barberton Greenstone Belt and Pilbara Craton 316
6.6.3 The High-Grade Metamorphic Terranes of the Eoarchean 322
6.7 Precambrian Earth: Synthesis of the Eoarchean Through Paleoproterozoic 328
6.8 Chapter Summary 332
References 335

1 Introduction and Overview

Iron has long been of great interest to geochemists given its high abundance and sensitivity to redox changes. Our goal in this chapter is to acquaint the reader with key concepts of Fe geochemistry and isotope geochemistry as a foundation for later chapters. Our short review of Fe geochemistry in Sect. 1.1 focuses on redox changes because these are responsible for some of the largest Fe isotope variations found in natural samples. The discussion here is necessarily brief, and we will revisit specific issues of Fe geochemistry in later chapters that are pertinent to that material. In Sect. 1.2 we introduce general concepts of isotope geochemistry, beginning with nomenclature and isotopic fractionation, and then illustrate some examples of how isotopic variations may be produced, such as equilibrium exchange and Rayleigh fractionation. The processes by which Fe isotopes are fractionated in nature, however, extend far beyond these simple models, as discussed in detail in later chapters. Finally, in Sect. 1.3 we provide the reader with a brief overview of subsequent chapters in the book.

1.1 Geochemistry of Fe

Iron is the most abundant redox-active metal in the Solar System. The unusually high abundance of Fe for its atomic number (Fig. 1.1) lies in the fact that it occurs at the end of the line in the sequence of stellar fusion and stability (e.g., Aller and McLaughlin 1965), in part reflecting the peak in binding energy that occurs for elements near mass 60 (e.g., Fewell 1995). On a cosmochemical basis, Fe belongs, to the so-called main group and has a 50% condensation temperature of ~1300 K. This group is between moderately volatile and refractory elements and includes other major elements, such as Mg and Si (e.g., Palme et al. 2014; Lee 2016). The abundance of Fe (normalized to Si) in the solar photosphere is identical to that in CI chondrites (e.g., Palme et al. 2014), indicating that the high Fe levels seen in cosmic abundances likely characterizes the bulk Earth. The electron configuration of zerovalent Fe is [Ar] $3d^6\ 4s^2$, and under oxidizing conditions, Fe^{2+} has a configuration of [Ar] $3d^6$, and Fe^{3+} has a configuration of [Ar] $3d^5$, the latter of which reflects a half-filled d shell. Iron, therefore, plays a major role in the electron flow of numerous reactions in the Earth.

Iron occurs as a wide variety of species. It may form an alloy, but it also binds with a range of anions. Following Goldschmidt's (1937) classification, Fe behaves as a siderophile element and is enriched in Earth's core. At high sulfide levels, it may also bind to S when it behaves as a chalcophile element, which can occur in the core, mantle, and crust. At high oxygen levels, Fe behaves as a lithophile element where, as a cation (Fe^{3+}, "ferric Fe"; Fe^{2+}, "ferrous Fe"), it is bound to O in a wide variety of silicates, oxide/hydroxides, and carbonates. Iron has a wide range of solubilities dependent on Eh, pH, and availability of binding ligands.

C. Johnson et al., *Iron Geochemistry: An Isotopic Perspective*, Advances in Isotope Geochemistry,
https://doi.org/10.1007/978-3-030-33828-2_1

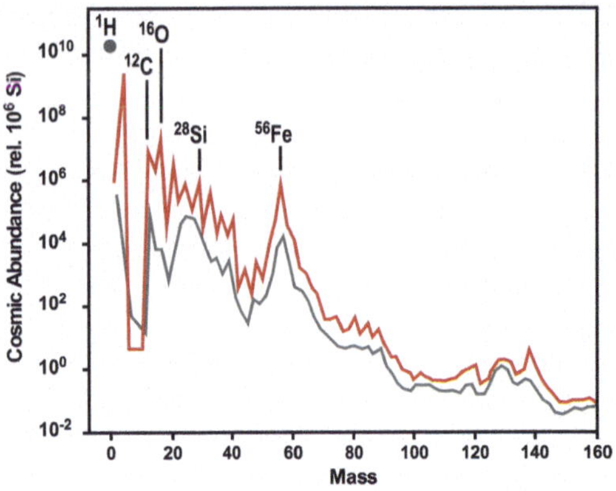

Fig. 1.1 Cosmic abundances based on the solar photosphere, normalized to 10^6 Si, up through mass 160. Even masses shown in red, odd masses in grey. The high enrichment of elements in the mass range of Fe reflects a binding energy peak and its effects on stellar nucleosynthesis. Also shown are the abundances of H, C, O, and Si, which are important for producing a range of major Fe-bearing minerals, including oxides, silicates, and carbonates. Adapted from Palme et al. (2014)

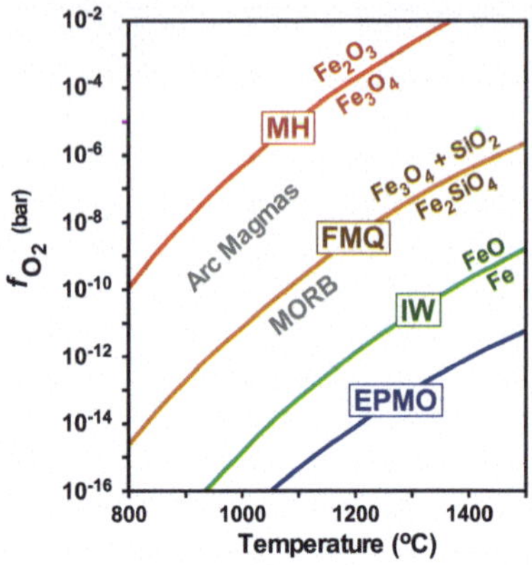

Fig. 1.2 Variations in redox state of Fe, as indicated by oxygen fugacity (f_{O_2}), relative to temperature, for a variety of mineral-O_2 buffers: iron-wüstite (IW, Eq. 1.2) shown in green curve, fayalite-magnetite-quartz (FMQ, Eq. 1.3) shown in brown curve, and magnetite-hematite (MH, Eq. 1.4) shown in red curve, reflecting progressively higher oxidation states of Fe. Blue curve labeled EPMO is an estimate for early planetary magma ocean where Fe^0 coexists with silicate melt. Most modern magmas lie within one log unit of FMQ, where mid-ocean ridge basalts (MORB) tend to lie below FMQ and arc magmas tend to lie above FMQ. Adapted from Behrens and Gaillard (2006)

Under oxidizing surface conditions, Fe is highly insoluble except at very low pH, but solubility may be enhanced by complexation with organic ligands (e.g., Schwertmann 1991). In contrast, at high-temperature, hydrothermal conditions, Fe solubility may be quite high, especially in chloride-bearing solutions (e.g., Chou and Eugster 1977).

1.1.1 Fe Redox

The redox state of Fe in high-temperature systems may be expressed by the oxygen fugacity, f_{O2}, through reactions such as:

$$2Fe^{2+} + {}^1\!/\!{}_2O_2 \rightarrow 2Fe^{3+} + O^{2-} \tag{1.1}$$

where O^{2-} represents structural O in oxides or silicates (e.g., Gaillard et al. 2003). More specifically, f_{O2} is commonly calculated from O_2-buffered mineral reactions (Fig. 1.2; e.g., Carmichael 1991; Kress and Carmichael 1991), which include, in increasing order of oxidation:

$$2(1-X)Fe^0 + O_2 \rightarrow 2Fe_{1-X}O \text{ (iron-wüstite; IW)} \tag{1.2}$$

$$3Fe_2SiO_4 + O_2 \rightarrow 2Fe_3O_4 + 3SiO_2 \text{ (fayalite-magnetite-quartz; FMQ)} \tag{1.3}$$

$$4Fe_3O_4 + O_2 \rightarrow 6Fe_2O_3 \text{ (magnetite-hematite; MH)} \tag{1.4}$$

An early Earth magma ocean in equilibrium with iron metal (Fe^0), would have had a very low f_{O2}, below the IW buffer (Fig. 1.2; e.g., Righter and Ghiorso 2012). For modern magmas, f_{O2} is commonly taken to reflect that of the source region(s) (e.g., Carmichael 1991). Among the variables for source region characteristics, water contents are perhaps the most important, and increasing water contents in magmas may increase the proportion of Fe^{3+} in the melt via charge coupling with H produced by water dissociation via the reaction:

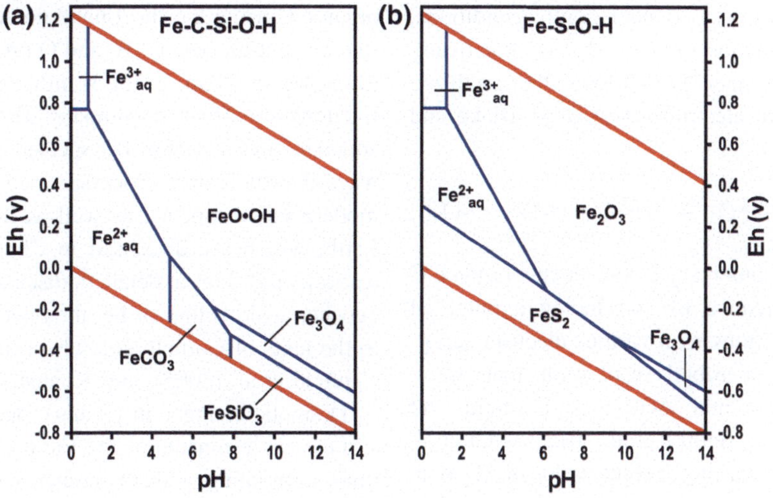

Fig. 1.3 Relations between redox potential (Eh, in volts) and pH for the Fe–C–Si–O–H system (**a**) and Fe–S–O–H system (**b**) at 25°C and 1 bar. Activities are: Fe: 10^{-6}, C: 10^{-3}, and Si: 10^{-3}. Phase boundaries bounded by stability limits of H_2O in red lines (upper boundary: H_2O–O_2, lower boundary: H_2O–H_2). Blue lines indicate phase boundaries in the region where H_2O is stable. Adapted from Brookins (1988)

$$H_2O \rightarrow H_2 + {}^1\!/_2 O_2 \tag{1.5}$$

Botcharnikov et al. (2005). The relatively oxidized nature of modern arc magmas is in part due to their relatively high H_2O contents (e.g., Kelley and Cottrell 2009). Charge balance between Na^+ or K^+ and Fe^{2+} and Fe^{3+} in alkaline magmas, particularly as a function of Fe pyroxene crystallization, may also change the proportion of Fe^{3+} in magmas (e.g., Markl et al. 2010). From a practical perspective, it can be safely assumed that the average proportion of Fe^{3+} in most magmas is $\sim$0.15, consistent with the observation that most magmas have f_{O2} within one log unit of FMQ (e.g., Bézos and Humler 2005) and the fact that hematite is not found as a phenocryst in volcanic rocks. Nevertheless, as discussed in Chap. 4, there are measurable changes in Fe isotope compositions in magmatic/high-temperature systems that can be ascribed to redox changes.

The reduced state of Fe in planetary interiors contrasts dramatically with the oxidation state of Fe in Earth's surface environments today. In Fig. 1.3 we show two Eh-pH diagrams for Fe, illustrating the equilibrium assemblages for C- and S-bearing systems in the presence of crystalline Fe^{3+}-oxide/hydroxide. At circum-neutral pH, Fe^{3+}-oxide/hydroxide is stable over a very wide range of Eh conditions. Only at low Eh is $FeCO_3$ (siderite), Fe_3O_4 (magnetite), and FeS_2 (pyrite) stable depending on C and S abundances. For Si-bearing systems, $FeSiO_3$ (Fe^{2+}-pyroxene) is stable at low Eh. Aqueous Fe^{3+}_{aq} is only stable under acidic conditions, and the phase field for Fe^{2+}_{aq} is smallest in the presence of Fe_2O_3 (hematite) relative to other Fe^{3+}-oxide/hydroxides. Importantly, the field for Fe^{2+}_{aq} expands significantly if poorly crystalline Fe^{3+}-hydroxides such as ferrihydrite are assumed, allowing co-existence of these components at circum-neutral pH under moderate Eh conditions. There is a large body of literature which shows that co-existence of Fe^{2+}_{aq} and Fe^{3+}-oxide/hydroxides induces auto-catalytic phase transformations to more stable Fe^{3+}-oxide/hydroxide, accompanied by extensive electron and Fe atom exchange (e.g., Cornell and Schwertmann 2003; Hansel et al. 2005; Handler et al. 2009; Yang et al. 2010). These effects turn out to be key in promoting Fe isotope equilibrium at low temperatures between Fe^{2+}_{aq} and Fe^{3+}-oxide/hydroxide, as discussed in Chap. 3.

A wide range of Eh conditions exist in modern surface environments. Oxidation of Fe^{2+}_{aq} by O_2 occurs rapidly at high Eh, circum-neutral pH conditions, on timescales of seconds to minutes

(Millero et al. 1987). Under acidic conditions, however, oxidation of $Fe^{2+}{}_{aq}$ is kinetically inhibited, but may be biologically catalyzed using O_2 as an electron donor (e.g., Blake and Johnson 2000):

$$2Fe^{2+} + {}^1\!/_2O_2 \rightarrow 2Fe^{3+} + H_2O \quad (1.6)$$

In many sedimentary environments organic C loading will reduce Eh, producing a range of diagenetic reactions with various electron acceptors through microbial respiration that are a function of Eh and ΔG (e.g., Canfield and Thamdrup 2009; Bethke et al. 2011; Arndt et al. 2013). With decreasing absolute value of ΔG, oxic respiration gives way to nitrate reduction, then manganese reduction, followed by iron reduction, sulfate reduction, and finally methanogenesis. These would produce authigenic minerals such as siderite in the case of reduction of Fe^{3+}, and pyrite in the case of reduction of sulfate. The corresponding terminal electron acceptor half reactions for iron and sulfate reduction would be:

$$Fe(OH)_3 + 3H^+ + e^- \rightarrow 2Fe^{2+} + 3H_2O \quad \left(\Delta G_r^o = -94.7\,kJ/e^-\right) \quad (1.7)$$

$$SO_4^{2-} + 9H^+ + 8e^- \rightarrow HS^- + 4H_2O \quad \left(\Delta G_r^o = -24.0kJ/e^-\right) \quad (1.8)$$

which in nature would be coupled to oxidation of organic C such as glucose via the half-reaction:

$$C_6H_{12}O_6 + 12H_2O \rightarrow 6HCO_3^- + 30H^+ + 24e^- \quad \left(\Delta G_r^o = 10.0\,kJ/e^-\right) \quad (1.9)$$

e.g., Arndt et al. (2013). Combining Eqs. 1.7 and 1.9, as well as Eqs. 1.8 and 1.9, shows that microbial Fe reduction should precede sulfate reduction in sediments that receive Fe^{3+}, sulfate, and organic C (C_{org}), as is commonly observed in modern marine sediments (e.g., Canfield et al. 1993; Thamdrup et al. 1994). Throughout the book, we will simplify such reactions with organic C using the short-hand notation "CH_2O" to represent glucose C, H, and O proportions. As discussed in Chap. 3, the significant Fe isotope fractionations that accompany the variety of aqueous and mineral Fe species provide key ways in which such processes may be traced in modern environments, as well as in the ancient Earth, which are discussed in Chaps. 5 and 6, respectively. For a detailed discussion of biogeochemical cycling of Fe, the reader is referred to the texts of Cornell and Schwertmann (2003), Canfield et al. (2005), and Konhauser (2006).

Dramatic changes in Eh have occurred in the surface environments of the Earth over geologic time, correlating with O_2 contents in the atmosphere. The Hadean or early Archean atmosphere was likely highly reduced, reflecting the reduced state of the planetary interior (e.g., Kasting et al. 1993). In the classic model, atmospheric O_2 contents remained very low up until ~2.3 Ga, where the first significant rise in O_2 (the "Great Oxidation Event", or GOE) was inferred from multiple lines of evidence (e.g., Holland 1984). An anoxic surface Earth would have stabilized $Fe^{2+}{}_{aq}$ in the oceans and produced a markedly different Fe biogeochemical cycle than that of the modern Earth. Evidence for an early "iron world" lies in abundant iron formation (IF) deposits of Archean and Paleoproterozoic age (e.g., Konhauser et al. 2017). Abundant $Fe^{2+}{}_{aq}$ in the early Earth likely affected biochemistry, including mediation of RNA folding and catalysis by ribozymes and proteins (e.g., Okafor et al. 2018). It is now clear, however, that O_2 levels were at least locally high significantly prior to the GOE (e.g., Lyons et al. 2014). Given the large changes that occurred in the Fe cycle over geologic time, a major effort has been expended on exploring the Fe isotope record of the Precambrian, as discussed in Chap. 6.

1.2 Stable Isotope Geochemistry

Approximately three-quarters of the elements in the Periodic Table have two or more stable isotopes, reflecting variations in neutron abundances. Stable isotopes may be thought of as a "third dimension" to the Periodic Table, where small differences in isotopic abundances for a given element may be used to constrain temperature, mass fluxes, fluid-rock interaction, biogeochemical processes, etc. Stable isotopes fractionate during kinetic processes, such as evaporation or diffusion, reflecting the high translational velocities of light isotopes. The most widely used aspect of stable isotope variations, however, is based on isotope fractionation between phases that are in physicochemical equilibrium. In this case, stable isotope partitioning between phases largely reflect differences in zero-point energies for isotopically substituted species, although in detail, complexities exist that are mass-independent or reflect non-stochastic distribution of rare isotopes. Importantly, common issues that face geochemical studies such as activity coefficients or variations in molar volumes are unimportant in all but the most extreme conditions for stable isotope studies. The basic thermodynamic framework for stable isotope geochemistry was laid out by Bigeleisen and Mayer (1947) and Urey (1947). More recent reviews of the general principles of isotopic fractionation can be found in O'Neil (1986b), Criss (1999), Schauble (2004), Blanchard et al. (2017), and Hoefs (2018), all of which provide distinct and valuable perspectives. Over seven decades of research has been dedicated to study of the stable isotope variations of H, C, N, O, and S, and in the last two decades there has been increased attention to elements that span the Periodic Table (Johnson et al. 2004b; Teng et al. 2017). Our focus here, of course, is on the stable isotope geochemistry of Fe.

1.2.1 Nomenclature

Stable isotope variations are generally on the order of 1% or less, particularly for heavy elements where the relative mass differences are small. The vast majority of isotopic data are reported in "delta notation", where the isotopic composition is cast as the deviation of an isotopic ratio relative to the same ratio in a standard (e.g., O'Neil 1986a):

$$\delta^{i}\mathrm{E_X} = \left[\left(\mathrm{R}^{i/j}_{\mathrm{X}} - \mathrm{R}^{i/j}_{\mathrm{STD}}\right)/\mathrm{R}^{i/j}_{\mathrm{STD}}\right]10^3 \quad (1.10)$$

where i and j are the specific isotopes used in ratio R of element E, X is the sample of interest, and STD is a standard reference material or reservoir. It is traditional to list isotope i in the δ value, and it is important to note the specific ratio $\mathrm{R}^{i/j}$ that is used. The units for the $\delta^{i}\mathrm{E_X}$ value are in parts per thousand or "per mil", which is commonly noted using the per mil sign (‰). In the case of Fe, the two most abundant isotopes are ^{56}Fe and ^{54}Fe, which, using Eq. 1.10, would produce:

$$\delta^{56}\mathrm{Fe_X} = \left[\left({}^{56}\mathrm{Fe}/{}^{54}\mathrm{Fe_X} - {}^{56}\mathrm{Fe}/{}^{54}\mathrm{Fe_{STD}}\right)/{}^{56}\mathrm{Fe}/{}^{54}\mathrm{Fe_{STD}}\right]10^3 \quad (1.11)$$

Traditionally, during development of stable isotope geochemistry of light elements, $\mathrm{R}^{i/j}$ was defined as the abundance ratio of the rare isotope to the major isotope (e.g., O'Neil 1986a), which corresponded to *heavy over light* masses, leading to a consistent nomenclature where a positive $\delta^{i}\mathrm{E_X}$ value refers to a sample that is relatively enriched in the heavy isotope (a high $\mathrm{R}^{i/j}$ ratio relative to the standard).

The vast majority of stable Fe isotope literature is reported relative to the IRMM-14 standard, and that practice is followed here. Some labs continue to report δ values relative to the average of igneous rocks, which is approximately equal to that of average crust. The advantage of defining δ^{56}Fe values relative to a major geologic reservoir such as average igneous rocks is this provides a useful interpretive context for discussing geologic processes. The disadvantage of using the average igneous rock reference frame is that, as discussed in Chap. 4, the Fe isotope compositions of igneous rocks vary. This is, however, a quandary that is similar to that faced in the choice of reference frames for other isotopic systems. For example, defining δ^{18}O values relative to Standard Mean Ocean Water (SMOW) continues to be used because of its useful geologic context, despite the fact that the δ^{18}O values of ocean water are quite variable. For stable Fe isotopes, the two reference frames have had a consistent offset since the beginnings of Fe isotope geochemistry, allowing one to freely move between reference frames regardless of publication date. On the IRMM-14 scale, the δ^{56}Fe value of the average of igneous rocks is +0.09‰ (Beard et al. 2003), identical to the average δ^{56}Fe value for Mid-Ocean-Ridge basalts (MORB; see Chap. 4). In the future, a different reference frame may be chosen for stable Fe isotopes because the original IRMM-14 standard is no longer readily available, and given the rapid increase in stable isotope research across the Periodic Table, policies are under development for establishing isotopic reference frames (e.g., Teng et al. 2017).

Some early Fe isotope data have been reported using different nomenclature. The ε value has been used, which is defined in parts per 10^4 (e.g., Zhu et al. 2000). Use of the ε value for reporting stable isotope data continues for other stable isotope systems (not to be confused with the use of ε to report stable isotope fractionation factors in the literature), but was not extensively used for Fe isotopes. Iron isotope data reported as ε values appear to vary by 10X relative to that reported as δ values, although the analytical uncertainty is also a factor of 10X greater for data reported as ε values. In addition, some publications have reported Fe isotope compositions as "F_{Fe}", defined as the ‰/amu deviation in the δ value (e.g., Dauphas et al. 2004). The community no longer uses ε or F_{Fe} for reporting Fe isotope compositions, but the reader should be aware of their existence in older literature.

The motivation for defining the $\delta^i E_X$ value as *rare isotope over major isotope* lies in the fact that the mathematical forms of mixing relations and other physical processes are greatly simplified in cases where the rare isotope *i* is very low in abundance, which leads to the simplification that the abundance of isotope *j* may be treated as invariant, particularly when the range in isotopic compositions is relatively restricted. For example, the exact mixing relation for a single element between components A and B which differ in their isotopic compositions is given by:

$$M_A/M_B = -(C_B/C_A)\left[(R^*_{MIX} - R^*_B)/(R^*_{MIX} - R^*_A)\right] \quad (1.12)$$

where M_A and M_B are the masses of components A and B, respectively, C_A and C_B are the concentrations of the element in components A and B, respectively, and R^* is defined as the ratio of the mass of the rare isotope over the total mass of the element (see Eq. 1.14 in Criss 1999). In the case of O, for example, where the abundances of ^{18}O and ^{16}O are 0.20% and 99.76%, respectively, R^* is very nearly equal to the ^{18}O/^{16}O ratio, allowing us to relate the exact mixing equation directly to the measured isotopic compositions. In such cases, we may further simplify Eq. 1.12 using δ notation as:

$$\delta^i E_{MIX} \approx \delta^i E_A f + \delta^i E_B (1 - f) \quad (1.13)$$

where *f* is the fraction of component A in the two-component mixture.

For stable Fe isotopes, defining the $\delta^i E_X$ value as *heavy over light isotope* does not correspond to *rare over major isotope*, which introduces error in the approximation of Eq. 1.13 for mixtures that involve very large (>10‰) isotopic contrasts. This may be a problem, for example, in experiments that use enriched Fe isotope tracers (see Chap. 3). For most natural systems, however, the range in Fe isotope compositions is limited, and the errors introduced by Eq. 1.13 are small. Despite the complexity of non-linear mixing relations and complexities that accompany a *heavy over light isotope* definition for the $\delta^i E_X$ value for stable Fe isotopes, the geochemical community generally follows this practice for all isotopic systems so that relative isotopic shifts remain consistent.

1.2.2 Isotopic Fractionation

Following standard practice (e.g., O'Neil 1986a), the isotopic fractionation factor between two substances A and B is defined as:

$$\alpha_{A-B} = R^{i/j}_A / R^{i/j}_B \quad (1.14)$$

which may be cast in terms of $\delta^i E$ values as:

$$\alpha_{A-B} = \left(1000 + \delta^i E_A\right) / \left(1000 + \delta^i E_B\right) \quad (1.15)$$

Note that α_{A-B} simply reflects the contrast in isotopic compositions between two substances and, in terms of physical processes, could reflect equilibrium or non-equilibrium partitioning of isotopes. If α_{A-B} is an equilibrium isotope fractionation factor, the principles of equilibrium thermodynamics allow the fractionation factor to be treated as an additive property that may be combined across multiple phases to obtain different combinations. For an isotope exchange reaction in which one atom is exchanged, α_{A-B} is equal to the equilibrium constant. Because α_{A-B} is very close to unity, generally on the order of 1.00X, we may take advantage of the relation that $10^3\ln(1.00X) \approx X$ which provides the useful relation:

$$10^3\ln\alpha_{A-B} \approx \delta^i E_A - \delta^i E_B \equiv \Delta^i_{A-B} \quad (1.16)$$

This allows us to describe isotopic fractionations by simply subtracting the $\delta^i E$ values of substances A and B. It is important to clearly define the order of reference between substances A and B and to keep the order of subtraction consistent, which, unfortunately, is not always the case in the literature. Assuming a fractionation of 10‰ ($\alpha_{A-B} = 1.010$), which is very large for stable Fe isotopes, an error of only 0.05‰ is introduced if the fractionation is described using Δ^i_{A-B} as compared to α_{A-B}, which is immaterial for Fe isotopes.

Examples of stable O and Fe isotope fractionations are shown in Fig. 1.4, cast in the traditional relation of $10^3\ln\alpha_{A-B}$ relative to $10^6/T^2$ (T in K). To a close approximation, the isotopic fractionation factor varies linearly relative to $1/T^2$, although for some low-temperature systems linear variations are more closely approximated by 1/T (e.g., O'Neil 1986b; see further discussion in Chap. 3). A slight curvature in fractionation factors relative to $1/T^2$, as shown in Fig. 1.4, reflects higher-order temperature variations that are dependent upon modeling or experimental results. Figure 1.4 illustrates the generally larger isotopic fractionations that are

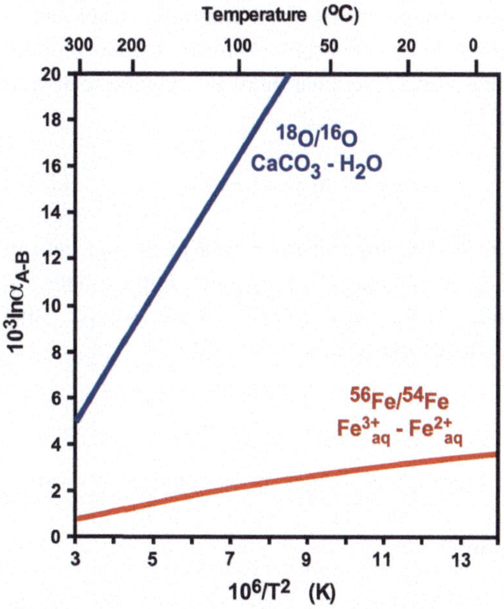

Fig. 1.4 Examples of isotopic fractionations for O and Fe, as cast in the traditional $10^3\ln\alpha_{A-B}$–$10^6/T^2$ diagram. The quantity $10^3\ln\alpha_{A-B}$ places the isotope fractionation factor α_{A-B} in units of per mil (‰). To a good approximation, many stable isotope fractionation factors vary linearly relative to $1/T^2$. Isotopic fractionations for relatively light elements, such as O, are generally higher than those of higher-mass elements such as Fe, as expected based on their relative mass differences. $CaCO_3$–H_2O curve for $^{18}O/^{16}O$ fractionations based on experiments from O'Neil et al. (1969). Fe^{3+}_{aq}–Fe^{2+}_{aq} curve for $^{56}Fe/^{54}Fe$ fractionations based on calculations of Rustad et al. (2010), which align with the experiments of Welch et al. (2003)

associated with light elements relative to heavy elements, where equilibrium isotope fractionations roughly scale to the relative mass difference over the square of the average mass (e.g., Johnson et al. 2004a). In addition, Fig. 1.4 shows one of the major controls on Fe isotope fractionation, Fe oxidation state, which changes the Fe–O bond lengths and hence vibrational frequencies (see Chap. 3).

For elements that have three or more isotopes, isotopic fractionations may be defined using two or more isotopic ratios. Assuming that isotopic fractionation occurs through a mass-dependent process, the extent of fractionation will be a function of the relative mass differences of the two isotope ratios. For example, assuming a simple harmonic oscillator for molecular motion, the isotopic fractionation of $R^{i/j}$ may be related to $R^{k/j}$ as:

$$\alpha_{k/j} = (\alpha_{i/j})^{Z} \quad (1.17)$$

where $Z = (m_i/m_k)[(m_k - m_j)/(m_i - m_j)]$, and m refers to the masses of the individual isotopes *i*, *j*, and *k* (e.g., Criss 1999). For Fe isotopes, this produces the relation:

$$\alpha_{57/54} = (\alpha_{56/54})^{1.4750} \quad (1.18)$$

Over small ranges in isotopic composition, Eq. 1.17 may be approximated by the linear form:

$$\delta^{k}E \approx [(k - j)/(i - j)]\delta^{i}E \quad (1.19)$$

where *i*, *j*, and *k* are integer masses. In the case of Fe, this relation would be:

$$\delta^{57}Fe \approx 1.5\delta^{56}Fe \quad (1.20)$$

These relations are important because some studies report Fe isotope data only in terms of $^{57}Fe/^{54}Fe$, although most studies now report both $^{56}Fe/^{54}Fe$ and $^{57}Fe/^{54}Fe$ ratios. Some studies report data in terms of $^{57}Fe/^{56}Fe$ ratios. All of these may be converted to $\delta^{56}Fe$ values via the relations above. In virtually all cases, Fe isotope variations in nature appear to follow mass-dependent relations, although there are some exceptions, as discussed in Chap. 3. In addition, there may be non-mass-dependent variations in Fe isotope compositions in extraterrestrial samples, as discussed in Chap. 4.

1.2.3 Processes that Produce Isotopic Variations

The occurrence of isotopic variations among natural samples indicates that geochemical reactions occur that are accompanied by isotopic fractionation, providing a powerful means for tracing processes associated with mass transfer. One of the simplest processes that produces isotopically distinct reservoirs would be slow reaction of substance A to B, where A and B remain open to complete isotopic exchange during the process, and where isotopic exchange occurs under equilibrium conditions. This is commonly referred to as *closed system equilibrium*, and the changes in isotopic compositions that occur may be defined by the exact relation:

$$\delta^{i}E_{B} = [\alpha_{B-A}\delta^{i}E_{SYS} + 1000f(\alpha_{B-A} - 1)]/[\alpha_{B-A} - \alpha_{B-A}f + f] \quad (1.21)$$

where α_{B-A} is the B–A fractionation factor, $\delta^{i}E_{SYS}$ is the $\delta^{i}E$ value for the total system (= composition when only A exists), and *f* is the fraction of A remaining (*f* = 1 when the system is entirely A) (see Eq. 3.19 in Criss 1999). Commensurate with the description of this model, α_{B-A} is an equilibrium isotope fractionation factor, as opposed to a kinetic isotope fractionation factor that would be pathway dependent. Note that Eq. 1.21 is simpler if cast in terms of α_{B-A} rather than α_{A-B}, as was defined in Eq. 1.14 above. If α_{B-A} is close to unity, Eq. 1.21 may be simplified to:

$$\delta^{i}E_{B} \approx \delta^{i}E_{SYS} + f(\alpha_{B-A} - 1)10^{3} \quad (1.22)$$

Criss (1999), which may be further simplified to:

$$\delta^{i}E_{B} \approx \delta^{i}E_{SYS} + f\Delta^{i}_{B-A} \quad (1.23)$$

using the approximation $\Delta^{i}_{B-A} \approx (\alpha_{B-A} - 1)10^{3}$. Following these simplifications, the corresponding $\delta^{i}E_{A}$ value, at a given *f*, is:

$$\delta^{i}E_{A} \approx \delta^{i}E_{B} - \Delta^{i}_{B-A} \quad (1.24)$$

Criss (1999). The *closed system equilibrium* model described by Eqs. 1.23 and 1.24 plot as straight lines in terms of $\delta^{i}E_{A}$ or $\delta^{i}E_{B}$ as a function of *f* (fraction of A remaining). In Fig. 1.5 we illustrate an example where Fe^{2+}_{aq} ("A") undergoes oxidation and precipitation to poorly crystalline Fe^{3+}-hydroxide ($Fe(OH)_3$) ("B"), where the

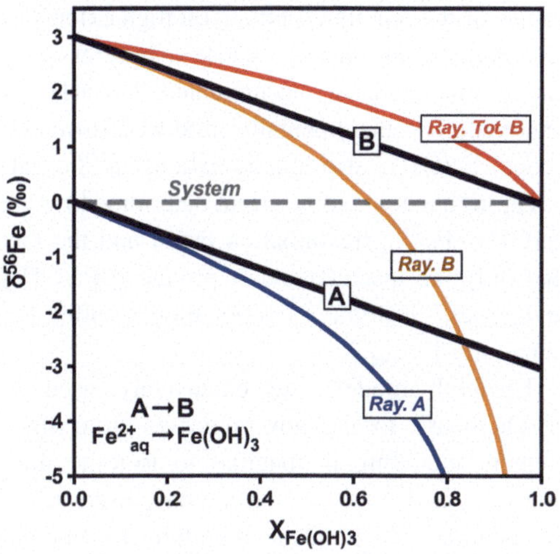

Fig. 1.5 Comparison of isotopic fractionations produced by *closed-system equilibrium* and *Rayleigh fractionation* processes where phase A is reacted to phase B, as a function of the proportion of B (X_B) that is produced using Eqs. 1.21 through 1.26 ($X_B = 1-f$). The example illustrated is oxidation of Fe^{2+}_{aq} (phase A) and precipitation of $Fe(OH)_3$ (phase B), relative to the proportion of $Fe(OH)_3$ produced ($X_{Fe(OH)3}$). Isotopic compositions are reported as $\delta^{56}Fe$ values, and the isotopic fractionation between $Fe(OH)_3$ and Fe^{2+}_{aq} is +3.0‰ ($\alpha_{Fe(OH)3-Fe2+aq}$ = 1.0030). The isotopic composition of the system is set to $\delta^{56}Fe = 0$ (dashed grey line). Solid black lines show the changes in $\delta^{56}Fe$ values for a *closed-system equilibrium* process, where isotopic equilibrium is continuously maintained between Fe^{2+}_{aq} and $Fe(OH)_3$. A Rayleigh process is illustrated in the colored curves using the same system isotopic composition and isotopic fractionation factor. The total (integrated) $Fe(OH)_3$ during reaction is shown in the red curve ("*Ray. Tot. B*"), and the instantaneous $Fe(OH)_3$ produced at any increment of reaction ($X_{Fe(OH)3}$) is shown in the brown curve ("*Ray. B*"). The instantaneous Fe^{2+}_{aq} is shown in the blue curve ("*Ray. A*"). At any value for $X_{Fe(OH)3}$, the isotopic difference between the brown "Ray. B" curve and the blue "Ray. A" curve is equal to the isotopic fractionation factor

system is initially composed only of Fe^{2+}_{aq} and has an initial $\delta^{56}Fe$ value of zero and where $10^3 \ln\alpha_{B-A} \approx \Delta^i_{B-A} = +3.0‰$. The fraction of B produced ($1-f$) is shown as $X_{Fe(OH)3}$. The first fraction of Fe $(OH)_3$ to form will have a $\delta^{56}Fe$ value of 3‰. As the reaction proceeds, the shifting mass balance of Fe^{2+}_{aq} and $Fe(OH)_3$ will require shifts in their $\delta^{56}Fe$ values while maintaining a constant isotopic fractionation between Fe^{2+}_{aq} and $Fe(OH)_3$. When the reaction is complete and the system is entirely Fe $(OH)_3$, mass balance requires that the $\delta^{56}Fe$ value of $Fe(OH)_3$ is the same as the initial $\delta^{56}Fe$ value of Fe^{2+}_{aq}. The importance of this illustration is that in cases where reactions go to completion, if there is no addition or loss of Fe from the system, there will be no net change in isotopic composition, even if the Fe isotope fractionation factor is large.

The isotopic-mass relations of the *closed system equilibrium* model indicate that interpretation of a given $\delta^{56}Fe$ value may vary greatly depending on isotopic mass balance. For example, as extensively discussed in Chap. 6, many workers point to the need of a high-$\delta^{56}Fe$ component in the ancient Earth to mass-balance the large number of highly negative $\delta^{56}Fe$ values measured for Precambrian rocks and minerals, yet the simple relations in Fig. 1.5 show this may not be the case; a $\delta^{56}Fe$ value of −2.5‰, for example, could be mass-balanced by a component that had a $\delta^{56}Fe$ value of 0.5‰ if that component was 80% of the system. A $\delta^{56}Fe$ value of −3‰ could be mass-balanced by a component that had a $\delta^{56}Fe$ value equal to that of average crust if it formed 99% of the system. The relations in Fig. 1.5 also show that extremely positive or negative $\delta^{56}Fe$ values can only be produced by components that represent only a small fraction of the total Fe pool that was involved in isotopic exchange. Finally, it is important to note that the relations described by the *closed system equilibrium* model are defined by ratios of the components but cannot directly constrain absolute sizes of the components involved in isotopic exchange. For example, as discussed in Chaps. 5 and 6, the absolute sizes of Fe components required to mass-balance Fe in the modern oceans are orders-of-magnitude smaller than those that must have been involved in Fe isotope mass balance in Fe^{2+}_{aq}-rich Archean oceans.

In the case of reactions where the products do not continue to exchange with other phases in the system, as might be the case during precipitation of a mineral from solution, *Rayleigh fractionation* may best describe the changes in δ^iE values for the individual components. In terms of isotopic fractionation, the well-known Rayleigh equation (Rayleigh 1902) is:

$$[R^{i/j}]/[R^{i/j}]_i = f^{(\alpha_{B-A}-1)} \quad (1.25)$$

where $[R^{i/j}]_i$ is the initial ratio $R^{i/j}$ (which may be defined for either A or B), and f is the fraction of A remaining. Cast in terms of $\delta^i E$ values, Eq. 1.25 becomes:

$$(1000+\delta^i E)/(1000+\delta^i E_i) = f^{(\alpha_{B-A}-1)} \quad (1.26)$$

where $\delta^i E$ may be defined for either A or B, and the subscript i refers to the initial $\delta^i E$ value.

The products of *Rayleigh fractionation* are effectively isolated from isotopic exchange with the rest of the system immediately upon formation. If the process occurs slowly, such that each increment of product B forms in isotopic equilibrium with the reactant A prior to isolation of B from the system, then α_{B-A} would be an equilibrium isotope fractionation factor. If, however, the process of formation of B is rapid, incremental formation of B may be out of isotopic equilibrium with A. In this case, α_{B-A} would be a kinetic isotope fractionation factor, which may be a function of reaction rates or other pathway-dependent processes. It is sometimes assumed that *Rayleigh fractionation* must be a kinetic process, but this does not have to be the case; a Rayleigh model simply means that the products are isolated from the reactants after the products are formed.

Because the product in *Rayleigh fractionation* is progressively isolated, the isotopic compositions may become more extreme at high extents of reaction as compared to those produced by a *closed system equilibrium* model (Fig. 1.5). For the example in Fig. 1.5, where Fe^{2+}_{aq} is oxidized, followed by precipitation of $Fe(OH)_3$, the total (integrated) $Fe(OH)_3$ must attain a $\delta^{56}Fe$ value equal to that of the system at the point of complete reaction ($X_{Fe(OH)3} = 1$), although at intermediate extents of reactions, the $\delta^{56}Fe$ value of the total solid is slightly higher than that of the *closed system equilibrium* model. In a *Rayleigh fractionation* model, the $\delta^{56}Fe$ values of Fe^{2+}_{aq} are determined by the isotopic fractionation between the incrementally produced $Fe(OH)_3$, not the total $Fe(OH)_3$, and hence extreme $\delta^{56}Fe$ values of the remaining Fe^{2+}_{aq} (at high extents of reaction) are produced because a progressively larger inventory of accumulated $Fe(OH)_3$ is removed. The incrementally produced (instantaneous) $Fe(OH)_3$ also attains extreme $\delta^{56}Fe$ values at high extents of reaction, determined by the $Fe(OH)_3$–Fe^{2+}_{aq} fractionation factor and the fact that only the instantaneously-produced $Fe(OH)_3$ undergoes isotopic fractionation with the remaining Fe^{2+}_{aq}.

Rayleigh models are extensively used in geochemistry to explain processes from trace element modeling in magmas to isotopic fractionation, and have been commonly used in the Fe isotope geochemical literature to explain processes ranging from oxidation of groundwater Fe^{2+}_{aq} to precipitation of Precambrian iron formations. As will be discussed in the chapters that follow, however, there are many cases of inappropriate application of Rayleigh models to Fe isotopes. For example, abundant experimental data show that isotopic equilibrium between Fe^{2+}_{aq} and Fe^{3+}-hydroxides may be maintained through coupled electron and atom exchange (see Chap. 3), and for such situations, a Rayleigh model should not be used. In addition, because a Rayleigh model is effectively a closed-system box model, it is not an appropriate choice to model open systems where there are input and output fluxes. As discussed in Chap. 6, for example, Rayleigh models have been inappropriately applied to the Precambrian oceans, whereas more appropriate models based on reaction transport concepts are better suited to marine systems. There are, however, clear cases where a Rayleigh model may be appropriate, but these are relatively few compared to its common use in the literature.

1.3 Overview of the Chapters

The $\delta^{56}Fe$ values for natural samples vary by ~8‰ (Fig. 1.6). Average upper crust, as represented by loess samples, has an average $\delta^{56}Fe$ value of 0.09‰, identical to the average of igneous rocks defined by Beard et al. (2003).

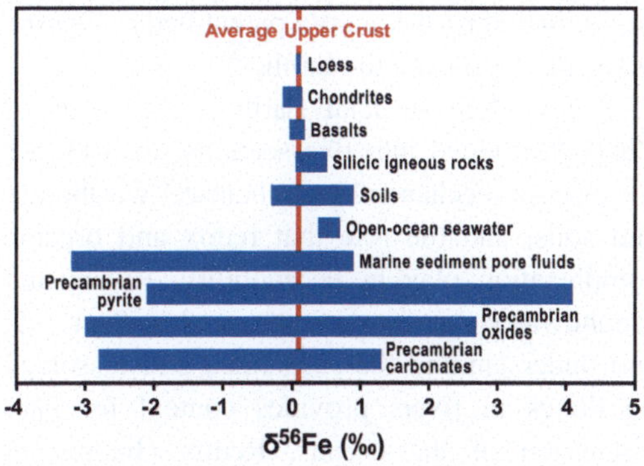

Fig. 1.6 Range in Fe isotope compositions for various fluids, rocks, and minerals. Isotopic data reported as $\delta^{56}Fe$ values relative to the IRMM-14 scale, a convention used throughout the book. The average $\delta^{56}Fe$ value for loess is identical to the average of igneous rocks from Beard et al. (2003), and may be taken to reflect average upper crust (vertical red line at $\delta^{56}Fe = 0.09‰$). High-temperature samples, including meteorites, basalts, and silicic igneous rocks have a restricted range in $\delta^{56}Fe$ values, generally <1‰. Soils, modern seawater, and pore fluids from modern marine sediments define a much larger range in $\delta^{56}Fe$ values, reflecting the relatively large Fe isotope fractionations that occur at low temperature. Also shown is the range in $\delta^{56}Fe$ values measured for minerals from Precambrian sedimentary rocks, including pyrite, oxides (hematite and magnetite), and carbonates, which define the largest range measured on Earth

High-temperature samples, including basalts and silicic igneous rocks, have a relatively restricted range in $\delta^{56}Fe$ values, <1‰, overlapping with the $\delta^{56}Fe$ values of meteorites. Although the range in $\delta^{56}Fe$ values for high-temperature samples is relatively small, reflecting the small isotopic fractionations that occur at high temperatures (Fig. 1.4), Fe isotopes provide key insights into processes from planetary accretion to hydrothermal alteration. Low-temperature samples define the largest range in $\delta^{56}Fe$ values, commensurate with the relatively large isotopic fractionations at low temperature. Minerals and fluids in low-temperature environments both define large ranges in $\delta^{56}Fe$ values, including pore fluids from modern marine sediments, as well as minerals from Precambrian marine sedimentary rocks. This large range reflects fluid-mineral fractionations as well as redox changes.

Chapter 2 discusses analytical methods for Fe isotope analysis, with a strong emphasis on the best practices required to ensure accurate and precise Fe isotope ratio measurements. In Sect. 2.2, the methods used to purify Fe in natural matrices by ion exchange chromatography are discussed. This is followed by a section on mass spectrometry methods that includes a historical perspective (Sect. 2.3.1). Early stable Fe isotope analysis was done by Thermal Ionization Mass Spectrometry (TIMS), which imposed serious limitations in terms of precision and sample throughput. With the advent of Multi-Collector Inductively Coupled Plasma Mass Spectrometry (MC-ICP-MS), precision and throughput increased dramatically, and methods used for Fe isotope analysis by MC-ICP-MS are discussed in Sects. 2.3.2 and 2.3.3. This section finishes with a discussion of the origin of instrumental mass bias and matrix effects (Sect. 2.3.4). The chapter ends with a discussion of in situ Fe isotope analysis techniques (Sect. 2. 4), a relatively new area of focus in Fe isotope analysis of natural samples.

As with any stable isotope system, a deep understanding of isotopic fractionation factors is required to fully understand natural systems. Chapter 3 discusses mass-dependent fractionation of Fe isotopes between different Fe bearing phases in both abiologic and biological systems at low-temperatures (<350 °C). In particular, this chapter makes comparisons between Fe isotope fractionation factors that have been experimentally measured with those calculated by either inverting spectroscopic data or obtained using ab initio methods. This comparison highlights a strong consistency between many calculated and experimentally measured Fe isotope fractionation factors, and we suggest that this provides strong evidence for the accuracy of experimentally measured fractionation factors (Sect. 3.1). The chapter includes an initial discussion of how one can calculate fractionation factors (Sect. 3.2), and is followed by discussion of methods used to measure fractionation factors by experimentation (Sect. 3.3). Section 3.4 compares experimentally measured fractionation factors with those derived by inversion of spectroscopic measurements or ab initio methods in abiological experiments. This section is broken down into a section that described fractionation factors between dissolved Fe species (Sect. 3.4.1) and between dissolved Fe and Fe-bearing minerals (Sect. 3.4.2). Section 3.5 discusses Fe isotope experiments in which organisms process Fe metabolically, and includes a section on Fe-oxidizing bacteria (Sect. 3.5.1), magnetotactic bacteria (Sect. 3.5.2), and iron-reducing bacteria (Sect. 3.5.3). The chapter ends with a discussion of a proposed self-consistent data set of β-values that are deemed to be accurate based on consistency between experiments and calculated Fe isotope fractionation factors.

In Chap. 4 we focus on processes that occur at high temperatures, including the formation of the planets and planetary cores, partial melting and metasomatism in the Earth's mantle, and the formation and differentiation of the crust. The chapter is subdivided in three parts: in parts one and two, Fe isotope compositions and variations of the major planetary (Sect. 4.1) and silicate Earth (Sect. 4.2) reservoirs are presented and discussed. The planetary reservoirs include different chondritic components and chondrite types, as well as materials from different planetary bodies. The terrestrial reservoirs include different rock types from the Earth's mantle, basalts from different tectonic settings, different material from the differentiated crust (e.g., different granitic rocks, rhyolites, etc.), metamorphic rocks, hydrothermally altered rocks and magmatic ores. Discussion of these suites is based on data compilations of the respective reservoirs and rock types from the literature. In the third part of the chapter (Sect. 4.3) Fe isotope fractionation during selected important high-temperature processes, including accretion, core formation, partial melting, differentiation of melts and diffusion-driven processes during mantle metasomatism and melt evolution are examined in more detail, combining information from experimental studies, ab initio modelling and natural observations. This subchapter closes with conclusions on how these different processes may explain the differences observed between the Fe isotope compositions of different planetary crusts, including the angrite parent body, 4 Vesta, Mars, the Moon and the Earth.

The modern surficial Earth is the focus of Chap. 5, divided into five sections. In Sect. 5.1 we discuss mechanical and chemical weathering and soils, and the role that redox and organic complexation play in Fe mobility. Rivers and groundwater chemistry are covered in Sect. 5.2, and understanding the suspended and dissolved Fe fluxes in rivers provides context for later discussion of the oceans. Redox changes in groundwater and terrestrial hydrothermal systems are also discussed, including processes that may produce large variations in Fe isotope compositions through oxidation of Fe^{2+}_{aq}. Continuing the theme of redox, Sect. 5.3 reviews Fe isotope data from redox-stratified water bodies, including lakes and the Black Sea, where redox is important, as well as the role of sulfide. Extensive discussion of marine sediments is contained in Sect. 5.4, where multiple processes are involved in biogeochemical cycling of Fe. Reactive Fe inventories in sediments are discussed, and interaction between Fe^{2+}_{aq} in pore fluids and reactive solid-phase Fe in sulfide-rich and sulfide-poor systems is shown to produce a very wide range of Fe isotope compositions. Sedimentary pyrite paragenesis is discussed in detail, as well as the concept of basin-scale transport of benthic Fe that is released by microbial diagenesis in marine sediments. Chapter 5 is concluded with a discussion of the Fe budget for the modern oceans, including the distribution and speciation of Fe, as well as using Fe isotopes and other geochemical data to trace Fe sources, including aeolian, riverine, benthic, and hydrothermal sources.

We conclude the book with an in-depth discussion of the ancient surface Earth in Chap. 6, the longest chapter in the volume, reflecting the very large Fe isotope database that is available for pre-Holocene samples. The chapter progressively steps back in time, beginning in Sect. 6.1 with a discussion of Cenozoic climatic changes that impacted changes in seawater chemistry, whose Fe isotope record is contained in Fe–Mn crusts. In Sect. 6.2 we look at Cretaceous anoxic events, which represent periods of very high C_{org} loading and development of large marine redox gradients. We concentrate on the detailed geochemical record that is available for the Cenomanian–Turonian OAE-2 event, where Fe isotope and other data demonstrate large changes in the Fe–S cycle, including geographically heterogenous distribution of benthic Fe shuttling and development of euxinic conditions. Section 6.3 begins our discussion of the Precambrian with an overview of broad geochemical changes from the Neoproterozoic to the Eoarchean. Specific to the Fe cycle, we review broad temporal changes in the Fe abundance and speciation in Precambrian sedimentary rocks. In addition, we highlight expected differences in the marine Fe cycle in the Precambrian relative to the discussion of the modern marine system in the previous chapter, particularly for hydrothermal and benthic Fe contributions. This section also contains a discussion of differences in marine sediment biogeochemical cycling in the low-oxygen environments of the Precambrian, which contrasts with those discussed in modern marine sediments in Chap. 5. The overview contained in Sect. 6.3 provides the context for our first specific discussion of the Precambrian rock record in Sect. 6.4, which addresses the Fe cycle in the late Precambrian that is recorded in Neoproterozoic clastic sedimentary rocks and iron formations (IFs), a time period of large changes in oxygen levels and biogeochemistry.

The remaining part of Chap. 6 is focused on the transition through the "Great Oxidation Event (GOE) of the early Paleoproterozoic, and continuing back to the earliest rock record. One of the largest efforts in research on Precambrian surface environments has been on the Paleoproterozoic-Neoarchean transition, which is covered in Sect. 6.5, and this section reflects nearly half of our discussion of the Precambrian. Stepping back in time through the GOE transition, changes in weathering across the GOE are discussed, both in terms of marine sedimentary rocks but also continental red beds and paleosols. We revisit key issues of Fe mass balance and oxidation of Fe^{2+}_{aq} in an anoxic world. The very large IFs deposited in this time period have been subject of great debate, and we discuss the relative roles of abiologic and biologic processes in their genesis, including data from Mn IFs that suggest very high Eh conditions. An important complementary record to IFs is found in coeval platform carbonates and shales, and these are used to constrain oxygen contents and the extent of benthic Fe shuttling. Moving deeper into time, Sect. 6.6 discusses the early Archean record, first concentrating on well-preserved Mesoarchean basins, and then turning to Paleoarchean continental margin sequences, which document the earliest record of oxygen increases and extensive microbial Fe cycling. The rock record older than 3.5 Ga exclusively exists as high-grade metamorphic terranes, and we evaluate the Fe isotope record for Eoarchean sequences in light of the effects of metamorphism, in an attempt to identify the earliest marine Fe cycles. We conclude the chapter in Sect. 6.7 with a synthesis of the Eoarchean through Paleoproterozoic, with an eye toward the relation between the surface Earth and solid-Earth processes, including the development of plate tectonics.

References

Aller LH, McLaughlin DB (1965) Stellar structure, vol 8. The University of Chicago Press, Stars and Stellar Systems

Arndt S, Jørgensen BB, LaRowe DE, Middelburg JJ, Pancost RD, Regnier P (2013) Quantifying the degradation of organic matter in marine sediments: a review and synthesis. Earth Sci Rev 123:53–86. https://doi.org/10.1016/j.earscirev.2013.02.008

Beard BL, Johnson CM, Skulan JL, Nealson KH, Cox L, Sun H (2003) Application of Fe isotopes to tracing the geochemical and biological cycling of Fe. Chem Geol 195(1–4):87–117

Behrens H, Gaillard F (2006) Geochemical aspects of melts: volatiles and redox behavior. Elements 2:275–280

Bethke CM, Sanford RA, Kirk MF, Jin Q, Flynn TM (2011) The thermodynamic ladder in geomicrobiology. Am J Sci 311(3):183–210. https://doi.org/10.2475/03.2011.01

Bézos A, Humler E (2005) The Fe3+/ΣFe ratios of MORB glasses and their implications for mantle melting. Geochim Cosmochim Acta 69(3):711–725. https://doi.org/10.1016/j.gca.2004.07.026

Bigeleisen J, Mayer MG (1947) Calculation of equilibrium constants for isotopic exchange reactions. J Phys Chem 13:261–267

Blake R, Johnson DB (2000) Phylogenetic and biochemical diversity among acidophilic bacteria that respire on iron. In: Lovley DR (ed) Environmental microbe-metal interactions. ASM Press, Washington, D.C

Blanchard M, Balan E, Schauble EA (2017) Equilibrium fractionation of non-traditional isotopes: a molecular modeling perspective. Rev Mineral Geochem 82 (1):27–63

Botcharnikov RE, Koepke J, Holtz F, McCammon C, Wilke M (2005) The effect of water activity on the oxidation and structural state of Fe in a ferro-basaltic melt. Geochim Cosmochim Acta 69(21):5071–5085. https://doi.org/10.1016/j.gca.2005.04.023

Brookins DG (1988) Eh-pH Diagrams for geochemistry. Springer

Canfield DE, Kristensen E, Thamdrup B (2005) Aquatic geomicrobiology, vol 48. Academic Press

Canfield DE, Thamdrup B (2009) Towards a consistent classification scheme for geochemical environments, or, why we wish the term 'suboxic' would go away. Geobiology 7(4):385–392. https://doi.org/10.1111/j.1472-4669.2009.00214.x

Canfield DE, Thamdrup B, Hansen JW (1993) The anaerobic degradation of organic matter in Danish coastal sediments: iron reduction, manganese reduction, and sulfate reduction. Geochim Cosmochim Acta 57:3867–3883

Carmichael ISE (1991) The redox states of basic and silicis magmas: a reflection of their source regions? Contrib Miner Petrol 106:129–141

Chou I-M, Eugster HP (1977) Solubility of magnetite in supercritical chloride solutions. Am J Sci 277:1296–1314

Cornell RM, Schwertmann U (2003) The iron oxides: structure, properties, reactions, occurences and uses. Wiley

Criss RE (1999) Principles of stable isotope distribution. Oxford University Press, USA

Dauphas N, van Zuilen M, Wadhwa M, Davis AM, Marty B, Janney PE (2004) Clues from Fe isotope variations on the origin of early Archean BIFs from Greenland. Science 306:2077–2080

Fewell MP (1995) The atomic nuclide with the highest mean binding energy. Am J Phys 63(7):653–658

Gaillard F, Pichavant M, Scaillet B (2003) Experimental determination of activities of FeO and Fe2O3 components in hydrous silicic melts under oxidizing conditions. Geochim Cosmochim Acta 67(22):4389–4409. https://doi.org/10.1016/s0016-7037(03)00376-4

Goldschmidt VM (1937) The principles of distribution of chemical elements in minerals and rocks. The seventh Hugo Müller Lecture, delivered before the chemical society on March 17th 1937. J Chem Soc (Resumed) (0):655–673. https://doi.org/10.1039/jr9370000655

Handler RM, Beard BL, Johnson CM, Scherer MM (2009) Atom exchange between aqueous Fe(II) and goethite: an Fe isotope tracer study. Environ Sci Technol 43:1102–1107

Hansel CM, Benner SG, Fendorf S (2005) Competing Fe (II)-induced mineralization pathways of ferrihydrite. Environ Sci Technol 39:7147–7153

Hoefs J (2018) Stable isotope geochemistry. Springer

Holland HD (1984) The chemical evolution of the atmosphere and oceans. Princeton University Press, Princeton

Johnson CM, Beard BL, Albarède F (2004a) Chapter 1. Overview and general concepts. In: Rosso JJ (ed) Geochemistry of non-tradtional stable isotopes. Reviews in mineralogy and geochemistry

Johnson CM, Beard BL, Albarède F (2004b) Geochemistry of non-traditional stable isotopes, vol 55. Reviews in mineralogy and geochemistry

Kasting JF, Eggler DH, Raeburn SP (1993) Mantle redox evolution and the oxidation state of the Archean atmosphere. J Geol 101:245–257

Kelley KA, Cottrell E (2009) Water and the oxidation state of subduction zone magmas. Science 325 (5940):605–607

Konhauser KO (2006) Introduction to geomicrobiology. Wiley-Blackwell

Konhauser KO, Planavsky NJ, Hardisty DS, Robbins LJ, Warchola TJ, Haugaard R, Lalonde SV, Partin CA, Oonk PBH, Tsikos H, Lyons TW, Bekker A, Johnson CM (2017) Iron formations: a global record of Neoarchaean to Palaeoproterozoic environmental history. Earth Sci Rev 172:140–177. https://doi.org/10.1016/j.earscirev.2017.06.012

Kress VC, Carmichael ISE (1991) The compressibility of silicate liquids containing Fe2O3 and the effect of composition, temperature, oxygen fugacity and pressure on their redox states. Contrib Miner Petrol 108:82–92

Lee C-T (2016) Geochemical classification of elements. In: Encyclopedia of engineering geology. Encyclopedia of earth sciences series, pp 1–5. https://doi.org/10.1007/978-3-319-39193-9_255-1

Lyons TW, Reinhard CT, Planavsky NJ (2014) The rise of oxygen in Earth's early ocean and atmosphere. Nature 506(7488):307–315. https://doi.org/10.1038/nature13068

Markl G, Marks MAW, Frost BR (2010) On the controls of oxygen fugacity in the generation and crystallization of peralkaline melts. J Petrol 51(9):1831–1847. https://doi.org/10.1093/petrology/egq040

Millero FJ, Sotolongo S, Izaguirre M (1987) The oxidation kinetics of Fe(II) in seawater. Geochim Cosmochim Acta 51:793–801

O'Neil JR (1986a) Appendix: terminology and standards In: Valley JW, Taylor HPJ, O'Neil JR (eds) Stable isotopes in high temperature geological processes, vol 16. Reviews in mineralogy, pp 561–570

O'Neil JR (1986b) Theoretical and experimental aspects of isotopic fractionation. In: Valley JW, Taylor HPJ, O'Neil JR (eds) Stable isotopes in high temperature geological processes, vol 16. Reviews in mineralogy, pp 1–40

O'Neil JR, Clayton RN, Mayeda TK (1969) Oxygen isotope fractionation in divalent metal carbonates. J Chem Phys 51:5547–5558

Okafor CD, Bowman JC, Hud NV, Glass JB, Williams LD (2018) Folding and catalysis near life's origin: support for Fe2+ as a dominant divalent cation. In: Prebiotic chemistry and chemical evolution of nucleic acids. Nucleic acids and molecular biology, pp 227–243. https://doi.org/10.1007/978-3-319-93584-3_8

Palme H, Lodders K, Jones A (2014) Solar system abundances of the elements. In: Treatise on geochemistry, pp 15–36. https://doi.org/10.1016/b978-0-08-095975-7.00118-2

Rayleigh L (1902) On the distillation of binary mixtures. Philos Mag Ser 6(4):521–537

Righter K, Ghiorso MS (2012) Redox systematics of a magma ocean with variable pressure-temperature gradients and composition. Proc Natl Acad Sci U S A 109(30):11955–11960. https://doi.org/10.1073/pnas.1202754109

Rustad JR, Casey WH, Yin Q-Z, Bylaska EJ, Felmy AR, Bogatko SA, Jackson VE, Dixon DA (2010) Isotopic fractionation of Mg2+ (aq), Ca2 + (aq), and Fe2 + (aq) with carbonate minerals. Geochim Cosmochim Acta 74(22):6301–6323. https://doi.org/10.1016/j.gca.2010.08.018

Schauble EA (2004) Applying stable isotope fractionation theory to new systems. In: Johnson CM, Beard BL, Albarède F (eds) Geochemistry of non-traditional stable isotopes, vol 56. Reviews in mineralogy and geochemistry

Schwertmann U (1991) Solubility and dissolution of iron oxides. Plant Soil 130:1–25

Teng FZ, Dauphas N, Watkins JM (2017) Non-traditional stable isotopes: retrospective and prospective. Rev Mineral Geochem 82(1):1–26

Thamdrup B, Fossing H, Jørgensen BB (1994) Manganese, iron, and sulfur cycling in a coastal marine sediment, Aarhus Bay, Denmark. Geochim Cosmochim Acta 58(23):5115–5129

Urey HC (1947) The thermodynamic properties of isotopic substances. J Chem Soc (Lond):562–581

Welch SA, Beard BL, Johnson CM, Braterman PS (2003) Kinetic and equilibrium Fe isotope fractionation between aqueous Fe(II) and Fe(III). Geochim Cosmochim Acta 67:4231–4250

Yang L, Steefel CI, Marcus MA, Bargar JR (2010) Kinetics of Fe(II)-catalyzed transformation of 6-line ferrihydrite under anaerobic flow conditions. Environ Sci Technol 44:5469–5475

Zhu X-K, O'Nions RK, Guo Y, Reynolds BC (2000) Secular variation of iron isotopes in North Atlantic deep water. Science 287:2000–2002

2 Analytical Methods

2.1 Introduction

Iron has four naturally occurring stable isotopes: ^{54}Fe (5.85%), ^{56}Fe (91.75%), ^{57}Fe (2.12%) and ^{58}Fe (0.28%) and its isotopic composition has been of interest since the invention of isotope ratio mass spectrometry. For example, Fe isotopes have been used to evaluate if extra-terrestrial Fe has the same isotope composition of terrestrial Fe (Valley and Anderson 1947), how Fe is cycled in the human body (Tang and Trassy 1986; Walczyk 1997), and the extent to which Fe is fractionated by geological processes (Beard and Johnson 1999; Bullen et al. 2001; Dixon et al. 1992). In this chapter, we discuss how Fe isotope analyses are conducted, with an emphasis on the methods used to ensure accurate Fe isotope analysis. The biggest analytical challenges for analysis of Fe that preserves naturally occurring mass-dependent isotopic fractionation include (1) correcting for instrumental mass bias, and (2) resolving Fe isotopes from isobaric interferences that can include elemental isobars such as ^{58}Ni on ^{58}Fe, or molecular isobars such as $^{40}Ar^{16}O$ on ^{56}Fe.

Prior to the invention of Ar plasma ionization mass spectrometers, thermal ionization mass spectrometry (TIMS) was used for Fe isotope analysis. TIMS analysis is challenging because of the poor thermal ionization efficiency of Fe, and because the instrumental mass bias continuously changes during an analysis as, for example, ^{54}Fe is more efficiently vaporized from the filament as compared to ^{56}Fe, such that during a typical analysis an initially low $^{56}Fe/^{54}Fe$ ratio becomes progressively higher as the analysis progresses. However, the magnitude of this mass bias is relatively small (typically less than 1% for the $^{56}Fe/^{54}Fe$ ratio) and follows an exponential or Rayleigh fractionation mass-bias law (Beard and Johnson 1999), making it possible to accurately correct for instrumental mass bias using a double-spike technique (Johnson and Beard 1999). In contrast, plasma ionization mass spectrometry is highly efficient at ionizing Fe, and although instrumental mass bias is large ($\sim$5–10% for the $^{56}Fe/^{54}Fe$ ratio), it is relatively constant because an isotopically homogenous liquid is aspirated into the plasma ionization source, eliminating the continuous change in instrumental mass bias associated with Fe distillation from a filament, as in TIMS analysis. This constant mass bias allows one to correct for instrumental mass bias by comparison to standards run at the same conditions.

To ensure that mass bias of samples and standards are the same and do not suffer from matrix effects (Sect. 2.3.4), Fe from a sample is purified using ion-exchange chromatography so that Fe from the sample has the same matrix as an ultrapure Fe standard. Additionally, this purification must be quantitative to avoid laboratory induced mass dependent fractionation (Sect. 2.2) associated with loss of Fe from the ion exchange columns. However, plasma ionization presents a major challenge to Fe isotope analysis because all Fe isotopes have interferences created by Ar molecular isobars that are formed in the Ar plasma. These argide

C. Johnson et al., *Iron Geochemistry: An Isotopic Perspective*, Advances in Isotope Geochemistry,
https://doi.org/10.1007/978-3-030-33828-2_2

interferences require special analysis procedures to ensure accurate Fe isotope analysis, as discussed in Sects. 2.3.2 and 2.3.3. Despite these challenges, there is excellent agreement among laboratories for the Fe isotope analysis of different standards, indicating that inter-laboratory comparisons are not an issue. Moreover, as the sensitivity and mass-resolution capabilities of mass spectrometers have improved, the precision of Fe isotope analysis has improved significantly with smaller amounts of Fe being used for a single analysis. Indeed, we are now able to make in situ Fe isotope analyses by laser ablation (LA) multi-collector inductively coupled plasma mass spectrometry (MC-ICP-MS) or by secondary ionization mass spectrometry (SIMS) (Sect. 2.4), allowing exploration of Fe isotope variations at the micron scale.

2.2 Iron Purification Methods

High-precision Fe isotope ratio measurements of geological materials typically require dissolution using mineral acids, separation of Fe from the rest of the matrix by ion-exchange chromatography, and analysis of the purified Fe using a magnetic-sector multi-collector mass spectrometer. Chemical purification is important because it allows removal of elemental isobars that include ^{54}Cr on ^{54}Fe and ^{58}Ni on ^{58}Fe. These elemental isobars are too close in mass to be resolved using modern high or pseudo-high mass resolution mass spectrometry methods (a resolving power of >74,000 would be required; Table 2.1). Additionally, a high concentration of matrix elements in the sample tends to decrease the Fe ionization efficiency and causes problems when correcting for instrumental mass bias (see Sect. 2.3.4). Most commonly, Fe is separated using anion exchange resin by taking advantage of the fact that ferric iron readily forms an anion chloro-complex $(FeCl_4)^-$ in high molarity HCl, whereas most other major-element cations do not form anion complexes. Thus, when the sample is loaded in high molarity HCl onto columns with anion exchange resin, $(FeCl_4)^-$ adsorbs onto the resin. Subsequent washing steps with high molarity HCl removes major-element cations, after which the Fe is quantitatively removed from the anion exchange resin using low molarity HCl (Strelow 1980) or with H_2O and HNO_3 (Schoenberg and von Blanckenburg 2005).

The distribution coefficient (a measure of how strongly an element is retained on ion-exchange resin as compared to the amount of the element in the eluting liquid) of ferric iron is positively correlated with HCl molarity (Fritz 2004; Kraus and Nelson 1956; Strelow 1980; Van der Walt et al. 1985a, b), and thus the higher the molarity of HCl the more strongly Fe is retained on the resin. In practice, however, concentrated HCl is not ideal because its density is greater than anion-exchange resin, which causes the resin to float. Therefore, HCl molarities that are appropriate for this step are typically up to $\sim$8 M. A variety of anion-exchange resins are suitable, including macroporous resins such as AG MP 1 and the analytical grade anion-exchange resins with cross linkages of 4 or greater (e.g., AG 1 $\times$ 4). In general, lower cross linkage resins are preferable because they promote sharper eluted element profiles, minimizing crossover of contaminants during the Fe elution step (Van der Walt et al. 1985a, b). However, resin type is

Table 2.1 Most common Fe isotope isobars

	^{54}Fe	^{56}Fe	^{57}Fe	^{58}Fe
Elemental isobars	^{54}Cr (74,032)			^{58}Ni (123,851)
Argide isobars	$^{40}Ar^{14}N$ (2,087)	$^{40}Ar^{16}O$ (2,501)	$^{40}Ar^{16}OH$ (1,915)	

The number in parentheses shows the resolving power required for mass analysis. Resolving power is defined as $\Delta m/m$; where Δm is the mass difference between the isobar and the Fe isotope of interest and m is the mass of the Fe isotope of interest. Most mass spectrometers are not able to resolve elemental isobars so one typically monitors for the ^{54}Cr and ^{58}Ni isobars using ^{52}Cr or ^{53}Cr and ^{60}Ni. If the interference is significant, one can correct for these by assuming a Cr or Ni isotope composition typically chosen based on what elements are in the matrix material, and if other elements besides Fe will be separated for isotopic analysis. The choice of resin and acid molarity is made by considering elemental distribution coefficients (K_d) that have been determined for a variety of anion exchange resins and elutants (e.g., Fritz 2004; Kraus and Nelson 1956). Additionally a number of methods have been developed to separate different metals from one another, making it possible to analyze a wide number of metal isotope ratios (e.g., Cu–Fe–Zn) from a single dissolution of a sample (e.g., Borrok et al. 2007; Yamakawa et al. 2009).

The separation of Fe from chemically complex samples needs to be quantitative because Fe loss during ion-exchange chromatography can result in mass-dependent isotope fractionation. For example, at equilibrium, there is a −0.3‰ fractionation in $^{56}Fe/^{54}Fe$ between the aqueous Fe complexes of $[FeCl_4]^-$ and a mixture of $[FeCl_2(H_2O)_4^+]$ and $[FeCl_3(H_2O)_3]$ (Hill et al. 2009) at 3 M HCl. Because the isotopic fractionation between these different ferric iron chloro-complexes is small at high chloride concentrations, substantial loss of Fe would be required to significantly bias the sample. For example, if 5% of the Fe is lost as a mixed $[FeCl_3(H_2O)_3]$–$[FeCl_2(H_2O)_4^+]$ complex and the remaining Fe is retained on the resin as $FeCl_4^-$, the sample would be biased by 0.015‰ in $^{56}Fe/^{54}Fe$. Larger fractionations (~3‰) are possible if a moderate or low molarity HCl (e.g., 2 M) is used, as shown by Anbar et al. (2000) and Roe et al. (2003). In contrast, if Fe is lost as a result of the presence of ferrous iron (Fe^{2+} does not form an anion chloro-complex) the isotopic composition of the sample could be more strongly biased because the fractionation between ferrous and ferric iron is an order of magnitude larger as compared to the fractionation between different ferric iron chloro-complexes (e.g., Welch et al. 2003). It is, therefore important to convert all the Fe in the sample to ferric iron prior to anion exchange chromatography.

Some of the initial ion-exchange chromatography studies noted that Fe can be auto-reduced on resin with low cross linkages, which could result in loss of any ferrous iron that is produced (Van der Walt et al. 1985b). Reduction of Fe on anion-exchange resin can be minimized by using a mixture of HCl and 0.001% H_2O_2 in order to keep all Fe as a ferric species (Van der Walt et al. 1985b). In practice, Fe is quantitatively oxidized in most samples during the dissolution process because oxidizing mineral acids are evaporated to dryness on a hot plate in atmosphere, and except for ion exchange resin, there is typically not any material in a geologic specimen that will reduce Fe once it is converted to ferric iron in the initial dissolution procedure. For samples with a high amount of organic material relative to Fe, care must be exercised to destroy organic ligands that can strongly bond with Fe creating low ion-exchange yields and Fe isotope fractionation.

Problematic samples, such as seawater, require specially developed methods due to low Fe concentration and high total dissolved solids. Appropriate methods have only recently been perfected (Conway et al. 2013; de Jong et al. 2007; John and Adkins 2010; Lacan et al. 2008), and typically require an initial pre-concentration step using a chelating resin, followed by purification using anion-exchange resin. Additionally, for these difficult, low-level samples (10's of nanograms of Fe per analysis), workers use a double Fe spike technique where an enriched isotope mixture composed of two Fe isotopes is added to the sample and allowed to equilibrate (Beard and Johnson 1999; Johnson and Beard 1999). The double-spike method permits determination of the primary Fe isotope composition of the sample, even if Fe loss occurs during Fe separation procedures.

If Fe isotope analysis is not conducted using the double-spike approach, ion-exchange column yields are measured to ensure quantitative recovery of Fe. This typically involves either analysis of Fe concentration using instrumentation such as ICP-MS or ICP-OES, or UV/Vis spectrophotometric methods. The ICP methods tend to have lower detection limits and better precision, while UV/Vis spectrophotometric techniques allow measurement of different Fe species. For example, the ferrozine method is based on the formation of a purple ferrous iron complex; where absorption at 562 nm is positively correlated with the ferrous iron concentration (Stookey 1970). Total Fe concentrations can then be determined by adding a reducing agent, such as hydrochloric hydroxylamine, to reduce all iron to ferrous iron and the concentration of ferric iron is determined by difference.

2.3 Mass Spectrometry

2.3.1 Beginnings of Fe Isotope Analysis

Prior to the wide adoption of multi-collector inductively-coupled-plasma mass-spectrometry (MC-ICP-MS) in the early 2000s, Fe isotope ratios were analyzed using numerous techniques. The first Fe isotope study was done in 1939 to determine the isotope composition of iron metal (Nier 1939). In 1947, a more complete study was conducted to evaluate the Fe isotope composition of terrestrial samples as compared to the Fe isotope composition of meteorites (Valley and Anderson 1947). These initial Fe isotope measurements were made by TIMS and no resolvable Fe isotope differences between terrestrial and extraterrestrial Fe could be detected outside of the analytical uncertainty of $\pm$10‰ in $^{56}Fe/^{54}Fe$ (1σ). Additional early Fe isotope investigations were conducted by both TIMS and SIMS to evaluate if there was excess ^{57}Fe or ^{58}Fe in extraterrestrial materials. These studies achieved a high-level of precision on $^{57}Fe/^{56}Fe$ and $^{58}Fe/^{56}Fe$ measurements ($\pm$0.3‰ and 1.5‰, respectively for measurements made by TIMS) by internally normalizing the data to a constant $^{56}Fe/^{54}Fe$ ratio to correct for instrumental mass bias. This technique allowed identification of small deviations in ^{58}Fe abundances in some small inclusions found in calcium aluminum inclusions (CAI) from chondritic meteorites (Hutcheon et al. 1987; Völkening and Papanastassiou 1989). This excess ^{58}Fe provided evidence of injection of extra solar material prior to solar system condensation (see Sect. 4.1 in Chap. 4 for a detailed discussion of extraterrestrial Fe isotope variations). Other analyses of micrometeorites in the form of type I deep sea spherules showed that Fe isotope variations were mass dependent but defined a range of 40‰ in $^{56}Fe/^{54}Fe$, reflecting vaporization processes as the micrometeorite entered the Earth's atmosphere (Herzog et al. 1999). Importantly, for all of these measurements, elemental isobars from ^{54}Cr and ^{58}Ni were monitored and corrected for by monitoring ^{52}Cr or ^{53}Cr and ^{60}Ni and assuming a $^{54}Cr/^{53}Cr$ or a $^{58}Ni/^{60}Ni$ ratio. The practices of monitoring and correcting for these interferences is followed by all other workers.

Additional, novel methods were developed to study Fe isotopes in biological systems using a variety of mass spectrometry techniques (Flory et al. 1993; Gotz and Heumann 1988; Stenberg et al. 2003; Ting and Janghorbani 1986; Turnlund and Keyes 1990; Walczyk 1997). These methods, however, were unable to correct for instrumental mass bias, limiting the precision to 1–10‰ in $^{56}Fe/^{54}Fe$.

The need to correct for instrumental mass bias and preserve naturally occurring, mass-dependent isotope fractionation led workers to evaluate methods to produce Fe species, such as $Fe(PF)_3$, with high vapor pressures that allowed analysis of Fe isotope compositions using a dual-inlet gas-source mass spectrometer (Taylor et al. 1993b), and correction for instrumental mass bias by comparing to an $Fe(PF)_3$ standard. Moreover, workers also gravimetrically prepared Fe isotope standards with a wide range of $^{56}Fe/^{54}Fe$ ratios from pure ^{56}Fe and ^{54}Fe spikes making it possible to analyze these standards by TIMS and develop empirically derived instrumental mass-bias corrections. These standards were then applied to the analysis of a sample of Fe metal in order to more accurately determine its Fe isotope composition (Taylor et al. 1993a). This methodology was applied in certifying the Fe isotope composition of the IRMM-014 standard (Taylor et al. 1992), and is similar to methods used for common TIMS Pb isotope analysis in which an instrumental mass-bias correction for samples is estimated by the analysis of NIST SRM-981 and SRM-982 standards. This empirically derived instrumental mass-bias correction method was applied to natural Fe samples to identify that some banded iron formations and Fe from microbial mats had Fe isotope compositions different from that of a Fe metal standard (Dixon et al. 1992, 1993). Moreover, this group reported that ^{57}Fe abundance for one sample was not mass-dependent relative to $^{56}Fe/^{54}Fe$ (Dixon et al. 1992). This methodology had an external reproducibility of ~1‰ in $^{56}Fe/^{54}Fe$.

In order to improve the precision of TIMS-based Fe isotope analysis, a double-spike method was developed to allow for a robust correction of instrumental mass bias (Beard and Johnson 1999; Johnson and Beard 1999). This double-spike method improved Fe isotope ratio measurements to a 2-standard deviation (2-SD) precision of 0.6‰ in $^{56}Fe/^{54}Fe$. The geologic samples analyzed using this TIMS double-spike method had mass-dependent variations in ^{54}Fe–^{56}Fe–^{57}Fe–^{58}Fe. Overall, this TIMS double-spike method was able to show that:

1. terrestrial and lunar rocks had overlapping Fe isotope composition (Beard and Johnson 1999),
2. low temperature geologic specimens, such as banded iron formations, had variable Fe isotope compositions (Beard and Johnson 1999),
3. that there was a 1.3‰ fractionation during the rapid precipitation of hematite from an aqueous Fe^{3+} solution (Skulan et al. 2002),
4. ferrous iron produced by microbial dissimilatory iron reduction had a lower $^{56}Fe/^{54}Fe$ ratio compared to the ferrihydrite that the microbes reduced (Beard et al. 1999),
5. magnetotactic bacteria under Fe-limited conditions did not produce an Fe isotope fractionation between the grown magnetosomes and aqueous Fe (Mandernack et al. 1999) and,
6. oxidized Fe produced by exposure of anoxic, Fe-rich groundwater to the atmosphere produced ferric iron that had a $^{56}Fe/^{54}Fe$ ratio greater than the aqueous ferrous iron (Bullen et al. 2001).

These first TIMS double-spike-based studies, coupled with new theoretical predictions of Fe isotope fractionation factors (Polyakov and Mineev 2000; Schauble et al. 2001) stimulated numerous other workers in the early 2000's to begin conducting Fe isotope studies using newly available MC-ICP-MS instruments.

2.3.2 Multi Collector Inductively Coupled Plasma Mass Spectrometry (MC-ICP-MS)

Today, the most common method for Fe isotope analysis is by MC-ICP-MS. This instrumentation benefits from a highly efficient ionization source coupled with a stable instrumental mass bias during the course of a sample analysis, as compared to TIMS, where the mass bias of the sample continuously changes as the sample is evaporated from the filament. The mass-bias stability of MC-ICP-MS allows for correction of instrumental mass fractionation by a sample-standard bracketing correction. This is similar to the method used by dual-inlet gas-source mass spectrometers in which δ values are calculated on a sample gas relative to a standard gas. Although the Ar plasma is an efficient ionization source, this ionization produces several molecular argon species that are isobaric with Fe isotopes. The most common argides include $^{40}Ar^{14}N$ which is isobaric on ^{54}Fe, $^{40}Ar^{16}O$ on ^{56}Fe and $^{40}Ar^{16}OH$ on ^{57}Fe. The $^{40}Ar^{16}O$ isobar at ^{56}Fe requires the highest resolving power at 2,500 (Table 2.1). This resolving power was greater than what could be achieved for multi-isotope collection in the first generation of commercially available MC-ICP-MS that included Nu Instruments *Nu Plasma*, the VG *Axiom*, and the Micromass *IsoProbe*.

The first MC-ICP-MS study (Belshaw et al. 2000) only reported $^{57}Fe/^{54}Fe$ measurements because it was not able to limit ArO from the mass spectrum. The formation of ArN in their study was minimized by using HCl as a diluting acid instead of HNO_3, and oxides (ArO and ArOH) were suppressed by using a desolvating nebulizer. The small argide isobars that occurred on ^{54}Fe and ^{57}Fe were corrected by using an on-peak-zero (OPZ) subtraction method in which HCl without Fe was inlet into the instrument to allow correction for the small amounts of $^{40}Ar^{14}N$ and $^{40}Ar^{17}O$ or $^{40}Ar^{16}OH$ that were

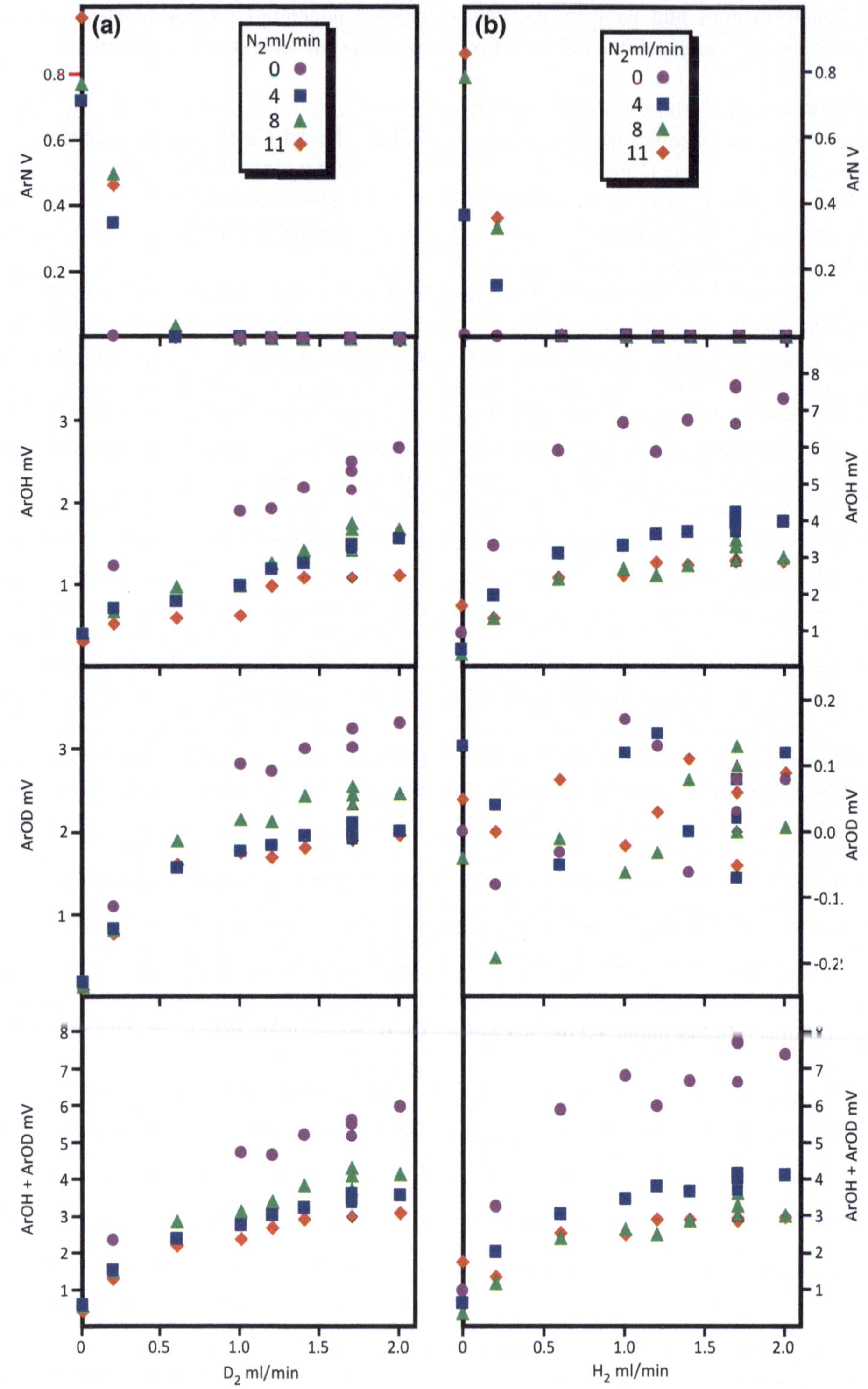
(a)
(b)
N2ml/min
0
4
8
11
ArN V
ArOH mV
ArOD mV
ArOH + ArOD mV
D2 ml/min
H2 ml/min
0
0.5
1.0
1.5
2.0

◀ **Fig. 2.1** Plot of argide isobars at masses 54, 57, and 58 as a function of N_2 injector gas flow rates and D_2 (**a**) or H_2 (**b**) collision gas flow rates. Addition of N_2 increases the $^{40}Ar^{14}N$ isobar at ^{54}Fe but addition of H_2 or D_2 can minimize production of $^{40}Ar^{14}N$. The argon hydroxide ($^{40}Ar^{16}OH$ or $^{40}Ar^{16}OD$) increases as a function of addition of H_2 or D_2, but production of the argon hydroxide can be minimized by addition of N_2 into the injector. Production of $^{40}Ar^{16}OH$ with D_2 in the collision cell implies that not all argon hydroxide is produced by reactions involving collision gas. For these tests, a Micromass *Isoprobe* was used with sample introduction using a Cetac *Aridus 2* desolvating nebulizer with a PFA concentric flow self-aspirating nebulizer. The instrument was operated with a resolving power of ~4,500 that was achieved by closing the defining slit aperture to 10% ion transmission and narrowing the entrance slit to the central axial Faraday collector, allowing full resolution of the argide isobars from the Fe mass spectrum. Each test was done by setting the N_2 flow of the *Aridus* to a set value and then tuning the instrument to maximum Fe sensitivity (the biggest change in tuning was a decrease in sweep gas with increasing N_2 flow rate). After establishing the best tuning conditions, a mass scan was done by aspiration of a 25 ppb Fe solution into the mass spectrometer and recording the ion intensity at ^{54}Fe, $^{40}Ar^{14}N$, ^{57}Fe, $^{40}Ar^{16}OH$, $^{58}Fe/^{58}Ni$, and $^{40}Ar^{16}OD$ (nickel was derived from the Ni sampler, skimmer, and extractor cones) over a range of H_2/D_2 collision gas flow rates. Additionally, a baseline was recorded at a mass 0.05 amu higher than the $^{40}Ar^{14}N$, $^{40}Ar^{16}OH$, and $^{40}Ar^{16}OD$ masses, which was subtracted from the ion intensities measured at nominal masses of 54, 57, and 58, respectively. Based on these findings, past Fe isotope work done at the University of Wisconsin was done using a H_2 collision gas with a flow rate of 1.8 to 2.2 ml/min and an N_2 flow rate of 8–11 ml/min with the analyte in a 0.1% HNO_3 matrix (Beard et al. 2003a, b). Although use of D_2 would limit production of $^{40}Ar^{16}OH$ it was not used because OPZ was sufficient to correct for the amount of $^{40}Ar^{16}OH$ produced, and D_2 is considerably more expensive as compared to H_2

present at masses 54 and 57, respectively. Other workers developed cool plasma techniques to limit production of argides coupled with OPZ subtraction to make Fe isotope measurements (Kehm et al. 2003). Additionally, instruments like the Micromass *IsoProbe* were equipped with a reaction cell, allowing workers to use a reactive gas, such as H_2, to significantly limit the production of argide isobars (Vogl et al. 2003). In practice, H_2, as a collision gas, reduces ArN and ArO signals to levels of intensity so they can be robustly subtracted from an ion signal using the OPZ subtraction method. However, use of H_2 collision gas produces $^{40}Ar^{16}OH$ which makes ^{57}Fe measurements more challenging. High levels of ArOH can be mitigated by adding N_2 to the plasma gases to limit oxide production, and increase instrument sensitivity in some cases (e.g., Lam and Horlick 1990). Of course, addition of N_2 promotes ArN production, which in turn requires high levels of H_2 collision gas to suppress (Fig. 2.1).

The production of argide isobars in reaction-cell based MC-ICP-MS has been explored using either H_2 or D_2 as a collision gas with N_2 addition to the injector using a Cetac *Aridus* desolvation nebulizer system. These tests are particularly germane because the next generation of MC-ICP-MS being developed (Nu Instruments *Sapphire*, and Thermo Fisher *Proteus*) includes a collision cell. Use of H_2 and D_2 allows evaluation of the formation of argon hydroxide (e.g., ArOH or ArOD) caused by reactions with the collision gas as compared to formation of argon hydroxide not associated with the collision gas. For example, if all argon hydroxide was produced by reactions with the collision gas, then introduction of D_2 instead of H_2 would produce an isobar at ^{58}Fe as $^{40}Ar^{16}OD$. The results of these tests, done at UW-Madison, are shown in Fig. 2.1, which plots the measured ion intensity after baseline subtraction of the Fe and argide peaks as a function of collision gas setting and N_2 injector gas flow rate. Addition of any N_2 to the injector produces a significant increase in the $^{40}Ar^{14}N$ production that requires ~1 ml/min H_2 or D_2 to minimize. At a given N_2 flow rate, a positive correlation exists between H_2/D_2 flow rate and the production of $^{40}Ar^{16}OH/D$, but the production of this isobar is minimized by using high N_2 flow rates. Overall, the total amount of $^{40}Ar^{16}OH$ produced under different H_2 and N_2 flow conditions is similar to the sum of $^{40}Ar^{16}OH$ and $^{40}Ar^{16}OD$ produced under different D_2 and N_2 flow conditions. Based on the D_2 collision gas tests, approximately 1/3 of the total $^{40}Ar^{16}OH/D$ is produced as $^{40}Ar^{16}OH$ and 2/3 as $^{40}Ar^{16}OD$, implying that not all argon hydroxide is a product of reactions with the collision gas. Approximately 1/3 is produced by other sources of H_2, such as the weak acid matrix of the solution aspirated into the instrument.

2.3.3 Modern MC-ICP-MS Using Pseudo High Mass Resolution Methods

The techniques that minimized the effects of argide isobars for the first generation MC-ICP-MS instruments, although successful, were not as robust as the method developed for the 2nd generation of MC-ICP-MS, such as the Thermo *Neptune/Neptune Plus* and the Nu Instruments *Plasma500/Plasma2/Plasma3*. These newer instruments used a technique frequently described as "pseudo high mass resolution". For this technique, the beam width is reduced by using a narrow defining slit (Fig. 2.2). The width of the collector slit is left constant (i.e., suitable for low mass resolution analyses). Interferences can be resolved by having the ion beam enter into the Faraday collector and only capturing the Fe signal, cutting off any contribution from argide mass interferences by adjusting the position of the Faraday collector or the quadratic "zoom" lenses (Weyer and Schwieters 2003) (Fig. 2.2c). A resolving power of m/Δm of >10,000 at the focal plane of the detector array can be achieved with this technique. As common mass interferences on Fe isotopes require only a resolution of ~2,500, this technique results in flat-top peak shoulders on which Fe isotopes can be reliably analyzed at high precision. A disadvantage of this technique is that, in multi-collection mode, parallel resolution of all interfering species is only possible if they appear at the same side of the ions of interest. This is true for all significant mass interferences on Fe isotopes, as well as for all molecular interferences on other first-row transition- or lighter metals (double-charged ions are exceptions that may cause problems in some cases). The reason is the decrease in relative nuclear binding energy of the elements until an atomic number of ~40, resulting in a generally lower mass of ions (e.g., ^{56}Fe) compared to molecular species of the same nominal mass (e.g., ^{40}Ar^{16}O). Another disadvantage of running at pseudo-high mass resolution with a narrow defining slit is a significant decrease in ion transmission, typically only ~10% of the ion is transmitted through the narrow slit. On the other hand, the physical elimination of the mass interferences allows

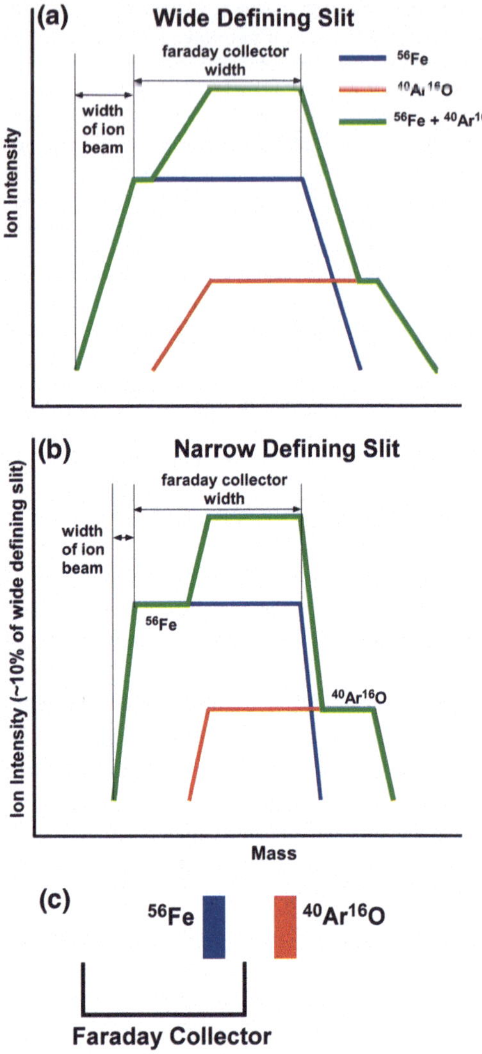

Fig. 2.2 Schematic plots of ion intensity vs mass for the Fe mass spectrum using a wide defining slit (**a**) and narrow defining slit (**b**), following the work of Weyer and Schwieters (2003). The complex peak shapes include ^{56}Fe (blue), ^{40}Ar^{16}O (red), and the green trace shows the combination of the Fe and argide mass spectrum. Use of a wide defining slit results in a wide ion beam that does not produce significant resolution between ^{56}Fe and ^{40}Ar^{16}O. Use of a narrow defining slit produces a flat- topped portion of the ^{56}Fe mass spectrum that is free of ^{40}Ar^{16}O. The pseudo high mass resolution technique allows the instrument to be configured so that only the flatted topped peak is allowed to enter into the faraday collector as shown in **c**. Note that because the atomic mass difference between ^{56}Fe and ^{40}Ar^{16}O is only 0.022 amu, the magnet and accelerating potential of the instrument must be very stable. Moreover, use of a narrow defining slit decreases the ion signal to approximately 10$ of the wide slit ion signal

optimization of all other instrument parameters to enhance ion transmission. Furthermore, these 2nd and 3rd generation instruments improved overall instrument sensitivity through the use of higher capacity vacuum pumps on the expansion chamber, and carefully designed cones to increase overall transmission. Indeed, these new instruments allowed workers to evaluate matrix effects that had a pronounced effect on the accuracy of Fe isotope ratio measurements (e.g., Dauphas et al. 2009; Schoenberg and von Blanckenburg 2005).

2.3.4 Matrix Effects and Instrumental Mass Fractionation Corrections

Instrumental mass fractionation in the ICP source is largely driven by collision between ions and Ar neutral atoms as the plasma is expanded through the sampler cone, and by space-charge effects caused after expansion of the plasma through the skimmer cone, resulting in the plasma losing charge neutrality (Douglas and Tanner 1998; Gillson et al. 1988). Collision effects result in preferential loss of light isotopes through ion scattering because light ions have a lower relative kinetic energy. The loss in charge neutrality results in the positive ions repelling one another, which leads to the lightest ions being preferentially lost relative to heavier ions. Thus, in all ICP-MS the measured analyte always has a "heavier" isotopic composition compared to its true isotopic composition for ratios where the numerator is the heavier mass. This is shown by studies that analyzed the isotopic composition of material plated onto sampler and skimmer cones during analysis; the material plated on the cones had a lighter isotopic composition compared to the material let into the mass spectrometer (Andren et al. 2004). In general, mass bias and sensitivity are a function of the atomic Z of the analyte, where mass bias is negatively correlated and sensitivity is positively correlated. For example, the heaviest elements tend to have the lowest amount of mass bias and highest sensitivity (as expressed in molar concentration) compared to lower atomic Z elements that display more mass bias and lower sensitivity. Overall, the magnitude of the mass bias in terms of percent difference from the true isotopic composition, normalized to mass difference, is a few percent for U and 5–10% for Fe depending on instrumentation type and configuration. An implication of these space-charge effects is that the ion beam is zoned isotopically with a higher proportion of heavy ions in the central section. Indeed, this zonation leads to the observation that the mass bias of an ion beam passing through a narrow defining slit has a heavier isotopic composition compared to an ion beam passing through a normal wider-sized slit. For example, Mg is an element that can be analyzed in either high- or low mass resolution modes because isobaric molecules on Mg are small and can be subtracted using the on-peak zero subtraction method. Using a Nu Instruments *NP2* coupled with a Cetac *Aridus 2* desolvating nebulizer, the Mg isotope composition of a solution was analyzed at both high- and low-mass resolution modes (Fig. 2.3). The mass bias is greater at high mass resolution. Moreover, instrumental stability is better at low mass resolution where bracketing standards define a narrower range in delta values as compared to those analyzed at high-mass resolution.

Not only does the expansion of the plasma through the expansion chamber result in an isotopically-zoned ion beam, it also results in non-spectral interferences or matrix effects (Albarède and Beard 2004). These happen because the mass bias of an analyte is a function of the composition of the analyte solution. One of the most prominent matrix effects is the change in instrument mass bias with analyte concentration (Fig. 2.4). Notably, the magnitude of this concentration matrix effect is typically less pronounced if the sample is introduced as a wet plasma as compared to aerosols produced using a desolvating nebulizer system (Fig. 2.4). The actual mechanism for moderation of matrix effects by water is uncertain. It may be a result of more consistent plasma loading for wet aerosols as compared to dry, or a result of changes in electron density of dry and wet plasmas sampled behind the skimmer cone as shown by Langmuir probe measurements (Alder et al. 1980; Browner and Long 1989; Caughlin and Blades 1987; Long and Browner 1988; Niu and Houk 1994; Niu et al. 1991; Ting and Janghorbani

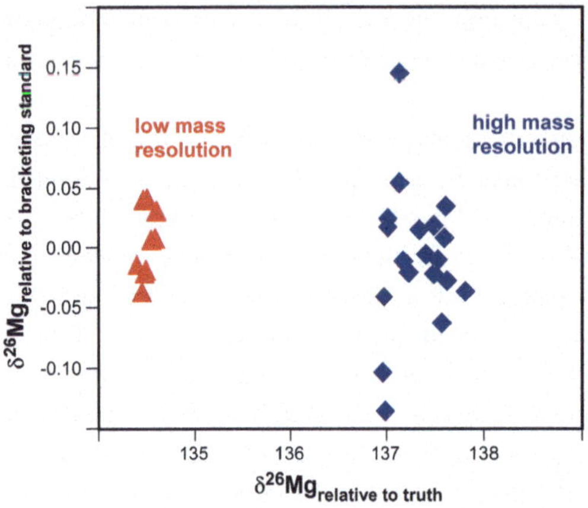

Fig. 2.3 Plot of the $\delta^{26}Mg/^{24}Mg$ ($\delta^{26}Mg$) relative to a bracketing standard (y-axis) versus the $\delta^{26}Mg$ relative to the true isotope composition of Mg (x-axis) for a Mg solution run under low- and high mass resolution (low mass resolution uses a 0.30 mm wide slit and the high mass resolution uses a 0.05 mm wide slit). The $\delta^{26}Mg$ value relative to truth is a measure of the instrumental mass bias whereas the $\delta^{26}Mg$ relative to the bracketing standard is a measure of the mass bias stability of the instrument. Mass bias is smaller and stability is better for low mass resolution settings, which can be explained by the isotopic zonation of an ion beam as it passes through the transfer optics of a MC-ICP-MS. Data were collected using a Nu Instruments NP2 housed at the University of Wisconsin-Madison. Solutions were aspirated using a Cetac Aridus 2 desolvating nebulizer. The same isotopic composition Mg solution was used for both the low- and high-mass resolution tests, but to keep the ion intensity similar (total ion intensity ~22 V) in the low and high mass resolution tests, a 0.1 ppm and a 1 ppm solution, respectively, were used

1986; Zhu and Browner 1988). In addition to this concentration matrix effect, the presence of other elements can affect mass bias. In general, matrix effects are greater for matrix elements that are higher in mass as compared to the analyte, and for matrix elements with low ionization potentials (Douglas and Tanner 1998; Gillson et al. 1988).

Matrix effects can lead to inaccurate Fe isotope results and considerable effort has been devoted to evaluate how matrix effects may impact the accuracy of Fe isotope measurements. It is important to note that matrix effects cannot generally be detected by deviation in mass-dependent relations among Fe isotope ratios. For most samples, ion exchange chromatography is sufficient to remove matrix elements. Additionally, matched bracketing spike to use has been explored for most of the periodic table, including a useful computer program

standards and sample Fe concentrations should minimize differential concentration matrix effects. To validate that these procedures yield accurate Fe isotope measurements, one typically prepares test solutions that mimic the chemical composition of a sample (or analyze the ion-exchange sample wash from a sample that had its Fe removed) with Fe of a known isotopic composition. After purifying Fe from these test samples by ion-exchange chromatography, a purified Fe isotope composition that matches the unadulterated Fe validates the accuracy of the methodology.

In addition to performing this validation, workers have also developed methods to improve instrumental mass-bias corrections. The double-spike method is the most robust, but requires accurate measurement of the abundance of ^{58}Fe, which is affected by an isobaric interference from ^{58}Ni (Finlayson et al. 2015). The ^{58}Ni isobar can be from Ni not removed from the sample during chemical purification or from sampler and skimmer cones if these are made of Ni. The double-spike method will work for any element that has four or more naturally occurring isotopes and for which enriched isotope tracers are available for at least two of the isotopes. A complete double-spike analysis consists of two analyses: one of the unspiked sample and one of the sample with a double spike added to it. Each analysis defines a mass-bias curve on a three-isotope ratio plot which has a common denominator. One can solve for the true isotopic composition by noting that the isotopic composition of the sample, spiked sample, and double spike define a straight line on a three-isotope ratio plot. Assuming an adequate mass-bias law (e.g. exponential law) and a measured isotopic composition of the double spike, one can explicitly calculate the true isotopic composition of the sample. Note that if the isotopic composition of the double spike is not perfectly known, this will not affect relative delta values if the same spike is used for the sample and standard. In practice, because natural Fe is rarely affected by non-mass dependent fractionation, one can assume that Fe exclusively follows a single fractionation curve, and thus only the analysis of the spiked sample is needed to infer the true isotope composition of sample. The choice of the best double to calculate isotope compositions using the double-spike method (Rudge et al. 2009).

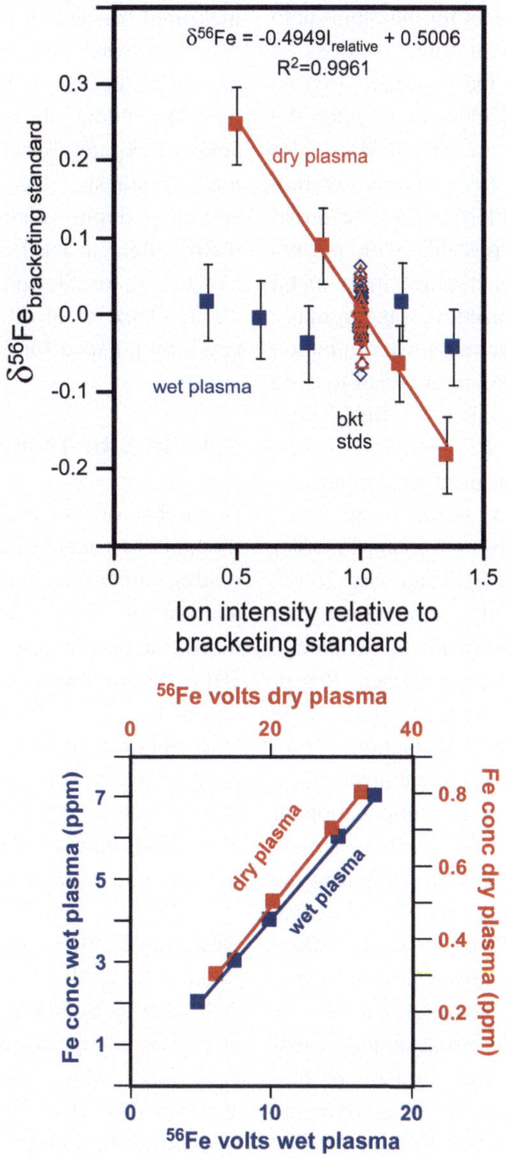

Fig. 2.4 Plot of the δ^{56}Fe normalized to a bracketing standard (y axis) versus the ion intensity relative to the bracketing standard of different Fe concentration solutions of the same Fe isotope composition. Iron isotope compositions were measured using a Nu Instruments NP2 housed at the University of Wisconsin-Madison, run is pseudo high mass resolution mode to resolve argides from the Fe mass spectrum ($\sim$10% ion transmission). The dry plasma solutions were inlet into the instrument using a Cetac *Aridus 2* desolvating nebulizer using a self-aspirating Savillex concentric flow nebulizer with an uptake rate of 50 μl/min. The wet plasma data were collected using a Glass Expansion Micromist nebulizer aspirating at $\sim$80 μl/min and a Glass Expansion Twisty spray chamber cooled to 7 °C. The δ^{56}Fe values of the variable concentration solutions are relative to the bracketing standard run immediately before and after the variable concentration solution, whereas the δ^{56}Fe values of the bracketing standards reflect a longer time extrapolation because variable concentration solutions were run in between every bracketing standard. Inset shows the Fe concentration (y-axis) of the wet (left) and dry (right) plasma conditions versus the ^{56}Fe voltage (x-axis) for the wet (bottom) and dry (top) plasma conditions. Dry plasma conditions result in much higher sensitivities and this is reflected in the fact that for dry plasma the bracketing standard was 600 ppb Fe and for wet plasma it was 5 ppm Fe

A second method to correct for mass bias is to add an element with two or more isotopes of similar atomic Z, and use the measured isotope composition of the added element to infer the instrumental mass bias for the analyte (Devulder et al. 2013; Marechal et al. 1999). This correction technique, sometimes referred as "external mass-bias correction," is possible with plasma source mass spectrometers because most metal elements are efficiently ionized. It was originally developed to correct for the mass discrimination of Pb isotopes using a Tl solution added to a Pb solution (e.g., Longerich et al. 1987; Rehkämper and Halliday 1998). For Fe isotope analyses, either Ni or Cu have been used as doping elements (Arnold et al. 2004; Kehm et al. 2003; Malinovsky et al. 2003; Poitrasson and Freydier 2005; Schoenberg and von Blanckenburg 2005). This methodology is slightly more robust than use of a sample-standard bracketing technique alone, assuming that the doping element (Cu or Ni) and the analyte (Fe) behave similarly to matrix-induced mass-bias fluctuation. Frequently, the doping method is combined with a standard bracketing method. A complication of this method is that analyte and doping element may display different mass discrimination in the plasma, as observed for Cu and Zn (Marechal et al. 1999). Similarly, Weyer et al. (2005) observed a slightly different mass discrimination for Fe and Cu. This offset, however, can also be corrected by sample standard bracketing. A disadvantage of using Cu to correct for Fe mass bias is that peak hopping, also referred to as "dynamic mode", is necessary in order to measure both Fe and Cu isotopes. In contrast, using Ni as a mass bias monitor avoids peak hopping because all Fe isotopes and at least 2 interference-free Ni isotopes can be collected simultaneously in the collector array (Malinovsky et al. 2003; Poitrasson and Freydier 2005). Accordingly, this method is also suitable for in situ isotope analyses with laser ablation (LA) MC-ICP-MS (Oeser et al. 2014), where short integration times are required. This elemental spike is co-aspirated into the instrument while the laser ablation takes place. A disadvantage of Ni is that $^{58}Fe/^{54}Fe$ isotope ratios cannot be accurately measured because of the large mass interference from ^{58}Ni when Ni cones are used. Measurement of $^{58}Fe/^{54}Fe$ is not important to evaluate mass-dependent effects, but it is important in the study of extraterrestrial materials that may record nucleosynthesis effects. Other methods, such as adding a doping element to create a constant matrix effect, as has been done by adding SO_4 or Mg to Si, for Si isotope analysis (Chen et al. 2017; Oelze et al. 2016), have not been successfully pursued for Fe isotopes.

2.4 In Situ Techniques

A number of studies have made use of in situ methods to analyze Fe-rich minerals (including oxides, sulfides, carbonates, and silicates), metals, and glasses. These studies used either secondary ionization mass spectrometry (SIMS) (Galic et al. 2017; Marin-Carbonne et al. 2011, 2014; Virtasalo et al. 2013; Whitehouse and Fedo 2007) or laser ablation (LA) coupled to MC-ICP-MS to analyze Fe isotope compositions of minerals (Collinet et al. 2017; Czaja et al. 2013; Graham et al. 2004; Hirata and Ohno 2001; Horn and von Blanckenburg 2007; Horn et al. 2006b; Kosler et al. 2005; Li et al. 2013; Nishizawa et al. 2010; Oeser et al. 2015, 2014, 2018; Sio et al. 2013; Steinhoefel et al. 2009a, b; 2010; Yoshiaki et al. 2012, 2015a, b). SIMS has been successfully used in the Fe isotope analysis of Fe sulfides (Galic et al. 2017; Marin-Carbonne et al. 2011, 2014; Virtasalo et al. 2013; Whitehouse and Fedo 2007). In contrast, Fe isotope analysis of iron oxides by SIMS is complicated by the variable instrumental mass bias associated with the orientation of the crystal lattice with respect to the primary ion beam (Kita et al. 2011). For example, the Fe isotope composition measured by SIMS of an isotopically homogenous magnetite sample defines a range of 0.6‰ in $^{56}Fe/^{54}Fe$. Moreover, there is a positive correlation between the measured Fe isotope composition and the Fe ion yield, which is interpreted to be a result of crystallographic channeling of the primary ion beam (Kita et al. 2011). This crystallographic effect is similar to that documented for O isotopes in magnetite (Huberty et al.

2010; Lyon et al. 1998) and in S isotopes in sphalerite (Kozdon et al. 2010), where it appears that the orientation of the high symmetry mineral with respect to the primary beam leads to ion channeling, resulting in production of higher secondary ion signals and, as a result, different instrumental mass-bias conditions. It is possible to minimize these orientation effects for O and S isotopes by lowering the potential difference between the primary and secondary ion beams (Huberty et al. 2010; Kita et al. 2011; Kozdon et al. 2010). These conditions allow for the precise and accurate measurement of O and S isotope compositions, albeit at lower secondary ion production rates. No tests have been performed to establish if a lower potential would minimize Fe isotope crystal orientation mass-bias effects.

Laser ablation techniques for Fe isotope analysis have focused on using two broad categories of lasers: those with nanosecond (ns, 10^{-9} s) pulse widths and those with femtosecond (fs, 10^{-15} s) pulse widths. The difference in pulse widths between these two laser types significantly impacts the amount of thermal damage the substrate undergoes during ablation. During laser ablation with a ns-laser, energy delivery by the pulsed phonons is long enough to allow for thermal diffusion within the target mineral, while the fs-laser pulse is too short for significant thermal diffusion, although considerable time still exists for laser-matter interaction (e.g., Russo et al. 2013). For example, following the treatment of Gonzales et al. (2007b), the heat affected zone can be approximated by the relationship

$$L = (a\tau)^{1/2}$$

where L is the length of the heat affected zone, τ is the laser pulse width and a is the thermal diffusivity. Figure 2.5 shows that the length of the heat affected zone for a 3 ns pulse width laser is two orders of magnitude greater than for a fs-laser with a 150 fs pulse width. On substrates with high to medium thermal diffusivities, such as Fe metal, Fe oxides, and Fe sulfides ablated by ns pulses, the size of the heat affected zone can reach several dozens of nanometers. Moreover, during ns-LA

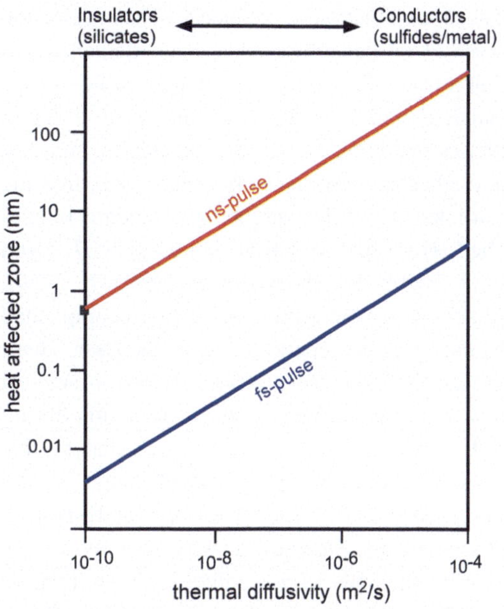

Fig. 2.5 Plot of the thermal diffusivity versus the length of the heat affected zone for a laser with a 150 fs pulse width and a 5 ns pulse width. Small thermal diffusivities which would be associated with most silicate minerals produce very small (nanometers) heat-affected zones for both fs- and ns-LA, but for minerals with higher thermal diffusivities such as iron oxides and sulfides, the heat-affected zone for ns-LA is large (microns). However, the heat-affected zone for all minerals is nearly two orders of magnitude lower for fs-LA as compared to ns-LA

of substrates that act as conductors, large particles are often produced by ejection of melted layers referred to as hydrodynamic sputtering (e.g., Hergenröder 2006), which leads to significant elemental and isotopic fractionation because these large particles (>1 μm in diameter) are only partially ionized in the ICP source (Jackson and Güther 2003). In contrast, fs-LA does not produce significant amounts of these large particles during ablation of conductors such as Fe metal (Gonzalez et al. 2007b). The differences in particle sizes and heat-affected zones is likely one of the reasons that Fe isotope analysis of Fe metal by ns-LA is inaccurate compared to fs-LA (Horn et al. 2006a; Kosler et al. 2005, 2006). Indeed, because the heat-affected zone size and corresponding melting of the target are minimized during fs-LA, the extent of isotopic fractionation linked to ablation is reduced below detection levels. Consequently, it has been demonstrated that it is possible to use Fe metal as an Fe isotope standard and to analyze a variety of different phases including Fe oxides, carbonates, sulfides, and silicates, with the only requirement that the ion intensity of the sample and standard is matched (Horn and von Blanckenburg 2007; Horn et al. 2006b; Steinhoefel et al. 2009a, b, 2010). Matching ion intensities is easily accomplished by changing the repetition rate of the laser. These observations led Horn et al. (2006b) to suggest that there was no Fe isotope fractionation during fs-LA and that fs-LA is a matrix-free analysis that does not require sample-standard matrix matching. Indeed, fs-LA for Fe isotope analysis stoichiometrically samples a wide variety of Fe bearing minerals over a range of laser fluences and wavelengths (d'Abzac et al. 2013, 2014; Zheng et al. 2017, 2018). However, additional fs-LA studies have shown that for Cu isotopes substrates are probably not sampled stoichiometrically at high fluences (Lazarov and Horn 2015), implying that one needs to carefully evaluate in situ methods for accuracy.

To better understand the origins of the differences in fs- and ns-LA for Fe isotope analysis, studies have characterized the Fe isotope composition and morphology of LA produced aerosol particles with different aerodynamic size distributions using both ns- and fs-LA and collection of particles with a cascade impactor (d'Abzac et al. 2013; 2014; Zheng et al. 2017). The cascade impactor used in these studies consists of a total of 12 stages, where the largest aerodynamic particles settle out on the top stages and lower stages catch progressively smaller particles. A final filter stage collects any remaining particles that did not settle out on the other impactor plates (Marple et al. 1991). Aerodynamic size reflects the diameter of the particle, as well as its shape and density. The relative Fe mass distribution of the aerodynamically sized particles generated by fs-LA is similar for all phases that have been studied including magnetite, hematite pyrite, pyrrhotite, and siderite (d'Abzac et al. 2013, 2014; Zheng et al. 2017) (Fig. 2.6). Morphological studies on these fs-LA produced particles using SEM and TEM show that all particles occur as either soft-bonded

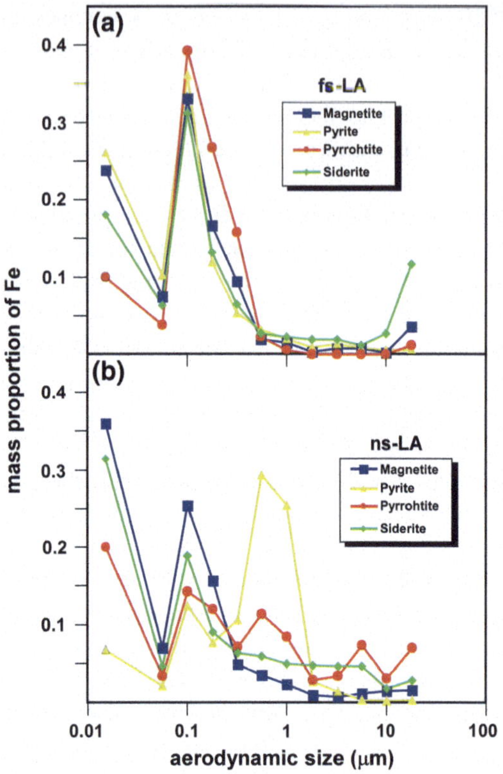

Fig. 2.6 Plot of proportion of Fe collected on each impaction stage of a cascade impactor relative to the total Fe collected as a function of aerodynamic size. The particle size distribution for particles produced by fs-LA for different phases are similar. In contrast, the particle size distribution for particles produced by ns-LA for different phases are a function of the phase that is ablated. The particle size distribution data are from the literature (d'Abzac et al. 2013; Zheng et al. 2017)

agglomerates composed of ~10 nm spheres with a wide range of sizes, or isolated spheres that range in size from 20 to 200 nm. This agglomerate and sphere morphology is produced by the ablation of all substrates investigated including oxides, sulfides, carbonates, silicates, phosphates, and native metals, as well as for a wide range of laser conditions including a range of laser wavelengths at ~800 nm, ~266, and ~198 nm (d'Abzac et al. 2012, 2013, 2014; Glaus et al. 2010; Gonzalez et al. 2007a, b; Zheng et al. 2017). In terms of their Fe isotope composition, the smallest aerodynamic sized particles produced by fs-LA have low δ^{56}Fe values as compared to larger aerodynamic size particles for all substrates analyzed (d'Abzac et al.

2013; Zheng et al. 2017). The magnitude of this fractionation is phase dependent, where the fractionation is greatest for magnetite (range of ~2‰), moderate for pyrrhotite (range of ~1‰), and not analytically resolvable for pyrite (d'Abzac et al. 2013). These isotopic fractionations are interpreted to be a result of unidirectional processes associated with condensation of the particles from the laser-produced plume. The degree of isotopic fractionation is controlled by condensation temperature and partial pressure of the different elements present (d'Abzac et al. 2013). Based on the relative Fe mass distribution and Fe isotope composition of the size-sorted particles, mass balance calculations indicate that fs-LA always samples the substrate stoichiometrically (d'Abzac et al. 2013, 2014; Zheng et al. 2017). The most important conclusion regarding these observations is that because the different sized particles have a range of Fe isotope compositions, accurate Fe isotopic analysis can only be achieved if all the particles are quantitatively carried to the ICP source, or, to a lesser extent, if samples and standards undergo the same level of particle loss through transport.

Similar studies were performed for particles produced by ns-LA using the same type of laser ablation cell and aerosol collection method as done for the fs-LA studies. The Fe mass distribution for particles produced by ns-LA of magnetite are similar to the distribution of particles produced by fs-LA (Fig. 2.6). However, for pyrrhotite, pyrite and siderite, ns-LA generated particles have a higher Fe mass proportion of large aerodynamic sized particles (Zheng et al. 2017) (Fig. 2.6). In terms of Fe isotope fractionation in size-sorted particles, ns-LA resulted in the smallest aerodynamic sized particles having relatively high δ^{56}Fe values and large particles with relatively low δ^{56}Fe values for both pyrite and pyrrhotite, but an opposite trend for siderite. This particle-size dependent Fe isotope fractionation appears to be also laser fluence-dependent; for magnetite at low fluence, the smallest particles have low δ^{56}Fe values compared to larger particles, but at higher fluences the smallest particles have high δ^{56}Fe values as compared to smaller particles. Moreover, based on mass-balance calculations, although ns-LA typically samples the substrate stoichiometrically, it is not the case at low fluence. The mass balance of the collected aerosol particles have a δ^{56}Fe value that is 0.75‰ less than the substrate (Zheng et al. 2017). Morphologically, particles produced by ns-LA are variable and reflect both mineralogy and laser conditions (Glaus et al. 2010; Gonzalez et al. 2007a, b; Zheng et al. 2017). For example, particles produced by ns-LA of magnetite at fluences below 1 J/cm^2 tend to occur solely as agglomerates whereas at higher fluences both spheres and agglomerates occur; at fluences of 4 J/cm^2 the spheres were very large ranging from 100's of nm up to 1 µm in size (Zheng et al. 2017). A wide array of particle types occur for sulfides, including hard bonded agglomerates and spheres up to several microns in size (Zheng et al. 2017).

Because of the different response (particle size distributions, particle morphology, Fe isotope fractionation) of substrates to ns-LA, Fe isotope studies conducted by ns-LA require careful matrix matching in terms of both mineralogy and mineral chemical composition (Sio et al. 2013; Zheng et al. 2018). Indeed, failure to match matrices by even a few wt% in major element composition can result in Fe isotope analyses with poor reproducibility (Zheng et al. 2018). In contrast, because the particles produced by fs-LA tend to have similar size distributions and morphology over a wide range of substrates and laser conditions, it is reasonable to suggest that fs-LA is free of matrix effects. The fs-LA analyses, performed in Hanover by Ingo Horn and co-workers (e.g., Horn and von Blanckenburg 2007; Horn et al. 2006b; Steinhoefel et al. 2009a, b, 2010) always have been performed at deep UV and wet plasma conditions. More recent work used a Ni solution as a mass bias monitor (Oeser et al. 2014, 2015). The use of a mass bias monitor for LA analyses has the additional advantage of allowing interference correction to be performed very precisely, by adding a mass bias estimate to the interference monitor, e.g. ^{53}Cr for the mass interference of ^{54}Cr on ^{54}Fe (Weyrauch et al. 2017). The latter is of particular advantage for LA analyses because no chemical purification is performed, and mass interferences of ^{54}Cr on ^{54}Fe can be significant. Furthermore, Oeser et al. (2014) used isotopically well-characterized glass standards, instead of Fe metal as reference material for sample-standard bracketing, for the analyses of isotopically zoned olivine and other silicates. This resulted in a more similar matrix in the plasma and had the advantage that little adjustment of the laser repetition rate was necessary to match intensities of the signal between sample and standard. Oeser et al. (2014) have shown that their method is insensitive to chemical variations among the analyzed silicates (such as variable SiO_2 or MgO) and is suitable to perform precise and accurate Fe (and Mg) isotope compositions of ~0.1‰. The latter has been demonstrated by the analysis of a variety of reference glasses with a wide range of compositions (komatiitic, basaltic, dioritic), both with conventional solution and fs-LA techniques. Although aspiration of a Ni solution has been shown to improve the precision and accuracy of Fe isotope analysis by fs-LA (Oeser et al. 2014), the fact that introduction of water or elemental solution effectively creates a wet plasma condition should not be ignored. Wet plasma tends to decrease matrix effects as compared to dry aerosol conditions and water addition to the laser ablation aerosol appears to have a similar effect. For example, co-aspiration of ~20 μl/min of water during LA analysis of magnetite it was found that accurate and precise Fe isotope compositions of magnetite that ranged from ~100% Fe_3O_4 to ~85% Fe_3O_4 could be made (Zheng et al. 2018). Although addition of water minimizes such matrix effects, it also decreases ion intensity; 20 μl/min water addition decreases Fe sensitivity by ~40%. It has also been demonstrated that by adding water to the LA aerosol, it is possible to use a pyrite standard to recover accurate and precise Fe isotope analyses of pyrrhotite (~10 μl/min water addition) and magnetite (~20 μl/min water addition) (Zheng et al. 2018).

Overall, fs-LA is the preferred method for Fe isotope analysis of phases that conduct heat, which includes iron oxides, sulfides, and metal. SIMS is a highly effective method for Fe isotope analysis of sulfides but suffers from a crystal orientation effect for Fe isotope analysis of high symmetry oxides such as magnetite. An effective method to correct for this phenomenon has not yet been investigated for SIMS analysis. Overall the spatial resolution of pyrite analysis by SIMS is better as compared to fs-LA, where a 2-σ precision of 0.17‰ in $^{56}Fe/^{54}Fe$ can be obtained on a volume of 200 μm^3 of pyrite (Galic et al. 2017), whereas a 2-σ precision of 0.15‰ ($^{56}Fe/^{54}Fe$) can be obtained on a volume of 1,500 μm^3 by fs-LA (Zheng et al. 2018). However, it should be noted that the fs-LA analyses have all been done on small turning radius MC-ICP-MS instruments that decrease the transmission of the ion beam by 90% in order to achieve a resolving power suitable to be free of argide isobars. If a large turning radius instrument such as the *Nu 1700* or a new generation MC-ICP-MS fitted with a collision cell was used, the spatial resolution of SIMS and fs-LA are likely to be comparable.

2.5 Summary

- Precise and accurate Fe isotope analysis that preserves naturally occurring, mass-dependent fractionation requires mass spectrometry methods that include correction for instrumental mass bias and removal of elemental and molecular isobars.
- Samples for mass spectrometric analysis using solution nebulization requires purification by ion-exchange chromatography to remove unwanted isobars and matrix elements. Because the sample can be isotopically fractionated, chemical purification must be quantitative.
- Initial MC-ICP-MS analysis methods minimize argide isobars such as $^{40}Ar^{14}N$ on ^{54}Fe, via cool plasma or reaction cells, and an on-peak zero subtraction for Fe mass analysis that preserved mass-dependent behavior between ^{54}Fe, ^{56}Fe, and ^{57}Fe. Modern ICP-MS analysis uses pseudo-high mass resolution

techniques to eliminate argide isobars. This pseudo-high mass resolution technique uses a narrow defining slit and adjusts the zoom optics or position of the Faraday collector so that only the Fe peaks enter into the Faraday collectors. However, these techniques decrease the ion signal, and only ~10% of the ion beam is typically transmitted through the narrow defining slit.

- Correction for instrumental mass bias during MC-ICP-MS analysis may use sample standard bracketing and assumes the mass bias of the sample is identical to the mass bias of the standard. Additional methods to improve on this approach include addition of either Cu or Ni and normalization to a constant Cu or Ni isotope ratio. The double-spike technique can also be used to correct for instrumental mass bias and if the sample is spiked before chemical purification the yield need not be quantitative, making the double-spike method for instrumental mass-bias correction superior for ultra-small samples in a complex matrix, such as the Fe isotope analysis of seawater.
- In MC-ICP-MS analysis, the light isotope is lost preferentially to the heavier isotope and the amount of mass-bias reflects what is present in the analyte solution. To avoid such matrix issues, the concentration of the sample and standard are matched and samples are chemically purified. In general, it is possible to minimize matrix effects by using wet plasma techniques as compared to dry plasma techniques.
- In situ Fe isotope analysis can be done using either SIMS or LA with MC-ICP-MS. SIMS analysis allows for superior spatial resolution for the Fe isotope analysis of pyrite, but suffers from orientation effects in high symmetry minerals such as magnetite. Precise and accurate Fe isotope analysis using lasers with a nanosecond-pulse width requires careful matrix matching between standards and samples. In contrast, lasers with a femtosecond pulse width do not require matrix matching between samples and standards if moist plasma techniques are utilized. For example, if ~20 µl/min of H_2O is co-aspirated during fs-LA analysis, a pyrite grain can be used as a standard for the analysis of magnetite samples.

References

Albarède F, Beard B (2004) Analytical methods for non-traditional isotopes. In: Johnson CM, Beard BL, Albarède F (eds) Geochemistry of non-traditional stable isotopes. Reviews in mineralogy & geochemistry, vol 55, pp 113–152. https://doi.org/10.2138/gsrmg.55.1.113

Alder JF, Bombelka RM, Kirkbright GF (1980) Electronic excitation and ionization temperature-measurements in a high-frequency inductively coupled argon plasma source and the influence of water-vapor on plasma parameters. Spectrochim Acta Part B At Spectrosc 35 (4):163–175. https://doi.org/10.1016/0584-8547(80)80063-2

Anbar AD, Roe JE, Barling J, Nealson KH (2000) Nonbiological fractionation of iron isotopes. Science 288(5463):126–128. https://doi.org/10.1126/science.288.5463.126

Andren H, Rodushkin I, Stenberg A, Malinovsky D, Baxter DC (2004) Sources of mass bias and isotope ratio variation in multicollector ICP-MS: optimization of instrumental parameters based on experimental observations. J Anal At Spectrom 19(9):1217–1224. https://doi.org/10.1039/b403938f

Arnold GL, Weyer S, Anbar AD (2004) Fe isotope variations in natural materials measured using high mass resolution multiple collector ICPMS. Anal Chem 76(2):322–327. https://doi.org/10.1021/ac034601v

Beard BL, Johnson CM (1999) High precision iron isotope measurements of terrestrial and lunar materials. Geochim Cosmochim Acta 63(11–12):1653–1660. https://doi.org/10.1016/s0016-7037(99)00089-7

Beard BL, Johnson CM, Skulan JL, Nealson KH, Cox L, Sun H (2003a) Application of Fe isotopes to tracing the geochemical and biological cycling of Fe. Chem Geol 195(1–4):87–117. https://doi.org/10.1016/s0009-2541(02)00390-x

Beard BL, Johnson CM, Von Damm KL, Poulson RL (2003b) Iron isotope constraints on Fe cycling and mass balance in oxygenated Earth oceans. Geology 31 (7):629–632. https://doi.org/10.1130/0091-7613(2003)031<0629:Iicofc>2.0.Co;2

Beard BL, Johnson CM, Cox L, Sun H, Nealson KH, Aguilar C (1999) Iron isotope biosignatures. Science 285(5435):1889–1892

Belshaw NS, Zhu XK, Guo Y, O'Nions RK (2000) High precision measurement of iron isotopes by plasma source mass spectrometry. Int J Mass Spectrom 197:191–195. https://doi.org/10.1016/s1387-3806(99)00245-6

Borrok DM, Wanty RB, Ridley WI, Wolf R, Lamothe PJ, Adams M (2007) Separation of copper, iron, and zinc from complex aqueous solutions for isotopic measurement. Chem Geol 242(3–4):400–414. https://doi.org/10.1016/j.chemgeo.2007.04.004

Browner RF, Long SE (1989) Comments on influence of water on ICP-reply. Spectrochim Acta Part B At Spectrosc 44(8):831–832. https://doi.org/10.1016/0584-8547(89)80083-7

Bullen TD, White AF, Childs CW, Vivit DV, Schulz MS (2001) Demonstration of significant abiotic iron isotope fractionation in nature. Geology 29(8):699–702. https://doi.org/10.1130/0091-7613(2001)029%3c0699:dosaii%3e2.0.co;2

Caughlin BL, Blades MW (1987) Effect of wet and dry nebulizer gas on the spatial-distribution of electron-density. Spectrochim Acta Part B At Spectrosc 42(1–2):353–360. https://doi.org/10.1016/0584-8547(87)80076-9

Chen XY, Lapen TJ, Chafetz HS (2017) Accurate and precise silicon isotope analysis of sulfur- and iron-rich samples by MC-ICP-MS. Geostand Geoanal Res 41(3):427–435. https://doi.org/10.1111/ggr.12158

Collinet M, Charlier B, Namur O, Oeser M, Medard E, Weyer S (2017) Crystallization history of enriched shergottites from Fe and Mg isotope fractionation in olivine megacrysts. Geochim Cosmochim Acta 207:277–297. https://doi.org/10.1016/j.gca.2017.03.029

Conway TM, Rosenberg AD, Adkins JF, John SG (2013) A new method for precise determination of iron, zinc and cadmium stable isotope ratios in seawater by double-spike mass spectrometry. Anal Chim Acta 793:44–52. https://doi.org/10.1016/j.aca.2013.07.025

Czaja AD, Johnson CM, Beard BL, Roden EE, Weiqiang L, Moorbath S (2013) Biological Fe oxidation controlled deposition of banded iron formation in the ca. 3770 Ma Isua Supracrustal Belt (West Greenland). Earth Planet Sci Lett 363:192–203

d'Abzac FX, Beard BL, Czaja AD, Konishi H, Schauer JJ, Johnson CM (2013) Iron isotope composition of particles produced by UV-femtosecond laser ablation of natural oxides, sulfides, and carbonates. Anal Chem 85(24):11885–11892. https://doi.org/10.1021/ac402722t

d'Abzac FX, Czaja AD, Beard BL, Schauer JJ, Johnson CM (2014) Iron distribution in size-resolved aerosols generated by UV-femtosecond laser ablation: influence of cell geometry and implications for in situ isotopic determination by LA-MC-ICP-MS. Geostand Geoanal Res 38(3):293–309

d'Abzac FX, Seydoux-Guillaume AM, Chmeleff J, Datas L, Poitrasson F (2012) In situ characterization of infra red femtosecond laser ablation in geological samples. Part B: the laser induced particles. J Anal At Spectrom 27(1):108–119. https://doi.org/10.1039/c1ja10154d

Dauphas N, Pourmand A, Teng FZ (2009) Routine isotopic analysis of iron by HR-MC-ICPMS: How precise and how accurate? Chem Geol 267(3–4):175–184. https://doi.org/10.1016/j.chemgeo.2008.12.011

de Jong J, Schoemann V, Tison JL, Becquevort S, Masson F, Lannuzel D, Petit J, Chou L, Weis D, Mattielli N (2007) Precise measurement of Fe isotopes in marine samples by multi-collector inductively coupled plasma mass spectrometry (MC-ICP-MS). Anal Chim Acta 589(1):105–119. https://doi.org/10.1016/j.aca.2007.02.055

Devulder V, Lobo L, Van Hoecke K, Degryse P, Vanhaecke F (2013) Common analyte internal standardization as a tool for correction for mass discrimination in multi-collector inductively coupled plasma-mass spectrometry. Spectrochim Acta Part B At Spectrosc 89:20–29. https://doi.org/10.1016/j.sab.2013.08.009

Dixon PR, Janecky DR, Perrin RE, Rokop DJ, Unkefer PL, Spall WD, Maeck R (1992) Unconventional stable isotopes-iron. Water-rock interaction, vols 1 and 2: Vol 1: low temperature environments; vol 2: moderate and high temperate environments

Dixon PR, Perrin RE, Rokop DJ, Maeck R, Janecky DR, Banar JP (1993) Measurement of iron isotopes (Fe-54, Fe-56, Fe-57, and Fe-58) in submicrogram quantities of iron. Anal Chem 65(15):2125–2130. 10.1021/ac00063a033

Douglas DJ, Tanner SC (1998) Fundamental considerations in ICP-MS. In: Montaser A (ed) Inductively coupled plasma mass spectrometry. Wiley-VCH, New York, pp 615–679

Finlayson VA, Konter JG, Ma L (2015) The importance of a Ni correction with ion counter in the double spike analysis of Fe isotope compositions using a Fe-57/Fe-58 double spike. Geochem Geophys Geosyst 16(12):4209–4222. https://doi.org/10.1002/2015gc006012

Flory DR, Miller LV, Fennessey PV (1993) Development of techniques for the isolation of iron from biological=material for measuremnet of isotope ratios by fast-atom-bombardment MASS-SPECTROMETRY. Anal Chem 65(23):3501–3504. https://doi.org/10.1021/ac00071a029

Fritz JS (2004) Early milestones in the development of ion-exchange chromatography: a personal`account. J Chromatogr A 1039(1–2):3–12. https://doi.org/10.1016/j.chroma.2003.12.068

Galic A, Mason PRD, Mogollon JM, Wolthers M, Vroon PZ, Whitehouse MJ (2017) Pyrite in a sulfate-poor Paleoarchean basin was derived predominantly from elemental sulfur: evidence from 3.2 Ga sediments in the Barberton Greenstone Belt, Kaapvaal Craton. Chem Geol 449:135–146. https://doi.og/10.1016/j.chemgeo.2016.12.006

Gillson GR, Douglas DJ, Fulford JE, Halligan KW, Tanner SD (1988) Nonspectroscopic interelement interferences in inductively coupled plasma mass-spectrometry. Anal Chem 60(14):1472–1474. https://doi.org/10.1021/ac00165a024

Glaus R, Kaegi R, Krumeich F, Gunther D (2010) Phenomenological studies on structure and elemental composition of nanosecond and femtosecond laser-generated aerosols with implications on laser ablation inductively coupled plasma mass spectrometry. Spectrochim Acta Part B At Spectrosc 65(9–10):812–822. https://doi.org/10.1016/j.sab.2010.07.005

Gonzalez JJ, Liu C, Wen SB, Mao X, Russo RE (2007a) Glass particles produced by laser ablation for ICPMS measurements. Talanta 73(3):577–582. https://doi.org/10.1016/j.talanta.2007.04.028

Gonzalez JJ, Liu CY, Wen SB, Mao XL, Russo RE (2007b) Metal particles produced by laser ablation for ICP-MS measurements. Talanta 73(3):567–576. https://doi.org/10.1016/j.talanta.2007.04.029

Gotz A, Heumann KG (1988) Iron isotope ratio measurements with the thermal ionization technique using a compact quadrupole mass-spectrometer. Int J Mass Spectrom Ion Process 83(3):319–330. https://doi.org/10.1016/0168-1176(88)80036-3

Graham S, Pearson N, Jackson S, Griffin W, O'Reilly SY (2004) Tracing Cu and Fe from source to porphyry: in situ determination of Cu and Fe isotope ratios in sulfides from the Grasberg Cu-Au deposit. Chem Geol 207(3–4):147–169. https://doi.org/10.1016/j.chemgeo.2004.02.009

Hergenröder R (2006) Hydrodynamic sputtering as a possible source for fractionation in LA-ICP-MS. J Anal At Spectrom 21(5):517–524. https://doi.org/10.1039/b600705h

Herzog GF, Xue S, Hall GS, Nyquist LE, Shih CY, Wiesmann H, Brownlee DE (1999) Isotopic and elemental composition of iron, nickel, and chromium in type I deep-sea spherules: implications for origin and composition of the parent micrometeoroids. Geochim Cosmochim Acta 63(9):1443–1457. https://doi.org/10.1016/s0016-7037(99)00011-3

Hill PS, Schauble EA, Shahar A, Tonui E, Young ED (2009) Experimental studies of equilibrium iron isotope fractionation in ferric aquo-chloro complexes. Geochim Cosmochim Acta 73(8):2366–2381. 10.1016/j.gca.2009.01.016

Hirata T, Ohno T (2001) In-situ isotopic ratio analysis of iron using laser ablation-multiple collector-inductively coupled plasma mass spectrometry (LA-MC-ICP-MS). J Anal At Spectrom 16(5):487–491. https://doi.org/10.1039/b100946j

Horn I, von Blanckenburg F (2007) Investigation on elemental and isotopic fractionation during 196 nm femtosecond laser ablation multiple collector inductively coupled plasma mass spectrometry. Spectrochim Acta Part B At Spectrosc 62(4):410–422. https://doi.org/10.1016/j.sab.2007.03.034

Horn I, Schoenberg R, von Blanckenburg F (2006a) Comment on "Analysis of Fe isotopes in sulfides and iron meteorites by laser ablation high-mass resolution multi-collector-ICP mass spectrometry" by J. Kosler, R.B. Pedersen, C. Kruber and P.J. Sylvester. J Anal At Spectrom 21(2):211–213. https://doi.org/10.1039/b504720j

Horn I, von Blanckenburg F, Schoenberg R, Steinhoefel G, Markl G (2006b) In situ iron isotope ratio determination using UV-femtosecond laser ablation with application to hydrothermal ore formation processes. Geochim Cosmochim Acta 70(14):3677–3688. https://doi.org/10.1016/j.gca.2006.05.002

Huberty JM, Kita NT, Kozdon R, Heck PR, Fournelle JH, Spicuzza MJ, Xu H, Valley JW (2010) Crystal orientation effects in delta O-18 for magnetite and hematite by SIMS. Chem Geol 276(3–4):269–283. https://doi.org/10.1016/j.chemgeo.2010.06.012

Hutcheon ID, Armstrong JT, Wasserburg GJ (1987) Isotopic studies of Mg, Fe, Mo, Ru and W in Fremdlinge from Allende refractory inclusions. Geochim Cosmochim Acta 51(12):3175–3192. https://doi.org/10.1016/0016-7037(87)90126-8

Jackson SE, Güther D (2003) The nature and sources of laser induced isotopic fractionation in laser ablation-multicollector-inductively coupled plasma-mass spectrometry. J Anal At Spectrom 18 (3):205–212. https://doi.org/10.1039/b209620j

John SG, Adkins JF (2010) Analysis of dissolved iron isotopes in seawater. Mar Chem 119(1–4):65–76. https://doi.org/10.1016/j.marchem.2010.01.001

Johnson CM, Beard BL (1999) Correction of instrumentally produced mass fractionation during isotopic analysis of Fe by thermal ionization mass spectrometry. Int J Mass Spectrom 193(1):87–99. https://doi.org/10.1016/s1387-3806(99)00158-x

Kehm K, Hauri EH, Alexander CMO, Carlson RW (2003) High precision iron isotope measurements of meteoritic material by cold plasma ICP-MS. Geochim Cosmochim Acta 67(15):2879–2891. https://doi.org/10.1016/s0016-7037(03)00080-2

Kita NT, Huberty JM, Kozdon R, Beard BL, Valley JW (2011) High-precision SIMS oxygen, sulfur and iron stable isotope analyses of geological materials: accuracy, surface topography and crystal orientation. Surf Interface Anal 43(1–2):427–431. https://doi.org/10.1002/sia.3424

Kosler J, Pedersen RB, Kruber C, Sylvester PJ (2005) Analysis of Fe isotopes in sulfides and iron meteorites by laser ablation high-mass resolution multi-collector ICP mass spectrometry. J Anal At Spectrom 20 (3):192–199. https://doi.org/10.1039/b412169d

Kosler J, Pedersen R, Kruber C, Sylvester P (2006) Comment on "Analysis of Fe isotopes in sulfides and iron meteorites by laser ablation high-mass resolution multi-collector-ICP mass spectrometry"—Reply. J Anal At Spectrom 21(2):214–216. https://doi.org/doi:10.1039/b512647a

Kozdon R, Kita NT, Huberty JM, Fournelle JH, Johnson CA, Valley JW (2010) In situ sulfur isotope analysis of sulfide minerals by SIMS: precision and accuracy, with application to thermometry of similar to 3.5 Ga Pilbara cherts. Chem Geol 275(3–4):243–253. 10.1016/j.chemgeo.2010.05.015

Kraus KA, Nelson F (1956) Anion exchange studies of the fission products. In: Proceedings of the first united conference of the peaceful uses of atomic energy, pp 113–125

Lacan F, Radic A, Jeandel C, Poitrasson F, Sarthou G, Pradoux C, Freydier R (2008) Measurement of the isotopic composition of dissolved iron in the open ocean. Geophys Res Lett 35(24). https://doi.org/10.1029/2008gl035841

Lam JWH, Horlick G (1990) A comparison of argon and mixed gas plasmas for inductively coupled plasma mass-spectrometry. Spectrochim Acta Part B At Spectrosc 45(12):1313–1325. https://doi.org/10.1016/0584-8547(90)80185-l

Lazarov M, Horn I (2015) Matrix and energy effects during in-situ determination of Cu isotope ratios by ultraviolet-femtosecond laser ablation multicollector inductively coupled plasma mass spectrometry. Spectrochim Acta Part B At Spectrosc 111:64–73. https://doi.org/10.1016/j.sab.2015.06.013

Li W, Huberty JM, Beard BL, Kita NT, Valley JW, Johnson CM (2013) Contrasting behavior of oxygen and iron isotopes in banded iron formations revealed by in situ isotopic analysis. Earth Planet Sci Lett 384:132–143

Long SE, Browner RF (1988) Influence of water on conditions in the inductively coupled agon pasma. Spectrochim Acta Part B At Spectrosc 43(12):1461–1471. https://doi.org/10.1016/0584-8547(88)80185-x

Longerich HP, Fryer BJ, Strong DF (1987) Determination of lead isotope ratios by inductively coupled plasma-mass spectrometry (ICP-MS). Spectrochim Acta Part B At Spectrosc 42(1–2):39–48. https://doi.org/10.1016/0584-8547(87)80048-4

Lyon IC, Saxton JM, Cornah SJ (1998) Isotopic fractionation during secondary ionisation mass spectrometry: crystallographic orientation effects in magnetite. Int J Mass Spectrom 172(1–2):115–122. https://doi.org/10.1016/s0168-1176(97)00143-2

Malinovsky D, Stenberg A, Rodushkin I, Andren H, Ingri J, Ohlander B, Baxter DC (2003) Performance of high resolution MC-ICP-MS for Fe isotope ratio measurements in sedimentary geological materials. J Anal At Spectrom 18(7):687–695. https://doi.org/10.1039/b302312e

Mandernack KW, Bazylinski DA, Shanks WC, Bullen TD (1999) Oxygen and iron isotope studies of magnetite produced by magnetotactic bacteria. Science 285 (5435):1892–1896. https://doi.org/10.1126/science.285.5435.1892

Marechal CN, Telouk P, Albarède F (1999) Precise analysis of copper and zinc isotopic compositions by plasma-source mass spectrometry. Chem Geol 156(1–4):251–273. https://doi.org/10.1016/s0009-2541(98)00191-0

Marin-Carbonne J, Rollion-Bard C, Luais B (2011) In-situ measurements of iron isotopes by SIMS: MC-ICP-MS intercalibration and application to a magnetite crystal from the Gunflint chert. Chem Geol 285(1–4):50–61. https://doi.org/10.1016/j.chemgeo.2011.02.019

Marin-Carbonne J, Rollion-Bard C, Bekker A, Rouxel O, Agangi A, Cavalazzi B, Wohlgemuth-Ueberwasser CC, Hofmann A, McKeegan KD (2014) Coupled Fe and S isotope variations in pyrite nodules from Archean shale. Earth Planet Sci Lett 392:67–79. https://doi.org/10.1016/j.epsl.2014.02.009

Marple VA, Rubow KL, Behm SM (1991) A microorifice uniform deposit impactor (MOUDI): description, calibration, and use. Aerosol Sci Tech 14:434–446

Nier AO (1939) The isotopic constitution of iron and chromium. Phys Rev 55:1143

Nishizawa M, Yamamoto H, Ueno Y, Tsuruoka S, Shibuya T, Sawaki Y, Yamamoto S, Kon Y, Kitajima K, Komiya T, Maruyama S, Hirata T (2010) Grain-scale iron isotopic distribution of pyrite from Precambrian shallow marine carbonate revealed by a femtosecond laser ablation multicollector ICP-MS technique: possible proxy for the redox state of ancient seawater. Geochim Cosmochim Acta 74 (9):2760–2778. https://doi.org/10.1016/j.gca.2010.02.014

Niu HS, Houk RS (1994) Langmuir prove measurements of the ion extraction process in inductively-coupled plasma-mass spectrometry. 1. Spatially-resolved determination of electron-density and electron-temperature. Spectrochim Acta Part B At Spectrosc 49(12–14):1283–1303. https://doi.org/10.1016/0584-8547(94)80109-6

Niu HS, Hu K, Houk RS (1991) Langmuir probe measurements of electron-temperature and electron-density behind the skimmer of an inductively coupled plasma mass-spectrometer. Spectrochim Acta Part B At Spectrosc 46(6–7):805–817. https://doi.org/10.1016/0584-8547(91)80082-e

Oelze M, Schuessler JA, von Blanckenburg F (2016) Mass bias stabilization by Mg doping for Si stable isotope analysis by MC-ICP-MS. J Anal At Spectrom 31(10):2094–2100. https://doi.org/10.1039/c6ja00218h

Oeser M, Dohmen R, Horn I, Schuth S, Weyer S (2015) Processes and time scales of magmatic evolution as revealed by Fe-Mg chemical and isotopic zoning in natural olivines. Geochim Cosmochim Acta 154:130–150. https://doi.org/10.1016/j.gca.2015.01.025

Oeser M, Ruprecht P, Weyer S (2018) Combined Fe-Mg chemical and isotopic zoning in olivine constraining magma mixing-to-eruption timescales for the continental arc volcano Irazu (Costa Rica) and Cr diffusion in olivine. Am Miner 103(4):582–599. https://doi.org/10.2138/am-2018-6258

Oeser M, Weyer S, Horn I, Schuth S (2014) High-precision Fe and Mg isotope ratios of silicate reference glasses determined in situ by femtosecond

LA-MC-ICP-MS and by solution nebulisation MC-ICP-MS. Geostand Geoanal Res 38(3):311–328. https://doi.org/10.1111/j.1751-908X.2014.00288.x

Poitrasson F, Freydier R (2005) Heavy iron isotope composition of granites determined by high resolution MC-ICP-MS. Chem Geol 222(1–2):132–147. https://doi.org/10.1016/j.chemgeo.2005.07.005

Polyakov VB, Mineev SD (2000) The use of Moessbauer spectroscopy in stable isotope geochemistry. Geochim Cosmochim Acta 64(5):849–865. https://doi.org/10.1016/S0016-7037(99)00329-4

Rehkämper M, Halliday AN (1998) Accuracy and long-term reproducibility of lead isotopic measurements by multiple-collector inductively coupled plasma mass spectrometry using an external method for correction of mass discrimination. Int J Mass Spectrom 181:123–133. https://doi.org/10.1016/s1387-3806(98)14170-2

Roe JE, Anbar AD, Barling J (2003) Nonbiological fractionation of Fe isotopes: evidence of an equilibrium isotope effect. Chem Geol 195(1–4):69–85. http://doi.org/10.1016/s0009-2541(02)00389-3

Rudge JF, Reynolds BC, Bourdon B (2009) The double spike toolbox. Chem Geol 265(3–4):420–431. https://doi.org/10.1016/j.chemgeo.2009.05.010

Russo RE, Mao XL, Gonzalez JJ, Zorba V, Yoo J (2013) Laser ablation in analytical chemistry. Anal Chem 85 (13):6162–6177. https://doi.org/10.1021/ac4005327

Schauble EA, Rossman GR, Taylor HP (2001) Theoretical estimates of equilibrium Fe-isotope fractionations from vibrational spectroscopy. Geochim Cosmochim Acta 65(15):2487–2497. https://doi.org/10.1016/S0016-7037(01)00600-7

Schoenberg R, von Blanckenburg F (2005) An assessment of the accuracy of stable Fe isotope ratio measurements on samples with organic and inorganic matrices by high-resolution multicollector ICP-MS. Int J Mass Spectrom 242(2–3):257–272. https://doi.org/10.1016/j.ijms.2004.11.025

Sio CKI, Dauphas N, Teng FZ, Chaussidon M, Helz RT, Roskosz M (2013) Discerning crystal growth from diffusion profiles in zoned olivine by in situ Mg-Fe isotopic analyses. Geochim Cosmochim Acta 123:302–321. https://doi.org/10.1016/j.gca.2013.06.008

Skulan JL, Beard BL, Johnson CM (2002) Kinetic and equilibrium Fe isotope fractionation between aqueous Fe(III) and hematite. Geochim Cosmochim Acta 66 (17):2995–3015. https://doi.org/10.1016/S0016-7037(02)00902-X

Steinhoefel G, Horn I, von Blanckenburg F (2009a) Matrix-independent Fe isotope ratio determination in silicates using UV femtosecond laser ablation. Chem Geol 268(1–2):67–73. https://doi.org/10.1016/j.chemgeo.2009.07.010

Steinhoefel G, Horn I, von Blanckenburg F (2009b) Micro-scale tracing of Fe and Si isotope signatures in banded iron formation using femtosecond laser ablation. Geochim Cosmochim Acta 73(18):5343–5360. https://doi.org/10.1016/j.gca.2009.05.037

Steinhoefel G, von Blanckenburg F, Horn I, Konhauser KO, Beukes NJ, Gutzmer J (2010) Deciphering formation processes of banded iron formations from the Transvaal and the Hamersley successions by combined Si and Fe isotope analysis using UV femtosecond laser ablation. Geochim Cosmochim Acta 74(9):2677–2696. https://doi.org/10.1016/j.gca.2010.01.028

Stenberg A, Malinovsky D, Rodushkin I, Andren H, Ponter C, Ohlander B, Baxter DC (2003) Separation of Fe from whole blood matrix for precise isotopic ratio measurements by MC-ICP-MS: a comparison of different approaches. J Anal At Spectrom 18(1):23–28. https://doi.org/10.1039/b210482b

Stookey LL (1970) Ferrozine-A new spetrophotometric reagent for iron. Anal Chem 42:779–781

Strelow FWE (1980) Improved separation of iron from copper and other elements by anion-exchange chromatography on a 4-percent cross-linked resin with high-concentrations of Hydrochloric-Acid. Talanta 27 (9):727–732. https://doi.org/10.1016/0039-9140(80)80166-4

Tang YQ, Trassy C (1986) Inductively coupled plasma-the role of water in axial excitation temperatures. Spectrochim Acta Part B At Spectrosc 41(1–2):143–150. https://doi.org/10.1016/0584-8547(86)80146-x

Taylor PDP, Maeck R, Debievre P (1992) Determination of the absolute isotopic composition and atomic-weight of a reference sample of natural iron. Int J Mass Spectrom Ion Process 121(1–2):111–125. https://doi.org/10.1016/0168-1176(92)80075-c

Taylor PDP, Maeck R, Hendrickx F, Debievre P (1993a) The gravimetric preperation of synthetic mixtures if iron isotopes. Int J Mass Spectrom Ion Process 128(1–2):91–97. https://doi.org/10.1016/0168-1176(93)87018-n

Taylor PDP, Valkiers S, Debievre P, Flegel U, Kruck T (1993b) Stable-isotope analysis of iron by gas-phase electron-impact mass-spectrometry. Anal Chem 65 (21):3166–3167. https:doi.org/10.1021/ac00069a036

Ting BTG, Janghorbani M (1986) Inductively coupled plasma mass-spectrometry applied to isotopic analysis of iron in human fecal matter. Anal Chem 58(7):1334–1340. https://doi.org/10.1021/ac00298a014

Turnlund JR, Keyes WR (1990) Automated-analysis of stable isotopes of Zinc, Copper, Iron, Calcium, and Magnesium by thermal ionization mass-spectrometry using double isotope-dilution from tracer studies in humans. J Micronutr Anal 7(2):117–145

Valley GE, Anderson HH (1947) A comparison of the abundance ratios of the isotopes of terrestrial and meteoritic iron. J Am Chem Soc 69:1871–1875

Van der Walt TN, Strelow FWE, Haasbroek FJ (1985a) Separation of Fe-52 from Chromium cyclotron targets on the 2-percent cross-linked anion-exchange resin AG1-X2 in hydrochloric-acid. Talanta 32(4):313–317. https://doi.org/10.1016/0039-9140(85)80086-2

Van der Walt TN, Strelow FWE, Verheij R (1985b) The influence of crosslinkage on the distribution coefficients and anion-exchange behavior of some elements in hydrochloric-acid. Solvent Extr Ion Exc 3 (5):723–740. https://doi.org/10.1080/07366298508918536

Virtasalo JJ, Whitehouse MJ, Kotilainen AT (2013) Iron isotope heterogeneity in pyrite fillings of Holocene worm burrows. Geology 41(1):39–42. https://doi.org/10.1130/g33556.1

Vogl J, Klingbeil P, Pritzkow W, Riebe G (2003) High accuracy measurements of Fe isotopes using hexapole collision cell MC-ICP-MS and isotope dilution for certification of reference materials. J Anal At Spectrom 18(9):1125–1132. https://doi.org/10.1039/b301812a

Völkening J, Papanastassiou DA (1989) Iron isotope anomalies. Astrophys J 347(1):L43–L46

Walczyk T (1997) Iron isotope ratio measurements by negative thermal ionisation mass spectrometry using FeF4-molecular ions. Int J Mass Spectrom Ion Process 161(1–3):217–227. https://doi.org/10.1016/s0168-1176(96)04532-6

Welch S, Beard B, Johnson C, Braterman P (2003) Kinetic and equilibrium Fe isotope fractionation between aqueous Fe (II) and Fe (III). Geochim Cosmochim Acta 67(22):4231–4250

Weyer S, Anbar AD, Brey GP, Munker C, Mezger K, Woodland AB (2005) Iron isotope fractionation during planetary differentiation. Earth Planet Sci Lett 240(2):251–264. https://doi.org/10.1016/j.epsl.2005.09.023

Weyer S, Schwieters J (2003) High precision Fe isotope measurements with high mass resolution MC-ICPMS. Int J Mass Spectrom 226(3):355–368. https://doi.org/10.1016/s1387-3806(03)00078-2

Weyrauch M, Oeser M, Bruske A, Weyer S (2017) In situ high-precision Ni isotope analysis of metals by femtosecond-LA-MC-ICP-MS. J Anal At Spectrom 32(7):1312–1319. https://doi.org/10.1039/c7ja00147a

Whitehouse MJ, Fedo CM (2007) Microscale heterogeneity of Fe isotopes in > 3.71 Ga banded iron formation from the Isua Greenstone Belt, Southwest Greenland. Geology 35(8):719–722. https://doi.org/10.1130/g23582a.1

Yamakawa A, Yamashita K, Makishima A, Nakamura E (2009) Chemical separation and mass spectrometry of Cr, Fe, Ni, Zn, and Cu in terrestrial and extraterrestrial materials using thermal ionization mass spectrometry. Anal Chem 81(23):9787–9794. https://doi.org/10.1021/ac901762a

Yoshiaki K, Nishizawa M, Sawaki Y, Ueno Y, Komiya T, Yamada K, Yoshida N, Hirata T, Wada H, Maruyama S (2012) In situ iron isotope analyses of pyrite and organic carbon isotope ratios in the Fortescue Group: metabolic variations of a Late Archean Ecosystem. Precambr Res 2012:169–193

Yoshiya K, Sawaki Y, Hirata T, Maruyama S, Komiya T (2015a) In-situ iron isotope analysis of pyrites in similar to 3.7 Ga sedimentary protoliths from the Isua supracrustal belt, southern West Greenland. Chem Geol 401:126–139. https://doi.org/10.1016/j.chemgeo.2015.02.022

Yoshiya K, Sawaki Y, Shibuya T, Yamamoto S, Komiya T, Hirata T, Maruyama S (2015b) In-situ iron isotope analyses of pyrites from 3.5 to 3.2 Ga sedimentary rocks of the Barberton Greenstone Belt, Kaapvaal Craton. Chem Geol 403:58–73. https://doi.org/10.1016/j.chemgeo.2015.03.007

Zheng XY, Beard BL, Johnson CM (2018) Assessment of matrix effects associated with Fe isotope analysis using 266 nm femtosecond and 193 nm nanosecond laser ablation multi-collector inductively coupled plasma mass spectrometry. J Anal At Spectrom 33 (1):68–83. https://doi.org/10.1039/c7ja00272f

Zheng XY, Beard BL, Lee S, Reddy TR, Xu HF, Johnson CM (2017) Contrasting particle size distributions and Fe isotope fractionations during nanosecond and femtosecond laser ablation of Fe minerals: implications for LA-MC-ICP-MS analysis of stable isotopes. Chem Geol 450:235–247. https://doi.org/10.1016/j.chemgeo.2016.12.038

Zhu GX, Browner RF (1988) Study of the influence of water-vapor loading and interface pressure in inductively coupled plasma mass-spectrometry. J Anal At Spectrom 3(6):781–789. https://doi.org/10.1039/ja9880300781

Fe Isotope Fractionation Factors

3

3.1 Introduction

The Fe isotope system is used in a variety of Earth and planetary science fields, including high- and low-temperature applications. We have a significant understanding of the controls on Fe isotope fractionation and rates of Fe isotope exchange between different Fe-bearing species. Various studies have characterized Fe isotope fractionation factors and isotope exchange kinetics by empirical methods, including experimental studies and analysis of well-characterized natural samples. Moreover, multiple theoretical methods have been used to calculate Fe isotope fractionation factors (see Sect. 3.2). Overall, these studies have found that the largest Fe isotope fractionations are associated with redox processes. For example, the fractionation between aqueous Fe^{3+} and aqueous Fe^{2+} is 3‰ in $^{56}Fe/^{54}Fe$ ratios at 20 °C. These studies have also revealed that the magnitude of Fe isotope redox fractionations is a function of the Fe mineral produced and the aqueous Fe speciation, and that Fe redox couples such as Fe oxides and aqueous Fe^{2+} have rapid Fe isotope exchange rates that result in near-complete Fe isotope exchange even at low temperatures.

It is important to note the excellent agreement between experiments and theoretical calculations, which make Fe isotope investigations highly versatile geochemical tools. Indeed, one of the main goals of this review is to highlight a set of calculated reduced partition function ratios (used to calculate the equilibrium isotope fractionation factor) for different aqueous Fe species and minerals that agree well with experimental studies and are accurate and internally consistent with one another (see Sect. 3.6). Extrapolation of these derived fractionation factors and reduced partition function ratios can be applied to a wide variety of geochemical reactions over a range of temperature conditions, making Fe isotope investigations a highly versatile geochemical tool.

One of the focuses of this chapter is equilibrium Fe isotope fractionation between fluids and minerals from low-temperature to hydrothermal conditions. For a discussion of Fe isotope fractionation factors appropriate for high temperature studies, suitable for investigation of igneous or metamorphic processes, see Chap. 4. Specifically, this chapter focuses on the identification of equilibrium Fe isotope fractionation factors (Sect. 3.4) and methods used to investigate Fe isotope fractionation (Sects. 3.2 and 3.3). This chapter also evaluates kinetic processes where a unidirectional process may occur rapidly, prohibiting equilibrium isotope exchange from occurring. In natural settings, these kinetic issues may be important during the oxidation or reduction of Fe by microbial processes (Sect. 3.5). Kinetic processes can also make inference of equilibrium fraction from experiments very challenging. Section 3.4 discusses methods used in Fe isotope exchange

C. Johnson et al., *Iron Geochemistry: An Isotopic Perspective*, Advances in Isotope Geochemistry,
https://doi.org/10.1007/978-3-030-33828-2_3

experiments to unravel kinetic processes from experiments designed to capture equilibrium Fe isotope fraction.

3.2 Deriving Fe Isotope Fractionation Factors from First Principles

For studies that calculate Fe isotope fractionations, one derives reduced partition function ratios, β-values, for different Fe-bearing species, A and B. These β-values are related to the α-value (the fractionation factor between two species; Eq. 1.14, Chap. 1) using the relation:

$$\alpha_{A-B} = \frac{\beta_A}{\beta_B} \quad \text{or} \quad 1000\ln\alpha_{A-B} = 1000ln\beta_A - 1000ln\beta_B \tag{3.1}$$

Reduced partition function ratios are classically calculated using differences in zero-point energies and a harmonic oscillator approximation, as shown by Bigeleisen and Mayer (1947). There are numerous other treatments for evaluating β-values using a harmonic approximation and the reader is referred to the following references for generalized discussions (Dove 1993; Kieffer 1982; Schauble 2004; Young et al. 2015; Criss 1999). The most recent review was prepared by Blanchard et al. (2017). In general, β-values reflect differences in the vibrational frequencies between two different isotopes of the same element in a molecule. These vibrational controls are a function of bond stiffness and four main Fe-bonding characteristics control Fe isotope fraction:

(1) The nature of the chemical bond. A substance with covalent-like bonds (such as pyrite) will concentrate heavy Fe isotopes as compared to a substance with ionic-like bonds (such as iron oxides).
(2) A substance with stronger bonds tends to concentrate heavy Fe isotopes as compared to a substance with weaker bonds. Notably, shorter bonds tend to be stronger than longer bonds.
(3) Substances with low coordination numbers tend to have short, strong bonds and thus concentrate heavy Fe isotopes as compared to substances with Fe that has a higher coordination number.
(4) Iron in a higher redox state tends to have stronger bonds and concentrates heavy Fe as compared to Fe in a lower oxidation state.

Vibrational frequencies can be measured by a variety of spectroscopic methods. Schauble et al. (2001) used Raman, infrared, and inelastic neutron scattering spectroscopic measurements of Fe salts and metal, combined with a Modified Urey-Bradley Force Field (MUBFF) model, to calculate β-values for a number of Fe species. Spectroscopic observations based on the second order Doppler (SOD) shift measured by Mössbauer spectroscopy is another method used to calculate β-values. The SOD shift is proportional to the vibrational kinetic energy of Mössbauer-sensitive ^{57}Fe, as shown by Polyakov (1997), and this vibrational kinetic energy can be used to constrain β-values. However, this method is probably not as robust as β-values derived from neutron resonant inelastic X-ray scattering (NRIXS; in the literature this is also abbreviated as INRXS), because determining the SOD shift by Mössbauer entails collecting Mössbauer spectra over a range of temperatures and making assumptions about how the SOD shift changes as a function of temperature. Moreover, an isomeric shift in the Mössbauer spectra can also be confused with the SOD shift, although isomeric shifts tend to be much smaller than SOD shifts (Polyakov et al. 2007). Polyakov et al. (2005, 2007) and Dauphas et al. (2012) derived equations to relate NRIXS spectra to β-values. In the derivation by Polyakov et al. (2005), the β-value is calculated from the partial phonon density of states (pDOS), whereas in the derivation by Dauphas et al. (2012) the moments of the scatter spectrum, S(E), are used. Mathematically, the derivations are equivalent, but the S(E)-based derivation by Dauphas et al. (2012) may be easier to use because errors are not correlated between the different energy channels of the detector used to collect the NRIXS spectrum.

Moreover, assessment of the data reduction is more straightforward (see Blanchard et al. 2015 for a discussion of these different treatments). In addition, because NRIXS only works with Mössbauer-sensitive isotopes, such as ^{57}Fe, workers have been able to increase signal to noise ratio in collected NRIXS spectra by synthesizing minerals with pure ^{57}Fe (Dauphas et al. 2012; Roskosz et al. 2015).

First-principles electronic structure calculations are also used to constrain β-values. Two widely used approaches include the Hartee-Fock (HF) method and Density Functional Theory (DFT). The HF methods makes simplifying assumptions regarding the interaction of electrons, whereas DFT calculates electronic energies from electron densities (e.g., Schauble 2004). For Fe isotope studies, the HF method generally allows for more rapid calculations, but the DFT method is typically preferred because it is better suited for molecules containing transition elements (Schauble 2004). Additionally, hybrid DFT models are used. For example, the Becke three-parameter Lee-Yang-Par (B3LYP) method is well suited for large molecules and is frequently used in the earth sciences (Schauble 2004). All of these methods use a set of orbital-like basis functions called a basis set. Commonly used basis sets include 6-31G(d), 6-311G(d) and Ahlrich's VTZ. These basis sets differ in the degrees of freedom with which the electronic orbitals can be simulated, and the 6-311G(d) and Ahrlich's VTS have more Gaussian functions which may allow them to more accurately predict β-values as compared to the 6-31G(d) basis set (Hill and Schauble 2008).

β-values are calculated over a range of temperatures allowing investigation of Fe isotope fractionation over a wide range of conditions. In general, β-values follow a linear $1/T^2$ relationship (T in K), although at low temperatures ($\sim$150 °C) for molecules with high-frequency vibrations, this relationship tends toward a 1/T behavior (Blanchard et al. 2017). Thus, a plot of 1000ln β versus $1/T^2$ produces a concave-upward curve. This concave behavior is pronounced for atoms bonded directly to hydrogen, but for simple oxides the curve is nearly straight, even at low temperatures (Blanchard et al. 2017; Clayton and Kieffer 1991). In practice, 1000lnβ-values as a function of temperature are typically reported as a third order polynomial as a function of $10^6/T^2$ or as a tabulated set of β-values over a range of temperatures. For Fe isotopes, the curvature of β-values as a function of temperature is minor. For example, if the β-values discussed in Sect. 3.6 are calculated by extrapolation of the β-value as a function of $1/T^2$ from the β-value calculated at 0 °C to zero at infinite temperature, there is little difference (at most 0.2‰) between the extrapolated β-values and those calculated as a function of temperature from 0 to 300 °C. The largest difference between the β-values as originaly calculated and those estimated using a $1/T^2$ linear fit is for hexaquo Fe^{3+} (0.2‰). For the Fe oxides the difference is smaller (<0.1‰). Note that, in the literature, β-values are reported as a function of $^{57}Fe/^{54}Fe$ ($\beta^{57}{}_{Fe}/^{54}{}_{Fe}$) or $^{56}Fe/^{54}Fe$ ($\beta^{56}{}_{Fe}/^{54}{}_{Fe}$). The following relation is used to convert between these different β-values:

$$\beta_{^{56}Fe/^{54}Fe} \approx 0.678 \times \beta_{^{57}Fe/^{54}Fe} \tag{3.2}$$

In this chapter, all β-values are reported as a function of $^{56}Fe/^{54}Fe$. An important observation is that all of these different methods are independent of one another, and if the β-values for a substance determined by these different methods are similar, it is likely that calculated β-values are accurate.

3.3 Experimental Methods for Measuring Fe Isotope Fractionation Factors

Experimental methods devised to evaluate Fe isotope fractionation factors have relied heavily on methods developed to study O isotope fractionation in the 1960s through the 1990s by Robert Clayton and his students/coworkers at the University of Chicago (e.g., O'Neil 1986). These approaches include:

(1) Three-isotope method to characterize isotope exchange kinetics and to extrapolate isotope fractionation factors to complete isotope exchange.
(2) Reversed isotope exchange experiments where the starting isotope composition of one phase is changed relative to the other phase to determine if the same fractionation factor is determined.
(3) Partial exchange method in which the isotopic contrast between the initial two phases is varied.
(4) Synthesis methods used if isotope exchange between product and reactant is sluggish.

These different techniques, as discussed in detail below, were developed to evaluate if an experiment reflected laboratory-induced issues associated with synthesis of a mineral that experienced limited isotope exchange, as compared to formation in a natural environment where reaction rates can be much slower, allowing the phase to form at equilibrium. These kinetic issues are particularly problematic at low temperatures where laboratory studies may result in more rapid rates of mineral production, as compared to natural settings, because the mineral synthesis typically requires oversaturation to induce mineral formation in realistic timescales for laboratory experiments. Shahar et al. (2017) presents a review of experimental methods to constrain non-traditional stable isotope fractionations.

Synthesis methods are employed for many low-temperature studies because isotope exchange can be very sluggish. In this method, reactants are added at concentrations at or slightly above the solubility of a mineral, and then the system is changed, typically by adding a seed crystal or changing the pH to induce the mineral to form. Synthesis methods are most commonly used in the study of isotope fractionation between aqueous species and carbonate minerals, which are difficult to produce unless the system is oversaturated (e.g., Wiesli et al. 2004). Synthesis methods are also used as a check on isotope exchange experiments by evaluating if the fractionation factor changes as a function of precipitation rate (Frierdich et al. 2014a, b; Mansor and Fantle 2019).

In situations where isotope exchange occurs between two species, one can conduct reversal experiments in which the Fe isotope composition of component A is varied so that in one experiment it has a $^{56}Fe/^{54}Fe$ ratio greater than component B and in another experiment it has a $^{56}Fe/^{54}Fe$ ratio less than component B. If the final fractionation factor between A and B is the same in both experiments, the measured fractionation is unlikely to result from laboratory induced kinetic affects. The reversal method has been utilized in a number of Fe isotope studies between aqueous Fe^{2+} and iron oxyhydroxides (Frierdich et al. 2014a, b, 2019). An important aspect of all these exchange experiments is that there should not be a significant mass transfer between A and B during the duration of the experiment. Additionally, it is critical to characterize the products and reactants of the exchange experiment in terms of both mineralogy and speciation of dissolved Fe components.

The partial exchange method for isotope experiments, developed by Northrop and Clayton (1966), requires performing a variety of identical exchange experiments between two components that only differ in their initial isotopic contrast. As shown by Northrop and Clayton (1966), on a plot of the fractionation factor measured at the end of the exchange experiment versus the difference between fractionation factor at the end and beginning of the experiment, the experiments should define a line in which the y-intercept represents the fractionation factor at complete exchange, and the slope is a function of the amount of exchange (Fig. 3.1). Steep slopes are associated with little isotope exchange; the fractionation factor (y-intercept) is poorly determined because the line is sub-parallel to the y-axis. In contrast, if extensive isotope exchange has occurred, the slope would be less, with a slope of −1 corresponding to complete exchange. This method was initially developed for O isotope studies of carbonates (Northrop and Clayton 1966). The partial exchange method was applied to determine the fractionation factor between aqueous Fe and pyrite at hydrothermal conditions, as illustrated in Fig. 3.1 (Syverson et al. 2013).

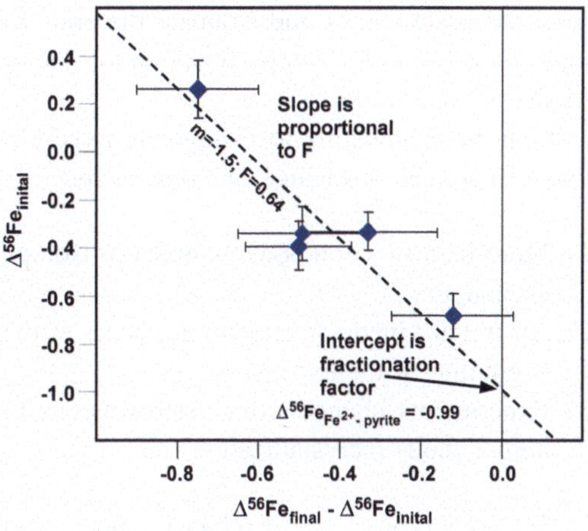

Fig. 3.1 Plot of $\Delta^{56}Fe_{aqueous\ Fe(II)–pyrite}$ from Fe isotope exchange experiments of Syverson et al. (2013). $\Delta^{56}Fe_{initial}$ is the difference between the $\delta^{56}Fe$ value of the starting ferrous chloride and the $\delta^{56}Fe$ value of the starting pyrite. $\Delta^{56}Fe_{final}$ is difference between the aqueous Fe and pyrite at the end of the experiment. The slope of the line defined by this experiment is proportional to the amount of exchange and the y-intercept is the $\Delta^{56}Fe_{aqueous\ Fe(II)–pyrite}$ at complete exchange. See Northrop and Clayton (1966) for derivation, figure modified from Syverson et al. (2013)

An extension of the partial exchange methodology is the three-isotope method, which rigorously documents the isotopic fractionation factor that would be measured at complete isotope exchange. To the degree that isotopic exchange occurs under equilibrium conditions, extrapolation produces the equilibrium isotope fractionation factor. This technique was initially developed by Matsuhisa et al. (1978) to study O isotope fractionation between quartz and water. The technique assesses the isotope exchange between two components that do not initially plot on the same mass-dependent isotope fractionation line because one component contains an enriched isotope relative to the other component (T = 0 in Fig. 3.2). Typically, a series of identical exchange experiments are performed for differing amounts of time, which leads to different amounts of isotopic exchange as shown by time points labeled T1-T3 in Fig. 3.2. At complete isotopic exchange, the two components will plot along a common "secondary" mass

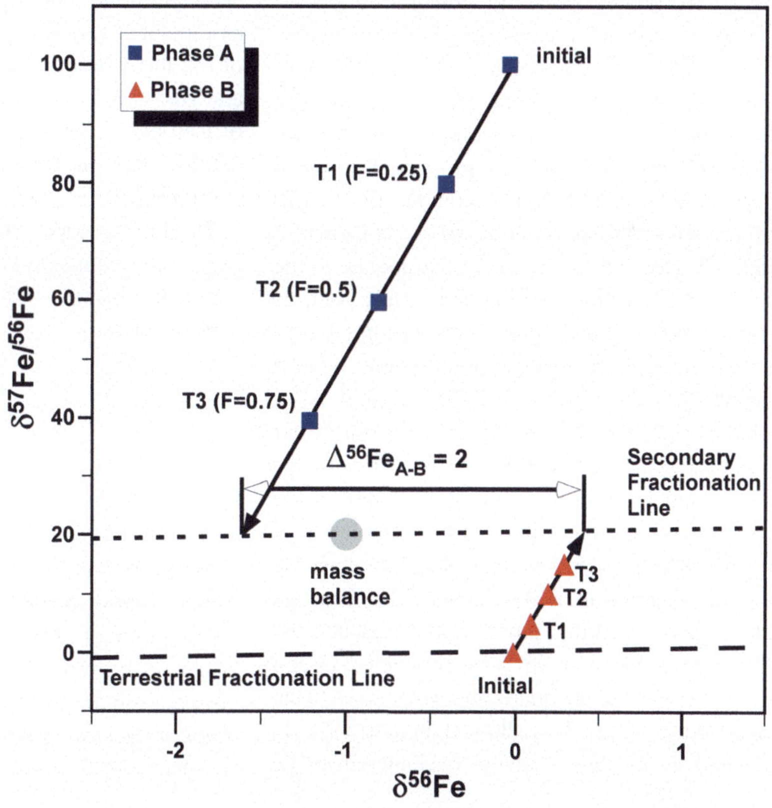

Fig. 3.2 Plot of $\delta^{56}Fe$ versus $\delta^{57}Fe/^{56}Fe$ for a typical Fe isotope exchange experiment between aqueous Fe^{2+} (phase A) and an iron oxide (phase B). In this example, phase A has been enriched by addition of a ^{57}Fe spike and phase B has a normal (non-enriched) Fe isotope composition. See text for discussion

fractionation line, which must pass through the isotopic composition of the system, as dictated by mass balance (Fig. 3.2). The slope of this secondary fractionation line is a function of the reduced masses of the two components and need not equal the slope of an inferred average terrestrial fractionation line (see Appendix 1 of Matsuhisa et al. 1978). For Fe isotopes, however, there is scant evidence for any resolvable differences in the mass-dependent slope of different Fe bearing species on a three isotope plot (except for some rare extraterrestrial samples and some magnetosomes; see Sect. 4.1 in Chap. 4, and Sect. 3.5.2), and the secondary fractionation line will be parallel to the terrestrial fractionation line. This stands in marked contrast to subtle but resolvable differences in the slope of different materials on a three-isotope oxygen isotope plot, where these differences in slope reflect different processes that produced O isotope exchange (e.g., Levin et al. 2014; Miller 2002; Rumble et al. 2007; Young et al. 2002). The difference in Fe isotope composition between the two components that lie on the "secondary" mass-fractionation line is the isotopic fractionation factor (Fig. 3.2). In the case of a simple two-component exchange that does not reach complete isotope exchange, one can extrapolate to the isotopic composition at complete exchange if the relative proportion of products and reactants is not varied in the time series experiments (Matsuhisa et al. 1978) as shown in Fig. 3.2. How far one needs to extrapolate to the secondary mass fractionation line largely controls the precision of the derived fractionation factor.

The choice of isotopic enrichment is also important. For example, in the original work conducted by Matsuhisa et al. (1978), the water was enriched in ^{18}O and the $\delta^{17}O$ values of the quartz and water were close to the equilibrium value. This required a significant extrapolation of the $\delta^{18}O$ to the equilibrium values based on the limited changes in $\delta^{17}O$ values. The Chicago group recognized this problem and repeated the experiments using water with enriched ^{17}O and the $\delta^{18}O$ values of the quartz and water close to the equilibrium values. This allowed them to robustly evaluate the kinetics of isotope exchange from the measured $\delta^{17}O$ values and evaluate the equilibrium quartz water O isotope fractionation factor (Matthews et al. 1983).

It can be challenging to extrapolate to conditions of complete exchange in cases where:

(1) More than two components undergo isotope exchange,
(2) An initial kinetic fractionation occurs at the beginning of the experiment, or
(3) Experiments change from heterogeneous to homogenous recrystallization.

This is because the exchange trajectories defined by the components may be curves rather than lines, and they may "over shoot" the secondary mass fractionation line (e.g., Beard et al. 2010; Frierdich et al. 2019; Matsuhisa et al. 1978). In such cases where a time series experiment defines a curve, a better practice may be to use only the last few time points to extrapolate to complete exchange (Beard et al. 2010; Frierdich et al. 2014a, b, 2019). Of course, if exchange has been minimal, significant uncertainty in the extrapolated values is likely. An alternative view of the three-isotope method is presented by Cao and Bao (2017), who suggest that if the kinetics of isotope exchange are not known a priori, the extrapolated fractionation factor may not be at equilibrium.

In three-isotope exchange experiments, the most common method to monitor the approach to complete isotope exchange is use of the expression:

$$F^t = \frac{\delta_a^t - \delta_a^i}{\delta_a^e - \delta_a^i} \tag{3.3}$$

where F^t is the fractional approach to mass balance at time t, δ_a^t is the Fe isotope composition of phase a at time t, δ_a^i is the initial Fe isotope composition of phase a, and δ_a^e is the Fe isotope composition of phase a at complete exchange. When using enriched isotope tracers, δ_a^e is essentially equivalent to the Fe isotope composition of the total system, which can be measured or calculated by mass balance because the small shifts in Fe isotope composition as a function of mass-dependent fractionation is small relative to the enriched Fe isotope composition. Figure 3.2 illustrates an example in which, at the start of the experiment, phase A has a $\delta^{57}Fe/^{56}Fe$ value of 100‰ and $\delta^{56}Fe$ value of 0‰, and phase B has a $\delta^{57}Fe/^{56}Fe$ and $\delta^{56}Fe$ value of 0‰. In this example, phase A has an Fe molar proportion of 0.2 and phase B, 0.8, and we also define that at complete exchange (F = 1) the $^{56}Fe/^{54}Fe$ fractionation between the two phases is $\Delta^{56}Fe_{a-b} = 2‰$.

If using the $\delta^{57}Fe/^{56}Fe$ value calculated from mass balance (20‰) instead of the $\delta^{57}Fe/^{56}Fe$ value that would reflect complete exchange (19.2‰), the F values calculated using the system mass-balance composition would be off by at most 1%. In contrast, in calculations using the non-enriched isotopes ($^{56}Fe/^{54}Fe$) and the mass-balance Fe isotope composition for δ^{e}_{a}, calculated F values would be in significant error (in this example F would be undefined because the denominator is 0). Note that an equivalent F equation can be written involving phase B; the F for each phase should match one another. In cases where the calculated F for each phase does not match, one needs to carefully evaluate if this is an artifact of the experiment caused by, for example, imperfect separation of nano-sized minerals from an aqueous solution. This internal check of analyzing both phases is an excellent method to evaluate the fidelity of the experimental method, but is not done in all studies. Note that use of the F parameter can also allow one to correct for a time series experiment in which multiple reactors may have some differences in their mass-balance values. It is often desirable to take the approach of sacrificing different reactors rather than sequentially sampling a single reactor, because the latter can produce large changes in mass balance. Using such a method, one could plot the $^{56}Fe/^{54}Fe$ ratio versus the F calculated for each component for the time series experiments and extrapolate to an F of 1, as compared to using the traditional three-isotope plot that requires the mass balance of the system for each time point to remain constant (e.g., Wu et al. 2011, 2012a).

Importantly, use of the F value as a measurement of complete isotope exchange cannot be used to identify the amount of atom exchange that has occurred between two phases (Gorski and Fantle 2017; Handler et al. 2014). The size of the different reservoirs and method of isotope exchange must be considered in order to calculate the amount of atom exchange. The end member models for isotope exchange are homogeneous and heterogeneous exchange. During homogeneous exchange, isotope equilibrium is maintained between the phases and both phases should be isotopically homogeneous. In the case of homogeneous exchange, the proportion of Fe atom exchange (AE) into phase a relative to the total amount of phase a at time t, $AE^{homogenous}_{a,t}$, as derived by Handler et al. (2014), is:

$$AE^{homogeneous}_{a,t} = \frac{N_b \times [\delta Fe^{i}_{b} - \delta Fe^{t}_{b}]}{N_a \times [\delta Fe^{t}_{b} - \delta Fe^{i}_{a}]} \quad (3.4)$$

where N is the moles of Fe in phase a and b, the δFe^{i} is the initial Fe isotope composition of phase a and b, and δFe^{t}_{b} is the Fe isotope composition of phase b at time t. Use of this equation requires that mass-dependent Fe isotope fractionation between a and b is negligible. When mass-dependent fractionation is significant (for example if one were conducting a study with little or no isotopic enrichment in one of the phases), an adjustment would need to be made for this mass-dependent fractionation (see Handler et al. 2014 for derivation).

During heterogeneous exchange, phases a and b do not isotopically re-equilibrate, thus one or both should be isotopically zoned. For a system that undergoes heterogeneous exchange, the proportion of atom Fe exchange in phase a at time t, for which mass-dependent fractionation is negligible, as derived by Frierdich et al. (2019), is:

$$AE_{a,t}^{heterogenous} = \frac{N_b}{N_a} \times ln\left(\frac{\delta Fe_a^i - \delta Fe_b^i}{\delta Fe_a^i - \delta Fe_b^t}\right) \quad (3.5)$$

A hybrid model has also been derived in which it is assumed that a combination of processes takes place during atom exchange (Joshi et al. 2017).

When the goal of isotope exchange experiments is to evaluate the isotope fractionation factor between components, the amount of atom exchange is not important because one can extrapolate to a secondary fractionation line without knowing the amount of atom exchange. However, if the goal is to understand recrystallization of, for example, iron hydroxides when exposed to aqueous ferrous iron, an understanding of these different models is critical because the amount of recrystallization affects the release of metals and nutrients absorbed onto the iron hydroxide. For example, the amount of atom exchange for each time point in Fig. 3.2 is shown in Table 3.1. The Fe mass balance of the system is largely contained in one phase, and the amount of atom exchange calculated using Eqs. 3.4 or 3.5 is very different as compared to the F value calculated using Eq. 3.3. The amount of atom exchange also differs depending on if a homogeneous or heterogeneous exchange model is used.

Gorski and Fantle (2017) proposed novel methods to evaluate homogenous and heterogeneous exchange by conducting experiments in which the mass balance of components A and B are varied over a wide range. Although these types of experiments have yet to be conducted, others (e.g., Zhou et al. 2018) have used partial leaching methods to identify if an iron hydroxide reacted with aqueous ferrous iron is isotopically zoned. Other researchers have performed reversed three-isotope exchange experiments using materials with different surface areas to evaluate these processes (Frierdich et al. 2019). In general, these studies have shown that during isotope exchange between aqueous Fe^{2+} and iron hydroxides, initially rapid exchange typically occurs when the iron hydroxide is first exposed to the Fe^{2+} and the aqueous Fe^{2+} is sorbed onto the surface of the iron hydroxide. This initial process probably occurs by heterogeneous exchange and likely involves significant kinetic isotopic fractionation. During later stages of these experiments, exchange is slower and appears to occur by homogeneous exchange (Frierdich et al. 2019), and equilibrium fractionation seems to erase, at least in part, the initial kinetic fractionations. These observations may help explain the observation that the morphology of goethite before and after long term ($\sim$ 14 days) exposure to aqueous Fe^{2+} appears identical (Handler et al. 2014), but that during the initial rapid reaction there are subtle differences in the length to width aspect ratio of the reacted goethite as compared to the initial and final goethite (Joshi and Gorski 2016).

3.3.1 The Role of Sorption in Isotope Exchange

An important observation from the Fe isotope exchange studies noted above between Fe^{2+} and iron hydroxides is that the process of sorption should not be viewed in isolation. Initial studies suggested that Fe isotope fractionation between aqueous Fe^{2+} and iron hydroxides was caused by sorption, but this was based solely on analyzing the aqueous Fe^{2+} isotope composition before and after exposure to iron oxyhydroxide (Icopini et al. 2004). In later studies, researchers realized that the process of sorption also led to atom and electron exchange between the aqueous Fe^{2+} and iron hydroxide (e.g., Crosby et al. 2005; Handler

Table 3.1 Calculated F values using different exchange models

Time	$AE_{a,t}^{homogenous}$	$AE_{a,t}^{heterogenous}$
T1 (F = 0.25)	0.063	0.056
T2 (F = 0.5)	0.169	0.129
T3 (F = 0.75)	0.385	0.233

et al. 2009; Mikutta et al. 2009). Therefore, one cannot view the sorption processes in isolation.

Of course, in abiological systems, sorption is important for atom exchange. Experiments conducted at low pH, where the negative charge of iron hydroxide is balanced by H^+ instead of by the sorption of other cations, no significant Fe isotope exchange occurs between aqueous Fe^{2+} and iron oxyhydroxides (Reddy et al. 2015). Moreover, the importance of both atom and electron exchange between Fe^{2+} and iron oxyhydroxides is illustrated by the fact that, if one uses enriched ^{57}Fe aqueous Fe^{3+} instead of Fe^{2+}, isotopic exchange is limited to only the surface atoms of the iron oxydyroxides (Poulson et al. 2005). In contrast, for many experiments that use ferrous iron and iron hydroxides, the amount of atom exchange is 90%, which far exceeds the number of surface atoms of the iron oxyhydroxide. Electron and atom exchange between aqueous Fe^{2+} and ferric substrates is not limited only to iron oxides; significant Fe exchange occurs between aqueous Fe^{2+} and ferric Fe clays (Neumann et al. 2015; Shi et al. 2016).

In exchange experiments, it is also critical to identify all components that undergo Fe isotope exchange. In the case of three-isotope exchange experiments between aqueous Fe^{2+} and goethite, Beard et al. (2010) conducted partial leaching studies that analyzed both Fe isotope compositions and determined ferrous and ferric iron concentrations using spectroscopic methods. They found at least four components with unique $^{56}Fe/^{54}Fe$ ratios. From lowest to highest $^{56}Fe/^{54}Fe$ ratios, these include: aqueous Fe^{2+}, sorbed Fe^{2+}, goethite, and an easily-leachable Fe^{3+} component. This latter component is thought to be the surface atoms of goethite which have different Fe bonding than the interior goethite atoms where all Fe is in octahedral coordination with oxygen. This is similar to the reactive ferric layer as identified in Crosby et al. (2005, 2007). The difference in Fe bonding of goethite surface atoms is thought to explain the larger fractionation between goethite and aqueous Fe^{2+} in experiments that use nanometer-sized goethite rods. These rods have a higher surface area as compared to experiments that use micrometer-sized goethite rods (Beard et al. 2010; Frierdich et al. 2014a).

3.4 Equilibrium Fractionation of Fe Isotopes: Working Toward a Unified Set of Fractionation Factors

As discussed above, Fe isotope fractionation factors are derived by empirical methods, by first-principles calculations, and by using spectroscopic measurements such as Mössbauer and NRIXS. Consistency between different methods is evidence that the inferred fractionation is likely to reflect equilibrium isotope partitioning and does not suffer from laboratory-induced fractionation or imperfections in assumed bonding or vibrational energies. Moreover, because equilibrium fractionation is essentially a thermodynamic property, these fractionation factors are additive and can be applied to unstudied systems. In other words, it is possible to use the β-values determined for a variety of aqueous Fe species to calculate the β-values for a specific chemical composition in which the aqueous Fe may have a variety of Fe species. For example, in a hydrothermal experiment with chloride present, the aqueous ferrous iron may include hexaquo Fe^{2+} and different ferrous iron chloro complexes (e.g., Saunier et al. 2011). To determine a composite β-value for the dissolved Fe, one could sum the product of the β-values and the mole fraction for the different aqueous Fe complexes. Similarly, it is possible to sum the fractionation measured between hexaquo ferrous and ferric iron ($\Delta^{56}Fe_{Fe(II)-Fe(III)} = -3.01‰$) (Welch et al. 2003) and hexaquo ferric iron and hematite ($\Delta^{56}Fe_{Fe(III)-hematite} = -0.15‰$) (Skulan et al. 2002) to calculate the fractionation between hexaquo Fe^{2+} and hematite ($\Delta^{56}{}_{Fe(II)-hematite} = -3.16‰$) at 20 °C (Wu et al. 2010).

3.4.1 Aqueous Fe Species

Some of the first investigations were devoted to understanding Fe isotope fractionation between different dissolved Fe species. These studies included both experiments, first-principles calculations, and vibrational data obtained from spectroscopy coupled with force field modelling using MUBFF (e.g., Johnson et al. 2002; Schauble et al. 2001; Anbar et al. 2005; Welch et al. 2003). These studies showed that significant fractionation between aqueous Fe^{2+} and Fe^{3+} should occur. Experimentally, investigating the isotopic fractionation between aqueous species is challenging because it is difficult to separate different Fe species from a single liquid. A number of methods have been employed to separate aqueous Fe species that make use of these characteristics:

(1) Fe^{3+} is insoluble at circumneutral pH (Dideriksen et al. 2008; Johnson et al. 2002; Welch et al. 2003).
(2) $Fe^{(III)}Cl_4^-$ dissolves into ether and ether is immiscible in aqueous liquids, allowing separation of $Fe^{(III)}Cl_4^-$ from other aqueous Fe species (Fujii et al. 2006; Hill et al. 2009, 2010).
(3) High molecular weight organic Fe species can be separated using dialysis bags with different molecular weight cut-offs (Morgan et al. 2010).
(4) Ferric iron forms anion complexes that absorb to anion-exchange resin, whereas ferrous iron and many ferric species are cations and do not absorb on to resin (Anbar et al. 2000; Matthews et al. 2001; Roe et al. 2003).

Identifying equilibrium fractionation between different Fe species requires the separation method to be faster than the isotope exchange rate between the Fe species. Welch et al. (2003) added $BaCO_3$ to low-pH mixed Fe^{3+}–Fe^{2+} solutions to rapidly neutralize the solution and cause Fe^{3+} to flocculate and precipitate as an iron hydroxide, allowing separation of the Fe^{3+} and Fe^{2+} components. They used $BaCO_3$ instead of $CaCO_3$ to avoid formation of $^{40}Ca^{16}OH$ during isotopic analysis, because this molecule is isobaric with ^{57}Fe and cannot be removed using a collision cell (see Sect. 2.2 in Chap. 2). It took about 1 s for the Fe^{3+} to flocculate for experiments done at 22 °C, and about 2 s for experiments done at 0 °C. Companion experiments using an enriched isotope tracer evaluated isotope exchange rates between Fe^{3+} and Fe^{2+}. Based on the rate of separation of Fe^{3+} from Fe^{2+} versus the isotope exchange rate between the species, it was possible to determine that there was approximately 30% Fe isotope exchange between the Fe^{3+} and Fe^{2+} components during precipitation for the 22 °C experiments and approximately 10% for the experiments done at 0 °C. This information was used to correct for isotope exchange during the separation method. These corrections produced the same $\Delta^{56}Fe_{Fe(III)-Fe(II)}$ over a range of Fe^{2+} to total Fe ratios. Overall, the researchers found that at 22 °C, $\Delta^{56}Fe_{Fe(III)-Fe(II)} = 3.01‰$ and at 0 °C it was 3.57‰. They conducted experiments over a range of Cl concentrations from 0 to 111 mM Cl, which had a minor effect on Fe speciation because some Fe formed chlorine complexes. These experiments measured no resolvable differences in the isotope fractionation factors as a function of Cl. In these experiments, the predominant Fe species were Fe^{3+} and Fe^{2} hexaquo species $Fe^{(III)}(H_2O)_6^{3+}$ and $Fe^{(II)}(H_2O)_6^{2+}$, and this fractionation factor is taken to represent the fractionation between hexaquo Fe^{3+} and Fe^{2+}.

For comparison, a number of workers have calculated β-values for $Fe^{(III)}(H_2O)_6^{3+}$ and $Fe^{(II)}(H_2O)_6^{2+}$ using a variety of methods (Anbar et al. 2005; Domagal-Goldman and Kubicki 2008; Fujii et al. 2014; Hill and Schauble 2008; Hill et al. 2010; Moynier et al. 2013; Ottonello and Zuccolini 2008, 2009; Rustad et al. 2010; Schauble et al. 2001). For the $Fe^{(III)}(H_2O)_6^{3+}$ species, workers calculated β-values ranging from 11.65 to 7.50 at 22 °C, while β-values at 22 °C ranged from 6.82–4.66 for the $Fe^{(II)}(H_2O)_6^{2+}$ species (Fig. 3.3a). The variations in the reduced partition function ratios for hexaquo Fe^{3+} is nearly as large as the range in $^{56}Fe/^{54}Fe$ ratios measured in nature. The calculated 1000lnα values from these β-values defines a narrower range (5.40–2.80), and many of these calculated 1000lnα values agree with the experiments of Welch et al. (2003) (Fig. 3.3b).

To better evaluate which hexaquo Fe β-values are most accurate, other experiments have used aqueous Fe species and minerals to determine the β-value for hexaquo Fe^{2+} and Fe^{3+} that consistently agrees with different experiments (Beard et al. 2010; Polyakov and Soultanov 2011;

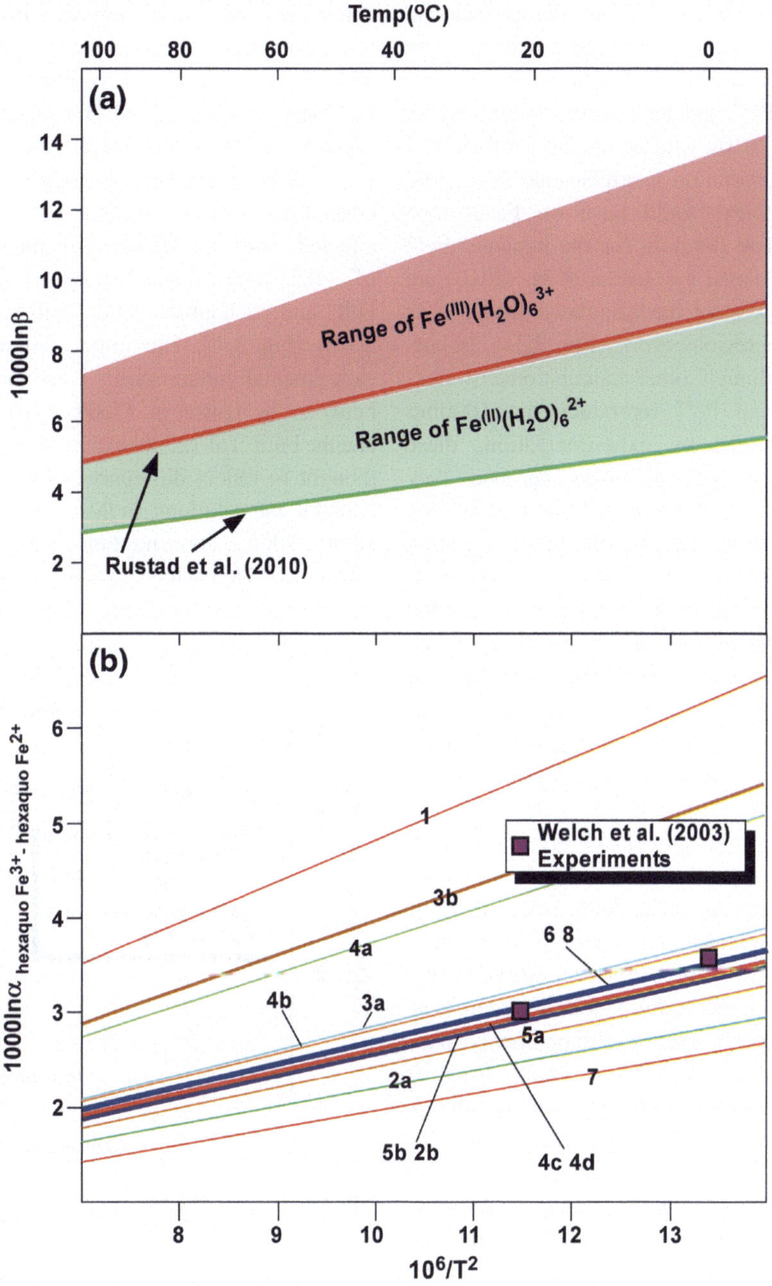

Fig. 3.3 **a** Plot of 1000lnβ values as a function of $10^6/T^2$ (T in Kelvin) for hexaquo Fe^{3+} (red field) and hexaquo Fe^{2+} (green field). The β-values calculated from the work of Rustad et al. (2010) are shown because these β-values agree well with aqueous Fe experiments and aqueous Fe mineral fractionation studies (Polyakov and Soultanov 2011; Rustad et al. 2010, see text for discussion). **b** Plot of $1000\ln\alpha_{\text{hexaquoFe(III)–hexaquoFe(II)}}$ versus $10^6/T^2$. The numbered curves correspond to the different studies and the letters to the different formulations used in the study to calculate the β-values. All curves but 1 are based on first-principles calculations; curve 1 used spectroscopy data to determined vibrational frequencies and a force field model to calculate β-values. **1** Schauble et al. (2001) MUBFF, **2a** Anbar et al. (2005) 1-DFT-PCM, **2b** 2-DFT, **3a** Ottonello and Zuccolini (2009) B3LYP (gas), **3b** PCM-HF-UAHF (solute), **4a** Hill and Schaulbe (2008) for Fe(III)and Hill et al. (2010) for Fe(II) UHF/6-31G(d), **4b** B3LYP/6-31G(d), 4C B3LYP/6311G(d), **4d** B3LYP/VTZ, **5a** Domgal-Goldman and Kubicki (2008) in vacuo; **5b** 2-IEFPCM, **6** Rustad et al. (2010) S_6, **7** Moynier et al. (2013), and **8** Fujii et al. (2014). The squares are the experimental results of Welch et al. (2003) done at 0 and 22 °C

Rustad et al. 2010). The β-values that provide the most consistent results between calculated and experimentally-measured fractionation factors for hexaquo Fe^{3+} and Fe^{2+} were determined by Rustad et al. (2010), who accounted for the effect that the outer hydration sphere around the central hexaquo complex would have on Fe isotope partitioning. The β-values for the hexaquo Fe^{2+} and Fe^{3+} calculated by Rustad et al. (2010) are among the lowest of the calculated β-values as compared to previous work (Fig. 3.3a). Importantly, even though other calculations of hexaquo Fe^{2+} and Fe^{3+} reproduce the 1000lnα values determined by experimentation, these calculations are *not accurate* because they exclude the effect of the outer hydration sphere. Indeed, because the effect of the outer hydration sphere is similar for Fe^{3+} and Fe^{2+}, a fortuitous error-cancelling occurred, making the calculated fractionation factor seem accurate if β-values are combined, even though the β-value is too large to be consistent with aqueous Fe-mineral fractionations. This issue is critical for applying β-values in an additive manner, which is possible only if they are accurate.

The effects that different Fe chlorine complexes have on Fe isotope fractionation has been experimentally studied (Fujii et al. 2006; Hill et al. 2009, 2010), and a number of different β-values have been calculated for these different species (Fujii et al. 2006, 2014; Hill and Schauble 2008; Hill et al. 2010). Iron isotope exchange experiments made use of the fact that the $Fe^{(III)}Cl_4^-$ complex dissolves into ether compounds, and ether is immiscible in aqueous solutions. This immiscibility allows separation of $Fe^{(III)}Cl_4^-$ from other aqueous Fe species simply by adding ether to an aqueous Fe solution. Immiscibility experiments using ether compounds have been conducted over a wide range of Cl concentrations to evaluate how Fe speciation of aqueous Fe solutions affects fractionation between mixed aqueous Fe^{3+} species and $Fe^{(III)}Cl_4^-$. These aqueous Fe species include different proportions of $Fe^{(III)}(H_2O)_6^{3+}$, $Fe^{(III)}(H_2O)_5Cl^{2+}$, $Fe^{(III)}(H_2O)_4Cl_2^+$, $Fe^{(III)}(H_2O)_3Cl_3$, and $Fe^{(III)}Cl_4^-$. In addition, a set of experiments was performed over a range of chloride and Fe^{2+} to total Fe ratios. These mixed Fe valance state experiments allowed extrapolation to conditions of pure ferrous iron, allowing researchers to investigate fractionations associated with Fe^{2+} chloride species, including $Fe^{(II)}(H_2O)_6^{2+}$, $Fe^{(II)}(H_2O)_5Cl^+$, and $Fe^{(II)}(H_2O)_4Cl_2$. In these experiments, Fe isotope fractionation of the mixed Fe^{2+} species was relatively insensitive to chloride concentration (Fig. 3.4), but Fe^{3+} species were affected with a decrease in the fractionation of ~0.25‰ per mole of chloride (Fujii et al. 2006; Hill and Schauble 2008; Hill et al. 2009, 2010) (Fig. 3.4). The minor differences in the experimentally-determined fractionation factors between the Hill et al. (2009, 2010) experiments and the Fujii et al. (2006) experiments (Fig. 3.4) are thought to reflect differences in the way Cl was added to the solutions. In the experiments by Fujii et al. (2006), concentrations were adjusted by addition of HCl which resulted in ionic strength

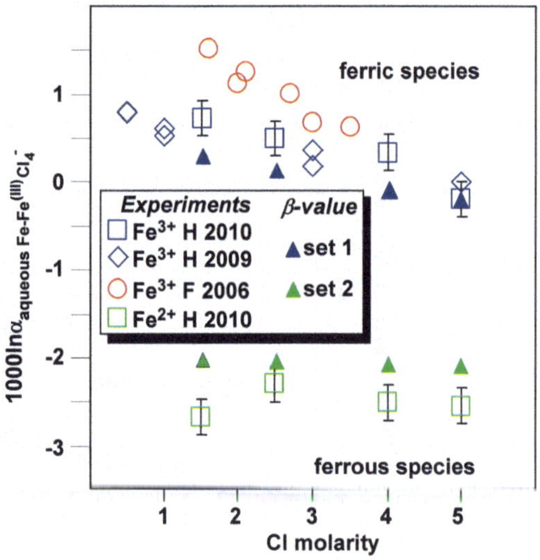

Fig. 3.4 Plot of $1000\ln\alpha_{\text{aqueous Fe–FeCl4}}$ versus Cl concentration. Aqueous Fe is ferrous for the green symbols and ferric for the blue and red symbols (exclusive of $FeCl_4^-$). The open symbols are experimental results for the measured fractionation factor between aqueous Fe and $FeCl_4^-$ as a function of Cl concentration (Fujii et al. 2006; Hill et al. 2009, 2010). The closed symbols are the calculated fractionations for the experiments conducted by Hill et al. (2010) using the Fe speciation reported in Hill et al. (2010) and the β-values for the Fe chloro complexes used the scaled UHF/6-31G(d) formulation (Hill and Schauble 2008; Hill et al. 2010) and for hexaquo Fe, the Rustad et al. (2010) β-values were used. Please see text for a discussion on scaling of β-values. β-values set 1 is for Fe^{3+} complexes and set 2 is for the Fe^{2+} species

variations in the different solutions, whereas the Hill et al. (2009, 2010) experiments kept the ionic strength of the solutions constant as a function of chloride concentration.

Calculated β-values for different Fe^{3+} chlorine complexes are consistent with the general trend of decreasing fractionation with increasing chloride concentration. However, all β-values for the $Fe^{(III)}(H_2O)_5Cl_2^{2+}$ reported by Hill and Schauble (2008) are higher than the Rustad et al. (2010) hexaquo Fe^{3+}, which is inconsistent with the experimental results. Therefore, following a method similar to that of Clayton and Kieffer (1991), here we scale the Fe^{3+} β-values of Hill and Schauble (2008) to be consistent with the Rustad et al. (2010) hexaquo and Fe^{3+} β-value. This correction was done by calculating a scaling factor based on the ratio of the hexaquo Fe^{3+} determined by Rustad et al. (2010) and the different hexaquo Fe^{3+} formulations of Hill and Schauble (2008).

There are four different formulations for the Fe^{3+} and Fe^{2+} β-values calculated by Hill and Schauble (2008) and Hill et al. (2010). One formulation used the unrestricted HF method (UHF) and the 6G-31(d) basis set, while three formulations used the B3LYP method and three different basis sets including 6-31G(d), 6-311G (d), and Ahrlich's VTZ. Thus the UHF/6G-31(d) formulation for all of the Fe^{3+} chloro complexes were scaled to the ratio of Rustad et al. (2010) hexaquo Fe^{3+} to the UHF/6-31G(d) hexaquo Fe^{3+}. The other formulations were scaled in a similar fashion. This scaling method resulted in good agreement between the fractionation factors calculated from the β-values and the experiments. Figure 3.4 shows the $1000\ln\alpha_{Fe(III)-Fe(III)Cl4}$ using the UHF/6-31G(d) formulation. The scaled β-value for the $Fe^{(III)}Cl_4^-$ complex is sub-equal to the unscaled Fe^{2+} chloro complex β-values. This is inconsistent with the experimental results that indicate that Fe^{2+} complexes should have $\delta^{56}Fe$ values that are 2.5‰ less than $Fe^{(III)}Cl_4^-$. Therefore the Fe^{2+} chloro complexes have been scaled in a fashion similar to that of the Fe^{3+} chloro complexes using the ratio of Rustad et al. (2010) Fe^{2+} hexaquo complex to that of the different Fe^{2+} hexaquo formulations calculated by Hill et al. (2010). The best fit between the Fe^{2+} chloro complex experiments and calculated β-values is based on the UHF/6-31G(d) formulations (Fig. 3.4). The other formulations predict a smaller fractionation between the different Fe^{2+} complexes and the $Fe^{(III)}Cl_4^-$ complex (not shown in Fig. 3.4). The good fit between the β-values calculated using the UHF/6-31G(d) formulation and the experiments is largely driven because this formulation predicts that the $Fe^{(III)}Cl_4^-$ complex should have a large β-value as compared to the $Fe^{(III)}(H_2O)_3Cl_3$ complex. In contrast, the B3LYP formulations using the different basis sets predict only a small increase for the $Fe^{(III)}Cl_4^-$ complex (6-31G(d) basis set) or no increase (6-311G(d) and Ahlrich's VTZ basis sets). The fact that the UHF/6-31G(d) formulation normalized to the Rustad et al. (2010) Fe^{2+} and Fe^{3+} hexaquo β-values provides the best fit to the experiments is somewhat surprising because the B3LYP model with the 6-311G(d) or Ahrlich's VTZ basis sets tend to be more accurate (Hill and Schauble 2008). However, the increase in the $Fe^{(III)}Cl_4^-$ β-value predicted using the UHF/6-31G(d) model is consistent with the general rule that molecules with lower coordination tend to have stronger bonds because Fe in the $Fe^{(III)}(H_2O)_3Cl_3$ complex is octahederally coordinated whereas Fe in the $Fe^{(III)}Cl_4^-$ complex is in tetrahedral coordination.

This comparison of calculated Fe^{3+} and Fe^{2+} chloro complex β-values to experiments has not considered combining β-values from different formulations. In fact, it is best not to combine β-values from different formulations to calculate fractionation factors unless one is able to verify the accuracy of the β-value (Blanchard et al. 2012) by comparison to experiments or independent calculation methods such as from NRIXS. Because more than two species are present in these experiments, it is difficult to evaluate the accuracy of any single β-value. Note that other workers (Fujii et al. 2014) have calculated β-values for some Fe chloro complexes that are very similar to those from Hill and

Schauble (2008), but it is impossible to distinguish which formulation provides the best fit to the experiments because not all studies calculated β-values for the $Fe^{(III)}Cl_4^-$ complex.

The effect that organic ligands complexed with aqueous Fe has on Fe isotope fractionation has also been investigated by experimentation and from first-principles calculations (Dideriksen et al. 2008; Domagal-Goldman et al. 2009; Fujii et al. 2014; Morgan et al. 2010; Moynier et al. 2013). In experiments conducted by Dideriksen et al. (2008), the Fe isotope fractionation between hexaquo Fe^{3+} and Fe^{3+} complexed with desferrioxamine-B (DFOB) (a siderophore that strongly binds Fe^{3+}), was $\Delta^{56}Fe_{FeDFOB\text{–}hexaquo\ Fe(III)}$ of 0.6‰ at room temperature (Fig. 3.5). Other experiments separated Fe^{3+} species complexed with organic ligands using dialysis membranes with different molecular weight cutoffs. These researchers identified a small but resolvable fractionation between Fe^{3+} complexed with DFOB and Fe^{3+} complexed with oxylate (Fig. 3.5), but no resolvable fractionation between Fe^{3+}–DFOB and Fe^{3+} complexed with EDTA (Morgan et al. 2010). A variety of β-values have been calculated for a host of aqueous Fe species complexed with organic ligands including Fe^{3+}–DFOB (Domagal-Goldman et al. 2009), ferrous and ferric Fe complexed with acetate (Ottonello and Zuccolini 2008), oxalate (Domagal-Goldman and Kubicki 2008; Domagal-Goldman et al. 2009), triscatecholate (Domagal-Goldman and Kubicki 2008), and citrate (Fujii et al. 2014; Moynier et al. 2013). The agreement between the Dideriksen et al. (2008) experiments and the β-values for Rustad et al. (2010) for Fe^{3+} hexaquo complex, and Domagal-Goldman et al. (2009) Fe^{3+}–DOB is excellent, as are the results for the Fe^{3+}–DFOB and the Fe^{3+} oxylate experiment done by Morgan et al. (2010) and the Domagal-Goldman et al. (2009) β-value for Fe^{3+}–DFOB and the Fe^{3+} oxylate reported by Domagal-Goldman and Kubicki (2008) 1-In vacuo formulation (Fig. 3.5).

Calculated β-values and these Fe organic ligand experiments predict Fe isotope fractionation between the Fe^{3+} species increases as the difference in their binding affinities increases, with higher $^{56}Fe/^{54}Fe$ ratios associated with Fe^{3+} bound to the organic ligand with a higher binding affinity (e.g., Morgan et al. 2010). This observation stands in contrast to leaching studies in which minerals leached by organic ligands produced leached ligand-bound Fe^{3+} with a lower $^{56}Fe/^{54}Fe$ as compared to the mineral (Brantley et al. 2004; Wiederhold et al. 2006). As noted by Morgan et al. (2010) these differences are thought to reflect kinetic effects in the leaching experiments because the Fe was not able to equilibrate with Fe in the minerals. In other leaching studies that used HCl or oxalic acid under anoxic conditions, Fe leached from rocks and minerals tended to have lower $^{56}Fe/^{54}Fe$ ratios as compared to the leached material, but that over time the leached Fe had a $^{56}Fe/^{54}Fe$ ratio closer to the Fe isotope composition of the substrate (Chapman et al. 2009; Kiczka et al. 2010). In all these leaching studies, the fractionation is interpreted to reflect kinetic effects, and the change in leached Fe isotope compositions with time is thought to represent partial

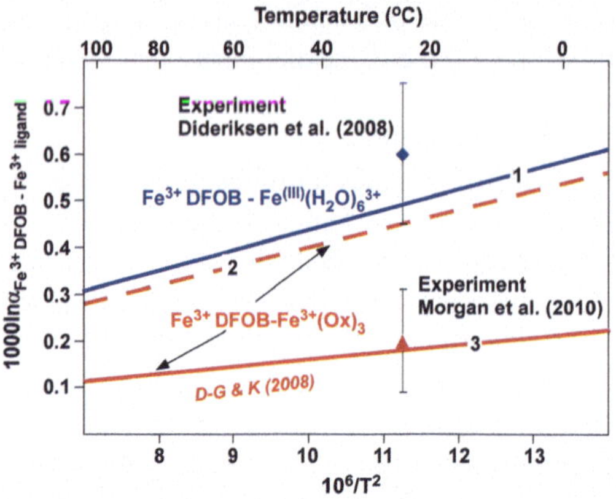

Fig. 3.5 Plot of $1000\ln\alpha_{Fe\ DFOB\text{–}Fe\ ligand}$ as a function of $10^6/T^2$ (T in Kelvin). The symbols are experiments showing the fractionation factor between Fe^{3+}–DFOB and hexaquo Fe^{3+} complex done by Dideriksen et al. (2008), and between Fe^{3+}–DFOB and Fe^{3+} oxalate ($Fe(Ox)_3$) by Morgan et al. (2010). The blue curve labeled 1 shows the calculated fractionation using the β-values for hexaquo Fe^{3+} by Rustad et al. (2010) and Domagal-Goldman et al. (2009) for Fe^{3+}–DFOB complex. The red curves show calculated fractionation using the Fe^{3+}–DFOB β-value by Domagal-Goldman et al. (2009) and the dashed curve labeled 2 uses the $Fe(Ox)_3$ β-value by Domagal-Goldman et al. (2009) and the solid curve labeled 3 uses the Fe $(Ox)_3$ β-value by Domagal-Goldman and Kubick (2008) 1-In vacuo formulation

leaching of a substrate with a high $^{56}Fe/^{54}Fe$ ratios that was generated by the initial leaching (Chapman et al. 2009) or perhaps a result of Fe isotope exchange between aqueous Fe^{2+} and ferric materials.

Several other Fe speciation experiments used ion-exchange resin to separate different Fe species (Anbar et al. 2000; Matthews et al. 2001; Roe et al. 2003). These experiments showed that Fe speciation can be important in causing Fe isotope fractionation. However, these experiments suffered from kinetic effects, and it was challenging to interpret the results under conditions that would be close to equilibrium. For example, Matthews et al. (2001) separated ferric iron in HCl from Fe^{2+} complexed with 2,2′-bipyridine using anion exchange resin and 6 M HCl, and they found that the Fe^{2+} 2,2′-bipyridine complex broke down during exposure to high molarity HCl. This breakdown caused partial exchange between $Fe^{(III)}Cl_4^-$ bound to the resin and other Fe components, causing significant kinetic fractionation. In these experiments, the Fe^{2+} complexed with 2,2′-bipyridine had a greater $^{56}Fe/^{54}Fe$ ratio as compared to the ferric components. This observation stands in marked contrast to most experiments and calculations that indicate ferric iron has a higher $^{56}Fe/^{54}Fe$ ratio compared to ferrous species. It is thought that the very high $^{56}Fe/^{54}Fe$ ratio of the Fe^{2+} 2,2′-bipyridine complex is caused by the strong covalent bonding between Fe^{2+} and the 2,2′-bipyridine molecule. In general the calculated β-values for Fe^{2+} complexed with organic ligands suggest that Fe^{2+} complexed with some organic ligands should have a lower or higher $^{56}Fe/^{54}Fe$ ratio as compared to hexaquo Fe^{2+} depending on the organic ligand. However, for all the Fe^{2+} organic ligands studied, including acetate, oxylate, trisctatecholate, citrate, and nicotianamine (Domagal-Goldman and Kubicki 2008; Fujii et al. 2014; Moynier et al. 2013; Ottonello and Zuccolini 2008), the calculated β-values are less than the β-value calculated form hexaquo Fe^{3+} by Rustad et al. (2010). Other calculations have been performed to evaluate the effects that deprotonation would have on aqueous Fe^{2+} species. These studies found that the $^{56}Fe/^{54}Fe$ ratio of deprotonated Fe^{2+} should increase as a function of deprotonation (Fujii et al. 2014; Ottonello and Zuccolini 2009). The deprotonation of Fe^{3+} species indicates that $^{56}Fe/^{54}Fe$ ratio of the deprotonated species tends to be higher or equal to that of the hexaquo Fe^{3+} complex (Domagal-Goldman et al. 2009; Fujii et al. 2014; Ottonello and Zuccolini 2009). Experiments performed by Balci et al. (2006) to evaluate the effects of deprotonated Fe^{2+} complexes were inconclusive regarding the amount of fractionation between hexaquo Fe^{2+} and deprotonated Fe^{2+} complexes.

Iron isotope experiments that explore how Fe is fractionated between liquid and vapor have been done by Syverson et al. (2014) and Wiesli et al. (2007). Syverson et al. (2014) evaluated Fe isotope fractionation during the phase separation of vapor from liquid under simulated mid-ocean ridge hydrothermal conditions. These experiments were performed over a range of Cl concentrations and pressures and temperatures. In all cases, the amount of Fe in the vapor was small, but it consistently had a $^{56}Fe/^{54}Fe$ ratio between 0.15 and 0.06‰ greater than coexisting liquid/halite. Other experiments studied the Fe isotope fractionation between liquid and vapor Fe pentacarbonyl. This study evaluated the utility of conducting Fe isotope studies by electron impact ionization of $Fe(CO)_5$ vapor using a miniature mass spectrometer that could be deployed on a Mars lander mission. Cabonylation reactions are also important pre-biotic chemistry reactions, and are an important industrial process used in producing pure iron. In these experiments, the liquid and vapor of $Fe(CO)_5$ were allowed to equilibrate for different amounts of time. At 21 °C there was no resolvable fractionation between liquid and vapor $Fe(CO)_5$ and at 0 °C, vapor had a 0.05‰ greater $^{56}Fe/^{54}Fe$ ratio as compared to liquid (Wiesli et al. 2007). In contrast to the minimal fractionation between liquid and vapor $Fe(CO)_5$, there were large fractionations between $Fe(CO)_5$ vapor that was partially decomposed to hematite by heating the vapor with a heat gun. In this partial decomposition experiment, the hematite had a $\delta^{56}Fe$ value that was 2.5‰ less than vapor at small amounts of decomposition. The fact that the solid had a lower $\delta^{56}Fe$ value as compared to the vapor illustrates that in a unidirectional process, the product concentrates the light isotope preferentially compared to the reactant regardless if it is a vapor or a solid.

3.4.2 Aqueous Fe Mineral Fractionation

A number of experiments involving aqueous Fe species and Fe minerals including Fe-bearing carbonates, phosphates, sulfides, oxides, and silicates have been performed over a range of temperatures. Numerous studies have also calculated β-values using first-principles calculations, NRIXS, and Mössbauer spectroscopy. Dauphas et al. (2017) tabulated a large set of β-values for different Fe minerals. This section highlights cases in which agreement between experimentally-measured fractionation factors and those calculated from β-values exists. As in the previous section, we suggest that agreement between experiments and calculated β-values is strong evidence that these experiments reflect equilibrium values, and that the β-values are accurate. Importantly, because many of these mineral aqueous Fe experiments have a common aqueous Fe speciation, it is possible to combine experiments to define mineral-mineral fractionations to help evaluate which calculated β-values produce consistent fractionation factors between minerals and fluids (Polyakov and Soultanov 2011; Rustad et al. 2010; Polyakov and Soultanov 2012). Moreover, because the β-values for many of these minerals were calculated using the independent methods of first-principles, NRIXS, and Mössbauer spectroscopy, it is possible to evaluate the accuracy of the calculated β-value.

3.4.2.1 Fe Carbonate and Phosphate

Several experiments have used synthesis methods to investigate Fe isotope fractionation between aqueous Fe and carbonates or phosphates. Croal et al. (2004) synthesized vivianite during the preparation of an aqueous Fe^{2+} minimal salts media used to grow anaerobic Fe-oxidizing bacteria. The addition of Fe^{2+} into the minimal salts media was done in an anaerobic chamber, and after standing for several hours a white precipitate formed with a $^{56}Fe/^{54}Fe$ ratio that was 0.3‰ greater than the ferrous chloride salt that was added to this minimal salt growth media. They extracted this precipitate from the growth media before it was inoculated with cells. Although the researchers did not characterize this precipitate mineralogically, based on the composition of the media, the precipitate had to be vivianite or a mixture of vivianite and siderite. If the precipitate contained any siderite, then the $\Delta^{56}Fe_{Fe(II)-vivianite}$ of −0.3‰ must be a minimum value because the Fe isotope fractionation between Fe^{2+} and siderite is in the opposite direction (Wiesli et al. 2004). There has been no work to evaluate β-values of solid Fe phosphates, but Fujii et al. (2014) report β-values for aqueous Fe complexed with phosphate. Relative to hexaquo Fe^{2+}, the β-values of these phosphate complexes are 0.9 to 0.1‰ greater, depending on the PO_4 complex. These β-values are consistent with the prediction that Fe phosphate should have a $^{56}Fe/^{54}Fe$ greater than hexaquo Fe^{2+} (Fig. 3.6).

Iron isotope fractionation between aqueous Fe^{2+} and siderite was studied by synthesizing siderite from aqueous Fe solutions that contained bicarbonate buffered to a pH of 4 by bubbling CO_2 gas through the mixture. Precipitation of the siderite was initiated by slowly increasing the pH by bubbling a mixture of Ar/H_2 into the solution instead of CO_2 (Wiesli et al. 2004). The fractionation factor was constrained by fitting a Rayleigh curve to a time series of aqueous Fe and solids separated by centrifugation and filtration. These experimental results were interpreted to reflect equilibrium conditions because it was thought that the precipitation rate was slow enough to allow the instantaneously-formed siderite to be in equilibrium with the aqueous Fe^{2+}. Reduced partition function ratios have been calculated for siderite from Mössbauer spectroscopy (Polyakov and Mineev 2000) and through density functional theory (Blanchard et al. 2009). The β-values determined from these studies are similar; at 22 °C the β-values determined by

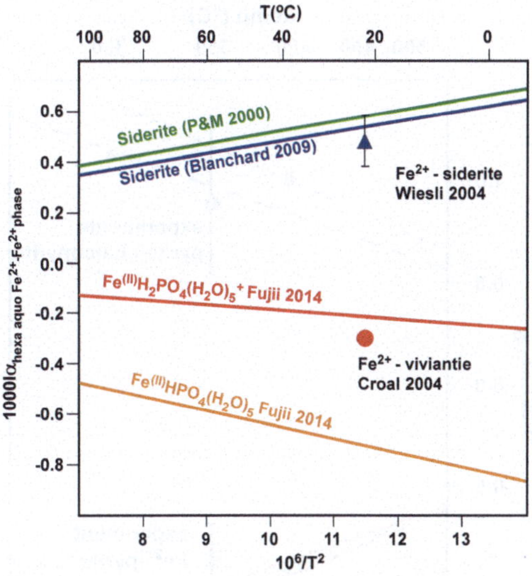

Fig. 3.6 Plot of $1000\ln\alpha_{hexaquo\ Fe(II)-Fe(II)\ phase}$ as a function of $10^6/T^2$ (T in Kelvin). The solid symbols show the results of synthesis experiments for forming siderite (blue) and vivianite (red) from Wiesli et al. (2004) and Croal et al. (2004), respectively. The blue and green solid lines show the calculated fractionation factors between hexaquo Fe^{2+} and siderite using the β-values for hexaquo Fe^{2+} by Rustad et al. (2010) and for siderite the β-value by Blanchard et al. (2009) derived from DFT (blue line) and Polyakov and Mineev (2000) from Mössbauer spectroscopy (green line). The red and orange curves show the calculated fractionation factors between hexaquo Fe^{2+} (Rustad et al. 2010) and different aqueous Fe^{2+}–PO_4 complexes calculated by Fujii et al. (2014) showing that Fe^{2+} phosphates tend to have lower $\delta^{56}Fe$ values as compared to hexaquo Fe^{2+}. No published studies have determined the β-value for the Fe phosphate vivianite

Blanchard et al. (2009) are 0.04‰ higher as compared to the value determined by Polyakov and Mineev (2000). Combining these siderite β-values with the hexaquo Fe^{2+} complex calculated by Rustad et al. (2010) results in $1000\ln\alpha_{Fe(II)-siderite}$ of 0.58 or 0.54‰ at 22 °C which agrees with the Wiesli et al. (2004) experiments that measured a $\Delta^{56}Fe_{Fe(II)-siderite}$ of 0.48 + 0.22‰ (Fig. 3.6). Substitution of Ca and/or Mg tends to decrease the β-values for Fe-bearing carbonates (Polyakov and Mineev 2000), suggesting, for example, that ankerite should have a lower $\delta^{56}Fe$ value than siderite at equilibrium.

3.4.2.2 Fe Sulfides

Several low- and high-temperature experimental studies have investigated Fe isotope fractionation between aqueous Fe species and Fe sulfides. Low-temperature experiments include synthesis and three-isotope exchange studies between aqueous Fe species and mackinawite from 2 to 35 °C, over a pH range from 4 to 8 (Butler et al. 2005; Guilbaud et al. 2010, 2011b; Wu et al. 2012a). Synthesis experiments in which pyrite was formed by the sulfidation of FeS were also performed at 40, 80, and 100 °C (Guilbaud et al. 2011a; Mansor and Fantle 2019). The goal of these low-temperature studies was to aid in the interpretation of sedimentary pyrite Fe isotope signatures. High-temperature studies that produced pyrite and chalcopyrite from aqueous Fe under hydrothermal conditions have been performed to evaluate Fe isotope fractionation that may occur during ore formation (Syverson et al. 2013, 2017) or in mid-ocean ridge hydrothermal settings.

The high-temperature pyrite study of Syverson et al. (2013) used the partial exchange approach developed by Northrop and Clayton (1966). The slope of the line corresponded to an F of 0.64 and the extrapolation to complete exchange was small. As such, the y-intercept determined in this study is well-constrained and defined a $\Delta^{56}Fe_{Fe(II)aq-pyrite}$ of -0.99 ± 0.29‰ at 350 °C (Syverson et al. 2013) (Fig. 3.1). These experiments were done with a chloride concentration of approximately ~1 M, and a combination of hexaquo Fe^{2+} and Fe^{2+} chlorine complexes is likely. In another hydrothermal experiment involving aqueous Fe and chalcopyrite in which an enriched isotope tracer was used to show Fe isotope exchange between aqueous Fe and chalcopyrite, a $\Delta^{56}Fe_{Fe(II)aq-chalcopyrite}$ of -0.09 ± 0.17‰ at 350 °C was measured (Syverson et al. 2017).

A number of reduced partition function ratio calculations for pyrite predict that pyrite should have high $^{56}Fe/^{54}Fe$ ratios compared to aqueous Fe^{2+}. Indeed, Polyakov and Mineev (2000) first noted that pyrite should have high $^{56}Fe/^{54}Fe$ ratios, comparable to many ferric rich minerals, and suggested that this was a result of the covalent nature of Fe–S bonding in pyrite. Pyrite β-values have been determined using Mössbauer spectroscopy (Blanchard et al. 2012; Polyakov and Mineev 2000; Polyakov et al. 2019; Polyakov and Soultanov 2011) and by DFT (Blanchard et al. 2009). At 350 °C, the β-values for pyrite range from 2.72 to 2.12. The best agreement between the experimental results of Syverson et al. (2013) for fractionation between pyrite and Fe^{2+}_{aq} is achieved by combining the β-value measured by Rustad et al. (2010) for aqueous Fe^{2+} and the pyrite β-values reported for DFT calculations by Blanchard et al. (2009) or from Mössbauer spectroscopy by Blanchard et al. (2012) or Polyakov et al. (2019) (lines labeled 1–3, Fig. 3.7). Because the Syverson et al. (2013) experiments contained chloride, it may be more appropriate to use β-values for ferrous chloro complexes. However, chloride concentration does not have a significant effect on isotopic fractionation for ferrous species, as noted above. If the pyrite and chalcopyrite experimental results of Syverson et al. (2013, 2017) are combined to avoid uncertainties in speciation (both experiments had similar amounts of chloride), a $\Delta^{56}Fe_{pyrite-chalcopyrite}$ of 0.9‰ is obtained, and this agrees perfectly with the β-values of pyrite by Blanchard et al. (2012) and Polyakov and Soultanov (2011) for chalcopyrite (Fig. 3.7). If the β-value of chalcopyrite reported in Dauphas et al. (2012) is used, a slightly worse fit is obtained as compared to the experimental results, although it is within experimental error. Both chalcopyrite β-values were determined from NRIXS.

In contrast, synthesis experiments done between 350 °C and 40 °C indicate that pyrite has a $\delta^{56}Fe$ value that is up to 0.5‰ greater than to 2.2‰ less than the aqueous ferrous iron component (Guilbaud et al. 2011a; Syverson et al. 2013; Mansor and Fantle 2019), a very large range. The magnitude of the Fe isotope fractionation between pyrite and aqueous Fe components is strongly controlled by precipitation rates, where at comparable temperatures (40–100 °C) the $\Delta^{56}Fe_{aqueousFe(II)-pyrite}$ varies from −0.5 to −2.5‰ over pyrite precipitation rates that range form 10^{-5} to 10^{-9} mol/L/s (Fig. 3.8). Mansor and

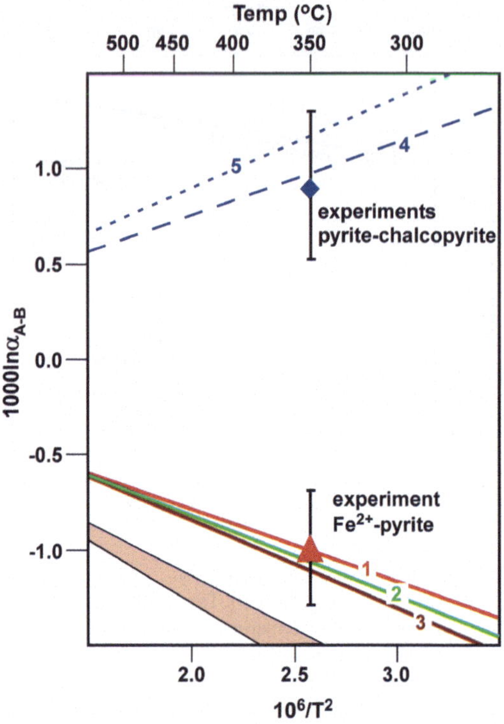

Fig. 3.7 Plot of 1000lnα as a function of $10^6/T^2$ (T in Kelvin) for aqueous Fe and/or Fe sulfides. The red symbol is the experimentally-measured fractionation between aqueous Fe^{2+} and pyrite determined by Syverson et al. (2013) and the 2-σ error bars. The solid lines show the calculated fractionation using the Rustad et al. (2010) hexaquo Fe^{2+} β-value and different pyrite β-values: Line **1**: Blanchard et al. (2009) derived from DFT, line **2**: Polyakov et al. (2019) derived from Mössbauer spectroscopy and line **3**: Blanchard et al. (2012) derived from Mössbauer spectroscopy. The red shaded field uses the pyrite β-values from Polyakov and Soultanov (2011), Polyakov et al. (2007), and Polyakov and Mineev (2000). The blue symbol is the fractionation factor between pyrite and chalcopyrite calculated by combining the aqueous Fe^{2+}–Fe sulfide fractionation determined for pyrite by Syverson et al. (2013) and chalcopyrite (Syverson et al. 2017), the error bar is 2-σ for the combined experiments. The pyrite and chalcopyrite experiments had similar Cl concentrations so we assume the Fe^{2+} speciation for the two experiments is the same for this calculation. The dashed and dotted lines show the fractionation between pyrite and chalcopyrite using the β-value determined by Blanchard et al. (2009) for pyrite and the chalcopyrite β-value determined by Polyakov and Soultanov (2011) for the dashed curve labeled 4, and the chalcopyrite β-value determined by Dauphas et al. (2012) for the dotted line. The β-values for the aqueous Fe have only been reported up to 300 °C so the β-values shown in this plot were calculated by using the value reported at 300 °C and linearly extrapolating as a function of $1/T^2$ to zero fractionation at infinite temperature

Fantle (2019) suggest that the variability in the fractionation factor as a function of precipitation is associated with changes in the magnitude of the kinetic fractionation factor as a function of pyrite precipitation rate, as well as differences in the amount of Fe isotope exchange that occurs after pyrite is formed. In their geochemical modeling of these different processes, they noted that isotope exchange rates would be greater at high temperatures, and this may be the origin for the smaller fractionation for the high temperature 300–350 °C pyrite synthesis studies of Syverson et al. (2013) that precipitated pyrite at a rate similar to the low temperature synthesis studies of Guilbaud et al. (2011a). Additionally, they noted that high surface area is likely to promote isotope exchange (Mansor and Fantle 2019).

How the rate of pyrite formation will affect kinetic Fe isotope fractionations is unknown. However, to the extent that pyrite should have high $^{56}Fe/^{54}Fe$ ratios, reflecting the covalent like bonding between Fe and S, fractionation between Fe^{2+} components and pyrite is likely to become less negative, and might even become positive, as pyrite formation rates decrease. Overall the magnitude of the fractionation between aqueous Fe^{2+} species and pyrite is large and reflects these rate-dependent processes. If one extrapolated the measured high temperature equilibrium fractionation to low temperatures, the prediction would be that pyrite should have a $\delta^{56}Fe$ value that is 5‰ greater than aqueous Fe^{2+} at 10 °C. In contrast, if exchange is limited and pyrite is precipitated rapidly, the fractionation changes sign, becoming slightly negative at intermediate precipitation rates to strongly negative at rapid pyrite precipitation rates. Based on such a wide range in fractionation factors, it is challenging to use the Fe isotope composition measured in

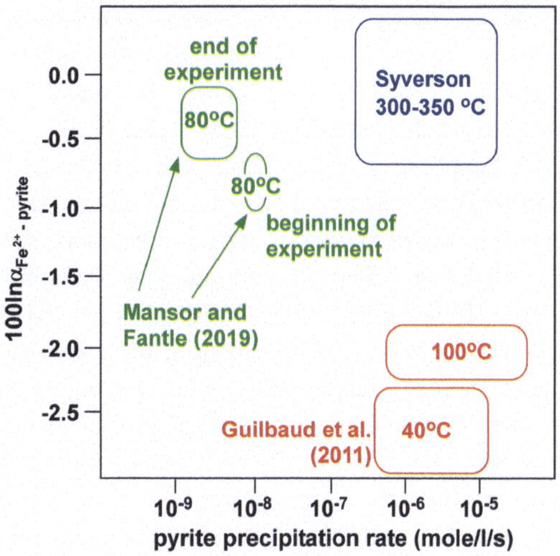

Fig. 3.8 Plot of the $1000ln\alpha_{Fe(II)-pyrite}$ measured in different pyrite synthesis experiments versus pyrite precipitation rate. The labeled fields show the results for the synthesis experiment done at 300–350 °C by Syverson et al. (2013) blue field, synthesis experiments done at 40 and 100 °C done by Guilbaud et al. (2011a) red field, and the synthesis experiments done by Mansor and Fantle (2019) at the beginning (more rapid precipitation) and end of the experiment, green fields. Modified from Mansor and Fantle (2019). See text for discussion

Table 3.2 Experimental results for three-isotope exchange experiments between aqueous Fe^{2+} species and mackinawite

Experiment	Predominant aqueous Fe speciation	pH	Temp	$\Delta^{56}Fe_{Fe(II)-mackinawite}$ at 100% exchange (‰)
Guilbaud et al. (2011b)	$Fe(H_2O)_6^{+2}$	4	2	−0.33 ± 0.12
Guilbaud et al. (2011b)	$Fe(H_2O)_6^{+2}$	4	25	−0.52 ± 0.16
Wu et al. (2012a)	$Fe(H_2O)_6^{+2}$	6	20	−0.36 ± 0.07
Wu et al. (2012a)	$Fe(H_2O)_6^{+2}$	7	20	−0.32 ± 0.29
Wu et al. (2012a)	$Fe(H_2O)_6^{+2}$	8	20	−0.32 ± 0.09
Wu et al. (2012a)	$Fe(H_2O)_6^{+2}$	7	35	−0.43 ± 0.26
Wu et al. (2012a)	$FeHS^+$	7	20	−0.64 ± 0.36
Wu et al. (2012a)	$FeHS^+$	7	35	−0.67 ± 0.23

pyrite as a proxy for the Fe isotope composition of seawater Fe. The use of pyrite for understanding Fe cycling in marine sediments is extensively discussed in Chaps. 5 and 6.

A number of experiments have been performed to constrain Fe isotope fractionation between mackinawite (a pyrite precursor; see Sect. 5.4.3 in Chap. 5) and aqueous Fe complexes (Butler et al. 2005; Guilbaud et al. 2010, 2011b; Wu et al. 2012a). Overall, these experiments agree exceptionally well among one another. Three-isotope exchange experiments between aqueous Fe^{2+} and nanoparticulate mackinawite have been performed over a pH range of 4–8 and from 2 to 35 °C. Under conditions of complete isotope exchange, there is no difference in the hexaquo Fe^{2+}-mackinawite fractionation factor from pH = 4 to pH = 8 (Table 3.2). The isotopic fractionation of Fe is larger in magnitude if there is more sulfide in the system, and an $FeHS^{+}$ complex is produced instead of hexaquo Fe.

In contrast to these three-isotope exchange studies, synthesis studies reveal that there are differences in Fe isotope exchange between hexaquo Fe^{2+} and mackinawite as a function of pH (Butler et al. 2005; Guilbaud et al. 2011b; Wu et al. 2012a). In synthesis time series experiments done at a pH of 4, the $\Delta^{56}Fe_{Fe(II)-mackinawite}$ did not reach the equilibrium Fe isotope fractionation as determined by the three-isotope exchange experiments (Butler et al. 2005; Guilbaud et al. 2011b; Wu et al. 2012a). Even after 38 days of aging, the $\Delta^{56}Fe_{Fe(II)-mackinawite}$ was only 0.1‰ (Wu et al. 2012a). In contrast, a $\Delta^{56}Fe_{Fe(II)-mackinawite}$ of −0.3‰ was measured at complete exchange. However, in similar synthesis experiments done at pH of 7, the final $\Delta^{56}Fe_{Fe(II)-mackinawite}$ was the same as the Fe isotope fractionation factor determined at complete isotope exchange by the three-isotope method experiments (Wu et al. 2012a). These differences in Fe isotope exchange rates between aqueous Fe and mackinawite at different pH may reflect aggregation of FeS nanoparticles. At low pH, particle aggregation keeps the inner portion of the nano particle aggregate from exchanging Fe isotopes with the fluid, but at higher pH, where aggregates do not form as easily, equilibrium is maintained. Because the pH is neutral to slightly alkaline in most natural settings where sedimentary pyrite would form, mackinawite should form at equilibrium and have a $\delta^{56}Fe$ value that is lower than aqueous Fe by 0.3–0.6‰ (at 20 °C) depending if the sulfide concentration is low or high, respectively.

One study reported a β-value for mackinawite (Polyakov and Soultanov 2011). This β-value combined with the Rustad et al. (2010) β-value for hexaquo Fe^{2+} would predict a $1000\ln\alpha_{Fe(II)-mackinawite}$ value of −2.82‰ at 20 °C, which is dramatically different as compared to the three-isotope exchange experiments that measured a value of −0.32‰. Because the $\Delta^{56}Fe_{hexaquoFe(II)-mackinawite}$ = −0.32‰ fractionation factor at 100% isotope exchange has been confirmed by two independent studies, it is suspected that the β-value determined for mackinawite is inaccurate. This inaccuracy perhaps reflects the quality of the source data used by Polyakov and Soultanov (2011) in the calculations—the Mössbauer measurements were made in 1965 (Bertaut et al. 1965), and it is also possible that the purity of the phase analyzed was not solely mackinawite, given the ease with which this mineral is oxidized. If the mackinawite β-value was scaled by multiplying it by a factor of 2.25 it would fit the experimental results.

3.4.2.3 Iron Oxides

Several studies have experimentally investigated Fe isotope fractionation between aqueous Fe and iron oxides and hydroxides. Many of these experiments use the fact that ferric-rich phases form a redox couple with aqueous Fe^{2+} that promotes significant amounts of Fe isotope exchange at low temperatures. This allows application of the three-isotope method to infer the iron isotope fractionation at complete isotope exchange. Moreover, both Si and O isotope exchange is promoted by Fe^{2+} atom and electron exchange with ferric iron, making this technique of Fe^{2+}-induced recrystallization of iron hydroxides a powerful method to investigate isotope exchange in a wide variety of systems and elements when Fe oxide/hydroxides are present (Frierdich et al. 2015; Zheng et al. 2016). To date, the crystalline Fe oxides, hematite, goethite and magnetite, and poorly crystalline ferrihydrite, and Fe–Si gels have been experimentally investigated. A number of β-values have been determined for the crystalline iron oxides, but no studies to determine reduced partition function ratios for ferrihydrite and Fe–Si gels. The lack of calculated β-values for these phases largely reflects the fact that the crystal structure of these phases are poorly known because they are x-ray amorphous.

Ferrihydrite-Amorphous Fe Gels

Studying Fe isotope exchange between aqueous Fe^{2+} and ferrihydrite is challenging because ferrihydrite is very reactive and will recrystallize to more stable iron oxides such as akaganeite, magnetite, or goethite upon exposure to aqueous Fe^{2+} (Boland et al. 2014; Hansel et al. 2005, 2003; Jang et al. 2003; Pedersen et al. 2005). However, ferrihydrite can be stabilized by either sorbing Si onto its surface or by coprecipitating ferrihydrite with either Si or organic C (Chen et al. 2015, 2016; Doelsch et al. 2003, 2000; Eusterhues et al. 2008; Henneberry et al. 2012; Zhou et al. 2018; Wu et al. 2011, 2012b). These co-precipitated gels, depending on the amount of Si or organic C that is used, do not recrystallize to more stable iron oxides but they do exchange significant amounts of Fe (ThomasArrigo et al. 2017, 2018; Wu et al. 2011, 2012b; Zhou et al. 2018). Experimentally-derived Fe isotope fractionation factors have been measured for ferrihydrite stabilized with sorbed Si and Fe–Si gels with a range of Fe to Si ratios; from these data, the fractionation factor has been extrapolated to pure ferrihydrite (Wu et al. 2011, 2012b) (Table 3.3). The change in the Fe isotope fractionation factor as a function of Fe to Si ratios is inferred to reflect differences in Fe bonding. The fractionation factor is least negative for an Fe:Si ratio of 1, which is thought to reflect the fact that Fe polymerization is at a minimum at this Fe to Si ratio. The resulting Fe–O bonds are longer and therefore weaker, resulting in less fractionation between hexaquo Fe^{2+} and the Fe–Si gel (Wu et al. 2012b). At higher Fe to Si ratios, Fe polymerization is increased, resulting in Fe–O bonds that are shorter/stronger, resulting in a larger magnitude fractionation between hexaquo Fe and the Fe–Si gel (Wu et al. 2012b).

Magnetite

Iron isotope fractionation between aqueous Fe^{2+} and magnetite has been studied using the three-isotope exchange method by Frierdich et al. (2014b), and a synthesis approach in which ferrihydrite was converted to magnetite by exposure to Fe^{2+} produced during bacterial dissimilatory iron reduction (Johnson et al. 2005), or by direct addition of Fe^{2+} to ferrihydrite (Frierdich et al. 2014b). In the synthesis method, a $\Delta^{56}Fe_{Fe(II)\text{–}magnetite}$ of −1.3‰ was determined by Johnson et al. (2005) and a value of −1.61‰ was determined by Frierdich et al. (2014b) (Fig. 3.9). In the three-isotope exchange experiments, a ^{57}Fe-enriched aqueous Fe^{2+} was reacted with magnetite and the approach to equilibrium was monitored using the $^{57}Fe/^{56}Fe$ ratios measured in both magnetite and aqueous Fe^{2+}. A total of four experiments were done in which the aqueous Fe^{2+} had different initial $^{56}Fe/^{54}Fe$ ratios that were greater, similar to, or less than the $^{56}Fe/^{54}Fe$ ratio of the magnetite. The exchange trajectories for the aqueous Fe^{2+} defined an initial linear trajectory up to an F of 0.66 that all converged to the same $^{56}Fe/^{54}Fe$ ratio. The final four time points that had $F > 0.66$ defined lines

Table 3.3 Three-isotope exchange experiments between aqueous Fe^{2+} and ferrihydrite and Fe–Si gels

Characteristics of ferric solid	$\Delta^{56}Fe_{Fe(II)\text{–}ferric\ solid}$ (‰)
Pure ferrihydrite corrected for Si	−3.2
Ferrihydrite with sorbed Si	−3.17 ± 0.08
Fe–Si coprecipitate Fe:Si 1:1	−2.58 ± 0.14
Fe–Si coprecipitate Fe:Si 1:2	−3.51 ± 0.20
Fe–Si coprecipitate Fe:Si 1:3	−3.99 ± 0.17

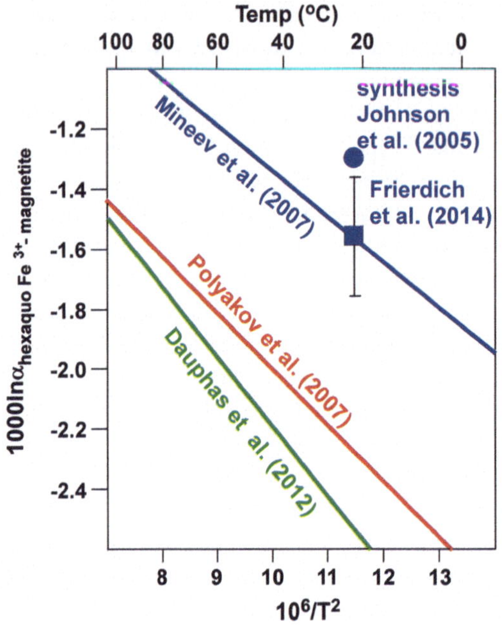

Fig. 3.9 Plot of $1000\ln\alpha_{\text{hexaquo Fe(II)–magnetite}}$ versus $10^6/T^2$ (T in Kelvin) showing the experimentally-measured fractionation from a synthesis experiment conducted by Johnson et al. (2005) and the combined results of a synthesis and three-isotope experiments conducted by Frierdich et al. (2014b). The 2-σ error bars are shown for the work of Frierdich et al. (2014b). For comparison, the lines show the calculated fractionation factors using the Rustad et al. (2010) hexaquo Fe^{2+} β-value and different magnetite β-values. The blue curve uses the magnetite β-value reported by Mineev et al. (2007), the red uses the Polyakov et al. (2007) β-value, and the green uses the Dauphas et al. (2012) magnetite β-value. Not shown is the magnetite β-value reported by Polyakov and Mineev (2000), which would plot at more negative $1000\ln\alpha$ values

with a different slope and the slope of these trajectories all converged to the same $^{56}Fe/^{54}Fe$ ratio that was lower as compared to the earlier time samples. The extrapolation of the last four time points to an F of 1 was taken as the equilibrium aqueous Fe^{2+} composition and produced a $\Delta^{56}Fe_{\text{hexaquo Fe(II)–magnetite}} = -1.56 \pm 0.2‰$. The magnetite in the time series did not display any changes in slope, because magnetite contained the majority of Fe in the experiments. The change in slope was inferred to represent either mixing differences between the A and B sites in the spinel structure, or a change from an initial kinetic isotope fractionation associated with rapid Fe isotope exchange followed by a slower equilibrium exchange. Because the exchange kinetics for pure magnetite and Co substituted magnetite are essentially the same (Gorski et al. 2012), the change from a kinetically- to equilibrium-driven exchange is the favored interpretation by Frierdich et al. (2014b). The kinetically-driven fractionation defined an aqueous Fe that was 0.4‰ greater than the aqueous Fe inferred to be in equilibrium with magnetite. The $\Delta^{56}Fe_{\text{Fe(II)–magnetite}}$ at complete isotope exchange was $-1.56 \pm 0.2‰$, which matches the synthesis experiments.

Reduced partition function ratios for magnetite have been calculated from Mössbauer spectra and NRIXS. At 20 °C, these magnetite $1000\ln\beta$-values span a 1‰ range from 6.39 to 7.38. Some of this range may reflect Mössbauer spectra collected on non-stoichiometric magnetite (Frierdich et al. 2014b; Polyakov et al. 2007). The best overall fit between the experimental results and the $1000\ln\alpha$ values use the Rustad et al. (2010) hexaquo Fe^{2+} and the magnetite β-values calculated by Mineev et al. (2007) from Mössbauer spectroscopy (Fig. 3.9).

Hematite

A number of independent experiments have investigated Fe isotope fractionation between aqueous Fe and hematite including:

1. Hydrothermal experiments that investigated fractionation between Fe^{3+} or Fe^{2+} Cl-rich fluids and hematite at 200 and 300 °C (Saunier et al. 2011),
2. Three-isotope reversed exchange experiments between hexaquo Fe^{2+} and hematite (Frierdich et al. 2019) done at 20 °C,
3. Simple exchange experiment between hexaquo Fe^{3+} and hematite (Skulan et al. 2002) done at 100 °C, and
4. Exchange experiment between Fe^{2+} and hematite (Wu et al. 2010) conducted at 20 °C.

All of these experiments consistently observed that the fractionation between aqueous Fe^{3+}-rich fluids and hematite is minimal, and the fractionation between Fe^{2+} and hematite is significant (Fig. 3.10). Of the low temperature experiments, the three-isotope exchange experiments conducted by Frierdich et al. (2019) provides the best picture of how Fe isotope exchange occurs between Fe^{2+} and hematite. In these experiments, an enriched ^{57}Fe aqueous Fe solution with a lower or higher $^{56}Fe/^{54}Fe$ ratio as compared to normal isotopic composition hematite was sampled at different times over 60 days. Hematite with three different surface areas was used to investigate how exchange rate correlates with mineral surface area, perhaps reflecting changes from heterogeneous to homogenous exchange. Overall, the degree of isotope exchange correlates with surface area, but in all experiments the final time points converged to a common $^{56}Fe/^{54}Fe$ ratio, allowing one to infer the Fe^{2+}-hematite Fe isotope fractionation factor at complete exchange. The $\Delta^{56}Fe_{Fe(II)–hematite}$ for the coarsest hematite grains (BET surface areas 6 m^2/g) was -3.14 ± 0.34‰, an intermediate size hematite (BET surface area of 32 m^2/g) was -2.61 ± 0.26‰, and the finest grain size hematite (BET surface area 60 m^2/g) was -2.77 ± 0.37‰. The cause for the differences in fractionation factor for the different hematite grain sizes is unknown; at the 2-σ level it is not possible to resolve these differences. The exchange experiment performed by Skulan et al. (2002) investigated the Fe isotope fractionation between a predominately hexaquo Fe^{3+} solution and hematite at 100 °C. Companion experiments using enriched ^{57}Fe tracers showed that Fe isotope exchange was controlled by surface area. From these experiments, a $\Delta^{56}Fe_{Fe(III)–hematite}$ of -0.1 ± 0.2‰ was inferred at 100 °C.

Hematite reduced partition function ratios have been determined by Polyakov et al. (2007) from Mössbauer spectra, by Dauphas et al. (2012) from NRIXS spectra, and by Blanchard et al. (2009) using density functional theory (Fig. 3.10). The $1000\ln\alpha_{aqueous\ Fe–hematite}$ values calculated using the Rustad et al. (2010) hexaquo Fe^{2+} and Fe^{3+} β-values and the Polyakov et al. (2007) and Dauphas et al. (2012) hematite β-values yield a good fit to the low temperature aqueous Fe hematite experiments (Frierdich et al. 2019; Skulan et al. 2002) (green lines labeled 4 and 5 and red lines labeled 1 and 2, respectively; Fig. 3.10a). In order to compare calculated fractionation factors to the aqueous Fe hematite fractionations determined using the hydrothermal experiments of Saunier et al. (2011), the reduced partition function ratios for the aqueous Fe needed to be adjusted for the mixed Fe^{3+}–Fe^{2+} solutions that contained chlorine complexes. For these calculations, the speciation reported by Saunier et al. (2011) was combined with the β-values of Rustad et al. (2010) for hexaquo Fe, Hill and Schauble (2008) for the Fe^{3+} chlorine complexes, and the Hill et al. (2010) for the Fe^{2+} chloro complexes using the scaled UHF/6-31G(d) formulation. There is agreement in ferric-rich experiments conducted at 200 °C and the $1000\ln\alpha_{Fe(III)–hematite}$ using the Fe^{3+} hexaquo value of (Rustad et al. 2010) and the hematite β-values of Dauphas et al. (2012) (Fig. 3.10b). However, this is not a fair comparison because the fluid in these experiments are inferred to be dominated by the $Fe^{(III)}Cl_3(H_2O)_3$ complex and about 1/3 of the Fe in the fluid is ferrous iron. This ferrous and ferric speciation makes this fluid β-value less than the β-values determined for hematite by (Dauphas et al. 2012), resulting in a $1000\ln\alpha_{aqueous\ Fe–hematite}$ of -0.5‰ at 200 °C (blue line labeled 2; Fig. 3.10b) as compared to the experiment that measured a $\Delta^{56}Fe_{aqueous\ Fe–hematite} = 0.01 \pm 0.05$‰. For the two ferrous-rich hydrothermal experiments conducted by Saunier et al. (2011), the $1000\ln\alpha_{aqueous\ Fe–hematite}$ values, calculated using the reported fluid speciation and the Dauphas et al. (2012) hematite β-value, are more negative by ~0.3‰ as compared to the experiments (blue lines 3 and 4; Fig. 3.10). It is unknown if the small (~0.4‰) misfit of the ferric-rich hematite hydrothermal experiments and the calculated $1000\ln\alpha$ values based on reduced partition function ratios is a result of uncertainty in the speciation calculations or the reduced partition function ratios for the Fe^{3+} chloro complexes. Overall, these experimental results provide strong support for the accuracy of either the Polyakov et al. (2007) or Dauphas et al. (2012) β-values for hematite and the β-values for the Fe^{2+} complexes reported by Hill et al. (2010).

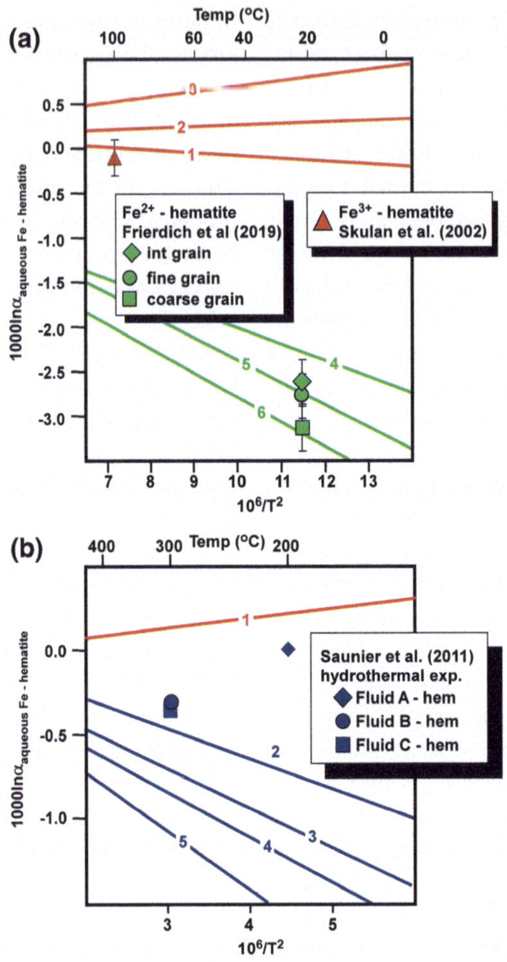

Fig. 3.10 Plot of $1000\ln\alpha_{\text{aqueous Fe–hematite}}$ versus $10^6/T^2$ (T in Kelvin). The symbols show the experimental results for different methods of producing hematite. In **a** the red diamond shows the acid hydrolysis formation of hematite from aqueous Fe^{3+} done by Skulan et al. (2002). The red lines show the calculated fractionation factors between hexaquo Fe^{3+} and hematite using the Rustad et al. (2010) hexaquo Fe^{2+} β-value and the Polyakov et al. (2007) hematite β-value for line 1, the Dauphas et al. (2012) hematite β-value for line 2, and the Blanchard et al. (2009) hematite β-value for line 3. The green symbols show the reversed three-isotope exchange experiment between aqueous Fe^{2+} and hematite with different grain sizes done by Frierdich et al. (2019). The green curves show the calculated fractionation factors between hexaquo Fe^{2+} using the Rustad et al. (2010) β-value and for line 4, the Blanchard et al. (2009) hematite β-value, line 5 the Dauphas et al. (2012) hematite β-value, and curve 6 the Polyakov et al. (2007) hematite β-value. **b** The blue symbols show the hydrothermal hematite precipitation experiments that used three different chloride-rich fluid compositions conducted by Saunier et al. (2011). The blue lines show the calculated fractionation between aqueous Fe and hematite. For all the lines shown in (**b**) the Dauphas et al. (2012) hematite β value was used. To calculate the β-value for the different fluids, the speciation was taken from the speciation calculation shown in Fig. 3 of Saunier et al. (2011). The β-value for the hexaquo Fe^{2+} and Fe^{3+} are from Rustad et al. (2010), the Fe^{3+} and Fe^{2+} chloro complexes are from the scaled Hill and Schauble (2008) and Hill et al. (2010) UHF/6-31G(d) formulations, respectively. The fluid A composition (line 2) is for the precipitation experiment that was predominately ferric iron (Saunier et al. 2011; Fig. 3.3b), fluid B (line 3) is for the experiment with moderate amounts of Fe^{3+} (Saunier et al. 2011; Fig. 3.3d), and fluid C (line 9) is for the experiment with minor Fe^{3+} (Saunier et al. 2011; Fig. 3f). The curves labeled 1 and 5 show the fractionation between hexaquo Fe^{3+} and hematite and hexaquo Fe^{2+} and hematite, respectively

Goethite

Two three-isotope exchange studies investigated the Fe isotope fractionation between hexaquo Fe^{2+} and goethite (Beard et al. 2010; Frierdich et al. 2014a). Both studies determined the same fractionation factor, where $\Delta^{56}Fe_{Fe(II)-goethite}$ = −1.06‰ for goethite micro-rods (BET surface area 45 m^2/g) and −1.22‰ for goethite nano-rods (BET surface area = 115 m^2/g). The study by Frierdich et al. (2014a) was motivated by the significant differences between the Beard et al. (2010) experimentally-determined fractionation factor as compared to the fractionation factor calculated by combining the Rustad et al. (2010) hexaquo Fe^{2+} complex with the β-value calculated for goethite by Dauphas et al. (2012) using NRIXS. These combined β-values predict a $1000\ln\alpha_{Fe(II)-goethite}$ of −4.9‰ at 22 °C as compared to the −1.2‰ value measured by Beard et al. (2010) for micro-goethite. The Dauphas et al. (2012) study suggested that the three-isotope exchange experiments, although extrapolated to 100% exchange, may have suffered kinetic effects because the time series trajectories for the aqueous Fe^{2+} experiments were curved and, therefore may not represent equilibrium fractionation. The Frierdich et al. (2014a) experimental investigation addressed these kinetic concerns by conducting three-isotope exchange experiments that were similar to those of Beard et al. (2010), but included reversal experiments in which the aqueous Fe^{2+} had $^{56}Fe/^{54}Fe$ ratios that were either greater, similar, or less than the $^{56}Fe/^{54}Fe$ ratios of the normal goethite. In these reversed experiments, the final time points of the aqueous Fe^{2+} converged to the same $^{56}Fe/^{54}Fe$ ratio at 100% exchange.

Additional exchange experiments were performed using goethite with coarser grain sizes or substituted with different cations. The Fe isotope exchange rates in these experiments were significantly slower as compared to the micro- and nano-rod goethite experiments, and they did not display any curvature on a plot of $^{56}Fe/^{54}Fe$ versus F. Although the amount of exchange in in these slow exchange experiments was less than in the micro- and nano-rod experiments, the extrapolation to 100% exchange for all of these slow exchange experiments yield the same $\Delta^{56}Fe_{Fe(II)-goethite}$ as the micro- and nano-rod three-isotope exchange experiments. Additionally, experiments conducted at 50 °C defined a lower magnitude fractionation ($\Delta^{56}Fe_{Fe(II)-nano\ goethite}$ = −0.99‰) as compared to experiments conducted at 22 °C ($\Delta^{56}Fe_{Fe(II)-nano\ goethite}$ = −1.22‰), consistent with an equilibrium process. Based on this body of evidence, Frierdich et al. (2014a) suggested that during rapid exchange in the beginning of the experiment there is a kinetic effect that is erased during slower exchange at the end of the experiment. They conclude that the fractionation factor extrapolated to 100% is the equilibrium fractionation factor between hexaquo Fe^{2+} and goethite.

As previously noted, agreement is poor between experiments and calculated Fe isotope fractionation that use the β-value of Rustad et al. (2010) for hexaquo Fe^{2+} and the Dauphas et al. (2012) β-value for goethite (Fig. 3.11). If one uses the β-value for goethite reported by Blanchard et al. (2015) based on the NRIXS data of Dauphas et al. (2012), but with a different background subtraction method introduced by Dauphas et al. (2014), agreement improves between the calculated and experimental results (Fig. 3.11). However, the best agreement between experiment and calculated fractionation factors uses the Rustad et al. (2010) hexaquo Fe^{2+} β-value and the β-value determined by Polyakov and Mineev (2000) constrained from Mössbauer spectroscopy (Fig. 3.11). It is unknown why the NRIXS-derived β-values for goethite predict such large fractionations whereas experiments always yield much smaller fractionations between aqueous Fe^{2+} and goethite.

Wiederhold et al. (2006) conducted long-term (300 day) ligand-promoted dissolution experiments of goethite using oxalic acid. In these time series experiments, the Fe isotope fractionation between Fe^{3+} oxalate and goethite had initial negative fractionation but over time became positive and stabilized to a constant $\Delta^{56}Fe_{Fe(III)\ Ox-goethite}$ of 0.33‰ after ~30 days (Fig. 3.11). Wiederhold et al. (2006) suggested this may represent the equilibrium fractionation between

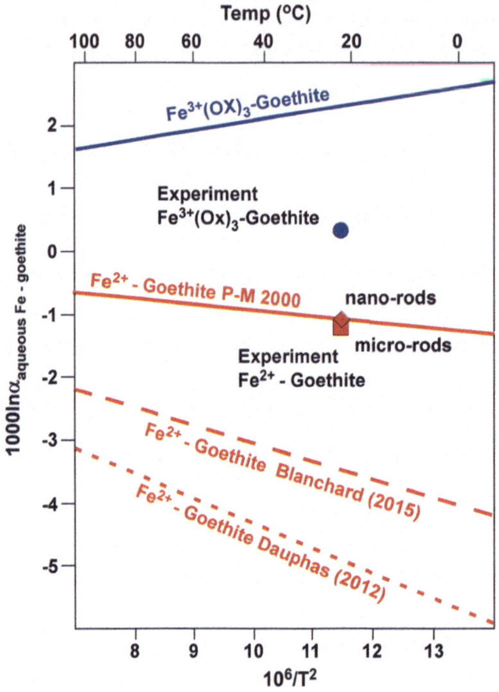

Fig. 3.11 Plot of $1000\ln\alpha_{aqueous\ Fe\text{–}goethite}$ versus $10^6/T^2$ (T in Kelvin). The red symbols show the experimental results for three-isotope exchange experiments between aqueous Fe^{2+} and goethite conducted by Beard et al. (2010) and Frierdich et al. (2014a) that used either nano-rod or micro-rod sized goethite. The blue symbol shows the long-term partial dissolution study of goethite using oxalic acid conducted by Wiederhold et al. (2006) that is inferred to be the equilibrium fractionation between ferric oxalate ($Fe^{III}(Ox)_3$) and goethite. The red curves show the fractionation factor between hexaquo Fe^{2+} using the β-value of Rustad et al. (2010) and different goethite β-values. The solid line uses the geothite β-value of Polyakov and Mineev (2000), the dashed line uses the Blanchard et al. (2015) goethite β-value, and the dotted line uses the Dauphas et al. (2012) goethite β-value. The blue line is the fractionation between aqueous $Fe^{III}(Ox)_3$ and goethite using the β-values of Domagal-Goldman and Kubicki (2008) 1-In vacuo and Polyakov and Mineev (2000), respectively

Fe^{3+} oxalate complexes and goethite. The reduced partition function ratio for Fe^{3+} oxalate has been calculated by Domagal-Goldman and Kubicki (2008) and Domagal-Goldman et al. (2009). The calculated β-values for Fe^{3+} oxalates reported in these studies are similar (0.4‰ spread at 0 °C) and Fig. 3.11 shows the $1000\ln\alpha_{Fe(III)oxalate\text{–}goethite}$ using the Polyakov and Mineev (2000) goethite β-value and the Domagal-Goldman and Kubicki (2008) β-value for Fe^{3+} oxalates. These reduced partition function ratios predict a much larger $1000\ln\alpha_{Fe(III)oxalate\text{–}goethite}$ as compared to the Wiederhold et al. (2006) experiments (Fig. 3.11). It is unknown if this disagreement reflects an issue with the reduced partition function ratios or if the Wiederhold et al. (2006) experiments truly reflect equilibrium conditions. However, considering that Fe isotope exchange between aqueous Fe^{3+} and ferric oxides tends to be limited to the oxide surface (Poulson et al. 2005; Skulan et al. 2002), reflecting the lack of electron exchange, it is likely that the Fe^{3+} oxalate only equilibrated with the goethite surface. Because of differences in Fe bonding between the surface and interior of goethite, the outermost surface of goethite is likely to have a higher $\delta^{56}Fe$ value as compared to the bulk goethite (Beard et al. 2010). If the surface of the goethite could be measured, it is likely that a larger fractionation between Fe^{3+} oxylate and the goethite surface would be measured, which would be more consistent with the calculated fractionation as shown in Fig. 3.11.

3.5 Biological Experiments

Biological experiments have constrained Fe isotope fractionation during microbial Fe oxidation and reduction. Additional experiments investigated how Fe uptake by different organisms produced Fe isotope fractionation. Overall, biological redox experiments show many similarities with abiological aqueous Fe^{2+} iron oxide exchange experiments because in both suites of experiments, the isotope exchange and fractionation that takes place between ferrous and ferric iron is being investigated. The main difference is that the abiological experiments are conducted under steady state Fe conditions with no net mass transfer of ferrous and ferric iron after the initial phase of Fe^{2+} sorption onto the iron oxide surface. In contrast, dissimilatory Fe reduction experiments start without any Fe^{2+}, but, as the bacteria grow, Fe^{2+} is produced and the ferric substrate surface is continuously recrystallized. Similarly, in Fe-oxidizing experiments, the experiment starts without any ferric material but its continuous production results in changes of the relative amounts of Fe^{2+} and Fe^{3+}. Thus, when comparing abiological and biological experiments, one must take into account that steady-state conditions are not achieved in biological experiments.

3.5.1 Fe Oxidizing Experiments

Iron-oxidizing experiments using bacteria with different metabolic pathways have been performed. Croal et al. (2004), Swanner et al. (2015), and Swanner et al. (2017) conducted Fe oxidation experiments using anoxygenic photosynthetic organisms. The study by Croal et al. (2004) used enrichment cultures obtained from freshwater settings. In the studies by Swanner et al. (2015) and Swanner et al. (2017), a marine photoferrotroph, *Rhodovulum iodosum,* was used. In the experiments conducted by Swanner et al. (2017) a growth media that contained 1 mM Si was used to simulate Archean conditions, whereas the Swanner et al. (2015) and Croal et al. (2004) experiments did not contain Si. Kappler et al. (2010) conducted Fe-oxidizing experiments in which the microbe *Acidovorax* sp. strain BoFeN1 oxidized Fe under anoxic conditions coupled to nitrate reduction. In the Kappler et al. (2010) experiments, the cells mineralized iron in the cell's periplasm, whereas in all the other experiments, iron mineralization was not associated with cells. Swanner et al. (2017) used the cyanobacteria *Synecochocus* which oxidized Fe by the production of O_2 during photosynthesis. Balci et al. (2006) used *A ferroxidins,* which is an autotrophic organism that thrives at low pH and oxidizes Fe.

Overall, these Fe-oxidation experiments found that oxidized Fe had a higher $\delta^{56}Fe$ value as compared to the reduced Fe, and the deduced fractionation between aqueous Fe^{2+} and ferric oxide was between 1.5 and 3‰. The origin of the Fe isotope fractionation in these oxidation experiments is thought to reflect an equilibrium fractionation between aqueous Fe^{2+} and aqueous Fe^{3+} followed by a kinetic fractionation associated with the precipitation of solid ferric oxide from the aqueous Fe^{3+}. Notably, in the Balci et al. (2006) experiments, where substantial amounts of aqueous ferric iron existed in solution because the experiments were done at a pH of ~2, the workers analyzed aqueous ferrous and ferric iron and found that the average $\Delta^{56}Fe_{Fe(III)-Fe(II)}$ was 3.2‰. This aqueous ferrous-ferric fractionation closely matches the Fe^{2+}–Fe^{3+} fractionation measured by Welch et al. (2003) in a strictly abiological experiment.

The unidirectional fractionation associated with the precipitation of ferric oxide from aqueous Fe^{3+} is thought to be a key feature of these experiments. Therefore, it is useful to divide these experiments according to their Fe oxidation rates. Figure 3.12 shows the proportion of Fe oxidation versus time for representative experiments from each study. Experiments are divided into three broad Fe oxidation rates based on the number of days it took to reach 50% oxidation:

(1) Rapid rate in which 50% oxidation occurred in 3 days,
(2) Intermediate rate in which 50% Fe oxidation occurred in 9 to 19 days, and
(3) Slow rate where the experiment reached 50% oxidation in 30 days or it never reached 50% oxidation over the duration of the experiment.

Using this oxidation rate classification, the Fe isotope composition measured for aqueous Fe^{2+} and precipitated ferric oxide is shown in Fig. 3.13, where from top to bottom, the rapid, intermediate, and slow rates are shown. For all of these studies, measured Fe isotope compositions have been recast such that the system Fe isotope composition has a $\delta^{56}Fe$ of 0‰, allowing direct comparison between all experiments. For each group, the calculated isotope composition for aqueous Fe^{2+} and solid ferric oxide based on either a Rayleigh fractionation model (curves) or a closed system model (straight lines) is shown. In the Rayleigh model, each aliquot of ferric oxide forms in equilibrium with the aqueous Fe^{2+} based on the assumed fractionation factor, but each aliquot is removed from the system and is

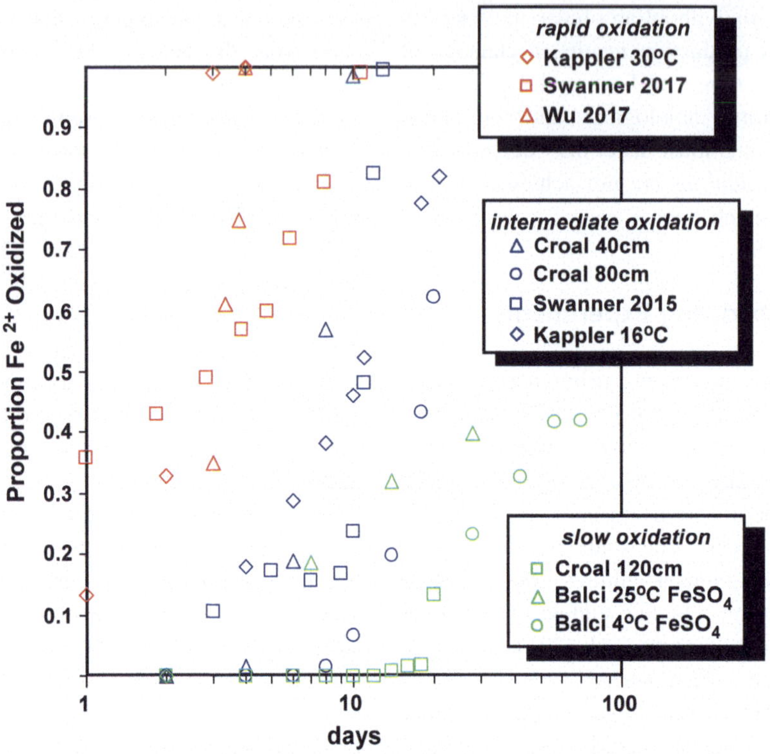

Fig. 3.12 Plot of the proportion of oxidized Fe^{2+} versus time in days (log scale) for bacterial Fe oxidation experiments. Experiments are divided into rapid, intermediate, and slow oxidation rates based on the number of days it took to reach 50% Fe oxidation. Rapid oxidation experiments include the Swanner et al. (2017), the Swanner et al. (2017) and Kappler et al. (2010) 30 °C experiment. The intermediate oxidation rate experiments include the Croal et al. (2004) experiments done at 40 and 80 cm away from a light source, the Swanner et al. (2015) experiment, and Kappler et al. (2010) 16 °C experiment. The slow oxidation experiments include the Croal et al. (2004) experiment done at 120 cm from a light source, and the Balci et al. (2006) 4 and 25 °C experiments

not able to re-equilibrate with the aqueous Fe^{2+}. In the closed system model, it is assumed that the aqueous Fe^{2+} and ferric oxide are able to continuously re-equilibrate. For most experiments, a Rayleigh model better predicts Fe isotope compositions for aqueous Fe^{2+} and ferric oxide. However, each experiment set requires a different fractionation factor. The rapid oxidation data are well fit using a $\Delta^{56}Fe_{Fe(II)\text{–ferric oxide}}$ of −2.5‰ (Fig. 3.13a), for intermediate oxidation rate a $\Delta^{56}Fe_{Fe(II)\text{–ferric oxide}}$ of −1.8‰ is shown (Fig. 3.13b), and for slow oxidation rate a $\Delta^{56}Fe_{Fe(II)\text{–ferric oxide}}$ of −1.5‰ is shown (Fig. 3.13c).

Past studies that investigated fractionation between aqueous Fe^{3+} and ferric oxides have shown that over long time scales (hundreds of days), the fractionation between aqueous Fe^{3+} and ferric oxide is small and is thought to reflect equilibrium conditions (Saunier et al. 2011; Skulan et al. 2002). In contrast, other studies have shown that if precipitation of aqueous Fe^{3+} occurs nearly instantaneously, no Fe isotope fractionation occurs (Johnson et al. 2002). This reflects the fact that all available aqueous Fe^{3+} is precipitated, so no net Fe isotope fractionation would occur. At intermediate precipitation rates, or times of equilibration (hours to weeks), the $\Delta^{56}Fe_{Fe(III)\text{–ferric oxide}}$ is positive and varies between 1.5 and 0.2‰, depending on precipitation rate and the mineralogy of the ferric oxide (Balci et al. 2006; Skulan et al. 2002). Indeed, if one considers only the precipitation experiments involving ferric chloride in which akaganeite precipitated (Balci et al. 2006), a negative

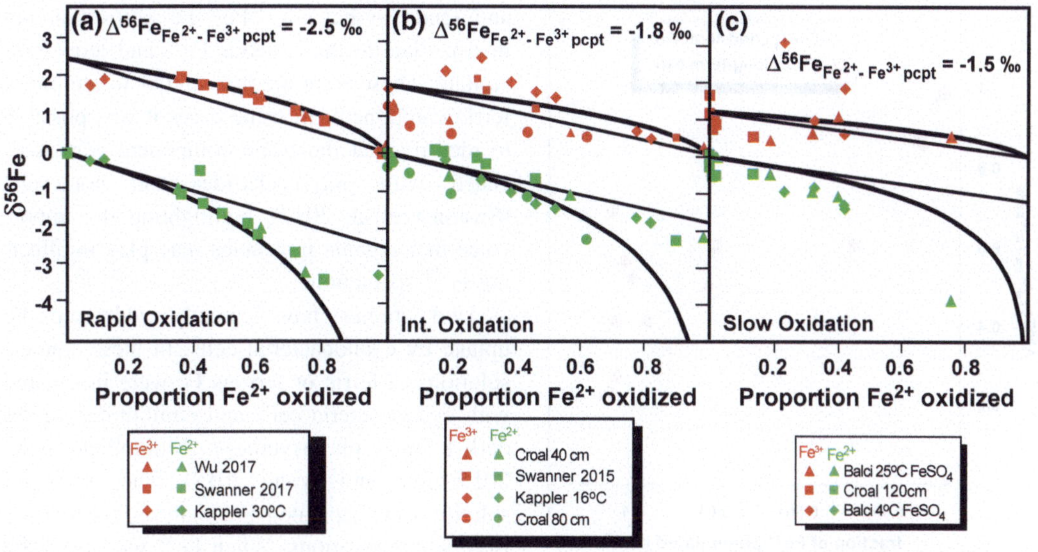

Fig. 3.13 Plot of δ^{56}Fe values for aqueous Fe^{2+} (green symbols) and Fe^{3+} precipitate (red symbols) versus the proportion of Fe^{2+} oxidized. All measured Fe isotope compositions are normalized to a system δ^{56}Fe of 0 to facilitate comparison between experiments. Only representative experiments from each study are plotted. **a** Rapid oxidation experiments include Ri-1 Swanner et al. (2017), exp 1 of Swanner et al. (2017) and the 30 °C bottle A experiment of Kappler et al. (2010). **b** Intermediate Fe^{2+} oxidation experiments include the Swanner et al. (2015) Ri3-1 experiment, the Kappler et al. (2010) 16 °C bottle A experiment and the Croal et al. (2004) 40 and 80 cm light distance experiments. **c** For slow oxidation, the experiments include the Croal et al. (2004) 120 cm light distance, the Balci et al. (2006) 25 °C Fe_SO_4 bottle B experiment, and the Balci et al. (2006) 4 ° C Fe_SO_4 experiment

correlation exists between the precipitation rate and $\Delta^{56}Fe_{Fe(III)\text{–ferric oxide}}$ (Fig. 3.14). In the abiological experiment conducted by Balci et al. (2006), a $\Delta^{56}Fe_{Fe(III)\text{–ferric oxide}}$ of 1‰ was measured for the slowest precipitation rates and 0.2‰ for faster precipitation rates (Fig. 3.14).

Based on these observations, we suggest that differences in the fractionation factor inferred for the different biological Fe oxidation studies reflect an initial ~3‰ fractionation between aqueous Fe^{3+} and aqueous Fe^{2+} that was effectively reduced because of a kinetic fractionation associated with precipitation of the ferric oxide. We infer that the kinetic fractionation is smaller for the most rapid Fe-oxidation experiments and is progressively larger for slower oxidation experiments, resulting in the $\Delta^{56}Fe_{Fe(II)\text{–ferric oxide}}$ being largest (least negative) for the most rapid oxidation experiments and progressively smaller (most negative) for the slower oxidation rate experiments. We note that there is also likely to be some isotopic re-equilibration between Fe^{2+} and ferric oxides, which would tend to move $\Delta^{56}Fe_{Fe(II)\text{–ferric oxide}}$ toward the equilibrium value of −3‰ if ferrihydrite or hematite are the main precipitates (Frierdich et al. 2019; Wu et al. 2011). Alternatively, if goethite was the main precipitate, equilibration would decrease the Fe^{2+}-precipitate fractionation reflecting the −1.2‰ fractionation between Fe^{2+} and goethite (Frierdich et al. 2014a). However, we note that in the Balci et al. (2006) low pH experiment, it is unlikely that exchange between Fe^{2+} and precipitate occurred because under acidic conditions, there is little sorption of Fe^{2+} onto the substrate, which limits isotopic exchange between aqueous Fe^{2+} and iron oxides (Reddy et al. 2015). In the other experiments, exchange between Fe^{2+} and the ferric precipitates is likely, but because the aqueous Fe^{2+} pool decreases with the duration of the experiment, the amount of exchange is likely to be minimal. However, in

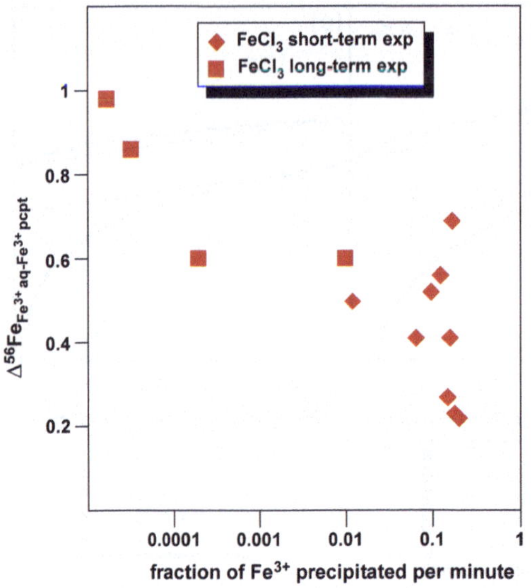

Fig. 3.14 Plot of the Fe isotope fractionation between aqueous ferric iron and a ferric precipitate as a function of the rate of abiological Fe precipitation from the experiments of Balci et al. (2006). The short-term experiments consisted of a ferric precipitate that formed from the oxidation of different ferric chloride stock solutions that ranged in pH from 2.22 to 3.53 over the course of 5 min. The long-term experiments allowed a ferric precipitate to form from a ferric chloride solution with a near constant 2.5 pH over the course of 14 days

cases in which Fe oxidation and oxide precipitation takes place in the periplasm, it is speculated that equilibrium isotope exchange between Fe^{2+} and the precipitate can be maintained (Kappler et al. 2010).

The other significant process that will affect the fractionation between Fe^{2+} and ferric oxide is complexation of Fe with organic ligands. Although the effects of organic ligands on Fe speciation was speculated upon by Croal et al. (2004) and Balci et al. (2006), it was not until later studies that the effects of organic ligands and Fe speciation were better evaluated. For example, studies conducted by Swanner et al. (2015, 2017) were able to identify both ferric and ferrous iron that were easily leached from the ferric oxide precipitate using H_2O followed by sodium acetate. The ferrous iron is thought to be a sorbed component, which has been identified in a number of other biological experiments (Crosby et al. 2005, 2007; Kappler et al. 2010) and tends to have a $^{56}Fe/^{54}Fe$ ratios that are intermediate to the aqueous Fe^{2+} and ferric precipitate. These were the first studies to identify a ferric component. In some cases it was possible to identify that the ferric component was associated with polysaccharides and not cells (Swanner et al. 2017), highlighting the importance that organic molecules may play in affecting Fe^{3+} speciation.

Two studies have directly addressed Fe uptake by cyanobacterial cells. In these studies, solutions of ferric or ferrous Fe were inoculated with cyanobacteria cells and equilibrated in the dark to limit photosynthesis (Mulholland et al. 2015; Sun and Wang 2018). The cells and solution were separated and analyzed for their Fe isotope compositions. When Fe^{3+} was used, the experiment was conducted at a pH of 3 to keep the aqueous Fe^{3+} from precipitating. The Fe isotope fractionation between aqueous Fe^{3+} and Fe^{3+} associated with cells was 0.97‰, and this fractionation was independent of the three different types of cyanobacteria used (Mulholland et al. 2015). In other experiments in which Fe^{2+} was used, the fractionation between the aqueous Fe^{2+} and iron associated with cells was typically larger as compared to the Fe^{3+} experiments (Mulholland et al. 2015). However, in some experiments, a fractionation in the opposite direction was reported (Sun and Wang 2018). One of the difficulties in interpreting experiments that used Fe^{2+} is that the Fe speciation was not determined, and so it is unknown if the Fe associated with cells was ferrous or ferric iron. In general, the authors of these studies suggest that the Fe associated with the cells is likely to have been oxidized and that this oxidation is the origin of the larger fractionation between aqueous Fe^{2+} and Fe associated with cells. Moreover, in the experiments conducted by Sun and Wang (2018) it is likely that some Fe-bearing phases precipitated in some of the long-duration experiments, making it very difficult to determine how much the measured $\delta^{56}Fe$ values of the aqueous Fe was changed by precipitation of an Fe-bearing mineral versus uptake of Fe by cells. Despite these uncertainties, it is clear that Fe associated with cyanobacteria cells has a $\delta^{56}Fe$ value that is greater than aqueous Fe. Notably, in the experiments conducted by Mulholland et al. (2015), workers were able to identify the different Fe ligands that may have been produced in the cellular-associated Fe, and they used the β-values reported by Fujii et al. (2014) for aqueous Fe species and Fe ligands to calculate $1000\ln\alpha_{Fe(III)-Fe(III)\ ligand}$ values. This comparison revealed a poor fit between experimental results and the calculated fractionation factor based on reduced partition ratios. The researchers concluded that the high symmetry Fe ligands used in the calculation of the reduced partition function ratios did not well approximate the multi-dentate Fe^{3+} complexes on cyanobacteria cells.

3.5.2 Magnetotactic Bacteria

Magnetic minerals such as magnetite and greigite can be formed intracellularly by magnetotactic bacteria. These magnetosomes give the cell a magnetic dipole that enables the cell to navigate up or down in a vertical chemical gradient using Earth's magnetic field. Two studies investigated Fe isotope fractionation during the formation of magnetosomes. In one of the first modern Fe isotope studies, Mandernack et al. (1999) found that the Fe isotope composition of magnetite from magnetosomes matched that of the aqueous Fe that was used in the growth media, within the uncertainty of the measurement method (+0.3‰). These experiments used two different magnetotactic bacteria, *Magnetovibrio blakmorei* MV-1 and *Magnetospirillum magnetotacticum* MS-1. Strain MV-1 synthesizes magnetite anaerobically and strain MS-1 produces magnetite as an obligate microaerophile and requires molecular O_2 to synthesize magnetite. For the MS-1 experiment, a growth medium containing ferric solution was used, whereas for the MV-1 strain both ferric and ferrous solutions were used. Considering the significant Fe isotope fractionation between ferrous and ferric iron, the lack of Fe isotope contrast between magnetite and ferrous and ferric solutions is surprising, if the magnetite formed under equilibrium conditions. In a more recent study using *Magnetospirillum magneticum* AMB-1 bacteria to produce magnetite, significant Fe isotope fractionation (~1‰ in $^{56}Fe/^{54}Fe$) was found between the Fe used in the growth media and the magnetosomes (Amor et al. 2016). In the AMB-1 experiments, the Fe mass balance of the system was accounted for by measuring the Fe content and Fe isotope composition of the used growth media, Fe from the magnetosomes, and remaining Fe from the lysis of the cells. The integration of these three components matched the Fe isotope composition of the starting growth media. Although Fe speciation was not measured, Amor et al. (2016) hypothesized that the Fe lysate was ferric iron, because the lysate Fe had a $\delta^{56}Fe$ value greater than the magnetite Fe in experiments that used either Fe^{2+} or Fe^{3+} in the growth media. As reported in Amor et al. (2016) Supplementary Table S1, the $\Delta^{56}Fe_{lysate\ Fe-magnetite}$ for the Fe^{3+} experiment in which lysate #1 and magnetite #1 were analyzed was 2.76 ± 0.20‰, and for the experiment in which Fe^{2+} was used in the growth media the $\Delta^{56}Fe_{lysate\ Fe-magnetite}$ for lysate and magnetite 1 was 2.19 ± 0.04‰ and 2.2 ± 0.29‰ for lysate and magnetite 2. The errors for these fractionation factors used the reported 2-SD in Table S1 of Amor et al. (2016) if only 1 analysis was reported, if two analyses were reported the average and standard deviation were calculated from the two analyses reported for magnetite #1 in the Fe^{3+} experiment and for magnetite #2 in the Fe^{2+} experiment. These experimentally-measured $\Delta^{56}Fe_{lysate\ Fe-magnetite}$ fractionations are larger than the $1000\ln\alpha$ value calculated from the Rustad et al. (2010) and Mineev et al. (2007) hexaquo Fe^{3+} and magnetite reduced partition function ratios, respectively. These experimentally-measured fractionations are larger compared to combining the Fe^{2+}-magnetite and Fe^{2+}–Fe^{3+} fractionations determined by Frierdich et al. (2014b) and Welch et al. (2003), respectively. However, if one considers that Fe^{3+} lysate is likely to be complexed with a strong organic ligand, a better fit to the experimental results would be obtained as shown by the $1000\ln\alpha$ calculated using Fe^{3+}–DFOB and magnetite β-values reported by Domagal-Goldman et al. (2009) and Mineev et al. (2007), respectively (Fig. 3.15).

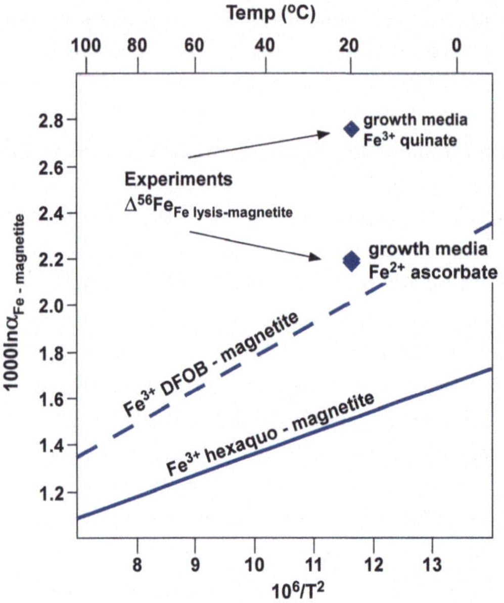

Fig. 3.15 Plot of $1000\ln\alpha_{Fe\text{–}magnetite}$ versus $10^6/T^2$. The symbols show the measured Fe isotope fractionation between Fe from the cell lysate and Fe from magnetite contained in magnetosomes (Amor et al. 2016). The experiment with the larger fractionation factor used Fe^{3+} quinate and the other experiments used Fe^{2+} ascorbate. It is thought that the Fe in the lysis is ferric iron (Amor et al. 2016). For comparison, the calculated fractionation factors between different aqueous Fe species and magnetite are shown. The solid blue curve used the Rustad et al. (2010) hexaquo Fe^{3+} β-value and the Mineev et al. (2007) magnetite β-value. The dashed blue line used the Domagal-Goldman et al. (2009) Fe^{3+}–DFOB β-value and the Mineev et al. (2007) magnetite β-value. See text for discussion

The cause for the minimal Fe isotope fractionation between the starting Fe isotope composition of the growth media that used either Fe^{3+} or Fe^{2+} and the magnetosome magnetite analyzed by Mandernack et al. (1999) is unknown. Johnson et al. (2005) suggest that these experiments do not reflect equilibrium fractionation, but rather reflect a kinetic process consistent with the lack of differences in the fractionation between starting Fe and magnetite for experiments done at 28 and 4 °C; at equilibrium, the lower temperature experiment should define a larger magnitude fractionation. Differences in fractionation between the Mandernack et al. (1999) and Amor et al. (2016) studies are not thought to be a function of metabolic pathways because AMB-1 and MS-1 produce magnetite using similar metabolic pathways as compared to MV-1. Amor et al. (2016) suggests that minimal Fe isotope fractionation found in the Mandernack et al. (1999) study may have occurred because of Fe limitation, and that all cellular Fe was converted to magnetite. In contrast, in the Amor et al. (2016) study, Fe occurred in the cell lysate, magnetosome, and in the growth media. In support of such a contention, Amor et al. (2016) noted that the Fe concentration in the starting growth media in the Mandernack et al. (1999) experiments used lower Fe concentrations as compared to the Amor et al. (2016) experiments. Regardless, it is important to note that the total proportion of Fe left in the growth media for the Mandernack et al. (1999) and Amor et al. (2016) experiments are similar. Because a complete Fe isotope mass balance of the experiments by Mandernack et al. (1999) was not made, it is difficult to evaluate these competing hypotheses.

In addition to the mass-dependent fractionations found in the experiments conducted by Amor et al. (2016), they also report mass-independent fractionations (MIF) for Fe isotope ratios that use ^{57}Fe as compared to isotope ratios that only contain ^{54}Fe, ^{56}Fe, and ^{58}Fe. They inferred that the MIF was caused by magnetic isotope effects, which results in MIF of odd-number isotopes (Buchachenko 2001). This study inferred that the MIF in the magnetosomes involved the intracellular formation of a small Fe^{2+} reservoir that interacted with the intracellular Fe^{3+} to produced magnetite with a MIF caused by the presence of the magnetic field imparted by the single domain magnetite grains in the magnetosomes. Amor et al. (2016) noted that the MIF is only associated with the Fe^{3+} growth media and suggest that the lack of a MIF in the Fe^{2+} growth media is thought to result in the MIF being diluted with transport of extracellular Fe^{2+} into the celll.

3.5.3 Fe Reducing Experiments

Microbial dissimilatory iron reduction (DIR) is an anaerobic respiration metabolism in which ferric iron acts as an electron acceptor and organic C or H_2 are electron donors (e.g., Lovley and Phillips 1986a, b). During DIR, some species (such as *Geobacter*) may need to be in contact with ferric material, whereas other species, such as *Schewenella,* do not need direct contact with the ferric substrate because they can use electron shuttle compounds such as humic substances (Lovley et al. 1991, 2011; Thamdrup 2000; Lovley et al. 2004). Additionally, these organisms can produce nanowire appendages that transfer electrons over greater distances such as through a biofilm (Lovley et al. 2011; Walker et al. 2018). These nanowires were discovered in the *Geobacter metallireducens* species by workers noting that the organism produced the appendage when grown on insoluble iron oxides but not when grown using a soluble chelated ferric iron (Childers et al. 2002).

Iron-reducing organisms can utilize a number of different ferric substances including aqueous iron, poorly crystalline iron oxides such as ferrihydrite, crystalline iron oxides such as hematite or goethite, and ferric-rich iron clays such as nontronite (Lovley 1991; Lovley et al. 2004; Roden 2003, 2006; Shi et al. 2016; Wu et al. 2017). The end product of DIR can be quite variable. This variability is largely controlled by the flux of the reduced iron that is produced, with secondary controls including the bacterial DIR species, the ferric substrate that is present, and the organic carbon oxidant (Roden and Urrutia 1999; Roden and Zachara 1996; Salas et al. 2009, 2010; Zachara et al. 2002). For example, during the coupled reduction of Fe and oxidation of organic carbon, both ferrous iron and CO_2 are produced, resulting in formation of Fe-rich carbonates such as siderite if the flux of Fe^{2+} is large or the ferric substrate is nonreactive with aqueous Fe^{2+}. If iron-reducing bacteria are grown on ferrihydrite, the aqueous Fe^{2+} produced will react with the ferrihydrite and recrystallize to more stable iron oxides such as goethite at low Fe^{2+} fluxes or magnetite if the flux of Fe^{2+} is greater (Zachara et al. 2002). In addition, other ferrous rich minerals such as green rust or vivianite or x-ray amorphous mineraloids such as ferrous hydroxides or Fe silica gels can form depending on the concentration of other anions in solution and/or the ferric iron substrate that is being reduced (Roden 2006; Salas et al. 2010; Zachara et al. 2002). To a large degree, the secondary minerals produced are governed by simple abiological interactions between aqueous Fe^{2+} and reactive iron oxides, although some organisms produce biofilms creating local microenvironments that allow formation of secondary minerals that are not in equilibrium with the conditions outside of the biofilm (Roden 2006; Salas et al. 2010).

All Fe isotope studies that have investigated bacterial DIR have found that the enzymatically-produced Fe^{2+} has a lower $^{56}Fe/^{54}Fe$ ratio compared to the ferric substrate (Beard et al. 1999, 2003; Crosby et al. 2005, 2007; Fortney et al. 2016; Icopini et al. 2004; Johnson et al. 2005; Percak-Dennett et al. 2011; Tangalos et al. 2010; Wu et al. 2009). Initial investigations suggested that this fractionation originated between pools of bacterial ligand-bound Fe^{2+} and Fe^{3+} (Beard et al. 1999, 2003; Johnson et al. 2004). Later studies suggested that the isotopically light aqueous Fe^{2+} originated because of sorption of Fe^{2+} onto the ferric hydroxide substrate (Icopini et al. 2004). However, in these initial studies, the workers did not isolate the Fe component that had the high $^{56}Fe/^{54}Fe$ ratios that is required to provide the mass balance to the aqueous Fe^{2+} with low $^{56}Fe/^{54}Fe$ ratios. Later studies that used partial leaching coupled with measuring Fe^{2+} and total Fe concentrations and Fe isotope compositions, identified that a reactive ferric iron layer formed from interfacial electron and atom exchange that occurred between bacterially-produced Fe^{2+} and the ferric substrate (Crosby et al. 2005, 2007).

Figure 3.16 is a carton illustrating the origin of Fe isotope variations caused by DIR. DIR bacteria pump electrons into the ferric substrate generating Fe^{2+}; in some cases the bacteria probably solubilize the ferric iron before reducing it. The Fe^{2+} produced is associated with the surface of the ferric substrate, and this generates

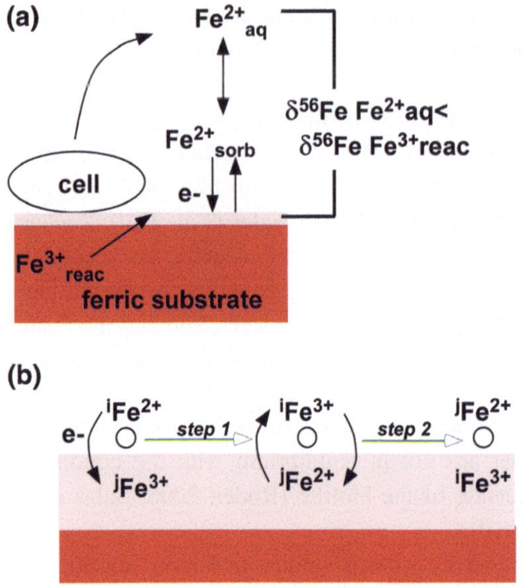

Fig. 3.16 Illustrative cartoon showing the origin of Fe isotope fractionation during dissimilatory Fe reduction caused by the formation of a reactive ferric iron layer by interfacial electron and atom exchange (based on the models presented by Crosby et al. (2005, 2007). **a** Illustration of bacterial dissimilatory iron reduction of a ferric substrate (not to scale). Aqueous Fe^{2+} is produced by electrons pumped from the bacteria, some of this $Fe^{2+}{}_{aq}$ is adsorbed onto the oxide surface ($Fe^{2+}{}_{sorb}$). The adsorbed Fe undergoes electron transfer and Fe^{2+}–Fe^{3+} atom exchange that produces a reactive ferric layer (lighter red shade). The reactive ferric layer has a δ^{56}Fe value greater than the initial substrate and this is the mass balance to the $Fe^{2+}{}_{aq}$ that has a δ^{56}Fe value lower than $Fe^{2+}{}_{sorb}$ and the ferric substrate. **b** An expanded view of the oxide surface, illustrating interactions between Fe isotopes i and j at the oxide surface. In this model, sorbed Fe^{2+} (atom i) transfers an electron to an Fe^{3+} (atom j) of the oxide. In the second step, isotopic exchange of atoms i and j occurs, as required by the Fe isotope changes observed in bacterial dissimilatory Fe-reducing experiments

interfacial electron and atom exchange in a fashion that is similar to abiologic Fe^{2+} recrystallization of ferric hydroxides. In the experiments conducted by Crosby et al. (2005, 2007) and Wu et al. (2009), pure strains of bacteria (*Geobacter sulfurreducens* strain PCA or *Shewanella putrefaciens* strain CN32) were grown on hematite or goethite with H_2 as the electron donor. These conditions resulted in limited (~4% maximum) reduction of the ferric material where the bulk of the Fe^{2+} was in the aqueous fraction (Fig. 3.17). Serial partial leaching of the ferric substrate using either sodium acetate (Crosby et al. 2005, 2007) or weak HCl (Wu et al. 2009) produced a leachate that was 96% Fe^{2+} that had a slightly higher δ^{56}Fe value as compared to the aqueous Fe^{2+}. For hematite, the sorbed Fe^{2+} was 0.3‰ greater than the aqueous Fe, and for goethite it was 0.9‰ greater than the aqueous Fe^{2+}. Following this weak leaching, the substrate was leached a second time in 0.5 M HCl, which produced a leachate that contained both Fe^{2+} and Fe^{3+}. The proportion of Fe^{3+} to total Fe in the 0.5 M HCl leach was variable, ranging from 8 to 83% ferric iron. The proportion of ferric iron removed in this leach was the lowest in the experiments with goethite, which likely reflects the fact that the sodium acetate leach of the goethite was not as effective as compared to the leaching of hematite. In all cases, the 0.5 M HCl leach had δ^{56}Fe values greater than the aqueous and sorbed Fe^{2+}. Assuming that the Fe^{2+} in the 0.5 M HCl leach was sorbed Fe^{2+} that was not fully removed by the weak HCl or sodium acetate, it is possible to calculate the Fe isotope composition of the endmember Fe^{3+} component by subtracting the Fe^{2+} component, assuming it has the same Fe isotope composition as the sorbed Fe^{2+}. This Fe^{3+} component, referred to as the reactive ferric component, had a δ^{56}Fe value that was greater than the starting hematite or goethite, indicating Fe isotope exchange. The calculated $\Delta^{56}{}_{Fe(II)aqueous–reactive\ ferric\ oxide}$ fractionation factor for both the goethite and hematite experiments was −2.95‰ by Crosby et al. (2007) and −2.70‰ for the Wu et al. (2009) experiment done at pH 7 without Si, which is similar to the conditions used by Crosby et al. (2007). These fractionations closely match the abiological $\Delta^{56}{}_{Fe(II)aqueous–hematite}$ = −2.9‰ or $\Delta^{56}{}_{Fe(II)}$

$_{aqueous-ferrihydrite}$ = −3.2‰ fractionation. Indeed, the abiological fractionation between aqueous Fe^{2+} and the reactive ferric layer identified in goethite by Beard et al. (2010) of −2.10 ± 0.48‰ is also similar. This similarity between the abiological and biological fractionations is why interfacial electron and atom exchange is now the preferred explanation for the origin of the low δ^{56}Fe aqueous Fe^{2+} produced by DIR. Based on isotope mass balance, the amount of reactive ferric iron can be calculated, and in these experiments, the reactive ferric component was similar to the number of hematite or goethite surface atoms.

In other experiments (Fortney et al. 2016; Percak-Dennett et al. 2011; Tangalos et al. 2010) that used bacterial DIR enrichment cultures, poorly crystalline ferric oxides, and acetate instead of H_2, the amount of Fe reduction was greatly increased, reaching levels of 68% reduction as compared to the small amount of Fe reduction in the Crosby et al. (2005, 2007) and Wu et al. (2009) experiments. The studies by Tangalos et al. (2010) and Fortney et al. (2016) used natural ferric oxides from an acid mine drainage site and a hydrothermal spring, respectively. In order to gain a better understanding of the types of material that might be produced during DIR in an Archean-age banded iron formation, the Percak-Dennett et al. (2011) study used a ferric silica gel and a growth media that simulated Archean seawater. In these high-Fe reducing experiments, the aqueous Fe^{2+} had the lowest δ^{56}Fe values. The adsorbed Fe^{2+} identified by partial leaching of the ferric substrate with weak molarity HCl had δ^{56}Fe values greater than those of the aqueous Fe^{2+}. Additionally, a reactive ferric component identified by 0.5 M HCl leaching and subtraction of any Fe^{2+} in the 0.5 M HCl leach had a δ^{56}Fe value that was typically greater than the starting crystalline ferric oxides. These are exactly the same Fe isotope compositions identified in the low Fe reducing experiments of Crosby et al. (2005, 2007) and Wu et al. (2009). However, in these high-Fe reduction experiments, the bulk of the Fe^{2+} was sorbed Fe and not the aqueous Fe^{2+} (Fig. 3.17).

As first noted by Percak-Dennett et al. (2011), differences in the amount of ferric oxide reduced, and the relative proportion of aqueous and sorbed Fe^{2+}, is likely to have a strong control on large basins with banded iron formations. For example, in an open basin, significant quantities of mM Fe^{2+} with low δ^{56}Fe values can be produced by small amounts of bacterial Fe reduction (Fig. 3.17). This small amount of DIR would be driven by the continuous supply of ferric oxides and organic carbon produced in the photic zone that settles through the water column and is utilized by DIR bacteria on the ocean floor. In contrast, in areas where the supply of ferric oxides is minimized by burial of the reactive ferric material, DIR is likely to produce a large amount of Fe^{2+} that is trapped on the ferric oxide. This trapped Fe^{2+} would result in diagenetic reactions that could produce magnetite or greenalite depending on the relative amount of silica to ferric oxide material. The application of stable Fe isotopes to tracing DIR in modern and ancient marine sediments is extensively discussed in Chaps. 5 and 6, respectively.

An important observation regarding the rate and amount of reduction by DIR is that it strongly correlates to the surface area of ferric substrate with only minor controls on ferric reduction rates as a function of the thermodynamic properties of the substrate (Roden 2003, 2006). Long-term reduction rates are also controlled by surface properties of the substrate and rates tend to decline as Fe^{2+} or other species adsorb onto the ferric substrate blocking electron transfer (Roden 2006; Roden and Urrutia 1999). In contrast, abiotic reductive dissolution is strongly controlled by the thermodynamic properties of the mineral (Roden 2006). An important observation in both the low-Fe and high-Fe

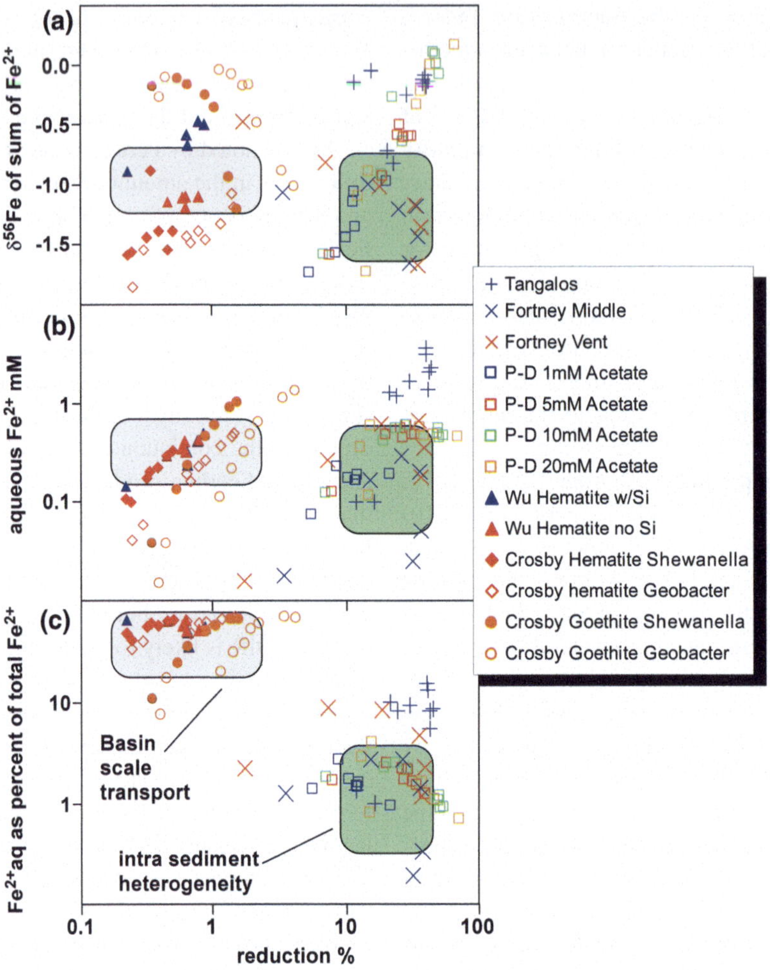

Fig. 3.17 **a** Plot of the $\delta^{56}Fe$ values of total Fe^{2+} produced during bacterial dissimilatory Fe reduction of ferric substrates as a function of the percent reduction of the substrate (log scale). The data are from bacterial DIR studies using pure and enrichment cultures. For pure cultures the circles are reduction of goethite (Crosby et al. 2007) by different bacterial DIR species, the diamonds are reduction of hematite (Crosby et al. 2007) by different bacterial DIR species, and the triangles are reduction of hematite at pH 7 (Wu et al. 2009) by *Geobacter* in the presence and absence of dissolved Si. For the enrichment cultures, the cross is the experiment reported by Tangalos et al. (2010) that used natural ferric oxides from an acid mine drainage site, the X's are for the work of Fortney et al. (2016) that used natural ferric oxides from a hydrothermal spring site, and the squares are from the reduction of Fe^{3+}-silica gels conducted by Percak-Dennett et al. (2011) that used different amounts of acetate to change growth conditions. **b** Plot of the aqueous Fe^{2+} (log scale) produced from the same bacterial DIR experiments shown in A versus the percent reduction of the ferric substrate. **c** Plot of the amount of aqueous Fe^{2+} as a function of total Fe^{2+} produced (log scale) from the same bacterial DIR experiments in A versus the percent reduction of the ferric substrate. See text for discussion

reduction experiments is that at higher amounts of Fe reduction, the Fe^{2+} begins to match the Fe isotope composition of the system. In the experiments that reduced a large amount of Fe the increase in the $\delta^{56}Fe$ of the Fe^{2+} is driven by simple mass balance because nearly all of the ferric oxide is consumed. In low Fe-reducing experiments, the increase in the $\delta^{56}Fe$ of the Fe^{2+} is thought to reflect the fact that DIR bacteria are beginning to reduce the reactive ferric substrate which has a $\delta^{56}Fe$ that is higher than that of the ferric substrate, and the bacterial DIR is being curtailed by the lower surface area of the crystalline hematite and goethite used in the experiments.

3.6 Preferred Set of β-Values

Table 3.4 presents preferred β-values for the wide variety of different Fe species discussed in this chapter. As discussed above, these reduced partition function ratios are consistent with experimental studies and are internally consistent among themselves. Therefore, the accuracy of the β-values is deemed to be high. Although this group of preferred β-values includes a host of oxide and sulfide minerals, as well as a variety of aqueous Fe organic and inorganic species, the list is not exhaustive. Because the number of calculated reduced partition function ratios far exceeds the number of experiments that have been conducted, it is worthwhile to develop some best practices to allow one to scale β-values for minerals or Fe species that have not been investigated experimentally. For reduced partition function ratios calculated by ab initio methods, one can scale untested β-values relative to one that has been tested. For example, Fujii et al. (2014) presents a host of reduced partition function ratios of aqueous Fe species calculated using density functional theory, including reduced partition function ratios for deprotonated aqueous ferric species such as $Fe^{(III)}OH(H_2O)_5^{+2}$, a species for which we have no experimental constraints. In order to scale the $Fe^{(III)}OH(H_2O)_5^{+2}$ β-value, one could calculate a scaled Fujii et al. (2014) $Fe^{(III)}OH(H_2O)_5^{+2}$ relative to the ratio of the Rustad et al. (2010) and Fujii et al. (2014) $Fe^{(III)}(H_2O)_6^{+3}$ β-values. Scaling β-values is based on the idea that calculated fractionation factors from reduced partition function ratios often closely match experimentally measured fractionation factors even if their accuracy is poor. For example, Fig. 3.3 shows that numerous formulations of the heaxaquo Fe^{3+} and Fe^{2+} β-values are internally consistent with experimental results (aquoeus-aqueous), but only the β-values of Rustad et al. (2010) are consistent with different aqueous-mineral fractionations. Although such scaling does not validate the accuracy of a scaled β-value, it would make the scaled β-value consistent with β-values for which the accuracy has been assessed. Attempting to scale reduced partition function ratios from NRIXS or Mössbauer spectroscopic data is more problematic because inaccurate β-values are likely to reflect imprecision in the input data, the Mössbauer or NRIXS spectra. Thus, best practice is to use such β-values as reported in the literature but recognize that they may be inaccurate.

The list of preferred β-values also includes values for mineraloids derived solely from experimentation, including ferrihydrite and Fe–Si gels. These experimentally-determined β-values were calculated based on the experimental results determined for hexaquo Fe^{2+} and the solid phase, and extrapolation to zero at infinite temperature on a plot of $1000\ln\alpha$ versus $1/T^2$ to determine the fractionation factor as a function of temperature. Note that the curvature of $1000\ln\beta$ versus $10^6/T^2$ is very minor for all of the investigated Fe minerals, so this is unlikely to introduce an error beyond ~0.1‰ over the temperature range from 0 to 300 °C. The β-value was then calculated using the difference between the Rustad et al. (2010) hexaquo Fe^{2+} reduced partition function ratio and the experimentally-determined $1000\ln\alpha$ values. Reduced partition function ratios have not been published for these phases, and calculating them would be challenging because they are x-ray amorphous with uncertain mineral structures, and thus we are forced to rely only on the experimental results. Nevertheless, it is important to have β-values available for these poorly-crystalline materials, given their great importance in natural environments (see Chap. 5).

Overall, the β-values for hexaquo Fe^{2+} and Fe^{3+} determined by Rustad et al. (2010) are considered to be highly accurate because they agree

Table 3.4 An internally consistent set of 1000lnβ-values that agree with experimental studies

Material	0 °C	25 °C	50 °C	100 °C	200 °C	300 °C	References
Aqueous Fe^{3+}							
$Fe^{(III)}(H_2O)_6$	9.04	7.69	6.62	5.05	3.20	2.21	1
$Fe^{(III)}(H_2O)_5Cl$	8.48	7.21	5.86	4.72	2.99	2.05	2
$Fe^{(III)}(H_2O)_4Cl_2$	8.12	6.90	5.64	4.50	2.85	1.96	2
$Fe^{(III)}(H_2O)_3Cl_3$	7.21	6.11	5.00	3.98	2.52	1.72	2
$Fe^{(III)}Cl_4$	7.70	6.52	5.41	4.22	2.65	1.82	2
$Fe^{(III)}(Ox)_3$	9.41	7.98	6.85	5.20	3.28	2.26	4
$Fe^{(III)}$–DFOB	9.62	8.15	*7.00*	5.31	3.35	2.30	5
Aqueous Fe^{2+}							
$Fe^{(II)}(H_2O)_6$	5.50	4.66	3.99	3.03	1.91	1.31	1
$Fe^{(II)}(H_2O)_5Cl$	5.42	4.59	3.78	2.98	1.87	1.29	3
$Fe^{(II)}(H_2O)_4Cl_2$	5.16	4.37	3.61	2.83	1.78	1.22	3
Minerals							
Siderite	4.87	4.11	3.51	2.65	1.66	1.130	6
Chalcopyrite	5.86	4.95	4.24	3.20	2.01	1.38	7
Pyrite	10.95	9.26	7.93	6.00	3.77	2.58	8
Magnetite	7.36	6.18	5.26	3.95	2.45	1.67	9
Goethite	6.77	5.71	4.88	3.68	2.30	1.57	10
Hematite	8.98	7.60	6.51	4.93	3.10	2.12	11
Ferrihydrite	9.24	7.76	6.60	4.95	3.08	2.10	This work
Gel Fe:Si 1:1	8.51	7.14	6.08	4.56	2.84	1.93	This work
Gel Fe:Si 1:2	9.60	8.06	6.86	5.14	3.20	2.18	This work
Gel Fe:Si 1:3	10.16	8.53	7.26	5.44	3.39	2.31	This work

References **1**: derived from DFT S6 formulation (Rustad et al. 2010). **2**: derived from UHF/6-31G(d) formulation (Hill and Schauble 2008) and scaled to the Rustad et al. (2010) hexaquo Fe^{3+}. The β-value at 50 °C was interpolated from the 25 and 100 °C values as a function of $1/T^2$. **3**: derived from UHF/6-31G(d) formulation (Hill et al. 2010) and scaled to the Rustad et al. (2010) hexaquo Fe^{2+}. The β-value at 50 °C was interpolated from the 25 and 100 °C values as a function of $1/T^2$. **4**: derived from DFT 1-in vacuo formulation (Domagal-Goldman and Kubicki 2008). **5**: derived from DFT (Domagal-Goldman et al. 2009). **6**: Average of the Mössbauer derived β-value reported in Polyakov and Mineev (2000) and the DFT derived β-value of Blanchard et al. (2009). **7**: Average of the NRIXS derived β-values (Dauphas et al. 2012; Polyakov and Soultanov 2011). The β-value from Dauphas et al. (2012) was calculated using their Eq. 12 with B_2 = 56,616 and a force constant of 146. **8**: The average of the β-values derived from Mössbauer spectroscopy (Blanchard et al. 2012; Polyakov et al. 2019) and DFT (Blanchard et al. 2009). **9**: The β-value was derived from Mössbauer spectroscopy (Mineev et al. 2007), **10**: The β-value was derived from Mössbauer spectroscopy (Polyakov and Mineev 2000). **11**: Average of the β-values derived from Mössbauer spectroscopy (Polyakov et al. 2007) and NRIXS (Dauphas et al. 2012). The β-value from Dauphas et al. (2012) was calculated using their Eq. 12 with B_2 = 54,915 and a force constant of 239

with numerous experiments involving different aqueous Fe species and aqueous-mineral experiments. Additionally, β-values for the minerals siderite, pyrite, and hematite are also considered to be highly accurate because very similar β-values for each mineral have been calculated using a combination of either Mössbauer spectroscopy, NRIXS, or by first-principles calculations using density functional theory. Moreover, these agree well with experimental results (Table 3.4).

3.7 Summary

- Fe isotope fractionation factors can be constrained from β-values determined by Mössbauer spectroscopy, NRIXS, and from first-principles calculations. A number of first-principles calculations are available, and the hybrid density functional theory B3LYP model is considered superior because it tends to model properties such vibrational frequencies and bond lengths that are consistent with measured values. These different methods for calculating β-values are independent of one another, providing a method to evaluate the accuracy of a β-value.
- Fe isotope fractionation factors are determined using experimental methods. These techniques include synthesis methods, the partial exchange approach, and the three-isotope method. The three-isotope method is considered the best experimental method because it allows evaluation of isotope exchange kinetics and the fractionation factor at complete exchange. The best method for investigating Fe isotope fractionation at low temperatures is a reversed three-isotope method in which the surface area of the solid material is varied to allow one to evaluate how isotope exchange kinetics may affect extrapolations to 100% exchange.
- The process of sorption of aqueous Fe^{2+} onto a ferric oxyhydroxide is important, but the isotopic fractionation between aqueous Fe^{2+} and ferric oxyhydroxides is driven by atom and electron exchange between the Fe^{2+} components and the ferric substrate, and not the sorption process itself. In abiological experiments, if Fe^{2+} is not allowed to sorb, because of low pH, for example, Fe isotope exchange between Fe^{2+} and the ferric substrate does not occur.
- Overall, there is good agreement between many of the calculated fractionation factors and experiments. The β-values for hexauqo Fe^{2+} and Fe^{3+} calculated by Rustad et al. (2010) agree well with experiments involving aqueous Fe^{2+} and Fe^{3+} and with aqueous-mineral experiments. These β-values are considered the best-constrained β-values for aqueous Fe species. The β-values determined for siderite, pyrite, and hematite are also considered accurate because they have been calculated by at least two of the three independent methods for calculating β-values, and these values agree well with experiments.
- Oxidation of ferrous iron by bacteria produces a ferric precipitate that has a $\delta^{56}Fe$ value that is greater than the aqueous Fe^{2+}. The magnitude of this fractionation is driven by the abiological fractionation between aqueous Fe^{2+} and Fe^{3+}, and a kinetic fractionation between aqueous Fe^{3+} and the ferric precipitate. Experiments that rapidly precipitate Fe result in a larger aqueous Fe^{2+}-precipitate fractionation factor as compared to experiments in which the Fe^{3+} is slowly precipitated.
- Magnetotactic bacteria produce intra-cellular magnetite. The Fe isotope fractionation between magnetite and the Fe source appears to be consistent with an equilibrium process between cellular Fe^{2+} complexed with an organic ligand and magnetite.
- Iron-reducing bacteria produce aqueous Fe^{2+} that has a $\delta^{56}Fe$ value that is less than the ferric substrate. The origin of this fractionation is driven by creation of a reactive ferric layer that has a high $\delta^{56}Fe$ value that provides the mass balance to the low $\delta^{56}Fe$ aqueous Fe^{2+}.
- Table 3.4 presents an internally consistent set of 1000lnβ-values that agree well with experimental results for aqueous Fe species and minerals.

References

Amor M, Busigny V, Louvat P, Gelabert A, Cartigny P, Durand-Dubief M, Ona-Nguema G, Alphandery E, Chebbi I, Guyot F (2016) Mass-dependent and -independent signature of Fe isotopes in magnetotactic bacteria. Science 352(6286):705–708. https://doi.org/10.1126/science.aad7632

Anbar AD, Jarzecki AA, Spiro TG (2005) Theoretical investigation of iron isotope fractionation between Fe (H2O)(3+)(6) and Fe(H2O)(2+)(6): implications for iron stable isotope geochemistry. Geochim Cosmochim Acta 69(4):825–837. https://doi.org/10.1016/j.gca.2004.06.012

Anbar AD, Roe JE, Barling J, Nealson KH (2000) Nonbiological fractionation of iron isotopes. Science 288(5463):126–128

Balci N, Bullen TD, Witte-Lien K, Shanks WC, Motelica M, Mandernack KW (2006) Iron isotope fractionation during microbially stimulated Fe(II) oxidation and Fe(III) precipitation. Geochim Cosmochim Acta 70(3):622–639. https://doi.org/10.1016/j.gca.2005.09.025

Beard BL, Handler RM, Scherer MM, Wu LL, Czaja AD, Heimann A, Johnson CM (2010) Iron isotope fractionation between aqueous ferrous iron and goethite. Earth Planet Sci Lett 295(1–2):241–250. https://doi.org/10.1016/j.epsl.2010.04.006

Beard BL, Johnson CM, Cox L, Sun H, Nealson KH, Aguilar C (1999) Iron isotope biosignatures. Science 285(5435):1889–1892. https://doi.org/10.1126/science.285.5435.1889

Beard BL, Johnson CM, Skulan JL, Nealson KH, Cox L, Sun H (2003) Application of Fe isotopes to tracing the geochemical and biological cycling of Fe. Chem Geol 195(1–4):87–117. https://doi.org/10.1016/s0009-2541(02)00390-x

Bertaut EF, Burlet P, chappert J (1965) Sur l'absence d'ordre magnetique dans la forme quadratique de FeS. Solid State Commun 3:335–338

Bigeleisen J, Mayer MG (1947) Calculation of equilibrium contants for isotopic exchange reactions. J Chem Phys 15:261–267

Blanchard M, Balan E, Schauble EA (2017) Equilibrium fractionation of non-traditional isotopes: a molecular modeling perspective. In: Teng FZ, Watkins J, Dauphas N (eds) Non-traditional stable isotopes, vol 82. Reviews in Mineralogy & Geochemistry, pp 27–63. https://doi.org/10.2138/rmg.2017.82.2

Blanchard M, Dauphas N, Hu MY, Roskosz M, Alp EE, Golden DC, Sio CK, Tissot FLH, Zhao J, Gao L, Morris RV, Fornace M, Floris A, Lazzeri M, Balan E (2015) Reduced partition function ratios of iron and oxygen in goethite. Geochim Cosmochim Acta 151:19–33. https://doi.org/10.1016/j.gca.2014.12.006

Blanchard M, Poitrasson F, Meheut M, Lazzeri M, Mauri F, Balan E (2009) Iron isotope fractionation between pyrite (FeS2), hematite (Fe2O3) and siderite (FeCO3): a first-principles density functional theory study. Geochim Cosmochim Acta 73(21):6565–6578. https://doi.org/10.1016/j.gca.2009.07.034

Blanchard M, Poitrasson F, Meheut M, Lazzeri M, Mauri F, Balan E (2012) Comment on "New data on equilibrium iron isotope fractionation among sulfides: Constraints on mechanisms of sulfide formation in hydrothermal and igneous systems" by VB Polyakov and DM Soultanov. Geochim Cosmochim Acta 87:356–359. https://doi.org/10.1016/j.gca.2012.01.048

Boland DD, Collins RN, Miller CJ, Glover CJ, Waite TD (2014) Effect of solution and solid-phase conditions on the Fe(II)-Accelerated transformation of ferrihydrite to lepidocrocite and goethite. Environ Sci Technol 48(10):5477–5485. https://doi.org/10.1021/es4043275

Brantley SL, Liermann LJ, Guynn RL, Anbar A, Icopini GA, Barling J (2004) Fe isotopic fractionation during mineral dissolution with and without bacteria. Geochim Cosmochim Acta 68(15):3189–3204. https://doi.org/10.1016/j.gca.2004.01.023

Buchachenko AL (2001) Magnetic isotope effect: nuclear spin control of chemical reactions. J Phys Chem A 105(44):9995–10011. https://doi.org/10.1021/jp011261d

Butler IB, Archer C, Vance D, Oldroyd A, Rickard D (2005) Fe isotope fractionation on FeS formation in ambient aqueous solution. Earth Planet Sci Lett 236(1–2):430–442. https://doi.org/10.1016/j.epsl.2005.05.022

Cao XB, Bao HM (2017) Redefining the utility of the three-isotope method. Geochim Cosmochim Acta 212:16–32. https://doi.org/10.1016/j.gca.2017.05.028

Chapman JB, Weiss DJ, Shan Y, Lemburger M (2009) Iron isotope fractionation during leaching of granite and basalt by hydrochloric and oxalic acids. Geochim Cosmochim Acta 73(5):1312–1324. https://doi.org/10.1016/j.gca.2008.11.037

Chen CM, Kukkadapu R, Sparks DL (2015) Influence of coprecipitated organic matter on Fe-(aq)(2+)-Catalyzed transformation of ferrihydrite: implications for carbon dynamics. Environ Sci Technol 49(18):10927–10936. https://doi.org/10.1021/acs.est.5b02448

Chen KY, Chen TY, Chan YT, Cheng CY, Tzou YM, Liu YT, Teah HY (2016) Stabilization of natural organic matter by short-range-order iron hydroxides. Environ Sci Technol 50(23):12612–12620. https://doi.org/10.1021/acs.est.6b02793

Childers SE, Ciufo S, Lovley DR (2002) Geobacter metallireducens accesses insoluble Fe(III) oxide by chemotaxis. Nature 416(6882):767–769. https://doi.org/10.1038/416767a

Clayton RN, Kieffer SW (1991) Oxygen isotopic thermometer calibrations. Spec Publ—Geochem Soc 3:3–10

Criss RE (1999) Principles of stable isotope distribution. Oxford University Press, New York, p 1999

Croal LR, Johnson CM, Beard BL, Newman DK (2004) Iron isotope fractionation by Fe(II)-oxidizing photoautotrophic bacteria. Geochim Cosmochim Acta 68(6):1227–1242. https://doi.org/10.1016/j.gca.2003.09.011

Crosby HA, Johnson CM, Roden EE, Beard BL (2005) Coupled Fe(II)-Fe(III) electron and atom exchange as a mechanism for Fe isotope fractionation during dissimilatory iron oxide reduction. Environ Sci Technol 39(17):6698–6704. https://doi.org/10.1021/es0505346

Crosby HA, Roden EE, Johnson CM, Beard BL (2007) The mechanisms of iron isotope fractionation produced during dissimilatory Fe(III) reduction by Shewanella putrefaciens and Geobacter sulfurreducens. Geobiology 5(2):169–189. https://doi.org/10.1111/j.1472-4669.2007.00103.x

Dauphas N, John SG, Rouxel OJ (2017) Iron isotope systematics. Rev Miner Geochem 82:415–510. https://doi.org/10.2138/rmg.2017.82.11

Dauphas N, Roskosz M, Alp EE, Golden DC, Sio CK, Tissot FLH, Hu MY, Zhao J, Gao L, Morris RV (2012) A general moment NRIXS approach to the determination of equilibrium Fe isotopic fractionation factors: Application to goethite and jarosite. Geochim Cosmochim Acta 94:254–275. https://doi.org/10.1016/j.gca.2012.06.013

Dauphas N, Roskosz M, Alp EE, Neuville DR, Hu MY, Sio CK, Tissot FLH, Zhao J, Tissandiere L, Medard E, Cordier C (2014) Magma redox and structural controls on iron isotope variations in Earth's mantle and crust. Earth Planet Sci Lett 398:127–140. https://doi.org/10.1016/j.epsl.2014.04.033

Dideriksen K, Baker JA, Stipp SLS (2008) Equilibrium Fe isotope fractionation between inorganic aqueous Fe (III) and the siderophore complex, Fe(III)-desferrioxamine B. Earth Planet Sci Lett 269(1–2):280–290. https://doi.org/10.1016/j.epsl.2008.02.022

Doelsch E, Masion A, Rose J, Stone WEE, Bottero JY, Bertsch PM (2003) Chemistry and structure of colloids obtained by hydrolysis of Fe(III) in the presence of SiO4 ligands. Colloids Surf a-Physicochem Eng Asp 217(1–3):121–128. https://doi.org/10.1016/s0927-7757(02)00566-6

Doelsch E, Rose J, Masion A, Bottero JY, Nahon D, Bertsch PM (2000) Speciation and crystal chemistry of iron(III) chloride hydrolyzed in the presence of SiO4 ligands. 1. An FeK-edge EXAFS study. Langmuir 16(10):4726–4731. https://doi.org/10.1021/la991378h

Domagal-Goldman SD, Kubicki JD (2008) Density functional theory predictions of equilibrium isotope fractionation of iron due to redox changes and organic complexation. Geochim Cosmochim Acta 72 (21):5201–5216. https://doi.org/10.1016/j.gca.2008.05.066

Domagal-Goldman SD, Paul KW, Sparks DL, Kubicki JD (2009) Quantum chemical study of the Fe(III)-desferrioxamine B siderophore complex-electronic structure, vibrational frequencies, and equilibrium Fe-isotope fractionation. Geochim Cosmochim Acta 73(1):1–12. https://doi.org/10.1016/j.gca.2008.09.031

Dove MT (1993) Introduction to lattice dynamics. Cambridge University Press, Cambridge, New York, p 1993

Eusterhues K, Wagner FE, Hausler W, Hanzlik M, Knicker H, Totsche KU, Kogel-Knabner I, Schwertmann U (2008) Characterization of ferrihydrite-soil organic matter coprecipitates by X-ray diffraction and mossbauer spectroscopy. Environ Sci Technol 42 (21):7891–7897. https://doi.org/10.1021/es800881w

Fortney NW, He S, Converse BJ, Beard BL, Johnson CM, Boyd ES, Roden EE (2016) Microbial Fe(III) oxide reduction potential in Chocolate Pots hot spring, Yellowstone National Park. Geobiology 14(3):255–275. https://doi.org/10.1111/gbi.12173

Frierdich AJ, Beard BL, Reddy TR, Scherer MM, Johnson CM (2014a) Iron isotope fractionation between aqueous Fe(II) and goethite revisited: new insights based on a multi-direction approach to equilibrium and isotopic exchange rate modification. Geochim Cosmochim Acta 139:383–398. https://doi.org/10.1016/j.gca.2014.05.001

Frierdich AJ, Beard BL, Rosso KM, Scherer MM, Spicuzza MJ, Valley JW, Johnson CM (2015) Low temperature, non-stoichiometric oxygen-isotope exchange coupled to Fe(II)-goethite interactions. Geochim Cosmochim Acta 160:38–54. https://doi.org/10.1016/j.gca.2015.03.029

Frierdich AJ, Beard BL, Scherer MM, Johnson CM (2014b) Determination of the Fe(II)(aq)-magnetite equilibrium iron isotope fractionation factor using the three-isotope method and a multi-direction approach to equilibrium. Earth Planet Sci Lett 391:77–86. https://doi.org/10.1016/j.epsl.2014.01.032

Frierdich AJ, Nebel O, Beard BL, Johnson CM (2019) Iron isotope exchange and fractionation between hematite (alpha-Fe2O3) a and aqueous Fe(II): a combined three-isotope and reversal-approach to equilibrium study. Geochim Cosmochim Acta 245:207–221. https://doi.org/10.1016/j.gca.2018.10.033

Fujii T, Moynier F, Blichert-Toft J, Albarede F (2014) Density functional theory estimation of isotope fractionation of Fe, Ni, Cu, and Zn among species relevant to geochemical and biological environments. Geochim Cosmochim Acta 140:553–576. https://doi.org/10.1016/j.gca.2014.05.051

Fujii T, Moynier F, Telouk P, Albarede F (2006) Isotope fractionation of iron(III) in chemical exchange reactions using solvent extraction with crown ether. J Phys Chem A 110(38):11108–11112. https://doi.org/10.1021/jp063179u

Gorski CA, Fantle MS (2017) Stable mineral recrystallization in low temperature aqueous systems: a critical review. Geochim Cosmochim Acta 198:439–465. https://doi.org/10.1016/j.gca.2016.11.013

Gorski CA, Handler RM, Beard BL, Pasakarnis T, Johnson CM, Scherer MM (2012) Fe atom exchange between aqueous Fe2 + and magnetite. Environ Sci Technol 46(22):12399–12407. https://doi.org/10.1021/es204649a

Guilbaud R, Butler IB, Ellam RM (2011a) Abiotic pyrite formation produces a large Fe isotope fractionation. Science 332(6037):1548–1551. https://doi.org/10.1126/science.1202924

Guilbaud R, Butler IB, Ellam RM, Rickard D (2010) Fe isotope exchange between Fe(II)(aq) and nanoparticulate mackinawite (FeSm) during nanoparticle growth. Earth Planet Sci Lett 300(1–2):174–183. https://doi.org/10.1016/j.epsl.2010.10.004

Guilbaud R, Butler IB, Ellam RM, Rickard D, Oldroyd A (2011b) Experimental determination of the equilibrium Fe isotope fractionation between Fe-aq(2 +) and FeSm (mackinawite) at 25 and 2 °C. Geochim

Cosmochim Acta 75(10):2721–2734. https://doi.org/10.1016/j.gca.2011.02.023

Handler RM, Beard BL, Johnson CM, Scherer MM (2009) Atom exchange between aqueous Fe(II) and goethite: an fe isotope tracer study. Environ Sci Technol 43(4):1102–1107. https://doi.org/10.1021/es802402m

Handler RM, Frierdich AJ, Johnson CM, Rosso KM, Beard BL, Wang CM, Latta DE, Neumann A, Pasakarnis T, Premaratne W, Scherer MM (2014) Fe (II)-Catalyzed recrystallization of goethite revisited. Environ Sci Technol 48(19):11302–11311. https://doi.org/10.1021/es503084u

Hansel CM, Benner SG, Fendorf S (2005) Competing Fe (II)-induced mineralization pathways of ferrihydrite. Environ Sci Technol 39(18):7147–7153. https://doi.org/10.1021/es050666z

Hansel CM, Benner SG, Neiss J, Dohnalkova A, Kukkadapu RK, Fendorf S (2003) Secondary mineralization pathways induced by dissimilatory iron reduction of ferrihydrite under advective flow. Geochim Cosmochim Acta 67(16):2977–2992. https://doi.org/10.1016/s0016-7037(03)00276-x

Henneberry YK, Kraus TEC, Nico PS, Horwath WR (2012) Structural stability of coprecipitated natural organic matter and ferric iron under reducing conditions. Org Geochem 48:81–89. https://doi.org/10.1016/j.orggeochem.2012.04.005

Hill PS, Schauble EA (2008) Modeling the effects of bond environment on equilibrium iron isotope fractionation in ferric aquo-chloro complexes. Geochim Cosmochim Acta 72(8):1939–1958. https://doi.org/10.1016/j.gca.2007.12.023

Hill PS, Schauble EA, Shahar A, Tonui E, Young ED (2009) Experimental studies of equilibrium iron isotope fractionation in ferric aquo-chloro complexes. Geochim Cosmochim Acta 73(8):2366–2381. https://doi.org/10.1016/j.gca.2009.01.016

Hill PS, Schauble EA, Young ED (2010) Effects of changing solution chemistry on Fe3+/Fe2+ isotope fractionation in aqueous Fe–Cl solutions. Geochim Cosmochim Acta 74(23):6669–6689. https://doi.org/10.1016/j.gca.2010.08.038

Icopini GA, Anbar AD, Ruebush SS, Tien M, Brantley SL (2004) Iron isotope fractionation during microbial reduction of iron: the importance of adsorption. Geology 32(3):205–208. https://doi.org/10.1130/g20184.1

Jang JH, Dempsey BA, Catchen GL, Burgos WD (2003) Effects of Zn(II), Cu(II), Mn(II), Fe(II), NO3-, or SO42- at pH 6.5 and 8.5 on transformations of hydrous ferric oxide (HFO) as evidenced by Mossbauer spectroscopy. Colloids Surf a-Physicochem Eng Asp 221 (1–3):55–68. https://doi.org/10.1016/s0927-7757(03)00134-1

Johnson CM, Beard BL, Roden EE, Newman DK, Nealson KH (2004) Isotopic constraints on biogeochemical cycling of Fe. In: Johnson CM, Beard BL, Albarede F (eds) Geochemistry of non-traditional stable isotopes, vol 55. Reviews in Mineralogy & Geochemistry, pp 359–408. https://doi.org/10.2138/gsrmg.55.1.359

Johnson CM, Roden EE, Welch SA, Beard BL (2005) Experimental constraints on Fe isotope fractionation during magnetite and Fe carbonate formation coupled to dissimilatory hydrous ferric oxide reduction. Geochim Cosmochim Acta 69(4):963–993. https://doi.org/10.1016/j.gca.2004.06.043

Johnson CM, Skulan JL, Beard BL, Sun H, Nealson KH, Braterman PS (2002) Isotopic fractionation between Fe(III) and Fe(II) in aqueous solutions. Earth Planet Sci Lett 195(1–2):141–153. https://doi.org/10.1016/s0012-821x(01)00581-7

Joshi P, Fantle MS, Larese-Casanova P, Gorski CA (2017) Susceptibility of Goethite to Fe2+ -Catalyzed Recrystallization over time. Environ Sci Technol 51 (20):11681–11691. https://doi.org/10.1021/acs.est.7b02603

Joshi P, Gorski CA (2016) Anisotropic morphological changes in goethtie during Fe^{2+}-catalyzed recrystallization. Environ Sci Technol 50:7315–7324

Kappler A, Johnson CM, Crosby HA, Beard BL, Newman DK (2010) Evidence for equilibrium iron isotope fractionation by nitrate-reducing iron(II)-oxidizing bacteria. Geochim Cosmochim Acta 74(10):2826–2842. https://doi.org/10.1016/j.gca.2010.02.017

Kiczka M, Wiederhold JG, Frommer J, Kraemer SM, Bourdon B, Kretzschmar R (2010) Iron isotope fractionation during proton- and ligand-promoted dissolution of primary phyllosilicates. Geochim Cosmochim Acta 74(11):3112–3128. https://doi.org/10.1016/j.gca.2010.02.018

Kieffer SW (1982) Thermodynamics and lattice vibrations of minerals, 5. Applications to phase equilibria, isotopic fractionation, and high-pressure thermodynamic properties. Rev Geophys Space Phys 20 (4):827–849. https://doi.org/10.1029/RG020i004p00827

Levin NE, Raub TD, Dauphas N, Eiler JM (2014) Triple oxygen isotope variations in sedimentary rocks. Geochim Cosmochim Acta 139:173–189. https://doi.org/10.1016/j.gca.2014.04.034

Lovley DR (1991) Dissimilatory Fe(III) and Mn(IV) reduction. Microbiol Rev 55(2):259–287

Lovley DR, Holmes DE, Nevin KP (2004) Dissimilatory Fe(III) and Mn(IV) reduction. In: Poole RK (ed) Advances in microbial physiology, vol. 49. Advances in Microbial Physiology, pp 219–286. https://doi.org/10.1016/s0065-2911(04)49005-5

Lovley DR, Phillips EJP (1986a) Availability of ferric iron for microbial reduction in bottom sediments of the fresh-water tidal Potomac river. Appl Environ Microbiol 52(4):751–757

Lovley DR, Phillips EJP (1986b) Organic-matter mineralization with reduction of ferric iron in anaerobic sediments. Appl Environ Microbiol 51(4):683–689

Lovley DR, Phillips EJP, Lonergan DJ (1991) Enzymatic versus nonenzymatic mechanisms for Fe(III) reduction in aquatic sediments. Environ Sci Technol 25 (6):1062–1067. https://doi.org/10.1021/es00018a007

Lovley DR, Ueki T, Zhang T, Malvankar NS, Shrestha PM, Flanagan KA, Aklujkar M, Butler JE, Giloteaux L, Rotaru AE, Holmes DE, Franks AE, Orellana R, Risso C, Nevin KP (2011) Geobacter: the microbe electric's physiology, ecology, and practical applications. In: Poole RK (ed) Advances in microbial physiology, vol 59. Advances in Microbial Physiology, pp 1–100. https://doi.org/10.1016/b978-0-12-387661-4.00004-5

Mandernack KW, Bazylinski DA, Shanks WC, Bullen TD (1999) Oxygen and iron isotope studies of magnetite produced by magnetotactic bacteria. Science 285 (5435):1892–1896. https://doi.org/10.1126/science.285.5435.1892

Mansor M, Fantle MS (2019) A novel framework for interpreting pyrite-based Fe isotope records of the past. Geochim Cosmochim Acta 253:39–62

Matsuhisa Y, Goldsmith JR, Clayton RN (1978) Mechanisms of hydrothermal crystallization of quartz at 250 °C and 15 kbar. Geochim Cosmochim Acta 42 (2):173–182. https://doi.org/10.1016/0016-7037(78)90130-8

Matthews A, Goldsmith JR, Clayton RN (1983) On the mechanisms and kinetics of oxygen isotope exchange in quartz and feldspars at elevated-temperatures and pressures. Geol Soc Am Bull 94(3):396–412. https://doi.org/10.1130/0016-7606(1983)94%3c396:Otmako%3e2.0.Co;2

Matthews A, Zhu XK, O'Nions K (2001) Kinetic iron stable isotope fractionation between iron (-II) and (-III) complexes in solution. Earth Planet Sci Lett 192 (1):81–92. https://doi.org/10.1016/s0012-821x(01)00432-0

Mikutta C, Wiederhold JG, Cirpka OA, Hofstetter TB, Bourdon B, Von Gunten U (2009) Iron isotope fractionation and atom exchange during sorption of ferrous iron to mineral surfaces. Geochim Cosmochim Acta 73(7):1795–1812. https://doi.org/10.1016/j.gca.2009.01.014

Miller MF (2002) Isotopic fractionation and the quantification of O-17 anomalies in the oxygen three-isotope system: an appraisal and geochemical significance. Geochim Cosmochim Acta 66(11):1881–1889. https://doi.org/10.1016/s0016-7037(02)00832-3

Mineev SD, Polyakov VB, Permyakov YV (2007) Equilibrium iron isotope fractionation factors for magnetite from Mossbauer spectroscopy and inelastic nuclear resonant X-ray scattering data. Geochim Cosmochim Acta 71(15):A669–A669

Morgan JLL, Wasylenki LE, Nuester J, Anbar AD (2010) Fe Isotope Fractionation during Equilibration of Fe-Organic Complexes. Environ Sci Technol 44 (16):6095–6101. https://doi.org/10.1021/es100906z

Moynier F, Fujii T, Wang K, Foriel J (2013) Ab initio calculations of the Fe(II) and Fe(III) isotopic effects in citrates, nicotianamine, and phytosiderophore, and new Fe isotopic measurements in higher plants. CR Geosci 345(5–6):230–240. https://doi.org/10.1016/j.crte.2013.05.003

Mulholland DS, Poitrasson F, Shirokova LS, Gonzalez AG, Pokrovsky OS, Boaventura GR, Vieira LC (2015) Iron isotope fractionation during Fe(II) and Fe (III) adsorption on cyanobacteria. Chem Geol 400:24–33. https://doi.org/10.1016/j.chemgeo.2015.01.017

Neumann A, Wu LL, Li WQ, Beard BL, Johnson CM, Rosso KM, Frierdich AJ, Scherer MM (2015) Atom exchange between aqueous Fe(II) and structural Fe in clay minerals. Environ Sci Technol 49(5):2786–2795. https://doi.org/10.1021/es504984q

Northrop DA, Clayton RN (1966) Oxygen-isotope fractionations in systems containing dolomite. J Geol 74 (2):174–196. https://doi.org/10.1086/627153

O'Neil JR (1986) Theoretical and experimental aspects of isotopic fractionation. Rev Mineral 16:1–40

Ottonello G, Zuccolini MV (2008) The iron-isotope fractionation dictated by the carboxylic functional: an ab-initio investigation. Geochim Cosmochim Acta 72(24):5920–5934. https://doi.org/10.1016/j.gca.2008.09.027

Ottonello G, Zuccolini MV (2009) Ab-initio structure, energy and stable Fe isotope equilibrium fractionation of some geochemically relevant H–O–Fe complexes. Geochim Cosmochim Acta 73(21):6447–6469. https://doi.org/10.1016/j.gca.2009.06.034

Pedersen HD, Postma D, Jakobsen R, Larsen O (2005) Fast transformation of iron oxyhydroxides by the catalytic action of aqueous Fe(II). Geochim Cosmochim Acta 69(16):3967–3977. https://doi.org/10.1016/j.gca.2005.03.016

Percak-Dennett EM, Beard BL, Xu H, Konishi H, Johnson CM, Roden EE (2011) Iron isotope fractionation during microbial dissimilatory iron oxide reduction in simulated Archaean seawater. Geobiology 9 (3):205–220. https://doi.org/10.1111/j.1472-4669.2011.00277.x

Polyakov VB (1997) Equilibrium fractionation of the iron isotopes: estimation from Mossbauer spectroscopy data. Geochim Cosmochim Acta 61(19):4213–4217. https://doi.org/10.1016/s0016-7037(97)00204-4

Polyakov VB, Clayton RN, Horita J, Mineev SD (2007) Equilibrium iron isotope fractionation factors of minerals: reevaluation from the data of nuclear inelastic resonant X-ray scattering and Mossbauer spectroscopy. Geochim Cosmochim Acta 71(15):3833–3846. https://doi.org/10.1016/j.gca.2007.05.019

Polyakov VB, Mineev SD (2000) The use of Mossbauer spectroscopy in stable isotope geochemistry. Geochim Cosmochim Acta 64(5):849–865. https://doi.org/10.1016/s0016-7037(99)00329-4

Polyakov VB, Mineev SD, Clayton RN, Hu G, Mineev KS (2005) Determination of tin equilibrium isotope fractionation factors from synchrotron radiation experiments. Geochim Cosmochim Acta 69(23):5531–5536. https://doi.org/10.1016/j.gca.2005.07.010

Polyakov VB, Osadchii EG, Voronin MV, Osadchii VO, Sipavina LV, Chareev DA, Tyurin AV, Gurevich VM, Gavrichev KS (2019) Iron and Sulfur isotope factors of pyrite: data from experimental mossbauer spectroscopy and heat capacity. Geochem Int 57 (4):369–383. https://doi.org/10.1134/s0016702919040098

Polyakov VB, Soultanov DM (2011) New data on equilibrium iron isotope fractionation among sulfides: Constraints on mechanisms of sulfide formation in hydrothermal and igneous systems. Geochim Cosmochim Acta 75(7):1957–1974. https://doi.org/10.1016/j.gca.2011.01.019

Polyakov VB, Soultanov DM (2012) Response to the comment by M Blanchard, F Poitrasson, M Me'heut, M Lazzeri, F Mauri, E Balan on "New data on equilibrium iron isotope fractionation among sulfides: constraints on mechanisms of sulfide formation in hydrothermal and igneous systems" published in Geochim. Cosmochim Acta 75:1957–1974. Geochim Cosmochim Acta 87:360–366

Poulson RL, Johnson CM, Beard BL (2005) Iron isotope exchange kinetics at the nanoparticulate ferrihydrite surface. Am Miner 90(4):758–763. https://doi.org/10.2138/am.2005.1802

Reddy TR, Frierdich AJ, Beard BL, Johnson CM (2015) The effect of pH on stable iron isotope exchange and fractionation between aqueous Fe(II) and goethite. Chem Geol 397:118–127. https://doi.org/10.1016/j.chemgeo.2015.01.018

Roden EE (2003) Fe(III) oxide reactivity toward biological versus chemical reduction. Environ Sci Technol 37(7):1319–1324. https://doi.org/10.1021/es026038o

Roden EE (2006) Geochemical and microbiological controls on dissimilatory iron reduction. CR Geosci 338(6–7):456–467. https://doi.org/10.1016/j.crte.2006.04.009

Roden EE, Urrutia MM (1999) Ferrous iron removal promotes microbial reduction of crystalline iron(III) oxides. Environ Sci Technol 33(11):1847–1853. https://doi.org/10.1021/es9809859

Roden EE, Zachara JM (1996) Microbial reduction of crystalline iron(III) oxides: Influence of oxide surface area and potential for cell growth. Environ Sci Technol 30(5):1618–1628. https://doi.org/10.1021/es9506216

Roe JE, Anbar AD, Barling J (2003) Nonbiological fractionation of Fe isotopes; evidence of an equilibrium isotope effect. Chem Geol 195(1–4):69–85. https://doi.org/10.1016/S0009-2541(02)00389-3

Roskosz M, Sio CKI, Dauphas N, Bi WL, Tissot FLH, Hu MY, Zhao JY, Alp EE (2015) Spinel-olivine-pyroxene equilibrium iron isotopic fractionation and applications to natural peridotites. Geochim Cosmochim Acta 169:184–199. https://doi.org/10.1016/j.gca.2015.07.035

Rumble D, Miller MF, Franchi IA, Greenwood RC (2007) Oxygen three-isotope fractionation lines in terrestrial silicate minerals: an inter-laboratory comparison of hydrothermal quartz and eclogitic garnet. Geochim Cosmochim Acta 71(14):3592–3600. https://doi.org/10.1016/j.gca.2007.05.011

Rustad JR, Casey WH, Yin QZ, Bylaska EJ, Felmy AR, Bogatko SA, Jackson VE, Dixon DA (2010) Isotopic fractionation of Mg2+ (aq), Ca2+ (aq), and Fe2+ (aq) with carbonate minerals. Geochim Cosmochim Acta 74(22):6301–6323. https://doi.org/10.1016/j.gca.2010.08.018

Salas EC, Berelson WM, Hammond DE, Kampf AR, Nealson KH (2009) The influence of carbon source on the products of dissimilatory iron reduction. Geomicrobiol J 26(7):451–462. https://doi.org/10.1080/01490450903060806

Salas EC, Berelson WM, Hammond DE, Kampf AR, Nealson KH (2010) The impact of bacterial strain on the products of dissimilatory iron reduction. Geochim Cosmochim Acta 74(2):574–583. https://doi.org/10.1016/j.gca.2009.10.039

Saunier G, Pokrovski GS, Poitrasson F (2011) First experimental determination of iron isotope fractionation between hematite and aqueous solution at hydrothermal conditions. Geochim Cosmochim Acta 75(21):6629–6654. https://doi.org/10.1016/j.gca.2011.08.028

Schauble EA (2004) Applying stable isotope fractionation theory to new systems. Rev Miner Geochem 55:65–111. https://doi.org/10.2138/gsrmg.55.1.65

Schauble EA, Rossman GR, Taylor HP (2001) Theoretical estimates of equilibrium Fe-isotope fractionations from vibrational spectroscopy. Geochim Cosmochim Acta 65(15):2487–2497. https://doi.org/10.1016/s0016-7037(01)00600-7

Shahar A, Elardo SM, Macris CA (2017) Equilibrium fractionation of non-traditional stable isotopes; an experimental perspective. Rev Miner Geochem 82:65–83. https://doi.org/10.2138/rmg.2017.82.3

Shi BJ, Liu K, Wu LL, Li WQ, Smeaton CM, Beard BL, Johnson CM, Roden EE, Van Cappellen P (2016) Iron isotope fractionations reveal a finite bioavailable Fe pool for structural Fe(III) reduction in nontronite. Environ Sci Technol 50(16):8661–8669. https://doi.org/10.1021/acs.est.6b02019

Skulan JL, Beard BL, Johnson CM (2002) Kinetic and equilibrium Fe isotope fractionation between aqueous Fe(III) and hematite. Geochim Cosmochim Acta 66 (17):2995–3015. https://doi.org/10.1016/s0016-7037(02)00902-x

Sun RY, Wang BL (2018) Iron isotope fractionation during uptake of ferrous ion by phytoplankton. Chem Geol 481:65–73. https://doi.org/10.1016/j.chemgeo.2018.01.031

Swanner ED, Bayer T, Wu W, Hao L, Obst M, Sundman A, Byrne JM, Michel FM, Kleinhanns IC, Kappler A, Schoenberg R (2017) Iron Isotope fractionation during Fe(II) oxidation mediated by the oxygen-producing marine cyanobacterium synechococcus PCC 7002. Environ Sci Technol 51(9):4897–4906. https://doi.org/10.1021/acs.est.6b05833

Swanner ED, Wu WF, Schoenberg R, Byrne J, Michel FM, Pan YX, Kappler A (2015) Fractionation of Fe isotopes during Fe(II) oxidation by a marine photoferrotroph is controlled by the formation of organic Fe-complexes and colloidal Fe fractions. Geochim Cosmochim Acta 165:44–61. https://doi.org/10.1016/j.gca.2015.05.024

Syverson DD, Borrok DM, Seyfried WE (2013) Experimental determination of equilibrium Fe isotopic fractionation between pyrite and dissolved Fe under hydrothermal conditions. Geochim Cosmochim Acta 122:170–183. https://doi.org/10.1016/j.gca.2013.08.027

Syverson DD, Luhmann AJ, Tan CY, Borrok DM, Ding K, Seyfried WE (2017) Fe isotope fractionation between chalcopyrite and dissolved Fe during hydrothermal recrystallization: an experimental study at 350 °C and 500 bars. Geochim Cosmochim Acta 200:87–109. https://doi.org/10.1016/j.gca.2016.12.002

Syverson DD, Pester NJ, Craddock PR, Seyfried WE (2014) Fe isotope fractionation during phase separation in the NaCl-H2O system: an experimental study with implications for seafloor hydrothermal vents. Earth Planet Sci Lett 406:223–232. https://doi.org/10.1016/j.epsl.2014.09.020

Tangalos GE, Beard BL, Johnson CM, Alpers CN, Shelobolina ES, Xu H, Konishi H, Roden EE (2010) Microbial production of isotopically light iron(II) in a modern chemically precipitated sediment and implications for isotopic variations in ancient rocks. Geobiology 8(3):197–208. https://doi.org/10.1111/j.1472-4669.2010.00237.x

Thamdrup B (2000) Bacterial manganese and iron reduction in aquatic sediments. In: Schink B (ed) Advances in microbial ecology, vol 16, pp 41–84

ThomasArrigo LK, Byrne JM, Kappler A, Kretzschmar R (2018) Impact of organic matter on Iron(II)-Catalyzed mineral transformations in ferrihydrite-organic matter coprecipitates. Environ Sci Technol 52(21):12316–12326. https://doi.org/10.1021/acs.est.8b03206

ThomasArrigo LK, Mikutta C, Byrne J, Kappler A, Kretzschmar R (2017) Iron(II)-Catalyzed iron atom exchange and mineralogical changes in iron-rich organic freshwater flocs: an iron isotope tracer study. Environ Sci Technol 51(12):6897–6907. https://doi.org/10.1021/acs.est.7b01495

Walker DJF, Adhikari RY, Holmes DE, Ward JE, Woodard TL, Nevin KP, Lovley DR (2018) Electrically conductive pili from pilin genes of phylogenetically diverse microorganisms. ISME J 12(1):48–58. https://doi.org/10.1038/ismej.2017.141

Welch SA, Beard BL, Johnson CM, Braterman PS (2003) Kinetic and equilibrium Fe isotope fractionation between aqueous Fe(II) and Fe(III). Geochim Cosmochim Acta 67(22):4231–4250. https://doi.org/10.1016/s0016-7037(03)00266-7

Wiederhold JG, Kraemer SM, Teutsch N, Borer PM, Halliday AN, Kretzschmar R (2006) Iron isotope fractionation during proton-promoted, ligand-controlled, and reductive dissolution of goethite. Environ Sci Technol 40(12):3787–3793. https://doi.org/10.1021/es052228y

Wiesli RA, Beard BL, Braterman PS, Johnson CM, Saha SK, Sinha MP (2007) Iron isotope fractionation between liquid and vapor phases of iron pentacarbonyl. Talanta 71(1):90–96. https://doi.org/10.1016/j.talanta.2006.03.026

Wiesli RA, Beard BL, Johnson CM (2004) Experimental determination of Fe isotope fractionation between aqueous Fe(II), siderite and "green rust" in abiotic systems. Chem Geol 211(3–4):343–362. https://doi.org/10.1016/j.chemgeo.2004.07.002

Wu LL, Beard BL, Roden EE, Johnson CM (2009) Influence of pH and dissolved Si on Fe isotope fractionation during dissimilatory microbial reduction of hematite. Geochim Cosmochim Acta 73(19):5584–5599. https://doi.org/10.1016/j.gca.2009.06.026

Wu LL, Beard BL, Roden EE, Johnson CM (2011) Stable iron isotope fractionation between aqueous Fe(II) and hydrous ferric oxide. Environ Sci Technol 45(5):1847–1852. https://doi.org/10.1021/es103171x

Wu LL, Beard BL, Roden EE, Kennedy CB, Johnson CM (2010) Stable Fe isotope fractionations produced by aqueous Fe(II)-hematite surface interactions. Geochim Cosmochim Acta 74(15):4249–4265. https://doi.org/10.1016/j.gca.2010.04.060

Wu LL, Druschel G, Findlay A, Beard BL, Johnson CM (2012a) Experimental determination of iron isotope fractionations among Fe-aq(2+)-FeSaq-Mackinawite at low temperatures: implications for the rock record. Geochim Cosmochim Acta 89:46–61. https://doi.org/10.1016/j.gca.2012.04.047

Wu LL, Percak-Dennett EM, Beard BL, Roden EE, Johnson CM (2012b) Stable iron isotope fractionation between aqueous Fe(II) and model Archean ocean Fe-Si coprecipitates and implications for iron isotope variations in the ancient rock record. Geochim Cosmochim Acta 84:14–28. https://doi.org/10.1016/j.gca.2012.01.007

Wu T, Griffin AM, Gorski CA, Shelobolina ES, Xu H, Kukkadapu RK, Roden EE (2017) Interactions between Fe(III)-oxides and Fe(III)-phyllosilicates during microbial reduction 2: natural subsurface sediments. Geomicrobiol J 34(3):231–241. https://doi.org/10.1080/01490451.2016.1174758

Young ED, Galy A, Nagahara H (2002) Kinetic and equilibrium mass-dependent isotope fractionation laws in nature and their geochemical and cosmochemical significance. Geochim Cosmochim Acta 66(6):1095–1104. https://doi.org/10.1016/s0016-7037(01)00832-8

Young ED, Manning CE, Schauble EA, Shahar A, Macris CA, Lazar C, Jordan M (2015) High-temperature equilibrium isotope fractionation of non-traditional stable isotopes: experiments, theory, and applications. Chem Geol 395:176–195. https://doi.org/10.1016/j.chemgeo.2014.12.013

Zachara JM, Kukkadapu RK, Fredrickson JK, Gorby YA, Smith SC (2002) Biomineralization of poorly crystalline Fe(III) oxides by dissimilatory metal reducing bacteria (DMRB). Geomicrobiol J 19(2):179–207. https://doi.org/10.1080/01490450252864271

Zheng X-Y, Beard BL, Reddy TR, Roden EE, Johnson CM (2016) Abiologic silicon isotope fractionation between aqueous Si and Fe(III)–Si gel in simulated Archean sea water; implications for Si isotope records in Precambrian sedimentary rocks. Geochim Cosmochim Acta 187:102–122. https://doi.org/10.1016/j.gca.2016.05.012

Zhou Z, Latta DE, Noor N, Thompson A, Borch T, Scherer MM (2018) Fe(II)-Catalyzed transformation of organic matter-ferrihydrite coprecipitates: a closer look using Fe isotopes. Environ Sci Technol 52(19):11142–11150. https://doi.org/10.1021/acs.est.8b03407

4 High-Temperature Fe Isotope Geochemistry

4.1 Iron Isotope Variations in the Solar System

Nucleosynthetic production of Fe in massive AGB stars (and in supernovae) generated five Fe isotopes, namely ^{54}Fe, ^{56}Fe, ^{57}Fe, ^{58}Fe, and ^{60}Fe, with long enough half-lives to be extant during the formation of the Solar System. Four of these five isotopes are stable and only ^{60}Fe decayed (β^-) with a half-life of ($t_{1/2}$ = 2.6 Ma) to ^{60}Ni. Variations in the occurrence of ^{60}Ni provide evidence that ^{60}Fe was indeed extant during the early Solar System, although it's initial abundance is debated (e.g. Mishra and Goswami 2014; Tang and Dauphas 2012). In this book, we focus on stable Fe isotopes and interested readers regarding the initial abundance of ^{60}Fe are referred to a recent review by Elliott and Steele (2017) and references therein. Fractionation of Fe isotopes during the processes of planet formation were recently reviewed by Dauphas et al. (2017) and Sossi et al. (2016a), with the former about stable Fe isotope systematics in general.

The first systematic study of Fe isotope compositions of terrestrial iron-bearing minerals and meteorites was performed by Valley and Anderson (1947). The result was that all investigated terrestrial and extraterrestrial material had identical Fe isotope compositions at analytical uncertainties of a $\sim$1% level. The motivation of this, and many following investigations, was to compare isotopic signatures of meteorites with those of terrestrial material in order to determine whether all material of the Solar System originated from a single source, or from several different sources. In the latter case, high-precision isotope analyses may help to elucidate the degree of mixing between different components during Solar System formation, and would be able to detect, for example, if the building blocks of Earth include material that was originally derived from the outer Solar System. Using Fe isotopes for the detection of nucleosynthetically diverse sources is promising, in theory, as not all Fe isotopes are produced by the same processes. While ^{54}Fe, ^{56}Fe, ^{57}Fe are essentially produced during nuclear burning, the production of the neutron-rich isotope ^{58}Fe requires nuclear fission (an s-process operating during He burning; the production of ^{60}Fe furthermore requires an r-process; (Woosley et al. 2002; Woosley and Weaver 1995).

Chondrites are the most primitive meteorites, as their parent bodies formed early in the Solar System and were not modified by subsequent planetary differentiation. Due to their importance in characterizing the potential building blocks of the terrestrial planets, both their chemical and isotopic composition have been investigated extensively. While the chemical composition of refractory elements in the Solar System has been overall well mixed (e.g. Palme and Jones 2003), high-precision isotope analyses reveal nucleosynthetic isotope anomalies for a number of elements at the 10–100 ppm level in many bulk meteorites, and considerably larger anomalies

C. Johnson et al., *Iron Geochemistry: An Isotopic Perspective*, Advances in Isotope Geochemistry,
https://doi.org/10.1007/978-3-030-33828-2_4

exist between meteoritic components (e.g. Brennecka et al. 2013; Burkhardt et al. 2011, 2012; Dauphas and Schauble 2016; Fischer-Gödde et al. 2015; Schönbächler et al. 2003). These isotope systematics indicate several nucleosynthetic sources and imperfect mixing in the protoplanetary disk, extending from the nanometer to the planetary body scale.

4.1.1 Chondrites and Chondritic Components

Chondrites display a homogeneous Fe isotope composition with no evidence of mass-independent Fe isotope variation at the bulk scale (Zhu et al. 2001; Craddock and Dauphas 2011), and this homogeneity also appears to extend to most chondritic components (although data for the minor isotope ^{58}Fe are barely reported). Exceptions are presolar SiC grains with a size of ≤ 1 μm, as can be found in some chondrites (e.g. Murchison) and which show variations in their Fe and Ni isotope composition on the order of several %. Isotope analyses of such grains have been performed either with nanoSIMS (Wang et al. 2013; Marhas et al. 2008) or with a resonance ionization mass spectrometer (CHILI at the University of Chicago; Kodolányi et al. 2018; Trappitsch et al. 2018). Very recently, coupled mass-independent and mass-dependent Fe isotope anomalies have also been detected in mineral separates of calcium-aluminum-rich inclusions (CAIs, the earliest dated condensates of the Solar System) (Shollenberger et al. 2019). Interestingly, CAI mineral separates exhibit excesses in ^{56}Fe (of up to 2 parts per ten thousand), rather than in the neutron-rich isotope, ^{58}Fe, as seen in neighboring elements with reported excesses in ^{48}Ca, ^{50}Ti, ^{54}Cr, and ^{64}Ni (e.g. Birck and Allègre 1984; Niederer and Papanastassiou 1984; Quitté et al. 2007; Niederer et al. 1980). This disparity indicates diverse nucleosynthetic sources for these elements of the Fe-peak of nucleosynthesis in the early Solar System. The large mass-dependent isotope fractionation (of up to 10‰ for δ^{56}Fe), observed for the same mineral separates, was likely generated by condensation-evaporation processes in the solar nebular. As Fe has a much lower condensation temperature than elements such as Ca, Al, and Ti, it is less abundant in high-temperature condensates like CAIs, and is therefore more susceptible to secondary enrichment. Accordingly, the correlated mass-dependent and mass-independent isotopic signatures in these CAI separates are likely the result of mixing primary isotopically anomalous CAI material with material of average Solar System composition that was added by secondary alteration. In this case, the observed mass-independent anomalies probably represent a minimum value for the isotopic difference between CAIs and average Solar System (chondrites). In any case, these investigations by Shollenberger et al. provide supporting evidence that the Solar System was not isotopically homogeneous at the time of the formation of its earliest solids (CAIs).

Most chondrites display a limited variation of mass-dependent Fe isotope fractionation (Craddock and Dauphas 2011; Dauphas et al. 2009; Hezel et al. 2010, 2018; Kehm et al. 2003; Moynier et al. 2007; Mullane et al. 2005; Needham et al. 2009; Poitrasson et al. 2005; Schoenberg and von Blanckenburg 2006; Wang et al. 2013, 2014a; Zhu et al. 2001; see Fig. 4.1). For chondrites, some interesting findings based on these investigations are worth noting: (1) the average of all chondrites compiled here (δ^{56}Fe = −0.011 ± 0.106‰, 2 SD, n = 146) is within analytical uncertainty identical to that of the most commonly used standard, IRMM-014 (or IRMM-643). (2) The average of the major chondrite groups, carbonaceous chondrites (CC), ordinary chondrites (OC), and enstatite chondrites (EC), are also within uncertainty of this standard value (with δ^{56}Fe = −0.008 ± 0.095‰, n = 47; δ^{56}Fe = −0.022 ± 0.092‰, n = 61 and δ^{56}Fe = 0.005 ± 0.135‰, n = 38, respectively). In this chapter, we will often report δ^{56}Fe values to three figures past the decimal to highlight subtle variations, although commonly analytical uncertainties are in the second decimal place. The uncertainties given here are *2 standard deviations (commonly referred to as "2SD" or "2σ")* of the individually analyzed meteorite splits (assuming a *Gaussian normal distribution*). The *2 standard error of the mean* (= $\sigma/\sqrt{n}$), combined with a *student-t* factor (below referred to as *95% confidence interval or "95% c.i."*) may be actually more suitable to evaluate if the mean value of a chondrite group differs from that of another group. Even at this much lower uncertainty level, the mean values for each group still overlap.

Low metal Enstatite chondrites (EL) show a significant variability in Fe isotopes ($-0.18 < \delta^{56}Fe < 0.18$), which is interpreted by Wang et al. (2014a) as the result of different mixing proportions of the major Fe carriers (metal, sulfides and silicates) during the complex impact history of these chondrites. Most of them are highly metamorphosed (EL6), resulting in thermal equilibration and isotope fractionation between the Fe-bearing minerals, in particular between metal (high $\delta^{56}Fe$) and troilite (low $\delta^{56}Fe$). Significant isotope fractionation, in particular between Fe-rich phases of highly metamorphosed OC and EC are confirmed by in situ Fe isotope analyses (Goldmann and Weyer 2013). Overall,

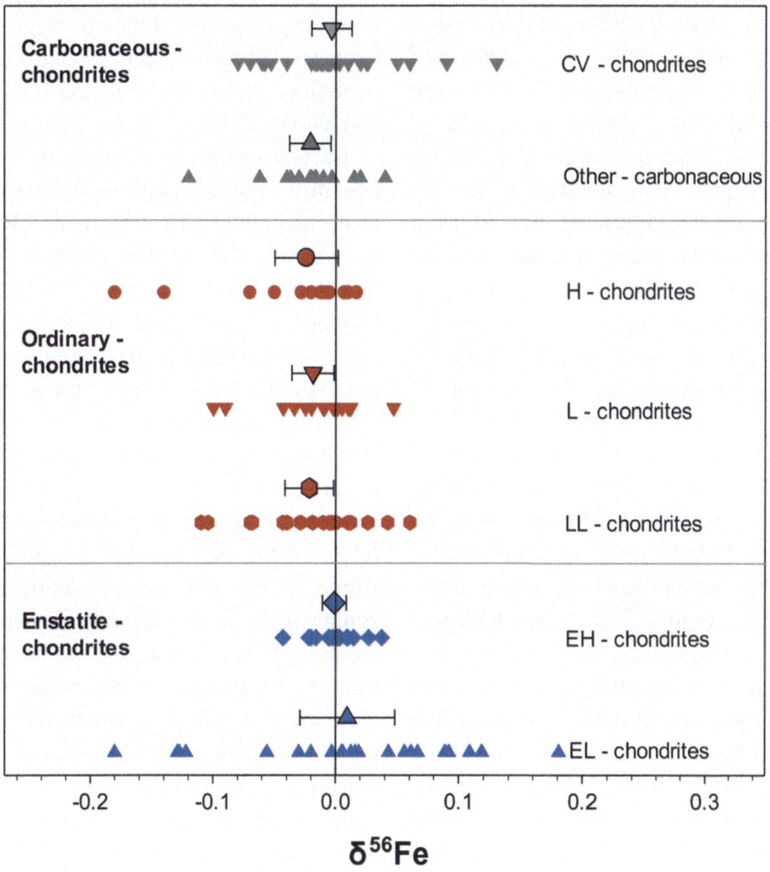

Fig. 4.1 Compilation of Fe isotope compositions of different chondrite groups, given in ‰ relative to IRMM-014 (as in all figures of this chapter, if not indicated otherwise), using data reported in Craddock and Dauphas (2011), Dauphas et al. (2009), Hezel et al. (2010, 2018), Kehm et al. (2003), Moynier et al. (2007), Mullane et al. (2005), Needham et al. (2009), Poitrasson et al. (2005), Schoenberg and von Blanckenburg (2006), Wang et al. (2013, 2014a), and Zhu et al. (2001). The individual symbols with error bars above each group indicates the group average with error bars representing the 2 standard error of the mean, combined with a student-t factor (95% confidence c.i.). The compilation combines data of the very early days of "high-precision" Fe isotope analyses (with typical uncertainties being in the range of 0.05 and 0.1‰), with those being conducted more recently (with typical uncertainties being in the range of 0.03 and 0.05‰)

the distribution and average δ^{56}Fe indicates that the building material of the Earth (as far as represented by chondrites) likely had a δ^{56}Fe value very close to 0.0‰.

There are also several investigations of chondritic components (such as chondrules and metals; Hezel et al. 2010, 2018; Mullane et al. 2005; Needham et al. 2009; Theis et al. 2008; Zhu et al. 2001), which show more significant variations in their stable Fe isotope composition than bulk chondrites. Stable isotope fractionation between chondritic components may in principle have been generated in the solar nebular (e.g., as a result of condensation-evaporation or diffusion processes), or on the parent body, as a result of equilibrium isotope fractionation between the components or also by kinetic processes (e.g., impact-driven). Of these processes, kinetic condensation or evaporation generates by far the largest isotope effects, followed by diffusion effects (e.g. Schauble 2004). Chondrules, the major component of many chondrites, consists of agglomerated dust material that was re-heated and molten by (an) event(s) very early in the Solar System, potentially in the context of proto-planetary accretion. Several studies have investigated the Fe isotope composition of chondrules (Hezel et al. 2010, 2018; Mullane et al. 2005; Needham et al. 2009; Zhu et al. 2001). They observed δ^{56}Fe value ranging from −1.33 to 0.69‰, which was interpreted to be generated by evaporation and re-condensation processes, rather than representing isotopic variations within the proto-planetary disc at the time of chondrule formation. Alteration on the parent body may also have played a role, but likely reduced, rather than increased isotope variations among chondrules. Notably, the average δ^{56}Fe value of all chondrules (−0.05 ± 0.04‰, at 95% conf.; n = 135) is indistinguishable from that of chondrites. There is also no difference between the average δ^{56}Fe of chondrules of carbonaceous and ordinary chondrites. These findings indicate that iron in the commonly Fe-poor (in particular type I) chondrules was either not lost by evaporation or that the iron vapor pressure around the chondrules was high enough to suppress kinetic fractionation (Dauphas et al. 2017). Bulk CAIs also display mass-dependent Fe isotope variations with a slightly larger range than chondrules (δ^{56}Fe from −1.71 to 1.32‰; Shollenberger et al. 2019; Mullane et al. 2005), which have also been interpreted to be generated by evaporation and re-condensation processes. Some larger variability in δ^{56}Fe values (on the order of −0.8 to 0.8‰) was also found in opaque phases of chondrites (such as metal and sulfides; Goldmann and Weyer 2013; Needham et al. 2009; Theis et al. 2008), which frequently control the Fe budget in chondrites (Hezel et al. 2010). These variations, however, essentially reflect isotope fractionation between kamacite, taenite and troilite during cooling of the chondrite parent body (Goldmann and Weyer 2013).

Exceptional groups among the carbonaceous chondrites are the metal-rich groups of the CR clan, including CB, CH, and CR chondrites. Specifically, CB and CH chondrites are particularly metal-rich (with up to 70% Fe in some CB chondrites) and appear to strongly genetically related as both groups have similar O and N isotope systematics and both groups contain metal grains that are chemically and isotopically zoned. Initial studies (Alexander and Hewins 2004; Zipfel and Weyer 2007) observed tremendous Fe isotopic zoning in a range of several ‰, with low δ^{56}Fe in the metal cores and about chondritic values at the rim, crudely anti-parallel to the zoning of Ni contents (and other refractory elements), which are high in the core and low in the rim. A recent study (Weyrauch et al. 2019) observed that isotopic zoning of the more refractory Ni in such grains is correlated to that of Fe. Unzoned metal grains show less Fe and Ni isotope variations, but still up to several ‰ in some grains. These chemical and isotopic signatures of CB and CH chondrite metal have been interpreted to be generated by condensation, likely as a result of a giant impact event in the early solar nebular, a few My after CAI formation. The latter was furthermore inferred by young chondrule ages, estimates of elevated gas/dust ratios and oxygen fugacities (Kleine et al. 2005; Meibom et al. 2001). The isotopic variations found in zoned metal grains, up to 8‰ in δ^{56}Fe values within a single metal grain, are among the largest naturally-generated Fe isotope variations yet found.

4.1.2 Differentiated Planetary Material

Several studies investigated and compared the Fe isotope composition of material from differentiated planetary bodies with each other and with that of chondrites (Barrat et al. 2015; Liu et al. 2010; Moynier et al. 2007; Poitrasson et al. 2004; Schoenberg and von Blanckenburg 2006; Sossi et al. 2016b; Wang et al. 2012, 2014a, b, 2015; Weyer et al. 2005; Wiesli et al. 2003). For Mars, Fe isotope analyses have been performed on a series of achondrites (siliceous meteorite samples from differentiated planetary bodies) assumed to come from Mars (Poitrasson et al. 2004; Sossi et al. 2016b; Wang et al. 2012; Stefan Weyer et al. 2005). Most of these samples are grouped as so-called SNC meteorites (Shergottites, Nakhlites and Chassigny), and have been identified to originate from Mars, using a variety of geochemical and isotope tracers (i.e. by comparison with direct analyses of the Mars surface material and atmosphere during Mars missions). From the Moon, Fe isotope analyses have been performed on a variety of rocks from the lunar crust, either from lunar meteorites or samples returned by Apollo missions, including high- and low-Ti mare basalts, Fe anorthosites, volcanic glass, KREEP-rich material (lunar crust enriched in highly incompatible elements), and highland breccia (Liu et al. 2010; Poitrasson et al. 2004; Wang et al. 2015; Weyer et al. 2005). Furthermore, several achondrites from planetary bodies of the asteroid belt have been analyzed, including HED samples (eucrites, diogenites and howardites), which are assumed to have originated from the asteroid 4 Vesta, as well as angrites, brachinites, ureilites and aubrites (Barrat et al. 2015; Poitrasson et al. 2004; Wang et al. 2012, 2014a, b; Weyer et al. 2005; Jordan et al. 2019). Finally, several iron meteorites (and pallasites) have been investigated, representing core material (and material with incomplete metal segregation) of early planetisimals (Chernonozhkin et al. 2016, 2017; Poitrasson et al. 2004, 2005; Weyer et al. 2005; Williams et al. 2006; Zhu et al. 2001).

The first few high-precision Fe isotope studies (Poitrasson et al. 2004, 2005; Weyer et al. 2005; Zhu et al. 2001) have already found isotopic differences between material from different planetary bodies, and also between different types of differentiated materials (Fig. 4.2). Poitrasson et al. (2004) observed that Lunar material and most igneous rocks from Earth are systematically isotopically heavier than chondrites, while achondrites from Mars, as well as those supposed to come from 4 Vesta, have chondritic Fe isotope compositions. They suggested that the heavier Fe isotope compositions observed for the investigated Lunar and terrestrial material are related to the widely accepted "giant impact" origin of the Moon, i.e. that the Moon (and to lesser degree the Earth) lost light isotopes as result of kinetic evaporation to space during the giant impact event. A problem in comparing bulk compositions of differentiated planets, however, is that the available material is scarce and none of it represents the bulk composition of the planets. The most common material available for investigation is that from planetary crust that was affected at least by two major differentiation events that may have generated Fe isotope fractionation.

The first differentiation event experienced by all differentiated planets by definition is the formation of a metal core. While core formation on early-formed planetisimals (within ≈1 Ma after CAI formation; Kleine et al. 2009) was driven by the energy provided by the decay of short-lived radionuclides (mostly ^{26}Al), late core formation during the final stage of terrestrial planet formation was driven by giant impact events (which may have lasted up to ≈100 Ma after CAI formation; Kleine et al. 2009). As most of the iron segregates to the core during these events, the cores rather than the silicate parts are more representative of the bulk planetary Fe isotope composition. However, from the currently existing planetary bodies, including the Moon and some achondrite parent bodies of the asteroid belt, no core material is available that could be

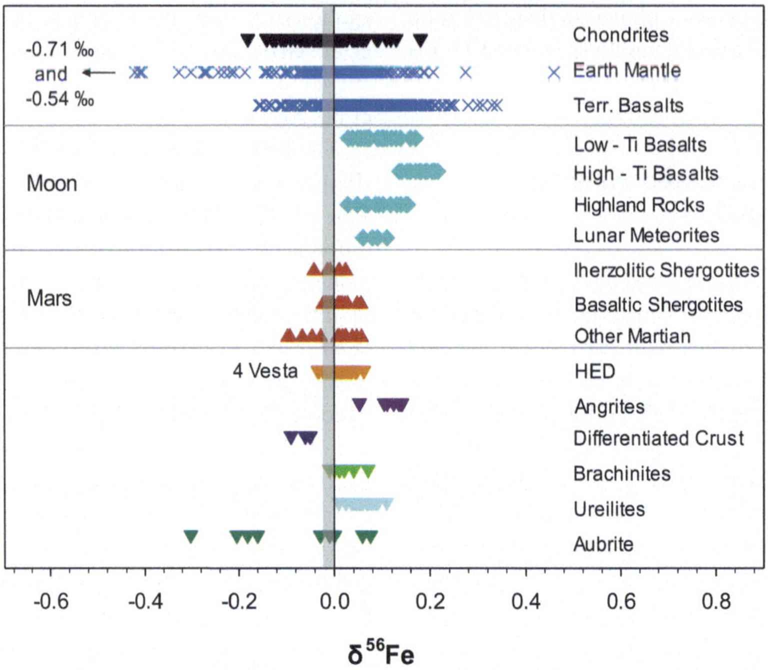

Fig. 4.2 Compilation of Fe isotope compositions of planetary materials, considering the studies of Barrat et al. (2015), Liu et al. (2010), Moynier et al. (2007), Poitrasson et al. (2004), Schoenberg and von Blanckenburg (2006), Sossi et al. (2016b), Wang et al. (2012, 2014a, b, 2015) Weyer et al. (2005), and Wiesli et al. (2003). Also shown for comparison is the range of all chondrites from Fig. 4.1, as well the ranges of all terrestrial mantle rocks and all terrestrial basalts, as shown in Figs. 4.3 and 4.4. The grey vertical bar represents the average of all chondrites (at 95% c.i.)

directly compared to the silicates. Investigation of iron meteorites, pallasites and aubrites shows that many of the metal phases appear to have heavier isotope composition compared to chondrites, indicating that some Fe isotope fractionation may have occurred during core formation (Poitrasson et al. 2005; Zhu et al. 2002; Jordan et al. 2019). This would infer, however, that the bulk planetisimals of the iron meteorites was either (1) systematically heavier than chondrites, (2) their planetary cores were very small or (3) their silicate portions were extremely light in order to satisfy mass balance (see modeling in Jordan et al. 2019). The latter is not very likely, considering the chondritic average Fe isotope composition of the silicate material from several planetary bodies, including Mars, Vesta (Poitrasson et al. 2004; Wang et al. 2012; Weyer et al. 2005).

A problem in using δ^{56}Fe values of iron meteorite in order to infer the bulk planetary Fe isotope composition is that they differentiated during cooling of the metal core into phases such as kamacite, taenite, troilite and schreibersite. This differentiation was still active at relatively low temperatures (potentially diffusive) and associated with large Fe isotope fractionation on the order of up to 1‰ (Chernonozhkin et al. 2017; Poitrasson et al. 2005; Weyer et al. 2005; Weyrauch et al. 2017; Williams et al. 2006). Accordingly, bulk Fe isotope compositions of irons (or pallasite metal) have to be taken with caution, as they simply may not represent the bulk core. For mass-balance reasons, the Fe isotope composition of kamacite may be more representative of the bulk composition of planetary cores. Indeed, kamacites display a narrower range (δ^{56}Fe values mostly between 0.0 and 0.2‰), however, are still slightly heavier than chondrites on average. To understand the meaning of these isotopic variations for the bulk Fe isotope composition of planets and isotope fractionation during planetary differentiation, the findings of experimental and theoretical studies have to be considered, which will be discussed further below (Sect. 4.3.2).

A second major fractionation event in all planets and planetisimals formed in the early solar system is partial melting of the silicate portion, associated with core formation and later magmatic events, that generated the planetary crust. It is widely accepted that crust formation on Earth results in sizable Fe isotope fractionation. Most terrestrial basalts display $\delta^{56}Fe$ values that are on the order of 0.05 to 0.1‰ higher than the average of mantle rocks, while highly differentiated crust commonly displays even higher $\delta^{56}Fe$ values, on the order of 0.1 to 0.5‰ (e.g. Dauphas et al. 2009, 2014; Heimann et al. 2008; Poitrasson and Freydier 2005; Schoenberg and von Blanckenburg 2006; Schuessler et al. 2009; Teng et al. 2008; Weyer et al. 2005, 2007; Weyer and Ionov 2007). Similarly, more detailed investigations of the lunar crust have revealed significant Fe isotope fractionation among different lunar crustal rocks, most notably between the low- and high-Ti mare basalts (Liu et al. 2010; Weyer et al. 2005). While the average $\delta^{56}Fe$ value of low-Ti mare basalts is indistinguishable from the average of all rocks from the lunar crust (0.089 ± 0.015‰, 95% c.i. vs. 0.101 ± 0.020‰, 95% c.i.), high-Ti mare basalts are systematically heavier (0.185 ± 0.015‰, 95% c.i.; see Fig. 4.2). Excluding the high-Ti basalts from the crustal average $\delta^{56}Fe$ value becomes 0.085 ± 0.011‰ (95% c.i.). Also excluded from the calculation of the average lunar crust are regolith samples, as Fe isotopes become systematically fractionated towards heavier composition (by ~0.1‰), as the result of space weathering (Moynier et al. 2007; Wiesli et al. 2003).

More recent investigations observed that, apart from the Moon and the Earth, crustal rocks of other achondrite parent bodies are also fractionated relative to chondrites (Barrat et al. 2015; Wang et al. 2012, 2014b; see Fig. 4.2). Among those, angrites display systematically higher $\delta^{56}Fe$ values (0.118 ± 0.025‰, 95% c.i.; Wang et al. 2012). Angrites have in common with the Earth that the mantle of the angrite parent body was more oxidized compared to e.g. Mars and Vesta (Jurewicz et al. 1991). Wang et al. (2012) also observed a tendency toward higher $\delta^{56}Fe$ values for eucrites (4 Vesta) from the so-called "Stannern Trend" (a group of eucrites with enriched trace element pattern; Barrat et al. 2007). However, if including analyses of earlier studies from Stannern-trend samples, they are indistinguishable from Main Group eucrites within uncertainties. Other groups of achondrites that have been investigated in more detail regarding their Fe isotope systematics include ureilites brachinites (both so-called primitive achondrites, sampling mantle residues of incompletely differentiated planetisimals), aubrites (differentiated enstatite meteorites), and feldspar-rich differentiated achondrites (Barrat et al. 2015; Wang et al. 2014a, b). Among these, aubrites and the feldspar-rich achondrites have $\delta^{56}Fe$ values that are slightly lower than that of chondrites. The light isotope composition of the feldspar-rich achondrites is in contrast to the heavier composition of highly differentiated crust on Earth (e.g. Heimann et al. 2008; Poitrasson and Freydier 2005). The variable $\delta^{56}Fe$ values of silicate material from different differentiated bodies is considered to be the result of condensation effects in the early solar nebular and variable Fe isotope fractionation during planetary differentiation, including core formation and partial melting in the silicate portion of the planets, as will be discussed in more detail below (Sect. 4.3).

4.2 The Silicate Earth

4.2.1 The Mantle and Its Minerals

The Fe isotope composition of the Earth's mantle has been investigated directly by analyses of a variety of mantle rocks, including lherzolites, harzburgites, dunites, wehrlites and pyroxenites, either as bulk rocks (e.g. An et al. 2017; Beard and Johnson 2004; Craddock et al. 2013;

Dauphas et al. 2009; Huang et al. 2011; Macris et al. 2015; Poitrasson et al. 2013; Schoenberg and von Blanckenburg 2006; Su et al. 2015; Weyer et al. 2005, 2007; Weyer and Ionov 2007; Williams et al. 2005; Williams and Bizimis 2014; Zhao et al. 2010, 2012, 2015) or mineral separates, such as olivine (ol), orthopyroxene (opx), clinopyroxene (cpx), spinel (spl), garnet (grt) or hydrous and other less abundant mantle minerals (e.g. An et al. 2017; Beard and Johnson 2004; Chen et al. 2015; Huang et al. 2011; Macris et al. 2015; Weyer et al. 2007; Weyer and Ionov 2007; Williams et al. 2004, 2005; Williams and Bizimis 2014; Zhao et al. 2010, 2012, 2015). Mantle rocks display a surprisingly large range in δ^{56}Fe values, varying from −0.71 to 0.28‰. However, >90% of the investigated samples fall in a much more limited range between −0.14 and 0.11‰ (Fig. 4.3). Although the δ^{56}Fe values of mantle rocks defines a large range, its average (0.002 ± 0.224‰, 2 SD; ± 0.014, 95% c.i., n = 249) is indistinguishable from that of chondrites (−0.010 ± 0.102‰, 2 SD; ± 0.009‰, 95% c.i., n = 146). The most analyzed mantle rocks are spinel peridotites (i.e. mantle rocks that origin from a depth of approximately 30–70 km), and only a few peridotites and pyroxenites from the garnet stability field have been studied (Beard and Johnson 2004; Poitrasson et al. 2013; Williams and Bizimis 2014). Accordingly, most analyzed mantle rocks sample the lithospheric mantle. However, the limited number of analyzed garnet peridotites reveals no systematic difference in δ^{56}Fe values compared to spinel peridotites. There is also neither a systematic difference between the Fe isotope composition of xenolitic and orogenic peridotites nor between the abyssal mantle rocks and the continental lithosphere. Recycling of subducted material fails to explain the large variations among mantle samples, for mass balance reasons (see e.g. Williams et al. 2005). Thus, processes within the mantle must be taken into account.

The large number of analyses allows a comparison between the Fe isotope composition of different mantle lithologies. Lherzolites (δ^{56}Fe = −0.011 ± 0.019‰ 95% c.i., n = 153), harzburgites (δ^{56}Fe = −0.026 ± 0.038‰ 95% c.i., n = 32) and dunites (δ^{56}Fe = 0.039 ± 0.042‰ 95% c.i., n = 17) are indistinguishable in their Fe isotope composition, although there is a weak tendency for harzburgites to be slightly lighter than lherzolites, and dunites being slightly heavier. Both lherzolites and harzburgites, however, suffer from significant outliers with very low δ^{56}Fe values, while, in contrast, some dunites have high δ^{56}Fe values (Fig. 4.3). In particular, for several lherzolites (and at least two harzburgites), most of the negative outlier samples are described to have suffered from strong metasomatism, associated with melt percolation and likely Fe diffusion (Poitrasson et al. 2013; Weyer and Ionov 2007; Zhao et al. 2012, 2015). Excluding these samples with the most negative δ^{56}Fe values, both average lherzolites and harzburgites become slightly heavier (0.013 ± 0.011‰ and −0.010 ± 0.019‰, 95% c.i.), and the heavier tendency of the lherzolites becomes more significant. In contrast to lherzolites and harzburgites, some dunites display high δ^{56}Fe, which may indicate that a different type of metasomatism, associated with enrichment in heavy Fe isotopes, affected at least some of the dunites. The only peridotite group with a distinctly different Fe isotope composition are wehrlites (with δ^{56}Fe = 0.104 ± 0.038‰ 95% c.i., n = 10; Weyer et al. 2007; Zhao et al. 2012). These mantle rocks also have been affected by melt percolation, which in their case, however, was associated with a reaction of the melt with opx, resulting in the generation of secondary cpx. Pyroxenites have also been observed to be enriched in heavy Fe isotopes (Williams and Bizimis 2014). However, pyroxenites from different localities show variable Fe isotope ratios. In average they are still slightly heavier than lherzolites (with δ^{56}Fe = 0.083 ± 0.066‰ 95% c.i., n = 16; Craddock et al. 2013; Macris et al. 2015; Poitrasson et al. 2013; Williams et al. 2005; Williams and Bizimis 2014).

In many studies, mantle minerals were measured together with bulk rocks, in order to investigate inter-mineral equilibrium or to better understand the interplay between mantle minerals and a melt or fluid. The δ^{56}Fe values in mantle minerals vary from −0.68 to 0.80‰ (An et al. 2017; Beard and Johnson 2004; Chen et al. 2015;

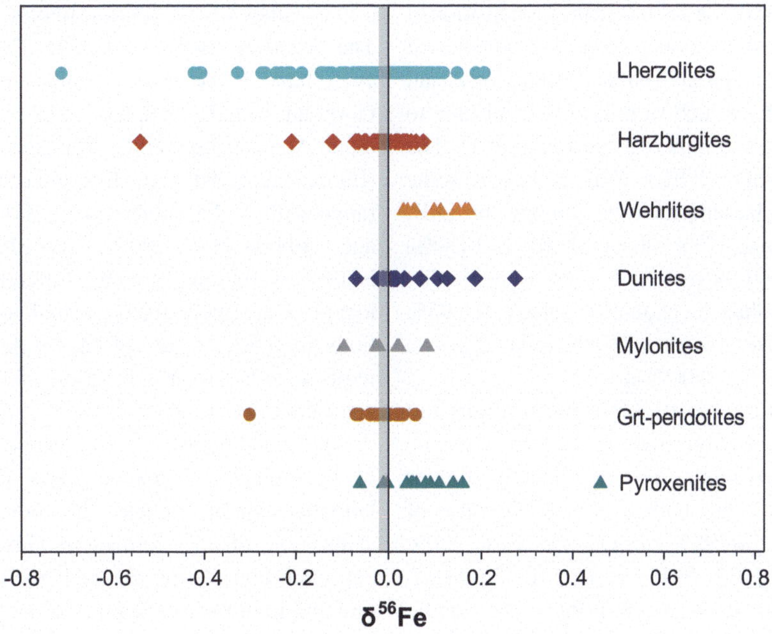

Fig. 4.3 Compilation of Fe isotope compositions of bulk rocks from the Earth's mantle, based on the studies of An et al. (2017), Beard and Johnson (2004), Craddock et al. (2013), Dauphas et al. (2009), Huang et al. (2011), Macris et al. (2015), Poitrasson et al. (2013), Schoenberg and von Blanckenburg (2006), Su et al. (2015), Weyer et al. (2005; 2007); Weyer and Ionov (2007), Williams et al. (2005), Williams and Bizimis (2014), Zhao et al. (2010, 2012; 2015). The grey vertical bar represents the average of all chondrites (at 95% c.i.)

Huang et al. 2011; Macris et al. 2015; Weyer and Ionov 2007; Williams et al. 2004, 2005; Williams and Bizimis 2014; Zhao et al. 2010, 2012, 2017). Their range is extended towards heavier Fe isotope compositions, compared to that of bulk rocks, which is essentially because of the overall heavy isotope signature of spinels (Williams et al. 2004, 2005; Zhao et al. 2015). Inter-mineral isotope fractionation occurs among mantle minerals with δ^{56}Fe values in the following order: grt $\leq$ ol opx $<$ cpx $<$ spl, which is in broad agreement with β-factor (reduced isotope partition function ratios; see Chap. 3) predictions, based on experimental work, synchrotron Nuclear Resonant Inelastic X-ray Scattering (NRIXS) or Moessbauer spectrometry (e.g. Dauphas et al. 2014; Polyakov et al. 2007; Roskosz et al. 2015; Sossi and O'Neill 2017). However, mantle minerals frequently show inter-mineral Fe-isotopic disequilibrium (Beard and Johnson 2004; Macris et al. 2015; Zhao et al. 2015), as also observed for other radiogenic and stable isotope systems (e.g. Tatsumoto et al. 1992). These findings support the hypothesis that many mantle rocks are affected by multi-stage metasomatic events, generated e.g. by diffusion in the presence of a melt or fluid. As diffusion at mantle temperatures (>1000 °C) is fast, disequilibrium among mantle minerals furthermore indicates that the final metasomatic event may have occurred shortly (a few years) before, or during the ascent of the samples to the surface.

4.2.2 Basalts and Komatiites

Several studies systematically investigated Fe isotope compositions of basaltic rocks (e.g. Beard et al. 2003a; Dauphas et al. 2009, 2010; Foden et al. 2018; Hibbert et al. 2012; Konter et al. 2016; McCoy-West et al. 2018; Nebel et al. 2013, 2014, 2015; Nebel et al. 2018; Peters et al. 2019; Schuessler et al. 2009; Teng et al. 2008, 2013; Weyer and Ionov 2007; Williams et al. 2018). Most of these studies focused either on the composition of the oceanic crust, including mid-ocean ridge basalt (MORB) and ocean island basalt (OIB), or on island arc basalt (IAB), which significantly contribute to the formation of continental crust (Fig. 4.4). The results, including MORB, OIB, IAB and some boninites and basaltic andesites among the IAB, reveal an average $\delta^{56}Fe$ value of 0.085 ± 0.130‰ (2 SD, n = 369) for modern basalts, which is distinctly heavier than the chondritic value at the 95% confidence level ($\delta^{56}Fe = 0.085 \pm 0.008$‰ vs. $\delta^{56}Fe = -0.010 \pm 0.009$‰).

Komatiites are mostly of Archean origin and generated at very high melting degrees. Their Fe isotope compositions are significantly lighter than modern basalts, with a mean $\delta^{56}Fe$ value of 0.010 ± 0.021‰ (95% c.i.; Dauphas et al. 2009; Hibbert et al. 2012; Nebel et al. 2015), indistinguishable from those of chondrites and mantle rocks ($\delta^{56}Fe = -0.001 \pm 0.014$‰, 95% c.i.). In detail, some "modern" komatiites from Gorgona island have a tendency towards slightly higher $\delta^{56}Fe$ values of ~0.05‰ (Hibbert et al. 2012). The Archean Alexo komatiites (Canada), however, have $\delta^{56}Fe$ values similar to those from Gorgona island (Dauphas et al. 2010), while other Archean komatiites from the Vetreny belt (Baltic shield) define the low end of the komatiite range with $\delta^{56}Fe$ values of ~−0.07‰, in average (Hibbert et al. 2012). Accordingly, the Fe isotope variations among different komatiite localities are more likely related to local variations in $\delta^{56}Fe$ values of the respective mantle regions, rather than a function of the komatiite age (Nebel et al. 2015).

The indistinguishable Fe isotope composition of komatiites and mantle rocks indicates that komatiites simply sampled the underlying mantle at the time of their formation. This might be as expected, considering the high melting temperature and high melting degrees of komatiites, minimizing potential isotope fractionation during partial melting (Dauphas et al. 2010). In contrast, the higher $\delta^{56}Fe$ values of modern basalts indicates

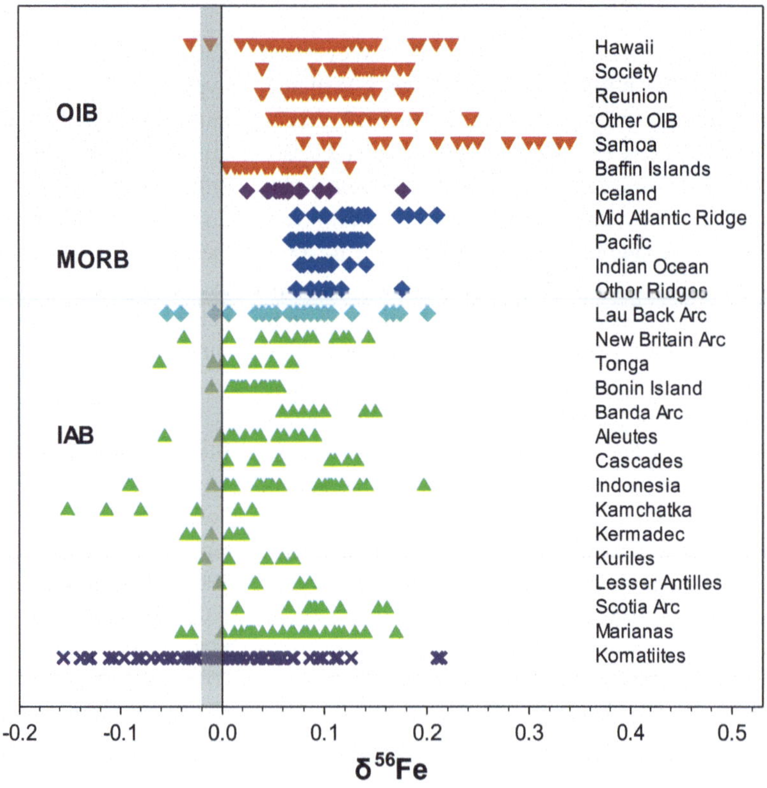

Fig. 4.4 Compilation of Fe isotope compositions of basalts, using data from the studies of Beard et al. (2003a), Dauphas et al. (2009, 2010), Foden et al. (2018), Hibbert et al. (2012), Konter et al. (2016), McCoy-West et al. (2018), Nebel et al. (2013, 2014, 2015, 2018), Peters et al (2019), Schuessler et al. (2009), Teng et al. (2008; 2013), Weyer and Ionov (2007), and Williams et al. (2018). The grey vertical bar represents the average of all chondrites (at 95% c.i.)

that Fe isotope fractionation took place during their formation, assuming that the analyzed mantle rocks provide the average Fe isotope composition of their mantle source (e.g. Dauphas et al. 2009; Schoenberg and von Blanckenburg 2006; Teng et al. 2008; Weyer et al. 2005; Weyer and Ionov 2007; Williams et al. 2005). Sub-dividing modern basalts corresponding to their tectonic setting, the mean δ^{56}Fe values for MORB and OIB are essentially indistinguishable (0.103 ± 0.008‰, 95% c.i., n = 76 and 0.111 ± 0.009‰, 95% c.i., n = 175, respectively; Beard et al. 2003a; Konter et al. 2016; McCoy-West et al. 2018; Peters et al. 2019; Schuessler et al. 2009; Teng et al. 2008, 2013; Weyer and Ionov 2007).

MORBs have relatively homogeneous δ^{56}Fe values (2 SD = ± 0.070‰), and there is no resolvable isotopic difference between the major basins of the Atlantic-, Pacific- and Indian Ocean. There is a weak tendency of E-MORB being slightly isotopically heavier than N-MORB (Sossi et al. 2016a). Included here are also data from Iceland, which actually represents a plume beneath the Mid-Atlantic Ridge. Not included are samples from the Lau Back-Arc Basin, which display in average slightly lower δ^{56}Fe value (0.072 ± 0.021‰, 2 SE, n = 31; Nebel et al. 2018). In contrast, OIBs have more variable δ^{56}Fe values (2 SD = ± 0.131‰), both among basaltic samples of the same locality, and among different localities. The lowest δ^{56}Fe values are found in Paleogene basaltic and picritic lavas and from the Baffin Islands (0.053 ± 0.017‰, 95% c.i., n = 17; McCoy-West et al. 2018), followed by basalts from the Hawaiian islands (0.096 ± 0.012‰, 95% c.i., n = 60; Beard et al. 2003a; Konter et al. 2016; Teng et al. 2008, 2013). The latter display δ^{56}Fe values indistinguishable from average MORB. Reunion and the Society islands have δ^{56}Fe of ≈0.11 and 0.13, respectively (Peters et al. 2019; Teng et al. 2013), which seems to be typical for many other ocean islands, of which only a few Fe isotope analyses are available. By far the heaviest Fe isotope composition have been observed for basalts from Samoa (δ^{56}Fe = 0.216 ± 0.044‰, 95% c.i., n = 17; Konter et al. 2016). Similarly high δ^{56}Fe values have only been observed for two samples, analyzed from the Easter Island (Weyer and Ionov 2007).

The variable Fe isotope compositions among different ocean islands support findings from komatiites (noted above) that different mantle regions likely have variable Fe isotope compositions, which is apparently reflected by ocean island basalts, more than by MORB. The reason may be that OIB sample deeper mantle sources and locally more distinct plume regions which are typically more heterogeneous in their chemical and isotopic composition (e.g. Hofmann 2003). This heterogeneity may include variable proportions of (isotopically heavier) pyroxenites among their source material (Williams and Bizimis 2014). Variable δ^{56}Fe values between different ocean islands or among basalts of the same island, however, may also be caused by variable melting degrees (which are lower in average for OIB as compared to MORB) and a variable degree of differentiation of the melts (e.g. Sossi et al. 2012), which will be discussed in more detail in Sect. 4.3.

In contrast to MORB and OIB, IAB have distinctly lighter Fe isotope compositions, with a mean δ^{56}Fe = 0.050 ± 0.008‰ (95% c.i., n = 183), and a spread of 2 SD = ± 0.110‰, including also some boninites from the Bonin Islands and New Caledonia (Dauphas et al. 2009; Foden et al. 2018; Nebel et al. 2013, 2015; Williams et al. 2018). The most comprehensive investigation of global arc Fe isotope compositions was recently performed by Foden et al. (2018), who observed the same mean δ^{56}Fe values as inferred by the compilation here. Similar to OIB, there are also significant variations among different arc systems. The lowest average δ^{56}Fe value of ≈−0.05‰ have been observed for Kamchatka (defined by only 6 samples), and the highest values of ≈0.1‰, i.e. similar to average MORB, for samples from the Scotia arc (Foden et al. 2018). Large isotopic variations, however, are also observed between different localities of the Indonesian arc systems, e.g. the Western Java (~−0.02‰) and the Eastern Sunda arc (~0.09‰), and also the Mariana arc system, shows an Fe isotopic spread of ±0.1‰ (2 SD). The average lighter Fe isotope composition of IAB compared to MORB has been either explained by a more depleted mantle source, compared to the depleted MORB mantle (DMM), or by suppressed isotope fractionation during the differentiation of IAB, compared to MORB and OIB magmas, which will be discussed in more detail in Sect. 4.3.

4.2.3 Differentiated Crust

While terrestrial basalts from various tectonic settings and geographic localities show only moderate Fe isotope fractionation relative to chondrites and the Earth's mantle, highly differentiated high-silica igneous rocks (e.g. granites, rhyolites) display distinctly heavier Fe isotope compositions, with δ^{56}Fe values of up to 0.6‰ (Du et al. 2017; Foden et al. 2015; He et al. 2017; Heimann et al. 2008; Poitrasson and Freydier 2005; Schoenberg and von Blanckenburg 2006; Schuessler et al. 2009; Sossi et al. 2012; Telus et al. 2012; Wawryk and Foden 2015; Xia et al. 2017; Xu et al. 2017; Zambardi et al. 2014). The average δ^{56}Fe values of all measured rocks above 60 wt% SiO_2 is ≈0.2‰, for both volcanic and plutonic rocks (n ≈ 360). The only exception among more evolved magmatic rocks, however, are carbonatites, which are associated with some intra-continental plumes and typically display negative δ^{56}Fe between −0.5 and 0‰ (Johnson et al. 2010).

Most of the studied differentiated rocks originate from the continental crust. Accordingly, they are commonly generated by a complex formation and differentiation history, which typically include two partial melting events, of which the first one occurred in the mantle (e.g. subduction zones) to form continental crust of basaltic to andesitic composition. Granitic rocks, which make up large parts of the upper continental crust, are typically formed by re-melting of existing crust, either in subduction zone settings (I-type granites) or in intra-continental settings (A-type granites, see e.g. Frost et al. 2001). In case of S-type granites, the source material for the second melting event consist of sediments, which represent a mix of initial crust formation with subsequent weathering. However, despite their complex and variable formation, all highly differentiated rocks define a common trend of increasing δ^{56}Fe values with increasing degree of differentiation, as expressed, for example, in increasing SiO_2 (Fig. 4.5) or decreasing MgO (Du et al. 2017; Foden et al. 2015; He et al. 2017; Heimann et al. 2008; Poitrasson and Freydier 2005; Schoenberg and von Blanckenburg 2006; Schuessler et al. 2009; Teng et al. 2008; Xia et al. 2017; Zambardi et al. 2014).

Accordingly, further statistical treatment of the average δ^{56}Fe values of differentiated rocks has little meaning and we rather focus on observations, based on their isotopic evolution during differentiation of the melt. There are a number of striking findings:

(1) Although different granite types (e.g. I-, A-, S- type or adakitic granitoids) display different ranges of Fe isotope fractionation (e.g. Foden et al. 2015 and Fig. 4.5), they display the same isotopic evolution with increasing SiO_2. The Fe isotopic evolution trends of volcanic differentiates from tholeiitic, alkaline or calc-alkaline basalts are also indistinguishable among each other and from that of plutonic differentiates.

(2) Independent on the initial magma chemistry, igneous rocks display essentially constant δ^{56}Fe values up to ≈65 wt% SiO_2. Above that, they show increasing δ^{56}Fe with values of ≈0.1 to 0.2‰ at 70 wt% SiO_2 and ≈ 0.1 to 0.3‰ at 75 wt% SiO_2. Extremely silicic rocks with $SiO_2 > 75$ wt% have highly variable and sometimes very high δ^{56}Fe (up to 0.64‰). The highest values of >0.46‰ have yet only been observed for rhyolites.

(3) Below $SiO_2 = 65$ wt%, the majority of plutonic—and also volcanic differentiates from a continental setting, display an initial δ^{56}Fe > 0.06‰, well above the average mantle value. I-type and adakitic diorites commonly have δ^{56}Fe values between 0.06 and 0.1‰, while A and S-type plutonic rocks with similarly low SiO_2 have more variable δ^{56}Fe values and, in average, higher δ^{56}Fe of up to 0.28‰. In contrast, tholeiitic volcanites from Iceland and many cretaceous alkaline volcanites from

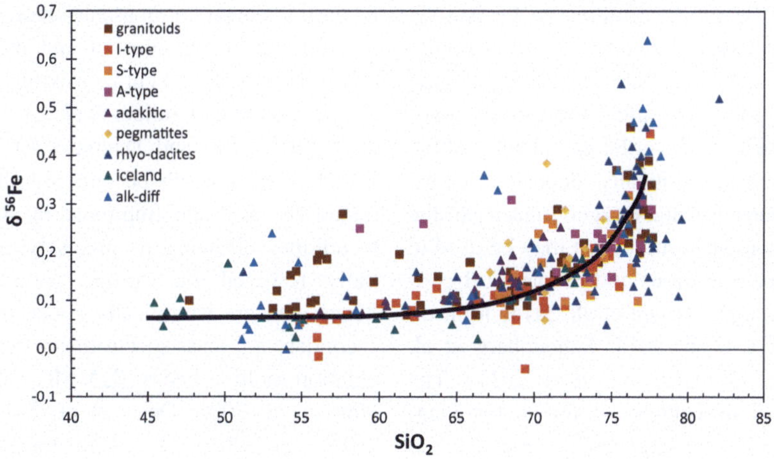

Fig. 4.5 Compilation of Fe isotope compositions of differentiated crustal rocks versus SiO_2 (in wt.%), considering the studies of Du et al. (2017); Foden et al. (2015), He et al. (2017), Heimann et al. (2008), Poitrasson and Freydier (2005), Schoenberg and von Blanckenburg (2006), Schuessler et al. (2009), Sossi et al. (2012), Telus et al. (2012), Xia et al. (2017), Zambardi et al. (2014). Black solid line represents a fit of increasing $\delta^{56}Fe$ with increasing SiO_2

China, display an offset towards lower $\delta^{56}Fe$ values between 0 and 0.1‰ for rocks with $SiO_2 < 65$ wt%. Nevertheless, the isotopic evolution during differentiation of these rocks with $SiO_2 > 70$ wt% is essentially indistinguishable from that of other igneous rocks.

Several mechanisms have been considered in order to explain the strong Fe isotope fractionation observed for highly evolved igneous rocks. Some studies investigated the role of fluid exsolution during fractional crystallization and the ascent of a granitic melt as a driving factor of Fe isotope fractionation (Poitrasson and Freydier 2005; Heimann et al. 2008; Telus et al. 2012; Xia et al. 2017). Other studies focused on the role of fractional crystallization (Dauphas et al. 2014; Du et al. 2017; Foden et al. 2015; He et al. 2017; Schoenberg and von Blanckenburg 2006; Schoenberg et al. 2009; Schuessler et al. 2009; Sossi et al. 2012; Teng et al. 2008; Wu et al. 2017; Xia et al. 2017). Furthermore, the effect of partial melting within the lower crust has been considered (Telus et al. 2012; Foden et al. 2015; He et al. 2017; Xia et al. 2017; Xu et al. 2017), or diffusion-driven processes, either between Fe-rich and Si-rich immiscible melts (Zhu et al. 2015), between the melt and magmatic minerals (Sossi et al. 2012), or by thermal diffusion (Telus et al. 2012; Zambardi et al. 2014). These mechanisms will be discussed in more detail in Sect. 4.3.5 and understanding their role is important in order to discuss the Fe isotope composition of the crust.

4.2.4 Magmatic Minerals

A number of studies investigated the Fe isotope composition of a variety of typical magmatic minerals, including olivine, opx, cpx, garnet, spinel, magnetite, ilmenite, pyrite, hornblende, biotite, alkali feldspar, calcite and others. These studies are essentially driven by two different motivations. The first one is the determination of equilibrium isotope fractionation between different minerals or between minerals and melt, in order to better understand the effect of fractional crystallization on Fe isotope fractionation during magma evolution (Chen et al. 2014; Dziony et al. 2014; Johnson et al. 2010; Liu et al. 2014; Schoenberg et al. 2009; Sossi et al. 2012; Teng et al. 2008; Weyer and Seitz 2012; Wu et al. 2017). Depending on the bonding environment of Fe in the minerals, which is essentially dependent on the redox state and coordination of Fe in the mineral, certain minerals favor either heavy or light isotopes, resulting in Fe isotope fractionation between the minerals. Accordingly, the order of minerals crystallizing during magma differentiation may also affect the isotopic evolution of the melt. As the order and identity of the crystallizing phases critically depends on the chemical composition and oxygen fugacity of the magma, the isotopic composition may be used to infer the redox conditions and/or chemical evolution of the magma (Chen et al. 2014; Johnson et al. 2010; Liu et al. 2014; Schoenberg et al. 2009; Sossi et al. 2012; Wu et al. 2017). The observed range of isotope variations for magmatic minerals is very large (δ^{56}Fe from −1.1 to 1.1‰; e.g. Dziony et al. 2014; Johnson et al. 2010; Wu et al. 2017). However, in many cases, the order of isotope fractionation among the minerals is only crudely in agreement with theoretical and experimental predictions, indicating disequilibrium between the minerals, generated by processes such as late-stage re-crystallization, interaction with a fluid, or diffusion-driven processes. Nevertheless, the investigation of the magmatic minerals revealed important insights into the role of fractional crystallization on magma evolution, discussed in more detail in Sect. 4.3.5.1.

The second motivation is driven by the finding that magmatic minerals are frequently not in isotopic equilibrium with each other or with a melt, as crudely represented by the volcanic bulk rocks (or mineral-free matrix, which may represent the melt at the final stage of fractional crystallization). Disequilibrium between melt and minerals develops as the melt undergoes compositional changes during magmatic evolution, resulting in chemical disequilibrium between minerals and melt, which is followed by diffusively-driven chemical exchange. As light isotopes diffuse faster than heavy isotopes, this process results in significant isotope effects (Oeser et al. 2015; Sio et al. 2013; Teng et al. 2008, 2011; Weyer and Seitz 2012), if the time is insufficient for complete re-equilibration between minerals and melt. These kinetic isotope effects may be an order of magnitude larger than equilibrium isotope effects at magmatic temperatures. In turn, the extent of isotopic disequilibrium may be used to constrain timescales between the onset of chemical disequilibrium and diffusion (e.g., driven by magma ascent or mixing), and the end of diffusion (e.g., Collinet et al. 2017; Oeser et al. 2015, 2018; Sio and Dauphas 2017; Sio et al. 2013). This research became significantly stimulated by new and improved in situ analytical techniques allowing to precisely measure spatially resolved Fe isotope compositions and hence measure isotopically zoned minerals with secondary ion mass spectrometry (SIMS) or laser ablation multi collector (LA-MC-) ICP-MS (e.g. Horn et al. 2006; Oeser et al. 2014; Sio et al. 2013; see Chap. 2). With these techniques, diffusion-driven zoning profiles can be precisely analyzed, allowing reconstruction of the duration of magma evolution. In turn, late diffusion-driven exchange between minerals and melt may also affect the melt composition itself. This will be discussed in more detail in Sect. 4.3.5.2.

4.2.5 Hydrothermal Products and Ores

Hydrothermal systems can be distinguished between those that occur at mid-ocean ridges, where seawater interacts with the hot and freshly-generated basaltic oceanic crust, and those occurring in subduction zones, where subduction-derived fluids interact with the overlying, frequently differentiated crust. While the latter are the host of many important metal ore deposits, the former have the advantage that the hydrothermal fluid, interacting with the crust can be sampled and analyzed directly for its isotopic composition (Beard et al. 2003b; Bennett et al. 2009; Rouxel et al. 2008, 2016, 2018; Severmann et al. 2004; Sharma et al. 2001). Coverage of MOR-associated hydrothermal systems in the book occurs in this section, as well as Sect. 5.5.4 in Chap. 5. In this chapter, focus is on the high-temperature portions of MOR hydrothermal systems, whereas coverage in Chap. 5 is more focused on interactions with seawater. Studies of MOR hydrothermal systems shows consistently low δ^{56}Fe values for the fluids, ranging between −0.7 and 0.1‰, which are thus fractionated by 0–0.8‰ relative to the average Fe isotope composition of fresh basalts (with $\delta^{56}Fe \sim 0.1‰$). Diffusive low-temperature hydrothermal fluids from along faults and fissures of the seafloor may have extremely negative $\delta^{56}Fe$ (as low as −2‰, Moeller et al. 2014; Rouxel et al. 2018). In subduction zones, fluids that are released from the subducted lithosphere are likely also enriched in light Fe isotopes, as a result of the breakdown of Fe^{3+}-rich minerals during subduction. This is indicated by the negative correlation of $\delta^{56}Fe$ values and $Fe^{3+}/\Sigma Fe$ observed in subduction related ophiolitic serpentinites from the Western Alps (Debret et al. 2016).

In a MOR setting, the Fe content of the fluid is determined by the amount of Fe that is leached from basalt during alteration of the oceanic crust and the amount of Fe that is precipitated from the fluid. The acidic properties of the hydrothermal fluids and reducing conditions within the oceanic crust at depth result in enhanced solubility of Fe, and substantial leaching of Fe from fresh basalt. This process results in the enrichment of light Fe isotopes in the fluid, either, because of the faster reaction kinetics of light Fe isotopes, or, because, at equilibrium between Fe^{2+} and Fe^{3+}, light Fe isotopes favor ferrous iron compounds (Polyakov and Mineev 2000; Rouxel et al. 2003), which are predominant in the fluid phase relative to the basalt. The preferential leaching of isotopically light Fe^{2+} during hydrothermal alteration from the oceanic crust is indicated by the positive correlation of Fe^{2+} and total Fe, as well as by the negative correlation of $\delta^{56}Fe$ values and total Fe frequently observed in altered oceanic crust (Fig. 4.6; Rouxel et al. 2003).

During ongoing interaction between hydrothermal fluids and the oceanic crust, the isotopically light fluids become saturated in iron sulfides. Furthermore, mixing with seawater results in more oxidizing conditions upward in the hydrothermal vent (within the chimney). Such conditions result in the precipitation of mostly Fe sulfides, which also affects the isotope composition of the fluid. The products, massive sulfides precipitated within the chimney of hydrothermal vents, show variable $\delta^{56}Fe$ values ranging between −2 and 0.3‰, as observed for the Lucky Strike hydrothermal field (Mid-Atlantic Ridge) and several hydrothermal vent sites along the East-Pacific Rise (Rouxel et al. 2004, 2008). The lowest and generally negative values have been observed for pyrite and marcasite of active vents, while chalcopyrite typically shows values around 0‰. The light Fe (as well as S-) isotope composition of the pyrites is thought to indicate kinetic isotope fractionation during pyrite precipitation. This is because in equilibrium with hydrothermal fluids, pyrite is predicted to be enriched in heavy isotopes. This is indicated from both experimental studies (Syverson et al. 2013), as well as from those, based on Moessbauer and Nuclear Resonant inelastic X-ray scattering (NRIXS) spectra (Polyakov et al. 2007; Polyakov and Soultanov 2011), and also from first-principles density functional theory (Blanchard et al. 2009). Chalcopyrite displays slightly higher $\delta^{56}Fe$ values than respective fluids (by ~0.14‰; Rouxel et al. 2008). These findings could indicate isotopic equilibrium with the fluids during its formation, as confirmed by experimental investigation (Syverson et al. 2013) and also in agreement with calculations from Moessbauer Spectra (Polyakov and Soultanov 2011). Interestingly, pyrite of inactive chimneys was observed to have $\delta^{56}Fe$ values similar to those of hydrothermal fluids, likely as a result of late-stage reworking (Rouxel et al. 2008).

Further Fe isotope fractionation occurs during the rise of the buoyant hydrothermal plume, as a result of the precipitation of isotopically light sulfides (of −0.89 to −0.73‰) and isotopically heavy Fe-oxide particles (of around −0.19‰; Bennett et al. 2009; Rouxel et al. 2016) during mixing of the hot, chemically reduced and acidic vent fluids with cold and oxygen-rich seawater. Even larger Fe isotope fractionation (−1.8 to 1.6‰) was observed during the precipitation of Fe oxide particles in low-temperature hydrothermal plumes (Moeller et al. 2014; Rouxel et al. 2018). The heavy Fe isotope signatures are explained by partial oxidation of the Fe in the vent fluid, likely mediated by Fe-oxidizing microbes. The very low $\delta^{56}Fe$ are explained by the very light Fe isotope composition of the low-temperature hydrothermal fluids or by additional isotope fractionation during previous precipitation of heavy Fe oxides.

Many types of metallic ore deposits are generated by the interaction of hydrothermal fluids,

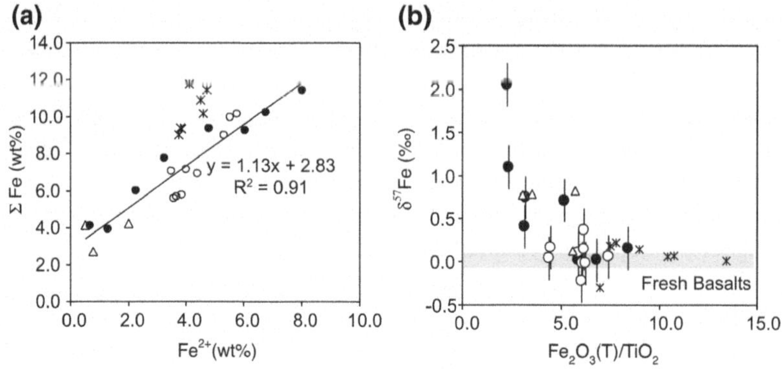

Fig. 4.6 **a** (left) Correlation between total iron (ΣFe and Fe^{2+}) of altered pillow basalts underlying Si–Fe hydrothermal deposits in MORB from different sites, indicating selective leaching of Fe^{2+}; **b** (right) negative correlation of δ^{57}Fe and total iron (normalized to TiO_2), indicating preferential leaching of light Fe isotopes (taken from Rouxel et al. 2003; their Fig. 6a and c)

which can be hot, up to 550 °C in active porphyry systems, with the overlying crust. Metals may come from the magma and/or from the crustal material that interacts with the fluid. Alteration of the crust commonly results in an increase of Fe isotope variability. As light Fe isotopes preferentially participate as Fe^{2+} into the fluid, altered rocks, e.g., granites, may preferentially display an increase of δ^{56}Fe values with increasing degree of alteration, which is also found in respective pyrite precipitates (Zhu et al. 2018). Minerals precipitating from late-stage fluids may show a range of δ^{56}Fe values (e.g., pyrites; Zhu et al. 2018), depending on the conditions (fO_2 and pH) and Fe isotope composition of the fluid, the mechanism of precipitation (i.e., kinetic or equilibrium), and in the latter case, dependent on equilibrium isotope fractionation factors. Wawryk and Foden (2015) also observed relatively heavy Fe isotope compositions for magnetite, pyrite, and chalcopyrite in granitic-hydrothermal Sn-W deposits, which they explained by the formation of these minerals from a fluid that equilibrated with a reduced magma and thus has not been fractionated by earlier-formed magnetite. However, the overall range observed in these mineralizations was large (between −1.0‰ for the lightest pyrrhotite and 1.6‰ for the heaviest pyrite), greatly exceeding the range usually observed for granites. In particular, the Fe isotope composition of chalcopyrite seems to be diagnostic for the oxidation state of the magma from which the hydrothermal fluids evolved, as suggested by Li et al. (2018a). These authors found relatively high δ^{56}Fe values (0.38–0.77‰) for magnetites from a porphory Cu–Au deposit (Duolong, Tibet, China), which formed early at temperatures of 550° to 480 °C from an oxidized magma. They found isotopically light chalcopyrite, which formed later at temperatures of 450° to 350 °C, with δ^{56}Fe values ranging from −0.6 to −0.3‰. The low δ^{56}Fe values of the chalcopyrite are interpreted to have formed by their precipitation from an evolved isotopically light fluid, due to prior magnetite formation.

A range of Fe isotope compositions (−1.5‰ < δ^{56}Fe < 0.9‰) has also been observed for mineralizations of the Black Forest (Germany), which are thought to have formed during interaction of saline Fe-rich hydrothermal fluids with both oxygenated- and CO_2-rich meteoric fluids (Markl et al. 2006), resulting in the formation of isotopically heavy hematite and light siderite. The range of isotopic heterogeneity was further enhanced during secondary low-temperature alteration. Significant Fe isotope variations have also been observed in a skarn-type Cu–Fe–Au deposit (Wang et al. 2011). Overall, the investigated skarn samples display significantly lighter Fe isotope compositions (by ~−1.2‰) than the quartz-monzodiorite stock (the likely Fe source), indicating that fluid exsolution results in the enrichment of light Fe isotopes. Further Fe isotope fractionation occurred during the evolution of the fluid, e.g. during magnetite precipitation. Similarly, Zhu et al. (2016) observed slightly lower δ^{56}Fe values for magnetite of a skarn-type deposit than for the associated diorites. They concluded that Fe was likely delivered by an isotopically light fluid.

Compared to the large Fe isotope variations observed among hydrothermally generated ore minerals, high-temperature ores show a more limited range of δ^{56}Fe values. Bilenker et al. (2016) observed for magnetite, the major Fe-bearing phase of two Chilean iron oxide apatite deposits, a range of δ^{56}Fe between 0.06 and 0.53‰, more similar to the isotopic range observed for granites. They used oxygen isotopes of coexisting magnetite and actinolite (and actinolite Fe numbers) to determine a formation temperature of 630° to 820 °C. With a similar approach, applying Fe isotopes and trace-element compositions of magnetites (which were all analyzed in situ with femtosecond Laser ablation MC-ICP-MS) and O isotopes, Günther et al. (2017) observed two different iron ore groups with two distinct types of magnetite in the Chagangnuoer and Zhibo iron ore districts. One group had a magmatic origin and was likely formed by fractional crystallization, resulting in a positive correlation of compatible- (V, Ni) and a negative correlation of incompatible elements with δ^{56}Fe in magnetite. The second group had a hydrothermal origin, associated with low δ^{56}Fe values (between −0.5 and 0‰) for magnetite, which likely formed by Fe-leaching from the ortho-magmatic ore body by a late stage fluid. These case studies show that Fe isotopes, combined with petrographic, geochemical and other isotopic tools, can be an important tool, for deciphering processes of magmatic and hydrothermal ore formation.

4.2.6 Metamorphic Rocks

Little work has been done on Fe isotope fractionation during metamorphism. Metamorphism, sensu stricto, only results in a re-equilibration of the chemical and isotopic composition of co-existing minerals. Accordingly, at a bulk rock scale, metamorphic rocks should mirror the Fe isotope composition of the precursor rocks, as observed for gneisses and amphibolites from the North China Craton (Zhu et al. 2018). At the mineral scale, this re-equilibration is the basis of all geothermo- and barometers, providing information on metamorphic conditions. It may even provide information on the cooling history and rates, if the closure temperature (the temperature at which no more diffusive exchange between minerals occurs) of a chemical/isotopic system is known. Accordingly, the Fe isotope fractionation between coexisting minerals in a metamorphic rock may be used as a thermometer (isotope fractionation is by far more temperature- than pressure sensitive). Even, at temperatures of several 100 °C, the extent of Fe isotope fractionation between coexisting minerals is still significant. For example, experimentally calibrated Fe isotope fractionation between magnetite and fayalite ($\Delta^{56}Fe_{Mag\text{-}Fa}$) is ~0.3‰ at 600 °C (Shahar et al. 2008). However, as Fe isotope fractionation is small compared to oxygen isotope fractionation, the limits of analytical precisions may limit the use of Fe isotopes as a thermometer. Hyslop et al. (2008) found correlated Δ^{56}Fe and Δ^{18}O fractionations between magnetite and silicate minerals in the Biwabik Iron Formation (Minnesota, USA), which underwent contact metamorphism by magmatic intrusion, indicating metamorphic re-equilibration of these minerals. A temperature decrease from likely >800 °C at the contact down to 375° C at a distance of 2.6 km from the contact, as determined by oxygen isotopes and mineral paragenesis, resulted in increasing $\Delta^{56}Fe_{Mag\text{-}sil}$ from ~0.2 to ~0.5‰ (this will be revisited in Sect. 6.6.3 in Chap. 6 in the discussion of high-grade iron formations).

In many cases, however, metamorphism is associated with mass flux (metasomatism), resulting in open-system behavior frequently associated with isotopic disequilibrium, as shown for mantle rocks (see above and, for example, Beard and Johnson 2004). Open system behavior is also assumed for the formation of eclogites in subduction zones. Williams et al. (2009) found correlated δ^{56}Fe values for garnet and pyroxene with $\Delta^{56}Fe_{cpx\text{-}grt}$ of ~0.28‰ in eclogites from the Kaalvallei and Bellsbank kimberlite pipes

(South Africa), indicating isotopic equilibrium between these minerals. Isotope fractionation between garnet and pyroxene is in broad agreement with that previously observed in mantle rocks (Beard and Johnson 2004). In contrast, Li et al. (2016) observed disequilibrium between garnet and omphacite of UHP (ultra-high-pressure) eclogites from the Bixiling and Dabie orogenic belt, with $\Delta^{56}Fe_{cpx\text{-}grt}$ ranging between 0.13 and 0.48‰. Omphacites from this locality display much larger variability in both $\delta^{56}Fe$ and $Fe^{3+}/\Sigma Fe$, likely as a result of omphacite alteration during retrograde metamorphism, while garnet remained essentially unaffected.

Several studies investigated the Fe isotope composition of subduction-related rocks on the bulk-rock scale (Debret et al. 2016; Inglis et al. 2017; Korh et al. 2017; Li et al. 2016). While Inglis et al. (2016) and Li et al. (2016) observed overall MORB-like Fe isotope signatures for the analyzed eclogites and metagabbros, Korh et al. (2016) observed higher $\delta^{56}Fe$ values for meta basites from Ile de Croix (France), ranging from 0.16 to 0.33‰, where systematically heavier values were measured for blueschists than for eclogites. The authors explain these variations to reflect a heterogeneous isotope composition of the protolith rather than isotope fractionation during metamorphism or metasomatism. In contrast, Inglis et al. (2016) observed a negative correlation of $\delta^{56}Fe$ values and fluid mobile elements, such as B, Rb or Sn, indicating the introduction of an isotopically light Fe component by fluids during subduction. Support for an isotopically light nature of such fluids may be provided by the negative correlation of $\delta^{56}Fe$ values and $Fe^{3+}/\Sigma Fe$ observed in subduction-related serpentinites, suggesting the release of isotopically light Fe-bearing fluids during serpentinite devolatilization (Debret et al. 2016).

4.3 Planetary Formation and Magmatic Processes

Basalts from different planetary bodies, including Mars, 4 Vesta, the Earth and the Moon, have different average Fe isotope compositions (e.g., Poitrasson et al. 2004; Weyer et al. 2005). These systematic variations may be explained by (1) a heterogeneous distribution of Fe isotopes in the early solar nebular, (2) Fe isotope fractionation during planetary accretion, (3) Fe isotope fractionation during the formation of the planetary cores, or (4) Fe isotope fractionation during silicate differentiation and the formation of the crust.

4.3.1 Planetary Accretion

There is increasing evidence from isotopic anomalies that the protoplanetary disc was heterogeneous (scenario 1) and not isotopically entirely mixed in the Solar nebular during the time of accretion in the early Solar system (e.g., Brennecka et al. 2013; Burkhardt et al. 2011). Systematic isotopic variations appear to exist between material accreted in the inner solar system, as sampled by Earth, and those accreted further outside, as sampled by the different types of chondrites (Fischer-Gödde et al. 2015; Fischer-Gödde and Kleine 2017). These variations are likely the result of incomplete mixing of source material (e.g. s- and r-process nuclides) in the protoplanetary disc. However, there is little evidence for systematic stable Fe isotope fractionation at planetary scales. Notably, all different types of chondrites (enstatite-, ordinary- and carbonaceous chondrites) have indistinguishable Fe isotope compositions (e.g. Craddock and Dauphas 2011). It is thus, unlikely that the Solar Nebula was heterogeneous regarding its stable Fe isotope composition.

A second possibility is that the accretion of the planets resulted in Fe isotope fractionation. It has been proposed by Poitrasson et al. (2004) that heavier Fe isotope signatures of terrestrial- and lunar crust, compared to that of Mars and Vesta, may be related to the impact origin of the Moon, resulting in volatilization-related loss of isotopically light Fe isotopes for both, Earth and Moon. However, other similar or more volatile elements in lunar rocks show little or weak evidence for evaporation-driven isotope fractionation (e.g., Cu, Li, Mg, Rb; Magna et al. 2006; Moynier et al. 2006, Pringle and Moynier 2017; Sedaghatpour et al. 2013; Seitz et al.

2006). For Si, higher $\delta^{30}Si$ values have been observed for the Earth and Moon, which however, have been commonly explained by Si isotope fractionation during Si partitioning into the (proto) Earth core during high-pressure core formation (Georg et al. 2007; Shahar et al. 2009; Hin et al. 2014). In contrast, Zambardi et al. (2013) proposed that some of the elevated $\delta^{30}Si$ observed for the Earth and Moon may be related to the Moon-forming impact event. This view has been challenged more recently by Dauphas et al. (2015), who proposed that isotope fractionation during condensation and early accretion was mostly responsible for the Si (which is similarly volatile as Fe) isotope variations observed between different planetary bodies and the Earth. These authors used the positive correlation of $\delta^{30}Si$ values and Mg/Si observed for different planetary bodies (including Mars, Vesta, Earth, Moon and the angrite parent body) to model this relationship, assuming equilibrium isotope fractionation between SiO gas and forsterite during forsterite condensation in the solar nebula, with subsequent mixing of forsterite with nebular gas. They assumed that Mg became essentially exhausted during nebular forsterite condensation, in order to explain why no Mg isotope fractionation is observed between the same planetary bodies (Sedaghatpour and Teng 2016). The new findings of Dauphas et al. (2015) set limits on the Si content in the Earth's core.

Sossi et al. (2016b) observed a correlation of $\delta^{56}Fe$ with $\delta^{30}Si$ values and Fe/Mn for basaltic crust of several planetary bodies and speculated that this may also be related to a reaction of $(Fe)_g$ with solids during condensation from the solar nebular (Sossi et al. 2016a). This would either presume partial condensation of Fe with equilibrium Fe isotope fractionation, similar as proposed for Si by Dauphas et al. (2015) and in contrast to Mg (for which no isotope fractionation between planets is observed; Sedaghatpour and Teng 2016). Alternatively (or additionally), isotope fractionation may have occurred during partial evaporation during accretion (for Earth and Moon potentially also during the giant impact; Poitrasson et al. 2004). Notably, such mechanisms appear to be plausible to explain the high $\delta^{56}Fe$ and $\delta^{30}Si$ values observed for angrites, which are highly depleted in volatile elements and for which the heavy isotope enrichment is not easy to explain otherwise. For other planetary bodies, however, a comparison as presented by Sossi et al. (2016b) is problematic, as they simply used the basaltic planetary crusts, for which $\delta^{56}Fe$ values is clearly best defined. They have not considered any other Fe isotope fractionation during planetary differentiation, including core formation and the formation of the crust. These processes will be discussed in the next sections. For example, for the Earth, the Fe isotope composition of the mantle and the bulk silicate Earth seems to be very close to chondritic, in contrast to the isotopically heavier composition of basaltic (oceanic) crust, indicating Fe isotope fractionation during crust formation. Thus, the correlations of $\delta^{56}Fe$ values observed by Sossi et al. (2016a) and their causality may need to be further elucidated. Nevertheless, some accretion-related Fe isotope variations among bulk planets may exist and accordingly, some planetary bodies may not have a chondritic Fe isotope composition.

4.3.2 Formation and Differentiation of Planetary Cores

Zhu et al. (2001) carried out one of the earliest studies of high-precision Fe isotope analyses and observed systematically higher $\delta^{56}Fe$ values for metal than for olivine, although this was based on only two pallasites. This observation was consistent in direction with a theoretical estimate based on Mössbauer spectroscopy (Polyakov and Mineev 2000), although at the high temperatures assumed for core formation on Earth ($\sim$3000–4000 °C; Wade and Wood 2005), only modest Fe isotope fractionation may be expected. Subsequently, several studies addressed the issue of potential Fe isotope fractionation between metal and silicates during the formation of planetary cores, either based on the analyses of natural samples, such as iron meteorites, pallasites, and ureilites (Barrat et al. 2015; Chernonozhkin et al. 2016; Poitrasson et al. 2005; Weyer et al. 2005; Williams et al. 2006; Jordan et al. 2019), or on experimental and theoretical approaches (Elardo and Shahar 2017; Hin et al. 2012; Liu et al. 2017; Poitrasson et al. 2009; Polyakov 2009; Roskosz et al. 2006; Rustad and Yin 2009; Shahar et al. 2015; Williams et al. 2012).

The studies on pallasites and iron meteorites report large Fe isotope variations, in particular between the different phases in iron meteorites or the metal part of pallasites, such as kamacite, taenite, troilite, and schreibersite (Chernonozhkin et al. 2016; Poitrasson et al. 2005; Weyer et al. 2005; Williams et al. 2006). Troilite displays significantly lower δ^{56}Fe values and taenite higher, relative to kamacite. Such isotope fractionation may occur during fractional crystallization of the metal core, as recently observed for Ru, using Ru stable isotope systematics (Hopp et al. 2018). However, Jordan et al. (2019) used a similar approach to explain the heavy Fe isotope signatures (δ^{56}Fe $\sim$ 0.04 to 0.13‰) they observed for IIIAB magmatic iron meteorites, simply assuming that the bulk planetary body had a chondritic Fe isotope composition. However, independent from the degree of crystallization, as estimated from HSE elements, they observed rather constant Fe isotope compositions, incompatible with the assumption of Fe isotope fractionation during core crystallization. Rather, as supported by recent studies applying spatially-resolved methods of Fe isotope analyses (Chernonozhkin et al. 2017; Weyrauch et al. 2017), the large Fe isotope fractionation between these phases was likely generated during late-stage (e.g., diffusion driven) re-equilibration at sub-solidus conditions. As such, the Fe isotope composition of metal phases may provide information on the cooling history of the metal core, rather than on the isotopic equilibrium during core formation.

For example, during cooling of Fe–Ni metal (taenite) a second metal phase (kamacite) with low Ni contents exsolves, resulting in Fe diffusion from taenite into kamacite and Ni diffusion in the opposite direction. As light isotopes diffuse faster than heavy isotopes, this process results in the enrichment of light Fe isotopes in kamacite (leaving high δ^{56}Fe values in taenite behind) and light Ni isotopes in taenite. This process of Fe–Ni exchange between taenite and kamacite proceeds until temperatures as low as $\sim$350–400 °C (Nichols et al. 2018), and results in Fe isotope fractionation of $\Delta^{56}Fe_{ta\text{-}ka}$ $\sim$ 0.3 to 0.8‰ (Poitrasson et al. 2005; Horn et al. 2006; Chernonozhkin et al. 2017; Weyrauch et al. 2017). Because of the lower abundance of Ni, the relative Ni flux is much higher than that of Fe, and isotope fractionation of Ni is even stronger than that of Fe (Chernonozhkin et al. 2017; Dauphas 2007; Weyrauch et al. 2017). Watson et al. (2016) experimentally determined the diffusivity ratio for two Fe or Ni isotopes during Fe-Ni exchange between different Fe-Ni alloys to be on the order of 1.005 to 1.006 for a Fe or Ni isotope ratio with 1 Dalton mass difference between the two isotopes. With this information, preserved diffusion-driven Fe and Ni isotope fractionation, which was generated during kamacite-taenite exsolution, may be used to estimate cooling rates of iron meteorites (Chernonozhkin et al. 2017; Dauphas 2007; Watson et al. 2016; see also Sect. 4.5.3.2 for more information about isotope diffusion chronometry).

The significant Fe isotope fractionation between metal (me) and troilite (tr) or schreibersite, but also the range of $\Delta^{56}Fe_{tr\text{-}m}$ $\sim$ −0.1 to −0.7‰ (Chernonozhkin et al. 2017; Weyer et al. 2005; Williams et al. 2006) may also indicate incomplete equilibration between these phases at sub-solidus conditions, as supported by a negative correlation of $\Delta^{56}Fe_{tr\text{-}me}$ and $\Delta^{65}Cu_{tr\text{-}me}$ (the latter is up to 9‰) observed by Williams and Archer (2011). In situ analyzed isotope systematics (with LA-MC-ICP-MS) of kamacite and adjacent troilite furthermore support sub-solidus Fe exchange between these phases at temperatures <900 °C (Chernonozhkin et al. 2017). The enrichment of light Fe isotopes in troilite may result in the overall enrichment of heavier Fe isotopes in the metal, depending on the abundance of troilite (Williams et al. 2006). Accordingly, super-chondritic δ^{56}Fe values, as frequently observed for metal in iron meteorites, may be balanced by troilite, rather than resulting from metal-silicate isotope fractionation during core formation (Hin et al. 2012; Williams et al. 2006). Schreibersite was even observed to be either isotopically heavier or lighter than metal, likely depending on whether it is in equilibrium or disequilibrium with the metal (Chernonozhkin et al. 2017).

Similar to the phases in iron meteorites (or within pallasite metal), olivine and metal in pallasites also show chemical and isotopic exchange at subsolidus conditions, as preserved in chemical and isotopic zoning in olivine in several cases (Miyamoto 1997; Chernonozhkin et al. 2017). Accordingly, the isotope fractionation observed between metal and olivine in pallasites (mostly between 0 and 0.2‰; Chernonozhkin et al. 2016, 2017; Poitrasson et al. 2005; Weyer et al. 2005) is unlikely to represent equilibrium Fe isotope fractionation produced by metal segregation. It rather represents either disequilibrium between metal and olivine, if diffusive exchange between metal phases (e.g. kamacite-taenite) occurred after closure of the metal-olivine exchange (Weyer et al. 2005), or re-equilibration of metal and olivine at much lower than magmatic temperatures (Chernonozhkin et al. 2017). In the latter case, the isotope composition of the pallasite olivines may indicate potential Fe isotope fractionation between metal and silicate. Combining the literature data (Zhu et al. 2002; Poitrasson et al. 2005; Weyer et al. 2005; Chernonozhkin et al. 2016, 2017) shows that pallasite olivines have an average δ^{56}Fe value of −0.08‰, likely the result of late stage re-equilibration at ~600° − 900 °C (Miyamoto 1997; Chernonozhkin et al. 2017). Assuming that about 80% of Fe is in metal and only 20% in silicate (si) (Buseck 1977), and that the pallasite parent body had a chondritic Fe isotope composition, this implies that pallasite metal should have an average δ^{56}Fe value on the order of 0.02‰, resulting in $\Delta^{56}Fe_{me\text{-}si}$ of ~0.1‰ at the re-equilibration temperatures given above (Chernonozhkin et al. 2016). At core formation temperatures of 2000–3000 °C, however, this would translate to very limited metal-silicate Fe isotope fractionation on the order of 0.01–0.02‰. Similarly, assuming a chondritic Fe isotope composition of the Earth and a δ^{56}Fe value of the silicate Earth of ~0.02‰ (Craddock et al. 2013; Weyer and Ionov 2007 and as inferred from the average mantle rocks from above), would also indicate a very small Fe isotope fractionation (though opposite in direction to pallasites) during core formation on Earth.

Jordan et al. (2019) investigated the Fe isotope compositions of metal and silicate in aubrites (achondrites closely related to enstatite chondrites) and observed $\Delta Fe_{metal\text{-}silicate}$ of 0.08 to 0.09‰. The aubrite metal was likely trapped in the mantle of the aubrite parent body during core formation at extremely reducing conditions, as indicated by the low FeO in silicate (mostly enstatite). From the Si content in the metal, they estimated a metal-silicate equilibration temperature of ~1150–1200 °C. The metal in the aubrites may have been less affected by sub-solidus re-equilibration, although the silicate metal equilibrium temperature for Si and Fe must not necessarily be identical. In contrast, ureilites, so-called unequilibrated achondrites of a C-rich parent body, display a slightly heavier than chondritic δ^{56}Fe value of 0.056‰. These superchondritic values are assumed to be the result of an S-rich metallic melt at low temperatures (<1100 °C, before melting of silicates), as furthermore indicated by a marked S depletion (Barrat et al. 2015).

According to the considerations above, metal isotope fractionation during core formation is likely small for many parent bodies. Many of the observed variations among metal phases and between metal and silicates may record sub-solidus re-equilibration, however, some isotope fractionation during metal segregation may not be excluded. Additional constraints have been provided by experimental studies and theoretical estimations of Fe isotope fractionation in the context of core formation. An overview of estimated Fe isotope fractionation based on (1) the analyses of natural samples (meteorites), (2) experimental approaches, and (3) theoretical modeling is provided in Fig. 4.7. Polyakov and Mineev (2000) realized, using Mössbauer spectroscopic data to determine the reduced isotopic partition function ratio (or β-factor), that native iron falls in between the range of β-factors determined for ferric and ferrous iron compounds. These findings are in accordance with the observation that pallasite metal appears to be slightly heavier than pallasite olivine. In agreement with the considerations above, the theoretical estimates suggest only minor isotope fractionation of <0.1‰ at potential core-formation temperatures for smaller planetary bodies (e.g. ~1600 °C) and negligible isotope fractionation of ~0.02‰ at estimated core-formation temperatures on Earth, assuming a deep magma ocean (e.g. 2900–3600 °C; Siebert et al. 2013) given the $1/T^2$ dependence of equilibrium isotope fractionation between two phases.

Metal-silicate Fe isotope fractionation was subsequently studied experimentally by Poitrasson et al. (2009) and Roskosz et al. (2006). The latter study used a technique of metal reduction in Pt capsules at 1500 °C and atmospheric pressures, and observed significant equilibrium isotope fractionation between metal and silicate ($\Delta^{56}Fe_{me\text{-}si}$) on the order of 0.4 ± 0.3‰ for $\delta^{56}Fe$ values. Poitrasson et al. (2009) performed experiments at slightly higher temperatures (2000 °C) and elevated pressures (1–7.7 GPa), which are more consistent with the conditions of core formation on planetary bodies. They observed no resolvable Fe isotope fractionation of $\Delta^{56}Fe_{me\text{-}si}$ ~ 0.03 ± 0.04‰. A similar finding was reported by Hin et al. (2012), who determined Fe isotope fractionation at low temperatures (1250–1300 °C and 0.5–1.3 GPa), with the addition of significant amounts of Sn or S to their metal-silicate mix in order to achieve more similar conditions to those assumed for metal segregation on small planetary bodies. They observed no resolvable Fe isotope fractionation ($\Delta^{56}Fe_{me\text{-}si}$ of ~0.01 ± 0.04‰).

Polyakov (2009) was the first who applied synchrotron (INRXS, also known as nuclear

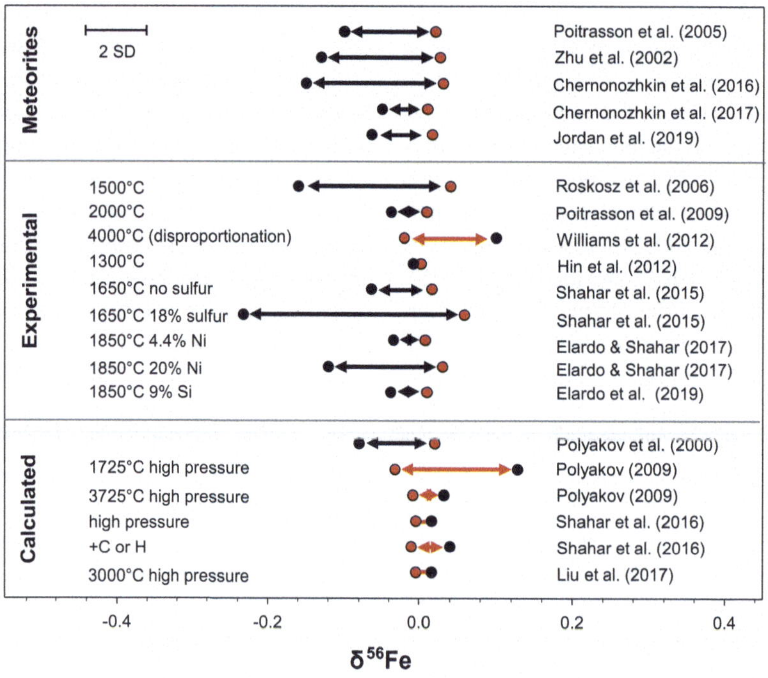

Fig. 4.7 Compilation of estimated Fe isotope fractionation (black arrows) between silicate (black dots) and metal (red dots) during planetary core formation from the investigation of meteorites (pallasites, aubrites and irons), from experimental determinations and as calculated from NRIXS or Mössbauer Spectra (see references in the legend of the figure). Note that (1) all absolute values for $\delta^{56}Fe$ of metal and silicates are arbitrary defined, assuming a chondritic bulk composition of a planetary body, consisting of 80% Fe in metal and 20% Fe in silicate; (2) natural samples have equilibrated and experiments have been performed at different P-T-conditions and metal compositions, resulting in different measured or calculated (from NRIXS spectra) Fe isotope composition. The temperature effect may simply be approximated with $\Delta^{56}Fe_{me\text{-}si}$ ~ $1/T^2$

resonant inelastic x-ray scattering—NRIXS) in order to determine the ^{57}Fe partial vibrational density states at high pressures (133 GPa), using a diamond anvil cell. He realized that Fe isotope fractionation between metal and silicate is pressure dependent. At high pressures, the compression of chemical bonds and increase of vibrational frequencies results in an increase of the isotope partitioning ratio (β factor; see Chap. 3). This effect is different for different chemical compounds. Based on his results, Polyakov (2009) calculated isotope fractionation between lower mantle minerals and metallic iron and revealed that at pressures >100GPa, both ferropericlase (Fe, Mg)O and post-perovskite (Fe, Mg)SiO_3 are enriched in heavy isotopes relative to metallic iron. This finding was interesting, as it may explain the heavier isotope composition of terrestrial basalts (by ~0.1‰) compared to those of basalts from Mars and Vesta, considering the much higher core-formation pressures in the Earth compared to smaller planetary bodies. However, metal—silicate isotope fraction determined for periclase ($\Delta^{56}Fe_{me\text{-}si}$ ~ −0.008‰ at 2000 K) is significantly smaller than that for post-perovskite ($\Delta^{56}Fe_{me\text{-}si}$ ~ −0.16‰ at 2000 K and ~ −0.041‰ at 4000 K), and thus a significant portion of Fe hosted by post-perovskite needs to be assumed. Note, for consistency reasons, we here report always metal-silicate isotope fractionation as $\Delta^{56}Fe_{me\text{-}si}$, which may differ in terms of the direction or the Fe isotope nomenclature (i.e., $\Delta^{57}Fe$ instead of $\Delta^{56}Fe$) in the original studies.

In a subsequent experimental study, Williams et al. (2012) assumed that a significant portion of iron in perovskite may be ferric (McCammon et al. 1997), generated by internal oxidation of the lower mantle, as a result of disproportionation of ferrous iron into ferric and metallic iron by perovskite crystallization during accretion of the Earth. They performed multi-anvil experiments, applying the "three isotope method" in order to extrapolate their result towards full isotopic exchange (see Chap. 3, and Shahar et al. (2008) and Williams et al. (2012) for high-temperature examples of the method), in which they mixed Fe^{3+}-bearing perovskite with Fe metal at 1850 K and 24 GPa. For their experiments, they observed significant Fe isotope fractionation between perovskite (pv) and metal of at least $\Delta^{56}Fe_{me\text{-}pv}$ = −1.02‰, which translates to $\Delta^{56}Fe_{me\text{-}pv}$ = −0.25‰, if assuming core-mantle boundary temperature of 4000 K. Assuming a higher $Fe^{3+}/\Sigma Fe$ of ~0.7 for perovskite (McCammon et al. 1997) instead of the perovskite with $Fe^{3+}/\Sigma Fe$ of 0.37 that was used for the experiments, $\Delta^{56}Fe_{me\text{-}pv}$ may further increase to ~−0.48‰. Accordingly, Williams et al. (2012) propose, similar to Polyakov (2009), an isotopically heavy mantle as a result of core formation, and they estimate that disproportionation of about 20% of the lower mantle with subsequent segregation of the light metallic Fe into the core may explain a bulk silicate Earth that is heavy by ~0.1‰ in $\delta^{56}Fe$ compared to chondrites and the silicate Mars or Vesta. They furthermore propose that such a mechanism likely occurred on the Proto-Earth, prior to the moon-forming "giant impact", as primitive lunar basalts are similarly heavy compared to terrestrial basalts.

A series of experimental studies have been performed that investigated metal-silicate Fe isotope fractionation, with a particular focus on the effect of metal composition and how these effects change with pressure and temperature of core formation. Shahar et al. (2015) investigated the effect of sulfur in metal and observed a significant increase of $\Delta^{56}Fe_{me\text{-}si}$ with increasing S in metal, reaching ~0.26‰ at 18% S (at 1650 °C). A similar effect was later on observed for Ni (Elardo and Shahar 2017) and Si (Elardo et al. 2019). The latter authors observed that all elements substituting for Fe in the metal structure, including Si, S and Ni, have a similar effect on metal-silicate Fe isotope fractionation, as they shorten and stiffen the bonds to their atomic neighbors, resulting in increasing $\Delta^{56}Fe_{me\text{-}si}$ with increasing substitutional impurities. Shahar et al. (2016) investigated the relative effect of different light-element substitution in metal on metal-silicate Fe isotope fractionation at high pressures of up to 60 GPa, by conducting high pressure INRXS experiments. They calculated isotope fractionation (β-factors) from the vibrational levels of different isotope-bearing structures for a larger pressure range. They found increasing β-factors with pressure for all investigated alloys and for pure Fe metal, which they explained by bond stiffening under compression. At a pressure of 60 GPa, which may be assumed for a magma ocean during Earth's core formation, they observed negative $\Delta^{56}Fe_{me\text{-}si}$ between all metal alloys and perovskite (bridgmanite). They furthermore observed a different pressure-dependence for different light elements, which are thought to explain the density deficit of the Earth's core, including H, C and O. According to their findings, the effect of those elements would be quite significant, in particular that of H and C. They concluded that H and C are unlikely to be the light elements in the Earth's core, as this would result in a significantly positive $\delta^{56}Fe$ value for the silicate Earth. Similar experiments, using high-pressure INRXS spectroscopy, have been performed by Liu et al. (2017) in order to investigate metal-silicate Fe isotope fractionation, with a focus on Earth's core formation. They investigated the pressure effect on isotope fractionation between Fe–S and Fe–Si alloys, which are important candidates for the light element in Earth's core, as well as Ni and a basaltic glass at pressures of up to 206 GPa. In accordance with previous studies, they observed negative $\Delta^{56}Fe_{me\text{-}si}$ between the investigated metal alloys, but with $\Delta^{56}Fe_{me\text{-}si}$ not exceeding 0.02‰ at high pressures and temperatures of >3500 K, which they assumed for core formation conditions on Earth.

4.3.3 Partial Melting on Earth and Other Planets

The Earth's mantle is relatively oxidized, as compared to other planetary bodies, and contains a significant amount of ferric iron (~3.7%; e.g., Canil and O'Neill 1996). The capability of Fe^{3+} to form stronger bonds and enrich heavy Fe isotopes, combined with its incompatible nature during partial melting, motivated researchers to explore the Fe isotope systematics of mantle rocks and minerals (e.g., Craddock et al. 2013; Huang et al. 2011; Schoenberg and von Blanckenburg 2006; Su et al. 2015; Weyer et al. 2005 and 2007; Weyer and Ionov 2007; Williams et al. 2004, 2005; Williams and Bizimis 2014). The aim of these studies was to investigate the suitability of Fe isotopes as a proxy for mantle redox conditions or to explain variations in $\delta^{56}Fe$ values among the silicate portions of planets.

Williams et al. (2004, 2005) observed large variations in $\delta^{56}Fe$ values between different mantle samples mostly from the continental lithosphere, at the mineral and bulk rock scale, that correlated negatively with indices for mantle depletion, such as Cr# in spinel (the molar ratio of Cr/(Cr + Al)), and Mg# of Fe–Mg minerals (the molar ratio of Mg/(Fe + Mg)). They furthermore observed a negative correlation between $\delta^{56}Fe$ values and $(Fe^{3+}/\Sigma Fe)_{spinel}$ and other indices for the oxygen fugacity of the mantle. Applying mass-balance constraints, the authors could exclude that the large variations in Fe isotopes where inherited from subducted material, such as ferromanganese crusts. They suggested that their findings may be the result of Fe isotope fractionation during partial melting, where the investigated mantle xenoliths represented the residue of such mantle melting. Interestingly, the most oxidized samples (being also the most depleted ones) show the lowest $\delta^{56}Fe$ values, which is not easy to explain, as ferric Fe is expected to result in enrichment rather than depletion of heavy Fe isotopes. Williams et al. suggested that those samples may have been extensively altered by oxidized fluids, resulting in high melting degrees and large isotope fractionation between melt and residue. Assuming, however, that mantle oxidation occurred before partial melting, this model requires that significant Fe^{3+} was retained (or buffered) during partial melting.

Weyer et al. (2005) studied lithospheric mantle samples that contained little evidence of metasomatic overprinting and observed a very limited range of $\delta^{56}Fe$ values (−0.024 to 0.065‰), with an average of 0.015‰. As this value is lower than the average of published values for basalts, they concluded that Fe isotopes fractionate during partial melting and the formation of the continental crust. A similar offset between mantle samples and basalts was also observed by Schoenberg and von Blanckenburg (2006). Weyer and Ionov (2007) measured a range of different mantle rocks, including samples that experienced variable degrees of depletion by partial melting, and others that experienced different types of metasomatic overprint. Similar to Williams et al. (2005), they observed a negative correlation of $\delta^{56}Fe$ values and Mg#, which was used as indicator for the degree of mantle depletion. They applied a simple Rayleigh melting model, assuming a constant partition coefficient D (= 0.9) for total Fe (including both Fe^{3+} and Fe^{2+}, assuming buffered $Fe^{3+}/\Sigma Fe$ during melting) between bulk rock and melt during partial melting and different isotope fractionation factors α in order to fit the modelled depletion curve to the trend they observed for the natural samples. The relatively broad array defined by all investigated samples could be fitted with α values between 1.001 to 1.003, corresponding to ε values (= $(\alpha - 1) * 1000$) of 0.1 to 0.3. They used this broad array to determine a value for the average Fe isotope composition of the primitive upper mantle (at Mg# = 0.894; McDonough and Sun 1995) of $\delta^{56}Fe \sim 0.02$‰, which is within uncertainties in agreement with the average value of the data compilation for mantle rocks from above.

With the aim to better define isotope fractionation between the upper mantle and MORB, Craddock et al. (2013) focused on abyssal peridotites and observed a limited range of $\delta^{56}Fe$ values and an almost identical average value for their investigated mantle rocks to that of Weyer and Ionov (2007), with $\delta^{56}Fe = 0.025$‰. From their findings, they conclude that the residual mantle itself should have experienced only limited Fe isotope fractionation during melt extraction, on the order of 0.01‰. This is in contradiction to the correlations of $\delta^{56}Fe$ values with indices of melt extraction observed by several studies (Su et al. 2015; Weyer and Ionov 2007; Williams et al. 2004, 2005; Williams and Bizimis 2014). Interestingly, Weyer and Ionov (2007) observed the strongest correlation for samples from localities with a relatively steep trend of $\delta^{56}Fe$ versus Mg#, indicating distinctly stronger isotope fractionation during partial melting (with $\varepsilon \sim 0.3$), as inferred from the difference between average mantle and MORB.

Williams et al. (2005) and Williams and Bizimis (2014), also modeled Fe isotope fractionation during partial melting, assuming partial melting of a peridotite sample with a typical upper mantle mineral composition. In their model, they assumed that Fe isotope fractionation during melting is driven by non-modal melting, i.e., Fe isotope fractionation between Fe^{3+}-bearing minerals, such as spinel and cpx and those containing only Fe^{2+}, such as olivine. As these minerals contribute differently to melt production, this results in Fe isotope fractionation between melt and mantle residue. They realized, however, that extreme mineral-melt isotope fractionation factors would be required to obtain the observed Fe isotope variations, and light Fe isotope compositions for samples which experienced high degree of partial melting (e.g. high Cr# in spinel or whole rock). Such high isotope fractionation is in contradiction with both the average isotope fractionation observed between a basaltic crust and the mantle and also with theoretical estimates of Fe isotope fractionation between silicate minerals at magmatic temperatures (Dauphas et al. 2014; Polyakov and Mineev 2000; Roskosz et al. 2015; Sossi and O'Neill 2017). In addition to depletion trends in peridotites, Williams and Bizimis (2014) also observed systematically higher $\delta^{56}Fe$ values for pyroxenites, which they propose may be responsible for some of the variations observed among basalts, in particular OIBs (e.g. Konter et al. 2016).

Similar observations of large variations in $\delta^{56}Fe$ values among mantle rocks correlating with mantle depletion were reported by Su et al. (2015), who focused on ophiolitic peridotites from Tibet, which are assumed to have been depleted by partial melting, presumably at a former ocean spreading center. From the large Fe isotope variations among the peridotites they modeled, this would correspond to a fractionated ocean crust with $\delta^{56}Fe$ value of ~ 0.3‰, significantly heavier than observed at present. $\delta^{26}Mg$ values for the same samples were also measured and correlated with $\delta^{56}Fe$ values. They inferred from these finding that Mg isotopes are also fractionated to some degree by partial melting, resulting in elevated $\delta^{26}Mg$ coupled with low $\delta^{56}Fe$, and during subduction-related metasomatic enrichment, results in low $\delta^{26}Mg$ coupled with high $\delta^{56}Fe$. Similar to other studies (Weyer and Ionov 2007; Williams et al. 2004, 2005; Williams and Bizimis 2014), however, the $\delta^{56}Fe$ depletion trends of the mantle rocks investigated by Su et al. (2015) indicate much stronger melting-related isotope fractionation, as that proposed by experimental and theoretical studies (Dauphas et al. 2014; Polyakov and Mineev 2000; Roskosz et al. 2015; Sossi and O'Neill 2017). Su et al. (2015) suggest that multiple melting events (with re-enrichment in between), may generate the observed isotope signatures, which would then imply a surprisingly correlated effect of partial melting and re-enrichment for $\delta^{56}Fe$ and $\delta^{26}Mg$ values, and indices of melt extraction, including Cr#, Mg#. Alternatively, metasomatic processes, resulting in chemical and isotopic disequilibrium may have been involved in producing the observed isotope compositions. Such processes may involve, for example, Fe–Mg exchange diffusion between the melt and the depleted mantle wall rocks, which would result in a negative correlation of $\delta^{56}Fe$ and $\delta^{26}Mg$ values, as observed by Su et al. (2015), and potentially also in a correlation of $\delta^{56}Fe$ and Mg#. These processes will be discussed in more detail in the Sect. 4.3.4.

Variably depleted (or enriched) mantle sources are also proposed to explain some of the variability in the Fe isotope compositions observed in basalts. In particular, the overall lower $\delta^{56}Fe$ values of IAB, as compared to MORB or OIB, has been attributed to mantle sources that have been depleted in heavy Fe isotopes by previous partial melting events (Dauphas et al. 2009; Foden et al. 2018; Nebel et al. 2013, 2015; Su et al. 2015; Williams et al. 2018). Dauphas et al. (2009) studied the effect of partial melting on $\delta^{56}Fe$ values by analyzing IAB that have been formed by different degrees of partial melting, using TiO_2 as a monitor for the melting degree. They observed MORB-like Fe isotope signatures for samples with high TiO_2 (low water flux and melting degree), and lower, almost chondritic $\delta^{56}Fe$ values for samples with low TiO_2 (high melting degree). They modeled the extent of Fe isotope fractionation between the melt and a primitive mantle residue ($\delta^{56}Fe$ = 0.02‰; Weyer and Ionov 2007), assuming an incompatible behavior of Fe^{3+} (with $D_{melt\text{-}residue}$ of about 0.1) with an initial mantle Fe^{3+}/Fe^{2+} Fe of 0.037 and an isotope fractionation between Fe^{3+} and Fe^{2+} of 0.3‰. The latter was estimated combining the theoretical/experimental findings from Polyakov and Mineev (2000), Schoenberg et al. (2009), and Schuessler et al. (2007). These assumptions would generate only moderate Fe isotope fractionation between melt and mantle residue, if the Fe^{3+} content in the mantle residue was not buffered during melting (Fig. 4.8a, Dauphas et al. 2009). Assuming, however, a buffered (constant Fe^{3+}) during melting in the mantle residue, their model would generate sizeable Fe isotope fractionation between melts and mantle residue on the order of 0.08‰, at low- to moderate-melting degrees (Fig. 4.8b, Dauphas et al. 2009). A constant Fe^{3+} content during melting was recently produced during experiments of dry peridotite partial melting, as a result of increasing compatibility of Fe^{3+} during melting according to changes in the mineralogy (Sorbadere et al. 2018). Such a mechanism may thus at least partially explain the elevated $\delta^{56}Fe$ values ($\sim$0.1‰) typically observed for MORB and OIB. The lower $\delta^{56}Fe$ values observed for many IAB was explained by Dauphas et al. (2009) by flux-driven, non-buffered melting at high melting degrees, resulting to limited Fe isotope fractionation, as Fe^{3+} rapidly decreases in the mantle residue during melting. They furthermore suggested that many IAB may originate from already pre-depleted sources (as also suggested by Nebel et al. 2013, 2015; Su et al. 2015; Williams et al. 2018; Foden et al. 2018), although non-buffered melting has actually little effect on the Fe isotope composition of the mantle residue (Craddock et al. 2013; Williams et al. 2005).

The above conclusions on Fe isotope fractionation during partial melting are also consistent with the lower $\delta^{56}Fe$ values, as compared to MORB, that are commonly observed for komatiites (Dauphas et al. 2010; Hibbert et al. 2012; Nebel et al. 2014). As komatiites are thought to have formed by very high degrees of partial melting (typically of 35–50%; Herzberg

1992), they may display only minor Fe isotope fractionation relative to the mantle source. This is in particular the case if it is assumed that Fe isotope fractionation is essentially driven by the fractionation between incompatible Fe^{3+} and Fe^{2+} and non-buffered melting (although buffering of Fe^{3+} may be unrealistic at such high melting degrees; Dauphas et al. 2010). Some komatiites still display isotope variations correlating with indicators for the degree of partial melting, such as Mg#, Cr# or V/Sc (Hibbert et al. 2012). As the compatibility of V is also sensitive to mantle redox conditions, the authors infer relatively oxidizing conditions during partial melting at fO_2 conditions equal to, and one log unit above, the fayalite-magnetite-quartz buffer (FMQ), taken to represent that of contemporary upper mantle (e.g., Frost and McCammon 2008). As these komatiites of different ages (ranging from the Neoarchean to Neoproterozoic) define similar trends of $\delta^{56}Fe$ and V/Sc, the authors conclude that the mantle was already oxidized to a similar extent in the Archean relative to today.

The potential extent of equilibrium isotope fractionation during partial melting has also been investigated in theoretical and experimental studies. The reduced partition function ratio (β-factor) of Fe in different phases can be determined using the second order Doppler (SOD) shift of Moessbauer spectroscopy (e.g., Polyakov and Mineev 2000) or calculated based on vibrational frequencies using quantum mechanical modeling (e.g., Schauble 2004), as discussed in Chap. 3. For solid Fe phases, significant improvement on the determination of the β-factors, based on the determination of the mean force constants for Fe bonds in Fe-bearing glasses and minerals, have been achieved with application of synchrotron nuclear resonance inelastic X-ray scattering (NIRXS; Dauphas et al. 2012, 2014; Polyakov et al. 2007; Roskosz et al. 2015). Findings from these investigations confirm significant Fe isotope fractionation between Fe^{3+}- and Fe^{2+}-bearing minerals on the order of 0.2–0.4‰ (Dauphas et al. 2014). The disadvantage of this method is that no mean force constants for Fe bonds in melts can be determined. Dauphas et al. (2014) tried to circumvent this problem by investigating glasses of different compositions. They also modeled Fe isotope fractionation during partial melting at conditions assumed for MORB formation with a mantle source $Fe^{3+}/\Sigma Fe$ of 0.043 and mineral-melt partition coefficients for Fe^{3+}- and Fe^{2+} in order to generate the typically observed $Fe^{3+}/\Sigma Fe$ of 0.16 for MORB at about 10% partial melting. They obtained $\Delta^{56}Fe_{melt\text{-}mantle}$ on the order of only 0.03‰, i.e., lower than the observed difference between average mantle and MORB (Fig. 4.9, Dauphas et al. 2014).

Roskosz et al. (2015) performed experiments in order to determine the mean force constants for olivine, cpx and spinel to determine β-factors for calculating Fe isotope fractionation factors. They observed a linear relationship of Fe isotope fractionation between olivine and spinel and $Fe^{3+}/\Sigma Fe$ in spinel, with essentially no Fe isotope fractionation for Fe^{3+}-free spinel and a maximum $\Delta^{56}Fe_{spinel\text{-}olivine}$ on the order of 0.2‰ for magmatic temperatures (in agreement with previous experimental determinations of Fe isotope fractionation between magnetite and olivine by Shahar et al. 2008). Sossi and O'Neill (2017) performed additional experiments with a focus on the effect of the bonding environment on Fe isotope fractionation. They observed significant Fe isotope fractionation between Fe^{2+} minerals, for which Fe occurs at different coordination numbers, including 8-fold (almandine), 6 fold (fayalite, ilmenite), and 4-fold (chromite, hercynite) coordination, with increasing β-factors for decreasing coordination numbers. In contrast, they observed little Fe isotope fractionation between minerals for which Fe occurs in the same coordination (such as fayalite and ilmenite). These findings indicate that Fe isotope fractionation between melts and a mantle residue may not only be driven by fractionation between Fe^{3+}- and Fe^{2+} (and the incompatible behavior of Fe^{3+}), but also on the coordination environment of Fe in the melt and minerals (Dauphas et al. 2014; Sossi and O'Neill 2017).

In a recent study, Yang et al. (2019) used Fe isotopes to address the question as to whether the depleted upper mantle may be non-representative of the bulk mantle, and with that (for mass-balance reasons) for the bulk silicate Earth. They

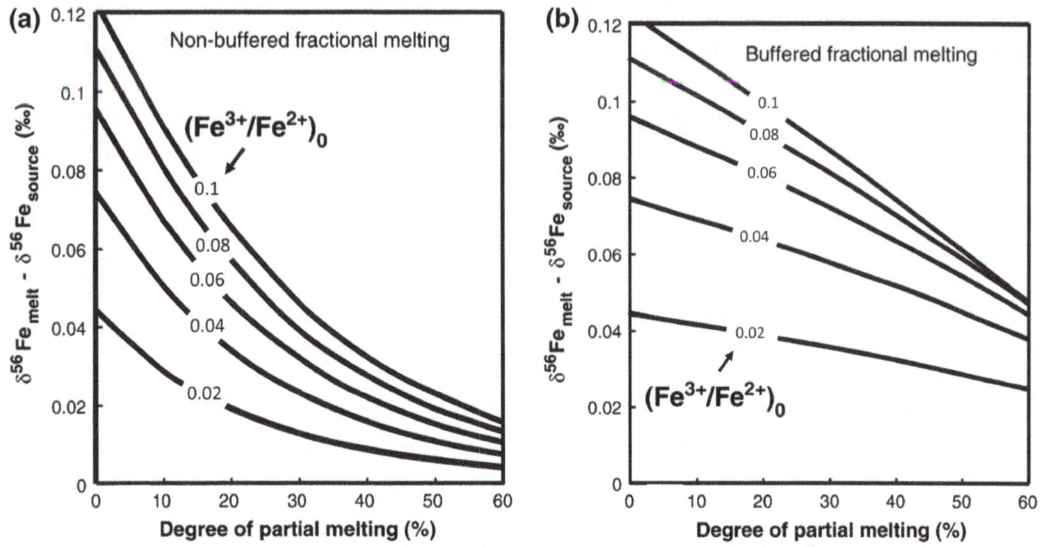

Fig. 4.8 Calculated melt-source iron isotopic fractionation as a function of degree of partial melting and initial Fe^{3+}/Fe^{2+} ratios. **a** (left) for non-buffered $Fe^{3+}/\Sigma Fe$ during fractional melting; **b** (right) for buffed fractional melting (modified from Dauphas et al. 2009; their Fig. 7)

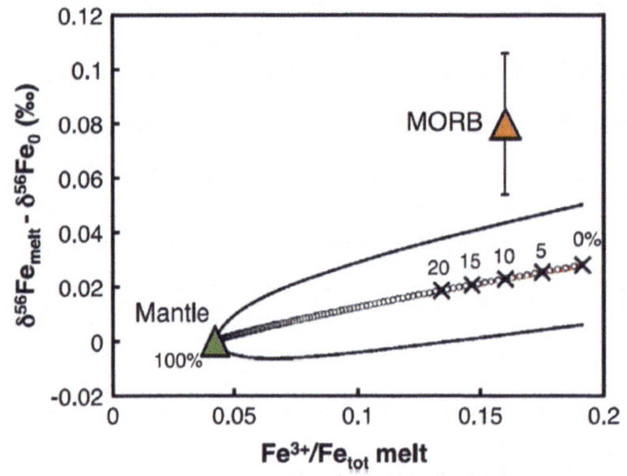

Fig. 4.9 Modeled Fe isotope and $Fe^{3+}/\Sigma Fe$ evolution during mantle melting in order to generate MORB. Orange triangle represents MORB with $Fe^{3+}/\Sigma Fe$ of 0.16 (Cottrell and Kelley 2011) and $\Delta^{56}Fe_{MORB\text{-}mantle} = 0.08$ (Craddock et al. 2013; Teng et al. 2013; Weyer and Ionov 2007); this book; adapted from Dauphas et al. 2014, their Fig. 7

determined Fe isotope fractionation between the lower mantle Fe minerals ferropericlase and bridgmanite based on high pressure experiments and analyses of the mean force constants with NRIXS. In agreement with earlier findings by Rustad and Yin (2009), they observed an increase in the force constant of ferropericlase and significant Fe isotope fractionation between ferropericlase and bridgmanite at lower mantle pressures. Modeling of the effect of melting in deep mantle plumes or magma ocean crystallization did not result in significant Fe isotope variations between mantle reservoirs, which furthermore have been continuously mixed with time (Yang et al. 2019). From these findings they conclude that primitive upper mantle is representative for the bulk silicate Earth in terms of its Fe isotope composition.

While thus far we focused on mantle melting, highly evolved rocks such as granites and rhyolites are frequently generated by partial melting of more mafic lower crustal material, which may result in isotope fractionation relative to the source region that may be different relative to mantle melting. These effects were modeled for trachytes and rhyodacitic rocks from the Hailar Basin, which are believed to have formed by partial melting of basaltic trachyandesites (Xia et al. 2017). For their model, with the software "rhyolite MELTS", they used the experimentally defined force constants for melts and minerals of Dauphas et al. (2014) and Roskosz et al. (2015), and generated melts with high Fe^{3+} and $\delta^{56}Fe$

values (i.e., using high force constants of the melt), in particular for a closed system and low melting degrees. Assuming that the igneous upper crust is commonly the result of two melting events (of mantle and lower crust), these results showed that these may also contribute to the variable and heavy Fe isotope compositions observed for some highly evolved rocks. Other processes controlling the Fe isotope composition of highly evolved rocks are discussed further in Sect. 4.3.5.

Iron isotope fractionation during partial melting, at variable source composition and conditions during melting, was also considered to explain the variable Fe isotope composition between the crust of different planetary bodies (see Fig. 4.2). Wang et al. (2012) analyzed a range of crustal rocks, including samples from Mars, howardites, eucrites, diogenites (from asteroid 4 Vesta), angrites, and one lunar meteorite. Similar to previous studies (Poitrasson et al. 2004; Weyer et al. 2005), they observed mostly chondritic Fe isotope compositions for samples from Mars and 4 Vesta. Angrites, however, displayed significantly heavier isotope compositions, similar to basalts on Earth. These findings may be explained by stronger Fe isotope fractionation during partial melting on the angrite parent body and the formation of the angrites, as compared to that during the formation of the crust on Mars and 4 Vesta. This scenario is supported by the more oxidized nature of the angrite parent body, as indicated by the occurrence of Fe^{3+} in angrites, while on Mars and 4 Vesta, Fe^{3+} is completely absent. Assuming Fe isotope fractionation during partial melting is mainly driven by the isotope fractionation between Fe^{3+} and Fe^{2+}, this may explain the similar δ^{56}Fe values of angrite and basalts on Earth. However, calculated fO_2 in angrites is only IW + 1 (Iron-Wüstite buffer + one log unit), i.e., 2–3 orders of magnitude lower than in the Earth's upper mantle (Wadhwa 2008). Furthermore, this scenario assumes a similar Fe isotope composition of the mantle of the angrite parent body to that of the Earth's mantle (which may not necessarily be the case—see discussion of Fe isotope fractionation during accretion and core formation above).

While variable redox conditions and Fe isotope fractionation between Fe^{3+} and Fe^{2+} minerals may explain some of the variations observed among planetary crusts, it cannot explain the heavy Fe isotope compositions observed for high-Ti mare basalts on the Moon (Liu et al. 2010; Poitrasson et al. 2004; Weyer et al. 2005), or the slightly higher δ^{56}Fe values observed in Stannern-group eucrites (Wang et al.2012), as partial melting of both the lunar mantle and 4 Vesta occurred under reducing conditions in the absence of Fe^{3+}. In contrast to the Earth's mantle, which has been extensively mixed and broadly homogenized by convection throughout Earth's history, rapid cooling resulted in a much shorter dynamic history of the Moon, producing significant heterogeneities in the lunar mantle. Such heterogeneities likely result from the differentiation and crystallization of the Lunar magma ocean (LMO) and were still preserved in the lunar mantle at the time of the formation of the lunar mare basalts. The heterogeneous lunar mantle source likely resulted in the different chemical signatures of high- and low-Ti mare basalts, and, furthermore, also in the different stable isotope signatures that have been generated during their formation. Apart from differences in δ^{56}Fe values, variable isotope signatures for oxygen, magnesium and titanium have been observed for these basalts (Liu et al. 2010; Millet et al. 2016, Sedaghatpour et al. 2013). While low-Ti basalts have δ^{26}Mg, δ^{49}Ti, and δ^{56}Fe values, very similar to terrestrial basalts, high-Ti mare basalts have more negative δ^{26}Mg and elevated δ^{49}Ti and δ^{56}Fe values. These isotope compositions are interpreted to originate from an ilmenite-rich mantle reservoir that formed during the later stages of magma ocean crystallization. Similarly, Wang et al. (2012) ascribed the slightly higher than chondritic δ^{56}Fe ($\sim$0.05‰) of Stannern-trend eucrites to the assimilation of ilmenite during re-melting of previously-formed crustal material, as indicated by, for example, the correlation of Fe isotopes and TiO_2.

The exact mechanism that formed the heavy Fe isotope signatures of the high-Ti mare basalts (and Stannern-trend-eucrites) is not quite clear. Indeed, heavy isotope compositions, as compared to bulk rocks and olivine, have been observed for ilmenite mineral separates from several low- and high-Ti mare basalts (Wang et al. 2015), indicating that heavy Fe of the high-Ti mare basalts could have been inherited from heavy ilmenite in the lunar mantle. However, a recent study from Sossi and O'Neill (2017) observed similar isotope fractionation factors for ilmenite and fayalite, suggesting that Fe isotope fractionation between VI-fold coordinated Fe minerals is insignificant. Instead, they suggest that a lower average coordination number of Fe in the melt, as compared to the mostly VI-fold coordinated Fe in mantle minerals, may result in sizable Fe isotope fractionation during fractional crystallization of the LMO, resulting in high δ^{56}Fe values of the residual melt at the time of ilmenite crystallization. In either case, the isotopic findings support previous evidence based on the chemical compositions that the mantle source of low-Ti mare basalts is more representative for the bulk composition of the Moon.

Wang et al. (2015) suggested that the deep lunar mantle might be isotopically light. This was based on the finding of a very light Fe isotope composition of a lunar dunite sample (72415), which belongs to the lunar Mg-suite of plutonic rocks (including dunites, troctolites, norites and gabbronorites), which are likely relicts of the earliest cumulates of LMO differentiation. In this case, the bulk Moon may easily have had a similar (or even lighter) Fe isotope composition than the Earth. It is, however, debated, if the Mg-suite rocks are products of the early magma ocean crystallization originating from the deep lunar mantle, or whether they represent later-formed cumulates originating from the shallow lunar mantle (Ryder 1991; Elardo et al. 2011).

4.3.4 Mantle Metasomatism

Systematic investigations on mantle rocks and their minerals revealed that (1) they display a surprisingly large range of δ^{56}Fe values (−0.7 to 0.8‰), considering the high mantle temperatures (>800 °C), and (2) that mantle minerals are frequently not (fully) equilibrated (e.g., Beard and Johnson 2004; Weyer and Ionov 2007; Williams et al. 2004, 2005 and see Sect. 4.2.1). Such large variations cannot be easily explained by partial mantle melting alone. Rather, they suggest that metasomatic interactions of the mantle rocks with a melt, percolating through the mantle, or a fluid, play an important role (Huang et al. 2011; Macris et al. 2015; Poitrasson et al. 2013; Su et al. 2015; Weyer and Ionov 2007; Williams and Bizimis 2014; Zhao et al. 2010, 2012, 2015, 2017). In addition, xenoliths may be penetrated during the ascent by the magmas that bring them up to the surface (e.g. Rudnick and Ionov 2007). Mantle metasomatism may have a strong effect on the chemical and isotopic composition of the mantle (e.g., Bodinier et al. 1990; Rudnick et al. 1993). As melts percolating through the mantle, disequilibrium is expected with the mantle rocks, and this frequently results in chemical reactions, or chemical diffusion of distinct elements between such melts and individual mantle minerals (Bodinier et al. 1990; Foley et al. 2013). These processes frequently result in isotopic disequilibria between mantle minerals (e.g. for Li and Fe; Beard and Johnson 2004; Rudnick and Ionov 2007), which may be preserved, if, for example, xenoliths were brought to the surface at relatively short timescales after mantle-metasomatism (or orogenic peridotites become cool and dry enough, with limited diffusive inter-mineral re-equilibration). However, even if mantle minerals re-equilibrated, large isotopic variations may exist between mantle rocks that were differently affected by mantle metasomatism.

In their investigation of mantle rocks from TOK (Siberia) and Tariat (Mongolia), Weyer and Ionov (2007) observed two distinct types of metasomatic processes (E1 and E2) at the bulk-rock scale, which could be distinguished by the combination of Fe isotopes and the chemical and mineralogical composition of the samples. Most samples from both suites experienced depletion by partial melting by ~20% (resulting in Mg# of around 91–92). Most peridotites from TOK and some from Tariat subsequently experienced metasomatic enrichment by the reaction with a Fe- and Ca-rich Si-undersaturated melt (defined as E1), resulting in the replacement of opx and the formation of secondary cpx and the formation of wherlites (Ionov et al. 2005). These

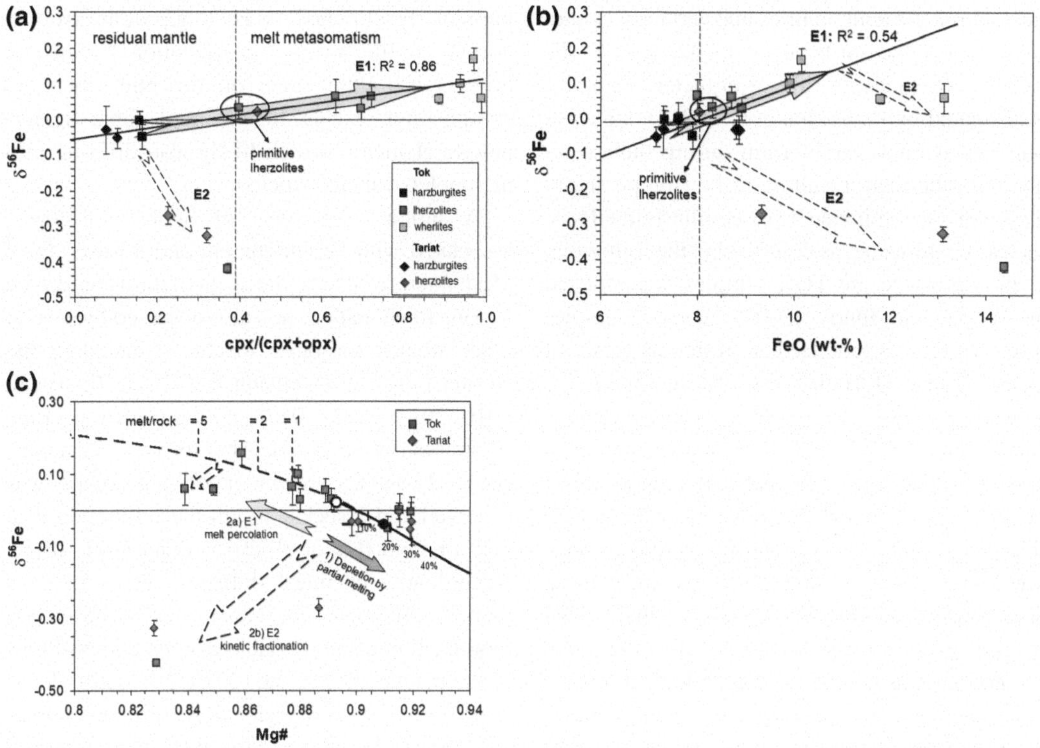

Fig. 4.10 Fe isotope composition of peridotites from TOK (Siberia) and Tariat (Mongolia) **a** relative to cpx/[cpx + opx], **b** relative to the FeO content and **c** relative to Mg#, indicating two different types of metasomatism: E1 = reaction with a melt and E2 = kinetic reaction with a melt or fluid (see text for more details). In (**c**) depletion of the peridotites was modeled as briefly described in Sect. 4.3.3; also modeled was the reaction of the peridotites with an Fe-rich melt at different melt/rock ratios (adapted from Weyer and Ionov 2007, their Figs. 2 and 4)

samples are characterized by a positive correlation of δ^{56}Fe values with cpx/(cpx + opx), CaO, FeO and several incompatible trace elements, and a negative correlation of δ^{56}Fe values with Mg# (Fig. 4.10a and b; Weyer and Ionov 2007). These geochemical signatures could be well explained and modelled by the reaction of a depleted mantle with an Fe-rich melt at relatively high melt-rock ratios (Fig. 4.10c; Weyer and Ionov 2007).

Similar reactions with a melt that affected the Fe isotope composition of depleted mantle rocks have been described for peridotites from the lithospheric mantle of eastern China (Huang et al. 2011; Zhao et al. 2010, 2012, 2015) and from the Kerguelen Plateau (Poitrasson et al. 2013). Depending on the mantle source that generated the melts, and the physio-chemical conditions of partial melting, the chemical and isotopic composition of the melt may have been quite variable, resulting in variable mineral reaction and also in a variable metasomatic overprinting of the initial Fe isotope composition of the mantle rocks. For example, Poitrasson et al. (2013) observed a shift towards significantly lighter Fe isotope compositions for those peridotites that reacted with a melt. In their case, the melt reaction resulted in the formation of secondary olivine, and the generation of dunites. Poitrasson et al. (2013) interpreted the different trends of their mantle rocks, as compared to that observed by Weyer and Ionov (2007), to be the result of the different products of the mineral-melt reactions. However, according to recent findings (Dauphas et al. 2014; Roskosz et al. 2015), equilibrium isotope fractionation between olivine, cpx and opx should be rather small at mantle temperatures and certainly cannot generate isotope shifts of up to 0.6‰ in δ^{56}Fe values. Alternatively, the Kerguelen samples, investigated by Poitrasson et al. (2013), have not achieved isotopic equilibrium during their reaction with the melt, resulting in Fe isotope signatures that are dominated by disequilibrium effects, such as Fe diffusion, as described in the following.

In contrast to the TOK samples, investigated by Weyer and Ionov (2007), several samples from Tariat, experienced a different type of metasomatism (E2) that resulted in low δ^{56}Fe values of ~ -0.3 to -0.4‰, and a negative correlation of δ^{56}Fe with FeO, a positive correlation of δ^{56}Fe with Mg#, and only a minor effect on the bulk chemical composition, except FeO (Fig. 4.10a and b; Weyer and Ionov 2007). Such low δ^{56}Fe values are difficult to explain with an equilibrium reaction with a melt, as this would require an extremely low δ^{56}Fe for the melt. Furthermore, as neither the mineralogical nor the bulk chemical compositions of such samples have changed significantly, this indicates a limited total mass flux between the metasomatic agent and the mantle. Such a small and selective mass flux is more likely to occur by chemical diffusion of Fe from a melt or fluid into the mantle at timescales that are too short to achieve chemical and isotopic equilibration (Poitrasson et al. 2013; Weyer and Ionov 2007). It has been shown in several studies, based on experimental approaches or natural observations, that chemical Fe–Mg exchange diffusion results in significant isotope effects of up to >1‰ in δ^{56}Fe values (Collinet et al. 2017; Oeser et al. 2015, 2018; Richter et al. 2003, 2009; Sio et al. 2013, 2018; Teng et al. 2011; see also Sect. 4.3.4.2). As light isotopes diffuse faster than heavy isotopes, this generally results in an enrichment of light isotopes in the direction of diffusion. Iron diffusion into the depleted mantle rocks of TOK and Tariat was likely driven by a gradient in the chemical potential between a melt (or fluid), that was highly enriched in Fe, and the mantle rocks. Notably, as depleted mantle rocks have generally high Mg#, they are depleted in Fe relative to most melts, even if assuming the melt is not particularly enriched in Fe (e.g., stems from a normal fertile mantle compositions). Such a process has little effect on the bulk chemical composition, except for Fe (and some incompatible elements, which likely also diffused into the depleted mantle rocks).

A similar enrichment of light Fe isotopes, associated with Fe enrichment and a lowering of Mg#, interpreted to be the result of diffusion-driven isotope fractionation, was also observed by several other studies for other localities, including the North China Craton (Huang et al. 2011; Zhao et al. 2010, 2012, 2015, 2017) and the Cameroon Line (Poitrasson et al. 2013). Also for other elements, such as Li and Ca, enrichment of light isotopes was observed and associated with the infiltration of a melt or fluid (Lai et al. 2015; Zhao et al. 2017). Accordingly, chemical diffusion seems to be a second important mechanism of metasomatic modification of metal stable isotope compositions of the mantle. Zhao et al. (2010, 2015) also determined the oxygen fugacity (fO_2) of their samples and observed relatively high fO_2 for the samples with the lowest δ^{56}Fe values, indicating that the metasomatic agent was relatively oxidizing. Furthermore, Su et al. (2015) and Zhao et al. (2012) measured the Mg isotope composition, along with the Fe isotope composition of their samples (Fig. 4.11; Su et al. 2015). They observed a strong negative correlation of δ^{56}Fe and δ^{26}Mg values, where samples with light Fe isotope enrichment had heavy Mg isotope compositions, which is indicative of an inter-diffusion origin of the observed isotopic variations. As (ferrous) Fe and Mg commonly share positions in the lattice of olivine and pyroxene, diffusion of both elements is commonly coupled, i.e. Fe diffusion into these minerals goes along with Mg diffusion outwards. Accordingly, diffusive Fe enrichment by Fe–Mg exchange results in low δ^{56}Fe values coupled with high δ^{26}Mg values in the affected mantle minerals. The slopes of δ^{56}Fe versus δ^{26}Mg, observed by Su et al. (2015) and Zhao et al. (2012), are crudely consistent with such a process.

Oeser et al. (2015), Sio et al. (2013), and Teng et al. (2011) observed, despite the higher relative mass difference for Mg, as compared to Fe isotopes, similar diffusion-driven isotope fractionation for δ^{56}Fe versus δ^{26}Mg variations in olivine. However, as mantle olivine (and pyroxene), including that analyzed by Zhao et al. (2012), commonly has high Mg# (of ≥ 90), diffusion-driven isotope fractionation for Fe is much higher than that for Mg, for mass balance reasons. Zhao et al. (2017) measured Ca isotopes along with Fe isotopes and observed a positive correlation of δ^{56}Fe and δ^{44}Ca values. Although, the interaction of Fe and Ca between melt and mantle minerals is not directly coupled, such as in the case of Fe–Mg exchange diffusion, this indicates that the light Ca isotope compositions are likely also the result of Ca diffusion from a Ca-rich melt into mantle minerals, such as for cpx, as successfully modeled by Zhao et al. (2017).

The common observation of diffusion-driven Fe isotope fractionation during mantle metasomatism raises the question as to which extent diffusion is involved in generating the isotope signature observed during partial mantle melting. As the melt generated typically occurs at depths on the order of 50 km (at MOR) or >100 km in subduction zones (Lee et al. 2009), it ascends through the depleted upper mantle (due to its lower density) through melt channels, a process that may take several years (Turner et al. 2001). During migration, both the mantle and the percolating melt are commonly affected by diffusive exchange and may re-equilibrate at shallower levels (Niu 1997). Depleted mantle rocks have high Mg#, reflecting depletion Fe and enrichment in Mg, relative to melt. Melts that have been generated by partial melting of a primitive or fertile mantle are in disequilibrium with the commonly depleted rocks in melt channels, which must result in Fe–Mg exchange diffusion. Assuming that melt ascent occurs for years or potentially much longer in the same melt channels (i.e., during multiple events), time is sufficient for a sizable Fe–Mg mass flux by diffusion, which would also affect the bulk Fe (and Mg-) isotope composition of the mantle minerals, including olivine, the dominant Fe–Mg host in the mantle.

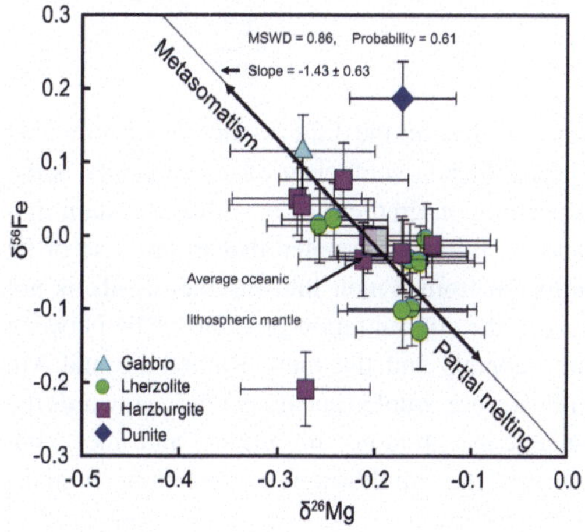

Fig. 4.11 δ^{56}Fe versus δ^{26}Mg of mantle rocks from the Purang ophiolite (Tibetian Plateau, China—adapted from Su et al. 2015, their Fig. 7)

The extent to which mantle minerals (we focus in the following on olivine for simplicity) are affected by diffusion depends on a variety of parameters. These include (1) the total interaction time of olivine with a melt (which may be the sum of several melt percolation events), (2) the concentration (or more precisely the chemical potential) gradient of the diffusing element between olivine and melt, (3) the olivine grain size, (4) the temperature, (5) the water content in the melt (e.g., in subduction zones), and (6) the oxygen fugacity (e.g., Chakraborty 1997; Dohmen and Chakraborty 2007). In most investigated mantle rocks, with the exception of those affected by E2 metasomatism described above, light Fe isotope compositions are commonly observed in samples with high fO_2 and high degrees of depletion (Poitrasson et al. 2013; Su et al. 2015; Weyer and Ionov 2007; Williams et al. 2004, 2005; Williams and Bizimis 2014; Zhao et al. 2010, 2012, 2015, 2017). These findings are in accord with a (at least partially) diffusive origin for the light Fe isotope signatures. According to Fick's first law, the diffusive flux is linearly correlated to the concentration gradient between to media:

$$J^{\alpha} = -D\frac{\partial C^{\alpha}}{\partial x}$$

where J^{α} denotes the diffusive flux of an atom α, D is the diffusion coefficient of α and dC/dx is the concentration gradient of α within a certain distance x. It has to be noted that in the case of Fe diffusion from a melt into olivine, dC/dx is not simply the concentration gradient of Fe between the minerals and the melt. Rather, Fe and Mg diffusion are coupled and do not move towards the same concentrations in olivine and melt, but towards the equilibrium composition. Furthermore, diffusion in the melt is fast compared to diffusion in olivine. Thus, the Fe–Mg exchange diffusion rate between olivine and melt will be essentially defined by the Fe–Mg exchange diffusion rate in olivine (Dohmen and Chakraborty 2007; Teng et al. 2011; Sio et al. 2013; Oeser et al. 2015), although, the melt can become locally depleted in Fe, in particular at low melt-rock ratios (Watson and Müller 2009). Accordingly, the more depleted the mantle rocks are (high Mg#), the stronger it is likely that disequilibrium between olivine and a percolating melt will exist, resulting in Fe diffusion from the melt into olivine, associated with the enrichment of light Fe isotopes in olivine. At the same time, depleted mantle rocks with high Mg# are enriched in Mg, resulting in Mg diffusion from olivine to the melt and the depletion of light Mg isotopes in olivine. Thus, Fe–Mg exchange diffusion may well explain the negative correlation of δ^{56}Fe and δ^{26}Mg values that was observed in olivine (Huang et al. 2011; Zhao et al. 2012) or at the bulk scale in depleted mantle rocks (Fig. 4.11; Su et al. 2015; Zhao et al. 2012). Because of the lower Fe concentration in olivine, as compared to that of Mg, a relatively larger diffusion-driven flux is associated with Fe, and therefore the effect of diffusion on Fe isotopes is somewhat larger than that on Mg isotopes.

In theory, Fe–Mg exchange diffusion between melts and depleted mantle rocks during melt ascent may also result in a stronger Fe isotope fractionation during mantle melting. Multiple melt extraction from the mantle wedge in subduction zones, which was previously enriched by metasomatic fluids from the subducted slap, may result in multiple diffusion events and significant enrichment of light Fe isotopes in the mantle residue. If such a mantle wedge is the source of subsequent arc production, this may contribute to the low δ^{56}Fe values that are commonly observed in IAB basalts (Dauphas et al. 2009; Foden et al. 2018; Nebel et al. 2013, 2015; Williams et al. 2018). Such a model was recently proposed by Foden et al. (2018), who observed that basalts from arc settings with a high assumed water flux display, on average, lower δ^{56}Fe values than those with lower fluid flux. This finding was interpreted to mirror the isotopic composition of the mantle source, which was assumed to be more depleted in heavy Fe isotopes in arc settings with high water flux, as the result of a more effective diffusion-driven pre-depletion history by multiple melting events.

Collectively, these findings imply that diffusive mass flux during mantle melting may represent an important process, which then may also modify the Fe and Mg isotope compositions of the melts themselves, assuming that melt ascent is slow enough for significant diffusion-driven Fe–Mg fluxes and fast enough to preserve isotopic disequilibrium. A similar process, such as diffusion-driven fractionation, may result from incomplete melt-rock equilibration during melting of the lower crust, as was recently proposed to account for the low Nb/Ta ratios frequently observed for the upper continental crust (Marschall et al. 2013). According to recent modeling results of Fe–Mg exchange between olivine and melt at temperatures of ~1300 °C, a time frame of only several months up to several decades would be required (Collinet et al. 2017; Oeser et al. 2015, 2018; Sio and Dauphas 2017; Sio et al. 2013, 2018). Accordingly, there may be a need to consider Fe–Mg exchange diffusion during

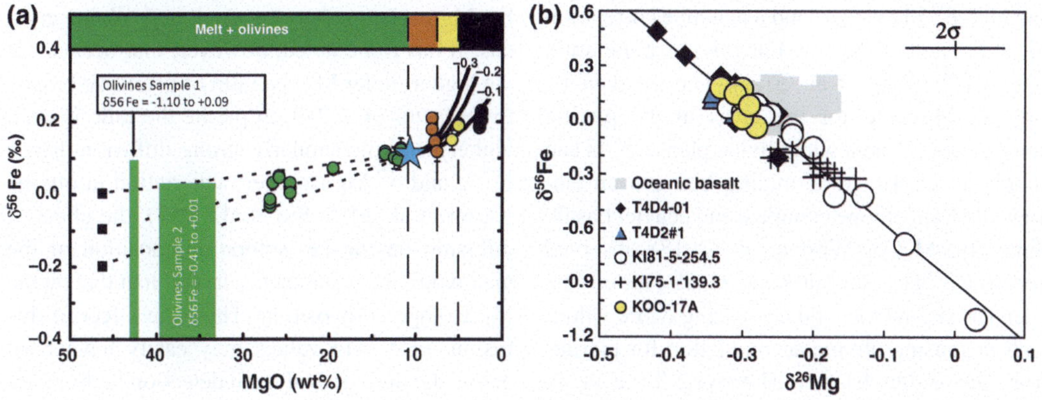

Fig. 4.12 **a** (left) Variation of δ^{56}Fe as a function of MgO from basalts from the Kilauea Iki lava lake (Hawaii). Samples with >11% MgO (green dots) represent mixtures of melt + olivine phenocrysts modeled with dashed lines. The blue star represents the most Mg-rich melt found for these samples. Samples with MgO < 11% experienced fractional crystallization of olivine (orange dots), plagioclase + augite (yellow dots) and oxides (black dots). Solid lines represent a Rayleigh fractionation model with ε values (isotope fractionation factors) during fractional crystallization. The green rectangles represent the ranges of measured δ^{56}Fe and estimated MgO for two drill core samples (KI81-5-254.5 and KI75-1-139.3, respectively). **b** (right) correlated Fe and Mg isotope compositions of olivine fragments of the same samples analyzed in Teng et al. (2008) (modified from Teng et al. 2008, their Fig. 4 and from Teng et al. (2011, their Fig. 1)

mantle melting when discussing isotope fractionation during silicate differentiation on Earth and other planets.

4.3.5 Differentiation of Melts

4.3.5.1 Fractional Crystallization and Fluid Exsolution

A variety of studies investigated the Fe isotope systematic of magmatic rocks, addressing isotopic fractionation during magmatic evolution by processes such as fractional crystallization or the exsolution of a fluid from highly evolved melts during late magmatic stages (Dauphas et al. 2014; Foden et al. 2015, 2018; Heimann et al. 2008; Konter et al. 2016; McCoy-West et al. 2018; Nebel et al. 2013, 2015, 2018; Poitrasson and Freydier 2005; Schoenberg et al. 2009; Schuessler et al. 2009; Sossi et al. 2012; Telus et al. 2012; Teng et al. 2008, 2011, 2013; Zambardi et al. 2014; Peters et al. 2019). While initial studies revealed limited variability and no systematic Fe isotope variations among basalts (Beard et al. 2003a; Poitrasson et al. 2004; Schoenberg and von Blanckenburg 2006; Weyer and Ionov 2007), those focusing on highly evolved rocks (with SiO_2 > 65–70 wt%; Heimann et al. 2008; Poitrasson and Freydier 2005; Schoenberg and von Blanckenburg 2006; Schuessler et al. 2009) observed significant variations and increasing δ^{56}Fe values with increasing differentiation (see Fig. 4.5 in Sect. 4.2.3).

Teng et al. (2008) also observed increasing δ^{56}Fe values with increasing differentiation for samples from a lava Lake of the Kilauea volcano (Hawaii), which cover a range from picritic compositions (with MgO up to 28 wt% and abundant olivine) towards differentiated composition (with <3 wt% MgO and olivine-free) (Fig. 4.12a, Teng et al. 2008). They also analyzed individual olivine phenocrysts, for which they observed highly variable but generally negative δ^{56}Fe values of −1.1 to 0‰. As shown in a subsequent study by Teng et al. (2011), the negative δ^{56}Fe values of the magmatic olivine was produced by Fe–Mg exchange diffusion, driven by chemical disequilibrium between early olivine crystals and the progressively evolving melt. During this process, isotopically light Fe diffuses into, and isotopically heavy Mg out of, the olivine crystals, resulting in olivine with low δ^{56}Fe values and high $\delta^{26}Mg$ values, and a negative correlation of $\delta^{56}Fe$ and $\delta^{26}Mg$ in the olivine population (Fig. 4.12b, Teng et al. 2011). The trend in Fe isotopes observed for the lavas in the picritic compositional range with olivine phenocrysts can simply be explained as mixing between an isotopically light olivine cumulate and heavier basalt. Similarly, McCoy-West et al. (2018) observed very low $\delta^{56}Fe$ values down to −0.8‰ for olivine phenocrysts and low and decreasing $\delta^{56}Fe$ values with increasing olivine accumulation for picrites from the Baffin Islands. However, Teng et al. (2008) also observed a negative correlation of $\delta^{56}Fe$ values and MgO for more evolved lavas, reaching $\delta^{56}Fe > 0.2$‰ for the most differentiated samples. This trend was modeled by Teng et al. (2008) by a Rayleigh-type fractionation with isotope fractionation factors between crystal and melt on the order of $\varepsilon \sim -0.3$ to −0.1‰. (Dauphas et al. 2014; Roskosz et al. 2015).

Looking in more detail at the data of Teng et al. (2008), however, there is some scatter towards high $\delta^{56}Fe$ values, in particular during fractional crystallization of olivine, which can barely be explained by equilibrium fractionation alone. Assuming α values of ≥ 0.1‰ also exceeds equilibrium fractionation factors, as estimated from NIRXS analyses for moderately differentiated melts at magmatic temperatures (Dauphas et al. 2014; Roskosz et al. 2015). Rather, the large, negatively coupled Mg and Fe isotopic variation in olivine clearly points towards kinetic isotope fractionation induced by Fe–Mg exchange diffusion between the melt and olivine (Teng et al. 2008, 2011). In theory, such a process should also cause Mg isotope fractionation in the melt. However, Teng et al. (2007) found no resolvable variation in $\delta^{26}Mg$ values among the Kilauea lavas at a bulk rock scale over a range of MgO contents, indicating limited Mg isotope fractionation during the differentiation of the melt. If Fe isotope fractionation of the lavas was diffusion-driven, this should also affect Mg isotope compositions, which should in that case become heavier for more differentiated samples which experienced significant diffusion-driven Fe–Mg exchange between melt and olivine crystals. It has to be noted, however, that because of the higher molar Mg–Fe ratio of primitive basalts (by a factor of ~2–4, in picrite the ratio is even higher), and the similarly strong diffusion-driven $\delta^{56}Fe$ and $\delta^{26}Mg$ isotope fractionation in olivine (Oeser et al. 2015; Sio et al. 2013), the effect of diffusion on the Fe isotope composition of the melt would be significantly larger than that on the Mg isotope composition. Thus, the effect of diffusion on $\delta^{26}Mg$ values may easily have been below the limit of analytical detection in the study of Teng et al. (2007).

In contrast to Teng et al. (2008), Schuessler et al. (2009) observed no Fe (and no Li) isotope fractionation for primitive basaltic to andesitic rocks from the Hekla volcano (Iceland). From this finding, they concluded that Fe isotope fractionation between melt and olivine is too small to generate any significant effect in the evolving melt. As diffusion-driven Fe and Li isotope fractionation in magmatic olivine are frequently coupled (Weyer and Seitz 2012), the absence of Li and Fe isotope fractionation in the basalts may be interpreted as insignificant diffusional Fe and Li exchange between olivine and melt. However, as Li is incompatible in olivine (with equilibrium partitioning Li_{ol}/Li_{melt} of ~1:5; Brenan et al. 1998), the effect of Li diffusion on isotopic variation in the melt may be small for mass balance reasons. For example, Tomascak et al. (1999) did not observe any Li isotope variation in the Kilauea Iki samples, which have been analyzed by Teng et al. (2007, 2008) for Mg and Fe isotopes. Also, Weyer and Seitz (2012) did not observe any effects at the bulk rock scale of their investigated basalts, despite large Li isotope variations in olivine. Still, a diffusion-driven effect on the Fe isotope composition of the Kilauea Iki lavas may be larger than that for the Hekla samples. Alternatively, other factors may play a role, such as fractional crystallization under closed- versus open-system conditions. Such a scenario was proposed by Sossi et al. (2012), who observed similar Fe isotope fractionation during magmatic evolution of basalts from Red Hill (Tasmania;

Australia) to that observed for Kilauea Iki lavas by Teng et al. (2008), which they explained by closed system magma evolution resulting in increasing $Fe^{3+}/\Sigma Fe$ in the melt during crystallization and potentially increasing mineral-melt Fe isotope fractionation factors. In contrast, lavas from the Hekla volcano may have evolved under more open system conditions, which resulted in a limited increase of Fe^{3+} and no detectable isotope fractionation.

For oceanic basalts, most studies assume an effect of isotope fractionation during fractional crystallization on the observed $\delta^{56}Fe$ values, in particular for OIB (Konter et al. 2016; McCoy-West et al. 2018; Teng et al. 2013; Peters et al. 2019). Olivine is commonly the dominant Fe carrier during early fractional crystallization, however, crystallization of Fe oxides, such as spinel, may also occur, and in some cases have a significant effect on $\delta^{56}Fe$ values, as equilibrium isotope fractionation between spinel and melt is assumed to be larger than that between olivine and melt. McCoy-West et al. (2018) proposed early crystallization of spinel in basalts from Baffin Island (with MgO < 21 wt%) by a strongly decreasing Cr# and a weakly positive correlation of $\delta^{56}Fe$ values and MgO, implying decreasing $\delta^{56}Fe$ values with increasing differentiation. They modeled this observation with $\Delta^{56}Fe_{olivine\text{-}melt}$ of −0.048‰ and $\Delta^{56}Fe_{spinel\text{-}melt}$ of 0.22‰ (Dauphas et al. 2014; Rosksoz et al. 2015), resulting in a total $\Delta^{56}Fe_{mineral\text{-}melt}$ of ~0.05‰ during fractional crystallization. Late fractional crystallization is typically accompanied with pyroxene fractionation, which also may have an effect on Fe isotopes, as may be indicated by a negative correlation of $\delta^{56}Fe$ values and CaO (Teng et al. 2013).

During fractional crystallization, melts frequently reach sulfide saturation, which also may have an effect on Fe isotope compositions (McCoy-West et al. 2018; Williams et al. 2018; Peters et al. 2019). Peters et al. (2019) observed a correlation for $\delta^{56}Fe$ and Pd/Ir, similar to that seen for $\delta^{56}Fe$ and Mg#, which they attributed to the crystallization of isotopically light sulfides. Williams et al. (2018) identified a crystallization sequence, combining major-, trace-element (e.g., V) and isotope signatures for basaltic and andesitic samples from the Mariana arc. According to their findings, crystallization started with olivine + pyroxene, resulting in increasing $\delta^{56}Fe$ values, followed by magnetite, resulting in a sharp drop of $\delta^{56}Fe$ values. Finally, sulfide oversaturation is indicated by a sharp decrease in Cu contents, which resulted in sulfide fractionation along with magnetite fractionation, resulting in a broadly constant $\delta^{56}Fe$ value during further melt evolution.

Isotope fractionation during fractional crystallization may explain some of the variability in Fe isotope compositions observed between MORB and OIB (see Fig. 4.4). However, this effect can barely exceed 0.1‰ in $\delta^{56}Fe$ values and is usually much smaller in most cases. Accordingly, the very heavy Fe isotope compositions observed for samples from the Samoa Islands and Koolau are difficult to explain by crystallization effects alone, and more likely indicate heterogeneities in the mantle source. Such mantle heterogeneities may have been generated during mantle metasomatism (e.g., Weyer and Ionov 2007; Zhao et al. 2010; Huang et al. 2011; see above) and/or reflect petrologic heterogeneities (e.g., melting of isotopically heavy pyroxene; Teng et al. 2013; Williams and Bizimis 2014). Fractional crystallization effects combined with a different (in this case more depleted) mantle source, could also explain the lower $\delta^{56}Fe$ values that are typically observed for IAB, as compared to MORB. Flux-driven multiple and extensive mantle melting events, potentially coupled with diffusion effects, typically result in an isotopically light mantle source for most IAB (Dauphas et al. 2009; Foden et al. 2018; Nebel et al. 2013, 2015; Williams et al. 2018). Subsequently, Fe isotope fractionation during early to intermediate differentiation of the melt by fractional crystallization may be suppressed in arcs, as compared to MOR settings, due to open system behavior and/or crystallization of magnetite in parallel to olivine (Foden et al. 2018; Nebel et al. 2015; Williams et al. 2018).

Highly differentiated igneous rocks display much more variable and on average heavier Fe isotope compositions than basalts, as noted above. There are essentially two major mechanisms discussed in order to explain the high $\delta^{56}Fe$ values observed for many highly evolved rocks, (1) fluid exsolution during magma differentiation (Poitrasson and Freydier 2005; Heimann et al. 2008; Telus et al. 2012), and (2) fractional crystallization (Dauphas et al. 2014; Du et al. 2017; Foden et al. 2015; He et al. 2017; Schoenberg and von Blanckenburg 2006; Schoenberg et al. 2009; Schuessler et al. 2007, 2009; Sossi et al. 2012; Teng et al. 2008; Wu et al. 2017; Xia et al. 2017). However, other effects, such as partial melting within the lower crust (Telus et al. 2012; Foden et al. 2015; He et al. 2017; Xia et al. 2017; Xu et al. 2017), or diffusion-driven processes between immiscible melts and/or minerals (Sossi et al. 2012; Zhu et al. 2015), as well as thermal diffusion (Telus et al. 2012; Zambardi et al. 2014), may also play a role. A remarkable feature of highly differentiated rocks that has to be explained by any of these mechanisms is the steep increase of $\delta^{56}Fe$ values up to ~ 0.6‰ for some highly evolved rocks with $SiO_2 > 75$ wt% (or MgO < 0.2 wt% or $FeO_{total} < 2$ wt%). This seems difficult to explain by fractional crystallization alone if assuming roughly constant fractionation factors between minerals and melt.

Heimann et al. (2008) observed a line of evidence indicating that late-stage magmatic fluid exsolution may play an important role in Fe isotope fractionation, including a positive correlation of $\delta^{56}Fe$ values with Rb concentrations and a negative correlation with Zr/Hf ratios, eruption temperatures, and Cl/F ratios in fluid inclusions. Low Cl/F indicate high volatile loss because of the preferential partitioning of Cl into vapor relative to F. As previously shown, Fe^{2+} has a high solubility in Cl-rich fluids (as $FeCl_4^{2}$-), particularly at low temperatures (e.g., Audétat and Pettke 2003). Such complexed Fe^{2+} should favor light Fe isotopes, in particular relative to magnetite, but also as compared to other Fe-bearing minerals (Polyakov et al. 2007; Schauble et al. 2001). Accordingly, the negative correlation of $\delta^{56}Fe$ values with temperature and Cl/F ratios in the fluid may indicate Fe isotope fractionation during effective Fe removal by fluid exsolution. Heimann et al. (2008) could successfully model Fe isotope fractionation on the order of what they observed for the investigated granitic rocks during release of 10% fluid at ~500 °C and a high magnetite biotite ratio (Fig. 4.13; Heimann et al. 2008). In some studies, Zn/Fe ratios and/or Zn isotopes were also measured along with Fe isotopes on some highly differentiated rocks (Telus et al. 2012; Xia et al. 2017). For some pegmatites, Telus et al. (2012) observed a moderate correlation of $\delta^{56}Fe$ values with $\delta^{66}Zn$ values and the dispersion of Zn/Fe. As Zn is assumed to be fluid mobile and Zn isotopes fractionate during mobilization, the authors assumed fluid mobilization to be one of the mechanisms generating high $\delta^{56}Fe$ values in these rocks.

Schuessler et al. (2009) also analyzed some highly differentiated volcanic rocks, including dacites and rhyolites from Iceland. While they did not observe significant Fe isotope fractionation from basaltic to andesitic magma evolution, the highly evolved rocks showed a clear increase in $\delta^{56}Fe$ values with increasing SiO_2 (and decreasing FeO_{total}). They furthermore measured Li isotopes, which are known to be highly fluid mobile and fractionate during fluid-rock interaction (Bouman et al. 2004; Li et al. 2018b). The absence of any resolvable Li isotope fractionation was interpreted by Schuessler et al. (2009) as evidence for a negligible role of fluids in the evolution of these rocks. In a parallel study (Schuessler 2008) revealed, using an experimental approach, that magnetite favors light isotopes in equilibrium with a rhyolitic melt that contains significant Fe^{3+} (with $\Delta^{56}Fe_{magnetite\text{-}melt}$ of ~−0.12 to −0.26‰). Using these experimental findings, Schuessler et al. (2009) could confirm the observed Fe isotope fractionation of the investigated Iceland rhyolites using a simple Rayleigh fractionation model with crystallization of titano-magnetite ($Fe^{2+}{}_{1+x}Fe^{3+}{}_{2-2x}Ti_xO_4$), which may expected to have a lower mineral-melt Fe isotope fractionation factor than magnetite, according to its lower content of Fe^{3+} (Sossi et al. 2012). Fractional crystallization with a very similar mineral-melt Fe isotope fractionation factor was also proposed to have generated high $\delta^{56}Fe$ values observed for two other high-SiO_2 rhyolitic rock suites from the Neoproterozoic volcanic-sedimentary basins in southern China and the Triassic Tu Le Basin in northern

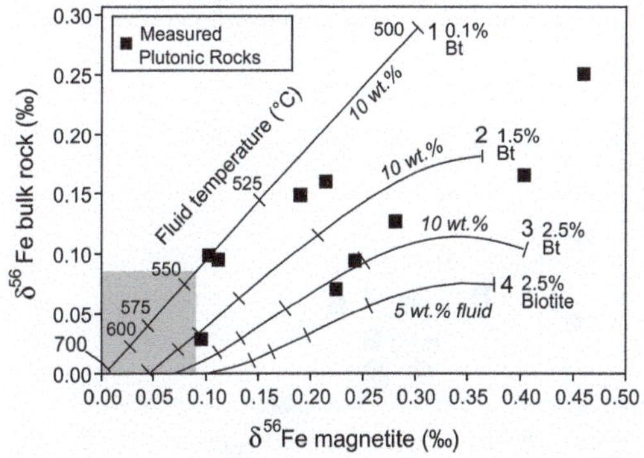

Fig. 4.13 Modeled $\delta^{56}Fe$ for magnetite and its bulk rock that result from the down-temperature interaction of a crystallizing granitoid with 10 wt% (curves 1–3) and 5 wt % (curve 4) exsolved fluid phase. Curves represent different magnetite and biotite (mgt, bt) modal abundances for a rock with 1.3% Fe content at 500 °C as follows: Tick marks represent temperature (°C) from 700 to 500 °C. Shown as squares are measured bulk-rock and magnetite compositions for granitoids from Questa (New Mexico, USA). The shaded field represents the mean $\delta^{56}Fe$ and 2 SD of crustal mafic- to intermediate-composition igneous rocks from Beard et al. (2003a) (adapted from Heimann et al. 2008, their Fig. 13)

Vietnam, respectively (Du et al. 2017). Similar to Schuessler et al. (2009), the authors also proposed that crystallization of titanomagnetite and ilmenite resulted in the removal of light Fe isotopes from the melt.

The finding that magnetite or at least titanomagnetite may have a negative mineral-melt Fe isotope fractionation factor was surprising, as magnetite is commonly enriched in heavy isotopes compared to most silicate minerals (e.g., Polyakov and Mineev 2000; Shahar et al. 2008). However, rhyolitic melts typically display enhanced abundances of alkaline elements (Na, K), resulting in enhanced stabilization of Fe^{3+} in the melt (Knipping et al. 2015). This may result in more negative mineral-melt Fe isotope fractionation factors, which is indicated by the correlation of $\delta^{56}Fe$ values and (Na + K)/Al or (Na + K)/(Ca + Mg), as frequently observed for granites (Foden et al. 2015; He et al. 2017; Sossi et al. 2012), supporting a melt compositional control on Fe isotope fractionation. Furthermore, Dauphas et al. (2014) measured the mean force constant of synthesized glasses of a range of basaltic to rhyolitic compositions with NIRXS, and observed that $Fe^{3+}/\Sigma Fe$ and structural conditions of the melt has a strong effect on equilibrium isotope fractionation during fractional crystallization. A particularly large change was observed in the coordination environment of Fe^{2+} in melts at about 75 wt% SiO_2, resulting in a large enhancement of the mean force constant between dacitic and rhyolitic melts. The authors used their findings to model the expected Fe isotope fractionation during fractional crystallization of an andesitic melt using the software "rhyolite MELTS", and could reproduce the steep increase of $\delta^{56}Fe$ values observed for highly differentiated rocks (Fig. 4.14a and b; Dauphas et al. 2014) (e.g.

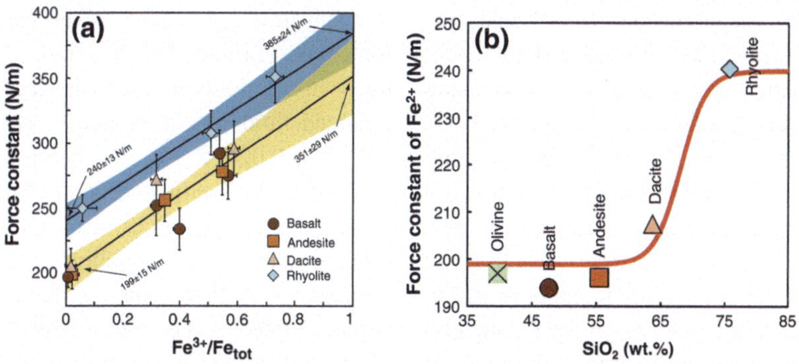

Fig. 4.14 **a** (left) Iron force constants in variably oxidized synthetic silicate glasses, including basalt, andesite, dacite, and rhyolite as a function of Fe^{3+}/Fe_{tot}.. note that all glasses can be extrapolated with one and the same linear array, while rhyolite plots on a separate array. **b** (right) Force constants of Fe^{2+} in olivine and the glasses from (**a**) relative to their SiO_2. Note the steep increase between dacite and rhyolite (adapted from Dauphas et al. 2014, their Figs. 5 and 9A)

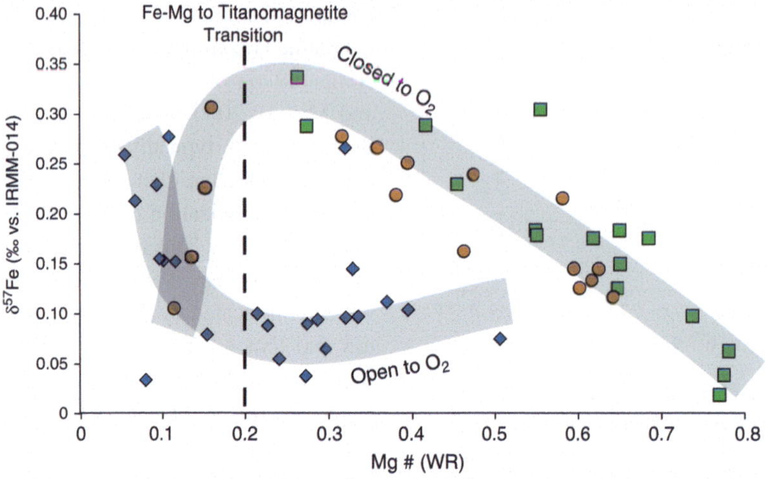

Fig. 4.15 δ^{57}Fe of co-genetic whole-rock samples from different tholeiitic suites: green squares = Kilauea Iki (Teng et al. 2008), blue diamonds = Hekla (Schuessler et al. 2009), orange circles = Red Hill (Sossi et al. 2012) plotted relative to their Mg#, characterizing two different differentiation trends, which are interpreted by Sossi et al. (2012) to depend on whether the system evolved open (Hekla) or closed (Red Hill, Kilauea Iki; modified from Sossi et al. 2012; their Fig. 7)

Heimann et al. 2008; Poitrasson and Freydier 2005; Schoenberg and von Blanckenburg 2006; Schuessler et al. 2009; Sossi et al. 2012; Telus et al. 2012 and see also compilation in Fig. 4.5).

In a field study of magmatic rocks from the Red Hill intrusion (Tasmania, Australia), Sossi et al. (2012) observed that the changes in δ^{56}Fe values with evolution during fractional crystallization seems to be coupled with the fO_2 evolution of the melt. They propose that the investigated magmatic suite evolved at initially relatively reducing conditions ($\sim$FMQ-1, reflecting one log unit lower than the fayalite-magnetite quartz buffered system) and then evolved as a closed system. These conditions resulted in progressively increasing δ^{56}Fe values during magma evolution until fO_2 and Fe^{3+} saturation in the melt became high enough for the crystallization of magnetite (at Mg# of $\sim$0.2), indicating significantly positive $\Delta^{56}Fe_{magnetite\text{-}melt}$ during magnetite crystallization (Fig. 4.15; Sossi et al. 2012). The authors speculated that the tholeiitic magmas at the Kilauea Iki lava lake (Teng et al. 2008) may have evolved under similar conditions, while those from Hekla (Iceland, Schuessler et al. 2009) evolved under more oxidizing, open-system conditions, resulting in early magnetite crystallization and delayed increase of δ^{56}Fe values.

According to this scenario, at Hekla, the change of the Fe bonding in the magnetite (which became more Ti-rich during melt evolution) and in the melt (which became more silicic and likely also more Fe^{3+}-rich at high SiO_2) may have resulted in an inversion of $\Delta^{56}Fe_{magnetite\text{-}melt}$ from initially positive to negative values, resulting in the steep increase of δ^{56}Fe values at high SiO_2. More oxidizing open-system conditions and early magnetite crystallization is also suggested to result in lower average δ^{56}Fe values observed for I-type granites (0.118 $\pm$ 0.023 2SE), which form in subduction zones in the presence of oxidized fluids, as compared to A-type granites (0.237 $\pm$ 0.024 2SE), which typically form in intracontinental settings (Foden et al. 2015; Sossi et al. 2012; mean values also include data from other studies; see compilation of Sect. 4.2.3 for references). S-type granites have on average slightly higher δ^{56}Fe values, which may indicate open system behavior during their formation (Foden et al. 2015). Notably, however, A-, S-, and I-type granites define essentially indistinguishable trends of δ^{56}Fe values versus SiO_2, and the higher average $\delta^{56}Fe$ values of A-type granites are mostly the result of their higher degree of differentiation (~73 wt% SiO_2 for A-type granites, as compared to 67 wt% for I-type granites).

In the context of the above discussion, it is remarkable that the increase in $\delta^{56}Fe$ values observed for the magmatic rocks (Kilauea Iki Lavas and Red Hill) investigated by Sossi et al. (2012) and Teng et al. (2008) occurs at relatively low SiO_2 contents. Although variables such as fO_2 and open versus closed system have a large effect on the Fe isotope evolution of the melt during fractional crystallization (see also modeling in Foden et al. 2015), high $\delta^{56}Fe$ values at low SiO_2 are difficult to explain. Assuming that fractional crystallization of olivine and pyroxene is mainly responsible for changes in $\delta^{56}Fe$ values with evolution of the melts, equilibrium Fe isotope fractionation factors between these minerals and a basaltic melt as proposed by Dauphas et al. (2014) and Roskosz et al. (2015) are too low to explain significant Fe isotope fractionation (>0.05‰) during early fractional crystallization. While Fe–Mg diffusion between olivine crystals and Kilauea Iki lava (Teng et al. 2011) likely also affects the Fe isotope composition of the melt to some extent, this is unlikely to play a role for granitic intrusions, although some Red Hills rocks display disequilibrium between pyroxene and magnetite (Sossi et al. 2012), indicating kinetic processes.

4.3.5.2 Isotope Diffusion Effects and Chronometry of Magmatic Systems

The mass difference between the isotopes of an element also results in isotope fractionation during diffusion, which is related to the different diffusivity of the isotopes. In general, light isotopes have higher diffusivities. The resulting isotope fractionation was described by Richter et al. (1999) with the form:

$$\frac{D_q}{D_r} = \left(\frac{M_r}{M_q}\right)^{\beta}$$

where *D* represents the diffusion coefficient and *M* the atomic mass of two isotopes *q* and *r* of an element. The exponent β is an empirical constant which depends on the diffusion medium, however, in solids and melts it is generally smaller than 0.5 (the velocity ratio of two isotopes in an ideal gas). In contrast to equilibrium isotope fractionation, diffusion-driven isotope fractionation is essentially temperature independent, although the diffusion rate is highly temperature dependent and a change in the diffusion mechanism (or structural changes) with temperature, may also result in a change of β (Richter et al. 2009b; Sio et al. 2018; Van Orman and Krawczynski 2015). We have already discussed the potential effect of chemical Fe–Mg exchange diffusion between the mantle and melts during their ascent to the surface on the Fe isotope composition of olivine and other mantle minerals, or even the composition of the melt itself (in Sect. 4.3.4). Similarly, chemical exchange diffusion may also take place during fractional crystallization of a melt in a magma chamber, such as Fe–Mg exchange diffusion between olivine crystals and the melt (Dauphas et al. 2010; Teng et al. 2011). As the melt evolves, chemical disequilibrium will develop between the melt and early-formed olivine phenocrysts or mantle-derived xenocrysts that have been brought from the mantle during ascent. Furthermore, a chemical gradient may also result in diffusion between the melt and the host rocks of the magma chamber or between two immiscible melts, such as a primitive basaltic mantle melt and a more differentiated (e.g., rhyolitic) melt that has already been differentiated earlier in the magma chamber. In this case, the large chemical gradient will result in diffusion of elements from the enriched-melt (e.g., basalt in Mg, Ca, Fe or pegmatite in Li) into the depleted melt/rock, which results in the enrichment of light isotopes in the latter (Chopra et al. 2012; Richter et al. 2003, 2009a; Teng et al. 2006; Wu et al. 2018). Beside chemical diffusion, thermal ("*Soret*") diffusion may act, as a result of a temperature gradient, such as that between a hot magma and the cooler wall-rocks of the magma chamber. Such a diffusion scenario will result in the enrichment of light isotopes at the hot end of the temperature gradient (e.g. Huang et al. 2011; Richter et al. 2009b; Zambardi et al. 2014).

In the following we will focus on Fe isotope fractionation during Fe–Mg exchange diffusion between olivine and melt during magma differentiation, which has recently been developed as an isotopic diffusion chronometer for magmatic evolution. As the melt becomes depleted in Mg and Ni during differentiation (crystallization of olivine), the resulting disequilibrium between melt and olivine provokes diffusion of Mg and Ni from olivine into the melt and of Fe (and other incompatible elements) from the melt into olivine. This commonly results in olivine grains that are chemically zoned in these elements and the Mg#. As diffusion coefficients for Fe–Mg exchange diffusion in olivine are relatively well understood (Dohmen and Chakraborty 2007), chemical zoning in these elements has already been established as a chronometer for magmatic evolution (e.g. Costa and Chakraborty 2004; Costa and Dungan 2005; Kahl et al. 2011). In a similar way, this method has also been used for other cations in olivine, (e.g., Kahl et al. 2011; Ruprecht and Plank 2013) or other magmatic minerals (e.g., Ca in plagioclase: Druitt et al. 2012 or Li in cpx and plagioclase: Coogan et al. 2005; Neukampf et al. 2019). Because of the low mass and high relative mass difference of the isotopes of the element Li, chemical diffusion results in particularly large isotope effects (Richter et al. 2003). Accordingly, initial approaches of potentially using isotopes for the estimation of magmatic time scales focused on Li isotopes (e.g., Beck et al. 2006; Jeffcoate et al. 2007; Gallagher and Elliott 2009).

The relatively large Fe isotope variations observed among magmatic olivines, and their covariation with other cations, such as Mg and Li, which could be modeled as a diffusion-driven process (Dauphas et al. 2010; Teng et al. 2008, 2011; Weyer and Seitz 2012), motivated researchers also to look for Fe (and Mg) isotopic zoning at the mineral scale. Improved analytical techniques using SIMS and LA-MC-ICP-MS (see Chap. 2), as well as micro-drilling of large olivine crystals, have shown correlated chemical (Mg#) and Fe–Mg isotopic zoning in olivine (Oeser et al. 2014, 2015; Sio et al. 2013). Sio et al. (2013) performed Fe and Mg isotope analyses with micro-drilling (and subsequent solution analyses after chemical purification), and compared results obtained using SIMS and LA-MC-ICP-MS, allowing comparison of Fe isotopic zoning at different spatial resolutions. They analyzed olivine grains from the Kilauea Iki lava lake, which previously have been investigated on a bulk gain scale. The advantage of this sample suite is that the cooling history of the lava lake is well characterized, which provides control on the modeled cooling rates. Oeser et al. (2014, 2015) developed a method applying femtosecond LA-MC-ICP-MS (Horn et al. 2006), enabling them to perform in situ Fe and Mg isotope analyses with a precision of $\sim$0.1‰ at a spatial resolution of 30–40 μm.

A significant finding of the in situ isotopic investigations of minerals is that a negatively correlated Mg–Fe isotopic zoning provides unambiguous evidence for Fe–Mg exchange diffusion, which can in cases be difficult to predict from chemical zoning alone. The latter also could have been generated during crystal growth in a melt that changed its chemical composition during the growth of the crystal. Such a scenario, however, would have only small effects on the isotopic composition of the melt (at least not for basaltic melt; see Sect. 4.3.4.1), where Fe isotope zoning of the crystal would be small (<0.1‰). In contrast, diffusion-driven Fe-isotope zoning can be larger by an order of magnitude (Oeser et al. 2015; Sio et al. 2013; Teng et al. 2011). Accordingly, using combined chemical and isotopic zoning is a very powerful approach of estimating magmatic timescales by diffusion chronometry. While isotopes serve for identification of the diffusion process, the typically better spatial resolution of EMPA Fe–Mg concentration profiles is better suited for a more precise modeling of the diffusion interval. Combined Fe–Mg isotope signatures can also serve to distinguish Fe–Mg exchange diffusion from *Soret* diffusion (Dauphas et al. 2010) or condensation-evaporation processes (Weyrauch et al. 2019), as only the first process results in anti-correlated Fe–Mg isotope variations, while the latter processes all result in correlated Fe–Mg signatures.

Combined Fe–Mg isotope signatures can also be used to unravel more complex aspects of magmatic evolution. As pointed out by Oeser et al. (2015), several different scenarios could be imagined just for the case of chemical and isotopic exchange between an olivine phenocrysts (or other magmatic Fe–Mg minerals) and an evolving melt, including (1) Fe–Mg inter-diffusion between an existing olivine and melt, (2) rapid diffusion-limited crystal growth, resulting in a boundary layer around the growing crystal, (3) Fe–Mg inter-diffusion between olivine zones from two distinct growth periods, (4) olivine growth and simultaneous Fe–Mg inter-diffusion, and (5) olivine dissolution and simultaneous Fe–Mg inter-diffusion (Fig. 4.16; Oeser et al. 2015).

In scenario (1), it is assumed that the olivine neither grows nor dissolves during diffusion. Strictly speaking this may be unrealistic, as the melt will change its chemical composition during magma differentiation, resulting in the chemical disequilibrium that triggers diffusion. Thus, chemical evolution of the melt during olivine growth is not unlikely, in theory, varying from compositions at which olivine grows to compositions of other mineral phases that form at the solidus, at which olivine may even dissolve. Investigations of a series of zoned olivine crystals from different basaltic rocks from different localities have shown that the zoning is often in line with the assumption that it was essentially generated solely by Fe–Mg inter-diffusion (Oeser et al. 2015, 2018; Sio and Dauphas 2017; Sio et al. 2013). As the Mg# of the melt commonly decreases during differentiation, Fe commonly diffuses into and Mg out of olivine. If the diffusion event is long enough, this process results in the enrichment of light Fe and heavy Mg isotopes in the olivine core (Fig. 4.17; Teng et al. 2011).

Commonly, however, the diffusion period is not long enough for light Fe (and heavy Mg) isotopes to reach the olivine core. In this case, olivine shows an unzoned Mg# plateau in its core and decreasing Mg# at the rims, associated with a symmetric valley of δ^{56}Fe values with the most negative values within the zoned rim (δ^{26}Mg are simply reversed to δ^{56}Fe). This kind of zoning has been described by Oeser et al. (2015) as "normal zoning" (Fig. 4.18; Oeser et al. 2015). It seems likely that other processes, such olivine growth or dissolution, are negligible or slow enough that they do not significantly alter the diffusion zoning.

The effect of scenario (2) on Fe and Mg isotope ratios of zoned olivine was not yet observed in nature (not even for komatiitic olivine; Dauphas et al. 2010), although modeled by Watson and Müller (2009). Very fast olivine growth would result in a depletion of compatible elements, such as Mg, and enrichment of incompatible elements in the melt adjacent to the olivine crystal (boundary layer). As Fe has an olivine-melt partition coefficient in basalts of ~ 1, the Fe concentration in the boundary layer would barely change (Sossi and O'Neill 2016). This would result in a chemical zoning of the olivine that looks similar to that of olivine-melt Fe–Mg exchange diffusion, with decreasing Mg# at the olivine rim. However, the Mg-depleted boundary layer would trigger Mg diffusion from the infinite melt reservoir into the boundary layer, resulting in light isotope compositions for compatible elements, which then would be incorporated into the fast-growing olivine (Watson and Müller 2009). This would result in very negative δ^{26}Mg values in olivine. In contrast, δ^{56}Fe values would show essentially no isotopic zoning. Thus, the negative δ^{26}Mg values and decoupled Fe–Mg isotopic zoning would be characteristic for such a process (Oeser et al. 2015).

In scenario (3), two growth stages of olivine with different Mg# are assumed, resulting in Fe–Mg exchange diffusion between the two zones. Such a scenario was used by Oeser et al. (2018), who applied their findings to estimate the rates and pathways of magma ascent of Irazu, currently most active volcano in Costa Rica. In agreement with a previous study by Ruprecht and Plank (2013), based on Ni zoning, they were able to construct a scenario in which olivine xenocrysts from a lower crustal cumulate were picked up by an ascending melt and transported to a magma chamber in the shallow crust. Differentiation of the melt resulted in additional olivine growth, either as separate olivine grains or rims around the existing xenolithic olivine. Subsequently, a second diffusion event

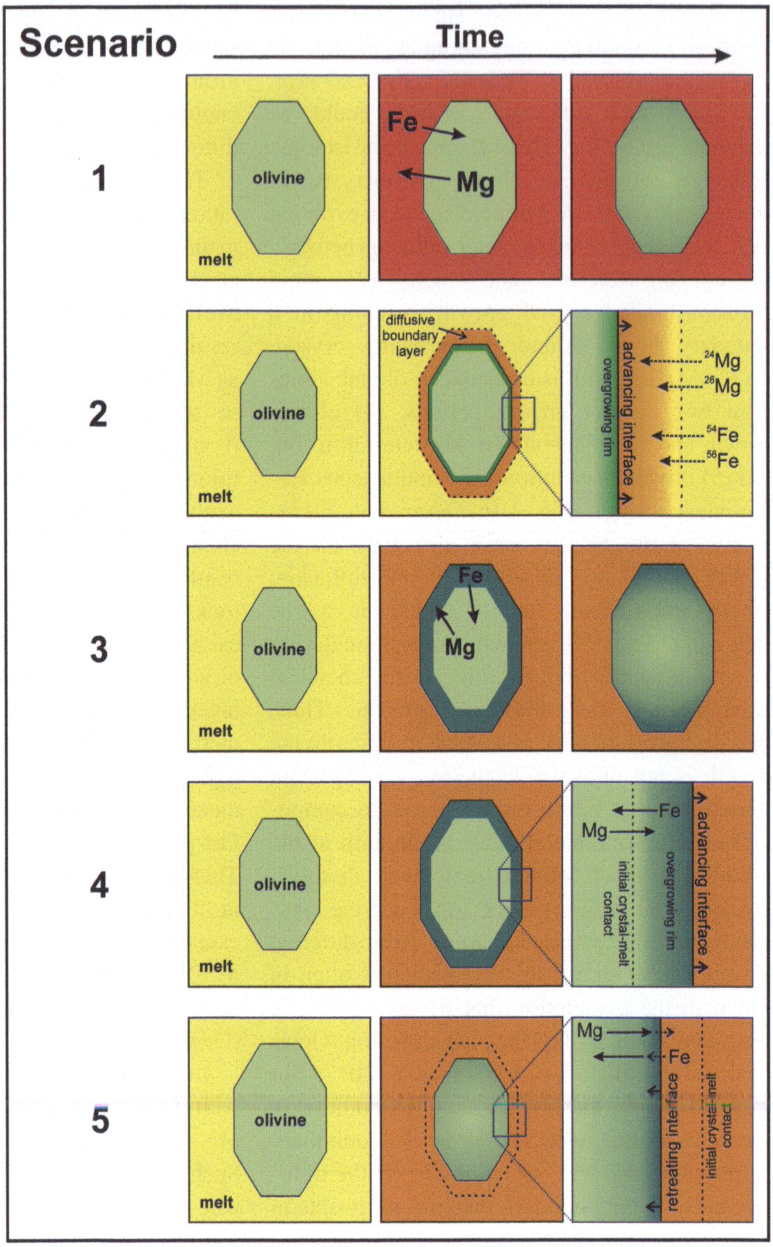

Fig. 4.16 Illustration of five different scenarios explored by Oeser et al. (2015) in order to theoretically investigate the combined effects of Fe–Mg inter-diffusion, crystal growth and dissolution on chemical and isotopic zoning of olivines. See text for further explanation. Adapted from Oeser et al. (2015), their Fig. 1; cartoon for scenario 1 is modified after Teng et al. (2011)

occurred in the shallow magma chamber and/or during magma ascent to the surface between the olivine rim and the surrounding melt. The latter was likely in disequilibrium with the olivine, because it mixed in the shallow magma chamber with a more differentiated melt and/or may also have been further differentiated during magma ascent to the surface. The different diffusion steps and their effects on the chemical (forsterite content) and Fe isotope composition are shown in Fig. 4.19 (Oeser et al. 2018).

The first diffusion event between the two olivine growth layers could be modeled assuming Fe–Mg inter-diffusion for 6 months (at a temperature of 1150 °C). While this model well described the observed chemical and isotopic

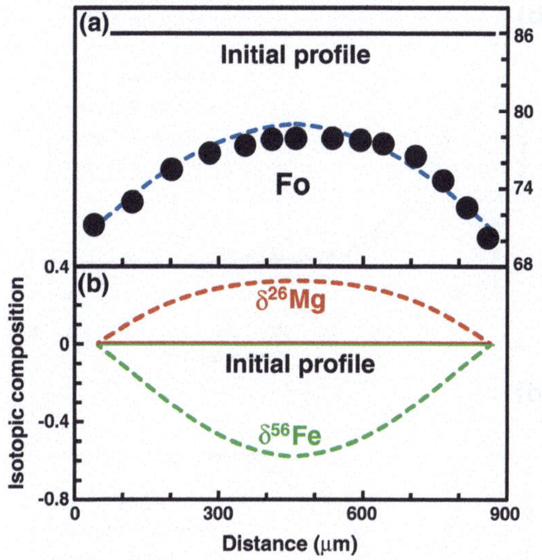

Fig. 4.17 **a** (upper panel) Measured Fo variation along the traverse in an olivine phenocryst from the Kilauea Iki lava lake (Hawaii) and theoretically reproduced diffusion curve (dotted line) by Mg–Fe inter-diffusion between olivines and melts. **b** (lower panel) Predicated Mg and Fe isotope fractionation along the traverse induced by the inter-diffusion. The solid lines are initial conditions and dotted curves are calculated spherical diffusion profiles (adapted from Teng et al. 2011, their Fig. 3)

zoning of the olivine core, it failed to fit the observed rim compositions (Fig. 4.19a and c). Rather, a second diffusion event between the olivine rim and a surrounding melt is necessary in order to reproduce the low Fo-content and δ^{56}Fe values measured for the olivine rim (Fig. 4.19b and d). In their model, Oeser et al. (2018) estimated a time scale of around 2.5 months for this latter event. Accordingly, these findings imply that magma ascent does not take longer than a year to pass from the deep crust to the surface, which indicates that supply of new magma from the deep crust through the plumbing system is possible during the active phase of the Irazu volcano.

Similar to scenario (1), scenario (3) is unlikely to occur sensu stricto, assuming two growth stages and diffusion only between the growth stages. Collinet et al. (2017) observed Fe–Mg isotopic zoning related to diffusion between zones of two different growth stages in magmatic olivines from Mars (i.e., from basaltic shergottites). However, the observed chemical and isotopic zoning could not be well reproduced with a model assuming separated olivine growth and diffusion stages. Rather, a much better fit with the observed chemical (including Fe–Mg, but also trace elements) and Fe–Mg isotopic zoning was achieved if assuming simultaneous olivine growth and Fe–Mg inter-diffusion (scenario 4 from above and Oeser et al. 2015). In addition, following Oeser et al. (2018), diffusion between the two olivine growth zones may have occurred already simultaneously to the growth of the olivine rim (as indicated by some trace element zoning patterns). In this case, their modelled duration of diffusion is slightly overestimated and the inferred magma ascent rates represent minimum estimates. Modeling of such a scenario is complex, as it includes a moving boundary between olivine and melt, and the respective zoning not only depends on the diffusion rate, but also on crystal growth rates (similarly for simultaneous olivine dissolution on the dissolution rates). If growth rates are high, relative to diffusion rates, this may result in asymptotic isotopic zoning at the crystal rim (i.e., if diffusion is limited to the rim region), with a maximum followed by a minimum in δ^{56}Fe values (from rim to core; Oeser et al. 2015). Sio and Dauphas (2017) successfully used a random search (or Monte Carlo inversion) approach to model a scenario of simultaneous diffusion and crystal growth for the zoned olivine crystals of the Kilauea Iki lava lake, for which diffusion acted long enough to affect the olivine cores (Sio et al. 2013).

Collinet et al. (2017) applied a similar model to explain data for two olivine megacrysts (700–800 μm in diameter) in shergottites from Mars. These olivines display surprisingly low δ^{56}Fe values of −0.6 to −1‰, combined with high δ^{26}Mg values of 0–0.2‰ (relative to the standard DSM-3), or ~1‰ lower and 0.4‰ higher, respectively, than the isotopic compositions reported for chondrites and bulk shergottites (see Sects. 4.1.1 and 4.1.2 for Fe, and Magna et al. 2017 and Teng et al. 2010 for Mg isotopes). Such fractionated Fe and Mg isotope compositions could be best reproduced by a model of slow crystal growth with

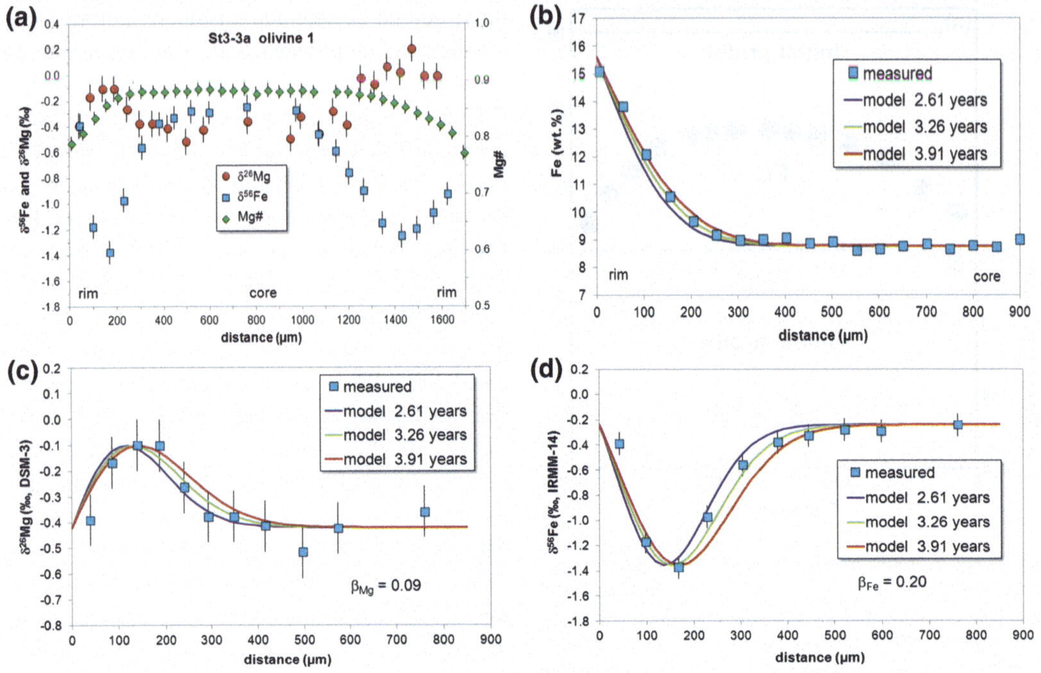

Fig. 4.18 **a** Fe–Mg chemical and isotopic profiles across a normally zoned olivine phenocryst (St3–3a olivine (1) from the Massif Central determined by fs-LA-MC-ICP-MS. **b–d** Simplified modeling of Fe and Fe and Mg isotope profiles during Fe–Mg exchange diffusion for the olivine shown in A for three different time spans. The diffusion coefficient D for Fe–Mg diffusion in olivine was calculated according to the parameterization of Dohmen and Chakraborty (2007). Model parameters: T = 1250 °C (constant; corresponding to 0.97 • T_{peak} and T_{peak} = 1570 K = 1297 °C); X_{Fe} = 0.165 (constant); fO_2 at NNO-2; P = 50 MPa (modified from Oeser et al. 2015, their Figs. 3a and 5 a, c and d)

simultaneous Fe–Mg exchange diffusion over several years between crystals and a melt, most likely to occur during slow cooling in a deeper magma chamber. This stage was followed by a second olivine growth stage, which was potentially initiated by additional magma supply, as well as formation of groundmass olivine. At the rims, the olivine megacrysts show a short (~100 μm) and steep anti-correlated Fe and Mg isotopic zoning pattern with essentially chondritic (or bulk shergottite) δ^{56}Fe and δ^{26}Mg values at the rim edge. This zoning can be best reproduced with a model assuming a second, much shorter (~100 days) diffusion event, with Fe–Mg exchange diffusion between the crystal rims and a surrounding melt. This latter diffusion event likely occurred during transport of the magma to its shallow site of emplacement (Collinet et al. 2017). According to these findings, magma ascent rates on Mars seem to be comparable to those on Earth (Rutherford 2008). These latter case studies show that combining information from chemical and Fe–Mg isotopic zoning may help to unravel complex crystallization and diffusion histories and contribute to a better understanding and timing of magma evolution on Earth and other planets.

In order to quantify diffusion-derived time scales, all parameters affecting diffusion rates in olivine must be defined, including temperature, pressure, and fO_2, as well as the chemical composition of the olivine (e.g. Chakraborty 2010). Diffusion in the melt is commonly considered to be infinitely fast (assuming moderate crystal growth rates; Watson and Müller 2009 and see scenario 2 above). Temperature has a large effect on diffusion in olivine and uncertainties in the determination of magmatic temperatures may bear the biggest source of uncertainty, where, for

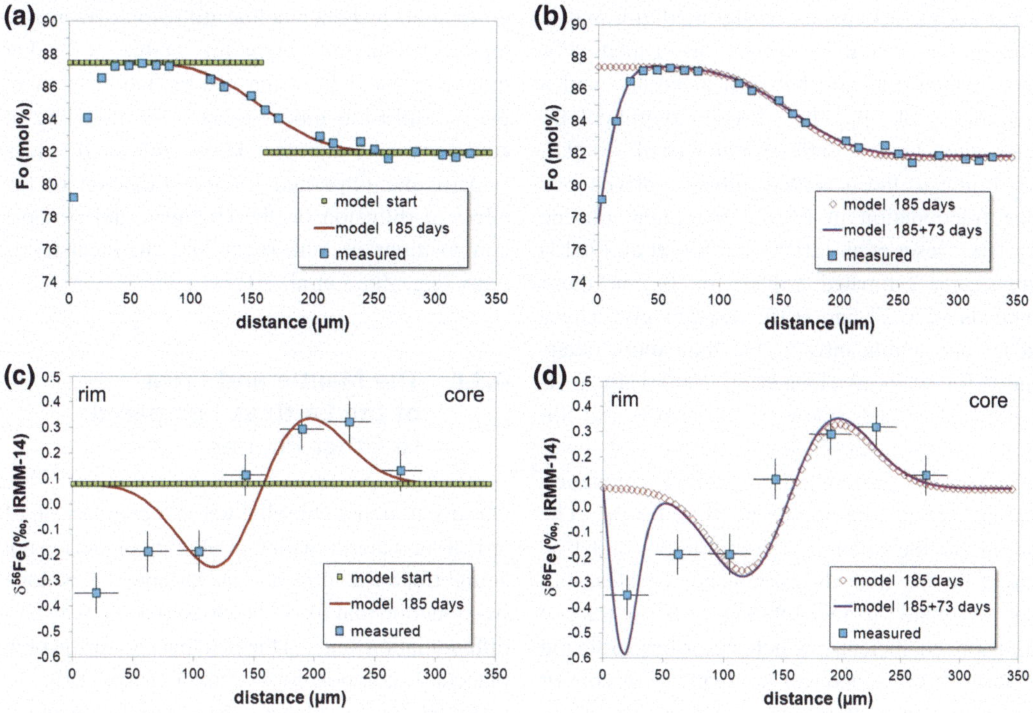

Fig. 4.19 Results of a stepwise diffusion modeling approach on the Fo content and the Fe and Mg isotope composition of a complexly zoned olivine (IZ-10–13 ol34). In **a** and **c** (left panels), only one diffusion episode between a second more Mg-rich outer olivine layer (distance 0–156 μm, assuming instantaneous growth) and the olivine core is assumed. In **b** and **d** (right panels), a second diffusion episode with diffusion of Fe into (and Mg out of) the olivine from a melt is assumed, which much better reproduce the low Fo content and relative enrichment of light Fe isotopes near the rim (modified from Oeser et al. 2018, their Fig. 9)

example, an uncertainty of 50 °C would result in an uncertainty of 2–3 in the estimated diffusion rate (e.g., Dohmen and Chakraborty 2007; Oeser et al. 2018). Commonly, diffusion in a cooling system is modeled, requiring the definition of a starting and a closure temperature (where diffusion becomes negligible). Pressure has only a minor effect on diffusion in the absence of volatiles. However, fO_2 is another important factor, where, for example, an increase in fO_2 by about one log unit results in a ~1.5 times faster diffusion in olivine (e.g., Dohmen and Chakraborty 2007; Chakraborty 2010). For diffusion in olivine, it has been shown by Dohmen and Chakraborty (2007) that the crystallographic orientation has a large effect on the diffusion rate; diffusion along the c axis, for example, is about a factor of six faster than diffusion along the a and b axes. Accordingly, as the chemical and isotopic olivine compositions are usually determined on two-dimensional sections, the crystallographic orientations of samples must be determined using methods such as elastic backscatter diffraction (EBSD, a method occasionally attached to SEM or electron microprobes), in order to precisely model diffusion rates. According to these uncertainties in the boundary conditions of diffusion, and also in the experimentally determined diffusion coefficients, modeled diffusion timescales commonly have uncertainties of half or even one order of magnitude.

Isotope fractionation during diffusion is commonly defined by the exponent β of Eq. 4.2 from above. It is important to note that this is a different β than that defining the reduced isotope partition that describes isotopic fractionation, but is here used to describe the distinct chemical bonding of an element relative to an ideal gas. As stated above, it depends on the medium, and is different between different minerals or melts. For Fe–Mg inter-diffusion between a basaltic and a rhyolitic melt, β_{Fe} has been experimentally determined to be $\sim$0.03 (Richter et al. 2009a). According to the observed diffusion-driven isotope fractionation in natural magmatic olivine crystals, Oeser et al. (2015) and Sio et al. (2013) empirically estimated values for β_{Fe} in these minerals of 0.27 and 0.16, respectively. These values are significantly higher than those determined for melts (Richter et al. 2009a). Interestingly, both studies, Oeser et al. (2015) and Sio et al. (2013), determined a factor of $\sim$2 lower values for β_{Mg}, defining Mg isotope fractionation during Mg diffusion in olivine, than for β_{Fe}. The reason for the different values for β_{Fe} and β_{Mg} during Fe–Mg inter-diffusion is not yet clear, but may be related to the different self- (or tracer-) diffusion coefficients, which describes diffusion in absence of a chemical gradient, in olivine of the same composition (e.g., Chakraborty 2010, and references therein). For the scenario of pure Fe–Mg exchange diffusion between an existing olivine and melt (or two different olivine growth zones), the precise knowledge of β_{Fe} and β_{Mg} is not essential for the estimation of diffusion timescales, which are determined from the width of chemical and isotopic zoning. Oeser et al. (2015) and Sio et al. (2013) simply fitted the isotope fractionation of the modeled diffusion profiles to match the naturally observed isotope fractionation to determine values for β_{Fe} and β_{Mg}. For the complex scenarios of simultaneous crystal growth and diffusion, precise knowledge of β values becomes more important (Collinet et al. 2017; Oeser et al. 2018; Sio and Dauphas 2017). Sio et al. (2018) performed an experimental study and determined β_{Fe} for different olivine compositions (with Fo > 83.4). For olivine with Fo < 88.8, they observed β_{Fe} equal to 0.16 for all crystal orientations, similar to the value observed by Oeser et al. (2015). However, for high Fo (>94), they observed an anisotropic behavior for β_{Fe} and a sharp decrease of β_{Fe} along the fast-diffusing c-axis (to a value as low as 0.03, similar to that previously observed for melt; Richter et al. 2009a). Certainly more work is required to better define diffusion-driven isotope fractionation, including studies of other minerals. Apart from the determination of more precise diffusion times scales from modeling of zoned crystals, a precise knowledge of β values is particular important for quantification of the effect of diffusion on the chemical and isotopic compositions of bulk rocks and chemical reservoirs (e.g. Zhao et al. 2017).

4.3.6 The Mantle and Crust of the Earth as Compared to Other Planets

Our discussion so highlighted that comparison of terrestrial material with that of other planets is not straightforward, due to (1) variable Fe isotope fractionation that likely occurred during planetary differentiation, and (2) the fact that chondritic bulk planetary compositions cannot necessarily be assumed for all planets. No mass-independent fractionation (MIF) of Fe isotopes between terrestrial or bulk meteorite samples have yet been observed. However, accretion-related mass-dependent variations between bulk planetary bodies may not be excluded, as might exist, for example, for the angrite parent body or even for the Earth (Poitrasson et al. 2004; Sossi et al. 2016b). We conclude from the literature and this review that Fe isotope variations observed among reservoirs on Earth can be explained by isotope fractionation during the differentiation of our planet and the formation of these reservoirs. Based on this working hypothesis, the above compilations of Earth and planetary reservoirs, and our discussion of mechanisms of Fe isotope fractionation, we will further consider the origin of Fe isotope variations among other planets.

As shown in several experimental and theoretical studies, there is the potential to generate Fe isotope fractionation during core formation. Furthermore, the extent of this isotope fractionation may significantly vary between different planetary bodies, considering the vastly different conditions assumed for their core formation and resulting different chemical and mineralogical compositions of their silicate and metal parts

(Elardo and Shahar 2017; Polyakov 2009; Shahar et al. 2015, 2016; Williams et al. 2012). At the high-temperature conditions expected to have prevailed during core formation on Earth, Fe isotope fractionation is likely small, i.e. <0.01 − 0.04‰ for $\Delta^{56}Fe_{silicate\text{-}metal}$ (Liu et al. 2017; Polyakov 2009). Notably, all studies, based on high-pressure experiments and analyses of the mean force constants with NRIXS propose a positive $\Delta^{56}Fe_{silicate\text{-}metal}$ (Liu et al. 2017; Polyakov 2009; Shahar et al. 2016). These results indicate that the silicate Earth may be marginally superchondritic in $\delta^{56}Fe$ values, although barely resolvable outside uncertainties.

With an average value of $\delta^{56}Fe = 0.00 \pm 0.02$‰, the depleted upper mantle is indeed identical to the chondritic value of $\delta^{56}Fe = -0.01 \pm 0.01$‰ (see compilation above). This still holds true if assuming that the Fe isotope composition of the mantle is better defined by non-metasomatized and relatively primitive mantle samples (defined, for example, by its Mg#), rather than using an average of all samples that may be biased by sampling. This may give a $\delta^{56}Fe$ value of 0.02 ± 0.02‰ (Craddock et al. 2013; Weyer and Ionov 2007) which is, however, still indistinguishable from the chondritic value. According to the findings of Yang et al. (2019), who investigated potential effects of partial melting in the deep mantle, we may conclude that the primitive upper mantle is representative of the bulk silicate Earth in terms of its Fe isotope composition. However, Williams et al. (2012) assume that disproportionation of ferrous iron into isotopically heavy ferric Fe + light metal, with subsequent re-equilibration of the metal with the core, may have generated a significantly superchondritic Fe isotope composition of the silicate Earth, assuming very high $Fe^{3+}/\Sigma Fe$ (McCammon et al. 1997). Accordingly, the Fe isotope composition of the silicate Earth might be slightly superchondritic as a result of core formation.

Core formation on smaller planetary bodies, such as Mars and Vesta, occurred at significantly lower temperatures (~1600 °C) and pressures (~6 GPa for Mars and <<1 GPa for Vesta) compared to those assumed for the Earth (Righter and Drake 1996). At such conditions, $\Delta^{56}Fe_{silicate\text{-}metal}$ was determined to be negative and its absolute value seems to increase with increasing quantities of light elements (such as O, Si and S), but also with increasing Ni in the metal core (Elardo and Shahar 2017; Elardo et al. 2019; Shahar et al. 2015, 2016, not confirmed by Liu et al. 2017). Notably, even if assuming significant (≥0.1‰) Fe isotope fractionation during core formation of smaller bodies, this would still barely explain the heavy Fe isotope composition frequently observed for iron meteorites based on mass balance. It may, however, influence the silicate portions of smaller planets, such as Mars. Indeed, a higher amount of S in the core is proposed for core formation under more oxidizing conditions, such as for Mars (Wänke and Dreibus 1988; Sanloup and Fei 2004). According to their findings, Shahar et al. (2016) and Elardo et al. (2019) propose that the mantles of Mars, Vesta, and even the Moon, should have negative $\delta^{56}Fe$ values. On the other hand, assuming isotopically light mantles for Mars and Vesta, additional Fe isotope fractionation during the formation of their crusts is required in order to balance Fe isotope fractionation during core formation and to achieve the observed chondritic Fe isotope composition for Martian and HED meteorites. Indeed, Sossi et al. (2016b) observed slightly elevated $\delta^{56}Fe$ values of ~0.03‰ for meteorites from the Martian crust with low olivine content, MgO, and high CaO, as compared to more primitive compositions. From these findings, they speculate that some Fe isotope fractionation during partial melting and magma evolution occurred, although likely less than commonly observed for oceanic basalts on Earth. The Martian mantle may have a slightly negative $\delta^{56}Fe$ value, although the limited number of analyzed lherzolitic shergottites have, within uncertainties, identical $\delta^{56}Fe$ values to basaltic shergottites. These processes are much better investigated on Earth, based on a much larger sample set.

The best studied, and most precisely known, high-temperature reservoir on Earth is the oceanic crust, represented essentially by MORB, which has a $\delta^{56}Fe$ value of 0.10 ± 0.01‰ (see compilation above; OIB also form parts of the oceanic crust, but with $\delta^{56}Fe$ = 0.11 ± 0.01‰, essentially identical to MORB). The Fe isotope composition of the continental crust is more difficult to define, as magmatic rocks from the continental crust show a significantly larger range of Fe isotope compositions. In particular, highly differentiated crust (with SiO_2 > 70 wt%) shows highly variable and on average much higher $\delta^{56}Fe$ values (see data compilation above). As discussed above, this is likely the result of fractional crystallization, fluid exsolution, partial melting, and kinetic processes (e.g. Dauphas et al. 2014; Foden et al. 2015; He et al. 2017; Heimann et al. 2008; Schuessler et al. 2009; Sossi et al. 2012; Telus et al. 2012; Xia et al. 2017; Zambardi et al. 2014). Such rocks represent a significant part of the upper crust, which has an average SiO_2 of ~67 wt% (e.g., Rudnick and Gao 2003). Using the simple relationship of $\delta^{56}Fe$ values and SiO_2, as shown in Fig. 4.5, the upper crust may have an average $\delta^{56}Fe$ value of ~0.11, which is, within uncertainties, indistinguishable from that of the oceanic crust (Foden et al. 2015; Teng et al. 2013).

Most modern continental crust is generated in subduction zones and island arc basalts and andesites have on average a lower $\delta^{56}Fe$ value than MORB (i.e., ~0.05‰ compared to ~0.1‰; Foden et al. 2018; Teng et al. 2013; see data compilation above). As the upper crust is often produced by partial melting of lower crust, elevated $\delta^{56}Fe$ values of the upper crust should be primarily a result of Fe isotope fractionation between the upper and lower crust. Accordingly, the upper and lower crust should display complementary Fe isotope compositions around the bulk crustal composition (Foden et al. 2015). However, because of the higher Fe content of the lower crust and overall high geochemical similarity of the lower crust with IAB (e.g., Hawkesworth and Kemp 2006), its Fe isotope composition may well be indistinguishable from that of island arcs. Large parts of mid- to early Archean continental crust was potentially not generated by modern-style subduction, but rather formed as thick plateaus above mantle plumes (e.g., Van Kranendonk 2010). Such crust may have a slightly different Fe isotope composition, which is, however, not well constrained. Dauphas et al. (2009) observed for the mafic lower crust from Isua (Greenland) an average $\delta^{56}Fe$ value of ~0.05‰, while four tonalities from the same locality yielded an average $\delta^{56}Fe$ value of ~0.09‰, which is very similar to modern style IAB and differentiated crustal rocks, respectively. According to these considerations, the bulk continental crust may have a slightly lower $\delta^{56}Fe$ values relative to the oceanic crust, more similar to that of island arcs (Foden et al. 2015).

The discussion above suggests there is a significant offset in $\delta^{56}Fe$ values between the upper mantle and the crust ($\Delta^{56}Fe_{crust\text{-}mantle}$ ~ 0.05 to 0.1‰), which, together with isotopic depletion trends in mantle peridotites (e.g., negatively correlated $\delta^{56}Fe$ and Mg#) and similar trends in magmatic rocks, indicate Fe isotope fractionation during partial melting and magma evolution (e.g., Dauphas et al. 2009; Teng et al. 2008; Weyer et al. 2005; Weyer and Ionov 2007; Williams et al. 2005). The question is what drives magmatic Fe isotope fractionation? Recent experimental findings, based on NIRXS synchrotron techniques, imply that equilibrium Fe isotope fractionation during partial melting may only explain about half of the Fe isotopic offset observed between average primitive mantle and MORB (i.e., ~0.02–0.04‰), depending on Fe^{3+} buffering in the mantle (Dauphas et al. 2014; Liu et al. 2017, see Fig. 4.9). Accordingly, kinetic processes very likely play a role to explain the Fe isotopic offset between mantle and crust on Earth. A likely process may be Fe–Mg exchange diffusion between a mantle melt and depleted mantle rocks (mostly with olivine, via interaction in melt channels), which should result in the enrichment of light Fe isotopes in the depleted mantle, with the most depleted mantle rocks having the lowest $\delta^{56}Fe$ values. Such a process would explain the steep slopes of $\delta^{56}Fe$ versus Mg# (or Cr#) commonly observed in depleted mantle rocks, which are not in line with equilibrium Fe isotope fractionation, as they would require unrealistically high

$\Delta^{56}Fe_{melt\text{-}mantle} \geq 0.3‰$ (Su et al. 2015; Weyer and Ionov 2007; Williams et al. 2005; Williams and Bizimis 2014). It would also be consistent with the negatively correlated $\delta^{26}Mg$ and $\delta^{56}Fe$ values that are commonly observed in such rocks (Huang et al. 2011; Su et al. 2015). Assuming the average interaction time of melt with olivine in depleted mantle rocks is long enough (e.g., a few weeks to several years, which is in the range of estimated magma ascent rates; Rutherford 2008), and providing a large interaction surface at a low melt/rock ratio ($\leq 10\%$), diffusive Fe–Mg exchange may also modify $\delta^{56}Fe$ and $\delta^{26}Mg$ values of the melt (for mass-balance reasons the effect on $\delta^{56}Fe$ is larger than that on $\delta^{26}Mg$). This process would than result in a heavier Fe isotope composition of the melt and may explain some of the offset and variability in the Fe isotope compositions observed for the Earth's mantle and crust.

Magmas may be further affected by Fe isotope fractionation during melt evolution, as a result of degassing and fractional crystallization. However, relying on recent experimental findings, equilibrium Fe isotope fractionation between melt and olivine or cpx should be insufficient to generate significant Fe isotope fractionation in basaltic to andesitic rocks (Dauphas et al. 2014). Nevertheless, correlations of $\delta^{56}Fe$ values and Mg# or MgO in evolving basaltic melts have been observed, indicating additional Fe isotope fractionation during melt evolution (e.g., Nebel et al. 2013, 2015; Teng et al. 2008). Nebel et al. (2013, 2015) proposed to correct for the effect of fractional crystallization assuming an Fe isotope fractionation ($\Delta^{56}Fe_{melt\text{-}mineral}$) $\sim 0.1‰$ (see also Sossi et al. 2012), which seems to be somewhat larger than that expected from experimental findings (Dauphas et al. 2014). On the other hand, the extremely fractionated and negatively correlated $\delta^{56}Fe$ and $\delta^{26}Mg$ values observed for olivine phenocrysts from basaltic rocks of the Kilauea Iki lava lake imply diffusive processes, which may also have slightly fractionated the $\delta^{56}Fe$ values of the basalts (Teng et al. 2011; see also McCoy-West et al. 2018 for similar findings). We may define an empirical isotope fractionation factor for fractional crystallization, although this would be expected to vary, as it is likely composed of two different mechanisms, including equilibrium- and diffusion-
driven isotope fractionation.

From other planetary bodies, we commonly have mostly crustal material available for analysis (e.g., from Mars, Moon, Vesta, the angrite parent body or some more differentiated feldspar-rich meteorites), which shows variable average $\delta^{56}Fe$ values between −0.08 and 0.12‰ (e.g., Poitrasson et al. 2004; Wang et al. 2012, 2014a, b; Weyer et al. 2005). Accordingly, terrestrial crustal material is on the heavy side of a range defined by the crustal material of other planets. Similar to Earth, Fe isotope fractionation during the formation of planetary crust may have occurred during partial melting and melt evolution, and for each of these processes, equilibrium and kinetic fractionation mechanisms may have been involved. For Mars, most crustal samples represent relatively primitive melts, which mainly experienced olivine crystallization. As the Mars mantle is expected to have no (or very little) Fe^{3+} in its source, insignificant ($\leq 0.01‰$; Dauphas et al. 2014) equilibrium Fe isotope fractionation may be expected. However, kinetic effects during partial melting and melt evolution cannot be excluded, as indicated, for example, by some very light Fe isotope compositions observed in basaltic shergottites, suggesting diffusion effects (Collinet et al. 2017). Accordingly, the Martian mantle may have either chondritic or slightly subchondritic $\delta^{56}Fe$ values, possibly as a result of core formation (Elardo and Shahar 2017). With similar arguments, bulk silicate Vesta may have a chondritic or slightly subchondritic Fe isotope composition. Angrites are more oxidized than basalts from Mars and Vesta, although still more reducing than terrestrial basalts, indicating that any equilibrium Fe isotope fractionation involved in their formation should be smaller than that during partial melting on Earth. Thus, the mantle of the angrite parent body may potentially have been slightly isotopically heavier than the Earth's mantle (i.e., superchondritic). According to experimental findings and modeling, this seems is not likely a result of core formation (Elardo and Shahar 2017; Liu et al. 2017). Potentially, the bulk angrite parent body had a superchondritic δ^{56}Fe value as a result of nebular condensation and accretion (Dauphas et al. 2015; Sossi et al. 2016b).

The lunar crust has overall similar Fe isotope compositions (δ^{56}Fe ~ 0.08 – 0.1‰ on average, see data compilation above) as crust on Earth, including low-Ti mare basalts (perhaps most similar to primitive basalts on Earth), highland rocks, and a variety of different lunar meteorites. Only a few lunar glass samples show slightly lower δ^{56}Fe values, and high-Ti mare basalts significantly higher δ^{56}Fe values (Liu et al. 2010; Moynier et al. 2006; Poitrasson et al. 2004; Weyer et al. 2005). As discussed above, there is a line of evidence that the latter are likely the result of partial melting of an ilmenite-rich mantle source and non-representative of the lunar crust. As the Moon is essentially Fe^{3+}-free, insignificant equilibrium Fe isotope fractionation can be expected during partial melting. Although some kinetic effects during melting are likely, as indicated by dunite samples with very low δ^{56}Fe values (Wang et al. 2015), the lunar mantle is unlikely to be lighter than the Earth's mantle and may have a superchondritic δ^{56}Fe values. Accordingly, accretion-related effects on the Fe isotope composition, either from early condensation in the nebular (Sossi et al. 2016b), or from the commonly accepted giant impact-origin of the Moon (Poitrasson et al. 2004), cannot be excluded. In contrast, an isotopically light lunar mantle, as proposed by Elardo and Shahar (2017), based on an experimental study of silicate-metal Fe isotope partitioning, is difficult to explain.

4.4 Summary

- Analyses of chondrites imply that the inner Solar System has a homogeneous stable Fe isotope composition of δ^{56}Fe = −0.01 ± 0.11‰ (2SD) relative to the commonly used reference materials (IRMM-014 or IRMM-643). Except for presolar grains and some components of calcium-aluminum-rich inclusions (CAIs), there is no evidence of mass-independent Fe isotope variations.
- Planetary crusts have variable Fe isotope compositions, mostly between δ^{56}Fe = – 0.1 and 0.2‰, that may partially arise from accretion-related non-chondritic bulk planetary Fe isotope compositions (e.g., for angrites). Most isotopic variations among planetary crusts, however, were likely generated during planetary differentiation, including core formation (and magma ocean crystallization), mantle melting, and crustal differentiation.
- The Fe isotope composition of the bulk Earth is likely essentially chondritic, assuming very minor Fe isotope fractionation at the very high temperatures (~3000–4000 K) of core formation on Earth, as indicated from experimental findings. This assumption is supported by most estimates for the average δ^{56}Fe value of the upper mantle of ~0.02‰, which is chondritic within uncertainties.
- Mantle rocks display variable Fe isotope compositions, ranging between δ^{56}Fe = −0.7 and 0.3‰, although >90% fall in a range between −0.14 and 0.11‰. These relatively large isotopic variations were likely generated by metasomatic processes, such as melt percolation and associated chemical diffusion effects.
- The oceanic crust has a δ^{56}Fe value of ~0.1 ± 0.1‰, isotopically heavier than the upper mantle by ~0.08‰, likely as a result of partial melting and fractional crystallization. This offset is slightly higher than that expected from experimental findings and may be partially driven by chemical diffusion-related processes.
- Ocean island basalts (OIB) display more variable Fe isotope compositions, as compared to MORB, most likely reflecting heterogeneous δ^{56}Fe values of their mantle sources.
- Island arc basalts (IAB) display isotopically lighter Fe isotope compositions than MORB and OIB, with δ^{56}Fe ~ 0.05 ± 0.1‰, most likely as a result of partial melting with high

melting degrees of previously-depleted (low $\delta^{56}Fe$) mantle sources.

- In contrast to the overall homogeneous Fe isotope composition of the oceanic crust, the continental crust is more heterogeneous in $\delta^{56}Fe$ values. In particular, highly evolved crustal rocks show exceptionally high $\delta^{56}Fe$ values up to 0.6‰ that correlate with SiO_2, most likely as a result of fractional crystallization, fluid exsolution, source effects, and diffusion-driven processes.
- The upper continental crust may be slightly heavier than the lower crust. The Fe isotope composition of the bulk continental crust is probably between that typically observed for IAB ($\delta^{56}Fe \sim 0.05$‰) and that observed for oceanic basalts ($\delta^{56}Fe \sim 0.1$‰.

References

Alexander CMO, Hewins RH (2004) Mass fractionation of Fe and Ni isotopes in metal in Hammadah al Hamra 237. Meteorit Planet Sci 39

An Y, Huang J-X, Griffin WL, Liu C, Huang F (2017) Isotopic composition of Mg and Fe in garnet peridotites from the Kaapvaal and Siberian cratons. Geochim Cosmochim Acta 200:167–185. https://doi.org/10.1016/j.gca.2016.11.041

Audétat A, Pettke T (2003) The magmatic-hydrothermal evolution of two barren granites: a melt and fluid inclusion study of the Rito del Medio and Canada Pinabete plutons in northern New Mexico (USA). Geochim Cosmochim Acta 67:97–121. https://doi.org/10.1016/S0016-7037(02)01049-9

Barrat JA, Rouxel O, Wang K, Moynier F, Yamaguchi A, Bischoff A, Langlade J (2015) Early stages of core segregation recorded by Fe isotopes in an asteroidal mantle. Earth Planet Sci Lett 419:93–100. https://doi.org/10.1016/j.epsl.2015.03.026

Barrat JA, Yamaguchi A, Greenwood RC, Bohn M, Cotten J, Benoit M, Franchi IA (2007) The Stannern trend eucrites: contamination of main group eucritic magmas by crustal partial melts. Geochim Cosmochim Acta 71:4108–4124. https://doi.org/10.1016/j.gca.2007.06.001

Beard BL, Johnson CM (2004) Inter-mineral Fe isotope variations in mantle-derived rocks and implications for the Fe geochemical cycle. Geochim Cosmochim Acta 68:4727–4743. https://doi.org/10.1016/j.gca.2004.04.023

Beard BL, Johnson CM, Skulan JL, Nealson KH, Cox L, Sun H (2003a) Application of Fe isotopes to tracing the geochemical and biological cycling of Fe. Chem Geol 195:87–117. https://doi.org/10.1016/s0009-2541(02)00390-x

Beard BL, Johnson CM, Von Damm KL, Poulson RL (2003b) Iron isotope constraints on Fe cycling and mass balance in oxygenated Earth oceans. Geology 31:629–632. https://doi.org/10.1130/0091-7613(2003)031%3c0629:iicofc%3e2.0.co;2

Beck P, Chaussidon M, Barrat JA, Gillet P, Bohn M (2006) Diffusion induced Li isotopic fractionation during the cooling of magmatic rocks: the case of pyroxene phenocrysts from nakhlite meteorites. Geochim Cosmochim Acta 70:4813–4825. https://doi.org/10.1016/j.gca.2006.07.025

Bennett SA, Rouxel O, Schmidt K, Garbe-Schönberg D, Statham PJ, German CR (2009) Iron isotope fractionation in a buoyant hydrothermal plume, 5°S Mid-Atlantic Ridge. Geochim Cosmochim Acta 73:5619–5634. https://doi.org/10.1016/j.gca.2009.06.027

Bilenker LD, Simon AC, Reich M, Lundstrom CC, Gajos N, Bindeman I, Barra F, Munizaga R (2016) Fe-O stable isotope pairs elucidate a high-temperature origin of Chilean iron oxide-apatite deposits. Geochim Cosmochim Acta 177:94–104. https://doi.org/10.1016/j.gca.2016.01.009

Birck JL, Allègre C (1984) Chronium isotopic anomalies in Allende refractory inclusions. Geophys Res Lett 11:943–946

Blanchard M, Poitrasson F, Méheut M, Lazzeri M, Mauri F, Balan E (2009) Iron isotope fractionation between pyrite (FeS2), hematite (Fe2O3) and siderite (FeCO3): A first-principles density functional theory study. Geochim Cosmochim Acta 73:6565–6578. https://doi.org/10.1016/j.gca.2009.07.034

Bodinier JL, Vasseur G, Vernieres J, Dupuy C, Fabries J (1990) Mechanisms of mantle metasomatism: geochemical evidence from the Lherz orogenic peridotite. J Petrol 31:597–628

Bouman C, Elliott T, Vroon PZ (2004) Lithium inputs to subduction zones. Chem Geol 212:59–79. https://doi.org/10.1016/j.chemgeo.2004.08.004

Brenan JM, Neroda E, Lundstrom CC, Shaw HF, Ryerson FJ, Phinney DL (1998) Behaviour of boron, beryllium, and lithium during melting and crystallization: constraints from mineral-melt partitioning experiments. Geochim Cosmochim Acta 62:2129–2141. https://doi.org/10.1016/S0016-7037(98)00131-8

Brennecka GA, Borg LE, Wadhwa M (2013) Evidence for supernova injection into the solar nebula and the decoupling of r-process nucleosynthesis. Proc Natl Acad Sci 110:17241–17246. https://doi.org/10.1073/pnas.1307759110

Burkhardt C, Kleine T, Dauphas N, Wieler R (2012) Origin of isotopic heterogeneity in the solar nebula by thermal processing and mixing of nebular dust. Earth

Planet Sci Lett 357–358:298–307. https://doi.org/10.1016/j.epsl.2012.09.048

Burkhardt C, Kleine T, Oberli F, Pack A, Bourdon B, Wieler R (2011) Molybdenum isotope anomalies in meteorites: constraints on solar nebula evolution and origin of the Earth. Earth Planet Sci Lett 312:390–400. https://doi.org/10.1016/j.epsl.2011.10.010

Buseck PR (1977) Pallasite meteorites-mineralogy, petrology and geochemistry. Geochim Cosmochim 41:711–740

Canil D, O'Neill HC (1996) Distribution of ferric iron in some upper-mantle assemblages. J Petrol 37:609–635

Chakraborty S (2010) Diffusion coefficients in olivine, wadsleyite and ringwoodite. Rev Mineral Geochemistry 72:603–639. https://doi.org/10.2138/rmg.2010.72.13

Chakraborty S (1997) Rates and mechanisms of Fe-Mg interdiffusion in olivine. J Geophys Res 102:12317–12331

Chen C, Su BX, Uysal I, Avci E, Zhang PF, Xiao Y, He YS (2015) Iron isotopic constraints on the origin of peridotite and chromitite in the Kizildağ ophiolite, southern Turkey. Chem Geol 417:115–124. https://doi.org/10.1016/j.chemgeo.2015.10.001

Chen LM, Song XY, Zhu XK, Zhang XQ, Yu SY, Yi JN (2014) Iron isotope fractionation during crystallization and sub-solidus re-equilibration: constraints from the Baima mafic layered intrusion, SW China. Chem Geol 380:97–109. https://doi.org/10.1016/j.chemgeo.2014.04.020

Chernonozhkin SM, Goderis S, Costas-Rodríguez M, Claeys P, Vanhaecke F (2016) Effect of parent body evolution on equilibrium and kinetic isotope fractionation: a combined Ni and Fe isotope study of iron and stony-iron meteorites. Geochim Cosmochim Acta 186:168–188. https://doi.org/10.1016/j.gca.2016.04.050

Chernonozhkin SM, Weyrauch M, Goderis S, Oeser M, McKibbin SJ, Horn I, Hecht L, Weyer S, Claeys P, Vanhaecke F (2017) Thermal equilibration of iron meteorite and pallasite parent bodies recorded at the mineral scale by Fe and Ni isotope systematics. Geochim Cosmochim Acta 217:95–111. https://doi.org/10.1016/j.gca.2017.08.022

Chopra R, Richter FM, Bruce Watson E, Scullard CR (2012) Magnesium isotope fractionation by chemical diffusion in natural settings and in laboratory analogues. Geochim Cosmochim Acta 88:1–18. https://doi.org/10.1016/j.gca.2012.03.039

Collinet M, Charlier B, Namur O, Oeser M, Médard E, Weyer S (2017) Crystallization history of enriched shergottites from Fe and Mg isotope fractionation in olivine megacrysts. Geochim Cosmochim Acta 207:277–297. https://doi.org/10.1016/j.gca.2017.03.029

Coogan LA, Kasemann SA, Chakraborty S (2005) Rates of hydrothermal cooling of new oceanic upper crust derived from lithium-geospeedometry. Earth Planet Sci Lett 240:415–424. https://doi.org/10.1016/j.epsl.2005.09.020

Costa F, Chakraborty S (2004) Decadal time gaps between mafic intrusion and silicic eruption obtained from chemical zoning patterns in olivine. Earth Planet Sci Lett 227:517–530. https://doi.org/10.1016/j.epsl.2004.08.011

Costa F, Dungan M (2005) Short time scales of magmatic assimilation from diffusion modeling of multiple elements in olivine. Geology 33:837–840. https://doi.org/10.1130/G21675.1

Cottrell E, Kelley KA (2011) The oxidation state of Fe in MORB glasses and the oxygen fugacity of the upper mantle. Earth Planet Sci Lett 305:270–282. https://doi.org/10.1016/j.epsl.2011.03.014

Craddock PR, Dauphas N (2011) Iron isotopic compositions of geological reference materials and chondrites. Geostand Geoanalytical Res 35:101–123. https://doi.org/10.1111/j.1751-908X.2010.00085.x

Craddock PR, Warren JM, Dauphas N (2013) Abyssal peridotites reveal the near-chondritic Fe isotopic composition of the Earth. Earth Planet Sci Lett 365:63–76

Dauphas N (2007) Diffusion-driven kinetic isotope effect of Fe and Ni during formation of the Widmanstätten pattern. Meteorit Planet Sci 42:1597–1613. https://doi.org/10.1111/j.1945-5100.2007.tb00593.x

Dauphas N, Craddock PR, Asimow PD, Bennett VC, Nutman AP, Ohnenstetter D (2009) Iron isotopes may reveal the redox conditions of mantle melting from Archean to Present. Earth Planet Sci Lett 288:255–267. https://doi.org/10.1016/j.epsl.2009.09.029

Dauphas N, John SG, Rouxel O (2017) Iron Isotope systematics. Rev Mineral Geochem 82:415–510. https://doi.org/10.2138/rmg.2017.82.11

Dauphas N, Poitrasson F, Burkhardt C, Kobayashi H, Kurosawa K (2015) Planetary and meteoritic Mg/Si and δ^{30}Si variations inherited from solar nebula chemistry. Earth Planet Sci Lett 427:236–248. https://doi.org/10.1016/j.epsl.2015.07.008

Dauphas N, Roskosz M, Alp EE, Golden DC, Sio CK, Tissot FLH, Hu MY, Zhao J, Gao L, Morris RV (2012) A general moment NRIXS approach to the determination of equilibrium Fe isotopic fractionation factors: application to goethite and jarosite. Geochim Cosmochim Acta 94:254–275. https://doi.org/10.1016/j.gca.2012.06.013

Dauphas N, Roskosz M, Alp EE, Neuville DR, Hu MY, Sio CK, Tissot FLH, Zhao J, Tissandier L, Médard E, Cordier C (2014) Magma redox and structural controls on iron isotope variations in Earth's mantle and crust. Earth Planet Sci Lett 398:127–140. https://doi.org/10.1016/j.epsl.2014.04.033

Dauphas N, Schauble EA (2016) Mass fractionation laws, mass-independent effects, and isotopic anomalies. Annu Rev Earth Planet Sci 44:709–783. https://doi.org/10.1146/annurev-earth-060115-012157

Dauphas N, Teng FZ, Arndt NT (2010) Magnesium and iron isotopes in 2.7 Ga Alexo komatiites: Mantle signatures, no evidence for Soret diffusion, and identification of diffusive transport in zoned olivine.

Geochim Cosmochim Acta 74:3274–3291. https://doi.org/10.1016/j.gca.2010.02.031

Debret B, Millet MA, Pons ML, Bouilhol P, Inglis E, Williams H (2016) Isotopic evidence for iron mobility during subduction. Geology 44:215–218. https://doi.org/10.1130/G37565.1

Dohmen R, Chakraborty S (2007) Fe-Mg diffusion in olivine II: point defect chemistry, change of diffusion mechanisms and a model for calculation of diffusion coefficients in natural olivine. Phys Chem Miner 34:409–430. https://doi.org/10.1007/s00269-007-0158-6

Druitt TH, Costa F, Deloule E, Dungan M, Scaillet B (2012) Decadal to monthly timescales of magma transfer and reservoir growth at a caldera volcano. Nature 482:77–80. https://doi.org/10.1038/nature10706

Du DH, Wang XL, Yang T, Chen X, Li JY, Li W (2017) Origin of heavy Fe isotope compositions in high-silica igneous rocks: a rhyolite perspective. Geochim Cosmochim Acta 218:58–72. https://doi.org/10.1016/j.gca.2017.09.014

Dziony W, Horn I, Lattard D, Koepke J, Steinhoefel G, Schuessler JA, Holtz F (2014) In-situ Fe isotope ratio determination in Fe–Ti oxides and sulfides from drilled gabbros and basalt from the IODP Hole 1256D in the eastern equatorial Pacific. Chem Geol 363:101–113. https://doi.org/10.1016/j.chemgeo.2013.10.035

El Korh A, Luais B, Deloule E, Cividini D (2017) Iron isotope fractionation in subduction-related high-pressure metabasites (Ile de Groix, France). Contrib Mineral Petrol 172:1–19. https://doi.org/10.1007/s00410-017-1357-x

Elardo SM, Draper DS, Shearer CK (2011) Lunar Magma Ocean crystallization revisited: Bulk composition, early cumulate mineralogy, and the source regions of the highlands Mg-suite. Geochim Cosmochim Acta 75:3024–3045. https://doi.org/10.1016/j.gca.2011.02.033

Elardo SM, Shahar A (2017) Non-chondritic iron isotope ratios in planetary mantles as a result of core formation. Nat Geosci 10:317–321. https://doi.org/10.1038/ngeo2896

Elardo SM, Shahar A, Mock TD, Sio CK (2019) The effect of core composition on iron isotope fractionation between planetary cores and mantles. Earth Planet Sci Lett 513:124–134. https://doi.org/10.1016/j.epsl.2019.02.025

Elliott T, Steele RCJ (2017) The isotope geochemistry of Ni. Rev Miner Geochem 82:511–542. https://doi.org/10.2138/rmg.2017.82.12

Fischer-Gödde M, Burkhardt C, Kruijer TS, Kleine T (2015) Ru isotope heterogeneity in the solar protoplanetary disk. Geochim Cosmochim Acta 168:151–171. https://doi.org/10.1016/j.gca.2015.07.032

Fischer-Gödde M, Kleine T (2017) Ruthenium isotopic evidence for an inner Solar System origin of the late veneer. Nature 541:525–527. https://doi.org/10.1038/nature21045

Foden J, Sossi PA, Nebel O (2018) Controls on the iron isotopic composition of global arc magmas. Earth Planet Sci Lett 494:190–201. https://doi.org/10.1016/j.epsl.2018.04.039

Foden J, Sossi PA, Wawryk CM (2015) Fe isotopes and the contrasting petrogenesis of A-, I- and S-type granite. Lithos 212–215:32–44. https://doi.org/10.1016/j.lithos.2014.10.015

Foley SF, Prelevic D, Rehfeldt T, Jacob DE (2013) Minor and trace elements in olivines as probes into early igneous and mantle melting processes. Earth Planet Sci Lett 363:181–191. https://doi.org/10.1016/j.epsl.2012.11.025

Frost BR, Barnes CG, Collins WJ, Arculus RJ, Ellis DJ, Frost CD (2001) A geochemical classification for granitic rocks. J Petrol 42:2033–2048

Frost DJ, McCammon CA (2008) The redox state of Earth's mantle. Annu Rev Earth Planet Sci 36:389–420. https://doi.org/10.1146/annurev.earth.36.031207.124322

Gallagher K, Elliott T (2009) Fractionation of lithium isotopes in magmatic systems as a natural consequence of cooling. Earth Planet Sci Lett 278:286–296. https://doi.org/10.1016/j.epsl.2008.12.009

Georg RB, Halliday AN, Schauble EA, Reynolds BC (2007) Silicon in the Earth's core. Nature 447:1102–1106. https://doi.org/10.1038/nature05927

Goldmann A, Weyer S (2013) In-situ determination of Fe Isotopes in Kamacite-, Taenite- and Troilite-phases of ordinary chondrites. In: Goldschmidt confrence abstract. https://doi.org/10.1180/minmag.2013.077.5.7

Günther T, Klemd R, Zhang X, Horn I, Weyer S (2017) In-situ trace element and Fe-isotope studies on magnetite of the volcanic-hosted Zhibo and Chagangnuoer iron ore deposits in the Western Tianshan, NW China. Chem Geol 453:111–127. https://doi.org/10.1016/j.chemgeo.2017.02.001

Hawkesworth CJ, Kemp AIS (2006) Evolution of the continental crust. Nature 443:811–817. https://doi.org/10.1038/nature05191

He Y, Wu H, Ke S, Liu SA, Wang Q (2017) Iron isotopic compositions of adakitic and non-adakitic granitic magmas: magma compositional control and subtle residual garnet effect. Geochim Cosmochim Acta 203:89–102. https://doi.org/10.1016/j.gca.2017.01.005

Heimann A, Beard BL, Johnson CM (2008) The role of volatile exsolution and sub-solidus fluid/rock interactions in producing high 56Fe/54Fe ratios in siliceous igneous rocks. Geochim Cosmochim Acta 72:4379–4396. https://doi.org/10.1016/j.gca.2008.06.009

Herzberg C (1992) Depth and degree of melting of komatiites. J Geophys Res 97:4521–4540

Hezel DC, Needham AW, Armytage R, Georg B, Abel RL, Kurahashi E, Coles BJ, Rehkämper M, Russell SS (2010) A nebula setting as the origin for bulk chondrule Fe isotope variations in CV chondrites. Earth Planet Sci Lett 296:423–433. https://doi.org/10.1016/j.epsl.2010.05.029

Hezel DC, Wilden JS, Becker D, Steinbach S, Wombacher F, Harak M (2018) Fe isotope composition of bulk chondrules from Murchison (CM2): constraints for parent body alteration, nebula processes and chondrule-matrix complementarity. Earth Planet Sci Lett 490:31–39. https://doi.org/10.1016/j.epsl.2018.03.013

Hibbert KEJ, Williams HM, Kerr AC, Puchtel IS (2012) Iron isotopes in ancient and modern komatiites: evidence in support of an oxidised mantle from Archean to present. Earth Planet Sci Lett 321–322:198–207. https://doi.org/10.1016/j.epsl.2012.01.011

Hin RC, Fitoussi C, Schmidt MW, Bourdon B (2014) Experimental determination of the Si isotope fractionation factor between liquid metal and liquid silicate. Earth Planet Sci Lett 387:55–66. https://doi.org/10.1016/j.epsl.2013.11.016

Hin RC, Schmidt MW, Bourdon B (2012) Experimental evidence for the absence of iron isotope fractionation between metal and silicate liquids at 1 GPa and 1250–1300 °C and its cosmochemical consequences. Geochim Cosmochim Acta 93:164–181. https://doi.org/10.1016/j.gca.2012.06.011

Hofmann AW (2003) Sampling mantle heterogeneity through oceanic basalts: isotopes and trace elements. In the mantle and the core. Treatise on Geochemistry

Hopp T, Fischer-Gödde M, Kleine T (2018) Ruthenium isotope fractionation in protoplanetary cores. Geochim Cosmochim Acta 223:75–89. https://doi.org/10.1016/j.gca.2017.11.033

Horn I, von Blanckenburg F, Schoenberg R, Steinhoefel G, Markl G (2006) In situ iron isotope ratio determination using UV-femtosecond laser ablation with application to hydrothermal ore formation processes. Geochim Cosmochim Acta 70:3677–3688. https://doi.org/10.1016/j.gca.2006.05.002

Huang F, Zhang Z, Lundstrom CC, Zhi X (2011) Iron and magnesium isotopic compositions of peridotite xenoliths from Eastern China. Geochim Cosmochim Acta 75:3318–3334. https://doi.org/10.1016/j.gca.2011.03.036

Hyslop EV, Valley JW, Johnson CM, Beard BL (2008) The effects of metamorphism on O and Fe isotope compositions in the Biwabik Iron Formation, northern Minnesota. Contrib to Mineral Petrol 155:313–328. https://doi.org/10.1007/s00410-007-0244-2

Inglis EC, Debtre B, Burton KW, Millet M-A, Pons M-L, Dale CW, Bouilhol P, Cooper M, Nowell GM, McCoy-West AJ, Williams HM (2017) The behavior of iron and zinc stable isotopes accompanying the subduction of mafic oceanic crust: a case study from Western Alpine ophiolites. Geochem Geophys Geosystems 2562–2579. https://doi.org/10.1002/2016gc006679

Ionov DA, Chanefo I, Bodinier JL (2005) Origin of Fe-rich lherzolites and wehrlites from Tok, SE Siberia by reactive melt percolation in refractory mantle peridotites. Contrib Mineral Petrol 150:335–353. https://doi.org/10.1007/s00410-005-0026-7

Jeffcoate AB, Elliott T, Kasemann SA, Ionov D, Cooper K, Brooker R (2007) Li isotope fractionation in peridotites and mafic melts. Geochim Cosmochim Acta 71:202–218. https://doi.org/10.1016/j.gca.2006.06.1611

Johnson CM, Bell K, Beard BL, Shultis AI (2010) Iron isotope compositions of carbonatites record melt generation, crystallization, and late-stage volatile-transport processes. Mineral Petrol 98:91–110. https://doi.org/10.1007/s00710-009-0055-4

Jordan MK, Tang HL, Kohl IE, Young ED (2019) Iron isotope constraints on planetesimal core formation in the early solar system. Geochim Cosmochim Acta 246:461–477. https://doi.org/10.1016/j.gca.2018.12.005

Jurewicz AJG, Mittlefehldt DW, Jones JH (1991) Partial melting of the allende (CV3) meteorite: implications for origins of basaltic meteorites. Science (80-) 252:695–698. https://doi.org/10.1126/science.252.5006.695

Kahl M, Chakraborty S, Costa F, Pompilio M (2011) Dynamic plumbing system beneath volcanoes revealed by kinetic modeling, and the connection to monitoring data: an example from Mt. Etna. Earth Planet Sci Lett 308:11–22. https://doi.org/10.1016/j.epsl.2011.05.008

Kehm K, Hauri EH, Alexander CMOD, Carlson RW (2003) High precision iron isotope measurements of meteoritic material by cold plasma ICP-MS. Geochim Cosmochim Acta 67:2879–2891. https://doi.org/10.1016/S0016-7037(03)00080-2

Kleine T, Mezger K, Palme H, Scherer E, Münker C (2005) Early core formation in asteroids and late accretion of chondrite parent bodies: Evidence from 182Hf-182 W in CAIs, metal-rich chondrites, and iron meteorites. Geochim Cosmochim Acta 69:5805–5818. https://doi.org/10.1016/j.gca.2005.07.012

Kleine T, Touboul M, Bourdon B, Nimmo F, Mezger K, Palme H, Jacobsen SB, Yin Q-Z, Halliday AN (2009) Hf–W chronology of the accretion and early evolution of asteroids and terrestrial planets. Geochim Cosmochim Acta 73:5150–5188. https://doi.org/10.1016/j.gca.2008.11.047

Knipping JL, Behrens H, Wilke M, Göttlicher J, Stabile P (2015) Effect of oxygen fugacity on the coordination and oxidation state of iron in alkali bearing silicate melts. Chem Geol 411:143–154. https://doi.org/10.1016/j.chemgeo.2015.07.004

Kodolányi J, Stephan T, Trappitsch R, Hoppe P, Pignatari M, Davis AM, Pellin MJ (2018) Iron and nickel isotope compositions of presolar silicon carbide grains from supernovae. Geochim Cosmochim Acta 221:127–144. https://doi.org/10.1016/j.gca.2017.05.029

Konter JG, Pietruszka AJ, Hanan BB, Finlayson VA, Craddock PR, Jackson MG, Dauphas N (2016) Unusual δ56Fe values in Samoan rejuvenated lavas generated in the mantle. Earth Planet Sci Lett 450:221–232. https://doi.org/10.1016/j.epsl.2016.06.029

Lai YJ, Pogge von Strandmann PAE, Dohmen R, Takazawa E, Elliott T (2015) The influence of melt infiltration on the Li and Mg isotopic composition of the Horoman Peridotite Massif. Geochim Cosmochim Acta 164:318–332. https://doi.org/10.1016/j.gca.2015.05.006

Lee CTA, Luffi P, Plank T, Dalton H, Leeman WP (2009) Constraints on the depths and temperatures of basaltic magma generation on earth and other terrestrial planets using new thermobarometers for mafic magmas. Earth Planet Sci Lett 279:20–33. https://doi.org/10.1016/j.epsl.2008.12.020

Li DY, Xiao YL, Li WY, Zhu X, Williams HM, Li YL (2016) Iron isotopic systematics of UHP eclogites respond to oxidizing fluid during exhumation. J Metamorph Geol 34:987–997. https://doi.org/10.1111/jmg.12217

Li J, Huang X-L, Wei G-J, Liu Y, Ma J-L, Han L, He P-L (2018a) Lithium isotope fractionation during magmatic differentiation and hydrothermal processes in rare-metal granites. Geochim Cosmochim Acta 240:64–79. https://doi.org/10.1016/j.gca.2018.08.021

Li JX, Qin KZ, Li GM, Evans NJ, Huang F, Zhao JX (2018b) Iron isotope fractionation during magmatic-hydrothermal evolution: a case study from the Duolong porphyry Cu-Au deposit, Tibet. Geochim Cosmochim Acta 238:1–15. https://doi.org/10.1016/j.gca.2018.07.008

Liu J, Dauphas N, Roskosz M, Hu MY, Yang H, Bi W, Zhao J, Alp EE, Hu JY, Lin J-F (2017) Iron isotopic fractionation between silicate mantle and metallic core at high pressure. Nat Commun 1–6. https://doi.org/10.1038/ncomms14377

Liu PP, Zhou MF, Luais B, Cividini D, Rollion-Bard C (2014) Disequilibrium iron isotopic fractionation during the high-temperature magmatic differentiation of the Baima Fe-Ti oxide-bearing mafic intrusion, SW China. Earth Planet Sci Lett 399:21–29. https://doi.org/10.1016/j.epsl.2014.05.002

Liu Y, Spicuzza MJ, Craddock PR, Day JMD, Valley JW, Dauphas N, Taylor LA (2010) Oxygen and iron isotope constraints on near-surface fractionation effects and the composition of lunar mare basalt source regions. Geochim Cosmochim Acta 74:6249–6262. https://doi.org/10.1016/j.gca.2010.08.008

Macris CA, Manning CE, Young ED (2015) Crystal chemical constraints on inter-mineral Fe isotope fractionation and implications for Fe isotope disequilibrium in San Carlos mantle xenoliths. Geochim Cosmochim Acta 154:168–185. https://doi.org/10.1016/j.gca.2015.01.024

Magna T, Hu Y, Teng FZ, Mezger K (2017) Magnesium isotope systematics in Martian meteorites. Earth Planet Sci Lett 474:419–426. https://doi.org/10.1016/j.epsl.2017.07.012

Magna T, Wiechert U, Halliday AN (2006) New constraints on the lithium isotope compositions of the Moon and terrestrial planets. Earth Planet Sci Lett 243:336–353. https://doi.org/10.1016/j.epsl.2006.01.005

Marhas KK, Amari S, Gyngard F, Zinner E, Gallino R (2008) Iron and nickel isotopic ratios in presolar sic grains. Astrophys J 689:622–645

Markl G, von Blanckenburg F, Wagner T (2006) Iron isotope fractionation during hydrothermal ore deposition and alteration. Geochim Cosmochim Acta 70:3011–3030. https://doi.org/10.1016/j.gca.2006.02.028

Marschall HR, Dohmen R, Ludwig T (2013) Diffusion-induced fractionation of niobium and tantalum during continental crust formation. Earth Planet Sci Lett 375:361–371. https://doi.org/10.1016/j.epsl.2013.05.055

McCammon C, Hutchison M, Harris J (1997) Ferric iron content of mineral inclusions in diamonds from Sao Luiz: a view into the lower mantle. Science (80-) 278:434–436

McCoy-West AJ, Fitton JG, Pons ML, Inglis EC, Williams HM (2018) The Fe and Zn isotope composition of deep mantle source regions: insights from Baffin Island picrites. Geochim Cosmochim Acta 238:542–562. https://doi.org/10.1016/j.gca.2018.07.021

McDonough WF, Sun SS (1995) The composition of the Earth. Chem Geol 120:223–253

Meibom A, Petaev MI, Alexander N, Keil K, Wood JA (2001) Growth mechanism and additional constraints on FeNi metal condensation in the solar nebula. J Geophys Res 106:797–801

Millet MA, Dauphas N, Greber ND, Burton KW, Dale CW, Debret B, Macpherson CG, Nowell GM, Williams HM (2016) Titanium stable isotope investigation of magmatic processes on the Earth and Moon. Earth Planet Sci Lett 449:197–205. https://doi.org/10.1016/j.epsl.2016.05.039

Mishra RK, Goswami JN (2014) Fe-Ni and Al-Mg isotope records in UOC chondrules: plausible stellar source of 60Fe and other short-lived nuclides in the early Solar System. Geochim Cosmochim Acta 132:440–457. https://doi.org/10.1016/j.gca.2014.01.011

Miyamoto M (1997) Chemical zoning of olivine in several pallasites. J Geophys Res 102:21613–21618

Moeller K, Schoenberg R, Grenne T, Thorseth IH, Drost K, Pedersen RB (2014) Comparison of iron isotope variations in modern and Ordovician siliceous Fe oxyhydroxide deposits. Geochim Cosmochim Acta 126:422–440. https://doi.org/10.1016/j.gca.2013.11.018

Moynier F, Albarède F, Herzog GF (2006) Isotopic composition of zinc, copper, and iron in lunar samples. Geochim Cosmochim Acta 70:6103–6117. https://doi.org/10.1016/j.gca.2006.02.030

Moynier F, Blichert-Toft J, Telouk P, Luck JM, Albarède F (2007) Comparative stable isotope geochemistry of Ni, Cu, Zn, and Fe in chondrites and iron meteorites. Geochim Cosmochim Acta 71:4365–4379. https://doi.org/10.1016/j.gca.2007.06.049

Mullane E, Russell SS, Gounelle M (2005) Nebular and asteroidal modification of the iron isotope composition of chondritic components. Earth Planet Sci Lett

239:203–218. https://doi.org/10.1016/j.epsl.2005.07.026

Nebel O, Arculus RJ, Sossi PA, Jenner FE, Whan THE (2013) Iron isotopic evidence for convective resurfacing of recycled arc-front mantle beneath back-arc basins. Geophys Res Lett 40:5849–5853. https://doi.org/10.1002/2013GL057976

Nebel O, Campbell IH, Sossi PA, Van Kranendonk MJ (2014) Hafnium and iron isotopes in early Archean komatiites record a plume-driven convection cycle in the Hadean Earth. Earth Planet Sci Lett 397:111–120. https://doi.org/10.1016/j.epsl.2014.04.028

Nebel O, Sossi PA, Bénard A, Wille M, Vroon PZ, Arculus RJ (2015) Redox-variability and controls in subduction zones from an iron-isotope perspective. Earth Planet Sci Lett 432:142–151. https://doi.org/10.1016/j.epsl.2015.09.036

Nebel O, Sossi PA, Foden J, Bénard A, Brandl PA, Stammeier JA, Lupton J, Richter M, Arculus RJ (2018) Iron isotope variability in ocean floor lavas and mantle sources in the Lau back-arc basin. Geochim Cosmochim Acta 241:150–163. https://doi.org/10.1016/j.gca.2018.08.046

Needham AW, Porcelli D, Russell SS (2009) An Fe isotope study of ordinary chondrites. Geochim Cosmochim Acta 73:7399–7413. https://doi.org/10.1016/j.gca.2009.08.034

Neukampf J, Ellis BS, Magna T, Laurent O, Bachmann O (2019) Partitioning and isotopic fractionation of lithium in mineral phases of hot, dry rhyolites: The case of the Mesa Falls Tuff, Yellowstone. Chem Geol 506:175–186. https://doi.org/10.1016/j.chemgeo.2018.12.031

Nichols CIO, Krakow R, Herrero-Albillos J, Kronast F, Northwood-Smith G, Harrison RJ (2018) Microstructural and paleomagnetic insight into the cooling history of the IAB parent body. Geochim Cosmochim Acta 229:1–19. https://doi.org/10.1016/j.gca.2018.03.009

Niederer FR, Papanastassiou DA (1984) Ca isotopes in refractory inclusions. Geochim Cosmochim Acta 48:1279–1293. https://doi.org/10.1016/0016-7037(84)90062-0

Niederer FR, Papanastassiou DA, Wasserburg GJ (1980) Endemic isotopic anomalies in titanium. Astrophys J 240:73–77

Niu Y (1997) Mantle melting and melt extraction processes beneath ocean ridges: evidence from abyssal peridotites. J Petrol 38:1047–1074

Oeser M, Dohmen R, Horn I, Schuth S, Weyer S (2015) Processes and time scales of magmatic evolution as revealed by Fe-Mg chemical and isotopic zoning in natural olivines. Geochim Cosmochim Acta 154:130–150. https://doi.org/10.1016/j.gca.2015.01.025

Oeser M, Ruprecht P, Weyer S (2018) Combined Fe-Mg chemical and isotopic zoning in olivine constraining magma mixing-to-eruption timescales for the continental arc volcano Irazú (Costa Rica) and Cr diffusion in olivine. Am Mineral 103:582–599. https://doi.org/10.2138/am-2018-6258

Oeser M, Weyer S, Horn I, Schuth S (2014) High-precision Fe and Mg isotope ratios of silicate reference glasses determined In Situ by Femtosecond LA-MC-ICP-MS and by solution nebulisation MC-ICP-MS. Geostand Geoanalytical Res 38:311–328. https://doi.org/10.1111/j.1751-908X.2014.00288.x

Palme H, Jones A (2003) Solar System abundance of the elements. Treatise Geochem 1:41–60

Peters BJ, Shahar A, Carlson RW, Day JMD, Mock TD (2019) A sulfide perspective on iron isotope fractionation during ocean island basalt petrogenesis. Geochim Cosmochim Acta 245:59–78. https://doi.org/10.1016/j.gca.2018.10.015

Poitrasson F, Delpech G, Grégoire M (2013) On the iron isotope heterogeneity of lithospheric mantle xenoliths: implications for mantle metasomatism, the origin of basalts and the iron isotope composition of the Earth. Contrib to Mineral Petrol 165:1243–1258. https://doi.org/10.1007/s00410-013-0856-7

Poitrasson F, Freydier R (2005) Heavy iron isotope composition of granites determined by high resolution MC-ICP-MS. Chem Geol 222:132–147. https://doi.org/10.1016/j.chemgeo.2005.07.005

Poitrasson F, Halliday AN, Lee DC, Levasseur S, Teutsch N (2004) Iron isotope differences between Earth, Moon, Mars and Vesta as possible records of contrasted accretion mechanisms. Earth Planet Sci Lett 223:253–266. https://doi.org/10.1016/j.epsl.2004.04.032

Poitrasson F, Levasseur S, Teutsch N (2005) Significance of iron isotope mineral fractionation in pallasites and iron meteorites for the core-mantle differentiation of terrestrial planets. Earth Planet Sci Lett 234:151–164. https://doi.org/10.1016/j.epsl.2005.02.010

Poitrasson F, Roskosz M, Corgne A (2009) No iron isotope fractionation between molten alloys and silicate melt to 2000 °C and 7.7 GPa: experimental evidence and implications for planetary differentiation and accretion. Earth Planet Sci Lett 278:376–385. https://doi.org/10.1016/j.epsl.2008.12.025

Polyakov VB (2009) Equilibrium iron isotope fractionation at core-mantle boundary conditions. Science (80-) 323:912–914. https://doi.org/10.1126/science.1166329

Polyakov VB, Clayton RN, Horita J, Mineev SD (2007) Equilibrium iron isotope fractionation factors of minerals: reevaluation from the data of nuclear inelastic resonant X-ray scattering and Mössbauer spectroscopy. Geochim Cosmochim Acta 71:3833–3846. https://doi.org/10.1016/j.gca.2007.05.019

Polyakov VB, Mineev SD (2000) The use of Mössbauer spectroscopy in stable isotope geochemistry. Geochim Cosmochim Acta 64:849–865. https://doi.org/10.1016/S0016-7037(99)00329-4

Polyakov VB, Soultanov DM (2011) New data on equilibrium iron isotope fractionation among sulfides: Constraints on mechanisms of sulfide formation in hydrothermal and igneous systems. Geochim Cosmochim Acta 75:1957–1974. https://doi.org/10.1016/j.gca.2011.01.019

Pringle EA, Moynier F (2017) Rubidium isotopic composition of the Earth, meteorites, and the Moon: evidence for the origin of volatile loss during planetary accretion. Earth Planet Sci Lett 473:62–70. https://doi.org/10.1016/j.epsl.2017.05.033

Quitté G, Halliday A, Meyer BS, Markowski A, Latkoczy C, Günther D (2007) Correlated Iron 60, Nickel 62, and Zirconium 96 in refractory inclusions and the origin of the solar system. Astrophys J 655:678–684. https://doi.org/10.1086/509771

Richter FM, Dauphas N, Teng FZ (2009a) Non-traditional fractionation of non-traditional isotopes: evaporation, chemical diffusion and Soret diffusion. Chem Geol 258:92–103. https://doi.org/10.1016/j.chemgeo.2008.06.011

Richter FM, Davis AM, DePaolo DJ, Watson EB (2003) Isotope fractionation by chemical diffusion between molten basalt and rhyolite. Geochim Cosmochim Acta 67:3905–3923. https://doi.org/10.1016/S0016-7037(03)00174-1

Richter FM, Liang Y, Davis AM (1999) Isotope fractionation by diffusion in molten oxides. Geochim Cosmochim Acta 63:2853–2861. https://doi.org/10.1016/S0016-7037(99)00164-7

Richter FM, Watson EB, Mendybaev R, Dauphas N, Georg B, Watkins J, Valley J (2009b) Isotopic fractionation of the major elements of molten basalt by chemical and thermal diffusion. Geochim Cosmochim Acta 73:4250–4263. https://doi.org/10.1016/j.gca.2009.04.011

Righter K, Drake MJ (1996) Core formation in Earth's moon, Mars, and Vesta. Icarus 124:513–529. https://doi.org/10.1006/icar.1996.0227

Roskosz M, Luais B, Watson HC, Toplis MJ, Alexander CMOD, Mysen BO (2006) Experimental quantification of the fractionation of Fe isotopes during metal segregation from a silicate melt. Earth Planet Sci Lett 248:851–867. https://doi.org/10.1016/j.epsl.2006.06.037

Roskosz M, Sio CKI, Dauphas N, Bi W, Tissot FLH, Hu MY, Zhao J, Alp EE (2015) Spinel-olivine-pyroxene equilibrium iron isotopic fractionation and applications to natural peridotites. Geochim Cosmochim Acta 169:184–199. https://doi.org/10.1016/j.gca.2015.07.035

Rouxel O, Dobbek N, Ludden J, Fouquet Y (2003) Iron isotope fractionation during oceanic crust alteration. Chem Geol 202:155–182. https://doi.org/10.1016/j.chemgeo.2003.08.011

Rouxel O, Fouquet Y, Ludden JN (2004) Subsurface processes at the lucky strike hydrothermal field, Mid-Atlantic ridge: evidence from sulfur, selenium, and iron isotopes. Geochim Cosmochim Acta 68:2295–2311. https://doi.org/10.1016/j.gca.2003.11.029

Rouxel O, Shanks WC, Bach W, Edwards KJ (2008) Integrated Fe- and S-isotope study of seafloor hydrothermal vents at East Pacific Rise 9–10°N. Chem Geol 252:214–227. https://doi.org/10.1016/j.chemgeo.2008.03.009

Rouxel O, Toner B, Germain Y, Glazer B (2018) Geochemical and iron isotopic insights into hydrothermal iron oxyhydroxide deposit formation at Loihi Seamount. Geochim Cosmochim Acta 220:449–482. https://doi.org/10.1016/j.gca.2017.09.050

Rouxel O, Toner BM, Manganini SJ, German CR (2016) Geochemistry and iron isotope systematics of hydrothermal plume fall-out at East Pacific Rise 9° 50′N. Chem Geol 441:212–234. https://doi.org/10.1016/j.chemgeo.2016.08.027

Rudnick RL, Gao S (2003) Composition of the continental crust. Treatise Geochem **3**:1–64

Rudnick RL, Ionov DA (2007) Lithium elemental and isotopic disequilibrium in minerals from peridotite xenoliths from far-east Russia: Product of recent melt/fluid-rock reaction. Earth Planet Sci Lett 256:278–293. https://doi.org/10.1016/j.epsl.2007.01.035

Rudnick RL, McDonough WF, Chappell BC (1993) Carbonatite metasomatism in the northern Tanzanian mantle: petrographic and geochemical characteristics. Earth Planet Sci Lett Sci 114:463–475

Ruprecht P, Plank T (2013) Feeding andesitic eruptions with a high-speed connection from the mantle. Nature 500:68–72. https://doi.org/10.1038/nature12342

Rustad JR, Yin QZ (2009) Iron isotope fractionation in the earth's lower mantle. Nat Geosci 2:514–518. https://doi.org/10.1038/ngeo546

Rutherford (2008) Magma ascent rates. Rev Mineral Geochem 69:241–271

Ryder G (1991) Lunar ferroan Anorthosites and Mare Basalts sources: the mixed connection. Geophys Res Lett 18:2065–2068

Sanloup C, Fei Y (2004) Closure of the Fe-S-Si liquid miscibility gap at high pressure. Phys Earth Planet Inter 147:57–65. https://doi.org/10.1016/j.pepi.2004.06.008

Schauble E, Rossman GR, Taylor HP (2001) Theoretical estimates of equilibrium Fe-isotope fractionations from vibrational spectroscopy. Geochim Cosmochim Acta 65:2487–2497

Schauble EA (2004) Applying stable isotope fractionation theory to new systems. Rev Mineral Geochem 55:65–111

Schoenberg R, von Blanckenburg F (2006) Modes of planetary-scale Fe isotope fractionation. Earth Planet Sci Lett 252:342–359. https://doi.org/10.1016/j.epsl.2006.09.045

Schoenberg R, Marks MAW, Schuessler JA, von Blanckenburg F, Markl G (2009) Fe isotope systematics of coexisting amphibole and pyroxene in the alkaline igneous rock suite of the Ilímaussaq Complex, South Greenland. Chem Geol 258:65–77. https://doi.org/10.1016/j.chemgeo.2008.06.023

Schönbächler M, Lee DC, Rehkämper M, Halliday AN, Fehr MA, Hattendorf B, Günther D (2003) Zirconium isotope evidence for incomplete admixing of r-process components in the solar nebula. Earth Planet Sci Lett 216:467–481. https://doi.org/10.1016/S0012-821X(03)00547-8

Schuessler JA (2008) Controls on stable iron isotope variations in magmatic systems. PhD theses, Hannover

Schuessler JA, Schoenberg R, Behrens H, von Blanckenburg F (2007) The experimental calibration of the iron isotope fractionation factor between pyrrhotite and peralkaline rhyolitic melt. Geochim Cosmochim Acta 71:417–433. https://doi.org/10.1016/j.gca.2006.09.012

Schuessler JA, Schoenberg R, Sigmarsson O (2009) Iron and lithium isotope systematics of the Hekla volcano, Iceland-evidence for Fe isotope fractionation during magma differentiation. Chem Geol 258:78–91. https://doi.org/10.1016/j.chemgeo.2008.06.021

Sedaghatpour F, Teng FZ (2016) Magnesium isotopic composition of achondrites. Geochim Cosmochim Acta 174:167–179. https://doi.org/10.1016/j.gca.2015.11.016

Sedaghatpour F, Teng FZ, Liu Y, Sears DWG, Taylor LA (2013) Magnesium isotopic composition of the Moon. Geochim Cosmochim Acta 120:1–16. https://doi.org/10.1016/j.gca.2013.06.026

Seitz HM, Brey GP, Weyer S, Durali S, Ott U, Münker C, Mezger K (2006) Lithium isotope compositions of Martian and lunar reservoirs. Earth Planet Sci Lett 245:6–18. https://doi.org/10.1016/j.epsl.2006.03.007

Severmann S, Johnson CM, Beard BL, German CR, Edmonds HN, Chiba H, Green DRH (2004) The effect of plume processes on the Fe isotope composition of hydrothermally derived Fe in the deep ocean as inferred from the Rainbow vent site, Mid-Atlantic Ridge, 36°14′N. Earth Planet Sci Lett 225:63–76. https://doi.org/10.1016/j.epsl.2004.06.001

Shahar A, Hillgren VJ, Horan MF, Mesa-Garcia J, Kaufman LA, Mock TD (2015) Sulfur-controlled iron isotope fractionation experiments of core formation in planetary bodies. Geochim Cosmochim Acta 150:253–264. https://doi.org/10.1016/j.gca.2014.08.011

Shahar A, Schauble EA, Caracas R, Gleason AE, Reagan MM, Xiao Y, Shu J, Mao W (2016) Pressure-dependent isotopic composition of iron alloys. Science (80-) 352:580–583

Shahar A, Young ED, Manning CE (2008) Equilibrium high-temperature Fe isotope fractionation between fayalite and magnetite: an experimental calibration. Earth Planet Sci Lett 268:330–338. https://doi.org/10.1016/j.epsl.2008.01.026

Shahar A, Ziegler K, Young ED, Ricolleau A, Schauble EA, Fei Y (2009) Experimentally determined Si isotope fractionation between silicate and Fe metal and implications for Earth's core formation. Earth Planet Sci Lett 288:228–234. https://doi.org/10.1016/j.epsl.2009.09.025

Sharma M, Polizzotto M, Anbar AD (2001) Iron isotopes in hot springs along the Juan de Fuca Ridge. Earth Planet Sci Lett 6036:1–13. https://doi.org/10.1016/j.engappai.2010.01.015

Shollenberger QR, Wittke A, Render J, Mane P, Schuth S, Weyer S, Gussone N, Wadhwa M, Brennecka GA (2019) Combined mass-dependent and nucleosynthetic isotope variations in refractory inclusions and their mineral separates. Geochim Cosmochim Acta 263:215–234. https://doi.org/10.1016/j.gca.2019.07.021

Siebert J, Bandro J, Antonangeli D, Ryerson FJ (2013) Terrestrial accretion under oxidizing conditions. Science (80-) 339:1194–1197

Sio CK, Roskosz M, Dauphas N, Bennett NR, Mock T, Shahar A (2018) The isotope effect for Mg-Fe interdiffusion in olivine and its dependence on crystal orientation, composition and temperature. Geochim Cosmochim Acta 239:463–480. https://doi.org/10.1016/j.gca.2018.06.024

Sio CKI, Dauphas N (2017) Thermal and crystallization histories of magmatic bodies by Monte Carlo inversion of Mg-Fe isotopic profiles in olivine. Geology 45:67–70. https://doi.org/10.1130/G38056.1

Sio CKI, Dauphas N, Teng FZ, Chaussidon M, Helz RT, Roskosz M (2013) Discerning crystal growth from diffusion profiles in zoned olivine by in situ Mg-Fe isotopic analyses. Geochim Cosmochim Acta 123:302–321. https://doi.org/10.1016/j.gca.2013.06.008

Sorbadere F, Laurenz V, Frost DJ, Wenz M, Rosenthal A, McCammon C, Rivard C (2018) The behaviour of ferric iron during partial melting of peridotite. Geochim Cosmochim Acta 239:235–254. https://doi.org/10.1016/j.gca.2018.07.019

Sossi PA, Foden JD, Halverson GP (2012) Redox-controlled iron isotope fractionation during magmatic differentiation: an example from the Red Hill intrusion, S. Tasmania. Contrib Mineral Petrol 164:757–772. https://doi.org/10.1007/s00410-012-0769-x

Sossi PA, Nebel O, Foden J (2016a) Iron isotope systematics in planetary reservoirs. Earth Planet Sci Lett 452:295–308. https://doi.org/10.1016/j.epsl.2016.07.032

Sossi PA, Nebel O, Anand M, Poitrasson F (2016b) On the iron isotope composition of Mars and volatile depletion in the terrestrial planets. Earth Planet Sci Lett 449:360–371. https://doi.org/10.1016/j.epsl.2016.05.030

Sossi PA, O'Neill HSC (2017) The effect of bonding environment on iron isotope fractionation between minerals at high temperature. Geochim Cosmochim Acta 196:121–143. https://doi.org/10.1016/j.gca.2016.09.017

Sossi PA, O'Neill HSC (2016) Liquidus temperatures of komatiites and the effect of cooling rate on element partitioning between olivine and komatiitic melt. Contrib Mineral Petrol 171:1–25. https://doi.org/10.1007/s00410-016-1260-x

Su BX, Teng FZ, Hu Y, Shi RD, Zhou MF, Zhu B, Liu F, Gong XH, Huang QS, Xiao Y, Chen C, He YS (2015) Iron and magnesium isotope fractionation in oceanic lithosphere and sub-arc mantle: perspectives from ophiolites. Earth Planet Sci Lett 430:523–532. https://doi.org/10.1016/j.epsl.2015.08.020

Syverson DD, Borrok DM, Seyfried WE (2013) Experimental determination of equilibrium Fe isotopic fractionation between pyrite and dissolved Fe under hydrothermal conditions. Geochim Cosmochim Acta 122:170–183. https://doi.org/10.1016/j.gca.2013.08.027

Tang H, Dauphas N (2012) Abundance, distribution, and origin of 60Fe in the solar protoplanetary disk. Earth Planet Sci Lett 359:248–263. https://doi.org/10.1016/j.epsl.2012.10.011

Tatsumoto M, Basu AR, Wankang H, Junwen W, Guanghong X (1992) Sr, Nd, and Pb isotopes of ultramafic xenoliths in volcanic rocks of Eastern China: enriched components EMI and EMII in subcontinental lithosphere. Earth Planet Sci Lett 113:107–128. https://doi.org/10.1016/0012-821x(92)90214-g

Telus M, Dauphas N, Moynier F, Tissot FLH, Teng F-Z, Nabelek PI, Craddock PR, Groat LA (2012) Iron, zinc, magnesium and uranium isotopic fractionation during continental crust differentiation: the tale from migmatites, granitoids, and pegmatites. Geochim Cosmochim Acta 97:247–265. https://doi.org/10.1016/j.gca.2012.08.024

Teng F-Z, Li W-Y, Ke S, Marty B, Dauphas N, Huang S, Wu F-Y, Pourmand A (2010) Magnesium isotopic composition of the Earth and chondrites. Geochim Cosmochim Acta 74:4150–4166. https://doi.org/10.1016/j.gca.2010.04.019

Teng FZ, Dauphas N, Helz RT (2008) Iron isotope fractionation during magmatic differentiation in Kilauea Iki lava lake. Science (80-) 320:1620–1622. https://doi.org/10.1126/science.1157166

Teng FZ, Dauphas N, Helz RT, Gao S, Huang S (2011) Diffusion-driven magnesium and iron isotope fractionation in Hawaiian olivine. Earth Planet Sci Lett 308:317–324. https://doi.org/10.1016/j.epsl.2011.06.003

Teng FZ, Dauphas N, Huang S, Marty B (2013) Iron isotopic systematics of oceanic basalts. Geochim Cosmochim Acta 107:12–26. https://doi.org/10.1016/j.gca.2012.12.027

Teng FZ, Mcdonough WF, Rudnick RL, Walker RJ, Sirbescu MLC (2006) Lithium isotopic systematics of granites and pegmatites from the Black Hills, South Dakota. Am Mineral 91:1488–1498. https://doi.org/10.2138/am.2006.2083

Teng FZ, Wadhwa M, Helz RT (2007) Investigation of magnesium isotope fractionation during basalt differentiation: implications for a chondritic composition of the terrestrial mantle. Earth Planet Sci Lett 261:84–92. https://doi.org/10.1016/j.epsl.2007.06.004

Theis KJ, Burgess R, Lyon IC, Sears DW (2008) The origin and history of ordinary chondrites: a study by iron isotope measurements of metal grains from ordinary chondrites. Geochim Cosmochim Acta 72:4440–4456. https://doi.org/10.1016/j.gca.2008.05.057

Tomascak PB, Tera F, Helz RT, Walker RJ (1999) The absence of lithium isotope fractionation during basalt differentiation: new measurements by multicollector sector ICP-MS. Geochim Cosmochim Acta 63:907–910. https://doi.org/10.1016/S0016-7037(98)00318-4

Trappitsch R, Stephan T, Savina MR, Davis AM, Pellin MJ, Rost D, Gyngard F, Gallino R, Bisterzo S, Cristallo S, Dauphas N (2018) Simultaneous iron and nickel isotopic analyses of presolar silicon carbide grains. Geochim Cosmochim Acta 221:87–108. https://doi.org/10.1016/j.gca.2017.05.031

Turner S, Evans P, Hawkesworth C (2001) Ultrafast source-to-surface movement of melt at island systematics. Science (80-) 292:1363–1366

Valley GE, Anderson HH (1947) A comparison of the abundance ratios of the isotopes of terrestrial and of meteoritic iron. J Am Chem Soc 69:1871–1875. https://doi.org/10.1021/ja01200a010

Van Kranendonk MJ (2010) Two types of Archean continental crust: plume and plate tectonics on early earth. Am J Sci 310:1187–1209. https://doi.org/10.2475/10.2010.01

Van Orman JA, Krawczynski MJ (2015) Theoretical constraints on the isotope effect for diffusion in minerals. Geochim Cosmochim Acta 164:365–381. https://doi.org/10.1016/j.gca.2015.04.051

Wade J, Wood BJ (2005) Core formation and the oxidation state of the Earth. Earth Planet Sci Lett 236:78–95. https://doi.org/10.1016/j.epsl.2005.05.017

Wadhwa M (2008) Redox conditions on small bodies, the Moon and Mars. Rev Mineral Geochem 68:493–510

Wang K, Day JMD, Korotev RL, Zeigler RA, Moynier F (2014a) Iron isotope fractionation during sulfide-rich felsic partial melting in early planetesimals. Earth Planet Sci Lett 392:124–132. https://doi.org/10.1016/j.epsl.2014.02.022

Wang K, Jacobsen SB, Sedaghatpour F, Chen H, Korotev RL (2015) The earliest Lunar Magma Ocean differentiation recorded in Fe isotopes. Earth Planet Sci Lett 430:202–208. https://doi.org/10.1016/j.epsl.2015.08.019

Wang K, Moynier F, Barrat JA, Zanda B, Paniello RC, Savage PS (2013) Homogeneous distribution of Fe isotopes in the early solar nebula. Meteorit Planet Sci 48:354–364. https://doi.org/10.1111/maps.12060

Wang K, Moynier F, Dauphas N, Barrat JA, Craddock P, Sio CK (2012) Iron isotope fractionation in planetary crusts. Geochim Cosmochim Acta 89:31–45. https://doi.org/10.1016/j.gca.2012.04.050

Wang K, Savage PS, Moynier F (2014b) The iron isotope composition of enstatite meteorites: implications for their origin and the metal/sulfide Fe isotopic fractionation factor. Geochim Cosmochim Acta 142:149–165. https://doi.org/10.1016/j.gca.2014.07.019

Wang Y, Zhu XK, Mao JW, Li ZH, Cheng YB (2011) Iron isotope fractionation during skarn-type metallogeny: a case study of Xinqiao Cu-S-Fe-Au deposit in the Middle-Lower Yangtze valley. Ore Geol Rev

43:194–202. https://doi.org/10.1016/j.oregeorev.2010.12.004
Wänke H, Dreibus G (1988) Chemical Composition and Accretion History of Terrestrial Planets. Philos Trans R Soc A 325:545–557. https://doi.org/10.1098/rsta.1988.0067
Watson EB, Müller T (2009) Non-equilibrium isotopic and elemental fractionation during diffusion-controlled crystal growth under static and dynamic conditions. Chem Geol 267:111–124. https://doi.org/10.1016/j.chemgeo.2008.10.036
Watson HC, Richter F, Liu A, Huss GR (2016) Iron and nickel isotope fractionation by diffusion, with applications to iron meteorites. Earth Planet Sci Lett 451:159–167. https://doi.org/10.1016/j.epsl.2016.06.030
Wawryk CM, Foden JD (2015) Fe-isotope fractionation in magmatic-hydrothermal mineral deposits: a case study from the Renison Sn-W deposit, Tasmania. Geochim Cosmochim Acta 150:285–298. https://doi.org/10.1016/j.gca.2014.09.044
Weyer S, Anbar AD, Brey GP, Münker C, Mezger K, Woodland AB (2007) Fe-isotope fractionation during partial melting on Earth and the current view on the Fe-isotope budgets of the planets (reply to the comment of F. Poitrasson and to the comment of B. L. Beard and C.M. Johnson on "Iron isotope fractionation during planetary. Earth Planet Sci Lett 256:638–646. https://doi.org/10.1016/j.epsl.2007.01.038
Weyer S, Anbar AD, Brey GP, Münker C, Mezger K, Woodland AB (2005) Iron isotope fractionation during planetary differentiation. Earth Planet Sci Lett 240:251–264. https://doi.org/10.1016/j.epsl.2005.09.023
Weyer S, Ionov DA (2007) Partial melting and melt percolation in the mantle: the message from Fe isotopes. Earth Planet Sci Lett 259:119–133. https://doi.org/10.1016/j.epsl.2007.04.033
Weyer S, Seitz HM (2012) Coupled lithium- and iron isotope fractionation during magmatic differentiation. Chem Geol 294–295:42–50. https://doi.org/10.1016/j.chemgeo.2011.11.020
Weyrauch M, Oeser M, Brüske A, Weyer S (2017) In situ high-precision Ni isotope analysis of metals by femtosecond-LA-MC-ICP-MS. J Anal At Spectrom 32:1312–1319. https://doi.org/10.1039/c7ja00147a
Weyrauch M, Zipfel J, Weyer S (2019) Origin of metal from CB chondrites in an impact plume-a combined study of Fe and Ni isotope composition and trace element abundances. GCA 246:123–137
Wiesli RA, Beard BL, Taylor LA, Johnson CM (2003) Space weathering processes on airless bodies: Fe isotope fractionation in the lunar regolith. Earth Planet Sci Lett 216:457–465. https://doi.org/10.1016/S0012-821X(03)00552-1
Williams HM, Archer C (2011) Copper stable isotopes as tracers of metal-sulphide segregation and fractional crystallisation processes on iron meteorite parent bodies. Geochim Cosmochim Acta 75:3166–3178. https://doi.org/10.1016/j.gca.2011.03.010
Williams HM, Bizimis M (2014) Iron isotope tracing of mantle heterogeneity within the source regions of oceanic basalts. Earth Planet Sci Lett 404:396–407. https://doi.org/10.1016/j.epsl.2014.07.033
Williams HM, Markowski A, Quitté G, Halliday AN, Teutsch N, Levasseur S (2006) Fe isotope fractionation in iron meteorites: new insights into metal-sulphide segregation and planetary accretion. Earth Planet Sci Lett 250:486–500. https://doi.org/10.1016/j.epsl.2006.08.013
Williams HM, McCammon CA, Peslier AH, Halliday AN, Teutsch N, Levasseur S, Burg JP (2004) Iron isotope fractionation and the oxygen fugacity of the mantle. Science (80-) 304:1656–1659. https://doi.org/10.1126/science.1095679
Williams HM, Nielsen SG, Renac C, Griffin WL, O'Reilly SY, McCammon CA, Pearson N, Viljoen F, Alt JC, Halliday AN (2009) Fractionation of oxygen and iron isotopes by partial melting processes: implications for the interpretation of stable isotope signatures in mafic rocks. Earth Planet Sci Lett 283:156–166. https://doi.org/10.1016/j.epsl.2009.04.011
Williams HM, Peslier AH, McCammon C, Halliday AN, Levasseur S, Teutsch N, Burg JP (2005) Systematic iron isotope variations in mantle rocks and minerals: the effects of partial melting and oxygen fugacity. Earth Planet Sci Lett 235:435–452. https://doi.org/10.1016/j.epsl.2005.04.020
Williams HM, Prytulak J, Woodhead JD, Kelley KA, Brounce M, Plank T (2018) Interplay of crystal fractionation, sulfide saturation and oxygen fugacity on the iron isotope composition of arc lavas: an example from the Marianas. Geochim Cosmochim Acta 226:224–243. https://doi.org/10.1016/j.gca.2018.02.008
Williams HM, Wood BJ, Wade J, Frost DJ, Tuff J (2012) Isotopic evidence for internal oxidation of the Earth's mantle during accretion. Earth Planet Sci Lett 321–322:54–63. https://doi.org/10.1016/j.epsl.2011.12.030
Woosley SE, Heger A, Weaver TA (2002) The evolution and explosion of massive stars. Rev Mod Phys 74:1015–1071. https://doi.org/10.1063/1.2943638
Woosley SE, Weaver TA (1995) The evolution and explosion of massive stars. II. Explosive hydrodynamics and nucleosynthesis. Astrophys J 101:181–235
Wu H, He Y, Bao L, Zhu C, Li S (2017) Mineral composition control on inter-mineral iron isotopic fractionation in granitoids. Geochim Cosmochim Acta 198:208–217. https://doi.org/10.1016/j.gca.2016.11.008
Wu H, He Y, Teng F-Z, Ke S, Hou Z, Li S (2018) Diffusion-driven magnesium and iron isotope fractionation at a gabbro-granite boundary. Geochim Cosmochim Acta 222:671–684. https://doi.org/10.1016/j.gca.2017.11.010
Xia Y, Li S, Huang F (2017) Iron and Zinc isotope fractionation during magmatism in the continental crust: Evidence from bimodal volcanic rocks from Hailar basin, NE China. Geochim Cosmochim Acta 213:35–46. https://doi.org/10.1016/j.gca.2017.06.018
Xu L-J, He Y, Wang S-J, Wu H, Li S (2017) Iron isotope fractionation during crustal anatexis: constraints from

migmatites from the Dabie orogen, Central China. Lithos 284–285:171–179. https://doi.org/10.1016/j.lithos.2017.04.005

Yang H, Lin J-F, Hu MY, Roskosz M, Bi W, Zhao J, Alp EE, Liu J, Liu J, Wentzowitch RM, Okuchi T, Dauphas N (2019) Iron isotopic fractionation in mineral phases from Earth's lower mantle: did terrestrial magma ocean crystallization fractionate iron isotopes? Earth Planet Sci Lett 506:113–122. https://doi.org/10.1016/j.epsl.2018.10.034

Zambardi T, Lundstrom CC, Li X, McCurry M (2014) Fe and Si isotope variations at Cedar Butte volcano; insight into magmatic differentiation. Earth Planet Sci Lett 405:169–179. https://doi.org/10.1016/j.epsl.2014.08.020

Zambardi T, Poitrasson F, Corgne A, Méheut M, Quitté G, Anand M (2013) Silicon isotope variations in the inner solar system: implications for planetary formation, differentiation and composition. Geochim Cosmochim Acta 121:67–83. https://doi.org/10.1016/j.gca.2013.06.040

Zhao X-M, Zhang H-F, Zhu X-K, Zhu B, Cao H (2015) Effects of melt percolation on iron isotopic variation in peridotites from Yangyuan, North China Craton. Chem Geol 401:96–110. https://doi.org/10.1016/j.chemgeo.2015.02.031

Zhao X, Zhang H, Zhu X, Tang S, Tang Y (2010) Iron isotope variations in spinel peridotite xenoliths from North China Craton: implications for mantle metasomatism. Contrib to Mineral Petrol 160:1–14. https://doi.org/10.1007/s00410-009-0461-y

Zhao X, Zhang H, Zhu X, Tang S, Yan B (2012) Iron isotope evidence for multistage melt–peridotite interactions in the lithospheric mantle of eastern China. Chem Geol 292–293:127–139. https://doi.org/10.1016/j.chemgeo.2011.11.016

Zhao X, Zhang Z, Huang S, Liu Y, Li X, Zhang H (2017) Coupled extremely light Ca and Fe isotopes in peridotites. Geochim Cosmochim Acta 208:368–380. https://doi.org/10.1016/j.gca.2017.03.024

Zhu B, Zhang HF, Zhao XM, He YS (2016) Iron isotope fractionation during skarn-type alteration: implications for metal source in the Han-Xing iron skarn deposit. Ore Geol Rev 74:139–150. https://doi.org/10.1016/j.oregeorev.2015.11.001

Zhu D, Bao H, Liu Y (2015) Non-traditional stable isotope behaviors in immiscible silica-melts in a mafic magma chamber. Sci Rep 5:1–10. https://doi.org/10.1038/srep17561

Zhu XK, Guo Y, O'nions RK, Young ED, Ash RD (2001) Isotopic homogeneity of iron in the early solar nebula. Nature 412:311–313

Zhu XK, Guo Y, Williams RJP, O'nions RK, Matthews A, Belshaw NS, Canters GW, De Waal EC, Weser U, Burgess BK, Salvato B (2002) Mass fractionation processes of transition metal isotopes. Earth Planet Sci Lett 200:47–62

Zhu ZY, Jiang SY, Mathur R, Cook NJ, Yang T, Wang M, Ma L, Ciobanu CL (2018) Iron isotope behavior during fluid/rock interaction in K-feldspar alteration zone–a model for pyrite in gold deposits from the Jiaodong Peninsula, East China. Geochim Cosmochim Acta 222:94–116. https://doi.org/10.1016/j.gca.2017.10.001

Zipfel J, Weyer S (2007) In situ analyses of Fe isotopes in zoned metall grains of Hammadah al Hamra 237. Lunar Planet, Sci Conf 38

5 The Modern Surficial World

The relatively large isotopic fractionations found for all stable isotope systems at low temperatures has attracted extensive interest in low-temperature environments, and stable Fe isotopes are no exception. The organization of this chapter starts with the continents, addressing weathering and soil-formation processes. We then move to terrestrial rivers and groundwater systems, followed by discussion of redox-stratified water bodies and their sediments, including lakes and the Black Sea. Next, we focus on modern marine sediments, which record extensive Fe biogeochemical cycling and authigenic mineral formation that is key to understanding the modern marine Fe budget. Significant space is allocated to discussion of modern lakes and marine sediments and their connections to water-column processes. As such, this chapter provides an essential framework for understanding the ancient rock record in Chap. 6. We finish this chapter with a detailed discussion of Fe isotopes in modern seawater and the constraints they impose on important Fe sources and cycles in the oceans, including dust, benthic, and hydrothermal components.

5.1 Weathering

5.1.1 Mechanical Weathering

Conceptually, we may consider mechanical weathering to be simple disaggregation of bedrock without net loss or gain of elemental constituents. Aerosols and loess deposits, as well as low-C shales, have been used to constrain the chemical composition of the upper crust based on the assumption that they represent a relatively non-fractionated sample of the upper crust (e.g., Taylor and McLennan 1985). This is not to say that these lithologies do not reflect elemental loss or gain, but simply that such changes do not significantly affect their bulk compositions. In terms of Fe isotope, this idea is supported by studies of aerosols and loess, where Beard et al. (2003) reported an average δ^{56}Fe value of +0.08 ± 0.08‰ for modern aerosols, and an average δ^{56}Fe value of +0.14 ± 0.04‰ for loess samples, which is essentially the same as average igneous rocks on the IRMM-14 scale. Non-soil Fe sources may deviate from the crustal average (Mead et al. 2013). Later studies of loess deposits have confirmed the restricted range in Fe isotope compositions, where Gong et al. (2016) report an average δ^{56}Fe value of 0.09 ± 0.03‰ for samples from the Chinese Loess Plateau. We may take this composition, then, as reflective of average upper crust on the IRMM-14 scale (see also Chap. 1).

5.1.2 Chemical Weathering

Chemical weathering produces soluble Fe, particularly if weathering involves organic acids that may complex Fe^{3+} or reduce Fe^{3+}-oxide/hydroxides (e.g., Jones 1998; Wilson 2004). In

C. Johnson et al., *Iron Geochemistry: An Isotopic Perspective*, Advances in Isotope Geochemistry,
https://doi.org/10.1007/978-3-030-33828-2_5

all studies to date, Fe that is soluble during weathering experiments, including those that assess plant-available Fe, has $\delta^{56}Fe$ values that are lower than that of the parent material (Wiederhold et al. 2006; Guelke and von Blanckenburg 2007; Chapman et al. 2009; Kiczka et al. 2010). The lowest $\delta^{56}Fe$ values for dissolved Fe are produced from dissolution of silicates and oxides by organic acids such as oxalate, which have been interpreted to record a kinetic isotope effect at low extents of dissolution relative to a small equilibrium isotope effect that dominates at greater extents of dissolution (Wiederhold et al. 2006; Chapman et al. 2009; Kiczka et al. 2010). In the case of interaction between aqueous Fe^{2+} and solid-phase Fe^{3+}, the observed isotopic changes may reflect equilibrium exchange between a reactive surface Fe layer and aqueous Fe (see discussion in Chap. 3), rather than a kinetic effect upon leaching of Fe.

Under oxidizing conditions, chemical weathering results in oxidation of Fe^{2+}-bearing silicates (Fritz 1988; Wilson 2004), which will produce Fe^{3+} oxides and hydroxides (Cornell and Schwertmann 2003). The $\delta^{56}Fe$ values of these secondary oxide/hydroxides tend to be higher than those of the parent material, reflecting Fe isotope fractionations expected between oxidized and reduced Fe species (see Chap. 3), although $\delta^{56}Fe$ values can be quite variable, possibly reflecting kinetic effects upon precipitation (Dideriksen et al. 2007, 2010). In contrast to dissolution by organic acids, proton-promoted dissolution of Fe^{3+} oxides and hydroxides does not fractionate Fe isotopes (Skulan et al. 2002; Wiederhold et al. 2006), although it may be difficult to draw an analogy between HCl leaching and natural weathering processes.

5.1.3 Soils

Biogeochemical weathering processes during pedogenesis may mobilize elements, including Fe, that eventually find their way into ground- and surface-water systems. The global average Fe content of soils is ~4 wt%, essentially identical to the crustal average, although soil Fe contents can range from <0.2 wt% to >50 wt%, indicating that soils may record extensive Fe loss or accumulation relative to average protoliths. Mobility of Fe in soils depends on interrelations among slope, drainage, and water flux, the latter of which determines infiltration of atmospheric oxygen, as well as the availability of organic matter and the nature of the bedrock. Climate imposes additional controls, particularly in terms of temperature effects on reaction rates (e.g., Chadwick et al. 2003). Iron mobilization in soils may occur via three mechanisms: (1) Fe^{2+}_{aq} produced by microbial reduction of Fe^{3+}-hydroxides, (2) Fe-organic complexes, which may occur under high Eh conditions as Fe^{3+}-C species, and (3) advective transport of nanoparticulates. Significant Fe isotope changes are expected for the first two processes, but in the last case, little Fe isotope effect is expected if Fe is mobilized by physical particle transport. There are numerous sources of information on soil classification and genesis, as well as biogeochemical processes that are associated with redox process, and we refer the reader to just a few (e.g., Stucki et al. 1988; Cornell and Schwertmann 2003; Sparks 2003; Weil and Brady 2017).

Eh–pH relations are strongly controlled by water saturation due to the limited solubility of O_2 in water. Well-drained soils have the highest Eh, seasonally saturated soils have intermediate Eh, and permanently or semi-permanently saturated (waterlogged) soils have the lowest Eh. In detail, Eh is a function of variables other than O_2, including the abundance of organic carbon and inorganic electron acceptors (e.g., nitrate, ferric Fe) and donors (e.g., sulfide, ferrous Fe). The Eh–pH relations for the redox couple Fe^{2+}_{aq}–$Fe(OH)_3$ may be taken as the dividing point between aerobic (oxidized) and anaerobic (reduced) soils. Soils that undergo seasonal wet-dry cycles tend to become acidic if labile electron acceptors such as nitrate and sulfate are lost. In addition, pH changes may occur through oxidation (proton producing) or reduction (proton consuming) reactions. In terms of Fe isotope variations, pH changes may affect the extent and rate of isotopic exchange and hence the likelihood that a soil may become heterogeneous in its

Fe isotope compositions, due to the strong control pH has on sorption (see Chap. 3).

A wide variety of soil types have been studied for Fe isotope variations. Inceptisols, which have minimal horizon development, are well drained, and reflect high Eh conditions, have limited variations in bulk-sample Fe isotope compositions (Yesavage et al. 2012). Studies of Cambisols, which are also relatively young soils, oxic, and well drained, with little re-distribution of Fe within the profile, have shown generally limited Fe isotope variations (Wiederhold et al. 2007b; Fekiacova et al. 2013). Podzols are characterized by vertical movement of Fe within the horizon, likely via organic-Fe^{3+} complexes under well-drained and high Eh conditions, and this is consistent with significant variations in bulk-sample Fe isotope compositions (Emmanuel et al. 2005; Wiederhold et al. 2007b). Soils that undergo seasonal or permanent water saturation, and hence have relatively low Eh, include albeluvisols and gleysols, and in some cases these soils record significant bulk-sample Fe isotope variations (Mansfeldt et al. 2011; Fekiacova et al. 2013; Akerman et al. 2014; Garnier et al. 2017). One study of andosols, which are soils formed on volcanic terranes, has shown bulk-sample Fe isotope variations across a climate gradient (Thompson et al. 2007). Ferralsols and laterites are soils that are highly enriched in ferric oxyhydroxides, reflecting deep weathering, high Eh, and low organic carbon content, and in some cases these are heterogeneous in bulk-sample Fe isotope compositions, but in other cases they have restricted Fe isotope variations (Liu et al. 2014; Li et al. 2017). Ultisols are examples of extreme weathering that is dominated by clay formation, and examples can be found that have significant bulk-sample Fe isotope variability, and others where Fe isotope compositions are restricted (Fantle and DePaolo 2004; Yesavage et al. 2012).

In Fig. 5.1 we classify soils that generally have bulk-sample δ^{56}Fe values that lie within ±0.2‰ of the crustal average (δ^{56}Fe = −0.11 to +0.29) as "low-mobility" soils, whereas soils that have bulk-sample δ^{56}Fe values which lie outside this range are termed "high-mobility soils". This definition is an isotopic one, which provides information that is distinct from that available based only on Fe contents. At the end of this section we will additionally consider Fe concentration mobility in the context of immobile elements such as Ti and Zr. We further divide soils into those that have Fe_2O_3 contents (as total Fe) <4 wt% ("low-Fe") and those that have higher Fe contents ("high-Fe"). Most soil profiles are less than 100 cm deep, and those that fall in the "low-mobility" group have no systematic variation in δ^{56}Fe values with depth or Fe content within the ±0.2‰ window (Fig. 5.1a). The average δ^{56}Fe value for low-mobility soils is +0.10‰, i.e., indistinguishable from that of average upper crust. Deeper soil profiles have been studied in ferralsols/laterites in Cameroon, Philippines, and China, and for the "low-mobility" group, there is no systematic change in δ^{56}Fe values with depth within the ±0.2‰ window (Fig. 5.1b), where the average δ^{56}Fe value is +0.06‰. In contrast, high-mobility soils (isotopic definition) have bulk-sample δ^{56}Fe values between −0.5 and +0.9‰, and tend to have Fe contents less than the crustal average, with the notable exception of an andosol from Hawaii (Fig. 5.1c), as well as ferralsols/laterites (Fig. 5.1d), in addition to a few other soils. As a group, "high-mobility" soils have slightly higher bulk-sample δ^{56}Fe values (δ^{56}Fe = +0.24), suggesting that, on average, the Fe that was lost had negative δ^{56}Fe values.

In detail, studies across several soil types in drainage profiles document bulk-sample Fe isotope variations that broadly correlate with soil redox conditions. For example, a traverse in Cameroon, from ferralsols on slopes to gleysols in wetlands, shows that the ferralsols have limited variations in Fe isotope compositions, whereas the water-saturated, low Eh gleysol samples have distinctly positive δ^{56}Fe values, indicating loss of reduced Fe that had low-δ^{56}Fe values (Poitrasson et al. 2008; Akerman et al. 2014). In a Hawaiian traverse, Thompson et al. (2007) noted a correlation of increasing bulk-sample δ^{56}Fe values with increasing rainfall and decreasing Eh, which they ascribed to increasing loss of low-δ^{56}Fe Fe under

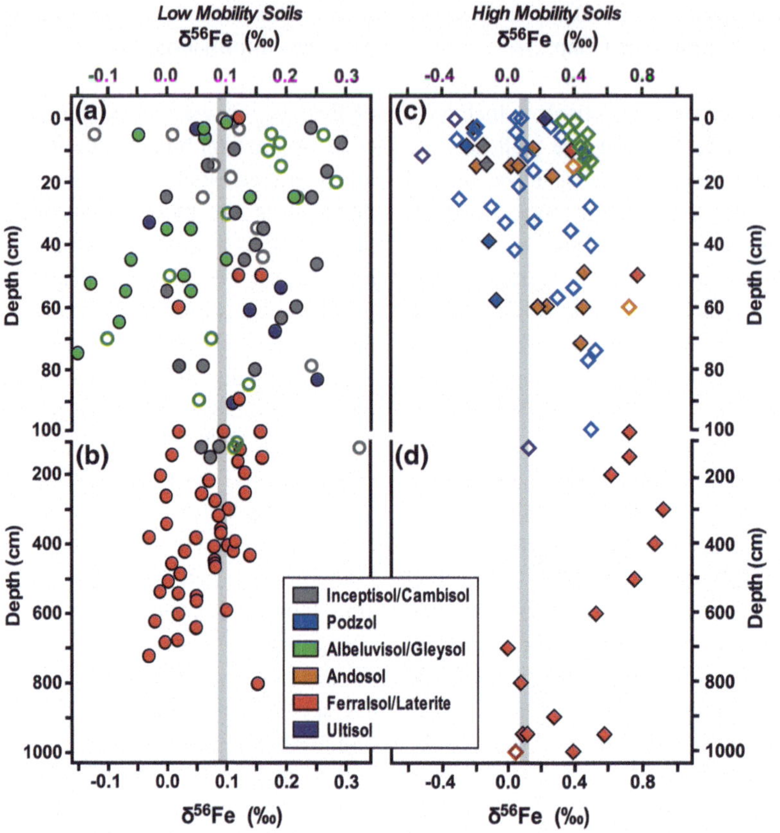

Fig. 5.1 Vertical changes in $\delta^{56}Fe$ values for bulk samples from soil profiles. Data for "low-mobility" (parts **a**, **b**) soils (circle symbols), as defined by soils that have $\delta^{56}Fe$ values that lie within 0.2‰ of average upper crust (vertical grey bars), from Wiederhold et al. (2007a, b), Poitrasson et al. (2008), Mansfeldt et al. (2011), Yesavage et al. (2012), Fekiacova et al. (2013), Liu et al. (2014), Garnier et al. (2017) and Li et al. (2017). Data for "high-mobility" (parts **c**, **d**) soils (diamond symbols), based on soils that have a range in $\delta^{56}Fe$ values that exceed 0.2‰ from that of average upper crust, from Fantle and DePaolo (2004), Emmanuel et al. (2005), Wiederhold et al. (2007b), Thompson et al. (2007), Akerman et al. (2014), and Liu et al. (2014). Lower panels show profiles for exceptionally deep profiles. Symbols are keyed to major soil types, where open symbols indicate low-Fe contents (<4 wt% as total Fe as Fe_2O_3) and filled symbols indicate high-Fe contents (>4 wt% as total Fe as Fe_2O_3)

successively more reducing conditions. In natural systems, production of low-$\delta^{56}Fe$ aqueous Fe^{2+} (Fe^{2+}_{aq}) that can be lost from a soil profile may occur through microbial dissimilatory iron reduction (DIR; e.g., Lovley et al. 2004) or ammonium oxidation (e.g., Yang et al. 2012). Such effects have been confirmed in experimental studies of soils under low-Eh conditions (Schuth et al. 2015); under high-Eh conditions, mobilized Fe (which exists as colloidal Fe^{3+}-hydroxides) had high $\delta^{56}Fe$ values, confirming the important control of Eh in determining the Fe isotope compositions of soils.

The reactive Fe component in soils has been investigated using various leaching methods to isolate the reactive Fe fraction. It has been long recognized that one of the safest leaching approaches that minimizes isotopic artifacts is proton-promoted partial dissolution, such as that obtained using low molarity (<1 M) HCl (Skulan et al. 2002; Wiederhold et al. 2006). Weak HCl leaching has been shown to target poorly crystalline Fe^{3+}-hydroxides, but could also include colloidal Fe, as well as organically bound Fe (Fantle and DePaolo 2004); unfortunately, Fe isotope studies of soils have not included Fe^{2+}–Fe^{3+} speciation or organic C measurements to help constrain the exact Fe components that have been extracted in weak HCl leaches. In some studies, several sequential leaches were done to isolate reactive Fe, well-crystalline Fe oxides, and silicate Fe (Wiederhold et al. 2007b).

The range in δ^{56}Fe values for 0.5 M HCl soil leaches is much larger than that seen in bulk-sample δ^{56}Fe values for the same soils (Fig. 5.2), supporting the idea that the most labile Fe components in soils are isotopically heterogeneous. Even for the "low-mobility" group, the range in δ^{56}Fe values for 0.5 M HCl leaches approaches 1‰ and the δ^{56}Fe values of the leaches are systematically lower than corresponding bulk-samples (Fig. 5.2a). For the "high-mobility" group, δ^{56}Fe values of the leaches vary greatly and are generally lower than those of the corresponding bulk-samples. In the majority of cases, reactive Fe extracted using weak HCl leaches have δ^{56}Fe values that are lower than those of the bulk Fe (Fig. 5.2). The low δ^{56}Fe values of reactive Fe has been generally taken to indicate reductive dissolution and reprecipitation (Wiederhold et al. 2007a, b) or organically bound Fe (Buss et al. 2010; Yesavage et al. 2012). Generally, soil leaches with low-δ^{56}Fe values have been inferred to reflect plant-available Fe in soils (Guelke et al. 2010). Specific to reactive Fe, if the interpretation that HCl extraction represents poorly crystalline Fe^{3+}-hydroxides is correct, then an oxidative step is required that preserves the low-δ^{56}Fe values of the reduced component (Yesavage et al. 2012; Liu et al. 2014). In a few studies, the δ^{56}Fe values for 0.5 M HCl leaches are higher than those of the bulk samples, most notably in the andosols of Hawaii. Thompson et al. (2007) argue that higher δ^{56}Fe values for reactive Fe indicate partial oxidation of reduced Fe, which may occur when low-Eh soils are periodically exposed to high-Eh conditions through redox cycling. A process of partial oxidation of reduced Fe might further increase isotopic variability in exported Fe from the residual reduced pool. Eh-driven Fe isotope effects are likely to be variable over the lifetime of soil development (Thompson et al. 2006).

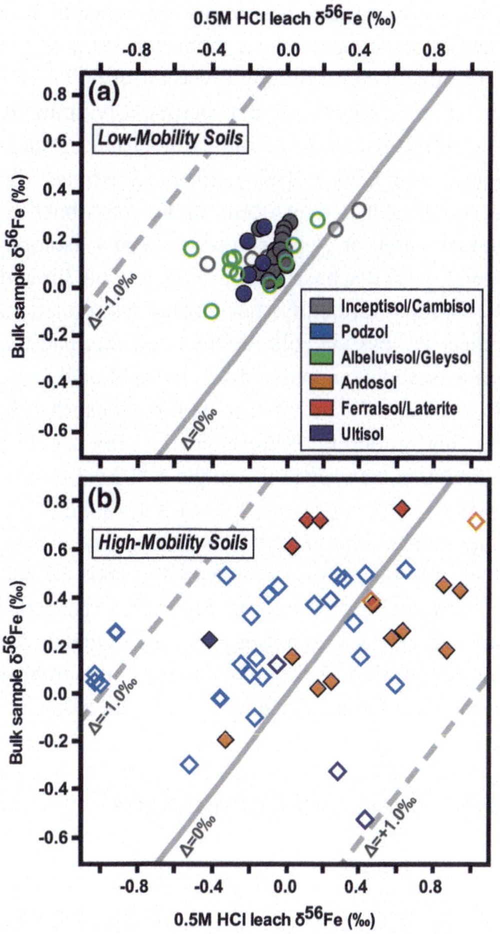

Fig. 5.2 Comparison of δ^{56}Fe values for bulk soil samples with those of 0.5 M HCl leaches of the same sample. Symbols same as in Fig. 5.1. For "low-mobility" soils (**a**), the range in δ^{56}Fe values for 0.5 M HCl leaches is relatively restricted, as compared to leaches of "high-mobility" soils (**b**), confirming that "high-mobility" soils have undergone Fe isotope exchange with reactive Fe components. As in Fig. 5.1, Fe mobility is based on an isotopic definition. Data from Fantle and DePaolo (2004), Wiederhold et al. (2007b), Thompson et al. (2007), Yesavage et al. (2012), and Liu et al. (2014)

Independent of Fe isotope studies, net Fe loss, based on Fe concentrations, are commonly expressed as the parameter tau (τ), through normalization to immobile elements such as Ti or Zr. For the "low-mobility" group, defined by a restricted range in bulk-sample δ^{56}Fe values, these soils vary greatly in terms of τ-based Fe loss (and in some cases, show modest gain), and yet have remained essentially invariant in δ^{56}Fe values (Fig. 5.3a); this effect is particularly dramatic for ferralsols/laterites from Cameroon, Philippines, and China. This contrast illustrates the decoupling that may occur in Fe mass-balance models that are concentration versus isotopic based. Soils that have a large range in τ, but limited Fe isotope variability, have been interpreted to reflect Fe loss via colloidal material that had the same isotopic compositions as the bulk soil (e.g., Li et al. 2017). In contrast, ferralsols/laterites in the "high-mobility" group have a broadly negative correlation between bulk-sample δ^{56}Fe values and τ (Fig. 5.3b), suggesting loss of a negative δ^{56}Fe component. Although more subtle and scattered, such a trend is consistent with the andosol and albeluvisol/gleysol soils in Fig. 5.3b). The relations in Fig. 5.3 do not provide any evidence for loss of a high-δ^{56}Fe component, a component that is important in the discussions of rivers below.

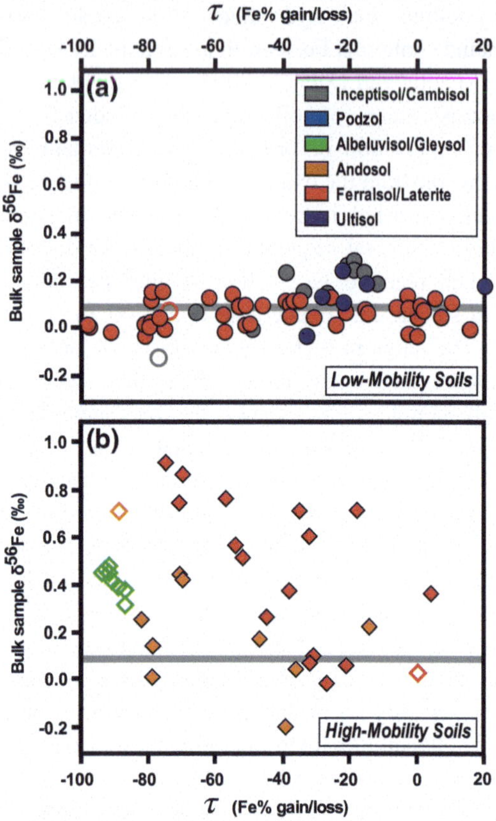

Fig. 5.3 Variations in bulk-sample δ^{56}Fe values for soils plotted in Fig. 5.1 for which % Fe gain/loss (expressed as positive or negative τ, respectively) has been calculated. Symbols and soils divisions/classifications, as well as data sources, same as in Fig. 5.1. Horizontal grey bars indicate the δ^{56}Fe value of average upper crust

5.2 Rivers and Groundwater

5.2.1 Rivers

Suspended sediment loads in rivers currently contribute a total element flux of ~1.8×10^{16} g/year to the oceans, a flux that was a factor of 2–10 less prior to development of farming (Milliman and Syvitski 1992). Suspended particulate matter (SPM) is generally defined as material retained on 0.22 or 0.45 μm filters; this size cutoff is similar to the definition used in the seawater literature (Sect. 5.5), where particulate Fe is defined as "pFe". Early isotopic studies suggested that the Fe isotope compositions of SPM were identical to those of average igneous rocks, aerosols, and loess (δ^{56}Fe = +0.08‰; Beard et al. 2003). The rivers analyzed by Beard et al. (2003) contained up to 50% Fe in non-silicate minerals, indicating a major component of poorly crystalline Fe^{3+}-hydroxides (Canfield 1997), suggesting little Fe isotope fractionation during modern oxidative weathering. Subsequent work has shown that δ^{56}Fe values for SPM range up to 2‰, although many rivers do have δ^{56}Fe values close to that of average upper crust (Fig. 5.4). Low-latitude rivers such as the Amazon, Maderia, and Solimoes rivers that have low organic contents tend to have δ^{56}Fe values similar to average upper crust, whereas the organic-rich Negro river has distinctly negative δ^{56}Fe values for SPM (Fig. 5.4). The Negro river drains podzol soils, and temporal variations indicate that during wet seasons δ^{56}Fe values are lowest, consistent with conditions expected to mobilize low-δ^{56}Fe Fe under low-Eh conditions (dos Santos Pinheiro et al. 2013, 2014; Poitrasson et al. 2014). It has been suggested that the δ^{56}Fe values for SPM is a function of river latitude (dos Santos Pinheiro et al. 2014), but consideration of all available data suggests no simple correlation.

The mid-latitude rivers that have been studied to date suggest a relatively restricted range in δ^{56}Fe values for SPM (Fig. 5.4). Despite the small range in Fe isotope compositions for SPM from mid-latitude rivers, Escoube et al. (2009) documented small changes due to mixing between seawater and fresh water in the North River estuary, which they inferred to reflect mixing of river-borne particles and lithogenic particles, but minimal Fe isotope fractionation during flocculation. A similar conclusion, based on other samples, was reached by Rouxel and Auro (2010). This interpretation contrasts with that of Bergquist and Boyle (2006b), who found that flocculation produced an Fe isotope fractionation of a few tenths per mil in δ^{56}Fe values. Finally, the last mid-latitude river shown in Fig. 5.4, the Aha Lake basin, has a distinctly different distribution in Fe isotope compositions, which Song et al. (2011) interpreted to reflect seasonal variability that records weathering of distinct mineralogy.

The range in δ^{56}Fe values for SPM appears to be greatest in high-latitude rivers (Fig. 5.4). Ingri et al. (2006) interpreted a positive correlation between δ^{56}Fe and SPM Fe content to reflect mixing between a low-δ^{56}Fe Fe-C colloidal component and a high-δ^{56}Fe Fe^{3+}-hydroxide component. Schroth et al. (2011) inferred the presence of low-δ^{56}Fe C complexes that were produced by decomposition of organic matter based on relations between δ^{56}Fe dissolved organic carbon (DOC). Later, Ilina et al. (2013) came to the opposite conclusion, using size-filtration analysis discussed in more detail below, where they inferred the Fe-C colloidal component to have very high δ^{56}Fe values, an effect superimposed upon moderately positive δ^{56}Fe values for SPM (Fig. 5.4). The largest Fe isotope variations for SPM in high-latitude rivers is found in the smallest rivers, including the

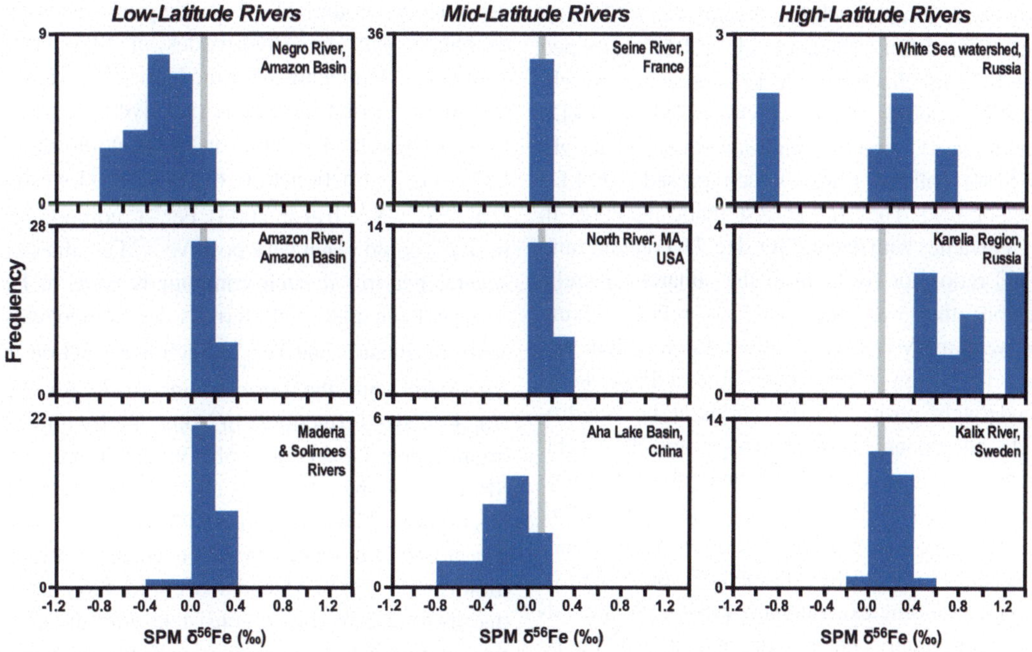

Fig. 5.4 Histograms of δ^{56}Fe values for Suspended Particulate Matter (SPM) for low, mid-, and high-latitude rivers. The widest range in δ^{56}Fe values for SPM is associated with the most organic-rich rivers. Data from Bergquist and Boyle (2006b), Ingri et al. (2006), Escoube et al. (2009, 2015), Song et al. (2011), Ilina et al. (2013), dos Santos Pinheiro et al. (2013), (2014), Chen et al. (2014), and Poitrasson et al. (2014). Vertical grey bars mark average upper crust

most positive $\delta^{56}Fe$ values yet measured for SPM. The origin of positive $\delta^{56}Fe$ values for SPM may reflect Fe^{3+}-hydroxides or isotopic exchange between SPM and colloidal Fe (Ilina et al. 2013; Escoube et al. 2015).

The total element flux of the dissolved load in rivers is currently <1% of that of the suspended load, or $\sim 10^{14}$ g/year (Moon et al. 2014). From an isotopic mass-balance perspective, this suggests that the dissolved component in rivers might be expected to be isotopically more variable than SPM. The definition of "dissolved", which we will define as "dFe", however, is variable, and commonly based on a filtration threshold that varies between <0.2 and <0.45 μm, depending on the study. As such, it is recognized that dFe in rivers may (likely?) include colloidal and nanoparticle components. Nine rivers have been analyzed for the Fe isotope compositions of dFe using a 0.2 or 0.45 μm threshold, and the majority of these have a restricted range in $\delta^{56}Fe$ values between −0.41 and 0.09 (Fig. 5.5a). In contrast, studies that have separated "dissolved" components into a variety of sizes have found much larger ranges in $\delta^{56}Fe$ values (Fig. 5.5b). Ilina et al. (2013) grouped sub-0.22 μm fractions into low-molecular weight (LMW) colloids (those which passed 1–10 kDa membranes) and high-molecular weight (HMW) colloids (those which passed 100 kDa membranes). They found systematic increases in $\delta^{56}Fe$ values with decreasing size, Fe contents, and Fe/C ratios for rivers from the subarctic Karelia region; this was interpreted to reflect mixing between a LMW C-rich colloidal component that had very high-$\delta^{56}Fe$ values, and a larger-sized moderately positive $\delta^{56}Fe$ colloidal Fe^{3+}-hydroxide component. Similar size-composition relations were observed by Escoube et al. (2015), also for organic-rich high-latitude rivers (Fig. 5.5b), highlighting the very large range in $\delta^{56}Fe$ values that may occur in the "dissolved" component of rivers, including exceptionally positive $\delta^{56}Fe$ values. Importantly, the broadly negative $\delta^{56}Fe$–Fe content relations cannot be explained by oxidation of Fe^{2+}_{aq} in the river samples, a process which would produce a positive relation between $\delta^{56}Fe$ values and Fe contents for aqueous Fe (see discussion in Sect. 5.2.2).

The possibility that organically bound dissolved Fe in rivers may have positive $\delta^{56}Fe$ values was first suggested by Bergquist and Boyle (2006b). This was confirmed in experimental studies of Fe isotope fractionation between siderophore complexes and Fe^{3+}_{aq}, where Dideriksen et al. (2008) inferred an equilibrium $^{56}Fe/^{54}Fe$ fractionation between Fe^{3+}-siderophore and Fe^{3+}_{aq} of +0.6‰ (see also Chap. 3). Importantly, using an enriched ^{57}Fe tracer, they showed that isotopic exchange was rapid, suggesting that an Fe^{3+}-C component in rivers would likely reflect equilibrium isotope fractionation. This work was extended to EDTA and oxalate by Morgan et al. (2010), who found that EDTA-bound Fe^{3+} is about 0.6‰ heavier in $^{56}Fe/^{54}Fe$ than Fe^{3+}_{aq}, and oxalate-bound Fe^{3+} was about 0.4‰ heavier in $^{56}Fe/^{54}Fe$ than Fe^{3+}_{aq}. Because the equilibrium Fe^{3+}_{aq}-ferrihydrite fractionation factor is near zero for $^{56}Fe/^{54}Fe$ (Chap. 3), these results suggest that if organically bound Fe^{3+} undergoes isotopic exchange with Fe^{3+}-hydroxides in SPM, this could be an explanation for the high $\delta^{56}Fe$ values measured for dFe in organic-rich rivers. In addition, when considered on top of the equilibrium Fe^{3+}_{aq}–Fe^{2+}_{aq} fractionation of ~+3‰ (Johnson et al. 2002; Welch et al. 2003), Fe^{3+}-C compounds in rivers could reach very positive $\delta^{56}Fe$ values of several per mil if such compounds were in Fe isotope exchange with ferrous Fe components such as silicates and Fe^{2+}_{aq}. Such a prediction is consistent with the observations made for the smallest sized fractions in some high-latitude, organic-rich rivers (Fig. 5.5b). Where co-existing SPM is available, in almost all cases dissolved Fe has higher $\delta^{56}Fe$ values than SPM (Fig. 5.6), and the largest differences in $\delta^{56}Fe$ values between dissolved Fe and SPM tend to occur for the smaller dissolved size fractions, consistent with the relations in Fig. 5.5. The major exception is the Seine river, where Chen et al. (2014) notes that the very low $\delta^{56}Fe$ values for dFe may reflect anthropogenic sources.

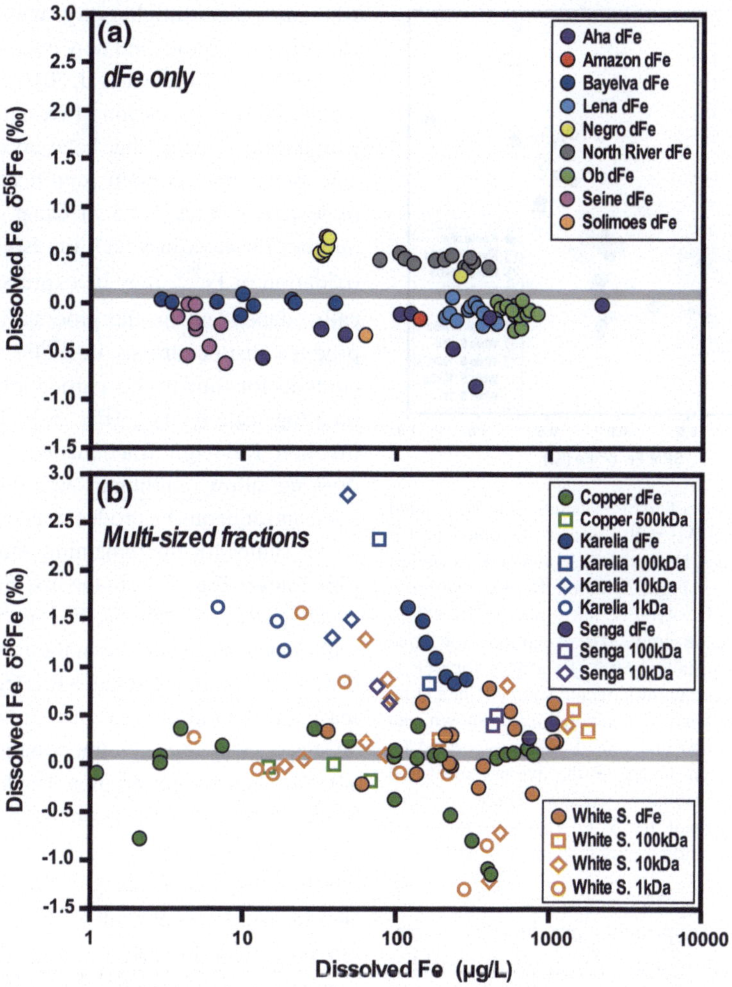

Fig. 5.5 Variations in δ^{56}Fe values of dissolved Fe in rivers relative to dissolved Fe contents of the sample. Where dissolved Fe is defined as dFe (filled circle symbols), this is the fraction that passed a 0.2 or 0.45 µm filter, depending on the study. In other cases, smaller size fractions were also separated based on different molecular-weight membranes, as noted. Panel **a** shows data where only dFe was available, whereas panel **b** shows samples where multiple size fractions were analyzed beyond dFe (open symbols: squares for 100 and 500 kDa, diamonds for 10 kDa, and circles for 1 kDa). The highest δ^{56}Fe values tend to be associated with the lowest molecular weight fractions. Horizontal grey bars indicate average upper crust. Data from Bergquist and Boyle (2006b), Escoube et al. (2009, 2015), Song et al. (2011), Ilina et al. (2013), dos Santos Pinheiro et al. (2014), Chen et al. (2014), and Zhang et al. (2015)

It cannot, however, be concluded that organically bound Fe must always have positive δ^{56}Fe values, given the fact that Fe taken up by plants has relatively low δ^{56}Fe values, and the fact that dissolution of silicate minerals by organic acids produces low δ^{56}Fe values (see Sect. 5.1.2). Ilina et al. (2013) observed that Fe^{3+} bound to HMW organic C tended to have lower δ^{56}Fe values than Fe bound to LMW compounds, and hence δ^{56}Fe values may be a function of the specific organic compound that is bound to Fe, an observation seen in experiments (Morgan et al. 2010). Nevertheless, the relations in Fig. 5.6 demonstrate that the dissolved riverine Fe load to the oceans has δ^{56}Fe values that are significantly higher than those of the suspended river load. The δ^{56}Fe

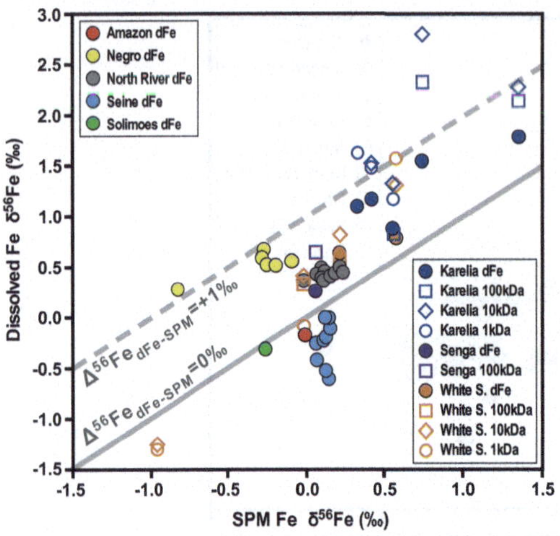

Fig. 5.6 Comparison of δ^{56}Fe values for Suspended Particulate Matter (SPM) with those of dissolved Fe for the same river samples. Definitions and symbols, as well as data sources, same as in Fig. 5.5, although colors may vary. In almost all cases, dissolved Fe has higher δ^{56}Fe values than SPM from the same sample. The exception is the Seine river, where the unusually low δ^{56}Fe values for dFe is interpreted to reflect an anthropogenic source (Chen et al. 2014). For reference, two fractionation lines are shown, one for $\Delta^{56}Fe_{dFe-SPM}$ = 0‰ and one $\Delta^{56}Fe_{dFe-SPM}$ = +1‰. Ignoring the data from the Seine, the average measured $\Delta^{56}Fe_{dFe-SPM}$ fractionation is +0.5‰. These relations document that, in general, the δ^{56}Fe values for dissolved Fe in rivers is significantly higher than those of SPM

values of these components are broadly correlated, meaning that a high δ^{56}Fe value for SPM is correlated with an even higher δ^{56}Fe value for dissolved Fe. Such observations will bear on discussions below on the Fe isotope inputs to the oceans (Sect. 5.5).

5.2.2 Groundwater and Terrestrial Hydrothermal Systems

Interaction between reduced, Fe^{2+}_{aq}-rich groundwaters and oxygenated environments produces a wide range in δ^{56}Fe values and Fe contents for the aqueous component, reflecting the Fe isotope fractionations associated with oxidation of Fe^{2+}_{aq} and precipitation of Fe^{3+}-hydroxides. Some groundwater systems may be in steady-state conditions relative to isotopic inputs and outputs, whereas others might vary with time due to changing reservoirs and may even reflect tectonic events (e.g., Schuessler et al. 2016). In this section, we divide our discussion of oxidation of reduced waters into systems that have no or little sulfide and those that contain significant sulfide. As reviewed in Sect. 1.2.3 in Chap. 1, there are a number of ways in which the isotopic effects of oxidation of Fe^{2+}_{aq} may be expressed mathematically, dependent on the process, and these may produce distinct trends in δ^{56}Fe values and Fe contents for aqueous Fe, particularly when plotted on a log scale for Fe_{aq} (Fig. 5.7). The commonly invoked Rayleigh fractionation process, which does not allow continuous equilibration between solid and aqueous Fe produces an extreme range in δ^{56}Fe values for the remaining aqueous Fe (solid pink lines in Fig. 5.7). In the first study of a spring system, where an anoxic, Fe_{aq}-rich spring mixed with ambient oxygen down the flow path, Bullen et al. (2001) suggested that a Rayleigh model fit at least the initial stages of oxidation, using a $\Delta^{56}Fe_{Ferrihydrite-Fe(II)aq}$ fractionation factor of +0.9‰. Deviations from a Rayleigh model at higher extents of oxidation were interpreted by Bullen et al. (2001) to reflect either "back-equilibration" or a change in isotopic fractionation factor. A recent study of Fe^{2+}_{aq} oxidation in a stream system similarly ascribed a decrease in Fe_{aq} contents and changes in δ^{56}Fe values to progressive oxidation of Fe^{2+}_{aq} (Ito et al. 2017).

In an attempt to model isotopic exchange between Fe^{2+}_{aq} and Fe^{3+}_{aq} during oxidation, as well as account for kinetic effects upon Fe^{3+}-hydroxide precipitation, Beard and Johnson (2004) proposed a "two-step" model, where Fe^{2+}_{aq} oxidizes to Fe^{3+}_{aq}, followed by precipitation as $Fe(OH)_3$. This was expanded upon by Dauphas and Rouxel (2006). The models differ in terms of the species that are in isotopic exchange and depend on the time it takes to attain isotopic equilibrium relative to the residence time of these species. The two-step model produces a significantly shallower slope than Rayleigh fractionation in Fig. 5.7, although it also cannot reproduce the curvature in Fe_{aq}–δ^{56}Fe space that is apparent

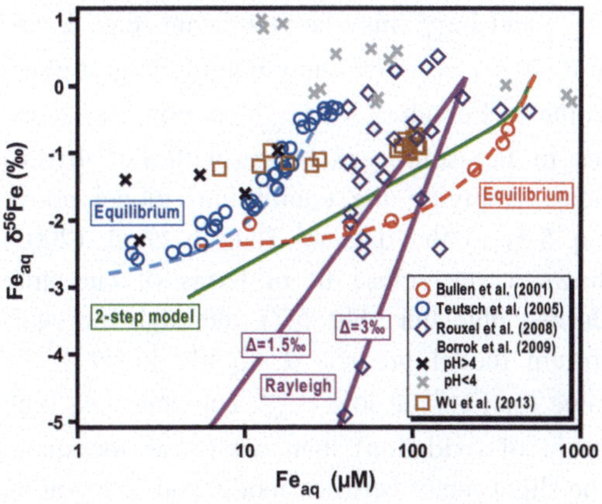

Fig. 5.7 Variations in δ^{56}Fe values of aqueous Fe relative to Fe_{aq} contents for non-sulfide groundwater systems, including moderate-temperature hydrothermal fluids. Aqueous fraction was obtained using a 0.2 or 0.45 μm filter, depending upon the study. In all cases, oxidation of Fe^{2+}_{aq} occurred, decreasing Fe_{aq} contents. It is important to note that Fe_{aq} speciation is not known for all studies, and may not consist only of Fe^{2+}_{aq}. Several Fe isotope fractionation models are illustrated for oxidation, including Rayleigh fractionation (shown in purple/pink lines for two $\Delta^{56}Fe_{Ppt-Aq}$ fractionation factors between $Fe^{3+}{}_{aq}$ hydroxide and Fe^{2+}_{aq} of +1.5 and +3.0‰, which are linear on the log-scale plots), a "two-step" model from Beard and Johnson (2004) (green line) that incorporates an equilibrium Fe^{3+}_{aq}–Fe^{2+}_{aq} fractionation factor of +3‰, with a +2‰ kinetic fractionation between Fe^{3+}_{aq} and Fe^{3+}-hydroxide upon precipitation, and a new equilibrium model that assumes equilibrium isotope exchange between Fe^{2+}_{aq} and $Fe(OH)_3$ and accounts for sorption in the Fe isotope mass balance (Eq. 5.5; see text), shown in dashed blue and red lines. Data sources are Bullen et al. (2001), Teutsch et al. (2005), Rouxel et al. (2008b), Borrok et al. (2009), and Wu et al. (2013)

in experimental and field data. The first part of the two-step oxidation model likely occurs under conditions of Fe isotope equilibrium, given the rapid rates of Fe isotope exchange between Fe^{2+}_{aq} and Fe^{3+}_{aq} (Johnson et al. 2002; Welch et al. 2003):

$$4Fe^{2+} + O_2 + 8OH^- + 2H_2O \rightleftharpoons 4Fe^{3+} + 12OH^- \quad (5.1)$$

The second step, precipitation of ferric oxide/hydroxide, may involve significant and variable kinetic isotope effects (Skulan et al. 2002):

$$Fe^{3+} + 3OH^- \rightarrow Fe(OH)_3 \quad (5.2)$$

Since this model was developed, however, it has become clear that coexistence of Fe^{2+}_{aq} and ferric oxyhydroxides may produce extensive Fe isotope exchange due to electron transfer reactions that are coupled to Fe atom exchange (e.g., Handler et al. 2009; see detailed discussion in Chap. 3). This exchange may maintain Fe isotope equilibrium between Fe^{2+}_{aq} and Fe^{3+}-hydroxide during the oxidation process:

$$Fe^{2+}_{aq} \rightleftharpoons Fe^{3+}_{solid} + e^- \quad (5.3)$$

We may therefore develop a third model that accounts for isotopic equilibration between aqueous Fe^{2+}, sorbed Fe^{2+}, and ferric hydroxide, and also assumes that the "solid" component that is separated from the fluid includes sorbed Fe^{2+} and ferric hydroxide. Such a model is particularly useful for studies of the aqueous component, as it accounts for removal of solid and sorbed Fe, which previous models have ignored.

First we define the isotopic mass balance for a closed system containing Fe^{2+}_{aq}, Fe^{2+}_{sorb}, and $Fe(OH)_3$:

$$\delta^{56}Fe_{system} = X_{Fe^{2+}_{aq}}\delta^{56}Fe_{Fe^{2+}_{aq}} + X_{Fe^{2+}_{sorb}}\delta^{56}Fe_{Fe^{2+}_{sorb}} + X_{Fe(OH)_3}\delta^{56}Fe_{Fe(OH)_3} \quad (5.4)$$

where $X_{Fe^{2+}_{aq}}$, $X_{Fe^{2+}_{sorb}}$, and $X_{Fe(OH)_3}$ are the mole fractions of Fe^{2+}_{aq}, Fe^{2+}_{sorb}, and $Fe(OH)_3$, respectively, and the δ^{56}Fe values are similarly defined.

The model uses the mass-balance relations in Wu et al. (2011b), where $X_{Fe^{2+}_{aq}}$, $X_{Fe^{2+}_{sorb}}$, and $X_{Fe(OH)_3}$ are 0.43, 0.08, and 0.49, respectively, and their estimated Fe^{2+}_{aq}-ferrihydrite and Fe^{2+}_{aq}–Fe^{2+}_{sorb} $^{56}Fe/^{54}Fe$ fractionations of −3.2‰ and −0.8‰, respectively. We thus derive an overall $^{56}Fe/^{54}Fe$ fractionation between Fe^{2+}_{aq} and solid (Fe^{2+}_{sorb} + ferrihydrite) of −2.6 to −2.8‰. Note that the total fractionation factor between aqueous and solid Fe cannot be derived by simply adding the Fe^{2+}_{aq}-ferrihydrite and Fe^{2+}_{aq}–Fe^{2+}_{sorb} fractionation factors as that would not account for the relative Fe balance in the aqueous and sorbed Fe and ferrihydrite. Substitution and re-arrangement produces the simple relation below, which assumes Fe isotope equilibrium is maintained Fe^{2+}_{aq}, Fe^{2+}_{sorb}, and $Fe(OH)_3$ during oxidation:

$$\delta^{56}Fe_{Fe^{2+}_{aq}} = \left(1 - X_{Fe^{2+}_{aq}}\right)\Delta^{56}Fe_{Fe^{2+}_{aq-solid}} + \delta^{56}Fe^{i}_{Fe^{2+}_{aq}} \tag{5.5}$$

where $\Delta^{56}Fe_{Fe^{2+}_{aq-soild}}$ is the $^{56}Fe/^{54}Fe$ fractionation between Fe^{2+}_{aq} and the total solid (Fe^{2+}_{sorb} + ferrihydrite) and $\delta^{56}Fe^{i}_{Fe^{2+}_{aq}}$ is the $\delta^{56}Fe$ value for the initial Fe^{2+}_{aq}, prior to oxidation. This third model provides an excellent fit to the data from Bullen et al. (2001) (red dashed line, Fig. 5.7), although it is not possible to test this in detail because the Fe isotope compositions of the sorbed component were not measured in that study. In addition, there is some uncertainty in the speciation of aqueous Fe in the dataset of Bullen et al. (2001), and the assumption that all aqueous Fe is Fe^{2+}_{aq} may not be correct. Nevertheless, the model supports the suggestion by Bullen et al. (2001) that isotopic "back reaction" (exchange) may have occurred in their system, although supporting experimental studies were not yet available at the time.

An in situ field experiment that involved oxygen injection into a reduced groundwater system by Teutsch et al. (2005) provides an extensive Fe isotope dataset to evaluate the isotopic changes that occur during oxidation of Fe^{2+}_{aq}. In this early study, Teutsch et al. (2005) do not consider exchange between Fe^{2+}_{aq} and ferrihydrite, as experimental studies were only just beginning that evaluated isotopic exchange kinetics between Fe^{2+}_{aq} and Fe^{3+}-hydroxides. Instead, they interpret the changes in $\delta^{56}Fe$ values for Fe^{2+}_{aq} to solely reflect the effect of sorption of Fe^{2+} to newly formed Fe^{3+}-hydroxides. Although the Fe isotope fractionation between Fe^{2+}_{sorb} and Fe^{2+}_{aq} may be significant (e.g., Beard et al. 2010), we now know that the largest effect on the $\delta^{56}Fe$ values for Fe^{2+}_{aq}, in non-S systems, lies in Fe isotope exchange with Fe^{3+}-hydroxides. Applying the equilibrium model above (Eq. 5.5) to the data of Teutsch et al. (2005) shows a very close fit in terms of curvature (dashed blue line; Fig. 5.7), although the equilibrium model predicts a slightly lower $\delta^{56}Fe$ value for Fe^{2+}_{aq} at low Fe^{2+}_{aq} contents (i.e., high extent of oxidation) than what was measured. The slight misfit between model and observation suggests that the model's experimentally derived proportions of Fe^{2+}_{aq}, Fe^{2+}_{sorb}, and $Fe(OH)_3$ (after Wu et al. 2011b) may not be appropriate for the groundwater system studied by Teutsch et al. (2005). If, for example, the proportions of Fe^{2+}_{aq} and Fe^{2+}_{sorb} were changed, as might occur with variations in ferrihydrite surface area, this will change the overall $\Delta^{56}Fe_{Fe2+aq-solid}$ fractionation factor. Broadly, however, the relations shown in the studies of Bullen et al. (2001) and Teutsch et al. (2005) suggest that equilibrium Fe isotope fractionation, where isotopic exchange occurs between Fe^{2+}_{aq} and ferrihydrite at circum-neutral pH, seems a reasonable explanation. An equilibrium model provides a much closer fit to the data than any Rayleigh model, which would produce linear trends in $\delta^{56}Fe$ values relative to the log of Fe_{aq} contents, and more extreme changes in $\delta^{56}Fe$ values for a given decrease in Fe^{2+}_{aq} contents (Fig. 5.7).

The data for other spring/groundwater systems do not fit any of the oxidation models discussed above (Rayleigh, two-step, or equilibrium) (Fig. 5.7). Rouxel et al. (2008b) measured a wide range in $\delta^{56}Fe$ values for Fe_{aq} in the subterranean estuary of Waquoit Bay, Cape Cod, USA. Such settings are of great interest because they often reflect mixing between reduced groundwaters and oxygenated seawater (e.g., Moore 1999). Rouxel et al. (2008b) identified two distinct sources of Fe: an anoxic freshwater/groundwater component that has high Fe^{2+}_{aq} and slightly positive to moderately negative $\delta^{56}Fe$ values that is interpreted to record DIR, and another low-Fe^{2+}_{aq} component (higher salinity/brackish composition) that has highly negative $\delta^{56}Fe$ values. They interpret these compositions to record oxidation of Fe^{2+}_{aq} and/or isotopic exchange between Fe^{2+}_{aq} and Fe^{3+}-hydroxides. The wide scatter in variations between Fe^{2+}_{aq} and $\delta^{56}Fe$ values (Fig. 5.7) suggests a mixture of processes, consistent with the proposal of Rouxel et al. (2008b) based on analysis of the Fe^{3+}-hydroxide-coated sands.

Mixing of distinct Fe sources is well documented in the alpine watershed studied by Borrok et al. (2009), where a low-Fe^{2+}_{aq} and low-$\delta^{56}Fe$ component derived from an anoxic fen of circum-neutral pH mixes with an acid-mine drainage component that has high Fe^{2+}_{aq} and high $\delta^{56}Fe$ values (Fig. 5.7). Although Borrok et al. (2009) interpret the low-Fe^{2+}_{aq} and low-$\delta^{56}Fe$ component to record oxidation of Fe^{2+}_{aq}, this could also reflect isotopic exchange between oxidized and reduced components in the fen environment. Borrok et al. (2009) suggest that the low-pH component, which has high-Fe_{aq} contents and high $\delta^{56}Fe$ values, was produced by dissolution of high-$\delta^{56}Fe$ oxyhydroxides that formed during oxidative weathering of pyrite. Importantly, Fe_{aq} speciation for the low-pH fluids indicate a high proportion of Fe^{2+}_{aq}, which is consistent with dissolution of oxyhydroxides under oxic and acidic conditions; the occurrence of a mixed Fe^{2+}–Fe^{3+} fluid will add complexity to the system, including scatter that reflects variable isotopic equilibration among aqueous and mineral components.

Isotopic compositions and aqueous Fe concentration relations for the circum-neutral Chocolate Pots hot spring (Yellowstone, USA) cannot be described by a Rayleigh or equilibrium model (Fig. 5.7). Despite oxidation of Fe^{2+}_{aq}, $\delta^{56}Fe$ values only mildly decrease with decreasing Fe^{2+}_{aq}. Wu et al. (2013) suggested this might be caused by changes in fractionation factors in the Si-rich hot spring, or the effects of DIR. In the latter case, reduction of high-$\delta^{56}Fe$ Fe-oxyhydroxides would provide a flux of high-$\delta^{56}Fe$ Fe^{2+}_{aq} into spring waters. Such high-$\delta^{56}Fe$ Fe-oxyhydroxides could be produced during the early stages of Rayleigh distillation, whereas much more negative Fe^{3+}–oxyhydroxides would be produced by extreme extents of oxidation of Fe^{2+}_{aq}. Alternatively, Wu et al. (2013) posit that Fe^{2+} sorption to Fe–Si precipitates may have an unusual Fe isotope fractionation factor. The Fe isotope data from Wu et al. (2013) remain perplexing in terms of a mechanistic model.

Sulfide-bearing groundwater systems have distinct trends among Fe^{2+}_{aq}, H_2S, and $\delta^{56}Fe$, reflecting a variety of kinetic and equilibrium processes and exchange between fluid, Fe-oxyhydroxides and sulfides (Fig. 5.8). Vertical profiles through the subterranean estuary at Indian River Lagoon (Florida, USA) identify three zones: a deep anoxic zone that has high Fe_{aq} contents and negative $\delta^{56}Fe$ values that record DIR, an intermediate zone of more moderate Fe_{aq} and $\delta^{56}Fe$, and a shallow zone that has elevated sulfide contents and high $\delta^{56}Fe$ values in Fe_{aq} that are interpreted to record precipitation of sulfides (Roy et al. 2012). Because the subterranean estuary intersects the marine realm, sulfide contents at the interface between terrestrial and marine interstitial fluids reach very high levels. Guo et al. (2013), in a study of groundwater from the Hetao Basin (Mongolia, China), combined Fe_{aq}, H_2S, Eh, and Fe isotope data to distinguish zones of DIR, oxidation of Fe^{2+}_{aq}, and sulfide (or siderite) precipitation. The DIR zone in the Hetao Basin was identified by elevated Fe^{2+}_{aq}, low H_2S, and low Eh, where $\delta^{56}Fe$ values are ~ -0.9‰. This low-$\delta^{56}Fe$ zone is similar to the Fe reduction zone noted by Roy et al. (2012). Based on the experimental studies of Fe^{2+}_{aq}–FeS_{aq}–FeS by Guilbaud et al. (2011a) and Wu et al. (2012), Fe_{aq} in equilibrium with FeS can be identified by its negative $\delta^{56}Fe$ values at relatively high H_2S and Fe_{aq} contents (Fig. 5.8).

Both Roy et al. (2012) and Guo et al. (2013) suggest that trends of increasing $\delta^{56}Fe$ for Fe_{aq} with decreasing H_2S and Fe_{aq} contents (Fig. 5.8) record precipitation of sulfides, which requires a

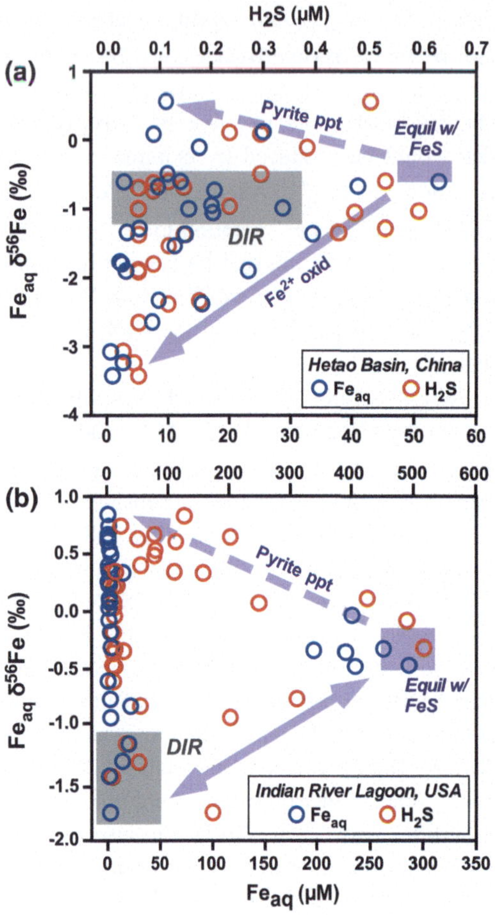

Fig. 5.8 $\delta^{56}Fe$ values for aqueous Fe as a function of Fe_{aq} and aqueous H_2S contents for two sulfide-bearing groundwater systems. Fe_{aq} obtained by passing through a 0.2 or 0.45 µm filter, depending upon the study. The Hetao Basin (Mongolia, China), shown in **a**, is a groundwater system that contains elevated As and sulfide (data from Guo et al. 2013). The Indian River Lagoon (Florida, USA), shown in **b**, is a sulfide-bearing subterranean estuary (data from Roy et al. 2012). Interpretive pathways in **a** adapted from Fig. 4 of Guo et al. (2013), and those in **b** based on endmembers defined by Roy et al. (2012); both **a** and **b** fix the FeS component based on the experiments of Guilbaud et al. (2011a) and Wu et al. (2012). Note that the placement of the end member boxes such as "DIR" varies between panels, as identified in the original studies

positive Fe_{aq}–sulfide $^{56}Fe/^{54}Fe$ fractionation factor (i.e., removal of low-$\delta^{56}Fe$ pyrite). Although a positive fractionation factor was measured between Fe^{2+}_{aq} and FeS by Butler et al. (2005) and Guilbaud et al. (2010), this is a kinetic isotope effect, and a negative fractionation factor is appropriate under equilibrium conditions (Guilbaud et al. 2011a; Wu et al. 2012). Therefore, as discussed by Guo et al. (2013), a trend of increasing $\delta^{56}Fe$ for Fe_{aq} and decreasing Fe_{aq} and H_2S likely reflects formation of pyrite under kinetic conditions (Fig. 5.8), which requires pyrite to have significantly lower $\delta^{56}Fe$ values than Fe^{2+}_{aq}. Such trends contrast with those produced by oxidation of Fe^{2+}_{aq}, which, as discussed above, produces decreasing $\delta^{56}Fe$ values for Fe_{aq} and decreasing Fe_{aq} (Fig. 5.8). The Fe isotope data for sulfide-bearing groundwaters, therefore, provide an understanding for when decreasing Fe_{aq} contents record oxidation or sulfide mineral formation, and they provide a hint that pyrite formation may be accompanied by significant kinetic Fe isotope fractionations; this later theme will become prominent in our discussions on pyrite below.

5.3 Redox-Stratified Water Bodies

A number of Fe isotope studies have investigated lakes and restricted marine settings. Lakes may have a wide range in $\delta^{56}Fe$ values for both water and sediment components. In some lakes, isotopic variability is not produced by redox gradients but instead is a function of seasonal input variations (e.g., Song et al. 2011). Subglacial lakes that contain highly saline brines have aqueous Fe that has very negative $\delta^{56}Fe$ values, in some cases recorded in Fe^{2+}_{aq}, but in others, in Fe^{3+}_{aq}, but these are relatively poorly understood from a mechanistic perspective (Mikucki et al. 2009; Proemse et al. 2017). Liu et al. (2015) measured lake water, pore fluids, and sediment extractions in Lake Tantare (Quebec, Canada), and found large Fe isotope fractionations among these components that they ascribed to DIR in the sediment. Highly negative $\delta^{56}Fe$ values in pore waters of Lake Kinneret (Sea of Galilee, Israel) in the zone of anaerobic oxidation of methane (AOM) led Sivan et al. (2011) to interpret such compositions to indicate microbial iron reduction coupled to AOM.

Below we focus our attention on two redox-stratified lakes (Nyos and Pavin) and the redox-stratified Black Sea, for which detailed Fe isotope profiles are available. In particular, the sulfide-rich Lake Pavin and Black Sea have strong redox gradients in the water column, and have been proposed to be modern analogs to Archean marine environments (e.g., Busigny et al. 2014; Rolison et al. 2018).

5.3.1 Lake Water

Figure 5.9 illustrates Fe isotope variations as a function of aqueous Fe for two redox-stratified lakes, Lake Nyos (Cameroon) and Lake Pavin (France). Both lakes have a permanently anoxic deep zone that has high Fe^{2+}_{aq} contents beneath a redoxcline (~100 m depth in Lake Nyos, ~60 m depth in Lake Pavin), a transition zone of ~1–10 m in the redoxcline, and an oxygen-rich upper layer where Fe^{2+}_{aq} contents are very low (e.g., Teutsch et al. 2009; Busigny et al. 2014). Despite markedly different water chemistry in the two lakes, particularly in terms of P abundances and Fe–P speciation, the $\delta^{56}Fe$–Fe_{aq} relations for both are very similar, where $\delta^{56}Fe$ values for Fe_{aq} decrease strongly with decreasing Fe_{aq} contents (Fig. 5.9). Although such relations are somewhat similar to those seen in groundwaters (Fig. 5.7), the range in Fe_{aq} contents in the lakes is much greater for a given range in $\delta^{56}Fe$ for Fe_{aq}. Teutsch et al. (2009) ascribe the Fe isotope variations to DIR across the redoxcline, whereas Busigny et al. (2014) interpret the isotopic compositions to record the effects of oxidation of upwelling Fe^{2+}_{aq}.

When viewed in the light of various redox and mixing processes, as well as currently known Fe^{2+}_{aq}—mineral fractionation factors, neither of the models proposed by Teutsch et al. (2009) and Busigny et al. (2014) fit the data. Although Teutsch et al. (2009) suggest mixing relations within the redoxcline, such a process cannot explain the entire range in Fe concentrations and isotopic compositions, given the strong mis-fit between the data and a simple mixing line

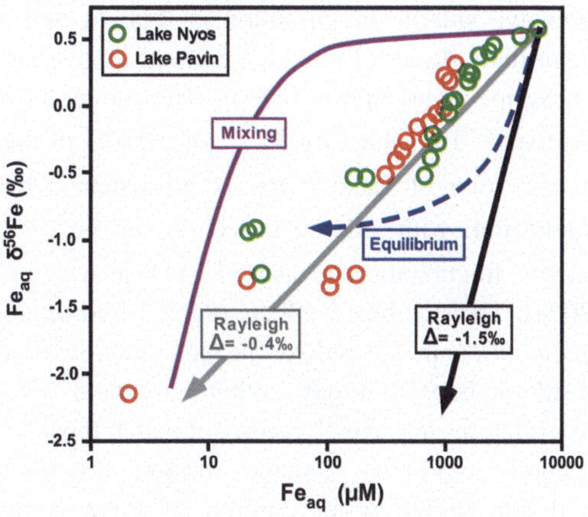

Fig. 5.9 $\delta^{56}Fe$ values for aqueous Fe as a function of Fe_{aq} contents for two redox-stratified lakes (Nyos and Pavin). Fe_{aq} obtained by passing through a 0.2 or 0.45 μm filter, depending upon the study. Data for Lake Pavin from Teutsch et al. (2009), and data from Lake Pavin from Busigny et al. (2014). Several Fe isotope evolution models are shown, including Rayleigh fractionation, with Fe^{2+}_{aq}–$Fe(OH)_3$ fractionation factors of −0.4 and −1.5‰ (linear grey and black lines, respectively), as well as the equilibrium fractionation model from Fig. 5.7 (dashed blue line). Simple mixing between high-Fe_{aq} and low-Fe_{aq} endmembers shown in solid purple curve

(Fig. 5.9). Similarly, oxidation of Fe^{2+}_{aq} under equilibrium conditions, as illustrated in Fig. 5.7, does not fit the data from either lake (Fig. 5.9). Oxidation of Fe^{2+}_{aq} via a Rayleigh process does not explain the data from the lakes using reasonable Fe^{2+}_{aq}–Fe^{3+}-oxyhydroxide fractionation factors (−1.5‰ in $^{56}Fe/^{54}Fe$ or lower). The linear relations in Fig. 5.9 may be explained by a Rayleigh process only if the Fe^{2+}_{aq}—mineral fractionation factor is ~−0.4‰, a fractionation factor that is quite small and not expected for oxidation of Fe^{2+}_{aq} to Fe^{3+}-hydroxides, at least if partial oxidation occurs (see Sect. 5.2.2).

The high CO_2 contents in Lake Nyos, which had a deadly CO_2 eruption in 1986, coupled with high Fe^{2+}_{aq} contents below the redoxcline, produces conditions where siderite is supersaturated in the anoxic zone (Bernard and Symonds 1989). Although Fe^{3+}-hydroxide particulates are found in the oxic layer, these are interpreted by some studies to have precipitated in situ, whereas extensive siderite precipitation is thought to be the primary driver of low Fe_{aq} contents above the redoxcline (Giggenbach 1990). The moderately positive $\delta^{56}Fe$ values for Fe^{2+}_{aq} of +0.6‰ in the anoxic zone of Lake Nyos is consistent with equilibrium with siderite based on the Fe^{2+}_{aq}—siderite fractionation measured by Wiesli et al. (2004). The combined effects of Fe^{3+}-hydroxide precipitation in the redoxcline and precipitation of siderite beneath it may explain why the $\delta^{56}Fe$–Fe_{aq} relations at Lake Nyos do not match those of a purely oxidizing system. Support for Fe^{2+} oxidation in the water column of Lake Nyos comes from the occurrence of microbial Fe^{2+} oxidation at several depths in Lake Nyos (Tiodjio et al. 2014), including those under aerobic and anaerobic conditions. The possibility that DIR is responsible for the highly negative $\delta^{56}Fe$ values, as proposed by Teutsch et al. (2009), might be independently tested via C isotope analysis of siderite or isolation of enrichment cultures of Fe^{3+}-reducing microbes in the water column. Importantly, without isotopic and mineralogic analysis of the suspended particles, it is difficult to further constrain the origin(s) of the wide range in $\delta^{56}Fe$ values at Lake Nyos, nor assess the possibility of Fe isotope exchange between Fe_{aq} and particulate Fe in the water column. In addition, because Fe_{aq} was sampled using a 0.45 μm filter (Teutsch et al. 2009), it is likely that the aqueous component contains colloidal or nanoparticulate Fe, which could complicate interpretation of Fe isotope compositions for Fe_{aq}.

Lake Pavin has been the subject of extensive geochemical study, where high authigenic Fe fluxes to the lake sediments (2–200 g m^{-2} yr^{-1}) are sustained by redox and biological cycling of spring-sourced Fe^{2+}_{aq}, as well as high input of organic carbon (C_{org}) and P (e.g., Schettler et al. 2007). The ferruginous nature of Lake Pavin has been used to propose it as a modern analog for Precambrian Fe cycling, potentially addressing the relative roles of water column processes versus sedimentary processes in explaining the wide range in Fe isotope compositions measured in Precambrian sulfides (e.g., Rouxel et al. 2005). Busigny et al. (2014) propose that the $\delta^{56}Fe$ values for Fe^{2+}_{aq} in the water column redoxcline are produced by partial oxidation of upwelling Fe^{2+}_{aq}. Further, the authors infer that these isotopic signatures are directly transferred to pyrite forming in the redoxcline with limited Fe isotope fractionation. The pyrites are envisioned to settle through the deep water and accumulate in the underlying sediment, where they are taken to represent a permanent isotope record of shallow-water redox cycling. Water column SPM is dominated by Fe^{3+}-hydroxides in the oxic zone. In contrast, vivianite and minor mackinawite (which we will refer to as FeS) are important Fe repositories in SPM in the anoxic zone, consistent with the supersaturated conditions with respect to these minerals beneath the redoxcline (Michard et al. 1994; Bura-Nakić et al. 2009; Cosmidis et al. 2014; Busigny et al. 2014, 2016). Although Busigny et al. (2014) draw an analogy between Fe isotope fractionation in Lake Pavin and Fe pathways for Precambrian sedimentary pyrite, pyrite was not found in SPM in Lake Pavin.

Busigny et al. (2016) modeled vertical variations in Fe^{2+}_{aq}, phosphate, and $\delta^{56}Fe$ at Lake Pavin to record the effects of Fe^{2+}_{aq} oxidation as well as vivianite formation, but there are some difficulties in their choices of Fe isotope fractionation factors. They assumed a −0.9‰ fractionation in $^{56}Fe/^{54}Fe$ between Fe^{2+}_{aq} and Fe-oxyhydroxide and a +0.5‰ fractionation in $^{56}Fe/^{54}Fe$ between Fe^{2+}_{aq} and vivianite. Although the former fractionation factor is supported by experimental studies (see Chap. 3) and draws its origins from Bullen et al. (2001), the vivianite fractionation factor they assume is opposite that reported in Croal et al. (2004), who inferred a Fe^{2+}_{aq}—vivianite fractionation factor of −0.3‰ in $^{56}Fe/^{54}Fe$ (see Chap. 3). Given the strong role vivianite has in Fe cycling in Lake Pavin, vivianite precipitation using a −0.3‰ (or −0.4‰) fractionation factor might explain part of

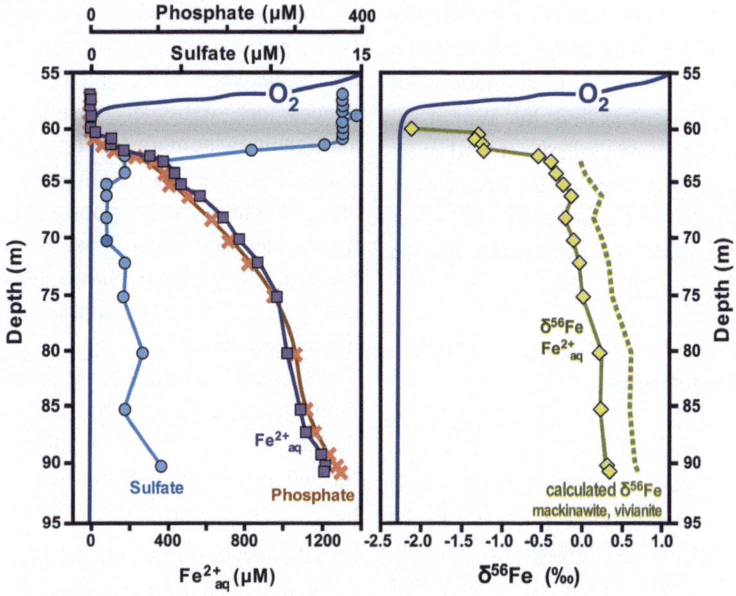

Fig. 5.10 Vertical variations in aqueous chemistry (left panel) and Fe isotope compositions (right panel) for the redox-stratified Lake Pavin (France). A redoxcline (grey shaded zone) at ~60 m depth is marked by a strong decrease in oxygen contents (dark blue curve; scaled from 0 to 200 μM). Just below the redoxcline, Fe_{aq} contents (purple squares) start to increase, and sulfate contents (blue circles) decrease, marking the top of an anoxic zone that has high Fe^{2+}_{aq} and increasing sulfide downward. Increasing Fe^{2+}_{aq} contents downward are matched by increases in phosphate contents (orange-pink "X" symbols). Large changes in δ^{56}Fe values for Fe^{2+}_{aq} occur immediately beneath, and within, the redoxcline. Below the redoxcline, elevated Fe^{2+}_{aq}, phosphate, and sulfide produces saturation in vivianite and mackinawite; the calculated equilibrium δ^{56}Fe values for a vivianite and/or mackinawite assemblage are indicated by the dark yellow dashed line, using the fractionation factors in Croal et al. (2004) and Wu et al. (2012). Chemical and isotopic data from Busigny et al. (2014)

the changes in Fe^{2+}_{aq} and δ^{56}Fe (Fig. 5.9), and in turn would decrease the importance of Fe^{3+}-hydroxide precipitation in determining the δ^{56}Fe values for Fe^{2+}_{aq}. Additional difficulties arise in modeling, where, for example, use of a fixed Fe isotope fractionation factor in a Rayleigh model cannot account for upwelling Fe^{2+}_{aq}, nor accommodate the variable mineral assemblage throughout the water column that is documented by studies of SPM in Lake Pavin (Cosmidis et al. 2014; Busigny et al. 2016). Similar to Lake Nyos, it is difficult to fully understand the δ^{56}Fe variations for Fe^{2+}_{aq} without detailed isotopic and mineralogic studies of co-existing SPM.

In detail, vertical variations in Fe^{2+}_{aq}, δ^{56}Fe, phosphate, sulfate, and oxygen at Lake Pavin (Fig. 5.10) suggest a more complex model than simple vivianite precipitation. As discussed by Busigny et al. (2016), although Fe^{2+}_{aq} and phosphate decrease upward toward the oxic/anoxic transition, their molar ratios in solution (Fe/P ~ 3–4) do not match those of vivianite (Fe/P = 1.5), suggesting that vivianite can only account for about half of the changes in Fe^{2+}_{aq}. Busigny et al. (2016) suggest other Fe–P components may be involved, such as sorption of P to Fe-oxyhydroxides, or Fe^{3+}-phosphates. The large decrease in sulfate contents beneath the oxic/anoxic transition, which is mirrored by increase in H_2S, suggests that FeS may also be important, as noted by other studies, although it must be recognized that Fe is in far greater excess than S in Lake Pavin. Given the circum-neutral pH of Lake Pavin waters and excess Fe over S, FeS is likely to be in Fe isotope equilibrium in the water column (e.g., Wu et al. 2012), indicating that FeS formed in the water column would have had near-zero to slightly positive $\delta^{56}Fe$ values (Fig. 5.10). Importantly, the Fe^{2+}_{aq}–FeS equilibrium fractionation of −0.3‰ for $^{56}Fe/^{54}Fe$ measured by Wu et al. (2012) is similar to that of vivianite (Croal et al. 2004), indicating that precipitation of both minerals would produce the same Fe isotope effect (indicated by the dark yellow dashed line in Fig. 5.10).

5.3.2 Lake Sediments

We continue our discussion of Lake Pavin to explore the connections between the water column and the underlying sediments. The strong Fe isotope gradient within, and immediately below, the redoxcline at Lake Pavin (Fig. 5.10) led Busigny et al. (2014) to propose that such isotopic signals would be recorded in the underlying sediments and therefore provide a permanent sedimentary record of redox cycling in the upper water column. The motivation for making this connection lies in the ancient rock record, which, if Fe isotope signals from water column are recorded in the sediments, indicates Fe isotope compositions of ancient sediments could be used to infer ancient water column processes. The isotopic pathways of authigenic and diagenetic mineral formation in Lake Pavin sediments are explored in Fig. 5.11, which allows testing this proposal, including the possibility that pyrite in the lake bottom sediments may record shallow water processes.

Comparison of the $\delta^{56}Fe$ values for pyrite, acid extractions, pore fluids, and lake water provide examples for Fe isotope equilibrium and disequilibrium. An important observation is that pore fluids from sediments in the redoxcline zone in Lake Pavin are not in Fe isotope equilibrium with lake water Fe^{2+}_{aq}, indicating diagenetic mineral formation in the sediment that is decoupled from processes in the water column in this zone. Comparison of the $\delta^{56}Fe$ values of HCl leaches with those of pyrite at Lake Pavin are highly variable (Fig. 5.11). There are limitations in our ability to interpret the Fe isotope data from HCl leaches from Lake Pavin, however, as Fe^{3+}–Fe^{2+} speciation was not determined, and correlations between $\delta^{56}Fe$ and Fe^{3+} and Fe^{2+} abundances are extremely useful in identifying authigenic Fe components (e.g., Severmann et al. 2006). Possible phases sampled by the HCl leaches would include poorly crystalline Fe^{3+}-hydroxides and amorphous Fe^{3+}-phosphates in the oxic zone, these phases plus vivianite and FeS in the redoxcline, and vivianite and FeS in the anoxic zones.

We may examine the relations between different reactive Fe pool by calculating the hypothetical $\delta^{56}Fe$ values of authigenic minerals, formed in isotopic equilibrium with pore water Fe^{2+}_{aq}, and comparing them with measured values for HCl leaches. Because the Fe^{2+}_{aq}-vivianite and Fe^{2+}_{aq}-mackinawite Fe isotope fractionation factors are similar, their calculated mineral isotope compositions are indistinguishable. For core samples obtained at 32 m, 60 m, and 65 m depths, these calculated vivianite/FeS compositions are reasonably close to the $\delta^{56}Fe$ values of the HCl leaches, although it is not clear if such comparisons are appropriate for the shallowest sediment cores in the oxic zone (Fig. 5.11). Busigny et al. (2014) interpret the $\delta^{56}Fe$ values of pyrite to reflect inheritance from FeS precursors, whose Fe isotope compositions were fixed in the redoxcline (60 m depth) through oxidation processes. The $\delta^{56}Fe$ values for most of the pore waters, HCl leaches, and pyrite from the oxic zone (32 m depth) are near zero, whereas the data for the oxic-anoxic transition (60 m depth) cluster about the FeS-pyrite marine sediment trend of Scholz et al. (2014b) (Fig. 5.11). Although Busigny et al. (2014) argue against an Fe isotope fractionation between FeS and pyrite during pyrite formation, because their hypothesis is that water column Fe isotope compositions are directly transferred to the sediment, it is striking that the sediment samples from the oxic-anoxic transition plot closely to the FeS-pyrite marine trend. Together with the observation that pore water and lake water are not in Fe isotope equilibrium, these two features suggest a kinetic Fe isotope fractionation during pyrite formation in

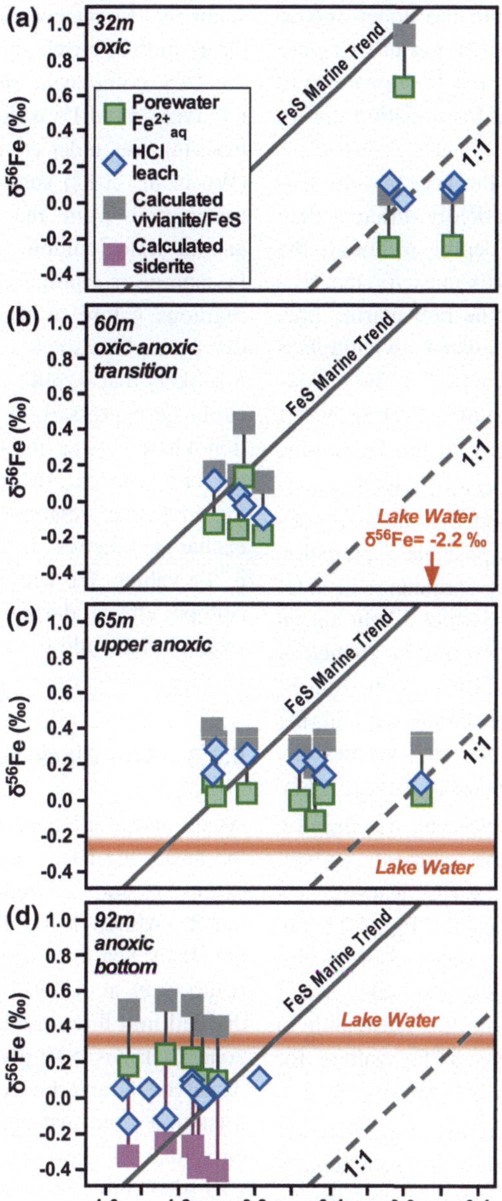

Fig. 5.11 Figure examines the relations between different reactive Fe pools in Lake Pavin sediments from four different water depth. Data from Busigny et al. (2014). The Fe isotope compositions of pore fluids and solid phase HCl leaches are shown relative to their concomitant pyrite δ^{56}Fe values. Porewater Fe was separated using a 0.2 μm filter. The HCl leach component was obtained via cold 1 M HCl for 3 h, followed by sequential extraction and isolation of the pyrite component by HNO_3. Minerals that might be extracted in the HCl leaches include vivianite, mackinawite, and siderite. Calculated δ^{56}Fe values for these minerals in equilibrium with the pore fluids are shown (Y-axis), using the following Fe isotope fractionation factors: vivianite (Croal et al. 2004), mackinawite (Wu et al. 2012), and siderite (Wiesli et al. 2004); tie lines aid in evaluating the possibility that one or more of these minerals may be reflected in the HCl leaches. The δ^{56}Fe values for lake water at the corresponding core depth, if available, are also noted by the horizontal shaded orange bar. The marine FeS trend is shown from Scholz et al. (2014b), which defines an average kinetic isotope fractionation between mackinawite and pyrite in marine sediments of ~+1‰. A 1:1 reference line (dashed) is also shown

sediments from this depth. In the upper anoxic zone, at 65 m depth, the data for pyrite-HCl-pore water scatter from a 1:1 line (no fractionation) to the FeS marine line (kinetic fractionation during pyrite formation).

The Fe isotope data for the lake bottom sediments (92 m depth) are markedly different than those from the shallower depth. Although the majority of measured $\delta^{56}Fe$ values for the HCl leaches plot near or above the FeS marine line, they do not agree with calculated vivianite/FeS isotope compositions. Calculated $\delta^{56}Fe$ values for vivianite/FeS in equilibrium with pore water Fe^{2+}_{aq} are far too high relative to the Fe isotope compositions of the HCl leaches, which is perplexing, given the likelihood that these phases were stable at this depth. A possible explanation for these relations lies in the abundance of siderite in the bottom sediments, which likely has an origin through microbial respiration (Schettler et al. 2007; Busigny et al. 2016). A mixture of vivianite, FeS, and siderite could explain the $\delta^{56}Fe$ values measured for the HCl leaches relative to those of the pore water Fe^{2+}_{aq} (Fig. 5.11). Such a proposal could be tested by determining C isotope compositions of the sediments, as well as inorganic C contents of the HCl leaches.

When all of the relations in Fig. 5.11 are considered, it is striking how isotopically heterogeneous the sediment and lake water components are in Lake Pavin. It is reasonable to ascribe the highly negative $\delta^{56}Fe$ values for Fe^{2+}_{aq} in the redoxcline to oxidation and formation of Fe^{3+}-oxyhydroxides. Directly transferring these Fe isotope compositions to sedimentary pyrite, however, is not supported by the discussion above. Such transfer must occur indirectly because pyrite is not found in SPM in Lake Pavin, and hence does not form in the water column. Moreover, it is important to keep in mind that the moles of Fe that have highly negative $\delta^{56}Fe$ values in the redoxcline are quite small relative to the very large Fe inventory in the anoxic zone, and it seems highly unlikely that transferring such a small (on a molar basis) negative-$\delta^{56}Fe$ signal through the Fe-rich anoxic section would be detectable in the bottom sediments because such a signal would likely be "lost" in the Fe-rich anoxic zone. In addition, for the FeS component in the water column, the relatively rapid Fe isotope exchange kinetics for mackinawite under circum-neutral pH conditions (Wu et al. 2012) suggests that any reactive Fe transported from the redoxcline through the anoxic water column would undergo extensive Fe isotope re-equilibration as it settled. The ferruginous nature of the water column indicates that pyrite formation is not Fe limited, and hence it is likely that significant Fe isotope fractionation could be expressed, in contrast to pyrite formation where Fe was limited. These issues highlight the importance of considering Fe isotope mass-balance relations, as well as isotopic exchange kinetics, in addition to variations in $\delta^{56}Fe$ values. We will extensively discuss the Fe isotope effects during pyrite formation in the sections that follow.

5.3.3 The Black Sea

As the world's largest permanently anoxic marine basin, the Black Sea has been the subject of extensive study of redox cycling of C, S, and metals. Although the depth of the redoxcline in the Black Sea may vary over decadal timescales (Murray et al. 1989), the water column can be divided into three zones: an upper (<80 m) oxic zone that has >10 μM oxygen and very low dissolved S and Fe contents, a 20–50 m-thick transition zone that contains the redoxcline and encompasses low oxygen and sulfide (both <10 μM) but relatively high (>10 nM) Fe_{aq} contents, and a deep (>115 m) euxinic zone that has high sulfide (up to 450 μM) and variable Fe_{aq} contents 100–400 nM), extending down to >2 km depth. Sulfur speciation in the water column is almost exclusively H_2S, where sulfide is produced by dissimilatory sulfate reduction (DSR) in the transition zone (e.g., Luther et al. 1991). In the oxic zone, Fe_{aq} consists of colloidal and organically complexed Fe, whereas Fe_{aq} consists of Fe^{2+}_{aq} and Fe–S colloids in the euxinic zone (Lewis and Landing 1991). Oxidation of sulfide occurs in the transition zone via oxygen and reduction of Fe^{3+}-hydroxides, whereas oxidation of Fe^{2+}_{aq} occurs from oxygen and nitrate reduction (Konovalov et al. 2006). Multiple lines of evidence indicate that enrichment of Fe_{aq} and particulate Fe in the deep euxinic zone is supplied by Fe shuttled from the shelf that has been liberated via DIR as well as sulfidization of Fe^{3+}-hydroxides (e.g., Wijsman et al. 2001b; Anderson and Raiswell 2004; Severmann et al. 2008), a concept that will be discussed in detail in Sect. 5.4.4.

A very large range in δ^{56}Fe values for aqueous Fe (<0.2 μm filter) was measured by Rolison et al. (2018) for two water column profiles in the deep parts of the Black Sea (Fig. 5.12). In addition to the large range in δ^{56}Fe values for Fe_{aq}, Fe_{aq} contents are exceptionally low, all less than 1 μM, relative to the groundwater and lakes discussed above. The peak in Fe_{aq} contents beneath the transition zone may in part reflect Fe that was shuttled from the shelf, which may largely be in the form of colloidal Fe^{3+}-hydroxides (e.g., Lenstra et al. 2019). Assuming that the Fe_{aq}-rich portions of the transition and euxinic zones have δ^{56}Fe values that largely reflect those of the Fe shuttled from the shelf (δ^{56}Fe = −0.7 to −1.2; Severmann et al. 2008; Rolison et al. 2018), the decreasing Fe_{aq} contents in the upper transition and oxic zones, at relatively constant δ^{56}Fe value, could be explained by complete oxidation of upwelling or laterally-transported Fe_{aq} (Fig. 5.12).

Rolison et al. (2018) focused their attention on the trend of increasing δ^{56}Fe values with decreasing Fe_{aq} contents in the euxinic zone and proposed that this reflected equilibrium fractionation between Fe_{aq} and pyrite with a very large $\Delta^{56}Fe_{Aq-Pyrite}$ fractionation factor of +2.75‰ in ^{56}Fe/^{54}Fe. The interpretation of equilibrium conditions is based an inverse relation between Fe_{aq} contents and δ^{56}Fe values (Fig. 5.12), as Fe_{aq} is converted to pyrite. Although Rolison et al. (2018) point to the experimental results of Guilbaud et al. (2011a) in support of their large proposed fractionation factor, it is important to note that Guilbaud et al.

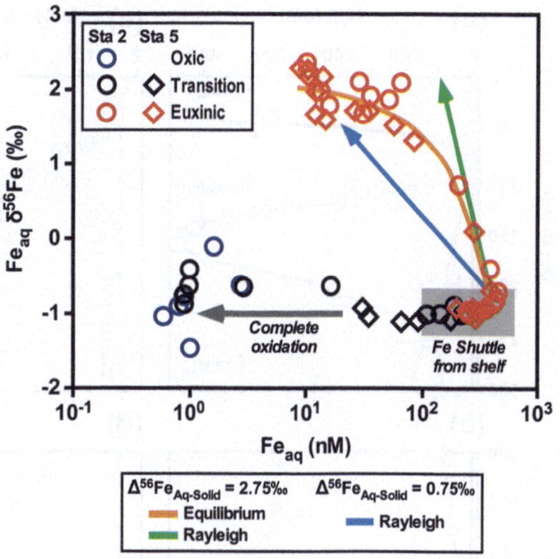

Fig. 5.12 Variations in δ^{56}Fe values for Fe_{aq} (defined as <0.2 μm) from two water column profiles from the deep portion of the Black Sea (Stations 2 and 5). Symbols divided into "oxic" (samples >10 μM oxygen, no sulfide), "transition" (<10 μM oxygen and sulfide), and "euxinic" (>10 μM sulfide, no oxygen). In Station 2, these boundaries correspond to 80 m and 115 m depths, and in Station 5, to 85 m and 130 m depths (note there are no data for the oxic zone at Station 5). Data from Rolison et al. (2018). The equilibrium fractionation line ($\Delta^{56}Fe_{Aq-Solid}$ = + 2.75‰; brown/orange line) shown from Rolison et al. (2018). Rayleigh fractionation lines shown for $\Delta^{56}Fe_{Aq-Solid}$ = +2.75‰ (green line) and $\Delta^{56}Fe_{Aq-Solid}$ = + 0.75‰ (blue). Samples from the oxic and transition zone, where Fe_{aq} decrease strongly at δ^{56}Fe ∼−1‰ is interpreted to reflect complete oxidation

(2011a) interpreted their results to indicate a large kinetic isotope fractionation upon pyrite formation from mackinawite. Rolison et al. (2018) rule out a kinetic or Rayleigh effect, which, as shown in Fig. 5.12, does not explain the topology of the Fe_{aq}–δ^{56}Fe relations, at least for a single Fe isotope fractionation factor.

The proposal by Rolison et al. (2018) for a very large equilibrium Fe isotope fractionation factor during pyrite precipitation in the water column contrasts with the model of Severmann et al. (2008) for the Black Sea, who call upon pyrite formation in the transition zone that records the δ^{56}Fe value of the shuttled Fe with only minor fractionation effects upon pyrite precipitation. The model of Rolison et al. (2018) may be evaluated through considering vertical

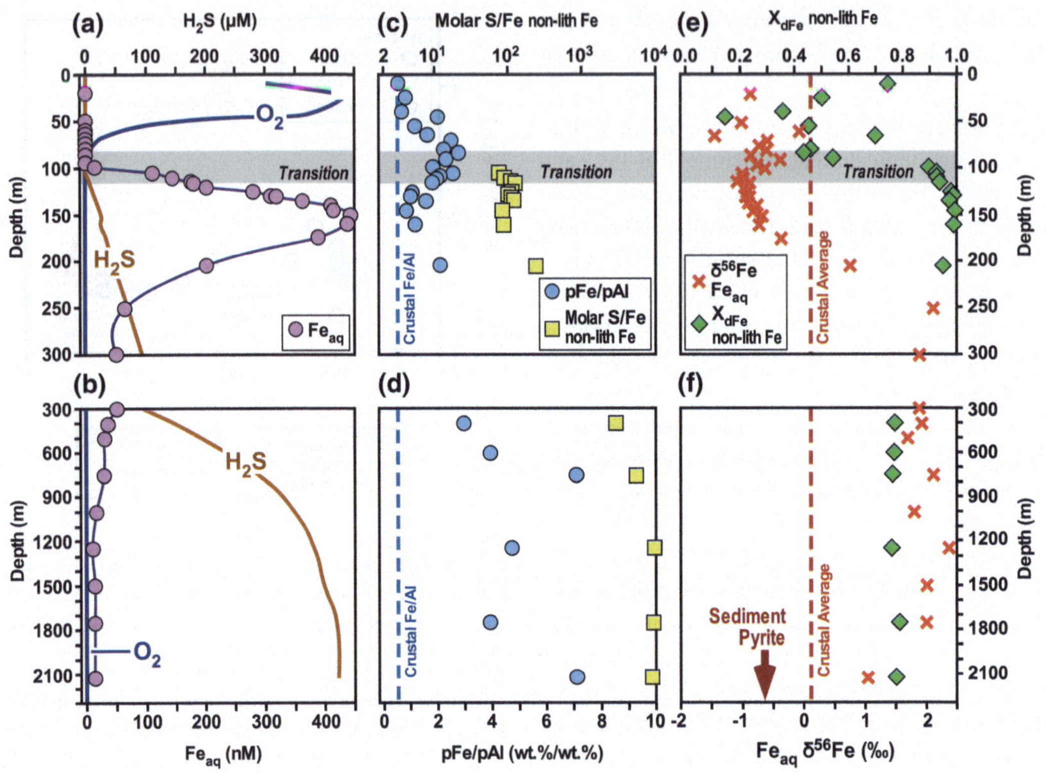

Fig. 5.13 Vertical profiles in water column chemistry and particulates for the western deep portions of the Black Sea. Data for dissolved components (O_2, H_2S, Fe_{aq}, and $\delta^{56}Fe$ for Fe_{aq} from Station 2 of Rolison et al. (2018), where aqueous components were determined on samples that passed a 0.2 μm filter. Scale for O_2 ranges from 0 to 350 μM. Parameters for particulate components, including Fe/Al ratios (defined as pFe/pAl), as well as "reactive" Fe components, including the mole fraction of dissolved Fe (X_{dFe}), and the molar S/Fe ratio of reactive Fe, additionally include data from Station BS3-2 from Lewis and Landing (1991), which lies in the same portion of the Black Sea as Station 2 of Rolison et al. (2018), ~150 km apart. The datasets were aligned using small (~10 m) adjustments to match water column Fe_{aq} contents from the two sources. Reactive particulate Fe calculated assuming a lithologic pFe/pAl ratio of 0.5 (as wt%/wt%), and this excess Fe is termed "non-lithologic". X_{dFe} is calculated from the measured dFe contents and the non-lithologic pFe contents. The average $\delta^{56}Fe$ value for sedimentary pyrite beneath the euxinic water column shown by vertical red arrow and is based on pyrite measured in euxinic samples by Severmann et al. (2008)

variations in the proportions of aqueous and solid components that may have undergone Fe isotope exchange. In Fig. 5.13 we combine the water column data from Rolison et al. (2018) with particulate data from the study of Lewis and Landing (1991) obtained from a similar setting and for which water column data allow cross-correlation. The combined dataset supports a model where Fe shuttled from the shelf is found in the high-Fe_{aq} portion of the upper euxinic zone (Fig. 5.13a, b) with an isotope composition corresponding with previous estimates of the basin shuttle Fe isotope compositions ($\delta^{56}Fe$ ~−1‰; Fig. 5.13e, f), and high Fe/Al particulates (defined as pAl/pFe) of the transition zone (Fig. 5.13c, d). The strong shift to high $\delta^{56}Fe$ values for Fe_{aq} occurs below 150 m in the euxinic zone, below the peak in Fe_{aq} contents, and maintains highly positive $\delta^{56}Fe$ values despite strong decreases in Fe_{aq} continuing down into the lower euxinic zone (Fig. 5.13a, b). The euxinic zone is characterized by particles that have very high Fe/Al enrichments, far higher than the average crustal ratio, indicating an enrichment in reactive particulate Fe (e.g., Anderson and Raiswell 2004).

An inherent assumption of the equilibrium mass-balance relation in the Black Sea water column, as proposed by Rolison et al. (2018), is that decreases in Fe_{aq} contents are matched by increases in solid-phase Fe (pFe). A comparison, however, of dissolved Fe (Rolison et al. 2018) and particulate Fe (Lewis and Landing 1991) suggests incongruent variations between Fe_{aq} and excess pFe: although Fe_{aq} contents decrease strongly from >400 nM to 10–20 nM with depth in the euxinic zone, it does not scale directly with an increase in the fraction of excess Fe contained in suspended particles. Excess Fe may be estimated from measured pFe/pAl compared to the lithogenic baseline, and together with dFe, this comprises the water column reactive Fe pool. If we define X_{dFe} as the fraction of Fe_{aq} that comprises the water column reactive Fe inventory, we see that X_{dFe} does not decrease from 1.0 to $\sim$0.1 (Fig. 5.13e, f), as assumed by Rolison et al. (2018) based on scaling Fe_{aq} contents, but only decreases to $\sim$0.8 in the euxinic zone. Such high values for X_{dFe} reflect the fact that particulate contents decrease more rapidly with depth than Fe_{aq} contents, recording particle settling. Full understanding of the highly positive $\delta^{56}Fe$ values for Fe_{aq} in the deep euxinic zone will require a detailed assessment of the co-existing particulate phases in the water column, including mineralogy and isotopic compositions, and it seems premature to propose a very large equilibrium Fe isotope fractionation for pyrite formation based on the data in Figs. 5.12 and 5.13. A detailed approach that considers the Fe phases and isotopic compositions in the dissolved and particulate phases is needed to help evaluate if pyrite of moderately negative $\delta^{56}Fe$ values, as measured in the euxinic sediments of the deep portion of the Black Sea (Fig. 5.13), can be produced from the highly positive $\delta^{56}Fe$ values for Fe_{aq} in the euxinic zone. Even with such data, understanding the dynamics between particle compositions, Fe_{aq} contents, and the effects of particle dissolution and sinking, is challenging. Nevertheless, assumptions of particulate compositions and mass-balance based on Fe_{aq} contents alone seem likely to be in error.

Production of pyrite in the water column of euxinic systems is generally thought to produce distinctive framboidal morphology, and such pyrite is generally accepted to be produced rapidly in the oxic/anoxic transition zone, followed by slow sinking (e.g., Muramoto et al. 1991; Wilkin et al. 1997; Cutter and Kluckhohn 1999; Lewis and Landing 1991). The molar S/Fe ratios of reactive aqueous and particulate components in the Black Sea water column reach very high levels, up to four-orders-of-magnitude higher than required for pyrite (Fig. 5.13c, d), indicating the extreme excess of S over Fe that is characteristic of euxinic environments. Although the euxinic zone is generally supersaturated with respect to pyrite precursors such as mackinawite and greigite, it is unclear how much framboidal pyrite is produced deep in the water column, given the strong limitation of Fe. It is therefore uncertain how such conditions could translate to an equilibrium Fe isotope effect at depth, given the very slow isotope exchange rates measured for pyrite (Fossing and Jørgensen 1990). We will explore the topic of pyrite paragenesis and its Fe isotope compositions in detail in Sect. 5.4.

5.4 Marine Sediments

5.4.1 Reactive Fe Inventories

The Fe redox processes discussed in Sects. 5.2 and 5.3, as well as Chap. 3, provide the framework required for interpreting Fe isotope variations in modern marine sediments. Here we return to some of the low-temperature Fe redox relations introduced in Sect. 1.1.1 in Chap. 1. Under oxygen-rich conditions, degradation ("mineralization") of organic carbon (C_{org}) occurs in the water column and sediment via the reaction:

$$CH_2O + O_2 \rightarrow H_2O + CO_2 \quad (5.6)$$

where C_{org} is represented as CH_2O using the short-hand glucose stoichiometry noted in Chap. 1. Even in oxic marine settings, however, sub-oxic sediment diagenesis is commonly the major pathway by which C_{org} is oxidized (C_{ox}) rather than through O_2 respiration, and anaerobic diagenetic processes include denitrification, and Mn, Fe, and sulfate reduction, and ammonia oxidation (Canfield et al. 1993a). For continental margin sediments, Canfield et al. (1993a, b) showed that oxidation of C_{org} via O_2 may comprise only a few percent of all C_{ox} pathways, and that DIR can, in some cases, be the dominant contributor to C_{ox}, even over sulfate reduction. DIR requires a source of both Fe^{3+} and organic carbon due to coupling Fe reduction to oxidation of C_{org}:

$$4Fe(OH)_3 + CH_2O + 8H^+ \rightarrow 4Fe^{2+}_{aq} + CO_2 + 11H_2O \quad (5.7)$$

Similarly, DSR requires C_{org}, as well as abundant sulfate:

$$SO_4^{2-} + 2CH_2O \rightarrow H_2S + 2HCO_3^- \quad (5.8)$$

Note that C_{ox} is the sum of all oxidation processes of C_{org}, both aerobic and anaerobic. Where bottom water O_2 is low, sulfide generation via sulfate reduction is commonly one of the most important pathways for C_{ox}, producing a variety of Fe–S aqueous and mineral species, which, of course, requires a supply of reactive Fe, as Fe^{2+} or Fe^{3+}, as well as sulfide (e.g., Thamdrup et al. 1994):

$$Fe^{2+}_{aq} + H_2S + 2OH^- \rightarrow FeS + 2H_2O \quad (5.9)$$

In Fig. 5.14 we illustrate two well-cited modern marine sediment sections where chemical extractions allowed characterization of all pools of Fe, including Fe^{2+}_{aq}, solid-phase Fe^{2+} that is not bound to sulfur (non-S $Fe(II)_{solid}$), poorly crystalline solid-phase Fe^{3+} or ferrihydrite ($Fe(III)_{solid}$), mackinawite (FeS), and pyrite (FeS_2). In the extraction scheme used, non-S Fe $(II)_{solid}$ is not magnetite but is interpreted to be a poorly crystalline Fe^{2+} solid and/or Fe^{2+} that is sorbed to ferric hydroxides. One sediment section is illustrated where C_{ox} is dominated by DIR in eastern Skagerrak, north of Denmark (Canfield et al. 1993a, b), and another where C_{ox} is dominated by DSR in Aarhus Bay, Denmark (Thamdrup et al. 1994). Using the proportions of reactive Fe phases determined through sediment extractions, and the appropriate Fe isotope fractionation factors discussed in Chap. 3, we can predict the equilibrium $\delta^{56}Fe$ values for species that were likely to have undergone isotopic exchange.

In the high iron-reduction case (Fig. 5.14a), high bottom water O_2 contents ($\sim$250 μM) are responsible for low pore water Fe^{2+}_{aq} contents in the upper sediment section via formation of Fe^{3+}-hydroxides by oxidation from O_2 that penetrates the upper sediment section (see Eqs. 5.1 and 5.2). Oxidation of Fe^{2+}_{aq} in near-surface sediments, therefore, in part drives the upward increase in $Fe(III)_{solid}$. Oxygen penetration into the sediment is dominated by bioturbation and bioirrigation, and may also occur via diffusion. In addition, oxidation of Fe^{2+}_{aq} in the shallow sediment sections may occur through MnO_2 reduction, as evidenced by pore water Mn^{2+} (data not shown). Deeper in the sediment, both pore water Fe^{2+}_{aq} and non-S $Fe(II)_{solid}$ contents increase in concert with decreases in $Fe(III)_{solid}$, indicating DIR. The limited C_{ox} that is associated with DSR in this example produces relatively small FeS and FeS_2 inventories. Importantly, although DIR is responsible for the majority of C_{ox} in this example, the majority of the Fe^{2+} inventory is contained in non-S $Fe(II)_{solid}$, and relatively little net escape of pore water Fe^{2+}_{aq} occurs to the open ocean due to high bottom water O_2 levels. Because of this, Canfield et al. (1993b) estimate that Fe (and Mn) is oxidized and reduced hundreds of times before ultimate burial in the sediment, indicating a very active Fe redox cycling regime near the sediment-water interface.

Figure 5.14c illustrates the reactive Fe inventories for the high-sulfate reduction case from Aarhus Bay, Denmark (Thamdrup et al. 1994). Although not obvious at the scale of Fig. 5.14, pore water Fe^{2+}_{aq} contents in the upper sediment section are one to two orders-of-magnitude higher in the Aarhus Bay section as compared to the eastern Skagerrak section (high iron reduction case), commensurate with the lower bottom water O_2 contents at Aarhus Bay ($\sim$50–90 μM), and

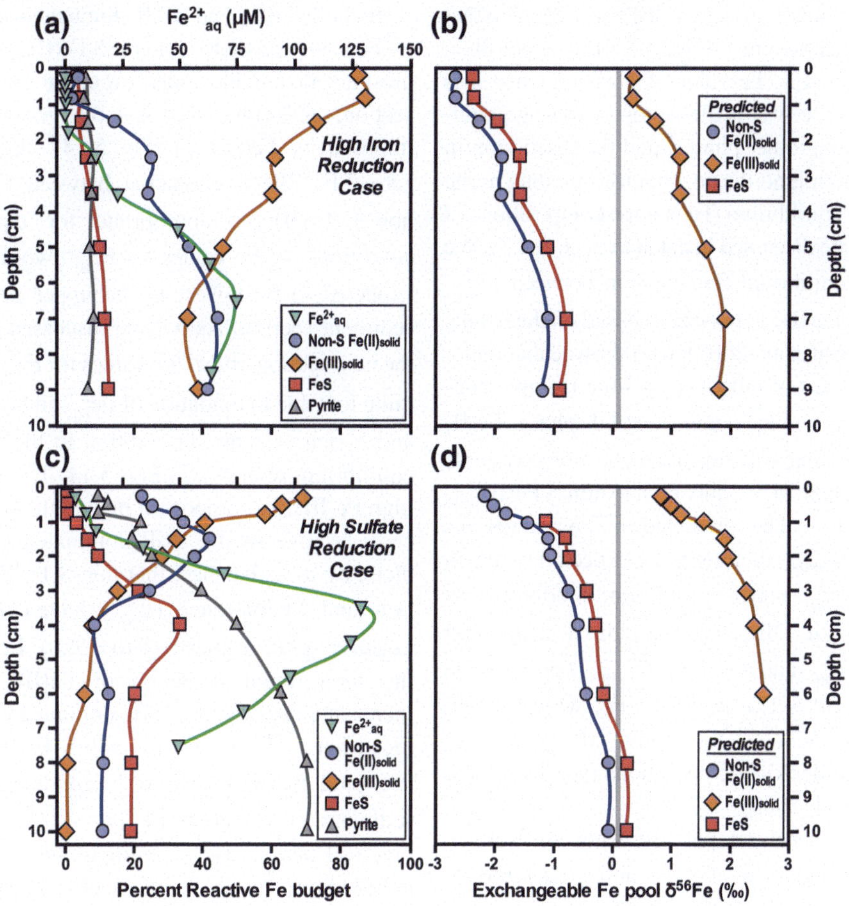

Fig. 5.14 Measured reactive Fe pools and predicted δ^{56}Fe values in modern marine sediment sections for **a**, **b** a high iron-reduction case (eastern Skagerrak, north of Denmark; Canfield et al. 1993a, b), and **c**, **d** a high-sulfate reduction case (Aarhus Bay, Denmark; Thamdrup et al. 1994). The high iron-reduction case sustains higher reactive Fe^{3+} and non-S solid Fe^{2+} relative to the high sulfate-reduction case, which produces larger quantities of FeS and pyrite. Predicted δ^{56}Fe values assume bulk sediment δ^{56}Fe = 0.09 (average upper crust; vertical grey bars in **b** and **d**), and uses the measured Fe mass balance for Fe pools expected to be in exchange equilibria: Fe^{2+}_{aq}, non-S $Fe(II)_{solid}$, and FeS, but not pyrite. The following Fe isotope fractionation factors are assumed: $\Delta^{56}Fe_{FeS-Fe(II)aq}$ = +0.3‰ and $\Delta^{56}Fe_{Fe(III)-Fe(II)}$ = +3‰

hence lower extents of Fe^{2+}_{aq} oxidation. This contrast must be considered when comparing the two sections. Nevertheless, oxidation of Fe^{2+}_{aq} in the Aarhus Bay locality in part explains the upward increase in $Fe(III)_{solid}$. At the Aarhus Bay locality, DSR is extensive below the first few cm of the sediment section, producing pore water H_2S, which reacts with Fe^{2+}_{aq}, non-S $Fe(II)_{solid}$, and $Fe(III)_{solid}$ to produce iron sulfides and S^0. Non-S $Fe(II)_{solid}$ is rapidly depleted in the zone of DSR, as is $Fe(III)_{solid}$. Fe^{2+}_{aq} and FeS contents peak at the same depth, suggesting solubility control of Fe^{2+}_{aq} by mackinawite. Pyrite abundance increases with depth and becomes the major Fe species below of few cm in the sediment section.

The Fe isotope compositions expected for Fe pools that are likely to undergo isotopic exchange, based on exchange kinetics discussed in Chap. 3, are $Fe(III)_{solid}$, non-S $Fe(II)_{solid}$, Fe^{2+}_{aq}, and FeS. Knowledge of the isotope fractionation factors allows predictions to be made of the δ^{56}Fe values expected for each component, assuming a bulk sediment $\delta^{56}Fe$ value equal to that of average upper crust. Figure 5.14 does not illustrate predicted $\delta^{56}Fe$ values for pyrite, since it is unlikely pyrite was in Fe isotope exchange equilibria with the other phases, and the discussion in Sect. 5.3 highlights the uncertainty regarding pyrite Fe isotope equilibrium (Fe isotope compositions of pyrite will be discussed in detail in Sect. 5.4.3). We take the equilibrium fractionation between Fe^{2+}_{aq} and non-S $Fe(II)_{solid}$ to be zero based on the results from Johnson et al. (2005), who showed that under moderate rates of DIR there is little isotopic contrast between Fe^{2+}_{aq} and a solid phase Fe(II) component that was not magnetic nor carbonate (which we take to be equivalent to non-S $Fe(II)_{solid}$ in Fig. 5.14). The predicted $\delta^{56}Fe$ values for non-S $Fe(II)_{solid}$, therefore, are expected to match those of Fe^{2+}_{aq} under equilibrium conditions. At high rates of DIR, however, the fractionation between Fe^{2+}_{aq} and non-S $Fe(II)_{solid}$ is positive, on the order of ~1‰ in $^{56}Fe/^{54}Fe$, which would produce $\delta^{56}Fe$ values for non-S $Fe(II)_{solid}$ that are very low under kinetic conditions (see Table 8 in Johnson et al. 2005).

At circum-neutral pH, the rates of isotopic exchange between Fe^{2+}_{aq} and mackinawite (FeS) are sufficiently high that isotopic equilibrium is expected (Wu et al. 2012), in contrast to the kinetic effects seen at low pH (Butler et al. 2005; Guilbaud et al. 2010, 2011b). We therefore assume that the equilibrium $^{56}Fe/^{54}Fe$ fractionation between FeS and Fe^{2+}_{aq} is +0.3‰, as discussed in Sect. 5.3. This fractionation is appropriate for aqueous Fe speciation that is dominated by hexaquo Fe^{2+} (Fe^{2+}_{aq}), but would increase to +0.6‰ or larger in sulfide-rich systems where the proportion of aqueous FeS (FeS_{aq}) is high (Wu et al. 2012); this would have the effect of increasing the $\delta^{56}Fe$ values for FeS relative to non-S $Fe(II)_{solid}$ in Fig. 5.14. Finally, the solid-phase sediment extractions in Fig. 5.14 identify the Fe^{3+} component as poorly crystalline ferric hydroxide (e.g., ferrihydrite), and we assume an equilibrium $Fe(III)_{solid}$–Fe^{2+}_{aq} fractionation of ~+3‰ for $^{56}Fe/^{54}Fe$ (Welch et al. 2003; Wu et al. 2011b).

In both DIR- and DSR-dominated examples, $\delta^{56}Fe$ values predicted for non-S $Fe(II)_{solid}$ are their most negative in the upper few cm of the sediment section, reflecting the isotopic mass balance imposed by $Fe(III)_{solid}$ (Fig. 5.14). Assuming a zero $^{56}Fe/^{54}Fe$ fractionation between Fe^{2+}_{aq} and non-S $Fe(II)_{solid}$, appropriate for equilibrium conditions, $\delta^{56}Fe$ values for Fe^{2+}_{aq} would also be expected to be lowest in the upper part of the sediment section. As will be discussed below in the next section, the $\delta^{56}Fe$ values for Fe^{2+}_{aq} may be quite low due to oxidation of Fe^{2+}_{aq} in both S-poor and S-rich systems, in addition to the predicted equilibrium relations in Fig. 5.14. The relatively high $Fe(III)_{solid}$ component in the DIR-dominated case, relative to the DSR-dominated case, predicts, by mass balance, that non-S $Fe(II)_{solid}$ and FeS (and, by inference, Fe^{2+}_{aq}), have consistently negative $\delta^{56}Fe$ values (−1 to −3‰). In contrast, the more rapid depletion of $Fe(III)_{solid}$ with increasing depth in the DSR-dominated case predicts that $\delta^{56}Fe$ values for non-S $Fe(II)_{solid}$ and FeS (and, by inference, Fe^{2+}_{aq}) approach the bulk sediment composition ($\delta^{56}Fe$ ~ 0‰). From a practical perspective, it is important to note that the predicted trends in Fig. 5.14b, d suggest that acid leaching that captures both solid Fe^{2+} and Fe^{3+} will have $\delta^{56}Fe$ values which should vary as a function of the proportion of ferrous and ferric Fe in the leach, where leaches that capture predominantly the Fe^{3+} fraction will have higher $\delta^{56}Fe$ values. This also underscores the importance that any study of authigenic minerals and leaches should include ferrous and ferric Fe determinations, which we will return to below. In anticipation of what might be permanently found in the rock record, assuming that pyrite inherits its Fe isotope compositions from solid Fe^{2+} precursors, it would be expected that the $\delta^{56}Fe$ values for pyrite would be most negative for sedimentary sections where DIR was dominant, which would be associated with the lowest degree of pyritization (DOP), whereas higher $\delta^{56}Fe$ values for pyrite would be expected where DSR was dominant and hence DOP was high. This conceptual model, however, ignores complexities from potential kinetic isotope effects upon pyrite formation, as well as the possibility that pyrite formation occurred through multiple diagenetic stages during lithification, involving different Fe reservoirs.

5.4.2 Pore Fluid-Sediment Interactions

The co-variations in δ^{56}Fe values for Fe^{2+}_{aq}, and Fe^{2+}_{aq} contents in modern marine sediments are quite distinct from those observed in groundwater systems or redox-stratified lakes, indicating fundamentally different processes in terms of Fe pathways than what we have seen earlier in the chapter. In the groundwater and lake systems, samples that have the highest aqueous Fe contents have near-zero or slightly positive δ^{56}Fe values, which are then subsequently modified by Fe oxidation and mineral precipitation, producing trends of decreasing δ^{56}Fe and aqueous Fe (Figs. 5.7 and 5.9). In contrast, pore fluids in modern marine sediments that have high Fe^{2+}_{aq} contents generally have much lower δ^{56}Fe values than high-Fe groundwaters and lakes (Fig. 5.15), with many data from low-sulfide systems clustering around a δ^{56}Fe value of ~ -1‰ for Fe^{2+}_{aq} (Fig. 5.15a). As show in Fig. 5.14, low-sulfide marine sediments, which are characterized by a low DOP, tend to be environments where reactive Fe is cycled by DIR. The most common DIR end member seems to be Fe^{2+}_{aq} contents of ~100–600 μM and δ^{56}Fe ~ −0.9 to −1.4‰, as seen in the Monterey Canyon, Eel River, Umpqua River, and Celtic Sea sites. Referring to Fig. 5.14, such Fe isotope compositions for Fe^{2+}_{aq}, as inferred from non-S $Fe(II)_{solid}$, would reflect Fe isotope exchange with reactive Fe^{3+}, consistent with a wealth of experimental studies on DIR discussed in Chap. 3. Because the sediments are a dynamic system, where DIR is sustained by influx of C_{org} and reactive Fe^{3+}, it would be expected that low-δ^{56}Fe Fe^{2+}_{aq} would be sustained as long as the delivery of C_{org} and Fe^{3+} was maintained.

Turning to the other low-DOP settings, Cape Margin pore fluids have a wide range of δ^{56}Fe values, but Fe^{2+}_{aq} contents are exceptionally low, extending into the nM range, in part reflecting the very low reactive Fe contents (Homoky et al. 2013). The Crozet data, although they are from low-DOP sediments, have δ^{56}Fe values near zero, which is interpreted to reflect a dominance of colloidal Fe particles and the effects of non-reductive dissolution of silicate minerals (Homoky et al. 2009, 2011). This is similar to the non-reductive dissolution observed in sediments near Papua New Guinea (Radic et al. 2011).

For high-DOP settings (Santa Barbara and Santa Monica basins and the North Sea), two examples have data that extend to high δ^{56}Fe at moderate Fe^{2+}_{aq} contents, a composition space generally not seen in the low-DOP settings (Fig. 5.15b). If we assume that this reflects equilibrium Fe isotope exchange with a solid that has near-zero δ^{56}Fe values, we may infer that this could reflect Fe isotope equilibrium between Fe^{2+}_{aq} and FeS, and we define this as an FeS endmember for high-DOP sediments, following the initial proposal by Severmann et al. (2006). The "FeS end member" reflects the importance of FeS in sediments that undergo high extents of sulfate reduction, which minimizes the importance of reactive Fe^{3+} in controlling the δ^{56}Fe values of Fe^{2+}_{aq} because reactive Fe^{3+} has been converted to Fe sulfide (Fig. 5.14). It is important to note that because the $\Delta^{56}Fe_{Fe(II)aq-FeS}$ fractionation is −0.3 to −0.6‰, depending upon FeS_{aq} contents (Wu et al. 2012) the trend of increasing δ^{56}Fe and decreasing Fe^{2+}_{aq} between the "FeS" and "DIR" components cannot be explained by precipitation of FeS from the DIR end member, but must instead represent mixing between aqueous Fe components.

In all cases, vertical profiles for the low-DOP data in Fig. 5.15a show decreasing δ^{56}Fe and Fe^{2+}_{aq} contents upward in the sediment section, where the lowest values are measured closest to the sediment-seawater interface. Such trends clearly reflect oxidation of pore fluids as they encounter oxygen from the overlying seawater (driven by bioirrigation and diffusion). This trend is well developed for the low-DOP settings of

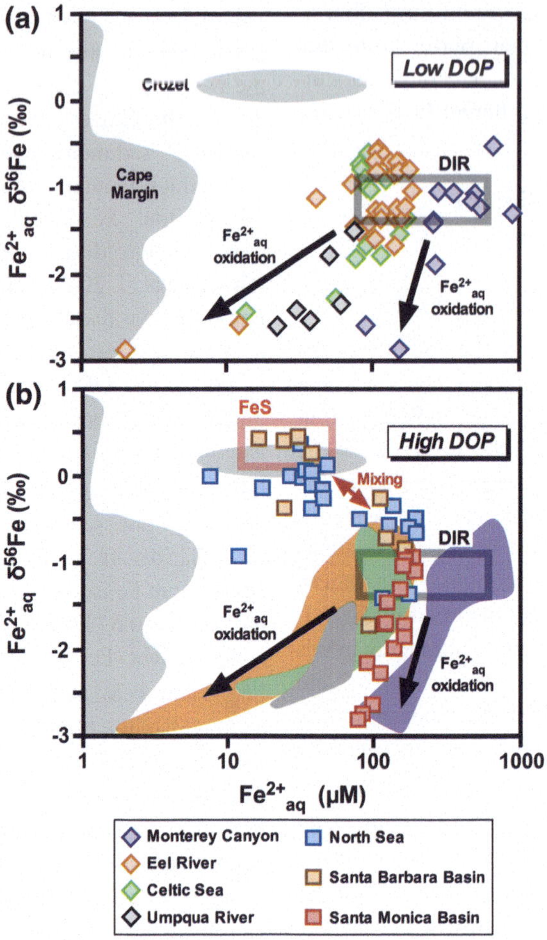

Fig. 5.15 $\delta^{56}Fe$ values and Fe^{2+}_{aq} contents for pore fluids from modern marine sediments. **a** shows data (diamonds) from sediments that have a low Degree of Pyritization (DOP), corresponding to cases where Dissimilatory Iron Reduction (DIR) dominates over Dissimilatory Sulfate Reduction (DSR) in terms of respiration of organic C (C_{org}). In all cases except the Crozet Plateau, the negative $\delta^{56}Fe$ values are interpreted to be produced by DIR, modified by Fe^{2+}_{aq} oxidation during pore fluid escape to the ocean. The data from Crozet are interpreted to reflect non-reductive dissolution of silicates. The data for the Cape Margin are plotted as a field because they extend to very low Fe^{2+}_{aq} contents (nM range); the wide range in $\delta^{56}Fe$ values is interpreted to reflect both DIR and non-reductive processes. Data sources: Crozet Plateau, Southern Ocean (Homoky et al. 2009), Cape Margin, Africa (Homoky et al. 2013), Monterey Canyon, California margin (Severmann et al. 2006), Eel River, California margin (Homoky et al. 2009; Severmann et al. 2010), Celtic Sea, NW European margin (Klar et al. 2017a), Umpqua River, California margin (Severmann et al. 2010). **b** shows data (squares) from sediment sections that have a high DOP, corresponding to examples where DSR is more important than DIR in terms of respiration of C_{org}. The moderately negative to positive $\delta^{56}Fe$ values are interpreted to reflect mixing between a DIR source and Fe^{2+}_{aq} in isotope equilibrium with FeS. In some cases (Santa Monica Basin), the trend of decreasing $\delta^{56}Fe$ and Fe^{2+}_{aq} is interpreted to additionally include oxidation of Fe^{2+}_{aq} of pore fluids as they emanate from the sediment into the overlying water column. Data sources: North Sea, German margin (Henkel et al. 2016), Santa Barbara Basin, California margin (Severmann et al. 2006), and Santa Monica Basin, California margin (Severmann et al. 2010). Fields of data from **a** also shown in **b** for reference

Monterey Canyon, Eel River, Celtic Sea, and Umpqua River. The slopes of these trajectories are variable, and scattered, likely reflecting differences in the net Fe isotope fractionation factor between Fe^{2+}_{aq} and Fe^{3+}-hydroxides, as well as the extent of oxidation (see, for example, Figs. 5.7 and 5.9). The shallowest pore fluids in all of the high-DOP suites also lie along a trend of decreasing Fe^{2+}_{aq} contents and $\delta^{56}Fe$ values that is indicative of oxidation of Fe^{2+}_{aq}, and this trend is particularly well developed for the Santa Monica Basin site (Fig. 5.15b). Importantly, the Fe^{2+}_{aq} oxidation trend is found even in pore waters from sediment sections that underlie low bottom water O_2 contents (~5 μM at Santa Monica), indicating that oxidation of Fe^{2+}_{aq} can still occur during escape of pore water Fe^{2+}_{aq} to the overlying ocean at low O_2 levels.

Three marine sediment profiles have Fe isotope data available for coexisting pore water Fe^{2+}_{aq} and acid-extractable solid-phase Fe: Monterey Canyon, Santa Barbara Basin, and the North Sea (Fig. 5.16), which allows an assessment of the likelihood that pore fluids and solid reactive Fe may be in isotope equilibrium. For the Monterey Canyon and Santa Barbara Basin datasets, extractions were done using 0.5 M HCl and Fe^{2+} and Fe^{3+} contents were additionally determined, which allows an assessment of the isotopic effects of extraction of a mixed Fe^{2+}–Fe^{3+} solid (Severmann et al. 2006); as noted earlier, such an approach is critical for understanding the solid phases that are sampled in such extractions, although ferrous-ferric Fe speciation is rarely done in Fe isotope studies of sediments. Extractions for the North Sea dataset did not include ferrous-ferric Fe speciation determinations, and we use the acetic acid extraction from that dataset in Fig. 5.16, which most likely records only Fe^{2+} phases such as FeS and non-S $Fe(II)_{solid}$ (Henkel et al. 2016). Broadly, we expect $\delta^{56}Fe$ values for aqueous and solid-phase Fe^{2+} to lie along a 1:1 line based on the fact that at equilibrium there should be little Fe isotope fractionation between Fe^{2+}_{aq} and non-S $Fe(II)_{solid}$, and the small (+0.3‰) fractionation between FeS and non-S $Fe(II)_{solid}$ (see discussion in Sect. 5.4.1). For solid-phase extractions that contain a mixture of Fe^{2+} and Fe^{3+}, an end member Fe^{2+} composition may be well estimated using the Fe isotope fractionation relations between ferrihydrite and non-S $Fe(II)_{solid}$, as well as accounting for the presence of FeS in the $Fe(II)_{solid}$ component (Johnson et al. 2005; Wu et al. 2011b, 2012).

The relations in Fig. 5.16a generally scatter about a 1:1 line using solid-phase Fe^{2+} (filled symbols), either measured in the acetic acid extracts or calculated for the pure $Fe(II)_{solid}$ using 0.5 M HCl extracts and ferrous-ferric Fe speciation. This indicates that, to a first approximation, equilibrium fractionation occurs between Fe^{2+}_{aq} and reactive solid-phase Fe^{2+}, a conclusion that will bear importantly on discussions below regarding sulfide paragenesis. This conclusion is particularly striking, given the ~3.5‰ range in $\delta^{56}Fe$ values that demonstrate a wide variety of Fe pathways is involved. Deviations from the 1:1 line may occur via kinetic isotope fractionation during rapid formation of non-S $Fe(II)_{solid}$, an increase in FeS_{aq} speciation, or rapid Fe^{2+}_{aq} oxidation, as indicated in Fig. 5.16. Such conclusions are well supported by experimental studies (Butler et al. 2005; Johnson et al. 2005; Wu et al. 2012). Extractions that contain significant Fe^{3+} also deviate from the 1:1 line, and the greatest deviations occur for samples that have the highest proportion of Fe^{3+} in the extracts (see dashed example tie lines between filled and open symbols in Fig. 5.16a), as expected based on the large Fe isotope fractionation between poorly crystalline Fe^{3+} and Fe^{2+}.

We may conclude that the relations in Fig. 5.16 allow us to define end member cases where FeS controls the $\delta^{56}Fe$ values for Fe^{2+}_{aq} (defined where the % Fe^{3+} in the extracts approaches zero) and those where DIR controls the $\delta^{56}Fe$ values for Fe^{2+}_{aq}, the latter of which is defined by high $Fe(III)_{solid}$ contents coupled with high Fe^{2+}_{aq} contents (see also Fig. 5.15). This is highly useful as it potentially provides an Fe isotope fingerprint for DIR- and DSR-dominated systems, in addition to the overall characteristics of Fe and S abundances. Samples that plot at

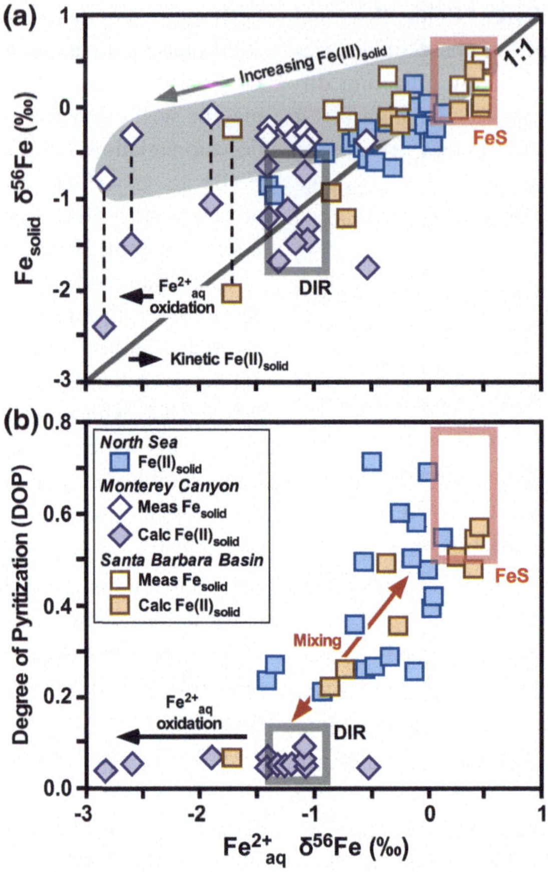

Fig. 5.16 Variations among δ^{56}Fe values for Fe^{2+}_{aq} and solid-phase acid-extractable solids and Degree of Pyritization (DOP) for modern marine sediments from the North Sea, Monterey Canyon, and Santa Barbara Basin. **a** shows the relation between δ^{56}Fe values of acid-extractable Fe_{solid} and Fe^{2+}_{aq}, supporting the suggestion that the FeS solid component has near-zero to slightly positive δ^{56}Fe values. **b** shows relations between δ^{56}Fe values for Fe^{2+}_{aq} and DOP, supporting identification of the FeS end member at high DOP. Symbols same as in Fig. 5.15, where diamonds note sediments that only contain low DOP values and squares note sediments that include high DOP values. For samples from the North Sea, data from HAc leaches are used and it is assumed all Fe_{solid} is Fe^{2+}. For samples from the Monterey Canyon and Santa Barbara Basin, HCl leaches are used that included Fe^{2+}–Fe^{3+} speciation measurements, and the measured solid extraction is shown in open symbols (Meas Fe_{solid}), which includes Fe^{2+} and Fe^{3+}. For these samples, an endmember $Fe(II)_{solid}$ δ^{56}Fe value is calculated for each point based on a Fe^{3+}–Fe^{2+} fractionation factor of +2.85‰ for $^{56}Fe/^{54}Fe$, which accounts for the presence of FeS in the $Fe(II)_{solid}$ component. Omission of the FeS component would increase the fractionation factor by 0.15‰. Samples that have very large corrections for the $Fe(II)_{solid}$ endmember, due to high proportions of Fe $(III)_{solid}$, shown with dashed tie lines. Isotopic compositions of the "DIR" and "FeS" end members in Fig. 5.15 also shown for their respective solid phase compositions. Data sources: North Sea, German margin (Henkel et al. 2016), Monterey Canyon and Santa Barbara Basin, California margin (Severmann et al. 2006)

lower δ^{56}Fe values than the DIR field belong to the trend of decreasing δ^{56}Fe values and decreasing Fe^{2+}_{aq} contents in Fig. 5.15, which is explained by oxidation of Fe^{2+}_{aq} at the top of the sediment section. Such samples are least likely to maintain Fe isotope equilibrium between Fe^{2+}_{aq} and $Fe(II)_{solid}$, and this is a likely explanation for their deviations from a 1:1 line in Fig. 5.16a despite a correction for Fe^{2+} and Fe^{3+}. As shown in Fig. 5.15, the δ^{56}Fe values of pore water Fe^{2+}_{aq}

that scatter between the DIR and FeS fields records mixing between these components, and Fig. 5.16b additionally shows that such mixing is well correlated with DOP, as would be predicted.

5.4.3 Solid-Phase Fe Components

In this section, we explore the Fe isotope implications of diagenetic processes that occur in modern marine sediments in the transition to lithified rocks, with a particular focus on Fe–S interactions and formation of diagenetic pyrite. In modern marine sedimentary deposits, approximately half of the C_{org} is respired via DSR, producing sulfide that ultimately finds its way into pyrite in the sedimentary rock record (e.g., Jørgensen 1982). Although in some modern marine environments, DIR is a more important pathway for C_{org} degradation (Fig. 5.14), the high sulfate content of modern seawater supports extensive sulfide production in most localities. This dominance of DSR would not be expected, however, in the early Earth, when seawater sulfate was much more limited (e.g., Habicht et al. 2002). Our goal here is to address the common observation in modern marine sediments that pyrite has significantly negative δ^{56}Fe values (e.g., Severmann et al. 2006, 2008; Fehr et al. 2008, 2010; Scholz et al. 2014b), although it should be noted that pyrite may have near-zero δ^{56}Fe values (Lin et al. 2017; Peng et al. 2017). Although syngentic (framboidal) pyrite is generally accepted to form in the water column (e.g., Wilkin and Barnes 1996), we focus here on diagenetic pyrite formation, which comprises the major portion of pyrite that occurs in modern marine sediments (e.g., Raiswell and Berner 1985).

At low temperatures (<100 °C), FeS (mackinawite) is an important precursor phase to pyrite formation (e.g., Morse et al. 1987; Rickard et al. 1995; Schoonen 2004), and five pathways have been identified:

$$\mathrm{FeS} + \mathrm{H_2S_{aq}} \rightarrow \mathrm{FeS_2} + \mathrm{H_2} \quad \text{"H}_2\text{S pathway"} \tag{5.10}$$

$$\mathrm{FeS} + \mathrm{S^0_{aq}} \rightarrow \mathrm{FeS_2} \quad \text{"S addition pathway"} \tag{5.11}$$

$$\mathrm{FeS} + \mathrm{S^{2-}_{aqX}} \rightarrow \mathrm{FeS_2} + \mathrm{S^{2-}_{aq[X-1]}} \quad \text{"Polysulfide pathway"} \tag{5.12}$$

$$\mathrm{FeS} + \mathrm{Fe(OH)_3} + 3\mathrm{H_2S_{aq}} \rightarrow 2\mathrm{FeS_2} + 3\mathrm{H_2O} + 3/2\mathrm{H_2} \quad \text{"Oxide sulfidation pathway"} \tag{5.13}$$

$$\mathrm{FeS} + 2\mathrm{H^+} \rightarrow \mathrm{FeS_2} + \mathrm{Fe^{2+}_{aq}} + \mathrm{H_2} \quad \text{"Fe loss pathway"} \tag{5.14}$$

The H_2S pathway was first recognized by Taylor et al. (1979) and involves H as the electron acceptor. In contrast, the S addition and polysulfide pathways involve S as the electron acceptor (e.g., Berner 1970; Luther 1991). Sulfidation of ferric hydroxides produces FeS that may further react to form pyrite (e.g., Rickard 1974; Pyzik and Sommer 1981). The Fe loss pathway produces Fe^{2+}_{aq} that further reacts with sulfide to form additional pyrite (e.g., Wilkin and Barnes 1996).

There is evidence that greigite, a mixed Fe^{2+}–Fe^{3+} sulfide (Fe_3S_4), may be an intermediate product in the conversion of mackinawite to pyrite (Wilkin and Barnes 1996; Benning et al. 2000). Greigite is found in SPM in the Black Sea water column (Lewis and Landing 1991; Cutter and Kluckhohn 1999), and in sulfidic sediments of the Santa Barbara Basin (Blanchet et al. 2009). Hunger and Benning (2007) demonstrated that the conversion of mackinawite to greigite to pyrite occurs in the solid-state via the reactions:

$$4\mathrm{FeS} \rightarrow \mathrm{Fe_3S_4} + \mathrm{Fe^{2+}_{aq}} + 2e^- \tag{5.15}$$

$$\mathrm{Fe_3S_4} + \mathrm{Fe^{2+}_{aq}} + 2e^- \rightarrow 2\mathrm{FeS_2} \tag{5.16}$$

Formation of greigite as an intermediate product couples Fe and S redox in the transition from mackinawite and pyrite. Unfortunately, we do not know the Fe isotope properties of greigite, although we might speculate that it would tend to have positive $\delta^{56}Fe$ values due to the presence of Fe^{3+} and covalent bonding (see Chap. 3). If, however, the transformation of mackinawite to greigite to pyrite occurs in the solid-state, the key Fe isotope fractionation step would be between mackinawite and pyrite and the isotopic properties of greigite would be immaterial. Once formed, it is not expected that pyrite would undergo Fe isotope exchange due to its high stability and very low solubility, a conclusion supported by enriched S isotope studies (Fossing and Jørgensen 1990).

Based on the relations in Fig. 5.16, if pyrite largely inherits the Fe isotope compositions of reactive Fe phases in FeS-dominated systems, pyrite is likely to have near-zero $\delta^{56}Fe$ values, whereas inheritance of the Fe isotope compositions of reactive Fe in DIR-dominated systems are likely to produce pyrite with negative $\delta^{56}Fe$ values. We will see below that some sites (e.g., Santa Barbara basin) have pyrite that cannot be explained by either of these end members. We recall our discussion earlier, however, where there are several lines of evidence that formation of sedimentary pyrite from FeS precursors may be associated with significant kinetic Fe isotope fractionation. Experimental studies by Guilbaud et al. (2011a) produced large kinetic Fe isotope fractionations between FeS and pyrite, between 2 and 3‰ in $^{56}Fe/^{54}Fe$ at 100 °C. As discussed in Czaja et al. (2012), such large fractionations at elevated temperatures were obtained at pyrite formation rates that were orders-of-magnitude faster than rates estimated for pyrite formation in natural diagenetic systems. Based on HCl extractions, ferrous-ferric Fe speciation, and pyrite analysis, Scholz et al. (2014b) suggested that the fractionation between FeS and pyrite was more modest in modern marine sediments, about 1‰ in $^{56}Fe/^{54}Fe$. This smaller fractionation, relative to experiments, would still be considered a kinetic fractionation, and hence could be variable in different marine sediment sections. For illustrative purposes, however, we will take the FeS–FeS_2 fractionation of Scholz et al. (2014b) to define the "FeS Marine Trend" in Fig. 5.17 (see also Fig. 5.11).

A number of key issues and Fe pathways are illustrated in Fig. 5.17. One of the most important is determination of Fe^{2+}–Fe^{3+} speciation in acid leaches. Using the calculated $\delta^{56}Fe$ value for the $Fe(II)_{solid}$ component in the HCl leaches, and assuming it to be equal to FeS, produces very large negative FeS-pyrite fractionations for low-DOP samples from Monterey Canyon and the Santa Barbara Basin. As noted in Sect. 5.4.2, however, such negative fractionations are associated with the upper sections of the sediment column that is unlikely to be in Fe isotope equilibrium due to rapid oxidation of Fe^{2+}_{aq} and/or production of Fe^{2+}_{aq} that has negative $\delta^{56}Fe$ values via DIR. In contrast to the fractionations observed in the Monterey Canyon samples, those measured between HCl leaches and pyrite for the low-DOP samples from the Baltic Sea have strongly positive fractionations (Fig. 5.17). If the leaches for the low-DOP Baltic Sea samples had high proportions of Fe^{3+}, this could explain the high fractionation at low-DOP, as is the case for the Monterey Canyon samples that were not corrected for Fe^{3+} contents (see open symbols in Fig. 5.16a). Unfortunately, this cannot be tested, as the Baltic Sea study did not perform ferrous-ferric Fe speciation on the HCl leaches. Because of this, firm inferences of FeS-pyrite fractionations for the Baltic Sea samples cannot be made, particularly for the low-DOP samples that may have significant Fe^{3+} contents in the HCl leaches.

At higher DOP levels, Fe^{3+} contents should be very low in HCl leaches, indicating that the leaches will recover the $\delta^{56}Fe$ values of FeS. This is shown by the close correspondence between the measured $^{56}Fe/^{54}Fe$ fractionations and those calculated for the $Fe(II)_{solid}$ component in the HCl leaches (compare open and closed symbols in Fig. 5.16a). In Fig. 5.17, many of these high-DOP samples, including those from the Santa Barbara Basin, Baltic Sea, and Peru Margin, scatter about an FeS-pyrite $^{56}Fe/^{54}Fe$ fractionation of $\sim 1‰$ ("FeS Marine Trend"), and in all cases are much lower than the experimental results from Guilbaud et al. (2011a), supporting the argument by Scholz et al. (2014b) that the large kinetic Fe isotope fractionations between FeS and pyrite measured experimentally by Guilbaud et al. (2011a) are not representative of natural systems. This observation also casts doubt on the proposal by Guilbaud et al. (2011a) that the entire range in $\delta^{56}Fe$ values of sedimentary pyrite can be simply explained by kinetic isotope effects during pyrite formation, with no role for Fe redox.

There are limited data for estimating FeS-pyrite Fe isotope fractionations at high extents of DOP, and those that are available may in part reflect unusual conditions. Although Fe isotope compositions were determined for acid extractions and Fe^{2+}_{aq} for high-DOP samples from the North Sea (Fig. 5.16; Henkel et al. 2016), Fe isotope measurements were not done on the pyrite fraction. In the case of the Oxygen Minimum Zone (OMZ) from the Peru margin, for which three FeS-pyrite pairs were measured, Scholz et al. (2014a) interpret these samples to have lost a low-$\delta^{56}Fe$ component via transport of Fe^{2+}_{aq} to the open ocean, which has the effect of shifting the Fe isotope compositions of pyrite in the OMZ to anomalously high $\delta^{56}Fe$ values; this produces an apparent near-zero $^{56}Fe/^{54}Fe$ fractionation between FeS and pyrite (Fig. 5.17). Alternatively, there may be something specific to the FeS-pyrite pathway during sulfide diagenesis in the OMZ of the Peru margin that produces no net Fe isotope fractionation. Despite the potential complexity of the Peru OMZ samples, several other samples from the Peru margin have fractionations that plot near the FeS Marine Trend. Regardless, none of the high-DOP samples plot near the experimentally determined FeS-pyrite fractionation factor of Guilbaud et al. (2011a), supporting the contention that these experimentally determined pyrite fractionations are not applicable to pyrite formation in marine sediments.

5.4.4 Benthic Fe Fluxes

Measured pore fluid Fe^{2+} contents in coastal sediments, as well as microbial incubation experiments, highlight the importance of DIR in modern marine sediments in terms of Fe redox cycling, as well as liberating Fe^{2+}_{aq} that may find its way into overlying seawater (Canfield et al. 1993a, b). Expanding on this, using benthic chambers ("landers"), Elrod et al. (2004) noted benthic Fe^{2+}_{aq} fluxes from marine sediments may be much higher than previously estimated using sediment pore fluid profiles due to oxidation in the shallow sediment section by bioirrigation, and they proposed that a flux of iron from continental shelf sediments was a major input of Fe to the oceans, traceable for 100's of km offshore. When compared to the estimated $\delta^{56}Fe$ values for fluxes that have been measured for their Fe isotope composition, either by benthic chambers or pore water compositions in the tops of sediment sections, there is a broad negative relation between benthic Fe flux and $\delta^{56}Fe$ (Fig. 5.18). Where Fe fluxes are highest, the $\delta^{56}Fe$ values are low, which is generally accepted to record a strong reductive dissolution signal (e.g., Homoky et al. 2016). This may reflect DIR, but reduction by sulfide may also liberate Fe^{2+}_{aq}. In contrast, sediments that produce small Fe fluxes have, in some cases, near-zero $\delta^{56}Fe$ values, which has been interpreted to record non-reductive Fe dissolution that has minimal isotopic fractionation, such as dissolution of silicate Fe (Homoky et al. 2013, 2016; Radic et al. 2011).

As shown in Figs. 5.15 and 5.16, the $\delta^{56}Fe$ values for Fe^{2+}_{aq} produced by reductive dissolution in the deeper portions of a sediment section are estimated to lie between -0.9 and $-1.4‰$. Reduction by sulfide may produce high Fe^{2+}_{aq} contents, but if the Fe isotope compositions of Fe^{2+}_{aq} are controlled by FeS, their $\delta^{56}Fe$ values

Table 5.1 Benthic Fe Fluxes

Location	C_{org} wt. %	Reactive Fe wt. %	DOP	C_{ox} mmol m^{-2} d^{-1}	Bottom water O_2 μM	Dissolved Fe μM m^{-2} d^{-1}	Dissolved Fe δ^{56}Fe ‰	Data sources
Black Sea		0.45	0.32		Nil			1
Peru Margin Transect—Shelf	3.63	0.45	0.50	8.4	<2	316		2
Peru Margin Transect—OMZ	10.01	0.09	0.66	3.5	<2	20		2
Peru Margin Transect—sub-OMZ	4.41	0.25	0.18	2.1	39.7	Nil		2
Santa Barbara Basin	2.30	0.41	0.39	1.9	3.5	490	−1.73	3
Santa Monica Basin	2.30	0.41	0.39	1.9	4.0	412	−3.25	4
Baltic Sea			0.30		20	100		5
Umpqua River, southern Oregon	1.21	0.40	Nil	16.3	65.5	56.3	−2.65	6
South China Sea		1.35	0.22		86.6			7
Monterey Canyon	0.35	0.58	0.06	11.0	100	11.0	−2.87	8
Eel River, northern California	0.94	0.60	Nil	12.0	103	17.8	−2.20	6
Cape Margin	1.52	0.17	Nil	4.2	200	0.17	0.23	9
Crozet Plateau	0.41	0.25	Nil	0.01	230	2.7	0.15	10
North Sea	1.20	0.63	0.47	16.1	233	12	−0.93	11
Celtic Sea—Late Spring	5.60	0.08	Nil	10	260	27.5	−2.45	12
Celtic Sea—Late Summer	5.60	0.07	Nil	10	260	3.8	−3.05	12

Sources
1—Eckert et al. (2013), Lyons and Severmann (2006), Severmann et al. (2008)
2—Scholz et al. (2011, 2014a, b)
3—Leslie et al. (1990), Severmann et al. (2006)
4—Leslie et al. (1990), Severmann et al. (2010)
5—Fehr et al. (2008, 2010)
6—Homoky et al. (2009), Severmann et al. (2010)
7—Lin et al. (2017)
8—Severmann et al. (2006)
9—Homoky et al. (2013)
10—Homoky et al. (2009)
11—Henkel et al. (2016)
12—Klar et al. (2017b)

should be much higher (Figs. 5.15 and 5.16). The range in Fe isotope compositions inferred for reductive dissolution is consistent with experimental studies of DIR discussed in Chap. 3 at moderate extents of reduction (ca. 10–40%). Such isotopic compositions are also expected given the inventories of reactive Fe^{2+} and Fe^{3+} in marine sediments where DIR is important (Fig. 5.14). As experimental studies have shown, at very small extents of reduction, DIR can produce δ^{56}Fe values for Fe^{2+}_{aq} as low as −3‰, approaching the maximum possible fractionation factor. Such extreme compositions would only be seen at the very low Fe^{2+}_{aq} contents produced by only a few % reduction. That many of the estimates for the Fe isotope composition of benthic Fe fluxes have lower δ^{56}Fe values (Table 5.1 and Fig. 5.18) than those estimated

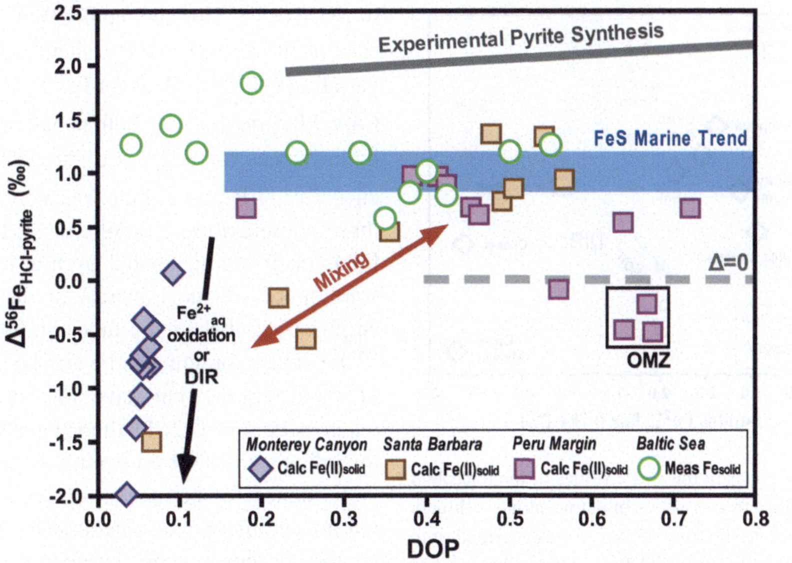

Fig. 5.17 Fe isotope fractionation between 0.5 M HCl extractions and pyrite ($\Delta^{56}Fe_{HCl-pyrite}$) relative to Degree of Pyritization (DOP) for modern marine sediments from Monterey Canyon and Santa Barbara Basin (California margin, USA), the Peru margin, and the Baltic Sea. This provides an estimate of the isotopic fractionation between reactive Fe, such as poorly crystalline Fe^{3+}-hydroxides or Fe^{2+}-sulfides such as mackinawite, and pyrite. Symbols for California margin samples same as in Figs. 5.15 and 5.16. HCl extractions of samples from the California and Peru margin includes Fe^{2+}–Fe^{3+} speciation, which allows calculation of the $\Delta^{56}Fe_{HCl-pyrite}$ fractionation for an end member $Fe(II)_{solid}$ component that is interpreted to be FeS, as describe in the caption to Fig. 5.16. Ferrous-ferric Fe speciation was not determined on the HCl extractions of samples from the Baltic Sea, and it is likely that low-DOP samples contain significant Fe^{3+}, as expected based on the high $\delta^{56}Fe$ values of the low-DOP samples. The experimentally determined FeS-pyrite fractionation determined by Guilbaud et al. (2011a) is shown by the dark grey line, which is interpreted to reflect a kinetic effect during conversion of FeS to pyrite. The blue band is the estimated fractionation between FeS and pyrite from Scholz et al. (2014b), which is also interpreted to reflect a kinetic fractionation, although much less than that obtained in experiments; this is defined as the "FeS Marine Trend" here and in other figures. Data from Severmann et al. (2006), Scholz et al. (2014b), Fehr et al. (2008, 2010)

for DIR require an additional process: oxidation of Fe^{2+}_{aq} near the sediment-seawater interface, and this is confirmed by the fact that, with a single exception, all pore fluid profiles for the locations in Fig. 5.18 trend toward negative $\delta^{56}Fe$ values at the top. If the effect of oxidation is accounted for in the flux values, the flux-weighted average $\delta^{56}Fe$ value for the benthic flux is −2.4‰ at 103 μM $Fe^{2+}_{aq}/m^2 \cdot d$ (Table 5.1).

The effect of oxidation of Fe^{2+}_{aq} as a function of C_{org} burial and bottom water O_2 is illustrated by high-DOP locations (Santa Monica and Santa Barbara basins, North Sea), as well as the seasonal comparison in the Celtic Sea (Fig. 5.18). Under conditions of high C_{org} loading and abundant sulfate, ambient O_2 will be low, and Fe^{2+}_{aq} contents will be relatively high in the top sections of sediments that undergo extensive sulfide diagenesis due to limited Fe^{2+}_{aq} oxidation at the sediment-seawater interface. This was noted in Sect. 5.4.1 and in the discussion of Fig. 5.14, and suggests that high-DOP samples might define a $\delta^{56}Fe$–Fe flux trend that lies at higher Fe fluxes than low-DOP sections (Fig. 5.18). For data from the Celtic Sea, samples taken in late spring were associated with high C_{org} loading, low bottom water O_2 contents, and low O_2 penetration into the sediment, leading to a relatively high Fe^{2+}_{aq} flux (Klar et al. 2017a). This

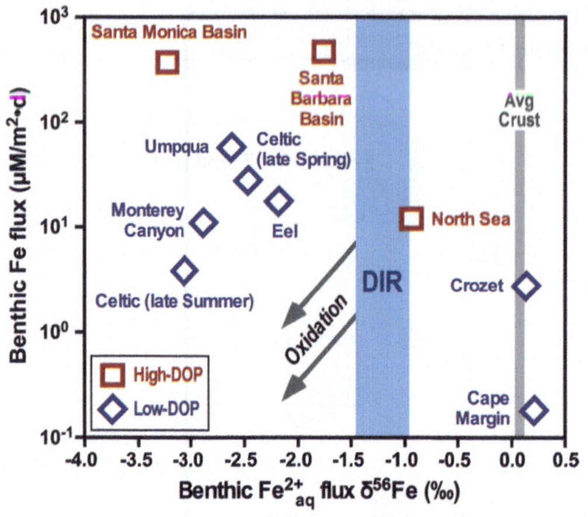

Fig. 5.18 Variations in the δ^{56}Fe values of the microbial benthic Fe^{2+}_{aq} flux from various modern marine sediment sections relative to Fe flux. Symbols are divided into sediments that have a high DOP (red squares; Santa Monica and Santa Barbara basins, California margin, and North Sea), where Fe fluxes are relatively high due to high C_{org} loading and low bottom water O_2 contents, and sediments that have a low-DOP (Umpqua and Eel rivers and Monterey Canyon, California margin, Celtic Sea, Crozet Plateau, and Cape Margin), where sulfide diagenesis is low and bottom water O_2 tends to be higher. The influence of seasonal changes in C_{org} and O_2 is seen in comparison of data from late spring (high C_{org}, low O_2) and late summer (low C_{org}, high O_2) in the Celtic Sea. Calculated Fe fluxes account for oxidation of pore fluid Fe^{2+}_{aq} in upper sediment sections, and the flux-isotopic trend of this effect is shown by the diagonal grey arrows. The estimated δ^{56}Fe values for DIR-generated Fe^{2+}_{aq} in modern marine sediments is given by the blue bar, and is taken from Figs. 5.15 and 5.16. Vertical grey line is the δ^{56}Fe value for average crust. Data sources given in Table 5.1

contrasts with samples taken in late summer, where lower C_{org} loading results in deeper O_2 penetration into the sediment section, producing a lower Fe^{2+}_{aq} flux to the ocean and lower δ^{56}Fe values for Fe^{2+}_{aq}. Such a comparison can be used to define relative Fe^{2+} oxidation trends (see arrows in Fig. 5.18), where a lower Fe^{2+}_{aq} flux is associated with lower δ^{56}Fe values, identical to oxidation trends seen in Figs. 5.7, 5.8, and 5.9.

The δ^{56}Fe values for Fe fluxes from two localities, Crozet and Cape Margin, are near zero (Fig. 5.18), and both localities are associated with sediments that contain relatively small proportions of reactive Fe (Homoky et al. 2009, 2013). In the case of the Crozet Plateau, there is little vertical variation in δ^{56}Fe values for Fe^{2+}_{aq}, although in the Cape Margin, there are both negative and positive δ^{56}Fe values for Fe^{2+}_{aq} at depth, reflecting influences by DIR and sulfide, respectively. Despite these complexities, Homoky et al. (2009, 2013) focus on fluxes generated by non-reductive dissolution of silicate minerals at these localities, emphasizing that such a flux will have near-zero δ^{56}Fe values for aqueous Fe and be characteristic of shelf regions that have low proportions of reactive Fe and C_{org} loading, hence minimizing the influence of DIR on benthic Fe fluxes.

Evidence for permanent enrichment of Fe in marine sediments via a benthic Fe flux is found in the proportion of extractable reactive Fe components and overall enrichment of Fe relative to background sedimentary baselines, commonly expressed in terms of Fe_{tot}/Al ratios, where total Fe contents (ferrous and ferric) are used (Canfield et al. 1996; Anderson and Raiswell 2004; Lyons and Severmann 2006). Marine basins that contain strong redox gradients such as the Black Sea, Cariaco Basin, and the Gotland Deep have represented textbook cases for studying transport of Fe released by reductive dissolution in shelf environments, followed by transport to the deep basin and sequestered in sediments under euxinic conditions. Studies by Fehr et al. (2008) of the Gotland Deep and Severmann et al. (2008) and Eckert et al. (2013) of the Black Sea documented a correlation of decreasing bulk sediment δ^{56}Fe values with increasing Fe_{tot}/Al ratios that were interpreted to record enrichment in low-δ^{56}Fe Fe in the euxinic portions of these basins (Fig. 5.19); this model has become known as the "benthic Fe redox shuttle" (see also Wijsman et al. 2001a). Severmann et al. (2008) noted that sediments deposited in the oxic/suboxic shelf of the Black Sea have slightly higher δ^{56}Fe values than average crust, and lower Fe_{tot}/Al ratios, features they interpreted to record loss of low-δ^{56}Fe Fe to the deep parts of the basin (Fig. 5.19). Through analysis of pyrite extractions, which may preferentially record the sulfidic trap for the transported Fe, as well as

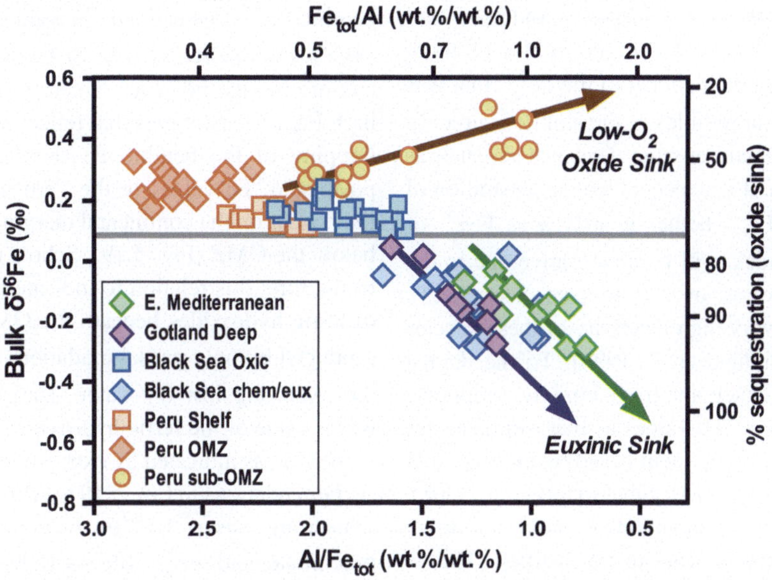

Fig. 5.19 Variations in bulk-sample Al/Fe_{tot} (Fe_{tot}/Al) (both as wt%/wt%) and $\delta^{56}Fe$ for modern marine sediments, illustrating the effects of the benthic Fe shuttle on sediment compositions. Fe_{tot} indicates that total Fe (ferrous and ferric) Fe is used. Scale for Al/Fe_{tot} is linear to allow extrapolation to a non-lithogenic component (Al/Fe_{tot} = 0). For reference, the Fe_{tot}/Al ratios for regional lithologic baselines are: Eastern Mediterranean (0.75), Gotland Deep (0.60), Black Sea (0.60), and Peru Margin (0.47); measured values that lie above these are interpreted to reflect Fe gain, whereas those that lie below record Fe loss. Samples that reflect euxinic conditions (sulfide sink) shown in diamonds, reflecting a 100% sequestration of low-$\delta^{56}Fe$ shuttled Fe, where variable Fe_{tot}/Al records different proportions of lithologic (clastic) and shuttled Fe. Samples that reflect oxic or sub-oxic conditions shown in squares, and record Fe loss via DIR. The sub-OMZ zone of the Peru margin records Fe accumulation via partial oxidation, and hence increasing Fe_{tot}/Al and increasing $\delta^{56}Fe$ (circle symbols); modeling the percent oxidation from the OMZ is shown on right-hand axis, assuming a Rayleigh model, an initial $\delta^{56}Fe$ value for Fe^{2+}_{aq} of −0.5‰, and an Fe^{3+} oxide-Fe^{2+}_{aq} fractionation factor of +1.2‰ for $^{56}Fe/^{54}Fe$ (see text). Data from Lyons and Severmann (2006), Fehr et al. (2008), Severmann et al. (2008), Scholz et al. (2011), (2014a), and Azrieli-Tal et al. (2014)

estimation of the $\delta^{56}Fe$ value of the enriched Fe component based on bulk-sample Fe_{tot}/Al ratios, Fehr et al. (2008) and Severmann et al. (2008) calculated the $\delta^{56}Fe$ values for the shuttled Fe to have been ~−1.1 to −1.6‰, which lies within the range measured directly from pore fluids in Fig. 5.18. Azrieli-Tal et al. (2014) studied organic- and sulfide-rich sediments (sapropels) from the Eastern Mediterranean and found a similar correlation of $\delta^{56}Fe$ and Fe_{tot}/Al (Fig. 5.19), which was also was interpreted to record a low-$\delta^{56}Fe$ benthic Fe shuttle.

There are a number of points between basin-scale transport and final sequestration of shuttled Fe where $\delta^{56}Fe$ values of shuttled Fe may be changed. Immediate oxidation of Fe^{2+}_{aq} released from reductive dissolution at the sediment-seawater interface is already accounted for in the Fe flux-$\delta^{56}Fe$ relations in Fig. 5.18, as noted above. If, however, the Fe^{2+}_{aq} that escapes the sediment under low bottom water O_2 later encounters O_2-rich water in another locality, additional ferric hydroxides would be produced distally from the source, and this could change $\delta^{56}Fe$ values of shuttled Fe as a function of the extent of oxidation (e.g., Chever et al. 2015). Complete oxidation, as is expected in high-O_2 seawater, will produce no change in $\delta^{56}Fe$ values of the original Fe^{2+}_{aq} (recall the discussion in Sect. 1.2.3 in Chap. 1). In this case, the shuttled Fe would be in the form of fine-grained particulate iron hydroxides that are advected to the deep basin via basin-wide transport (Anderson and Raiswell 2004), and the δ^{56}Fe values of these particles would be identical to the Fe^{2+}_{aq} that was released from pore fluids at the shuttle source. In basins that contain euxinic deep waters, such as the Black Sea, it is expected that sequestration of the shuttled Fe, whether it arrives as Fe^{2+}_{aq} or ferric hydroxides, will be quantitative due to the very high excess of sulfide (see Sect. 5.3.3), based on the very high reactivity of these species to sulfide (Canfield et al. 1996). To the degree that pyrite extractions from euxinic sediments have the same δ^{56}Fe values as that estimated for the shuttled Fe component based on bulk Fe_{tot}/Al ratios (Fehr et al. 2008; Severmann et al. 2008), the assumption of quantitative sequestration in euxinic systems appears to be justified. This is illustrated in Fig. 5.19, where the inferred shuttled Fe component in all of the euxinic basins studied to date has a δ^{56}Fe value of ~−1‰, based on extrapolation of δ^{56}Fe–Al/Fe_{tot} relations in Fig. 5.19 to Al/Fe_{tot} = 0 (the pure shuttle component). This Fe isotope composition lies in the range of pyrite extracts from the same sediments, consistent with a model of 100% sequestration of the shuttled Fe in the euxinic sink. A similar δ^{56}Fe–Fe_{tot}/Al (or Al/Fe_{tot}) relation is expected where Fe^{2+}_{aq} escapes the shelf sediment in a low-O_2 layer, such as beneath the chemocline, followed by transport in the low-O_2 layer, followed by sequestration directly into the euxinic layer (Lyons and Severmann 2006).

Markedly different results are obtained if only a small proportion of shuttled Fe is sequestered into the rock record because of significant Fe isotope fractionation between aqueous and solid phase Fe during partial oxidation (see again the discussion in Sect. 1.2.3 in Chap. 1). As noted earlier, the high productivity Peru margin produces a thick OMZ that intersects the continental margin, where production of Fe^{2+}_{aq} via reductive dissolution escapes with a high efficiency into the OMZ, followed by transport within the OMZ (Scholz et al. 2011; Noffke et al. 2012). Sediments in the OMZ have Fe_{tot}/Al ratios that are lower than the terrigenous input, and slightly positive δ^{56}Fe values (Fig. 5.19) are interpreted to indicate loss of benthic Fe^{2+}_{aq} that had a δ^{56}Fe value of ~−0.4 to −0.5‰ (Scholz et al. 2014a, b). Unlike the case for euxinic basins, however, when we consider the high Fe_{tot}/Al sediments that reflect accumulation/trapping of the benthic Fe shuttle, these have positive δ^{56}Fe values in the Peru margin where they occur on the continental margin immediately below the OMZ (Fig. 5.19). Scholz et al. (2014a, b) interpret this relation to indicate accumulation of ferric hydroxides beneath the OMZ that were produced by only partial oxidation of Fe^{2+}_{aq} from the overlying low-O_2 water. They estimate the δ^{56}Fe value of the ferric hydroxides to have been ~+0.7‰, as indicated by extrapolation of δ^{56}Fe–Al/Fe_{tot} relations in Fig. 5.19 to Al/Fe_{tot} = 0. This in turn suggests a ^{56}Fe/^{54}Fe fractionation between precipitate and Fe^{2+}_{aq} of ~+1.2‰, which lies within the range of that seen during partial oxidation of Fe^{2+}_{aq} in natural systems (e.g., Beard and Johnson 2004). The high inferred δ^{56}Fe value of the ferric hydroxides, relative to that inferred for the Fe^{2+}_{aq} source, indicates that oxidation and precipitation of the ferric hydroxides that settled from the OMZ involved perhaps ~20% oxidation. In Fig. 5.19, this is directly cast as the % sequestration of the shuttled Fe on the right-hand vertical scale, highlighting the contrast between the open ocean shuttle of the Peru margin (10's of % sequestration from the shuttle) and the redox shuttle in enclosed euxinic basins (100% sequestration from the shuttle).

5.5 The Fe Budget of the Modern Oceans

5.5.1 Seawater Fe

Iron contents in modern seawater are exceedingly low, nM or lower (e.g., Conway and John 2014; Conway et al. 2016; Gordon et al. 1982; Johnson et al. 1997). In part this reflects the fundamental instability of Fe^{2+}_{aq} under oxygenated conditions at circum-neutral pH, as well as the very low solubility of Fe^{3+} minerals (e.g., Millero et al. 1987). Given such low concentrations, it has long been recognized that Fe is a critical limiting micronutrient in many parts of the oceans (Martin and Gordon 1988; Martin and Fitzwater 1988), with the potential to dramatically affect CO_2 and climate via its control on primary productivity (Martin 1990). A key focus of studies of Fe in the oceans is its form. Discussion since the earliest studies has had to wrestle with operational definitions of Fe species (Gordon et al. 1982), and we have touched on this above in the discussion of river water. For studies of seawater, "dissolved" Fe, dFe, is commonly defined as that which passes a 0.2 μm filter (or 0.4 μm in some studies), and particulate Fe (pFe) would be sizes larger than this. Soluble Fe (sFe) has been defined as that which passes a 0.02 μm filter (Wu et al. 2001; Bergquist and Boyle 2006a; Bergquist et al. 2007), or sometimes a 0.03 μm filter or 10 kDa membrane (e.g., Fitzsimmons et al. 2015), and sFe may be considered what is "truly" dissolved Fe. Colloidal Fe (cFe) is defined as an intermediate size, often as the difference between dFe and sFe. Importantly, cFe is not expected to settle in the water column, in contrast to pFe.

Surface waters are generally the most depleted in dFe, often in the tenths of a nM or lower, yet, based on solubility of Fe^{3+}-hydroxides alone, Fe^{2+}_{aq} contents should be in the 10's of nM (Fox 1988). This discrepancy suggests a control other than simply ferrihydrite solubility. Extensive studies of vertical profiles of dFe show that the deeper oceans are relatively enriched in dFe, in the nM range, producing a depth profile that is correlated (or anti-correlated) to those species produced by biology for O and N. This characteristic profile shape for Fe has been interpreted to record the combined effect of biological Fe uptake in the shallow ocean, in addition to remineralization and particle scavenging below the thermocline (e.g., Martin and Gordon 1988; Johnson et al. 1997; Moore and Braucher 2008; Boyd and Ellwood 2010). Voltammetry has demonstrated that large proportions, in some cases virtually all, dFe consists of Fe^{3+} that is bound to organic ligands (e.g., Rue and Bruland 1995). Based on titration, the strongest ligands are defined as L_1, with L_2, L_3, etc., defining progressively weaker ligand binding strengths; the proportions of these ligands are variable with water depth, and in the case of strongly bound L_1, these tend to be most abundant in shallow waters and are thought to include siderophore-like compounds (e.g., Boiteau et al. 2016; Boyd and Ellwood 2010; Hunter and Boyd 2007). An abundance of L_2 (or L_3) ligands in deeper water may reflect degradation of organic matter upon sinking, including re-mineralization processes.

In the deep oceans, including the OMZ, increases in dFe with depth are thought to reflect dissolution and re-mineralization of sinking biologic or lithogenic particles (Martin and Gordon 1988; Johnson et al. 1997). Correlations between dFe and parameters that measure C_{org} re-mineralization, such as Apparent Oxygen Utilization (AOU), are broadly consistent with such a model, although in detail, scatter in such correlations suggest complexity in Fe:C ratios for biologic utilization or other factors, including ligand binding (e.g., Johnson et al. 1997; Hatta et al. 2015). An average residence time for Fe in the oceans is commonly cited as $\sim 10^2$ yr (e.g., Bruland et al. 1994), although the vertical variations in dFe in the oceans suggests changes in the residence time of Fe, on the order of 10^{-1} yr in the shallow oceans to $\geq 10^2$ yr in the deep oceans (Bergquist and Boyle 2006a; Hayes et al. 2015). Some modeling, however, suggests that even in the deep oceans, the residence time of Fe may be as short as $\sim 10^1$ yrs (e.g., Moore and Braucher 2008).

Key areas of research on Fe in seawater have been the bioavailability of Fe, exchange between dissolved and particulate pools, and enrichment or depletion in particles relative to lithologic inputs. Among the commonly measured dissolved quantities (dFe, cFe, and sFe), these often have distinct vertical distributions in the water column, suggesting different residence times, complexation, and biological uptake/release (Wu et al. 2001; Bergquist and Boyle 2006a; Bergquist et al. 2007). Particulate Fe tends to increase with depth along with dFe and can comprise the dominant inventory of Fe in seawater (e.g., Sherrell and Boyle 1992; Bruland et al. 1994). The biological inventory of Fe in particulates, as well as the bioavailability of particulate Fe, has been extensively studied via partial leaching methods (e.g., Tovar-Sanchez et al. 2003; Tang and Morel 2006; Berger et al. 2008; Rauschenberg and Twining 2015). Given the high proportion of poorly crystalline Fe^{3+}-hydroxides that may comprise riverine and dust particles (see Sect. 5.2.1), much of the focus of leaching has been on removing surface-bound Fe, especially on biogenic particulates. One leaching protocol, a combination of EDTA and oxalate involves non-redox, ligand-promoted dissolution (Tang and Morel 2006), and this has been successfully applied to Fe isotope studies (Revels et al. 2015b). A non-reductive leaching protocol is particularly important when poorly crystalline Fe^{3+}-hydroxides are present to avoid isotopic artifacts, given the catalytic nature of electron exchange in terms of isotopic exchange in Fe(II)–Fe(III) systems (see Chap. 3).

In the discussion below, we will refer to ligand-promoted partial leaching of particulate Fe as "L–pFe", following its use in the seawater literature. Such leaching approaches have also been combined with elemental analysis of Fe and Al for comparison to lithogenic sources of particles. In the North Atlantic, for example, pFe/pAl and L-pFe/pAl ratios that far exceed those of average crust have been interpreted by Milne et al. (2017) to record biologic surface uptake (and release at depth) of Fe, although dFe–AOU correlations indicate additional sources of dFe to the deep oceans. Only recently have studies of dFe and pFe included ferrous-ferric Fe speciation measurements and synchrotron X-ray methods to fully identify the nature of Fe (e.g., Ohnemus and Lam 2015; Sedwick et al. 2015; Heller et al. 2017). From the isotopic perspective, fully understanding the mineralogy of particles is critical, given the sensitivity of Fe isotope fractionation to redox state and mineralogy (Chap. 3).

5.5.2 Riverine and Aeolian Sources

Continental delivery of Fe to the oceans may occur via several pathways, the evidence for which lies in elevated dFe in proximity to continental shelves relative to the open oceans (see, for example, review by Moore and Braucher 2008). Riverine input of Fe largely occurs in continental shelf regions, and its influence is seen in elevated coastal primary productivity (e.g., Wetz et al. 2006). Re-suspension of shelf sediments is also a source of Fe in coastal environments (e.g., Johnson et al. 1999). As discussed in Sect. 5.2.1, the majority of SPM or pFe in rivers has a restricted range in $\delta^{56}Fe$ values relative to average upper crust, and hence Fe isotopes are relatively insensitive to tracing pFe input via rivers. The range in $\delta^{56}Fe$ values for dFe in rivers, however, is much larger (Fig. 5.5), which has raised the possibility that such input could be traced in seawater. This is particularly true considering the origin of positive $\delta^{56}Fe$ values for dFe in shallow seawater, given the highly positive values for the smallest size fractions of dFe in rivers (Fig. 5.5b). The difficulty with a major source of riverine dFe to the oceans lies in the fact that extensive loss of dFe occurs in estuaries via flocculation, where perhaps up to 90% Fe loss occurs (Boyle et al. 1977). In contrast, Poulton and Raiswell (2005) highlight the importance of re-release of particulate Fe as particles transition though salinity gradients, perhaps limiting net Fe loss to ~40%. Nevertheless, significant Fe loss in estuaries has led many workers to conclude that a "continental" input of Fe to the oceans comes largely from aerosols and dust, rather than riverine material (e.g., Moore and Braucher 2008). In addition, benthic Fe sources associated with release of Fe from detrital Fe^{3+}-hydroxides via reductive dissolution is recognized as a source of Fe to the oceans that might be thought of as an "indirect" continental source. Not all workers, however, accept a small direct riverine flux of Fe to the oceans (e.g., Labatut et al. 2014).

It is generally thought that aeolian dust is a major source of Fe input into the surface open oceans far from continents (e.g., Jickells et al. 2005). As noted in Sect. 5.1.1, the Fe isotope composition of loess and aerosols is very homogeneous and is taken to be identical to that of average upper crust at $\delta^{56}Fe = +0.09 \pm 0.10$‰ (Beard et al. 2003; Gong et al. 2016). If this is representative of the dust flux to the oceans, then deviations from this composition must reflect Fe isotope fractionation between pFe and other Fe pools in the oceans. A potential complication comes from possible anthropogenic sources, which have been posited to have negative δ^{56}Fe values (Majestic et al. 2009; Mead et al. 2013; Kurisu et al. 2016); see also data from the Seine River, Fig. 5.6). Using size-sorting and leaching methods, Kurisu et al. (2016) suggested a several per mil range in δ^{56}Fe values might be produced by anthropogenic sources, a remarkably wide range of Fe isotope compositions, although it is important to keep in mind that the mass of Fe in their highly negative δ^{56}Fe components was exceedingly small. Nevertheless, as anthropogenic Fe sources continue to be delivered to the surface oceans, it is possible that this may be seen in the Fe isotope compositions of the surface ocean over time, given the very short residence time of Fe. Use of Fe isotopes to track anthropogenic Fe delivery to the oceans seems a fruitful research avenue, especially as it may relate to climate change. For our purposes, however, we will assume that dust delivery has a homogenous Fe isotope composition identical to that of average upper crust.

The North Atlantic Ocean receives the highest dust input of the ocean basins. In Fig. 5.20 vertical profiles for δ^{56}Fe values of pFe, dFe, and L-pFe are illustrated for stations and water depths from North Atlantic GEOTRACES transect GA03 that are considered to be dominated by dust input (Conway and John 2014), far from the continental margin and hence have minimal riverine or benthic Fe input. In addition, concentration profiles are shown for O_2, sFe, dFe, L_I–Fe, L-pFe, pFe, and particulate Fe/Al (pFe/pAl). The relative proportions of sFe, dFe, and pFe vary greatly with depth, corresponding with important changes in biogeochemical cycling of Fe (Hatta et al. 2015; Jenkins et al. 2015b). This includes dFe drawdown in the chlorophyll maximum, followed by an increase with depth as O_2 decreases, reflecting dissolution of pFe. Although both L-pFe and pFe increase with depth, particulate Fe/Al ratios decrease below the regional crustal average below the chlorophyll maximum (Lam et al. 2015), indicting net transport of pFe into solution. The highest ratios of L-pFe to pFe are found in the chlorophyll maximum, likely reflecting biological uptake (Revels et al. 2015a). The concentration of L_I–Fe varies with depth, broadly correlated with sFe and dFe, but, importantly, is always higher than dFe, indicating excess Fe-binding ligands relative to Fe (Buck et al. 2015). Because dFe in this profile has essentially zero Fe^{2+} (Sedwick et al. 2015), dFe is most likely organically complexed Fe^{3+}.

Despite significant vertical concentration changes in the concentrations of Fe species, the δ^{56}Fe values for pFe, dFe, and L–pFe are relatively constant with depth (Fig. 5.20), although there may be a subtle anti-correlation between the δ^{56}Fe values of dFe and the proportion of dFe and sFe. The average δ^{56}Fe values for pFe, dFe, and L-pFe are +0.14‰, +0.55‰, and −0.27‰, respectively. The Fe isotope composition of pFe lies within error of that of average upper crust, as expected for particulates from this profile (Revels et al. 2015a). For dFe, near-surface samples that included determination of sFe suggest that the soluble Fe phase has δ^{56}Fe values that are as much as 1‰ higher than dFe, indicating significant Fe isotope fractionation between the soluble and colloidal components that are collectively measured as dFe (Fitzsimmons et al. 2015). In terms of variations in the δ^{56}Fe values of L-pFe, Revels et al. (2015a) observed a broad correlation between δ^{56}Fe values and the fraction that L-pFe that comprises total particulate Fe, leading them to infer a pure labile component that has a δ^{56}Fe value of ~−0.5‰. Their inferred labile end member is significantly lower than the average measured δ^{56}Fe value of ~−0.27‰, which suggest that the L-pFe data in Fig. 5.20 reflects a mixture between lithogenic Fe and labile Fe. The uncertainty of mixtures of components is commonplace when employing leaches on natural samples, and in the case of the oxalate-EDTA leaching method, results might be variable depending upon particulate composition.

The average ^{56}Fe/^{54}Fe fractionation between dFe and L-pFe in Fig. 5.20 is ~+0.8‰. If we consider the possibility that the δ^{56}Fe values for

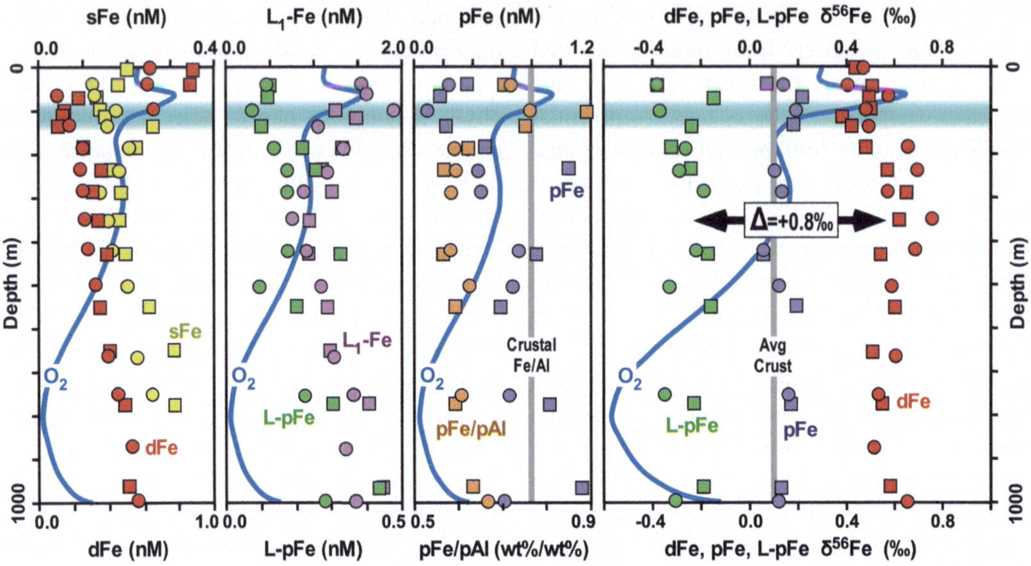

Fig. 5.20 Vertical profiles through upper 1000 m of Stations 12 (circles) and 14 (squares) for North Atlantic Ocean GEOTRACES transect GA03 where dFe is thought to be largely sourced to dissolution of dust particles (Conway and John 2014). Blue curve is dissolved O_2 (scaled from 150–250 μM) and horizontal blue band at ~100 m marks chlorophyll maximum (Jenkins et al. 2015b). Colored text labels in panels correspond with color of symbols: dissolved Fe (dFe; red symbols) and soluble Fe (sFe; yellow symbols) defined as <0.2 and <0.02 μm, respectively (from Conway and John 2014; Fitzsimmons et al. 2015). L_1–Fe (pink symbols) is defined as high-stability-constant organic ligands that pass a 0.2 μm filter (from Buck et al. 2015). L-pFe (green symbols) is ligand-leachable (EDTA + oxalate) Fe from the particulate fraction (>0.2 μm; from Revels et al. 2015a). For particulates, Fe contents shown (pFe, purple symbols; from Revels et al. 2015a), as well as Fe_{tot}/Al ratio (pFe/pAl, tan symbols; from Lam et al. 2015); regional crustal average shown as grey bar is that determined from aerosols collected on the westernmost sampling of the same cruise (Shelley et al. 2015) to minimize contribution from North Africa. δ^{56}Fe values for dFe (red symbols), pFe (purple symbols), and L-pFe (green symbols) from Conway and John (2014) and Revels et al. (2015a). The average δ^{56}Fe values for these components are: dFe = +0.55‰, pFe = +0.14‰, L-pFe = −0.27‰. This produces an average ^{56}Fe/^{54}Fe fractionation between dFe and L-pFe of +0.8‰. (shown as black double arrow). The average δ^{56}Fe value of pFe is identical to that of average upper crust (grey bar) within error

sFe is up to +1.5‰ and that an end member labile component has a δ^{56}Fe value as low as −0.5‰, the ^{56}Fe/^{54}Fe fractionation between a Low-Molecular Weight (LMW) Fe^{3+}-C component (sFe) and labile particulate Fe (L-pFe) could be as large as ~+2‰. Although this may seem quite an extreme isotopic fractionation, it lies within the range measured between LMW Fe^{3+}-C and total pFe (SPM) in rivers (Fig. 5.6). This reinforces the concept that Fe^{3+} bound to LMW C_{org} can have very high δ^{56}Fe values, as discussed in Sect. 5.2.1. Given the correlation between ligand strength and Fe isotope fractionation (Dideriksen et al. 2008; Morgan et al. 2010), it seems reasonable to interpret the positive δ^{56}Fe values for dFe and sFe to reflect an organically bound "L_1-type" or "LMW" or "siderophore-like" Fe^{3+} compound, or mix of compounds (e.g., L_1, L_2, L_3). More difficult is identifying what the L-pFe or "labile particulate" component is, or if it contains mixtures of components. It is not clear if L-pFe represents poorly crystalline Fe^{3+}-hydroxides, altered silicates, or a biological component of Fe or some combination of these. If, for example, L-pFe reflects poorly crystalline Fe^{3+}-hydroxides such as ferrihydrite, the expected fractionation between dFe and L-pFe would be about +0.6‰ for ^{56}Fe/^{54}Fe, based on combining the experiments of Welch et al. (2003), Dideriksen et al. (2008) and Wu et al. (2011b). If, however, L-pFe represents an Fe^{2+}-bearing altered silicate, a larger $^{56}Fe/^{54}Fe$ fractionation might be expected, based on the fact that Fe^{2+} silicate minerals will have lower $\delta^{56}Fe$ values than Fe^{3+}-hydroxides (see Chap. 3).

Regardless of the uncertainty in identifying the exact components involved, the relative constancy in $\delta^{56}Fe$ values for dFe and L-pFe where dust is the primary source of Fe (Fig. 5.20) suggests that their Fe isotope compositions are determined by isotopic exchange between dissolved and particulate Fe as part of the dynamics of Fe cycling in the upper water column. The possibility that such effects might reflect equilibrium isotopic exchange is supported by estimates of particle residence times of $\sim 10^1$–10^2 days for the profiles in Fig. 5.20 (Ohnemus and Lam 2015), which is much longer than the Fe isotope exchange kinetics measured for sFe, cFe, and pFe using enriched Fe isotope tracers in incubated seawater samples (Hurst and Bruland 2007). Rates of isotopic exchange and particle residence times, of course, might be significantly different in deeper water sections.

Seawater profiles in the upper 1000 m for stations from the GEOTRACES GP16 transect in the Eastern Pacific Ocean that are influenced by a strong dust component offer an opportunity to compare data from the North Atlantic GA03 transect shown in Fig. 5.20. In comparison to the North Atlantic profiles, the Eastern Pacific profiles record larger depletions in O_2 and more extensive influences from hydrothermal and continental shelf sources, as well as different trends in particle dissolution, Fe oxidation kinetics, and organic complexation (Buck et al. 2018; Jenkins et al. 2018; Moffett and German 2018; Peters et al. 2018; Resing et al. 2015). These issues contribute to a larger variability in Fe concentration and isotopic data from profiles that should be dominated by dust influence. Data from Stations 18, 20, 21, 23, 25, 26, 28, and 30 from GEOTRACES GP16, which are far from influence from the Peru margin and likely to be shallow enough to have minimal hydrothermal input (John et al. 2018), have average $\delta^{56}Fe$ values for dFe, pFe, and L-pFe of +0.43‰, +0.19‰, and −0.20‰, respectively (Fitzsimmons et al. 2017; John et al. 2018; Marsay et al. 2018). These are broadly similar to those from the North Atlantic GA03 transect (Fig. 5.20), although the range is more scattered for the Eastern Pacific data. The Eastern Pacific Ocean data therefore support a similar role for dust and Fe exchange between dissolved and particulate Fe in the upper 1000 m as seen in the North Atlantic Ocean, although the vertical profile trends are not as clear as those in Fig. 5.20.

Other studies have measured a range of $^{56}Fe/^{54}Fe$ fractionations between dFe and pFe, reflecting greater complexity in Fe cycling relative to the profiles in Fig. 5.20. Radic et al. (2011) and Labatut et al. (2014) studied dFe–pFe pairs from the western equatorial Pacific Ocean, where $^{56}Fe/^{54}Fe$ fractionations varied from −0.14 to +0.43‰, with an average of +0.17‰. They proposed that the modestly positive $\delta^{56}Fe$ values for dFe were produced by non-reductive dissolution of pFe, as well as uptake of low-$^{56}Fe/^{54}Fe$ Fe from phytoplankton. Ellwood et al. (2015) studied the effects of phytoplankton blooms on the Fe isotope compositions of dFe and pFe, where they measured an average dFe–pFe $^{56}Fe/^{54}Fe$ fractionation of −0.17‰ during the early stages of phytoplankton blooms, which they proposed to record photochemical and other inorganic fractionation effects. In contrast, the average dFe–pFe fractionation measured during the late stages of phytoplankton blooms was +0.43‰, accompanied by a decrease in dFe and increase in pFe, which Ellwood et al. (2015) interpreted to reflect biological uptake of low-$^{56}Fe/^{54}Fe$ Fe. In a study of the strongly stratified Baltic Sea, where shallow oxic conditions give way to euxinic conditions with increasing depth, Staubwasser et al. (2013) measured highly variable but consistently positive dFe–pFe $^{56}Fe/^{54}Fe$ fractionations in the non-sulfidic sections of the water column, which they took to indicate variable kinetic effects during Fe^{2+} oxidation. When considered in light of the above discussion of the GEOTRACES results, however, a positive dFe–pFe fractionation is similar to that expected between ligand-bound dFe and pFe and may therefore not reflect oxidation isotope effects. In addition, Staubwasser et al. (2013) measured negative dFe–pFe fractionations in the euxinic zone, which they postulated records a microbial Fe^{2+}

source. Alternatively, that the average measured dFe–pFe $^{56}Fe/^{54}Fe$ fractionation in the sulfide zone was −0.3‰, which matches that of the equilibrium Fe^{2+}_{aq}−FeS fractionation (see Sects. 5.3.1 and 5.4.2), suggests that Fe–S mineral formation may determine the $\delta^{56}Fe$ values in the euxinic zone, rather than microbial processes.

5.5.3 Benthic Sources

As discussed in Sect. 5.4.4, benthic Fe fluxes are primarily a function of bottom water O_2, as well as the extent of carbon respiration (C_{ox}) (Dale et al. 2015; Elrod et al. 2004). The highest Fe fluxes are associated with low O_2 levels, which reflects the more limited oxidation of Fe^{2+}_{aq} in the surface sediment layer under low-O_2 conditions. Benthic Fe fluxes vary by several orders of magnitude in marine sediments, reflecting fundamental differences in C_{ox} and O_2, but also in the methods employed (benthic chambers versus modeling pore water profiles), as well as the relative contribution to C_{ox} by DIR and DSR (Dale et al. 2015; Homoky et al. 2016). In Fig. 5.21 we illustrate the covariation of benthic Fe fluxes and bottom water O_2, as well as C_{ox}, using the model of Dale et al. (2015), for modern marine sediment sections for which Fe isotope compositions have been measured (Fig. 5.18 and Table 5.1). For illustration purposes, an Fe flux-O_2 relation for sediments with C_{ox} = 5 mmol m^{-2} d^{-1}, γ = 400 μmol m^{-2} d^{-1} is shown in Fig. 5.21. As defined by Dale et al. (2015), γ is the maximum possible benthic Fe flux, and this scales with the amount of highly reactive Fe in the sediment section. Many of the sections studied for Fe isotopes scatter about an exponential curve (blue line). Relative to the study of Elrod et al. (2004), this C_{ox} is slightly higher. In their global compilation, Dale et al. (2015) used a lower γ of 170 μmol m^{-2} d^{-1}, which they noted did not fit the data for the California margin, a possible explanation of which would be the higher reactive Fe contents for California margin sediments relative to the global average (Severmann et al. 2010). This suggests that the benthic

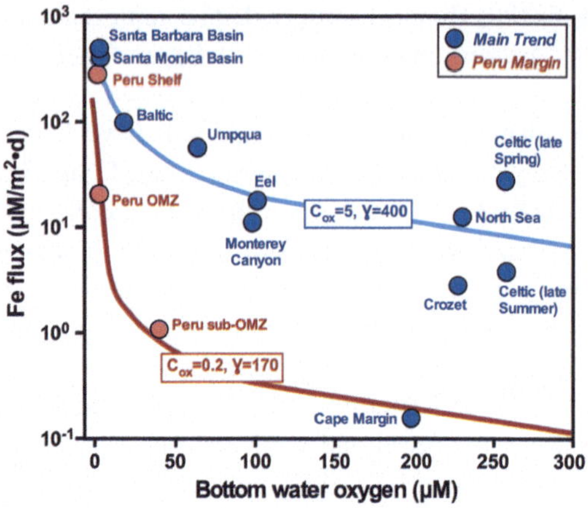

Fig. 5.21 Variations in benthic Fe flux and bottom water oxygen for localities for which Fe isotope data are available (see Fig. 5.18; data sources listed in Table 5.1). Blue curve approximately represents relations for the main trend (blue symbols) using the model of Dale et al. (2015), where carbon oxidation rate (C_{ox}) set to 5 mmol m^{-2} d^{-1}, and γ, the maximum Fe flux (proportion of reactive Fe abundance) set to 400 μmol m^{-2} d^{-1}. Red curve approximately represents relations for the Peru margin (red symbols), where very high rates of organic C burial suggest a lower C_{ox} (C_{ox} = 0.2 mmol m^{-2} d^{-1}) and reactive Fe abundances are lower (γ = 170 μmol m^{-2} d^{-1}) than the California margin; the steeper decline with oxygen also accounts for consumption of Fe^{2+}_{aq} via oxidation that is coupled to nitrate reduction (see text)

Fe flux is not only a function of O_2 and C_{ox} but of reactive Fe as well.

In addition, a second trend is illustrated in Fig. 5.21 for sediments off the Peru and Cape margins that uses a much lower C_{ox} (red line). In the case of the Peru margin sediments, C_{org} burial is exceptionally high (Scholz et al. 2011; Noffke et al. 2012), suggesting a low C_{ox}, although the estimated C_{ox} values from previous studies of the Peru margin are higher than that illustrated in Fig. 5.21. In addition, Fe^{2+}_{aq} in the Peru margin sediments is consumed via oxidation (under low-O_2 conditions) coupled to unusually high rates of nitrate reduction, strongly limiting the export of Fe^{2+}_{aq} to the open ocean (Scholz et al. 2016). Although the parameters in the model of Dale et al. (2015) cannot explicitly accommodate nitrate reduction, the choice of parameters for the Peru curve in Fig. 5.21 are meant to accommodate such effects. For our purposes, it is sufficient to simply illustrate the very rapid decline in benthic flux with increasing depth and increasing O_2 that occurs on the Peru margin (Noffke et al. 2012).

The most important conclusion to be drawn from Fig. 5.21 is that benthic Fe fluxes are highest when bottom water O_2 is low, the rate of organic carbon oxidation (C_{ox}) is high (indicating, in part, significant DIR), and the quantity of reactive Fe (γ) is high (required for DIR). These parameters logically follow Eq. 5.7, where DIR couples reduction of Fe^{3+} to oxidation of organic carbon. In addition, the effect of S cycling is important in its role in reductive dissolution, where high rates of DSR consume reactive Fe and suppress DIR. In S-rich systems, benthic Fe fluxes may remain quite high if bottom water O_2 is low, suppressing oxidation of Fe^{2+}_{aq} at the sediment-water interface. It is no accident, therefore, that for the localities plotted in Fig. 5.21 where bottom water O_2 is less than ~30 μM, all have very high DOP. Where bottom water O_2 is higher than ~30 μM, only one section has a high DOP (North Sea). Under conditions of low sulfide and low oxygen, as might exist in the Precambrian, benthic Fe fluxes would be expected to be much greater than the maximums shown in Fig. 5.21, assuming no limitation in supply of C_{org} and reactive Fe^{3+}, because reaction between Fe^{3+}-hydroxides and sulfides will be suppressed, as will be discussed in Chap. 6.

The influence of a benthic Fe source on seawater Fe isotope compositions is best seen in marine basins that have limited water exchange with the open oceans. We have discussed the Black Sea in Sect. 5.3.3 above as part of our discussion of redox-stratified water bodies. Here we highlight results from the extensional basins of the California continental borderland, which are not as strongly redox-stratified as the Black Sea, nor euxinic, yet have restricted deep-water exchange with the open oceans such that the water column is highly influenced by benthic Fe sources. Rates of DIR exceed those of DSR in near-surface sediments of the California borderland basins, producing a high benthic Fe flux, where sulfide production is dominant only in deeper portions of the sediment (Leslie et al. 1990). This vigorous Fe cycle in turn affects availability of nutrients such as P, where P is sequestered from solution during Fe^{3+}-hydroxide formation, but released during DIR (McManus et al. 1997).

The water columns of the Santa Barbara and San Pedro basins have high dFe contents and low $\delta^{56}Fe$ values (Fig. 5.22) relative to the very low dFe and high $\delta^{56}Fe$ values that are characteristic of the surface ocean far from continental margins (Fig. 5.20). The gradient of increasing dFe contents with depth in the Santa Barbara and San Pedro basins is very high, commensurate with the strong decrease in oxygen contents, reaching a minimum of 30 nM in the silled region of the Santa Barbara Basin, where dFe increases to very high levels. Although $\delta^{56}Fe$ values define a slight gradient in the Santa Barbara Basin above the sill, they become highly negative within the silled zone, reaching as low as ~−4‰ just above the sediment/water interface. Vertical gradients in dFe and $\delta^{56}Fe$ for the San Pedro Basin are more muted, changing only slightly in the silled region. Above the silled region, the vertical gradients in $\delta^{56}Fe$ values for seawater are distinct between the Santa Barbara and San Pedro basins.

The high dFe contents and negative $\delta^{56}Fe$ values in the silled zones of the Santa Barbara and San Pedro basins are attributed to a redox-driven benthic flux from the sediment, followed by modification in the upper water column through mixing and oxidation (John et al. 2012). Sediment pore water profiles in the Santa Barbara basin show that the shallow pore waters have very high Fe^{2+}_{aq} contents and negative $\delta^{56}Fe$ values, characteristics that are produced by reductive dissolution, although the shallowest samples record the effects of Fe^{2+}_{aq} oxidation (decreasing Fe^{2+}_{aq} and $\delta^{56}Fe$). Deeper in the sediment section, a decrease in Fe^{2+}_{aq} and increase in $\delta^{56}Fe$ values for Fe^{2+} reflects an increasing control by FeS (Severmann et al. 2006), as discussed in Sect. 5.4.2 (see also Figs. 5.15 and 5.16). Pore water profiles in the sediment from the Santa Monica Basin are more scattered and show less influence by FeS, as well as higher Fe^{2+}_{aq} contents than those observed in the Santa Barbara sediment (Fig. 5.22). Data from benthic chambers in the Santa Monica and San Pedro basins show very negative $\delta^{56}Fe$ values for the benthic Fe flux (Fig. 5.22), which extend to lower $\delta^{56}Fe$ values than pore water Fe^{2+}_{aq}, reflecting oxidation upon emanation from the sediment (Severmann et al. 2010). Although bottom water oxygen contents are low, oxidation of pore water Fe^{2+}_{aq} still occurs, driven by the effects of sediment bioturbation. The similarities in benthic chamber results from the Santa Monica and San Pedro basins suggests that the sediment

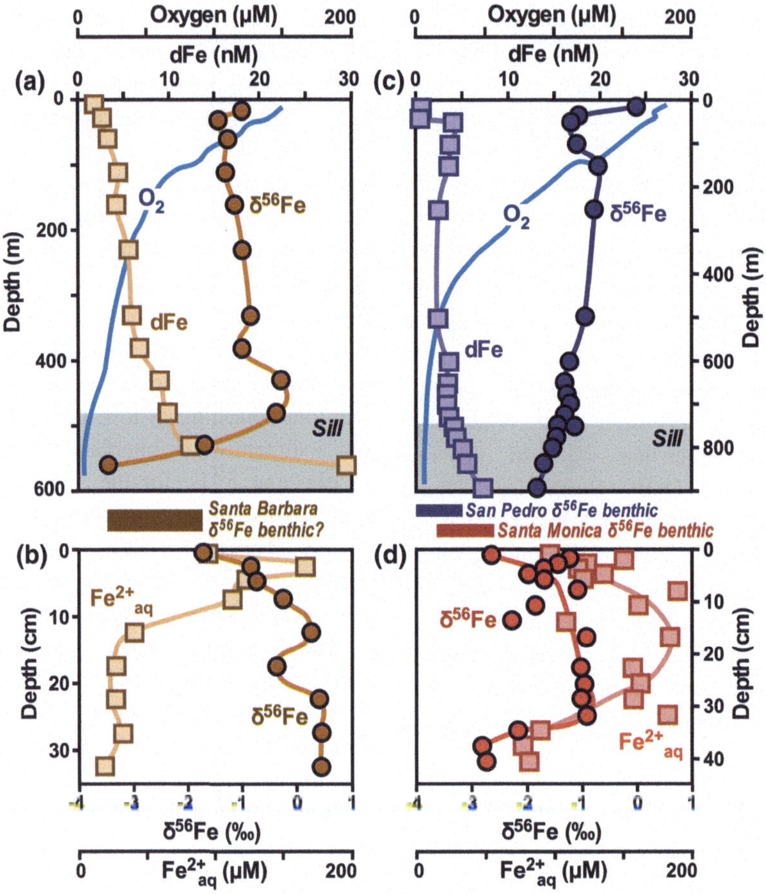

Fig. 5.22 Vertical seawater and sediment profiles for California Borderland locations that are heavily influenced by a DIR-driven benthic Fe flux due to restricted water exchange with the open oceans, accompanied by the effects of Fe^{2+}_{aq} oxidation and sulfide diagenesis. **a**—water column O_2, dFe, and $\delta^{56}Fe$ data for the Santa Barbara basin (John et al. 2012). **b**—pore water data for Fe^{2+}_{aq} contents and $\delta^{56}Fe$ values for sediment from the Santa Barbara basin (Severmann et al. 2006); see also Figs. 5.15, 5.16, and 5.17. **c**—water column O_2, dFe, and $\delta^{56}Fe$ data for the San Pedro basin (John et al. 2012). **d**—pore water data for Fe^{2+}_{aq} contents and $\delta^{56}Fe$ values for sediment from the Santa Monica basin, which lies adjacent to the San Pedro basin and has similar geochemical characteristics (Severmann et al. 2010); see also Fig. 5.15. Measured $\delta^{56}Fe$ values for sediment Fe^{2+}_{aq} flux from benthic incubation chambers shown as horizontal bars between **c** and **d** from the San Pedro and Santa Monica basins (Severmann et al. 2010). Estimated range of $\delta^{56}Fe$ values for benthic Fe^{2+}_{aq} flux shown for Santa Barbara basin between **a** and **b** is speculative and placed between Fe isotope compositions measured for upper-most pore fluids and deepest basin water. Grey bands in **a** and **c** mark silled zones, below which water exchange with the open oceans is highly restricted

sections may also be comparable. No pore water isotope data have been obtained from the San Pedro basin, and hence we compare the Santa Monica sediment to the San Pedro water column. That the deepest water column samples in the San Pedro Basin do not have δ^{56}Fe values that are as negative as the benthic chamber measurements was interpreted by John et al. (2012) to reflect a kinetic isotope effect upon precipitation of Fe^{3+}-hydroxides, as well as water mass mixing. Although there are no benthic chamber measurements from the Santa Barbara Basin, we infer a very negative δ^{56}Fe value based on the deepest water column data in the silled zone (Fig. 5.22), although this must remain speculative. It is worth noting that the deep water δ^{56}Fe value for dFe is lower than the δ^{56}Fe value inferred for the Santa Barbara Basin benthic flux based on pore water data (Fig. 5.18). Nevertheless, the data from the California borderland basins indicate that the redox driven benthic Fe flux has highly negative δ^{56}Fe values, where the lowest values reflect the additional effects of Fe^{2+}_{aq} oxidation.

The benthic Fe shuttle operates in a complex manner in the high-productivity Peru margin, reflecting open exchange with the ocean and the effects of intersection of the OMZ with the continental slope, and we have discussed some of these features above. Noffke et al. (2012) measured the highest benthic Fe fluxes in the shallow shelf ($\sim$100 m), which decrease an order of magnitude in the mid-shelf ($\sim$250–600 m). As noted in Sect. 5.4.4, Scholz et al. (2014a, b) infer that the net benthic flux on the Peru margin has an average δ^{56}Fe value of ~ -0.5‰, although it is important to note that this is calculated based on sediment mass balance (Fig. 5.19) and not based on measured pore waters or benthic chambers. Chever et al. (2015) made the first Fe isotope measurements of dFe and pFe from the Peru margin at 6°S, and, importantly, measured ferrous-ferric Fe speciation, where they determined δ^{56}Fe values as low as -1.3‰ for dFe in the OMZ and showed that dFe was nearly entirely Fe^{2+}. This is a more negative δ^{56}Fe value for the benthic Fe flux than that inferred by Scholz et al. (2014a, b) based on sediment data, but this contrasts reflects records of benthic flux on different timescales, where the sediment data reflect an integrated signal spanning 10^1–10^2 yrs. In all cases, Chever et al. (2015) measured higher δ^{56}Fe values for pFe, which was dominated by Fe^{3+}, both within and outside of the OMZ. John et al. (2018) measured less negative δ^{56}Fe values for dFe on the Peru margin at 12°S as part of the GEOTRACES GP16 transect, which is one degree south of the transect studied by Scholz et al. (2014a, b).

The enrichment of pFe that is dominated by Fe^{3+} in the Peru OMZ poses a problem—how does oxidation of benthic Fe^{2+}_{aq} by O_2 in an anoxic water column occur? The solution is oxidation by anaerobic pathways, such as nitrate reduction under low-O_2 conditions (Scholz et al. 2016; Heller et al. 2017). Straub et al. (1996) was one of the first to document anaerobic oxidation of Fe^{2+}_{aq} coupled to nitrate reduction, and they suggested this could be an important pathway for formation of ferric hydroxides where oxygen contents are low and nitrate was high:

$$10Fe^{2+}_{aq} + 2NO_3^- + 24H_2O \rightarrow 10Fe(OH)_3 + N_2 + 18H^+ \qquad (5.17)$$

In Fig. 5.23 we explore the interrelations among dFe, pFe, nitrate, and oxygen and their relations to δ^{56}Fe values for Station 1 of the GEOTRACES GP16 transect off the Peru margin. The low levels of dFe and NO_3^-, variable levels of Fe^{2+} for dFe, and high pFe/pAl in the upper OMZ supports a model for oxidation of Fe^{2+}_{aq} via NO_3^- reduction and formation of Fe^{3+}-rich particulates. Iron speciation measurements and X-ray analysis confirms that pFe is dominated by Fe^{3+}-hydroxides in the OMZ (Heller et al. 2017). Assuming an average δ^{56}Fe value for dFe on the shelf of ~ -0.5‰, the relatively high δ^{56}Fe values for dFe at Station 1, which overlies the Peru trench, is surprising if production of Fe^{3+}-hydroxides and particle settling occurs within the OMZ, because such particles should have higher δ^{56}Fe values than dFe, as shown by Chever et al. (2015). Insight into this quandary may come from Heller et al. (2017), who showed that although both dFe and

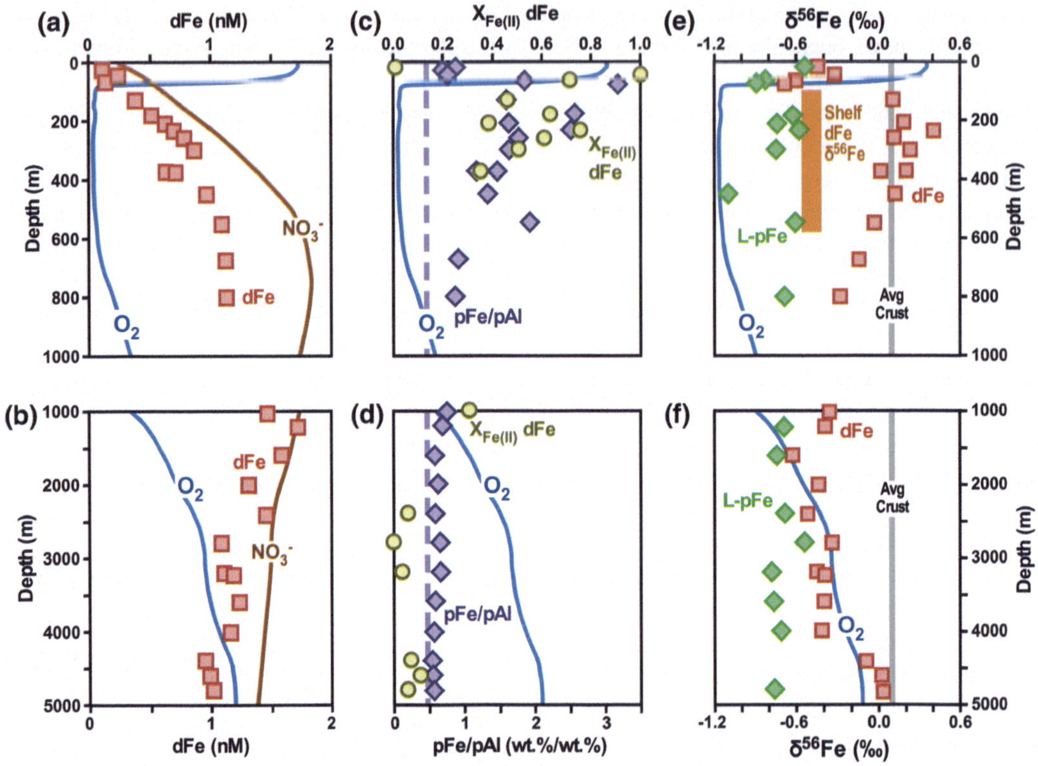

Fig. 5.23 Vertical seawater profiles for Station 1 of GEOTRACES transect GP16 off the Peru margin, which lies ~200 km west of the shelf. Oxygen curve (blue) scales from 0 to 300 μM in all panels. The Oxygen minimum Zone (OMZ) lies between 80 and 500 m at this location. Panels **a**, **b**: Dissolved Fe (dFe, red squares) and nitrate (brown curve) abundances increase with depth within the OMZ; nitrate curve scales from 0 to 50 μM (data from Heller et al. 2017; Peters et al. 2018). Panels **c**, **d**: Fe(II) abundance for dFe ($X_{Fe(II)}$ dFe) shows that dFe is largely Fe^{2+} in the upper OMZ, but this decreases with depth, and dFe below the OMZ is mostly Fe^{3+} (Heller et al. 2017). Particulate Fe/Al ratios (pFe/pAl) are very high in the OMZ, much greater than the background lithologic ratio (vertical dashed purple line), but lie near the background ratio below the OMZ (Lam et al. 2018; Lee et al. 2018). Particulates in the Peru OMZ, especially those in the upper OMZ, are enriched in Fe^{3+}-hydroxides, which are interpreted to have formed through Fe^{2+}_{aq} oxidation coupled to nitrate reduction (Heller et al. 2017). Panels **e**, **f**: For Fe isotope data (John et al. 2018; Marsay et al. 2018), $\delta^{56}Fe$ values for dFe in the OMZ at Station 1 are significantly higher than those on the shelf, which have an average $\delta^{56}Fe$ value of −0.5‰ (Chever et al. 2015; John et al. 2018), indicating an increase in $\delta^{56}Fe$ values for dFe during transport westward within the OMZ. The convergence of $\delta^{56}Fe$ values for ligand-leachable pFe (L-pFe) and dFe below the OMZ is interpreted to reflect reversible particle scavenging between solid and dissolved Fe during sinking from the OMZ (John et al. 2018; Marsay et al. 2018). Note that no data are plotted below 5000 m, due to the effects of sediment incorporation from the Peru trench (John et al. 2018). The $\delta^{56}Fe$ value of average upper crust is shown as vertical grey bar

pFe decrease within the OMZ with increasing distance from the Peru margin, the rate of dFe decrease is much higher. It is possible, although speculative, that the relatively high $\delta^{56}Fe$ values for dFe in the OMZ in Fig. 5.23 may reflect in part a Fe^{3+}-hydroxide cFe component in the dFe fraction. Such effects have been modeled by Chever et al. (2015). This is supported by the observation that dFe in the OMZ has $X_{Fe(II)}$ as low as 0.4, indicating a significant Fe^{3+} component. The rapid drop off in dFe with distance from the margin observed by Heller et al. (2017) would make the dFe component susceptible to the presence of any high-$\delta^{56}Fe$ cFe.

Below the OMZ, the trend of steadily decreasing $\delta^{56}Fe$ values for dFe is also unexpected for a model where Fe^{3+}-hydroxides form in the OMZ followed by settling, which should provide a high-δ^{56}Fe pFe flux into the sub-OMZ zone, which in turn would tend to increase the δ^{56}Fe values for dFe. Accumulation of high-δ^{56}Fe Fe^{3+}-hydroxides beneath the OMZ is supported by the Fe isotope relations observed by Scholz et al. (2014a, b) and shown in Fig. 5.19. Because the low-δ^{56}Fe dFe occurs beneath the OMZ, at mid-shelf depths (1–4 km), where the benthic Fe flux has been independently shown to be negligible, John et al. (2018) do not interpret such Fe isotope compositions to reflect a benthic source. Instead, John et al. (2018) and Marsay et al. (2018) point to the coincidence of δ^{56}Fe values for dFe and L-pFe beneath the OMZ (Fig. 5.23), suggesting that this reflects Fe isotope exchange between dFe and pFe during particle settling, which they term "reversible scavenging". Such a model implies that the equilibrium Fe isotope fractionation between dFe and L-pFe is zero, which is not known, and is likely sample dependent. In their reversible scavenging model, the negative δ^{56}Fe values for the benthic Fe flux is recorded in L-pFe in the OMZ, and particle settling and Fe isotope exchange between L-pFe and dFe results in a decrease in δ^{56}Fe values for dFe beneath the OMZ.

5.5.4 Hydrothermal Sources

There is an enormous amount of literature on marine hydrothermal systems, including detailed discussions on the geochemistry of hydrothermal processes (e.g., German and Von Damm 2003; German and Seyfried 2014), magmatic drivers (e.g., Lowell et al. 2013), relations to massive sulfide deposits (e.g., German et al. 2016), the magnitude of global hydrothermal systems (e.g., Beaulieu et al. 2015), hydrothermal plume processes (e.g., Sander and Koschinsky 2011), and, ultimately, the hydrothermal contribution of Fe to seawater (e.g., Holmes et al. 2017). Although it was initially assumed that marine hydrothermal vents contributed little to ambient seawater Fe contents due to rapid precipitation of Fe-sulfides and Fe^{3+}-hydroxides near the vent, correlations between excess ^{3}He anomalies, indicating magmatic sources, and dFe in the oceans indicate hydrothermal transport of dFe for perhaps thousands of km from the source (e.g., Jean-Baptiste et al. 2004; Wu et al. 2011a; Saito et al. 2013). The relatively recent recognition of long-scale transport of hydrothermal Fe may occur through complexation of Fe via organic matter entrained in hydrothermal vent fluids (e.g., German et al. 2015; Sander and Koschinsky 2011; Toner et al. 2009), as well as nanoparticulate Fe minerals, especially sulfides (e.g., Yücel et al. 2011). Long-distance transport of nanoparticulates from hydrothermal plumes is important, because, as discussed above, the common size cut-off for dFe is 0.2 or 0.4 μm, indicating that dFe may contain a substantial nanoparticulate component.

Just as bottom water O_2 contents affect benthic Fe fluxes (Fig. 5.21), deep water O_2 contents affect the oxidation rates of hydrothermal Fe^{2+}_{aq} (Fig. 5.24a). Oxidation of hydrothermal Fe^{2+}_{aq} occurs during mixing and dilution with ambient seawater, lowering the temperature and raising pH, which, together with exposure to dissolved oxygen, rapidly produces Fe^{3+}-hydroxides (e.g., Millero et al. 1987; Field and Sherrell 2000). The large range in O_2 contents among the ocean basis produces a wide range in oxidation half-lives ($t_{1/2}$) for hydrothermal vent fluids, ranging from minutes to hours. At these timescales, Fe isotope equilibrium is expected between Fe^{2+}_{aq} and Fe^{3+}_{aq} (Welch et al. 2003). As has been shown by studies of groundwater oxidation (Fig. 5.7), it is possible that both equilibrium and kinetic Fe isotope effects may occur during precipitation of Fe^{3+}-hydroxides during oxidation of hydrothermal Fe^{2+}_{aq}, but in all cases it is expected that the δ^{56}Fe values of the Fe^{3+} precipitates would either be positive (partial oxidation) or equal to that of the pure hydrothermal vent fluid (complete oxidation; see discussion in Chap. 1). Removal of Fe^{3+}-hydroxides via settling would produce a decrease in δ^{56}Fe values for the remaining aqueous Fe (Fig. 5.7).

Estimates for the δ^{56}Fe values of the pure hydrothermal Fe^{2+}_{aq} component are based on

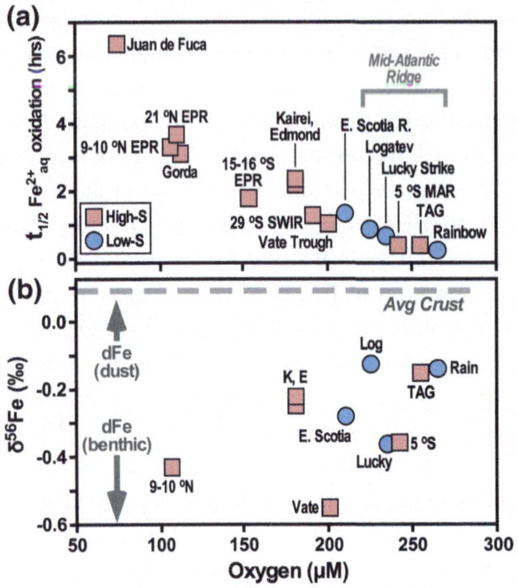

Fig. 5.24 Fe^{2+}_{aq} oxidation rate half-lives, $t_{1/2}$, **a** and $\delta^{56}Fe$ values for end-member hydrothermal vent fluids **b** relative to bottom-water oxygen contents. Symbols divide hydrothermal systems into high- and low-S contents based on molar Fe/S ratios <10 and >10, respectively. High-S systems are all basalt hosted, whereas low-S systems include ultramafic- and basalt-hosted systems. EPR: East Pacific Rise; SWIR: Southwest Indian Ridge. The broad decrease in oxidation rates (plotted as $t_{1/2}$) with increasing oxygen contents follows that shown by Millero et al. (1987). The fastest oxidation rates occur in the Atlantic Ocean, whereas oxidation rates are significantly slower in the Pacific Ocean. End-member $\delta^{56}Fe$ values for dFe in vent fluids calculated by subtracting the seawater component based on Mg contents. There is a weak positive correlation between $\delta^{56}Fe$ and bottom water oxygen contents. Also shown are the directions $\delta^{56}Fe$ values would plot for dFe produced by dissolution of dust, as well as where $\delta^{56}Fe$ values for a benthic component would plot (both plot off scale in the figure; see Figs. 5.18 and 5.20). Data in **a** largely from Field and Sherrell (2000), with additional data from Statham et al. (2005), Bennett et al. (2009), Lough et al. (2017), and Nasemann et al. (2018). Additional sources for data in **b** from Beard et al. (2003), Severmann et al. (2004), Rouxel et al. (2008a), (2016), and S. Severmann, C. Johnson, and B. Beard (unpub.)

subtracting ambient seawater contributions based on Mg contents (although for Fe, such contributions are small), and these estimates produce a range in $\delta^{56}Fe$ between −0.5 and −0.1‰ (Fig. 5.24b). These are all significantly lower than that of average upper crust or dFe that reflect Fe isotope exchange with dust particles (Fig. 5.20). There is no clear correlation between host composition (basaltic versus ultramafic) and $\delta^{56}Fe$ values for hydrothermal fluids. The end member vent fluids all have $\delta^{56}Fe$ values much greater than those estimated for benthic Fe sources (Fig. 5.18). There is a weak correlation between $\delta^{56}Fe$ values for end member vent fluids and seawater O_2 contents (Fig. 5.24b), and, as discussed next, this may reflect differences in the role of Fe sulfides.

Precipitation of hydrothermal Fe^{2+}_{aq} primarily produces Fe-metal-sulfides (e.g., FeS, FeS_2, $CuFeS_2$) and Fe^{3+}-hydroxides, the proportions of which vary dependent on the Fe/S ratios of the deep fluids, interactions with ambient seawater, and the O_2 contents of seawater. Basalt-hosted systems, which tend to occur in fast-spreading systems, often, but not always, have low Fe/S ratios, producing early and extensive Fe-sulfide precipitation (e.g., German and Seyfried 2014); for ridge systems in the Pacific Ocean, which has relatively low deep-water O_2 contents (Fig. 5.24 a), oxidation of Fe^{2+}_{aq} is suppressed relative to early Fe-sulfide formation. Production of nanoparticulate Fe-sulfides (10^1–10^2 nm in size), including pyrite, is now recognized as a major pathway for producing high-dFe plumes from hydrothermal systems. Even under high-O_2 conditions, oxidation half-lives ($t_{1/2}$) for sulfides may be on the order of 10^2 days, which is several orders-of-magnitude slower than oxidation of Fe^{2+}_{aq} (Fig. 5.24a). Such slow rates of oxidation of Fe-sulfides allows these particles to travel 10^3–10^4 km (Gartman and Luther 2014; Gartman et al. 2014). In contrast, ridge systems that are ultramafic-hosted have high Fe/S ratios and tend to occur in slow-spreading systems; when coupled to high seawater O_2 contents (Fig. 5.24a), Fe^{2+}_{aq} oxidation is early, producing large amounts of Fe^{3+}-hydroxide particles. As will be discussed at the end of this section, despite extensive Fe precipitation in near-vent settings, hydrothermal Fe may still be detected far from the vent.

In Fig. 5.25, we compare three marine hydrothermal plumes in terms of their particulate mineralogy (Fe sulfides relative to Fe^{3+}-hydroxides) and Fe isotope compositions. The Rainbow hydrothermal vent site on the Mid-Atlantic Ridge (MAR) is a large ultramafic-hosted system that has very high molar Fe/S ratios of ~24 and a relatively short Fe^{2+}_{aq} oxidation half-life ($t_{1/2}$) of ~15 min (Douville et al. 2002; German et al. 2010). Although Fe sulfides precipitate first from the vent fluids, the high Fe/S ratios produce a very high proportion of Fe^{3+}-hydroxide precipitation in the buoyant portion of the hydrothermal plume (Fig. 5.25; Edmonds and German 2004). In one of the first detailed Fe isotope studies of a marine hydrothermal system, Severmann et al. (2004) determined the Fe isotope compositions of vent fluids, plume particles, and Fe-rich sediments beneath the plume. They noted that with the exception of near-vent particles that had positive $\delta^{56}Fe$ values indicative of partial Fe^{2+}_{aq} oxidation, the majority of the data has a restricted range of Fe isotope compositions that has $\delta^{56}Fe$ values equal to, or slightly higher than, the vent fluids. This consistency was seen in sediments beneath the plume, providing a record back ~16 ka.

The MAR hydrothermal vent systems at ~5°S have similar $t_{1/2}$ for Fe^{2+}_{aq} oxidation as at Rainbow, but include basalt-hosted systems that are more sulfide rich, as reflected in molar Fe/S ratios near unity (German et al. 2008). Thermodynamic modeling demonstrates extensive early Fe-sulfide precipitation (Klevenz et al. 2011). Dissolved Fe contents measured down plume are much higher than those expected for such short $t_{1/2}$, and the dFe pool is thought to consist of both nanoparticulate sulfides as well as Fe^{3+}–C complexes (Bennett et al. 2008). The molar fraction of FeS and $Fe(OH)_3$ in pFe varies greatly, extending from the Rainbow cluster to a pure FeS component, that is well correlated with changes in $\delta^{56}Fe$ values for pFe (Bennett et al. 2009; Fig. 5.25). Relative to end-member vent fluid compositions, Bennett et al. (2009) proposed a $^{56}Fe/^{54}Fe$ fractionation between Fe^{2+}_{aq} and FeS of ~+0.5‰. This is slightly higher than the contrast in $\delta^{56}Fe$ values measured for sulfides from hydrothermal chimneys and vent fluids by

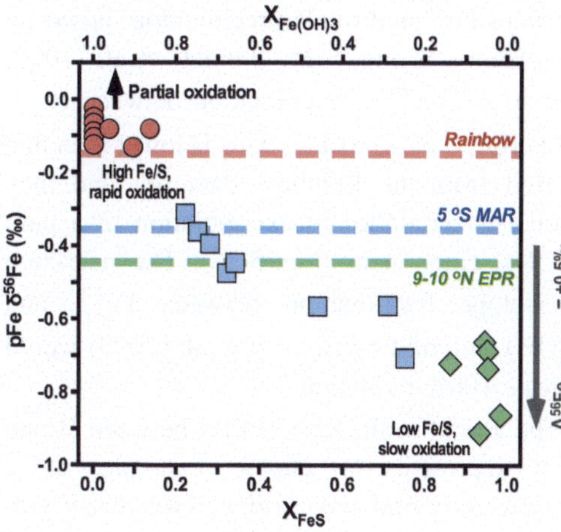

Fig. 5.25 $\delta^{56}Fe$ values for pFe from the buoyant portions of three hydrothermal plumes relative to the proportion of Fe-sulfide and -hydroxide in the particulate fraction. Rainbow (red symbols) is an ultramafic-hosted vent system with a very high Fe/S ratio on the Mid-Atlantic Ridge (MAR), whereas the 5°S MAR system (blue symbols) is basalt hosted with a low Fe/S ratio; both systems have similar ambient seawater O_2 (Fig. 5.24). The vent system at 9–10°N on the East Pacific Rise (EPR; green symbols) has the lowest Fe/S ratios and low seawater O_2 (Fig. 5.24). Color-coded horizontal dashed lines indicate the $\delta^{56}Fe$ values for the pure vent end member, corrected based on Mg contents. pFe is based on a 1 μm filter for Rainbow, 0.4 μm filter for 5°S MAR, and for 9–10°N, two values are plotted (averages of sediment traps, corrected for basalt particulate contributions), one <1 mm and one >1 mm (both give similar results). The mole fraction of FeS assumes two components for Fe, FeS and $Fe(OH)_3$, calculated based on P contents (see discussion in Bennett et al. 2009), and the results are consistent when checked against X-ray studies of the particles (Rouxel et al. 2016). The pure FeS end member is consistent with a kinetic $^{56}Fe/^{54}Fe$ fractionation of +0.5‰ between Fe^{2+}_{aq} and FeS (see discussion in Rouxel et al. 2016), show on right side of panel. Data from Severmann et al. (2004), Edmonds and German (2004), Bennett et al. (2009), and Rouxel et al. (2016)

Rouxel et al. (2008a). This order of fractionation, however, is opposite in sign to that measured in experiments under equilibrium conditions, but similar to what has been measured under kinetic conditions (Butler et al. 2005; Wu et al. 2012). In addition, this fractionation is opposite in sign to that measured between Fe^{2+}_{aq} and pyrite at hydrothermal temperatures (300–350 °C), where Syverson et al. (2013) measured an equilibrium $^{56}Fe/^{54}Fe$ fractionation of −1.0‰. Turning to the effects of Fe^{3+}-hydroxide precipitation, based on the relations in Fig. 5.25, Bennett et al. (2009) inferred a $^{56}Fe/^{54}Fe$ fractionation between Fe^{2+}_{aq} and $Fe(OH)_3$ of ~−0.1‰. This is lower than that inferred from the Rainbow data set, and may involve uncertainties in extrapolation to a pure $Fe(OH)_3$ component (Fig. 5.25). The near-zero Fe isotope fractionation between Fe^{2+}_{aq} and $Fe(OH)_3$ inferred by Bennett et al. (2009) would indicate 100% oxidation.

The East Pacific Rise (EPR) between 9 and 10°N represents a classic example of a fast-spreading mid-ocean ridge. Two recent volcanic eruptions have occurred in this section of the EPR, allowing comparison of vent chemistry and plume characteristics before and after eruption (Fornari et al. 2012). As is typical for the Pacific Ocean, $t_{1/2}$ for Fe^{2+}_{aq} oxidation is longer than for the Atlantic Ocean, on the order of 3–4 h (Fig. 5.24a). This, coupled with low Fe/S ratios, indicates that the Fe budgets of EPR vent fluids are commonly dominated by sulfide precipitation. Detailed Fe isotope studies of vent fluids, chimney sulfides, and particulates have been done by Rouxel et al. (2008a, 2016). Importantly, these studies document the mineralogy of plume particulates using both chemical compositions and x-ray spectroscopy. Rouxel et al. (2016) highlight the important role Fe/S ratios play in determining the $\delta^{56}Fe$ values for dFe in the buoyant portions of hydrothermal plumes, reflecting the opposing isotopic effects of Fe^{3+}-hydroxide and Fe-sulfide precipitation. They infer a kinetic fractionation in $^{56}Fe/^{54}Fe$ between Fe^{2+}_{aq} and FeS of ~+0.4‰, which is essentially identical to that inferred by Bennett et al. (2009) for the MAR at 5°S (Fig. 5.25).

If we assume that pyrite formation in hydrothermal plumes occurs via an FeS precursor (Eqs. 5.10–5.14), the observation of an ~+0.5‰ fractionation in $^{56}Fe/^{54}Fe$ between Fe^{2+}_{aq} and FeS/pyrite, which is ~1.5‰ higher than the inferred equilibrium fractionation between Fe^{2+}_{aq} and pyrite at hydrothermal temperatures, indicates strong kinetic effects (e.g., Rouxel et al. 2016). The low-pH experiments of Butler et al. (2005) and Guilbaud et al. (2010, 2011b) demonstrate a kinetic $^{56}Fe/^{54}Fe$ fractionation between Fe^{2+}_{aq} and FeS up to +1.2‰ during precipitation of FeS. Wu et al. (2012) showed that Fe isotope exchange in the system Fe^{2+}_{aq}–FeS is inhibited at low pH, but at circum-neutral pH, isotopic exchange is enhanced, allowing isotopic equilibrium on the order of 10^1 days. Given the low-pH and very rapid precipitation rates in hydrothermal fluids upon mixing with ambient seawater, it is probably no surprise that the $\delta^{56}Fe$ values of sulfides are not in isotopic equilibrium with the fluids (Rouxel et al. 2016). This is markedly different than the case of marine sediments, where isotopic equilibrium between Fe^{2+}_{aq} and FeS readily occurs (Sect. 5.4). Collectively, these results indicate that dispersion of hydrothermal Fe to the open oceans, if it occurs as nanoparticulate Fe sulfides, will produce $\delta^{56}Fe$ values for dFe that are moderately negative, approaching −1‰ (Fig. 5.25). In contrast, the relatively fast Fe isotope exchange rates between Fe^{2+}_{aq} and ferrihydrite (Wu et al. 2011b), coupled with the slower rates of precipitation of Fe^{3+}-hydroxides relative to sulfides in hydrothermal vent systems, suggests that Fe isotope fractionation involving Fe^{3+}-hydroxides may involve some equilibrium effects, in addition to kinetic processes. Regardless, for Fe^{3+}-hydroxides, there is no evidence that kinetic isotope effects produce a reversal in fractionation relative to equilibrium, as is seen in the sulfides, allowing confidence in inferring that the $\delta^{56}Fe$ values of Fe^{3+}-oxyhydroxides will be equal to or higher than those of the vent fluids.

The TAG (Trans-Atlantic Geotraverse project) hydrothermal site at 26°N on the MAR is a basalt-hosted system (molar Fe/S = 1.5) that has the same short $t_{1/2}$ as the MAR 5°S system discussed above. Early extensive Fe sulfide precipitation is well documented in the vent fluids and initial plume, whereas pFe from the neutrally buoyant part of the plume consists almost entirely of Fe^{3+}-hydroxides, indicating later Fe $(OH)_3$ precipitation (German et al. 1991; Rudnicki and Elderfield 1993). Station 16 of the GEOTRACES GA03 transect was located above the TAG site, and in Fig. 5.26, vertical geochemical variations are illustrated for the deep-water section of that station. The plume at this station, as shown by excess ^3He (δ^3He), is located between 200 and 600 m above the vent, indicating that excess ^3He reflects sampling of the neutrally buoyant part of the plume. For the dissolved components (Fig. 5.26a), excess ^3He is correlated with very large increases in dFe, a factor of 100 times greater than at shallow water levels. Importantly, the increase in dFe is accompanied by a decrease in the proportion of Fe^{3+} in dFe ($X_{Fe(III)}$), from $X_{Fe(III)} \sim 1$ that is typical of shallow water dFe, to $X_{Fe(III)} \sim 0$. This apparent increase in Fe^{2+} in dFe is interpreted by Sedwick et al. (2015) to reflect, at least in part, nanoparticulate pyrite, most likely in the cFe size fraction (0.02–0.2 μm; Fitzsimmons et al. 2015). Recognition that this highly negative $\delta^{56}Fe$ signal in dFe likely reflects nanoparticulate Fe-sulfides and not "unoxidized" Fe^{2+}_{aq} is important, as it reconciles large abundances of Fe^{2+} in dFe despite the fast Fe^{2+}_{aq} oxidation rates (Fig. 5.24). In addition, the plume contains high levels of organically bound Fe^{2+} in the dFe size fraction (Buck et al. 2015). The $\delta^{56}Fe$ values for dFe decrease from values near those in shallow waters (see Fig. 5.20), to highly negative values in the plume, down to −1.5‰, and such values are significantly lower than those of TAG vent fluids (Severmann et al. 2004).

For the particulate components (Fig. 5.26b), pFe/pAl ratios increase from the background lithologic ratio of ~0.5 to ~60 in the plume, indicating very large increases in authigenic Fe (average crustal Fe_{tot}/Al is ~0.5). This is

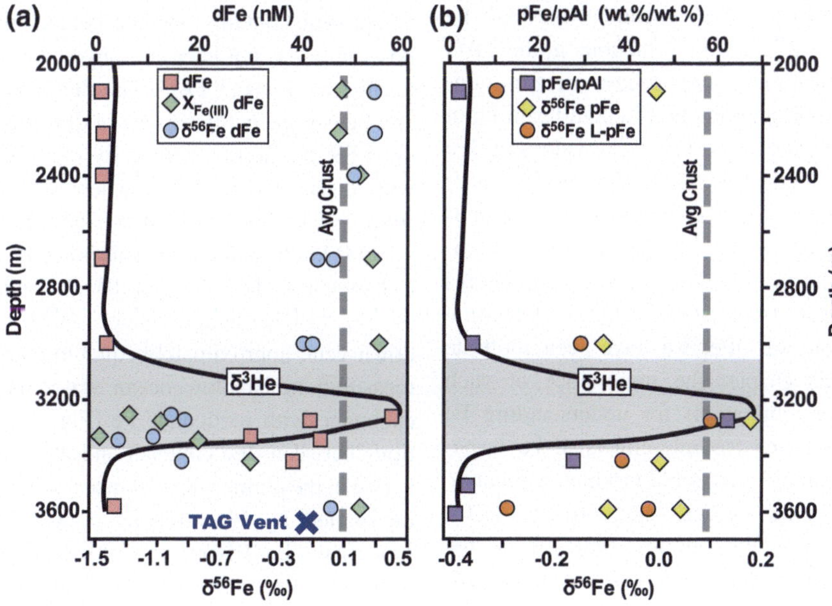

Fig. 5.26 Vertical water column geochemical profiles above the TAG hydrothermal vent, MAR from Station 16 of the GEOTRACES GA03 transect. The TAG plume (horizontal grey field) is identified by excess mantle ^3He (scale ranges from $\delta^3He = 5$ to 25 in both **a** and **b**). Vertical dashed grey lines in both panels show the $\delta^{56}Fe$ value of average upper crust. **a** In the plume, dFe markedly increases, and the fraction of Fe(III) in dFe, $X_{Fe(III)}$, decreases to near-zero, and $\delta^{56}Fe$ values for dFe strongly decrease ($X_{Fe(III)}$ scales from 0 to 1.0 in **a**). Blue "X" marks $\delta^{56}Fe$ value for TAG hydrothermal vent fluids. **b** pFe/pAl ratios strongly increase over the crustal average (pFe/pAl ~ 0.5) in the plume. $\delta^{56}Fe$ values for pFe and L-pFe are highly variable, ranging from slightly positive values within the plume, to values that scatter down to −0.3‰. Data sources are: δ^3He—Jenkins et al. (2015a); dFe and $\delta^{56}Fe$ for dFe—Conway and John (2014); $X_{Fe(III)}$ for dFe—Sedwick et al. (2015); $\delta^{56}Fe$ for TAG vent fluids—Severmann et al. (2004); pFe/pAl—Lam et al. (2015); $\delta^{56}Fe$ for pFe and L-pFe—Revels et al. (2015b)

correlated with strong increases in absolute concentrations of pFe and L-pFe (Lam et al. 2015; Revels et al. 2015a). X-ray spectroscopy identifies ferrihydrite as the major Fe phase in the particulate fraction (Ohnemus and Lam 2015), generally consistent with previous studies, but in strong contrast to the finding noted above that the dFe size fraction contains abundant Fe^{2+}-sulfides and Fe^{2+}-organics. The $\delta^{56}Fe$ values for pFe are near zero above the plume at ~2000 m water depth, and for L-pFe, $\delta^{56}Fe$ values are ~−0.3‰, similar to the Fe isotope compositions for pFe and L-pFe seen in the surface ocean (see Fig. 5.20). The $\delta^{56}Fe$ values for pFe and L-pFe in the plume, however, are anomalously high, equal to or greater than those of average upper crust, and greater than those of the TAG vent fluids. Beneath the plume, $\delta^{56}Fe$ values for pFe and L-pFe scatter between those of the plume and those similar to the TAG vent fluids. In toto, the relations in Fig. 5.26 document the high Fe enrichment in dFe and pFe fractions in the TAG hydrothermal plume, supporting the recent consensus that hydrothermal Fe is a significant input into the modern oceans. The $\delta^{56}Fe$ values of dFe in the TAG plume may extend to very negative values, reflecting high contents of nanoparticulate Fe sulfides and Fe^{2+} bound to organic complexes; identification of these components critically relies on ferrous-ferric Fe speciation determinations on dFe; we have seen multiple times in this chapter the importance of such speciation determinations for understanding Fe pathways and their isotopic effects. If Fe^{2+} species remain stable, this could produce a negative $\delta^{56}Fe$ flux to the oceans. Alternatively, if, for example, low-$\delta^{56}Fe$ sulfides settle from the plume, the remaining dFe might become isotopically heavy.

Hydrothermal plumes from the fast-spreading EPR can be exceptionally far-traveled, extending for >2,500 km based on excess ^3He anomalies (Resing et al. 2015; Jenkins et al. 2018). Vertical profiles above the EPR vent from the GEOTRACES GP16 transect show many similarities to the profiles above the TAG plume. Enrichment in dFe in the buoyant portion of the EPR plume correlate with enrichments in pFe and L-pFe, although it is noteworthy that both dissolved and particulate Fe sink compared to the ^3He and Mn plumes, suggesting exchange between particulate and dissolved phases. X-ray spectroscopy and trace-element data indicate that these enrichments are due to the presence of Fe^{3+}-hydroxides of all size ranges, including those contained in the dFe component (Fitzsimmons et al. 2017; Ho et al. 2018; Hoffman et al. 2018; Marsay et al. 2018). This is consistent with the expectation of early Fe–S precipitation near the vent in the basalt-hosted system, and the relatively slow Fe^{2+}_{aq} oxidation rates for the EPR. In addition, organic-bound Fe^{3+} is detected in both dFe and pFe enrichments in the plume, interpreted to be sourced to degraded biomass that was entrained in vent fluids (Fitzsimmons et al. 2017; Buck et al. 2018). Virtually no sulfides are detected in pFe in the plume above the GEOTRACES GP16 vent, although it is possible that, as in the TAG plume, the dFe component contains nanoparticulate pyrite; this possibility has not been investigated at the GP16 locality as it has at TAG. The dFe concentration and Fe isotope compositions of dFe and L-pFe in the EPR plume (as defined by high-^3He) varies with distance (Fig. 5.27), where dFe decreases from a peak of ~10 nM above the vent to <1 nM ~3800 km from the vent. This is accompanied by a change in $\delta^{56}Fe$ values for dFe above the vent of ~−0.2 to 0‰, increasing to ~+0.6 to +0.7‰ down-vent, approximately equal to compositions measured for shallow ocean water that reflects exchange with particulate Fe (Fig. 5.20). Variations between dFe contents and $\delta^{56}Fe$ values of L-pFe in the plume with distance from the vent are interpreted by Fitzsimmons et al. (2017) and Marsay et al. (2018) to record mixing between a near-vent nanoparticulate Fe^{3+}-hydroxide component that has $\delta^{56}Fe$ values equal to the vent fluids, with a high-$\delta^{56}Fe$ Fe^{3+}-bound C_{org} component that becomes dominant in the distal parts of the plume and is effectively identical to shallow open-ocean Fe. Although precipitation of low-$\delta^{56}Fe$ pyrite might be expected to produce some increase in $\delta^{56}Fe$ values in the remaining dFe as the plume evolves, Fitzsimmons et al. (2017) and Marsay et al. (2018) argue against such

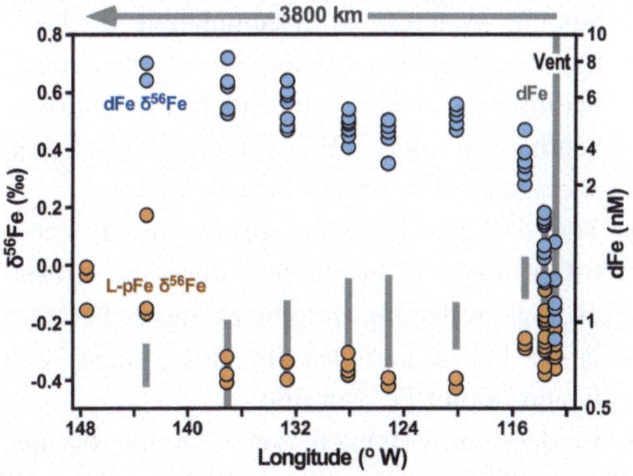

Fig. 5.27 Evolution of dFe concentration and Fe isotope compositions for dFe and L-pFe for the EPR hydrothermal plume sampled by the GEOTRACES GP16 transect. Only water column samples that occur in the plume (defined by δ^3He $\geq$ 32) are plotted. The vent is at Station 18 (112.8°W long), and the most distal station is 34 (147.5°W long), reflecting a distance of ~3,800 km for the transect shown. The relations shown are interpreted to reflect mixing between a dFe-rich plume component at the vent locality (δ^{56}Fe ~−0.2 to 0‰) and a high-δ^{56}Fe Fe^{3+}-C component distal from the vent that is similar to that of open-ocean shallow seawater (see also Fig. 5.20). The δ^{56}Fe values for L-pFe at the vent reflect Fe^{3+}-hydroxides produced by oxidation of vent fluids, whereas L-pFe more distal from the vent reflects exchange with Fe^{3+}-C species in the dFe component. Data from Fitzsimmons et al. (2017), Marsay et al. (2018), and John et al. (2018)

a model, noting that the distal, low-dFe plume component is controlled by the same Fe^{3+}-C component that determines the modestly positive δ^{56}Fe values of shallow open-ocean seawater.

The discussion above highlights the important role Fe–S, Fe^{3+}-hydroxide, and Fe-C_{org} species play in the Fe isotope evolution of marine hydrothermal plumes. Due in part to kinetic isotope effects upon Fe sulfide precipitation (see also Sect. 5.4.3), loss of Fe-sulfides (FeS and pyrite) from the plume will tend to push the remaining Fe to higher δ^{56}Fe values, whereas loss of Fe^{3+}-hydroxides will have the opposite effect, where the remaining Fe in the plume will shift to lower δ^{56}Fe values. In addition to the discussion above, the interplay between sulfide and Fe^{3+}-hydroxide precipitation has been highlighted in recent Fe isotope studies of the East Scotia Ridge (Klar et al. 2017b; Lough et al. 2017), Vanuatu back-arc (Nasemann et al. 2018), and Loihi (Rouxel et al. 2018) hydrothermal systems. But, a simple "sulfide versus hydroxide" model may be incorrect if the dFe component is dominated by nanoparticulate Fe-sulfides—in this case, in terms of Fe fluxes to the open oceans, the role of sulfide is not one of early near-vent precipitation, but forming the actual species of dFe that is transported far from the source. Key to understanding the Fe isotope effects of hydrothermal transport to the open oceans therefore requires a full characterization of the Fe species contained in the dFe and pFe components. Determining the conditions that favor early stage sulfide precipitation versus late-stage formation of nanoparticulate Fe-sulfides in the buoyant plume is important to understanding the Fe isotope compositions of hydrothermal plumes. To date, such comprehensive approaches have only been done at a few localities.

5.6 Summary

- Aerosols and loess have δ^{56}Fe values within error of average crust (δ^{56}Fe = +0.09‰), indicating no changes during mechanical weathering.
- The δ^{56}Fe values of bulk soils range from −0.5 to +0.9‰, significantly more variable than the crustal average. Acid leaches of soils tend to have δ^{56}Fe values that are lower than the bulk soil, suggesting that the labile component is isotopically light.
- Suspended particulate material (SPM) in rivers has a wide range in δ^{56}Fe values of up to 2‰, although SPM for many rivers cluster about the crustal average. Deviations from the crustal average for SPM reflect the presence of Fe-C species and Fe^{3+}-hydroxides.
- Dissolved Fe in rivers (dFe, generally <0.2 to <0.45 μm in size) has δ^{56}Fe values that vary greatly, with the largest variations in the smallest dFe fractions (1–100 kDa), reflecting variable contributions from high-δ^{56}Fe components inferred to be Fe^{3+}-bound organics, and components that have moderately positive δ^{56}Fe values that reflect nanoparticulate Fe^{3+} oxides/hydroxides.
- Oxidation of Fe^{2+}_{aq} in modern groundwater or terrestrial hydrothermal systems produces

Fe^{3+}-hydroxide precipitates that have $\delta^{56}Fe$ values equal to or greater than the source fluids, producing low $\delta^{56}Fe$ values in the remaining Fe^{2+}_{aq} pool. $\delta^{56}Fe$–Fe_{aq} trends can follow Rayleigh or equilibrium trends, dependent on oxidation rates and extent of oxidation.

- Redox-stratified lakes can have wide ranges in $\delta^{56}Fe$ values for aqueous Fe, reflecting a complex interplay among precipitating minerals. Low $\delta^{56}Fe$ values for Fe_{aq} may be produced by oxidation of Fe^{2+}_{aq} and formation of Fe^{3+}-hydroxides, or, under non-redox conditions, low $\delta^{56}Fe$ values for Fe_{aq} are produced if vivianite and/or mackinawite is formed. High $\delta^{56}Fe$ values for Fe_{aq} may be produced during siderite precipitation.
- Sedimentary pyrite, in lake or marine systems, is unlikely to form in Fe isotope equilibrium with aqueous Fe. Negative $\delta^{56}Fe$ values for sedimentary pyrite reflect variable kinetic isotope effects upon transformation from Fe–S precursors (which may initially form in Fe isotope equilibrium with Fe_{aq}) to pyrite via a wide variety of pathways. There is no evidence that pyrite formed in the water column is in Fe isotope equilibrium with Fe_{aq}.
- In terms of Fe isotope compositions for pore water Fe_{aq}, modern marine sediments define two end members: those dominated by microbial dissimilatory iron reduction (DIR) and reductive dissolution by H_2S have very negative $\delta^{56}Fe$ values for Fe_{aq}, recording exchange with Fe^{3+} hydroxides, whereas those dominated by microbial sulfate reduction have moderately positive $\delta^{56}Fe$ values for Fe_{aq}, reflecting exchange with FeS. Pore water Fe^{2+}_{aq} that undergoes oxidation from seawater O_2 will show decreasing $\delta^{56}Fe$ values with decreasing Fe_{aq} contents.
- Microbial iron reduction on continental shelves is supported by detrital Fe^{3+}-hydroxides and organic C from primary productivity. This provides a large benthic flux of Fe to the open oceans that is characterized by highly negative $\delta^{56}Fe$ values, and the magnitude of the flux inversely scales with O_2 contents. Traps for benthic Fe fluxes may be provided by sulfide in euxinic systems, or accumulation of Fe^{3+}-hydroxides in non-euxinic, low-oxygen zones. Sediments that have been enriched in Fe from a benthic flux have Fe/Al ratios higher than average crust or regional clastic inputs.
- The $\delta^{56}Fe$ values for surface ocean water (dFe) tend to be higher than the crustal average, reflecting isotopic exchange between labile Fe in particles (including dust) and ligand-bound Fe^{3+} in dFe.
- In deep, low-oxygen zones in the oceans, particularly regions restricted from water exchange with the open ocean, the $\delta^{56}Fe$ values for dFe can become quite negative, reflecting large inputs from benthic Fe fluxes generated by microbial iron reduction.
- The $\delta^{56}Fe$ values of marine hydrothermal plumes is strongly dependent on mineral precipitation processes, including the balance between Fe-sulfide and Fe^{3+} hydroxide formation, which tend to drive $\delta^{56}Fe$ values for dFe in opposite directions. For plumes that contain abundant nanoparticulate pyrite in the dFe component, these may be a source for low-$\delta^{56}Fe$ Fe to the open oceans.

References

Akerman A, Poitrasson F, Oliva P, Audry S, Prunier J, Braun J-J (2014) The isotopic fingerprint of Fe cycling in an equatorial soil–plant–water system: the Nsimi watershed, South Cameroon. Chem Geol 385:104–116. https://doi.org/10.1016/j.chemgeo.2014.07.003

Anderson RF, Raiswell R (2004) Sources and mechanisms for the enrichment of highly reactive iron in euxinic Black Sea sediments. Am J Sci 304:203–233

Azrieli-Tal I, Matthews A, Bar-Matthews M, Almogi-Labin A, Vance D, Archer C, Teutsch N (2014) Evidence from molybdenum and iron isotopes and molybdenum–uranium covariation for sulphidic bottom waters during Eastern Mediterranean sapropel S1 formation. Earth Planet Sci Lett 393:231–242. https://doi.org/10.1016/j.epsl.2014.02.054

Beard BL, Handler RM, Scherer MM, Wu L, Czaja AD, Heimann A, Johnson CM (2010) Iron isotope fractionation between aqueous ferrous iron and goethite. Earth Planet Sci Lett 295(1–2):241–250. https://doi.org/10.1016/j.epsl.2010.04.006

Beard BL, Johnson CM (2004) Fe isotope variations in the modern and ancient earth and other planetary bodies. Rev Miner Geochem 55:319–357

Beard BL, Johnson CM, Von Damm KL, Poulson RL (2003) Iron isotope constraints on Fe cycling and mass balance in oxygenated Earth oceans. Geol Soc Am 31 (7):629–632

Beaulieu SE, Baker ET, German CR (2015) Where are the undiscovered hydrothermal vents on oceanic spreading ridges? Deep Sea Res Part II 121:202–212. https://doi.org/10.1016/j.dsr2.2015.05.001

Bennett SA, Achterberg EP, Connelly DP, Statham PJ, Fones GR, German CR (2008) The distribution and stabilisation of dissolved Fe in deep-sea hydrothermal plumes. Earth Planet Sci Lett 270(3–4):157–167. https://doi.org/10.1016/j.epsl.2008.01.048

Bennett SA, Rouxel O, Schmidt K, Garbe-Schönberg D, Statham PJ, German CR (2009) Iron isotope fractionation in a buoyant hydrothermal plume, 5°S Mid-Atlantic Ridge. Geochim Cosmochim Acta 73 (19):5619–5634. https://doi.org/10.1016/j.gca.2009.06.027

Benning LG, Wilkin RT, Barnes HL (2000) Reaction pathways in the Fe-S system below 100C. Chem Geol 167:25–51

Berger CJM, Lippiatt SM, Lawrence MG, Bruland KW (2008) Application of a chemical leach technique for estimating labile particulate aluminum, iron, and manganese in the Columbia River plume and coastal waters off Oregon and Washington. J Geophys Res 113. https://doi.org/10.1029/2007jc004703

Bergquist BA, Boyle EA (2006a) Dissolved iron in the tropical and subtropical Atlantic Ocean. Glob Biogeochem Cycles 20(1):n/a-n/a. https://doi.org/10.1029/2005gb002505

Bergquist BA, Boyle EA (2006b) Iron isotopes in the Amazon River system: weathering and transport signatures. Earth Planet Sci Lett 248(1–2):54–68. https://doi.org/10.1016/j.epsl.2006.05.004

Bergquist BA, Wu J, Boyle EA (2007) Variability in oceanic dissolved iron is dominated by the colloidal fraction. Geochim Cosmochim Acta 71(12):2960–2974. https://doi.org/10.1016/j.gca.2007.03.013

Bernard A, Symonds RB (1989) The significance of siderite in the sediments from Lake Nyos, Cameroon. J Volcanol Geoth Res 39:187–194

Berner RA (1970) Sedimentary pyrite formation. Am J Sci 268:1–23

Blanchet CL, Thouveny N, Vidal L (2009) Formation and preservation of greigite (Fe3S4) in sediments from the Santa Barbara Basin: implications for paleoenvironmental changes during the past 35 ka. Paleoceanography 24(2):n/a-n/a. https://doi.org/10.1029/2008pa001719

Boiteau RM, Mende DR, Hawco NJ, McIlvin MR, Fitzsimmons JN, Saito MA, Sedwick PN, DeLong EF, Repeta DJ (2016) Siderophore-based microbial adaptations to iron scarcity across the eastern Pacific Ocean. Proc Natl Acad Sci USA 113(50):14237–14242. https://doi.org/10.1073/pnas.1608594113

Borrok DM, Wanty RB, Ian Ridley W, Lamothe PJ, Kimball BA, Verplanck PL, Runkel RL (2009) Application of iron and zinc isotopes to track the sources and mechanisms of metal loading in a mountain watershed. Appl Geochem 24(7):1270–1277. https://doi.org/10.1016/j.apgeochem.2009.03.010

Boyd PW, Ellwood MJ (2010) The biogeochemical cycle of iron in the ocean. Nat Geosci 3(10):675–682. https://doi.org/10.1038/ngeo964

Boyle EA, Edmond JM, Sholkovitz ER (1977) The mechanism of iron removal in estuaries. Geochim Cosmochim Acta 41:1313–1324

Bruland KW, Orians KJ, Cowen JP (1994) Reactive trace metals in the stratified central North Pacific. Geochim Cosmochim Acta 58(15):3171–3182

Buck KN, Sedwick PN, Sohst B, Carlson CA (2018) Organic complexation of iron in the eastern tropical South Pacific: results from US GEOTRACES Eastern Pacific Zonal Transect (GEOTRACES cruise GP16). Mar Chem 201:229–241. https://doi.org/10.1016/j.marchem.2017.11.007

Buck KN, Sohst B, Sedwick PN (2015) The organic complexation of dissolved iron along the US. GEOTRACES (GA03) North Atlantic section. Deep Sea Res Part II 116:152–165. https://doi.org/10.1016/j.dsr2.2014.11.016

Bullen TD, White AF, Childs CW, Vivit DV, Schulz MS (2001) Demonstration of significant abiotic iron isotope fractionation in nature. Geol Soc Am 29 (8):699–702

Bura-Nakić E, Viollier E, Jézéquel D, Thiam A, Ciglenečki I (2009) Reduced sulfur and iron species in anoxic water column of meromictic crater Lake Pavin (Massif Central, France). Chem Geol 266(3–4):311–317. https://doi.org/10.1016/j.chemgeo.2009.06.020

Busigny V, Jézéquel D, Cosmidis J, Viollier E, Benzerara K, Planavsky NJ, Albéric P, Lebeau O, Sarazin G, Michard G (2016) The Iron wheel in Lac Pavin: interaction with phosphorus cycle. In: Lake Pavin, pp 205–220. https://doi.org/10.1007/978-3-319-39961-4_12

Busigny V, Planavsky NJ, Jézéquel D, Crowe S, Louvat P, Moureau J, Viollier E, Lyons TW (2014) Iron isotopes in an Archean ocean analogue. Geochim Cosmochim Acta 133:443–462. https://doi.org/10.1016/j.gca.2014.03.004

Buss HL, Mathur R, White AF, Brantley SL (2010) Phosphorus and iron cycling in deep saprolite, Luquillo Mountains, Puerto Rico. Chem Geol 269 (1–2):52–61. https://doi.org/10.1016/j.chemgeo.2009.08.001

Butler IB, Archer C, Vance D, Oldroyd A, Rickard D (2005) Fe isotope fractionation on FeS formation in ambient aqueous solution. Earth Planet Sci Lett 236 (1–2):430–442. https://doi.org/10.1016/j.epsl.2005.05.022

Canfield DE (1997) The geochemistry of river particulates from the continental USA: major elements. Geochim Cosmochim Acta 61(16):3349–3365

Canfield DE, Jørgensen BB, Fossing H, Glud R, Gundersen J, Ramsing NB, Thamdrup B, Hansen JW, Nielsen LP, Hall POJ (1993a) Pathways of organic carbon oxidation in three continental margin sediments. Mar Geol 113:27–40

Canfield DE, Lyons TW, Raiswell R (1996) A model for iron deposition to euxinic Black Sea sediments. Am J Sci 296:818–834

Canfield DE, Thamdrup B, Hansen JW (1993b) The anaerobic degradation of organic matter in Danish coastal sediments: Iron reduction, manganese reduction, and sulfate reduction. Geochim Cosmochim Acta 57:3867–3883

Chadwick OA, Gavenda RT, Kelly EF, Ziegler K, Olson CG, Elliott WC, Hendricks DM (2003) The impact of climate on the biogeochemical functioning of volcanic soils. Chem Geol 202(3–4):195–223. https://doi.org/10.1016/j.chemgeo.2002.09.001

Chapman JB, Weiss DJ, Shan Y, Lemburger M (2009) Iron isotope fractionation during leaching of granite and basalt by hydrochloric and oxalic acids. Geochim Cosmochim Acta 73(5):1312–1324. https://doi.org/10.1016/j.gca.2008.11.037

Chen J-B, Busigny V, Gaillardet J, Louvat P, Wang Y-N (2014) Iron isotopes in the Seine River (France): Natural versus anthropogenic sources. Geochim Cosmochim Acta 128:128–143. https://doi.org/10.1016/j.gca.2013.12.017

Chever F, Rouxel OJ, Croot PL, Ponzevera E, Wuttig K, Auro M (2015) Total dissolvable and dissolved iron isotopes in the water column of the Peru upwelling regime. Geochim Cosmochim Acta 162:66–82. https://doi.org/10.1016/j.gca.2015.04.031

Conway TM, John SG (2014) Quantification of dissolved iron sources to the North Atlantic Ocean. Nature 511(7508):212–215. https://doi.org/10.1038/nature13482

Conway TM, John SG, Lacan F (2016) Intercomparison of dissolved iron isotope profiles from reoccupation of three GEOTRACES stations in the Atlantic Ocean. Mar Chem 183:50–61. https://doi.org/10.1016/j.marchem.2016.04.007

Cornell RM, Schwertmann U (2003) The iron oxides: structure, properties, reactions, occurences and uses. 2nd edn. Wiley-VCH Verlag GmbH & Co. KGaA, Weinheim. https://doi.org/10.1002/3527602097

Cosmidis J, Benzerara K, Morin G, Busigny V, Lebeau O, Jézéquel D, Noël V, Dublet G, Othmane G (2014) Biomineralization of iron-phosphates in the water column of Lake Pavin (Massif Central, France). Geochim Cosmochim Acta 126:78–96. https://doi.org/10.1016/j.gca.2013.10.037

Croal LR, Johnson CM, Beard BL, Newman DK (2004) Iron isotope fractionation by Fe(II)-oxidizing photoautotrophic bacteria 1 1Associate editor: D E canfield. Geochimica et Cosmochimica Acta 68(6):1227–1242. https://doi.org/10.1016/j.gca.2003.09.011

Cutter GA, Kluckhohn RS (1999) The cycling of particulate carbon, nitrogen, sulfur, and sulfur species (iron monosulfide, gregite, pyrite, and organic sulfur) in the water columns of Framvaren Fjord and the Black Sea. Mar Chem 67:149–160

Czaja AD, Johnson CM, Yamaguchi KE, Beard BL (2012) Comment on "Abiotic pyrite formation produces a large Fe isotope fractionation". Science 335:538

Dale AW, Nickelsen L, Scholz F, Hensen C, Oschlies A, Wallmann K (2015) A revised global estimate of dissolved iron fluxes from marine sediments. Glob Biogeochem Cycles 29:691–707. https://doi.org/10.1002/2014GB005017

Dauphas N, Rouxel O (2006) Mass spectrometry and natural variations of iron isotopes. Mass Spectrom Rev 25(4):515–550. https://doi.org/10.1002/mas.20078

Dideriksen K, Baker JA, Stipp SLS (2008) Equilibrium Fe isotope fractionation between inorganic aqueous Fe(III) and the siderophore complex, Fe(III)-desferrioxamine B. Earth Planet Sci Lett 269(1–2):280–290. https://doi.org/10.1016/j.epsl.2008.02.022

Dideriksen K, Christiansen BC, Baker JA, Frandsen C, Balic-Zunic T, Tullborg E, Mørup S, Stipp SLS (2007) Fe-oxide fracture fillings as a palæo-redox indicator: structure, crystal form and Fe isotope composition. Chem Geol 244(1–2):330–343. https://doi.org/10.1016/j.chemgeo.2007.06.027

Dideriksen K, Christiansen BC, Frandsen C, Balic-Zunic T, Mørup S, Stipp SLS (2010) Paleo-redox boundaries in fractured granite. Geochim Cosmochim Acta 74(10):2866–2880. https://doi.org/10.1016/j.gca.2010.02.022

dos Santos Pinheiro GM, Poitrasson F, Sondag F, Cochonneau G, Vieira LC (2014) Contrasting iron isotopic compositions in river suspended particulate matter: the Negro and the Amazon annual river cycles. Earth Planet Sci Lett 394:168–178. https://doi.org/10.1016/j.epsl.2014.03.006

dos Santos Pinheiro GM, Poitrasson F, Sondag F, Vieira LC, Pimentel MM (2013) Iron isotope composition of the suspended matter along depth and lateral profiles in the Amazon river and its tributaries. J S Am Earth Sci 44:35–44. https://doi.org/10.1016/j.jsames.2012.08.001

Douville E, Charlou JL, Oelkers EH, Bienvenu P, Jove Colon CF, Donval JP, Fouquet Y, Prieur D, Appriou P (2002) The rainbow vent fluids (36 14′N, MAR): the influence of ultramafic rocks and phase separation on trace metal content in Mid-Atlantic Ridge hydrothermal fluids. Chem Geol 184:37–48

Eckert S, Brumsack H-J, Severmann S, Schnetger B, März C, Fröllje H (2013) Establishment of euxinic conditions in the Holocene Black Sea. Geology 41(4):431–434. https://doi.org/10.1130/g33826.1

Edmonds HN, German CR (2004) Particle geochemistry in the Rainbow hydrothermal plume Mid-Atlantic Ridge. Geochimica et Cosmochimica Acta 68(4):759–772. https://doi.org/10.1016/s0016-7037(03)00498-8

Ellwood MJ, Hutchins DA, Lohan MC, Milne A, Nasemann P, Nodder SD, Sander SG, Strzepek R, Wilhelm SW, Boyd PW (2015) Iron stable isotopes track pelagic iron cycling during a subtropical phytoplankton bloom. Proc Natl Acad Sci USA 112(1):E15–20. https://doi.org/10.1073/pnas.1421576112

Elrod VA, Berelson WM, Coale KH, Johnson KS (2004) The flux of iron from continental shelf sediments: a missing source for global budgets. Geophys Res Lett 31(12):n/a-n/a. https://doi.org/10.1029/2004gl020216

Emmanuel S, Erel Y, Matthews A, Teutsch N (2005) A preliminary mixing model for Fe isotopes in soils. Chem Geol 222(1–2):23–34. https://doi.org/10.1016/j.chemgeo.2005.07.002

Escoube R, Rouxel OJ, Pokrovsky OS, Schroth A, Max Holmes R, Donard OFX (2015) Iron isotope systematics in Arctic rivers. CR Geosci 347(7–8):377–385. https://doi.org/10.1016/j.crte.2015.04.005

Escoube R, Rouxel OJ, Sholkovitz E, Donard OFX (2009) Iron isotope systematics in estuaries: the case of North river, Massachusetts (USA). Geochim Cosmochim Acta 73(14):4045–4059. https://doi.org/10.1016/j.gca.2009.04.026

Fantle MS, DePaolo DJ (2004) Iron isotopic fractionation during continental weathering. Earth Planet Sci Lett 228(3–4):547–562. https://doi.org/10.1016/j.epsl.2004.10.013

Fehr MA, Andersson PS, Hålenius U, Gustafsson Ö, Mörth C-M (2010) Iron enrichments and Fe isotopic compositions of surface sediments from the Gotland Deep. Baltic Sea. Chemical Geology 277(3–4):310–322. https://doi.org/10.1016/j.chemgeo.2010.08.014

Fehr MA, Andersson PS, Hålenius U, Mörth C-M (2008) Iron isotope variations in Holocene sediments of the Gotland Deep Baltic Sea. Geochimica et Cosmochimica Acta 72(3):807–826. https://doi.org/10.1016/j.gca.2007.11.033

Fekiacova Z, Pichat S, Cornu S, Balesdent J (2013) Inferences from the vertical distribution of Fe isotopic compositions on pedogenetic processes in soils. Geoderma 209–210:110–118. https://doi.org/10.1016/j.geoderma.2013.06.007

Field MP, Sherrell RM (2000) Dissolved and particulate Fe in a hydrothermal plume at 9 45′N, East Pacific Rise: slow Fe (II) oxidation kinetics in Pacific plumes. Geochim Cosmochim Acta 64(4):619–628

Fitzsimmons JN, Carrasco GG, Wu J, Roshan S, Hatta M, Measures CI, Conway TM, John SG, Boyle EA (2015) Partitioning of dissolved iron and iron isotopes into soluble and colloidal phases along the GA03 GEOTRACES North Atlantic Transect. Deep Sea Res Part II: Top Stud Oceanogr 116:130–151. https://doi.org/10.1016/j.dsr2.2014.11.014

Fitzsimmons JN, John SG, Marsay CM, Hoffman CL, Nicholas Sarah L, Toner BM, German CR, Sherrell RM (2017) Iron persistence in a distal hydrothermal plume supported by dissolved–particulate exchange. Nat Geosci 10(3):195–201. https://doi.org/10.1038/ngeo2900

Fornari D, Von Damm K, Bryce J, Cowen J, Ferrini V, Fundis A, Lilley M, Luther G, Mullineaux L, Perfit M, Meana-Prado MF, Rubin K, Seyfried W, Shank T, Soule SA, Tolstoy M, White S (2012) The East Pacific Rise between 9°N and 10°N: twenty-five years of integrated, multidisciplinary oceanic spreading center studies. Oceanography 25(1):18–43. https://doi.org/10.5670/oceanog.2012.02

Fossing H, Jørgensen BB (1990) Isotope exchange reactions with radiolabeled sulfur compounds in anoxic seawater. Biogeochemistry 9:223–245

Fox LE (1988) The solubility of colloidal ferric hydroxide and its relevance to iron concentrations in river water. Geochim Cosmochim Acta 52:771–777

Fritz SJ (1988) A comparative study of gabbro and granite weathering. Chem Geol 68:275–290

Garnier J, Garnier JM, Vieira CL, Akerman A, Chmeleff J, Ruiz RI, Poitrasson F (2017) Iron isotope fingerprints of redox and biogeochemical cycling in the soil-water-rice plant system of a paddy field. Sci Total Environ 574:1622–1632. https://doi.org/10.1016/j.scitotenv.2016.08.202

Gartman A, Findlay AJ, Luther GW (2014) Nanoparticulate pyrite and other nanoparticles are a widespread component of hydrothermal vent black smoker emissions. Chem Geol 366:32–41. https://doi.org/10.1016/j.chemgeo.2013.12.013

Gartman A, Luther GW (2014) Oxidation of synthesized sub-micron pyrite (FeS 2) in seawater. Geochim Cosmochim Acta 144:96–108. https://doi.org/10.1016/j.gca.2014.08.022

German CR, Bennett SA, Connelly DP, Evans AJ, Murton BJ, Parson LM, Prien RD, Ramirez-Llodra E, Jakuba M, Shank TM, Yoerger DR, Baker ET, Walker SL, Nakamura K (2008) Hydrothermal activity on the southern Mid-Atlantic Ridge: Tectonically- and volcanically-controlled venting at 4–5°S. Earth Planet Sci Lett 273(3–4):332–344. https://doi.org/10.1016/j.epsl.2008.06.048

German CR, Campbell AC, Edmond JM (1991) Hydrothermal scavenging at the Mid-Atlantic Ridge: modification of trace element dissolved fluxes. Earth Planet Sci Lett 107:101–114

German CR, Legendre LL, Sander SG, Niquil N, Luther GW, Bharati L, Han X, Le Bris N (2015) Hydrothermal Fe cycling and deep ocean organic carbon scavenging: model-based evidence for significant POC supply to seafloor sediments. Earth Planet Sci Lett 419:143–153. https://doi.org/10.1016/j.epsl.2015.03.012

German CR, Petersen S, Hannington MD (2016) Hydrothermal exploration of mid-ocean ridges: where might the largest sulfide deposits be forming? Chem Geol 420:114–126. https://doi.org/10.1016/j.chemgeo.2015.11.006

German CR, Seyfried WEJ (2014) Treatise on geochemistry (2nd edn). Ref Modul Earth Syst Environ Sci 8:191–233. https://doi.org/10.1016/b978-0-08-095975-7.00607-0

German CR, Thurnherr AM, Knoery J, Charlou JL, Jean-Baptiste P, Edmonds HN (2010) Heat, volume and chemical fluxes from submarine venting: a synthesis of results from the Rainbow hydrothermal field, 36°N MAR. Deep Sea Res Part I 57(4):518–527. https://doi.org/10.1016/j.dsr.2009.12.011

German CR, Von Damm KL (2003) Treatise on geochemistry (Hydrothermal Processes), vol 6

Giggenbach WF (1990) Water and gas chemistry of Lake Nyos and its bearing on the eruptive process. J Volcanol Geoth Res 42:337–362

Gong Y, Xia Y, Huang F, Yu H (2016) Average iron isotopic compositions of the upper continental crust: constrained by loess from the Chinese Loess Plateau. Acta Geochimica 36(2):125–131. https://doi.org/10.1007/s11631-016-0131-5

Gordon RM, Martin JH, Knauer GA (1982) Iron in north-east Pacific waters. Nature 299:611–612

Guelke M, von Blanckenburg F (2007) Fractionation of stable iron isotopes in higher plants. Environ Sci Technol

Guelke M, von Blanckenburg F, Schoenberg R, Staubwasser M, Stuetzel H (2010) Determining the stable Fe isotope signature of plant-available iron in soils. Chem Geol 277(3–4):269–280. https://doi.org/10.1016/j.chemgeo.2010.08.010

Guilbaud R, Butler IB, Ellam RM (2011a) Abiotic pyrite formation produces a large Fe isotope fractionation. Science 332(6037):1548–1551. https://doi.org/10.1126/science.1202924

Guilbaud R, Butler IB, Ellam RM, Rickard D (2010) Fe isotope exchange between Fe(II)aq and nanoparticulate mackinawite (FeSm) during nanoparticle growth. Earth Planet Sci Lett 300(1–2):174–183. https://doi.org/10.1016/j.epsl.2010.10.004

Guilbaud R, Butler IB, Ellam RM, Rickard D, Oldroyd A (2011b) Experimental determination of the equilibrium Fe isotope fractionation between and FeSm (mackinawite) at 25 and 2 °C. Geochim Cosmochim Acta 75(10):2721–2734. https://doi.org/10.1016/j.gca.2011.02.023

Guo H, Liu C, Lu H, Wanty RB, Wang J, Zhou Y (2013) Pathways of coupled arsenic and iron cycling in high arsenic groundwater of the Hetao basin, Inner Mongolia, China: An iron isotope approach. Geochim Cosmochim Acta 112:130–145. https://doi.org/10.1016/j.gca.2013.02.031

Habicht KS, Gade M, Thamdrup B, Berg P, Canfield DE (2002) Calibration of sulfate levels in the Archean Ocean. Science 298

Handler RM, Beard BL, Johnson CM, Scherer MM (2009) Atom exchange between Aqeuous Fe(II) and Goethite: an Fe isotope tracer study. Environ Sci Technol 43:1102–1107

Hatta M, Measures CI, Wu J, Roshan S, Fitzsimmons JN, Sedwick P, Morton P (2015) An overview of dissolved Fe and Mn distributions during the 2010–2011 US GEOTRACES north Atlantic cruises: GEOTRACES GA03. Deep Sea Res Part II 116:117–129. https://doi.org/10.1016/j.dsr2.2014.07.005

Hayes CT, Fitzsimmons JN, Boyle EA, McGee D, Anderson RF, Weisend R, Morton PL (2015) Thorium isotopes tracing the iron cycle at the Hawaii Ocean time-series station ALOHA. Geochim Cosmochim Acta 169:1–16. https://doi.org/10.1016/j.gca.2015.07.019

Heller MI, Lam PJ, Moffett JW, Till CP, Lee J-M, Toner BM, Marcus MA (2017) Accumulation of Fe oxyhydroxides in the Peruvian oxygen deficient zone implies non-oxygen dependent Fe oxidation. Geochim Cosmochim Acta 211:174–193. https://doi.org/10.1016/j.gca.2017.05.019

Henkel S, Kasten S, Poulton SW, Staubwasser M (2016) Determination of the stable iron isotopic composition of sequentially leached iron phases in marine sediments. Chem Geol 421:93–102. https://doi.org/10.1016/j.chemgeo.2015.12.003

Ho P, Lee J-M, Heller MI, Lam PJ, Shiller AM (2018) The distribution of dissolved and particulate Mo and V along the US GEOTRACES East Pacific Zonal Transect (GP16): the roles of oxides and biogenic particles in their distributions in the oxygen deficient zone and the hydrothermal plume. Mar Chem 201:242–255. https://doi.org/10.1016/j.marchem.2017.12.003

Hoffman CL, Nicholas SL, Ohnemus DC, Fitzsimmons JN, Sherrell RM, German CR, Heller MI, Lee J-m, Lam PJ, Toner BM (2018) Near-field iron and carbon chemistry of non-buoyant hydrothermal plume particles, Southern East Pacific Rise 15°S. Mar Chem 201:183–197. https://doi.org/10.1016/j.marchem.2018.01.011

Holmes TM, Chase Z, van der Merwe P, Townsend AT, Bowie AR (2017) Detection, dispersal and biogeochemical contribution of hydrothermal iron in the ocean. Mar Freshw Res 68(12). https://doi.org/10.1071/mf16335

Homoky WB, Hembury DJ, Hepburn LE, Mills RA, Statham PJ, Fones GR, Palmer MR (2011) Iron and manganese diagenesis in deep sea volcanogenic sediments and the origins of pore water colloids. Geochim Cosmochim Acta 75(17):5032–5048. https://doi.org/10.1016/j.gca.2011.06.019

Homoky WB, John SG, Conway TM, Mills RA (2013) Distinct iron isotopic signatures and supply from marine sediment dissolution. Nat Commun 4:2143. https://doi.org/10.1038/ncomms3143

Homoky WB, Severmann S, Mills RA, Statham PJ, Fones GR (2009) Pore-fluid Fe isotopes reflect the extent of benthic Fe redox recycling: evidence from continental shelf and deep-sea sediments. Geology 37 (8):751–754. https://doi.org/10.1130/g25731a.1

Homoky WB, Weber T, Berelson WM, Conway TM, Henderson GM, van Hulten M, Jeandel C, Severmann S, Tagliabue A (2016) Quantifying trace element and isotope fluxes at the ocean-sediment boundary: a review. Philos Trans A Math Phys Eng Sci 374 (2081). https://doi.org/10.1098/rsta.2016.0246

Hunger S, Benning LG (2007) Greigite: a true intermediate on the polysulfide pathway to pyrite. Geochem Trans 8:1. https://doi.org/10.1186/1467-4866-8-1

Hunter KA, Boyd PW (2007) Iron-binding ligands and their role in the ocean biogeochemistry of iron. Environ Chem 4(4). https://doi.org/10.1071/en07012

Hurst MP, Bruland KW (2007) An investigation into the exchange of iron and zinc between soluble, colloidal, and particulate size-fractions in shelf waters using low-abundance isotopes as tracers in shipboard incubation experiments. Mar Chem 103(3–4):211–226. https://doi.org/10.1016/j.marchem.2006.07.001

Ilina SM, Poitrasson F, Lapitskiy SA, Alekhin YV, Viers J, Pokrovsky OS (2013) Extreme iron isotope fractionation between colloids and particles of boreal and temperate organic-rich waters. Geochim Cosmochim Acta 101:96–111. https://doi.org/10.1016/j.gca.2012.10.023

Ingri J, Malinovsky D, Rodushkin I, Baxter DC, Widerlund A, Andersson P, Gustafsson Ö, Forsling W, Öhlander B (2006) Iron isotope fractionation in river colloidal matter. Earth Planet Sci Lett 245(3–4):792–798. https://doi.org/10.1016/j.epsl.2006.03.031

Ito A, Otake T, Shin K-C, Ariffin KS, Yeoh F-Y, Sato T (2017) Geochemical signatures and processes in a stream contaminated by heavy mineral processing near Ipoh city, Malaysia. Appl Geochem 82:89–101. https://doi.org/10.1016/j.apgeochem.2017.05.007

Jean-Baptiste P, Fourré E, Charlou J-L, German CR, Radford-Knoery J (2004) Helium isotopes at the Rainbow hydrothermal site (Mid-Atlantic Ridge, 36° 14′N). Earth Planet Sci Lett 221(1–4):325–335. https://doi.org/10.1016/s0012-821x(04)00094-9

Jenkins WJ, Lott DE, German CR, Cahill KL, Goudreau J, Longworth B (2018) The deep distributions of helium isotopes, radiocarbon, and noble gases along the US GEOTRACES East Pacific Zonal Transect (GP16). Mar Chem 201:167–182. https://doi.org/10.1016/j.marchem.2017.03.009

Jenkins WJ, Lott DE, Longworth BE, Curtice JM, Cahill KL (2015a) The distributions of helium isotopes and tritium along the US GEOTRACES North Atlantic sections (GEOTRACES GAO3). Deep Sea Res Part II 116:21–28. https://doi.org/10.1016/j.dsr2.2014.11.017

Jenkins WJ, Smethie WM, Boyle EA, Cutter GA (2015b) Water mass analysis for the US GEOTRACES (GA03) North Atlantic sections. Deep Sea Res Part II 116:6–20. https://doi.org/10.1016/j.dsr2.2014.11.018

Jickells TD, An ZS, Andersen KK, Baker AR, Bergametti G, Brooks N, Cao JJ, Boyd PW, Duce RA, Hunter KA, Kawahata H, Kubilay N, laRoche J, Liss PS, Mahowald N, Prospero JM, Ridgwell AJ, Tegen I, Torres R (2005) Global iron connections between desert dust, ocean biogeochemistry, and climate. Science 308

John SG, Helgoe J, Townsend E, Weber T, DeVries T, Tagliabue A, Moore K, Lam P, Marsay CM, Till C (2018) Biogeochemical cycling of Fe and Fe stable isotopes in the Eastern Tropical South Pacific. Mar Chem 201:66–76. https://doi.org/10.1016/j.marchem.2017.06.003

John SG, Mendez J, Moffett J, Adkins J (2012) The flux of iron and iron isotopes from San Pedro Basin sediments. Geochim Cosmochim Acta 93:14–29. https://doi.org/10.1016/j.gca.2012.06.003

Johnson CM, Roden EE, Welch SA, Beard BL (2005) Experimental constraints on Fe isotope fractionation during magnetite and Fe carbonate formation coupled to dissimilatory hydrous ferric oxide reduction. Geochim Cosmochim Acta 69(4):963–993. https://doi.org/10.1016/j.gca.2004.06.043

Johnson CM, Skulan JL, Beard BL, Sun H, Nealson KH, Braterman PS (2002) Isotopic fractionation between Fe(III) and Fe(II) in aqeuous solutions. Earth Planet Sci Lett 195:141–153

Johnson KS, Chavez FP, Friedrich GE (1999) Continental-shelf sediment as a primary source of iron for coastal phytoplankton. Nature 398

Johnson KS, Gordon MR, Coale KH (1997) What controls dissolved iron concentrations in the world ocean? Mar Chem 57:137–161

Jones DL (1998) Organic acids in the rhizosphere—a critical review. Plant Soil 205:25–44

Jørgensen BB (1982) Mineralization of organic matter in the sea bed—the role of sulphate reduction. Nature 296:643–645

Kiczka M, Wiederhold JG, Frommer J, Kraemer SM, Bourdon B, Kretzschmar R (2010) Iron isotope fractionation during proton- and ligand-promoted dissolution of primary phyllosilicates. Geochim Cosmochim Acta 74(11):3112–3128. https://doi.org/10.1016/j.gca.2010.02.018

Klar JK, Homoky WB, Statham PJ, Birchill AJ, Harris EL, Woodward EMS, Silburn B, Cooper MJ, James RH, Connelly DP, Chever F, Lichtschlag A, Graves C (2017a) Stability of dissolved and soluble Fe(II) in shelf sediment pore waters and release to an oxic water column. Biogeochemistry 135(1–2):49–67. https://doi.org/10.1007/s10533-017-0309-x

Klar JK, James RH, Gibbs D, Lough A, Parkinson I, Milton JA, Hawkes JA, Connelly DP (2017b) Isotopic signature of dissolved iron delivered to the Southern Ocean from hydrothermal vents in the East Scotia Sea. Geology 45(4):351–354. https://doi.org/10.1130/g38432.1

Klevenz V, Bach W, Schmidt K, Hentscher M, Koschinsky A, Petersen S (2011) Geochemistry of vent fluid particles formed during initial hydrothermal fluid-seawater mixing along the Mid-Atlantic Ridge. Geochem Geophys Geosyst 12(10):n/a-n/a. https://doi.org/10.1029/2011gc003704

Konovalov SK, Murray JW, Luther GW, Tebo BM (2006) Processes controlling the redox budget for the oxic/anoxic water column of the Black Sea. Deep Sea Res Part II 53(17–19):1817–1841. https://doi.org/10.1016/j.dsr2.2006.03.013

Kurisu M, Takahashi Y, Uematsu M (2016) Very low isotope ratio of iron in fine aerosol related to its contribution to the surface ocean. J Geophys Res: Atmos 121(11). https://doi.org/10.1002/2016jd024957

Labatut M, Lacan F, Pradoux C, Chmeleff J, Radic A, Murray JW, Poitrasson F, Johansen AM, Thil F (2014) Iron sources and dissolved-particulate interactions in the seawater of the Western Equatorial Pacific, iron isotopes perspectives. Glob Biogeochem Cycles 28:1044–1065. https://doi.org/10.1002/2014GB004928

Lam PJ, Lee J-M, Heller MI, Mehic S, Xiang Y, Bates NR (2018) Size-fractionated distributions of suspended particle concentration and major phase composition from the US GEOTRACES Eastern Pacific Zonal Transect (GP16). Mar Chem 201:90–107. https://doi.org/10.1016/j.marchem.2017.08.013

Lam PJ, Ohnemus DC, Auro ME (2015) Size-fractionated major particle composition and concentrations from the US GEOTRACES North Atlantic Zonal Transect. Deep Sea Res Part II 116:303–320. https://doi.org/10.1016/j.dsr2.2014.11.020

Lee J-M, Heller MI, Lam PJ (2018) Size distribution of particulate trace elements in the US GEOTRACES Eastern Pacific Zonal Transect (GP16). Mar Chem 201:108–123. https://doi.org/10.1016/j.marchem.2017.09.006

Lenstra WK, Hermans M, Séguret MJM, Witbaard R, Behrends T, Dijkstra N, van Helmond NAGM, Kraal P, Laan P, Rijkenberg MJA, Severmann S, Teacă A, Slomp CP (2019) The shelf-to-basin iron shuttle in the Black Sea revisited. Chem Geol 511:314–341. https://doi.org/10.1016/j.chemgeo.2018.10.024

Leslie BW, Hammond DE, Berelson WM, Lund SP (1990) Diagenesis in anoxic sediments from the California continental borderland and its influence on iron, sulfur, and magnetite behavior. J Geophys Res 95(B4). https://doi.org/10.1029/jb095ib04p04453

Lewis BL, Landing WM (1991) The biogeochemistry of manganese and iron in the Black Sea. Deep Sea Res Part A Oceanogr Res Pap 38:S773–S803. https://doi.org/10.1016/s0198-0149(10)80009-3

Li M, He Y-S, Kang J-T, Yang X-Y, He Z-W, Yu H-M, Huang F (2017) Why was iron lost without significant isotope fractionation during the lateritic process in tropical environments? Geoderma 290:1–9. https://doi.org/10.1016/j.geoderma.2016.12.003

Lin Z, Sun X, Lu Y, Strauss H, Xu L, Gong J, Teichert BMA, Lu R, Lu H, Sun W, Peckmann J (2017) The enrichment of heavy iron isotopes in authigenic pyrite as a possible indicator of sulfate-driven anaerobic oxidation of methane: Insights from the South China Sea. Chem Geol 449:15–29. https://doi.org/10.1016/j.chemgeo.2016.11.032

Liu K, Wu L, Couture R-M, Li W, Van Cappellen P (2015) Iron isotope fractionation in sediments of an oligotrophic freshwater lake. Earth Planet Sci Lett 423:164–172. https://doi.org/10.1016/j.epsl.2015.05.010

Liu S-A, Teng F-Z, Li S, Wei G-J, Ma J-L, Li D (2014) Copper and iron isotope fractionation during weathering and pedogenesis: insights from saprolite profiles. Geochim Cosmochim Acta 146:59–75. https://doi.org/10.1016/j.gca.2014.09.040

Lough AJM, Klar JK, Homoky WB, Comer-Warner SA, Milton JA, Connelly DP, James RH, Mills RA (2017) Opposing authigenic controls on the isotopic signature of dissolved iron in hydrothermal plumes. Geochim Cosmochim Acta 202:1–20. https://doi.org/10.1016/j.gca.2016.12.022

Lovley DR, Holmes DE, Nevin KP (2004) Dissimilatory Fe(III) and Mn(IV) Reduction. In. Advances in microbial physiology, pp 219–286. https://doi.org/10.1016/s0065-2911(04)49005-5

Lowell RP, Farough A, Hoover J, Cummings K (2013) Characteristics of magma-driven hydrothermal systems at oceanic spreading centers. Geochem Geophys Geosyst 14(6):1756–1770. https://doi.org/10.1002/ggge.20109

Luther GW (1991) Pyrite synthesis via polysulfide compounds. Geochim Cosmochim Acta 55:2839–2849

Luther GW, Church TM, Powell D (1991) Sulfur speciation and sulfide oxidation in the water column of the Black Sea. Deep Sea Res Part A Oceanogr Res Pap 38:S1121–S1137. https://doi.org/10.1016/s0198-0149(10)80027-5

Lyons TW, Severmann S (2006) A critical look at iron paleoredox proxies: new insights from modern euxinic marine basins. Geochim Cosmochim Acta 70(23):5698–5722. https://doi.org/10.1016/j.gca.2006.08.021

Majestic BJ, Anbar AD, Herckes P (2009) Stable isotopes as a tool to apportion atmospheric iron. Environ Sci Technol 43:4327–4333

Mansfeldt T, Schuth S, Häusler W, Wagner FE, Kaufhold S, Overesch M (2011) Iron oxide mineralogy and stable iron isotope composition in a Gleysol with petrogleyic properties. J Soils Sediments 12(1):97–114. https://doi.org/10.1007/s11368-011-0402-z

Marsay CM, Lam PJ, Heller MI, Lee J-M, John SG (2018) Distribution and isotopic signature of ligand-leachable particulate iron along the GEOTRACES GP16 East Pacific Zonal Transect. Mar Chem 201:198–211. https://doi.org/10.1016/j.marchem.2017.07.003

Martin JH (1990) Glacial-Interglacial CO2 Change: the iron hypothesis. Paleoceanography 5(1):1–13

Martin JH, Fitzwater SE (1988) Iron deficiency limits phytoplankton growth in the north-east Pacific subarctic. Nature 331:341–343

Martin JH, Gordon RM (1988) Northeast Pacific iron distributions in relation to phytoplankton productivity. Deep Sea Res 35(2):177–196

McManus J, Berelson WM, Coale KH, Johnson KS, Tammy E (1997) Phosphorus regeneration in continental margin sediments. Geochim Cosmochim Acta 61(14):2891–2907

Mead C, Herckes P, Majestic BJ, Anbar AD (2013) Source apportionment of aerosol iron in the marine environment using iron isotope analysis. Geophys Res Lett 40(21):5722–5727. https://doi.org/10.1002/2013gl057713

Michard G, Viollier E, Jézéquel D, Sarazin G (1994) Geochemical study of a crater lake: Pavin Lake, France—Identification, location and quantification of the chemical reactions in the lake. Chem Geol 115:103–115

Mikucki JA, Pearson A, Johnston DT, Turchyn AV, Farquhar J, Schrag DP, Anbar AD, Priscu JC, Lee PA (2009) A contemporary microbially maintained subglacial ferrous "Ocean". Science 324

Millero FJ, Sotolongo S, Izaguirre M (1987) The oxidation kinetics of Fe(II) in seawater. Geochim Cosmochim Acta 51:793–801

Milliman JD, Syvitski JPM (1992) Geomorphic/Tectonic control of sediment discharge to the ocean: the importance of small mountainous rivers. J Geol 100:525–544

Milne A, Schlosser C, Wake BD, Achterberg EP, Chance R, Baker AR, Forryan A, Lohan MC (2017) Particulate phases are key in controlling dissolved iron concentrations in the (sub)tropical North Atlantic. Geophys Res Lett 44:2377–2387

Moffett JW, German CR (2018) The US GEOTRACES Eastern Tropical Pacific transect (GP16). Mar Chem 201:1–5. https://doi.org/10.1016/j.marchem.2017.12.001

Moon S, Chamberlain CP, Hilley GE (2014) New estimates of silicate weathering rates and their uncertainties in global rivers. Geochim Cosmochim Acta 134:257–274. https://doi.org/10.1016/j.gca.2014.02.033

Moore JK, Braucher O (2008) Sedimentary and mineral dust sources of dissolved iron to the world ocean. Biogeosciences 5:631–656

Moore WS (1999) The subterranean estuary: a reaction zone of ground water and sea water. Mar Chem 65:111–125

Morgan JLL, Wasylenki LE, Nuester J, Anbar AD (2010) Fe isotope fractionation during equilibration of Fe-Organic complexes. Environ Sci Technol 44:6095–6101

Morse J, Millero F, Cornwell J, Rickard D (1987) The chemistry of the hydrogen sulfide and iron sulfide systems in natural waters. Earth-Sci Rev 24:1-42. https://doi.org/10.1016/0012-8252(87)90046-8

Muramoto JA, Honjo S, Fry B, Hay BJ, Howarth RW, Cisne JL (1991) Sulfur, iron and organic carbon fluxes in the Black Sea: sulfur isotopic evidence for origin of sulfur fluxes. Deep Sea Res Part A Oceanogr Res Pap 38:S1151–S1187. https://doi.org/10.1016/s0198-0149(10)80029-9

Murray JW, Jannasch HW, Honjo S, Anderson RF, Reeburgh WS, Top Z, Friederich GE, Codispoti LA, Izdar E (1989) Unexpected changes in the oxic/anoxic interface in the Black Sea. Nature 338

Nasemann P, Gault-Ringold M, Stirling CH, Koschinsky A, Sander SG (2018) Processes affecting the isotopic composition of dissolved iron in hydrothermal plumes: a case study from the Vanuatu back-arc. Chem Geol 476:70–84. https://doi.org/10.1016/j.chemgeo.2017.11.005

Noffke A, Hensen C, Sommer S, Scholz F, Bohlen L, Mosch T, Graco M, Wallmann K (2012) Benthic iron and phosphorus fluxes across the Peruvian oxygen minimum zone. Limnol Oceanogr 57(3):851–867. https://doi.org/10.4319/lo.2012.57.3.0851

Ohnemus DC, Lam PJ (2015) Cycling of lithogenic marine particles in the US GEOTRACES North Atlantic transect. Deep Sea Res Part II 116:283–302. https://doi.org/10.1016/j.dsr2.2014.11.019

Peng X, Guo Z, Chen S, Sun Z, Xu H, Ta K, Zhang J, Zhang L, Li J, Du M (2017) Formation of carbonate pipes in the northern Okinawa Trough linked to strong sulfate exhaustion and iron supply. Geochim Cosmochim Acta 205:1–13. https://doi.org/10.1016/j.gca.2017.02.010

Peters BD, Jenkins WJ, Swift JH, German CR, Moffett JW, Cutter GA, Brzezinski MA, Casciotti KL (2018) Water mass analysis of the 2013 US GEOTRACES eastern Pacific zonal transect (GP16). Mar Chem 201:6–19. https://doi.org/10.1016/j.marchem.2017.09.007

Poitrasson F, Cruz Vieira L, Seyler P, dos Santos Márcia, Pinheiro G, Santos Mulholland D, Bonnet M-P, Martinez J-M, Alcantara Lima B, Resende Boaventura G, Chmeleff J, Dantas EL, Guyot J-L, Mancini L, Martins Pimentel M, Ventura Santos R, Sondag F, Vauchel P (2014) Iron isotope composition of the bulk waters and sediments from the Amazon river Basin. Chem Geol 377:1–11. https://doi.org/10.1016/j.chemgeo.2014.03.019

Poitrasson F, Viers J, Martin F, Braun J-J (2008) Limited iron isotope variations in recent lateritic soils from Nsimi, Cameroon: implications for the global Fe geochemical cycle. Chem Geol 253(1–2):54–63. https://doi.org/10.1016/j.chemgeo.2008.04.011

Poulton SW, Raiswell R (2005) Chemical and physical characteristics of iron oxides in riverine and glacial meltwater sediments. Chem Geol 218(3–4):203–221. https://doi.org/10.1016/j.chemgeo.2005.01.007

Proemse BC, Murray AE, Schallenberg C, McKiernan B, Glazer BT, Young SA, Ostrom NE, Bowie AR, Wieser ME, Kenig F, Doran PT, Edwards R (2017) Iron cycling in the anoxic cryo-ecosystem of Antarctic Lake Vida. Biogeochemistry 134(1–2):17–27. https://doi.org/10.1007/s10533-017-0346-5

Pyzik AJ, Sommer SE (1981) Sedimentary iron monosulfides: kinetics and mechanism of formation. Geochim Cosmochim Acta 45:687–698

Radic A, Lacan F, Murray JW (2011) Iron isotopes in the seawater of the equatorial Pacific Ocean: new constraints for the oceanic iron cycle. Earth Planet Sci Lett 306(1–2):1–10. https://doi.org/10.1016/j.epsl.2011.03.015

Raiswell R, Berner RA (1985) Pyrite formation in euxinic and semi-euxinic sediments. Am J Sci 285:710–724

Rauschenberg S, Twining BS (2015) Evaluation of approaches to estimate biogenic particulate trace metals in the ocean. Mar Chem 171:67–77. https://doi.org/10.1016/j.marchem.2015.01.004

Resing JA, Sedwick PN, German CR, Jenkins WJ, Moffett JW, Sohst BM, Tagliabue A (2015) Basin-scale transport of hydrothermal dissolved metals across the South Pacific Ocean. Nature 523(7559):200–203. https://doi.org/10.1038/nature14577

Revels BN, Ohnemus DC, Lam PJ, Conway TM, John SG (2015a) The isotopic signature and distribution of particulate iron in the North Atlantic Ocean. Deep Sea Res Part II 116:321–331. https://doi.org/10.1016/j.dsr2.2014.12.004

Revels BN, Zhang R, Adkins JF, John SG (2015b) Fractionation of iron isotopes during leaching of natural particles by acidic and circumneutral leaches and development of an optimal leach for marine particulate iron isotopes. Geochim Cosmochim Acta 166:92–104. https://doi.org/10.1016/j.gca.2015.05.034

Rickard DT (1974) Kinetics and mechanism of the sulfidation of geothite. Am J Sci 274:941–952

Rickard D, Schoonen M, Luther G (1995) Chemistry of iron sulfides in sedimentary environments. Geochem Trans Sediment Sulfur 612. https://doi.org/10.1021/bk-1995-0612.ch009

Rolison JM, Stirling CH, Middag R, Gault-Ringold M, George E, Rijkenberg MJA (2018) Iron isotope fractionation during pyrite formation in a sulfidic Precambrian ocean analogue. Earth Planet Sci Lett 488:1–13. https://doi.org/10.1016/j.epsl.2018.02.006

Rouxel O, Shanksiii W, Bach W, Edwards K (2008a) Integrated Fe- and S-isotope study of seafloor hydrothermal vents at East Pacific Rise 9–10°N. Chem Geol 252(3–4):214–227. https://doi.org/10.1016/j.chemgeo.2008.03.009

Rouxel O, Sholkovitz E, Charette M, Edwards KJ (2008b) Iron isotope fractionation in subterranean estuaries. Geochim Cosmochim Acta 72(14):3413–3430. https://doi.org/10.1016/j.gca.2008.05.001

Rouxel O, Toner B, Germain Y, Glazer B (2018) Geochemical and iron isotopic insights into hydrothermal iron oxyhydroxide deposit formation at Loihi Seamount. Geochim Cosmochim Acta 220:449–482. https://doi.org/10.1016/j.gca.2017.09.050

Rouxel O, Toner BM, Manganini SJ, German CR (2016) Geochemistry and iron isotope systematics of hydrothermal plume fall-out at East Pacific Rise 9° 50′N. Chem Geol 441:212–234. https://doi.org/10.1016/j.chemgeo.2016.08.027

Rouxel OJ, Auro M (2010) Iron isotope variations in coastal seawater determined by multicollector ICP-MS. Geostand Geoanal Res 34:135–144

Rouxel OJ, Bekker A, Edwards KJ (2005) Iron isotope constraints on the archean and paleoproterozoic ocean redox state. Science 307:1088–1091

Roy M, Rouxel O, Martin JB, Cable JE (2012) Iron isotope fractionation in a sulfide-bearing subterranean estuary and its potential influence on oceanic Fe isotope flux. Chem Geol 300–301:133–142. https://doi.org/10.1016/j.chemgeo.2012.01.022

Rudnicki MD, Elderfield H (1993) A chemical model of the buoyant and neutrally buoyant plume above the TAG vent field, 26°N, Mid-Atlantic Ridge. Geochim Cosmochim Acta 57:2939–2957

Rue EL, Bruland KW (1995) Complexation of iron(III) by natural organic ligands in the Central North Pacific as determine by a new competitive ligand equilibration/adsorptive cathodic stripping voltammetric method. Mar Chem 50:117–138

Saito MA, Noble AE, Tagliabue A, Goepfert TJ, Lamborg CH, Jenkins WJ (2013) Slow-spreading submarine ridges in the South Atlantic as a significant oceanic iron source. Nat Geosci 6(9):775–779. https://doi.org/10.1038/ngeo1893

Sander SG, Koschinsky A (2011) Metal flux from hydrothermal vents increased by organic complexation. Nat Geosci 4(3):145–150. https://doi.org/10.1038/ngeo1088

Schettler G, Schwab MJ, Stebich M (2007) A 700-year record of climate change based on geochemical and palynological data from varved sediments (Lac Pavin, France). Chem Geol 240(1–2):11–35. https://doi.org/10.1016/j.chemgeo.2007.01.003

Scholz F, Hensen C, Noffke A, Rohde A, Liebetrau V, Wallmann K (2011) Early diagenesis of redox-sensitive trace metals in the Peru upwelling area—response to ENSO-related oxygen fluctuations in the water column. Geochim Cosmochim Acta 75(22):7257–7276. https://doi.org/10.1016/j.gca.2011.08.007

Scholz F, Löscher CR, Fiskal A, Sommer S, Hensen C, Lomnitz U, Wuttig K, Göttlicher J, Kossel E, Steininger R, Canfield DE (2016) Nitrate-dependent iron oxidation limits iron transport in anoxic ocean regions. Earth Planet Sci Lett 454:272–281. https://doi.org/10.1016/j.epsl.2016.09.025

Scholz F, Severmann S, McManus J, Hensen C (2014a) Beyond the Black Sea paradigm: The sedimentary fingerprint of an open-marine iron shuttle. Geochim Cosmochim Acta 127:368–380. https://doi.org/10.1016/j.gca.2013.11.041

Scholz F, Severmann S, McManus J, Noffke A, Lomnitz U, Hensen C (2014b) On the isotope composition of reactive iron in marine sediments: redox shuttle versus early diagenesis. Chem Geol 389:48–59. https://doi.org/10.1016/j.chemgeo.2014.09.009

Schoonen M (2004) Mechanisms of sedimentary pyrite formation. https://doi.org/10.1130/0-8137-2379-5.117

Schroth AW, Crusius J, Chever F, Bostick BC, Rouxel OJ (2011) Glacial influence on the geochemistry of riverine iron fluxes to the Gulf of Alaska and effects of deglaciation. Geophys Res Lett 38(16):n/a-n/a. https://doi.org/10.1029/2011gl048367

Schuessler JA, Kämpf H, Koch U, Alawi M (2016) Earthquake impact on iron isotope signatures recorded in mineral spring water. J Geophys Res: Solid Earth 121 (12):8548–8568. https://doi.org/10.1002/2016jb013408

Schuth S, Hurraß J, Münker C, Mansfeldt T (2015) Redox-dependent fractionation of iron isotopes in suspensions of a groundwater-influenced soil. Chem Geol 392:74–86. https://doi.org/10.1016/j.chemgeo.2014.11.007

Sedwick PN, Sohst BM, Ussher SJ, Bowie AR (2015) A zonal picture of the water column distribution of dissolved iron(II) during the US. GEOTRACES North Atlantic transect cruise (GEOTRACES GA03). Deep Sea Res Part II 116:166–175. https://doi.org/10.1016/j.dsr2.2014.11.004

Severmann S, Johnson CM, Beard BL, German CR, Edmonds HN, Chiba H, Green DRH (2004) The effect of plume processes on the Fe isotope composition of hydrothermally derived Fe in the deep ocean as inferred from the Rainbow vent site, Mid-Atlantic Ridge, 36°14′N. Earth Planet Sci Lett 225(1–2):63–76. https://doi.org/10.1016/j.epsl.2004.06.001

Severmann S, Johnson CM, Beard BL, McManus J (2006) The effect of early diagenesis on the Fe isotope compositions of porewaters and authigenic minerals in continental margin sediments. Geochim Cosmochim Acta 70(8):2006–2022. https://doi.org/10.1016/j.gca.2006.01.007

Severmann S, Lyons TW, Anbar A, McManus J, Gordon G (2008) Modern iron isotope perspective on the benthic iron shuttle and the redox evolution of ancient oceans. Geology 36(6). https://doi.org/10.1130/g24670a.1

Severmann S, McManus J, Berelson WM, Hammond DE (2010) The continental shelf benthic iron flux and its isotope composition. Geochim Cosmochim Acta 74 (14):3984–4004. https://doi.org/10.1016/j.gca.2010.04.022

Shelley RU, Morton PL, Landing WM (2015) Elemental ratios and enrichment factors in aerosols from the US-GEOTRACES North Atlantic transects. Deep Sea Res Part II 116:262–272. https://doi.org/10.1016/j.dsr2.2014.12.005

Sherrell RM, Boyle EA (1992) The trace metal composition of suspended particles in the oceanic water column near Bermuda. Earth Planet Sci Lett 111:155–174

Sivan O, Adler M, Pearson A, Gelman F, Bar-Or I, John SG, Eckert W (2011) Geochemical evidence for iron-mediated anaerobic oxidation of methane. Limnol Oceanogr 56(4):1536–1544. https://doi.org/10.4319/lo.2011.56.4.1536

Skulan JL, Beard BL, Johnson CM (2002) Kinetic and equilibrium Fe isotope fractionation between aqueous Fe(III) and hematite. Geochim Cosmochim Acta 66 (17):2995–3015

Song L, Liu C-Q, Wang Z-L, Zhu X, Teng Y, Liang L, Tang S, Li J (2011) Iron isotope fractionation during biogeochemical cycle: information from suspended particulate matter (SPM) in Aha Lake and its tributaries, Guizhou China. Chemical Geology 280(1–2):170–179. https://doi.org/10.1016/j.chemgeo.2010.11.006

Sparks DL (2003) Enviromental soil chemistry. Academic Press. https://doi.org/10.1016/b978-0-12-656446-4.x5000-2

Statham P, German C, Connelly D (2005) Iron (II) distribution and oxidation kinetics in hydrothermal plumes at the Kairei and Edmond vent sites, Indian Ocean. Earth Planet Sci Lett 236(3–4):588–596. https://doi.org/10.1016/j.epsl.2005.03.008

Staubwasser M, Schoenberg R, von Blanckenburg F, Krüger S, Pohl C (2013) Isotope fractionation between dissolved and suspended particulate Fe in the oxic and anoxic water column of the Baltic Sea. Biogeosciences 10(1):233–245. https://doi.org/10.5194/bg-10-233-2013

Straub KL, Benz M, Schink B, Widdel F (1996) Anaerobic, nitrate-dependent microbial oxidation of ferrous iron. Appl Environ Microbiol 1458–1460 (1996)

Stucki JW, Goodman BA, Schwertmann U (1988) Iron in soils and clay minerals. Nato science series C, vol 217, 1st edn. Springer, Netherlands. https://doi.org/10.1007/978-94-009-4007-9

Syverson DD, Borrok DM, Seyfried WE (2013) Experimental determination of equilibrium Fe isotopic fractionation between pyrite and dissolved Fe under hydrothermal conditions. Geochim Cosmochim Acta 122:170–183. https://doi.org/10.1016/j.gca.2013.08.027

Tang D, Morel FMM (2006) Distinguishing between cellular and Fe-oxide-associated trace elements in phytoplankton. Mar Chem 98(1):18–30. https://doi.org/10.1016/j.marchem.2005.06.003

Taylor P, Rummery TE, Owen DG (1979) Reactions of iron monosulfide solids with aqueous hydrogen sulfide up to 160 °C. J Inorg Nucl Chem 41:1683–1687

Taylor SR, McLennan SM (1985) The continental crust: its composition and evolution. Blackwell Scientific Publications

Teutsch N, Schmid M, Müller B, Halliday AN, Bürgmann H, Wehrli B (2009) Large iron isotope fractionation at the oxic–anoxic boundary in Lake Nyos. Earth Planet Sci Lett 285(1–2):52–60. https://doi.org/10.1016/j.epsl.2009.05.044

Teutsch N, von Gunten U, Porcelli D, Cirpka OA, Halliday AN (2005) Adsorption as a cause for iron isotope fractionation in reduced groundwater. Geochim Cosmochim Acta 69(17):4175–4185. https://doi.org/10.1016/j.gca.2005.04.007

Thamdrup B, Fossing H, Jørgensen BB (1994) Manganese, iron, and sulfur cycling in a coastal marine sediment, Aarhus Bay Denmark. Geochimica et Cosmochimica Acta 58(23):5115–5129

Thompson A, Chadwick OA, Rancourt DG, Chorover J (2006) Iron-oxide crystallinity increases during soil redox oscillations. Geochim Cosmochim Acta 70 (7):1710–1727. https://doi.org/10.1016/j.gca.2005.12.005

Thompson A, Ruiz J, Chadwick OA, Titus M, Chorover J (2007) Rayleigh fractionation of iron isotopes during pedogenesis along a climate sequence of Hawaiian basalt. Chem Geol 238(1–2):72–83. https://doi.org/10.1016/j.chemgeo.2006.11.005

Tiodjio RE, Sakatoku A, Nakamura A, Tanaka D, Fantong WY, Tchakam KB, Tanyileke G, Ohba T, Hell VJ, Kusakabe M, Nakamura S, Ueda A (2014) Bacterial and archaeal communities in Lake Nyos (Cameroon, Central Africa). Sci Rep 4:6151. https://doi.org/10.1038/srep06151

Toner BM, Fakra SC, Manganini SJ, Santelli CM, Marcus MA, Moffett JW, Rouxel O, German CR, Edwards KJ (2009) Preservation of iron(II) by carbon-rich matrices in a hydrothermal plume. Nat Geosci 2(3):197–201. https://doi.org/10.1038/ngeo433

Tovar-Sanchez A, Sañudo-Wilhelmy SA, Garcia-Vargas M, Weaver RS, Popels LC, Hutchins DA (2003) A trace metal clean reagent to remove surface-bound iron from marine phytoplankton. Mar Chem 82(1–2):91–99. https://doi.org/10.1016/s0304-4203(03)00054-9

Weil RR, Brady NC (2017) The nature and properties of soils, 15th edn. Pearson

Welch SA, Beard BL, Johnson CM, Braterman PS (2003) Kinetic and equilibrium Fe isotope fractionation between aqueous Fe(II) and Fe(III). Geochim Cosmochim Acta 67(22):4231–4250. https://doi.org/10.1016/s0016-7037(03)00266-7

Wetz MS, Hales B, Chase Z, Wheeler PA, Whitney MM (2006) Riverine input of macronutrients, iron, and organic matter to the coastal ocean off Oregon, USA, during the winter. Limnol Oceanogr 51 (5):2221–2231

Wiederhold JG, Kraemer SM, Teutsch N, Borer PM, Halliday AN, Kretzschmar R (2006) Iron isotope fractionation during proton-promoted, ligand-controlled, and reductive dissolution of goethite. Environ Sci Technol 40:3787–3793

Wiederhold JG, Teutsch N, Kraemer SM, Halliday AN, Kretzschmar R (2007a) Iron isotope fractionation during pedogenesis in redoximorphic soils. Soil Sci Soc Am J 71(6). https://doi.org/10.2136/sssaj2006.0379

Wiederhold JG, Teutsch N, Kraemer SM, Halliday AN, Kretzschmar R (2007b) Iron isotope fractionation in oxic soils by mineral weathering and podzolization. Geochim Cosmochim Acta 71(23):5821–5833. https://doi.org/10.1016/j.gca.2007.07.023

Wiesli RA, Beard BL, Johnson CM (2004) Experimental determination of Fe isotope fractionation between aqueous Fe(II), siderite and "green rust" in abiotic systems. Chem Geol 211(3–4):343–362. https://doi.org/10.1016/j.chemgeo.2004.07.002

Wijsman JWM, Middelburg JJ, Heip CHR (2001a) Reactive iron in Black Sea Sediments: implications for iron cycling. Mar Geol 172:167–180

Wijsman JWM, Middleburg JJ, Herman PMJ, Bottcher ME, Heip CHR (2001b) Sulfur and iron speciation in surface sediments along the northwestern margin of the Black Sea. Mar Chem 74:261–278

Wilkin RT, Arthur MA, Dean WE (1997) History of water-column anoxia in the Black Sea indicated by pyrite framboid size distribution. Earth Planet Sci Lett 148:517–525

Wilkin RT, Barnes HL (1996) Pyrite formation by reaction of iron monosufides with dissolved inorganic and organic sulfur species. Geochim Cosmochim Acta 60(21):4167–4179

Wilson MJ (2004) Weathering of the primary rock-forming minerals: processes, products and rates. Clay Miner 39(03):233–266. https://doi.org/10.1180/0009855043930133

Wu J, Boyle E, Sunda W, Wen L-S (2001) Soluble and colloidal iron in the oligotrophic North Atlantic and North. Pac Sci 293:847–849

Wu J, Wells ML, Rember R (2011a) Dissolved iron anomaly in the deep tropical–subtropical Pacific: evidence for long-range transport of hydrothermal iron. Geochim Cosmochim Acta 75(2):460–468. https://doi.org/10.1016/j.gca.2010.10.024

Wu L, Beard BL, Roden EE, Johnson CM (2011b) Stable iron isotope fractionation between aqueous Fe(II) and hydrous ferric oxide. Environ Sci Technol 45 (5):1847–1852. https://doi.org/10.1021/es103171x

Wu L, Brucker RP, Beard BL, Roden EE, Johnson CM (2013) Iron isotope characteristics of hot springs at chocolate pots, Yellowstone National Park. Astrobiology 13(11):1091–1101. https://doi.org/10.1089/ast.2013.0996

Wu L, Druschel G, Findlay A, Beard BL, Johnson CM (2012) Experimental determination of iron isotope fractionations among Fe^{2+}_{aq}–FeS_{aq}–mackinawite at low temperatures: Implications for the rock record. Geochim Cosmochim Acta 89:46–61. https://doi.org/10.1016/j.gca.2012.04.047

Yang WH, Weber KA, Silver WL (2012) Nitrogen loss from soil through anaerobic ammonium oxidation coupled to iron reduction. Nat Geosci 5(8):538–541. https://doi.org/10.1038/ngeo1530

Yesavage T, Fantle MS, Vervoort J, Mathur R, Jin L, Liermann LJ, Brantley SL (2012) Fe cycling in the Shale Hills Critical Zone Observatory, Pennsylvania: an analysis of biogeochemical weathering and Fe isotope fractionation. Geochim Cosmochim Acta 99:18–38. https://doi.org/10.1016/j.gca.2012.09.029

Yücel M, Gartman A, Chan CS, Luther GW (2011) Hydrothermal vents as a kinetically stable source of iron-sulphide-bearing nanoparticles to the ocean. Nat Geosci 4(6):367–371. https://doi.org/10.1038/ngeo1148

Zhang R, John SG, Zhang J, Ren J, Wu Y, Zhu Z, Liu S, Zhu X, Marsay CM, Wenger F (2015) Transport and reaction of iron and iron stable isotopes in glacial meltwaters on Svalbard near Kongsfjorden: from rivers to estuary to ocean. Earth Planet Sci Lett 424:201–211. https://doi.org/10.1016/j.epsl.2015.05.031

6 The Ancient Earth

Much of Chap. 5 focused on the fluid envelope of various surface environments in the modern Earth, where Fe redox changes, organic complexation, and mineral precipitation produce large changes in Fe isotope compositions of fluids and minerals. As we step into the ancient Earth, we lose the ability to measure fluids directly, yet the fluid envelope remained a key component of the Fe biogeochemical cycle in Earth's past, especially in an anoxic Earth when large quantities of Fe^{2+}_{aq} existed in the oceans. Our understanding of ancient fluids, therefore, depends on inferences from the rock (mineral) record, which requires application of the large array of mineral-fluid Fe isotope fractionation factors that is reviewed in Chap. 3. In addition, as the record of the past becomes incomplete, with missing pieces of the system, we cannot view Fe isotope compositions in isolation, but must also consider key mass-balance relations and bring a multitude of other geochemical data to the mix, including Fe speciation and other isotopic systems.

In the first part of this chapter, we discuss the Cenozoic Fe isotope record of seawater as recorded in Fe–Mn crusts, extending what we have discussed at the end of Chap. 5. Next, we turn to an example of ancient redox cycling, Cretaceous ocean anoxic events, which show similarities, and differences, in the marine Fe cycle as compared to that discussed in Chap. 5, under conditions of a fully oxygenated atmosphere. The majority of the chapter, however, is devoted to the Precambrian, stepping progressively further back into the rock record. Over 4,000 Fe isotope analyses are available for Precambrian rocks and minerals, reflecting the considerable attention the Precambrian Fe cycle has received. The distinct nature of the Fe biogeochemical cycle in the Precambrian, relative to that of the Phanerozoic, necessitates a broad comparative discussion before we focus on specific localities. In the end, it should become clear that the Fe isotope record for Precambrian rocks documents large changes in the Fe cycle as a function of atmospheric and ocean chemistry, evolution of the biosphere, and the solid Earth.

6.1 The Cenozoic Marine System

Large climatic and tectonic changes have occurred in the Cenozoic, and these are well recorded in the geochemical characteristic of marine sedimentary rocks. This includes potential changes in erosion rates and lithologies, the dust load to the oceans, primary productivity, burial of organic carbon, pyrite genesis, and mid-ocean ridge hydrothermal activity, any of which might be recorded in changes in the Fe isotope compositions of seawater as a function of age and location. We start this section with a

C. Johnson et al., *Iron Geochemistry: An Isotopic Perspective*, Advances in Isotope Geochemistry,
https://doi.org/10.1007/978-3-030-33828-2_6

brief review of climatic and tectonic changes in the Cenozoic before turning our attention to the record of Cenozoic seawater that is contained in Fe–Mn crusts.

6.1.1 Global Changes in the Cenozoic

The literature on Cenozoic climate is enormous, and we focus here on proxies that are expected to be most closely tied to processes that may affect changes in the Fe isotope composition of seawater, including $\delta^{13}C$ and $\delta^{18}O$ in benthic foraminifera, $\delta^{34}S$ values of marine sulfate, $^{87}Sr/^{86}Sr$ ratios of seawater, burial rates of organic carbon (C_{org}), changes in pCO_2 of the atmosphere, and oceanic heat flow (Fig. 6.1). The C and O isotope compositions of benthic foraminifera remain one of the most important means for understanding the relations between temperature, pCO_2, and the presence of continental-scale ice sheets during the Cenozoic (e.g., De Vleeschouwer et al. 2017; Westerhold et al. 2018; Zachos et al. 2001). Species-normalized $\delta^{18}O$ values of benthic foraminifera (e.g., Pearson 2012) record the effects of ice volume and temperature, where deconvolving the two requires an independent proxy for temperature (e.g., Cramer et al. 2011). The lowest $\delta^{18}O$ values occur during warm periods such as the late Paleocene thermal maximum and early Eocene climatic optimum, whereas increases in $\delta^{18}O$ values mark glaciation events, including O isotope event 1 (Oi-1) at the Eocene-Oligocene boundary that marks the initiation of major Antarctic glaciation, the middle-late Miocene development of the east Antarctic ice sheet, and the late Pliocene to Pleistocene initiation of northern hemisphere glaciation (Fig. 6.1). Although most studies assume a single $\delta^{18}O$ curve for the deep oceans, inter-ocean variability for $\delta^{13}C$ values can be significant, reflecting localized C budgets for C_{org} and carbonate (e.g., Cramer et al. 2009). Major changes in the $\delta^{13}C$ values for deep-water marine carbonate (shown in Fig. 6.1 for the Pacific Ocean) occur across the late Paleocene to early Eocene warm periods, as well as the cooling periods of the late Miocene to Pleistocene (Fig. 6.1), reflecting complex interrelations between C_{org} burial, primary productivity, carbonate compensation depth (CCD), and pCO_2 (e.g., Lyle et al. 2008).

The Eocene saw a large change in the $\delta^{34}S$ values of marine sulfate, as recorded in barite or carbonate-associated sulfate (CAS) in benthic foraminifera (Fig. 6.1) that has been interpreted in several ways. Kurtz et al. (2003) suggest that the increase in $\delta^{34}S$ values record an increase in pyrite burial, accompanied by a decrease in C_{org} burial due to respiration via microbial sulfate reduction. In contrast, Wortmann and Paytan (2012) argue against a change in pyrite burial, but infer changes in marine sulfate contents due to increased aerosol loading to the oceans, and later modeling by Algeo et al. (2015) supports a control on $\delta^{34}S$ values by marine sulfate levels. In terms of possible Fe isotope changes, an increase in pyrite burial would tend to shift pore waters in marine sediments to positive $\delta^{56}Fe$ values relative to other processes such as DIR (see Figs. 5.15 and 5.16 in Chap. 5), but changes in seawater sulfate contents, by itself, would not affect $\delta^{56}Fe$ values. Leavitt et al. (2013) emphasized a control by sulfate reduction rates rather than sulfate contents, the former being controlled more by C_{org} deposition rates rather than sulfate abundance. Recently, Rennie et al. (2018) compared the $\delta^{34}S$ values of foraminifera CAS with barite and concluded that the shift in S isotope compositions in the Eocene reflected a change in the location of pyrite burial from the shelf to the deep ocean, where C_{org} would become limiting in the deep ocean, accompanied by a shift of reactive Fe delivery from the shelf to the deep ocean. This proposal would favor a model where pore fluids in deep marine settings would tend to have high-$\delta^{56}Fe$ values (see Sect. 5.4.2 in Chap. 5).

Changes in the isotopic compositions of long-residence time elements such as Sr have been widely used to infer changes in the contributions from oceanic (mantle) and continental sources, and the mid-Cenozoic increase in seawater $^{87}Sr/^{86}Sr$ ratios is classically interpreted to record uplift of the Himalayas and Tibetan Plateau (e.g., Richter et al. 1992), which in turn has

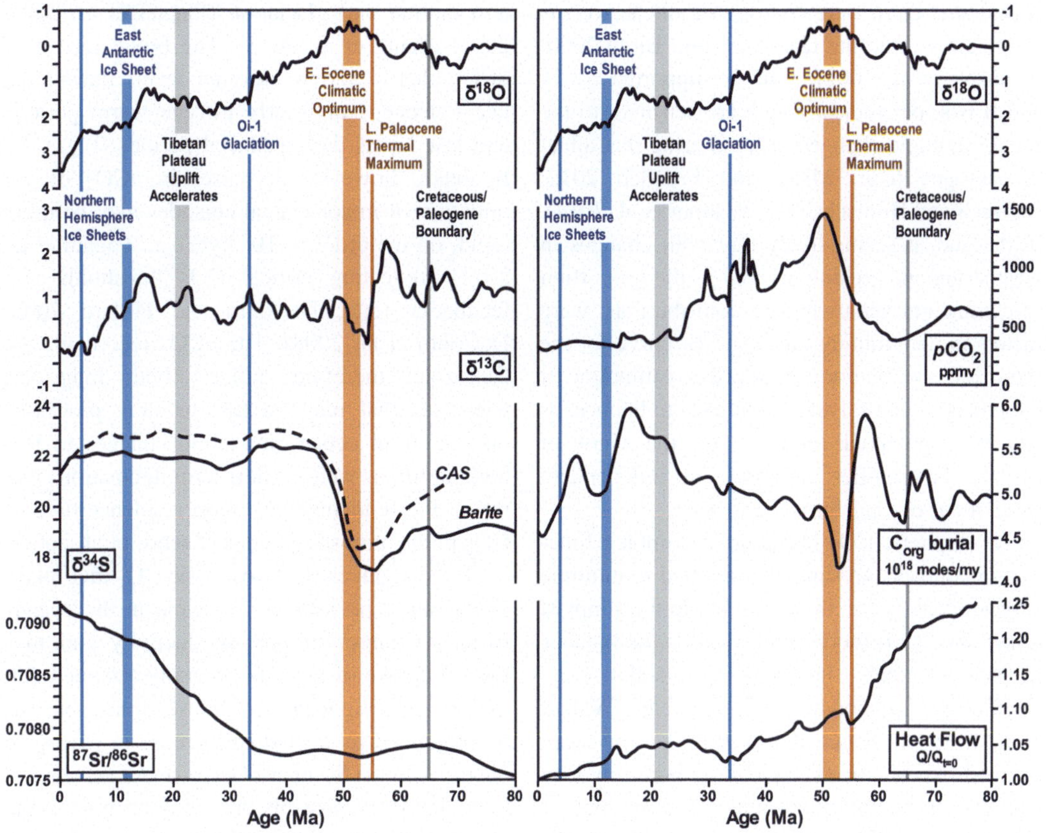

Fig. 6.1 Temporal variations in paleoenvironmental proxies for the latest Cretaceous through Cenozoic. Oxygen isotope ($\delta^{18}O$ on PDB scale, in ‰) curve for benthic foraminifera based on global compilation by Cramer et al. (2009); $\delta^{18}O$ shown for both left and right panels to aid comparison (note the scale convention of increasing $\delta^{18}O$ downward to emphasize down for cooling and up for warming). Carbon isotope ($\delta^{13}C$ on PDB scale, in ‰) curve for marine carbonate from the Pacific Ocean (Cramer et al. 2009). Sulfur isotope ($\delta^{34}S$ on CDT scale, in ‰) curves shown for marine sulfate based on barite (Paytan et al. 1998) and carbonate-associated sulfate (CAS) in benthic foraminifera (Rennie et al. 2018). Marine $^{87}Sr/^{86}Sr$ curve from McArthur et al. (2001). Atmospheric pCO_2 curve from of Zhang et al. (2013) for 0–40 Ma, aligned with curve of Kent and Muttoni (2013) for 40–80 Ma. Curve for burial of organic carbon (C_{org}) for 0–65 Ma from Kurtz et al. (2003), and for 65–80 Ma, extrapolated using same approach based on $\delta^{13}C$ values for marine carbonate. Total oceanic heat flow curve (relative to modern) from Becker et al. (2009), using their "M08" model. Major climatic and tectonic changes for Cenozoic noted as vertical bars (blue for glacial events, brown/orange for warm events), adapted from Zachos et al. (2001). A number of these climatic and biogeochemical changes could affect the Fe cycle of the oceans, which in turn might be detected by changes in Fe isotope compositions of seawater

been correlated to drawdown of atmospheric CO_2 through increased silicate weathering (e.g., Raymo and Ruddiman 1992). Inflections in the $^{87}Sr/^{86}Sr$-time curve in the Cenozoic have been interpreted to reflect increased erosion rates, such as the high slope between ~30 and 15 Ma (Fig. 6.1). If true, such changes might be accompanied by increased reactive Fe and C_{org} fluxes to continental shelves, which could affect processes such as the benthic Fe shuttle that provides high fluxes of negative $\delta^{56}Fe$ Fe to the oceans (see for example, Sects. 5.4.4 and 5.5.3 in Chap. 5). An alternative is that the $^{87}Sr/^{86}Sr$ ratio of the continental (riverine) flux has changed over the Cenozoic, and this possibility finds support in studies of depositional fans adjacent to the Himalaya (Derry and France-Lanord 1996). Changes in the Sr isotope compositions would be expected if there were changes in silicate versus carbonate weathering (Jacobson and Blum 2000; Jacobson et al. 2002). This is supported by Li and B isotope studies, which are sensitive to the same lithologic changes in weathering that affect Sr isotopes (e.g., Misra and Froelich 2012; Raitzsch and Hönisch 2013; Wanner et al. 2014). In the end, it seems likely that both changes in weathering and erosion, as well as the proportion of continental versus mantle contributions, were factors in determining the $^{87}Sr/^{86}Sr$ curve for the Cenozoic oceans. For our purposes, estimation of the potential detrital and C_{org} fluxes to the oceans and their potential influence on the vigor of benthic Fe shuttles, requires an independent assessment of C_{org} burial rates.

Atmospheric CO_2 levels are a complex function of sources and sinks that are tied to multiple geochemical cycles, including Fe. In the simplest terms, the "CO_2 thermostat" records the balance in sources (e.g., igneous) and sinks (e.g., weathering) to moderate climate (e.g., Walker et al. 1981). In detail, this picture is more complicated, where, for example, C_{org} burial may be a sink, or its respiration or mineralization may be a source (e.g., France-Lanord and Derry 1997). A single lithology such as basalt, may be both a source (e.g., igneous CO_2) and a sink, the latter dependent on the compositional effects of weathering on CO_2 drawdown (e.g., Dessert et al. 2003; Li et al. 2016), as well as the paleogeography of weatherable terranes (e.g., Kent and Muttoni 2013). Weathering of ultramafic terranes is a particularly strong potential CO_2 sink (Reusch 2011). Sedimentary sulfides may also play a complex role, where oxidation of pyrite, when coupled to carbonate dissolution, can increase CO_2 (Torres et al. 2014). The balance between sources and sinks is a critical one, given the small inventory of CO_2 in the atmosphere, which would be depleted in only a few m.y. in the absence of adequate sources (e.g., Li and Elderfield 2013).

Alkenone-based pCO_2 reconstructions, although subject to significant uncertainties, have provided a detailed record of atmospheric CO_2 levels for the Cenozoic that suggests correlations between inferred climate changes, including temperature and glaciation (Pagani et al. 2005, 2011; Zhang et al. 2013). The broad record of pCO_2 clearly shows elevated levels during the late Paleocene and early Eocene warm periods and low levels during glacial periods (Fig. 6.1). In detail, however, decreases in pCO_2 during initiation of major glacial episodes may be quite rapid, on the order of 10^2 kyr (e.g., Pagani et al. 2011), reflecting critical CO_2 thresholds and feedbacks (e.g., DeConto and Pollard 2003; DeConto et al. 2008). The pCO_2 record for the Cenozoic, therefore, reflects both long-term processes that may be tied to slow processes ranging from ocean circulation changes to tectonic uplift, as well as short-term fluctuations that reflect the fundamentally short residence time of CO_2 in the atmosphere (e.g., Zachos et al. 2001).

Rapid changes in atmospheric CO_2 that might significantly tie into the Fe cycle in the oceans include changes in paleoproductivity and biological uptake of Fe. It is generally accepted that the Eocene-Oligocene "ice house" transition was accompanied by a marked increase in paleoproductivity in the Southern Ocean and a decrease at lower latitudes, possibly driven by opening of the Drake Passage and attendant changes in nutrient availability (e.g., Moore et al. 2014; Ladant et al. 2018). Such increases are envisioned to have drawn down atmospheric CO_2 while simultaneously thermally isolating the Antarctic continent and hence facilitating development of ice sheets. These changes were accompanied by a marked drop in the CCD, well expressed in the Pacific Ocean, reflecting an increase in particulate scavenging and deepening of the thermocline (Lyle et al. 2008). In addition to changes in nutrient abundance, it has been proposed that an increase in dust delivery to the Southern Ocean from southern hemisphere deserts may have spurred an increase in paleoproductivity (Kohn et al. 2015). The mid-Cenozoic also saw an increase in large volcanic eruptions in the circum-Pacific arcs, an "ignimbrite flare-up", and this has been invoked as a means for increasing paleoproductivity that would have drawn down CO_2 and induced global cooling across the Eocene-Oligocene boundary (Cather et al. 2009; Jicha et al. 2009). Such a model builds on an extensive literature that documents the Fe fertilization potential of volcanic ash that is deposited in seawater (e.g., Frogner et al. 2001; Duggen et al. 2010).

The proportion of C_{org} and inorganic C (e.g., carbonate) exerts a strong control on atmospheric CO_2 and O_2, and this has been explored over geologic time using the $\delta^{13}C$ values of marine carbonate and isotopic mass-balance relations (e.g., Garrels and Lerman 1981). Drawdown of CO_2 via primary productivity may be expressed by the classic photosynthesis reaction:

$$H_2O + CO_2 + h\nu \rightarrow CH_2O + O_2 \quad (6.1)$$

where CH_2O represents an idealized sugar and is analogous to "C_{org}". As discussed in Sect. 5.4 in Chap. 5, oxidation ("mineralization" in some literature) of C_{org} occurs during settling of biogenic particles in the water column via:

$$CH_2O + O_2 \rightarrow H_2O + CO_2 \quad (6.2)$$

as well as via respiration during microbial sulfate and Fe^{3+} reduction, among other pathways:

$$SO_4^{2-} + 2CH_2O \rightarrow H_2S + 2HCO_3^{-} \quad (6.3)$$

$$4Fe(OH)_3 + CH_2O + 8H^+ \rightarrow 4Fe^{2+}{}_{aq} + CO_2 + 11H_2O \quad (6.4)$$

In considering the balance between C_{org} burial and weathering as mechanisms for the decline in atmospheric CO_2 since the mid-Miocene, researchers have pointed to the decline in $\delta^{13}C$ values of marine carbonate (Fig. 6.1) as supporting a decline in C_{org} burial, possibly reflecting tectonically-driven increased silicate weathering (e.g., Raymo and Ruddiman 1992). Recent modeling suggests that only a small increase in silicate weathering is required to explain both the decrease in CO_2 contents and changes in the marine Li, Sr, and Os isotope records (Caves et al. 2016). It is now recognized that simple isotopic mass balance models that assume a constant $^{13}C/^{12}C$ fractionation factor between organic and inorganic C is incorrect (e.g., Derry and France-Lanord 1996; Kump and Arthur 1999). This has produced a C_{org} burial curve for the Cenozoic that is more complicated than a direct scaling to $\delta^{13}C$ values (Fig. 6.1).

Consideration of both C and S isotope records demonstrates changing C_{org} and pyrite burial over the Cenozoic, with large changes through the late Paleocene to early Eocene warm periods (Fig. 6.1), and this lead Kurtz et al. (2003) to suggest very high C_{org} burial relative to that of pyrite during the Paleocene peak in $\delta^{13}C$ values, possibly indicating extensive C_{org} burial in terrestrial environments. This was followed by a strong decrease in the ratio of C_{org}/pyrite burial across the Paleocene-Eocene boundary, suggesting extensive pyrite burial in transient euxinic environments. C_{org} burial broadly increased from the warm periods of the Eocene-Paleocene to the Miocene, accelerating during increased rates of the Tibetan Plateau uplift, followed by a decrease since the mid-Miocene. These changes would be expected to produce a wide range in $\delta^{56}Fe$ values of the oceans, where, for example, increasing C_{org} burial in the middle Cenozoic, coupled to increasing weathering and erosion, might increase the extent and vigor of benthic Fe shuttles near continental margins. The large changes in C_{org} and pyrite burial through the Paleocene-Eocene warm periods might be expected to produce wide ranges in $\delta^{56}Fe$ values of seawater, given the opposite isotopic effects of pyrite formation and DIR (see also Sect. 5.4 in Chap. 5).

Finally, we consider potential changes in marine hydrothermal activity, given the clear signal of hydrothermal Fe to the oceans (see Sect. 5.5.4 in Chap. 5). A number of studies have concluded that the rate of plate creation has not varied significantly since the mid-Mesozoic (e.g., Rowley 2002), suggesting that marine hydrothermal activity has been roughly constant, a conclusion that finds support in geochemical data for foraminifera (e.g., Delaney and Boyle 1986 EPSL). Such results have figured prominently in models of atmospheric CO_2 evolution, where igneous CO_2 sources have been taken to be constant over the Cenozoic. Plate creation rates have been re-investigated, however, additionally considering lost (subducted) volumes of oceanic crust, and this approach highlights a non-constant plate creation rate over the Cenozoic (e.g., Conrad and Lithgow-Bertelloni 2007). In Fig. 6.1 we use the total oceanic heat flow model of Becker et al. (2009), which shows a steady decrease from the late Cretaceous until the Paleocene-Eocene warm period, followed by a gentle decline to current values since then. Interestingly, the periods of high heat flow in the early Cenozoic do not correlate with periods of high pCO_2, highlighting the importance of considering the full picture of sources and sinks. If we consider the Fe isotope signals of hydrothermal fluids, where hydrothermal dFe can have very low δ^{56}Fe values (see Fig. 5.26 in Chap. 5), we might expect such signals to be more common in seawater from the Cretaceous through the Paleocene-Eocene warm period, but less common afterwards. Such signals are likely to be localized, however, given the rapid changes in δ^{56}Fe values for dissolved Fe that occur with distance from the vent (see Fig. 5.27 in Chap. 5).

6.1.2 Fe–Mn Crusts as Archives of Paleo-Seawater Compositions

For metals, Fe–Mn crusts provide the most complete record of changes in seawater chemistry and isotopic compositions over geologic time (e.g., Frank 2002). Iron and Mn in crusts reflect scavenging from the water column as colloids, precipitating at very slow rates of $\sim$1–15 mm/m.y. In terms of Fe, we will assume that hydrogenous Fe–Mn crusts sample dFe from seawater (see Sect. 5.5.1 in Chap. 5), given the very low clastic components in hydrogenous crusts. The temporal record found in Fe–Mn crusts is dependent on the residence time of specific elements in seawater and their responses to changes in relative fluxes. For radiogenic isotopes, the residence times in seawater vary greatly: Pb (10^1–10^2 yrs), Nd (10^2–10^3 yrs), Hf (10^3 yrs), Os (10^4 yrs), and Sr (10^6 yrs). Relative to Fe, Pb is closest in its residence time. Only for Os and Sr are the isotopic compositions homogeneous for the modern ocean, indicating that Os and Sr isotopes are responsive largely to long-term changes in sources, whereas Pb, Nd, and Hf are expected to show isotopic provinciality in the oceans at any given time. The efficiency of particle scavenging near hydrothermal vents for REEs, Hf, and Pb lead to minimal hydrothermal influence far from marine hydrothermal systems, indicating that Nd, Hf, and Pb isotope compositions of seawater will be most sensitive to crustal inputs. Similarly, Os isotope variations in seawater record a major component of continental material, with the additional complexity that the high Os abundances of meteoritic material makes Os isotopes also sensitive to extraterrestrial particle fluxes. In the open ocean, far from continental margins, continental input occurs primarily as dust and aerosols.

Comparison of the Nd and Pb isotope compositions of Fe–Mn crusts from the Atlantic, Pacific, and Indian oceans shows that the isotopic provinciality of these oceans has been maintained for at least the past 20 m.y. (O'Nions et al. 1998; Frank et al. 1999). Large temporal changes in Nd and Pb isotopes to less radiogenic and more radiogenic compositions, respectively, in the last 3 m.y. in the North Atlantic Ocean record an increasing influence of Archean terranes in Canada and Greenland upon closure of the Panama gateway, as well as changes in erosion and sediment transport upon initiation of northern hemisphere glaciation (Burton et al. 1997; Reynolds et al. 1999). An extensive dataset for Fe–Mn crusts exists for the Pacific Ocean, documenting both latitudinal and depth changes with time in Nd, Hf, and Pb isotope compositions that are tied to events including closure of the Panama gateway and initiation of the Antarctic Circumpolar Current (ACC) via opening of the Tasman gateway and Drake Passage (e.g., Lee et al. 1999; Ling et al. 1997; van de Flierdt et al. 2004a, b).

The short residence time of Pb has made this element of particular interest, but interpretations have been variable. Christensen et al. (1997) noted correlations between Pb isotope changes and benthic foraminifer δ^{18}O record, suggesting that Pb isotope variations record changes in climate and weathering. In contrast, Klemm et al. (2007) argued that the relative homogeneity of Pb isotope compositions for Fe–Mn crusts at any particular time reflected delivery through volcanic aerosols that became homogenized in the upper atmosphere. More recently Chen et al. (2013) document significant Pb isotope heterogeneity for the central North Pacific, and they interpret three distinct crustal Pb sources as (1) Andean volcanism >45 Ma, (2) hydrothermal activity 45–20 Ma, and (3) Asian dust <20 Ma. These three sources were identified based on, respectively, anomalously high $^{207}Pb/^{204}Pb$, unusually low $^{206}Pb/^{204}Pb$, and intermediate $^{206}Pb/^{204}Pb$ that matches that of average upper crust ($\sim$18.6–18.7).

Fe–Mn crusts have also been used to infer redox conditions of Cenozoic marine settings using Mo and U isotope compositions. Siebert et al. (2003) documented invariant $^{98}Mo/^{95}Mo$ ratios for the Atlantic and Pacific oceans over the last 60 m.y., which suggests no significant changes in the proportions of oxic, suboxic, and euxinic sinks during the Cenozoic, at the time resolution of the crusts studied. Such an interpretation is consistent with $^{238}U/^{235}U$ ratios for Fe–Mn crusts, which are invariant over the Cenozoic, indicating no large changes in marine redox (Goto et al. 2014; Wang et al. 2016a), within the $\sim$3 m.y. time periods covered by individual analyses. On shorter timescales, however, data from ODP core show significant Mo isotope excursions across the Late Paleocene thermal maximum and Eocene thermal maximum 2, leading Dickson and Cohen (2012) to conclude that these events correspond to expansions of local euxinic sinks; this is broadly consistent with C and S isotope modeling (Kurtz et al. 2003; Yao et al. 2018) that suggests transient euxinic events during the Paleocene-Eocene warm period.

6.1.3 Fe Isotope Variations in Cenozoic Seawater

In a survey of surface layers of 37 Fe–Mn crusts from the Atlantic, Pacific, and Indian oceans, Levasseur et al. (2004) found significant Fe isotope heterogeneity, where $\delta^{56}Fe$ values ranged from −0.8 to −0.1‰. There were no correlations between $\delta^{56}Fe$ values and Nd or Pb isotope compositions, suggesting no control by crustal composition, contrary to the conclusions of an early study by Zhu et al. (2000). In fact, Levasseur et al. (2004) re-analyzed the outer three samples of crust BM1969.05 from the North Atlantic Ocean studied by Zhu et al. (2000), but they could not reproduce the data from the 2000 study. Levasseur et al. (2004) found no correlation between $\delta^{56}Fe$ values and the ratio of hydrothermal input to dust input for specific oceans, as proposed in the two-component mixing model of Beard et al. (2003), which Levasseur et al. (2004) took to indicate that hydrothermal Fe is not likely far-traveled in the open ocean. This conclusion is not supported by studies of hydrothermal plumes that show hydrothermal Fe, particularly in the form of nanoparticulates, may be transported far from the vent (see Sect. 5.5.4 in Chap. 5). In contrast, the Fe isotope signal of dFe from hydrothermal plumes may quickly move toward that of ambient seawater with distance from the vent (see Fig. 5.27 in Chap. 5). At the time of the 2000–2004 studies, there were no direct measurements available of the Fe isotope compositions of seawater, so it was unknown if there were variations in Fe isotope fractionation between crust and seawater, but the data suggested that the Fe isotope composition of seawater might be variable, a conclusion that is clear from the discussion in Sect. 5.5 of Chap. 5.

With the availability of direct measurements of the Fe isotope compositions of seawater (see Sect. 5.5 in Chap. 5), it has become clear that the $\delta^{56}Fe$ values for Fe–Mn crusts are offset from those of seawater. Although Chu et al. (2006) argued that the short residence time of Fe in seawater should result in no effective fractionation between seawater and crust, the relations in Fig. 6.2 suggest that this is not the case. Horner et al. (2015) complied crust-seawater pairs using the data from Levasseur et al. (2004) and Conway and John (2014), the latter of which represents seawater from the GEOTRACES GA03 transect. This produces an average $^{56}Fe/^{54}Fe$ fractionation between Fe–Mn crust and seawater

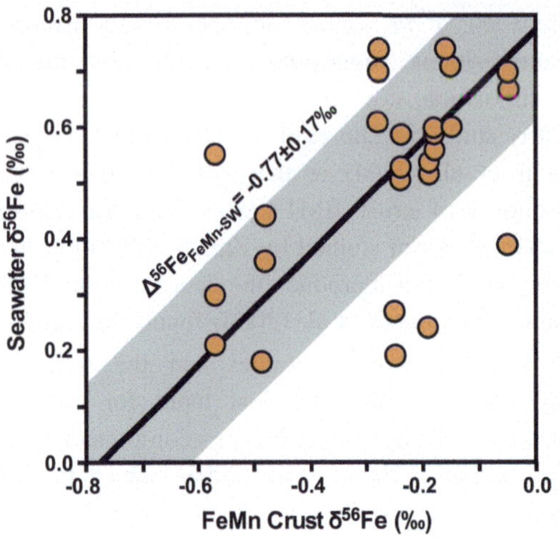

Fig. 6.2 Comparison of δ^{56}Fe values of surface layers of Fe–Mn crusts with δ^{56}Fe values estimated for proximal seawater, based on compilation by Horner et al. (2015). Fe–Mn crusts and seawater taken from the North Atlantic, including GEOTRACES transect GA03, but excluding data near the TAG hydrothermal site. Average ^{56}Fe/^{54}Fe fractionation between Fe–Mn crust and seawater ($\Delta^{56}Fe_{FeMn\text{-}SW}$) is −0.77 ± 0.17‰ (1SD, shown in grey band); this is identical to that calculated by Horner et al. (2015), although they report a smaller error, using the 2SE. The scatter suggests that in detail various factors may influence $\Delta^{56}Fe_{FeMn\text{-}SW}$, possibly including crust chemistry, disequilibrium effects, or errors in inferring seawater compositions, as well as differences in the timescale of samples taken from Fe–Mn crusts ($\leq 10^6$ yrs) relative to the residence time of Fe in seawater ($\leq 10^2$ yrs)

of −0.77 ± 0.17‰ (1SD). As shown in Fig. 6.2, there is significant scatter in this relation, possibly indicating errors in inferring local seawater compositions, but also possibly other factors that could influence the crust-seawater fractionation factor, such as composition. In addition, Fe–Mn crusts could record variable contributions from different depths of the water column. It also must be kept in mind that mm-scale sampling of Fe–Mn crusts will reflect timescales of 10^5–10^6 yrs, many orders of magnitude longer than the residence time of Fe in seawater. Nevertheless, this relation allows an estimation of δ^{56}Fe values for ancient seawater based on the Fe–Mn crust record.

Detailed Fe isotope records are available for five Fe–Mn crusts that range in age from 5 to 10 Ma, including the North Atlantic, Central and Southern Pacific, and the Western Pacific oceans (Fig. 6.3). This age range spans significant changes in climate, as discussed above in Sect. 6.1.1 and illustrated in Fig. 6.3 using the δ^{18}O curve for benthic foraminifera. In addition, we compare variations in Pb isotopes for the same (or comparable) crusts, given the similarly short residence time of Pb and Fe. Using the average Fe isotope fractionation factor in Fig. 6.2, calculated δ^{56}Fe values for seawater range from −0.1 to +0.7‰. The upper bound of this overlaps with δ^{56}Fe values for the open surface ocean that are controlled by exchange between ligand-bound Fe^{3+} in solution and dust particles (see Fig. 5.20 in Chap. 5), and the lower bound of which lies at the upper end for Mg-free endmembers of marine hydrothermal fluids (see Fig. 5.24 in Chap. 5).

North Atlantic crust BM1969.05 (Fig. 6.3a) lies on the Cretaceous New England seamount chain at 1929 m water depth and is located ~400 km from the North American shelf-slope break. This crust has been the subject of Nd and Pb isotope studies (Burton et al. 1997; O'Nions et al. 1998). As noted above, the outer three samples were re-analyzed for Fe isotopes but the initially determined δ^{56}Fe values could not be reproduced (Zhu et al. 2000; Levasseur et al. 2004), and it is uncertain what confidence can be assigned to the temporal variations in Fe isotope compositions for this crust at >1 Ma age. Within the scatter in δ^{56}Fe values for the crust (which likely reflects the fact that the initial analyses were done early in the development of Fe isotope geochemistry), there is no correlation between Fe isotope compositions and climate or changes in crustal weathering and sediment transport, as expressed in changes in O and Pb isotope compositions. The moderately positive δ^{56}Fe values of ~+0.1 to +0.4‰ inferred for seawater from crust BM1969.05 are similar to those measured for seawater dFe at the same depths (1–2 km) for stations 6 and 8 from the GEOTRACES GA03 transect by Conway and John (2014). The Fe isotope compositions for dFe are interpreted to reflect non-reductive release of Fe from shelf and slope sediments. Given the tectonic stability of this crust location over the last 6 m.y., this seems to be the most likely Fe source for this crust over that time period.

Two crusts from the Southern and Central Pacific Ocean have been studied, SPG-2 and 28DSR-9 (Fig. 6.3b). Crust SPG-2 lies in the South Pacific Ocean on Cretaceous-age oceanic crust, 1800 km east of the Tonga trench at 5126 m water depth (Marcus et al. 2015). It is therefore very far from large continental land masses or marine hydrothermal systems, and its calculated seawater δ^{56}Fe values of ~+0.6‰ directly overlap the range estimated for the open surface ocean. As shown by vertical seawater profiles far from the influence of hydrothermal or continental shelf sediments, such compositions may be maintained into the deep ocean through Fe isotope exchange between ligand-bound Fe^{3+} and sinking particles (e.g., Conway and John 2014). Crust 28DSR-9 lies in the equatorial Central Pacific Ocean, 1400 km NNE of Fiji at 3033 m water depth (Chu et al. 2006). Similar to crust SPG-2, crust 28DSR-9 lies far from continental or hydrothermal influences. The highly homogenous Fe isotope composition of this crust over the last 10 m.y. is best explained by a consistent fractionation between dFe and particles that is fixed by a consistent dust delivery such that the proportion of dFe and pFe has been relatively constant. This is supported by the restricted range in ^{206}Pb/^{204}Pb ratios for these two crusts (with the possible exception of the youngest time point for crust 28DSR-9). Importantly, the small differences in Pb isotope compositions between the crusts could indicate slightly different dust sources that may also indicate different relative particle loading levels, although the Pb isotope data for SPG-2 are inferred from another locality. Small differences in dust loading and dFe-pFe mass balances could explain small differences in the inferred δ^{56}Fe values for seawater. Nevertheless, the homogeneity of Fe isotope compositions with time is remarkable and document that Fe sources to the open oceans may remain constant over time periods of ~10^1 m.y.

The effects of Fe isotope provinciality are well illustrated by two crusts from the Western Pacific Ocean in the Izu-Bonin back-arc (Fig. 6.3c). Crust D105-5AB is located 120 km west of the volcanic front at a water depth of 2120 m, whereas crust D97-1 is located 70 km to the SSW at ~400 m greater water depth (Chu et al. 2006). For crust D105-5AB, δ^{56}Fe values have been homogeneous for the last 5 m.y.; Pb isotopes have been somewhat variable, following the general trend of increasing ^{206}Pb/^{204}Pb ratios that has been ascribed to Asian dust (Chen et al. 2013), although the range in Pb isotope compositions is quite small. The inferred seawater δ^{56}Fe values for crust D105-5AB are not quite as positive as those expected for the open surface ocean, likely reflecting additional components given its proximity to the Izu-Bonin arc and Asian continent relative to the Central and South Pacific crusts. Isotopic variations for crust D97-1 are quite different, where periods of low-δ^{56}Fe excursions occurred at ~3 Ma, ~1 Ma, and the Holocene. Chu et al. (2006) noted that these low-δ^{56}Fe excursions correlated with increased abundances of V, Ni, Mn, Cu, and Zn, which they ascribed to pulses of hydrothermal input that were captured by the crust. That crust D97-1 recorded a significantly different Fe isotope history than nearby crust D105-5AB may reflect the greater depth of crust D97-1, which increased the likelihood that it would intersect back-arc hydrothermal pulses. The Pb isotope evolution of crust D97-1 does not record these pulses, lending support to the model that Pb isotopes are most sensitive to continental influences (e.g., dust, aerosol) rather than hydrothermal components. This highlights the decoupling that may occur between Fe and Pb isotopes in seawater, despite their similarly short residence times.

The longest record of seawater Fe isotope composition during the Cenozoic is provided by a single Fe–Mn crust, CD92-2, extending back to almost 80 Ma (Fig. 6.4). This crust is from the

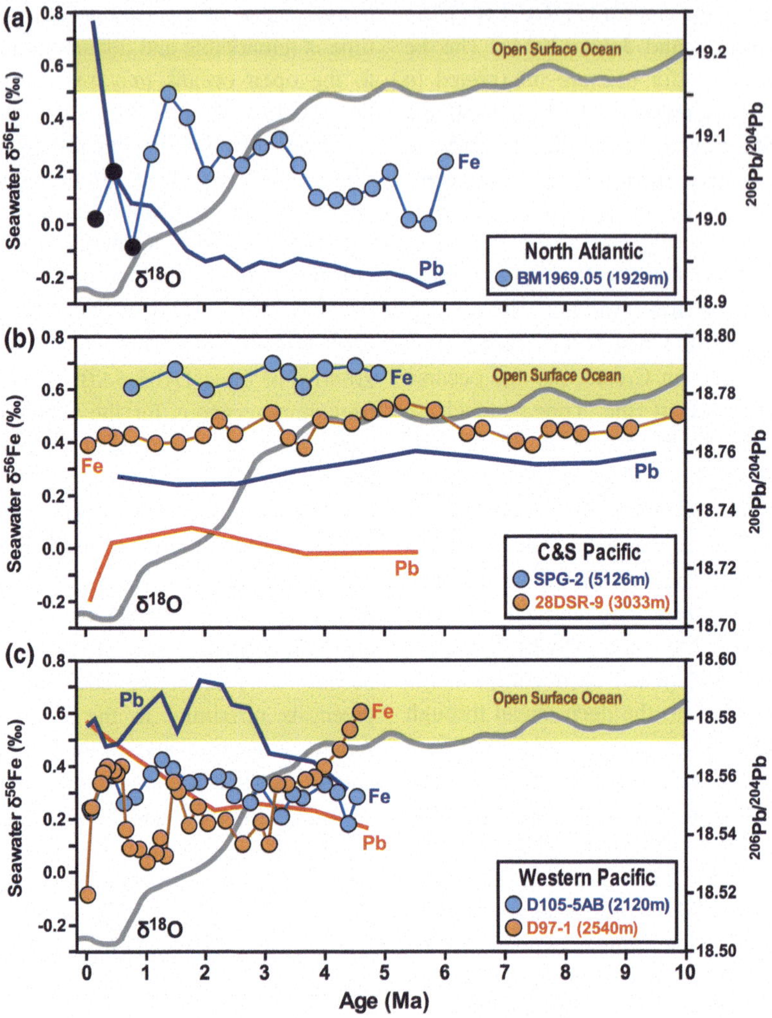

Fig. 6.3 Temporal variations for Fe (symbols + lines) and Pb (lines only) isotope compositions for Fe–Mn crusts that provide a record over the last 10 m.y. Ages for these crusts are well constrained by ^{10}Be geochronology. Measured δ^{56}Fe values converted to seawater values using $\Delta^{56}Fe_{FeMn\text{-}SW} = -0.77$‰ (see Fig. 6.2). Current water depths for each crust indicated in legend. Also shown in yellow band is the range in δ^{56}Fe values for open surface ocean based on the dominant range of values shown for dFe in Fig. 5.20 (Chap. 5), which is interpreted to be controlled by Fe isotope exchange with particles, and likely reflects the maximum δ^{56}Fe values attained in seawater. Grey curve in all three panels is δ^{18}O curve for benthic foraminifera from Fig. 6.1, scaled from +2.0‰ (top of graphs) to +3.5‰ (bottom of graphs), indicating initiation of global cooling and northern hemisphere glaciation beginning ~3 Ma. **a**: North Atlantic crust BM1969.05 (Zhu et al. 2000; Levasseur et al. 2004); outer three layers from the original 2000 data set, which suggested very high δ^{56}Fe values, could not be reproduced by Levasseur et al. (2004), and filled black circles indicate the values obtained by the 2004 study; this adds uncertainty for the Fe isotope data on samples >1 Ma in age. **b**: Central and South Pacific crusts SPG-2 (Marcus et al. 2015) and 28DSR-9 (Chu et al. 2006). Lead isotope data plotted for SPG-2 uses data for "Nova D137-01" crust from van de Flierdt et al. (2004a), which is the closest crust that has Pb isotope data and is a likely reasonable approximation given the relatively consistent Pb isotope compositions for this region of the Pacific Ocean in this age range (van de Flierdt et al. 2004a). **c**: Western Pacific crusts D105-5AB and D97-1 from the Izu-Bonin back-arc (Chu et al. 2006)

Central Pacific Ocean on the Karin Ridge/Line Islands at a water depth of ~2 km, 1400 km west of Hawaii. Basalts from the Karin Ridge have ages between 81 and 86 Ma, whereas younger ages of 68–73 Ma occur on a cross-trending seamount chain south of Johnston Atoll ~150 km WSW of the CD92-2 locality (Davis et al. 2002). Crust growth began shortly after seamount formation, and plate reconstructions suggest an original location of the seamount at ~10 °N at ~1 km water depth, followed by NW translation to its current position (Klemm et al. 2007; Meynadier et al. 2008; Horner et al. 2015). Its location suggests minimal continental margin inputs or hydrothermal inputs from the East Pacific Rise, although local inputs from the nearby 68–73 Ma seamounts is possible. Crust CD29-2 has been subject to extensive previous studies of various isotopic systems, including, Ni (Gall et al. 2013), Nd (Ling et al. 1997), Hf (Lee et al. 1999), Os (Klemm et al. 2005), Tl (Nielsen et al. 2009), Pb (Christensen et al. 1997), and U (Wang et al. 2016a). Because the crust history extends beyond that accessible to ^{10}Be geochronology, the timescale of the crust >10 Ma has been a challenge. Frank et al. (1999) assumed Co precipitation rates remained constant with time, whereas Klemm et al. (2005) used Os isotopes to determine ages relative to the Os isotope curve for seawater. This timescale was further modified by Nielsen et al. (2009), who used Tl isotopes and the Tl isotope curve for seawater.

Calculated δ^{56}Fe values for seawater from crust CD29-2 span a remarkably wide range, from −0.4 to +2.3‰ (Fig. 6.4). Very few seawater δ^{56}Fe values overlap those expected for the open surface ocean where exchange with dust determines the Fe isotope composition of dFe, which led Horner et al. (2015) to emphasize that the majority of the Fe must have come from various deep ocean sources. The notably positive δ^{56}Fe values, which are significantly higher than any yet measured for modern seawater (e.g., Conway and John 2014), were interpreted by Horner et al. (2015) to reflect hydrothermal Fe that had undergone extensive sulfide precipitation. As shown in Fig. 5.25 in Chap. 5, MOR hydrothermal systems that have high proportions of FeS particulates have highly negative δ^{56}Fe values for particulate Fe (pFe), which would move the remaining dFe to positive δ^{56}Fe values. Lower δ^{56}Fe values were interpreted to possibly reflect Fe released from sediments or hydrothermal fluids that has undergone oxidation.

Looking at the broad climactic, tectonic, and volcanic record of the Cenozoic, as well as considering the drivers of Fe isotope compositions of modern seawater (Sect. 5.4 in Chap. 5), suggests some alternative interpretations of the Fe isotope record for Cenozoic seawater provided by crust CD29-2 (Fig. 6.4). The relatively low δ^{56}Fe values between 75 and 64 Ma seem best explained by hydrothermal fluids emanating from the seamount chain south of the Johnston Atoll, given its proximity and age range. This time period reflects relatively high total oceanic heat flow (Fig. 6.4b), commensurate with establishment of a number of hot spot or volcanic chains in the Pacific Ocean. Lead isotope compositions, however, vary greatly over this time period, where ^{207}Pb/^{206}Pb ratios steadily increase with decreasing age (Fig. 6.4c), suggesting major changes in dust sources, but this is not apparently recorded in the δ^{56}Fe values of the crust, which may have been dominated by local hydrothermal sources. δ^{56}Fe values shift markedly positive at the Cretaceous-Paleogene boundary, although looking at the broad changes in the marine record over this boundary (Fig. 6.1) does not immediately identify a reason for the sudden Fe isotope shift. The moderately positive δ^{56}Fe values during the Paleocene overlap those that are typical of the open surface ocean, and the middle to late Paleocene is also a time period of little change in Pb isotope compositions for the crust, suggesting a stable source of dust.

Large changes in δ^{56}Fe values occur through the late Paleocene thermal maximum and the early Eocene climactic optimum, with values both below and above those of the open surface ocean (Fig. 6.4). The change from low to high δ^{56}Fe values aligns with the marked increase in δ^{34}S values for seawater sulfate (Fig. 6.1) and the commensurate decline and recovery of C_{org} burial rates that corresponds to development of

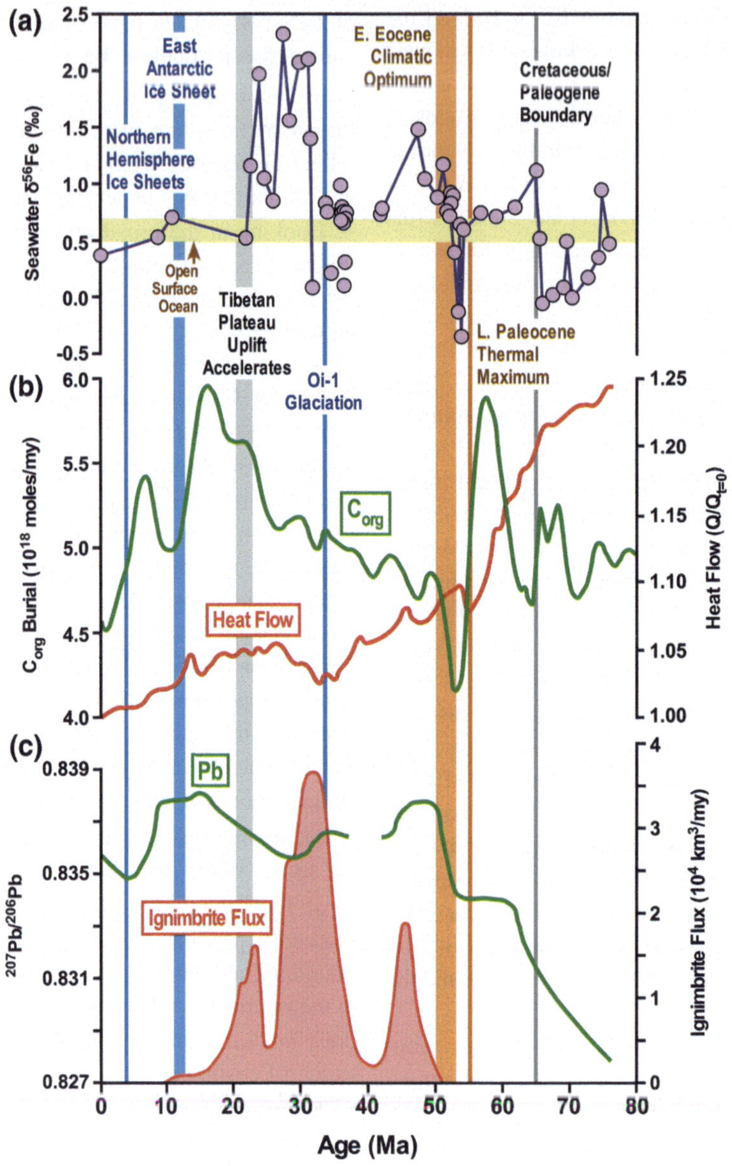

Fig. 6.4 Temporal variations in isotopic compositions of Central Pacific Fe–Mn crust CD29-2, compared to major climactic and tectonic events and other indicators from Fig. 6.1. Ages for Fe and Pb isotope trends for CD29-2 are calculated using the Tl-based timescale of Nielsen et al. (2009). **a** Measured δ^{56}Fe values (Horner et al. 2015) converted to seawater values using $\Delta^{56}Fe_{FeMn\text{-}SW} = -0.77$‰ (see Fig. 6.2). Yellow band as in Fig. 6.3, noting seawater δ^{56}Fe values for the open surface ocean as controlled by dust exchange. **b** Variations in organic carbon (C_{org}) burial (green curve) and total oceanic heat flow (red curve), from Fig. 6.1. **c** $^{207}Pb/^{206}Pb$ variations (green curve) for crust CD29-2 (Christensen et al. 1997), re-cast into the Tl-based timescale of Nielsen et al. (2009). Lead isotope data plotted as $^{207}Pb/^{206}Pb$ here because this ratio has significantly higher precision for the sample sizes analyzed as compared to $^{206}Pb/^{204}Pb$ (e.g., Fig. 6.3), allowing illustration of fine-scale changes with time. Also shown is the ignimbrite flux curve for "supervolcanic" eruptions from the Sierra Madre Occidental from Cather et al. (2009). The marked positive δ^{56}Fe values for seawater between ~35 Ma and 20 Ma correspond to the period of "ignimbrite flare-up" from the Sierra Madre Occidental

transient euxinic conditions, as discussed above. Might, for example, the steep rise in δ^{56}Fe values record an increase in pyrite burial, or a change in pyrite burial locations from the shelf to the deep oceans, as noted earlier? As extensively discussed in Sect. 5.4.3 in Chap. 5, pyrite in modern marine sediments commonly have negative δ^{56}Fe values, and extensive pyrite burial during the Paleocene and Eocene warm periods might have shifted the Fe isotope compositions for seawater dFe to highly positive δ^{56}Fe values. Such a model is distinct from that proposed by Horner et al. (2015), who envisioned pyrite to only play a role during precipitation from hydrothermal fluids. Although the Paleocene and Eocene warm periods correspond with a rapid increase in $^{207}Pb/^{206}Pb$ ratios for crust CD29-2, which seem likely to record a change in weathering and dust sources, the very high δ^{56}Fe values were not likely to reflect dust (see Sect. 5.5 in Chap. 5).

The period between the Paleocene-Eocene warm intervals and acceleration of the Tibetan Plateau uplift is accompanied by decreases in pCO_2, initiation of Antarctic glaciation (Oi-1), large increases in burial of C_{org}, and a continued decline in total oceanic heat flow, as discussed above in Sect. 6.1.1 (see also Fig. 6.1), but it was also a time of major ignimbrite "supervolcano" eruptions in the circum-Pacific arcs, especially in the Sierra Madre Occidental (Fig. 6.4). Although the increase in C_{org} burial might be expected to increase the vigor of a benthic Fe shuttle, crust CD29-2 has been too remote from continental margins to likely have seen the effects microbial sources of Fe from continental margin sediments. The remote location of the CD29-2 site, however, should have made it highly sensitive to changes in dust flux. The relatively low primary productivity of the Central Pacific Ocean, as discussed above, might have made the Fe isotope record of crust CD29-1 responsive to Fe fertilization if dust loads were increased.

We offer here a new proposal that the unusually positive δ^{56}Fe values for crust CD29-2 between $\sim$35 Ma and $\sim$20 Ma reflected phytoplankton blooms through Fe fertilization, where biological uptake of low-δ^{56}Fe Fe (Ellwood et al. 2015), followed by settling of biogenic Fe particles, produced exceptionally high δ^{56}Fe values for dFe. That this process produced more extreme Fe isotope compositions than that seen in the modern oceans might be expected, considering the enormous volume of ash dispersal that occurred during the Oligocene and Miocene ignimbrite flare-up, producing dust loads far greater than seen in the Holocene. Transport of the Fe isotope signal from such Fe fertilization events to the deep oceans, to be sequestered into Fe–Mn crusts, likely would have been enhanced by the depressed CCD during this time interval (see Sect. 6.1.1). If one wanted to continue the speculation, the drawdown of atmospheric CO_2 that accompanied the initi-

ation of Antarctic glaciation across the Eocene-Oligocene boundary might have been facilitated by the ignimbrite-induced Fe fertilization, as proposed by Cather et al. (2009) and Jicha et al. (2009), a proposal that finds some support in the Fe isotope record of crust CD29-2. In detail, the Fe isotope shift to positive values occurs slightly after this transition (Fig. 6.4), although uncertainties in timescale calibrations for Fe–Mn crusts >10 Ma in age remain significant.

The proposal that the unusually positive δ^{56}Fe values in the Oligocene-Miocene sections of crust CD29-2 record an ignimbrite flare-up ash fertilization event(s) contrasts with that proposed by Horner et al. (2015), who called upon extensive sulfide precipitation from hydrothermal fluids. The difficulty with invoking a hydrothermal source include the fact that heat flow was relative low during this time (Fig. 6.4), and crust CD29-2 has always remained >4,000 km from the East Pacific Rise (EPR). As discussed in Sect. 5.5.4 in Chap. 5, there is little evidence in the modern oceans that a hydrothermal Fe isotope signal for dFe travels such distances (see also Fig. 5.27 in Chap. 5). Moreover, there is no evidence for local hydrothermal venting of this age in the Karin Ridge/Line Islands. Finally, considering the down-vent evolution of hydrothermal plumes from the EPR (e.g., Fig. 5.27 in Chap. 5), δ^{56}Fe values do not exceed those for the open surface ocean, reflecting instead a progressive dilution with ambient seawater with increasing distance from the vent. There is no evidence that distal hydrothermal plumes may attain the δ^{56}Fe values >1‰. In contrast, support for an influence from the Oligocene-Miocene ignimbrite flare-up can be found in the Pb isotope data from crust CD29-2. Although relatively few Pb isotope analyses have been reported for the silicic units of the Sierra Madre Occidental, those that have show relatively non-radiogenic ^{206}Pb/^{204}Pb ratios, as expected for the relatively young age of the basement beneath the province (Ferrari et al. 2007). The ^{207}Pb/^{206}Pb ratios for the Sierra Madre Occidental (~0.836) exactly match those measured in crust CD29-2 at this time (Fig. 6.4; McDowell et al. 1999), consistent with a model where very large ignimbrite eruptions produced high ash loads into the upper atmosphere, followed by deposition in the equatorial Pacific Ocean (Cather et al. 2009).

We conclude this section with a few comments about what future Fe isotope studies of Fe–Mn crusts might detect. Hydrothermal venting events are likely to have been local features that do not influence the large global climate changes that have occurred in the Cenozoic. In contrast, models for large changes in C_{org} burial might be tested through search for evidence of changes in benthic Fe shuttles, but this will require studies of crusts that lay close to continental margins. Similarly, a pyrite burial signal for Fe isotopes would be expected during and after the Paleocene and Eocene warm periods, where there was a large excursion in seawater δ^{34}S values (see Fig. 6.1), perhaps variable on a regional scale and strongest near continental margins. Specific to the proposal here that a Oligocene-Miocene ignimbrite flare up provided Fe fertilization to the oceans and attendant CO_2 drawdown, this will require studies of additional Fe–Mn crusts that are far from continental margins and contain a record >10 Ma, which in turn will rely on development of detailed geochronology timescales.

6.2 Cretaceous Anoxic Events

Oceanic anoxic events (OAEs) are defined by periods of extensive organic carbon burial, reflected in rapid changes in δ^{13}C values for C_{org} and marine carbonates (Fig. 6.5; e.g., Jenkyns 2010). Although many OAEs have been proposed, several of them have been considered to be of global significance, including OAE-1A at the early Aptian and OAE-2 at the Cenomanian-Turonian boundary. There is broad consensus that extensive burial of C_{org} during OAEs reflects high primary productivity that was driven by enhanced nutrient delivery, including P, as well as an enhanced Fe cycle (e.g., Van Cappellen and Ingall 1994; Leckie et al. 2002; Handoh and Lenton 2003; Mort et al. 2007b; Trabucho Alexandre et al. 2010; Kraal et al. 2010; Monteiro et al. 2012; Tessin et al. 2016; Owens et al. 2018; Sweere et al. 2018; Zeng et al. 2018). Burial of C_{org} during OAEs was superimposed on the broad "Greenhouse World" of the Mesozoic (e.g., Takashima et al. 2006). Specific to the Cretaceous, this was a period of relatively high seafloor-spreading rates, oceanic heat flow, and marine hydrothermal activity (e.g., Delaney and Boyle 1986; Becker et al. 2009; Seton et al. 2009), as well as relatively high sea-level stand (e.g., Müller et al. 2008). Temperatures were relatively high, reflecting high pCO_2, yet pO_2 levels were relatively low in comparison to earlier times in the Phanerozoic, reflecting the broadly antithetic relation between CO_2 and O_2 (e.g., Berner 2006; Hong and Lee 2012; Friedrich et al. 2012; Poulsen et al. 2015).

There is extensive literature on potential drivers for the Cretaceous OAEs as recorded in various geochemical proxies (e.g., Jenkyns 2010). A compilation of Cretaceous sea-surface temperatures based on the TEX_{86} thermometer (O'Brien et al. 2017) suggests warming periods preceding OAE-1A and OAE-2 (Fig. 6.5), possibly related to volcanic CO_2 inputs (see discussion in Jenkyns 2018). Changes in the hydrologic cycle, weathering, and nutrient delivery have been proposed prior to the onset of OAEs (e.g., van Helmond et al. 2014), and these are marked by prominent inflections in the ^{87}Sr/^{86}Sr seawater curve (Fig. 6.5) that have been interpreted to record

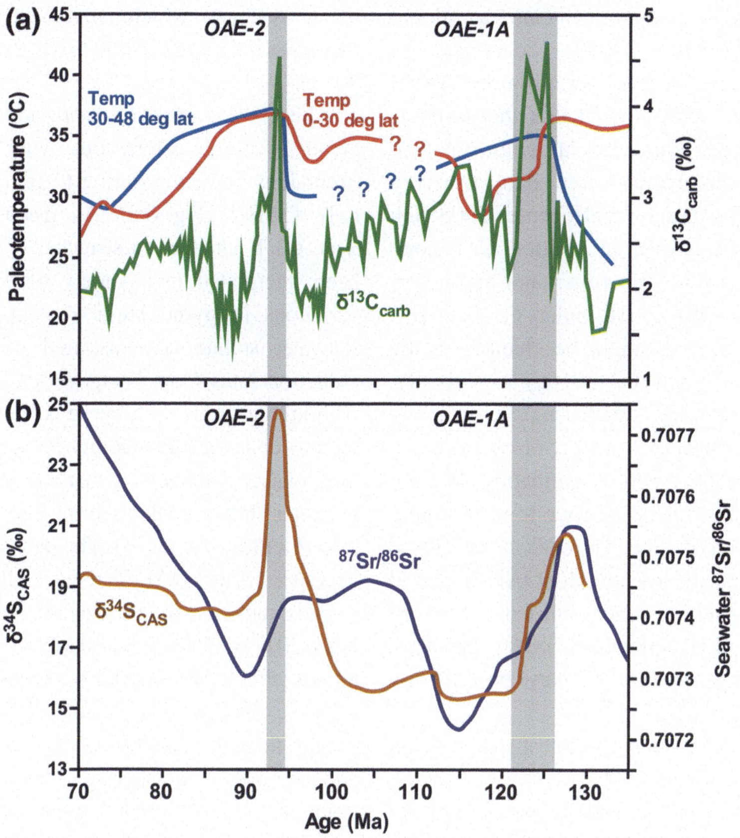

Fig. 6.5 Temporal marine changes through the major part of the Cretaceous highlighting associations with global Oceanic Anoxic Events (OAE) OAE-1A and -2 (vertical grey bands; time width reflects absolute age uncertainties for events, not duration of events). **a** Paleotemperatures for surface ocean (not calibrated to present-day) from compilation of TEX_{86} thermometer data, divided by low (red curve) and intermediate (blue curve) latitudes (Jenkyns 2018). Also shown is average curve (green) for $\delta^{13}C$ values (PDB scale) for marine carbonates (Takashima et al. 2006), with modification for the early Cretaceous from Mills et al. (2017). **b** Compilation curve for $\delta^{34}S$ values (CDT scale) for seawater sulfate (brown; CAS and barite) from Gomes et al. (2016) and Mills et al. (2017), as modified from the barite curve (Paytan et al. 2004) to account for very large (and heterogenous) fluctuations about OAE-2. Does not include data that suggest locally high $\delta^{34}S$ values between 115 and 100 Ma (Gomes et al. 2016; Mills et al. 2017). Compilation curve for seawater $^{87}Sr/^{86}Sr$ ratios (purple) from Jones and Jenkyns (2001), modified for early Cretaceous from Mills et al. (2017). OAE-1A and OAE-2 are marked by large positive $\delta^{13}C$ excursions for marine carbonate, which are preceded by rapid warming, especially at mid-latitudes, followed by long-term cooling. Large changes in $\delta^{34}S$ values occur about OAE-1A and OAE-2 that may record large shifts in seawater sulfate contents and/or pyrite burial. $^{87}Sr/^{86}Sr$ ratios broadly increase during times preceding OAE-1A and OAE-2, followed by declines across the events, which have been interpreted to reflect changes in weathering and igneous inputs

enhanced hydrothermal input to the oceans during OAEs (Jones and Jenkyns 2001). Detailed records, however, indicate an increase in $^{87}Sr/^{86}Sr$ ratios for seawater immediately before OAE-1A and OAE-2, accompanied by a decrease in $^{44}Ca/^{42}Ca$ ratios, which suggests an increase in continental weathering preceding the OAEs (Blättler et al. 2011), although Ca isotope changes could reflect changes in the fractionation factor (Du Vivier et al. 2015a). Detailed Li isotope studies across OAE-2 suggest an increase in continental weathering (Pogge von Strandmann et al. 2013), and a similar conclusion was reached by Lechler et al. (2015) for OAE-1A. Osmium isotope variations clearly indicate a global (although regionally variable) decrease in $^{187}Os/^{188}Os$ ratios for seawater prior to OAEs, reflecting a large hydrothermal flux that is interpreted to have been sourced to large igneous province (LIP) emplacement, followed by recovery within the OAE (Turgeon and Creaser 2008; Du Vivier et al. 2014, 2015b). In addition, Cr isotope data have been taken to record a major hydrothermal flux during OEA-2 (Holmden et al. 2016), although an alternative might be changes in the extent of euxinia (Wang et al. 2016b). Importantly, evidence for increased weathering immediately preceding OAEs from Li, Ca, and Sr isotopes is not seen in the Os isotope record, presumably reflecting the much greater Os mass balance toward mantle inputs relative to continental Os. Collectively, these results suggest that the trigger(s) for OAEs included increased weathering and nutrient delivery accompanied by LIP activity and pCO_2 increases, followed by expansion of primary productivity during the OAE.

Several lines of evidence suggest major redox changes leading up to OAEs, as well as redox changes within OAEs. Trace-element ratios that are sensitive to redox (e.g., I/Ca, V/Al, Cr/Al), as well as the presence of euxinia (e.g., Mo/Al), suggest development of local and regional oxygen-deficient marine environments as early as 1 m.y. prior to OAEs, followed by variable shift to euxinic environments within OAEs (e.g., van Helmond et al. 2014; Owens et al. 2016, 2017a; Zhou et al. 2015, 2017). Such conclusions are supported by Fe speciation determinations for "highly reactive" (acid-extractable) and sulfide-bound Fe (Owens et al. 2017a). Detailed Tl isotope studies across the Toarcian and Cenomanian-Turonian OEAs indicate a change in Tl sinks prior to the OAEs, indicating expansion of anoxic marine environments (Ostrander et al. 2017; Owens et al. 2017b; Them et al. 2018). Variable $\delta^{98}Mo$ values in several OAE-2 sections suggest a short residence time for Mo, indicating significant drawdown in euxinic environments on local and regional scales (Westermann et al. 2014; Dickson et al. 2017; Goldberg et al. 2016). Increased anoxia during OAE-2 was also inferred based on changes in $\delta^{238}U$ values, which also suggests transient increases in oxygen within OAE-2 (Clarkson et al. 2018).

The $\delta^{34}S$ values for seawater sulfate varied greatly over the Cretaceous, with large positive excursions occurring during OAE-1A and particularly OAE-2 (Fig. 6.5). As discussed above in Sect. 6.1.1, changes in seawater $\delta^{34}S$ values have been interpreted in a variety of ways, including variations in pyrite burial (extent and location), seawater sulfate contents, and microbial sulfate reduction rates (correlating with C_{org} abundance). Although a single $\delta^{34}S$ curve is shown in Fig. 6.5, there is considerable viability for any given time in the Cretaceous, indicating that S residence times and seawater sulfate contents were low, perhaps a few mM (Gomes et al. 2016). The peaks in $\delta^{34}S$ values for OAE-1A and OAE-2 are generally interpreted to record a rapid rise in seawater sulfate contents. Mills et al. (2017) point to the similarity in S and Sr isotope records after OAE-1A (Fig. 6.5), which they suggest is best explained by an increase in sulfate contents due to increased delivery via hydrothermal sources and continental weathering. The rise in $^{87}Sr/^{86}Sr$ ratios at ~ 115 Ma is inferred to record increased evaporite deposition; large variability in $\delta^{34}S$ values not reflected in the general curve in Fig. 6.5 supports a model of relatively low sulfate contents leading up to OAE-2. Adams et al. (2010) suggest the increase in $\delta^{34}S$ values before OAE-2 may be explained by increased sulfate contents from volcanism, supported, for example by Os isotope data (above), which then increased microbial sulfate reduction that accompanied increases in nutrient delivery; this would "set the stage" for development of euxinic zones during OAE-2. Owens et al. (2013) model the very high $\delta^{34}S$ values during OAE-2 as a shift to euxinic and high sulfate conditions. The decrease in $\delta^{34}S$ following OAE-2 records the contraction of euxinic conditions and decreased pyrite burial in their model.

6.2.1 Cenomanian-Turonian OAE-2

We focus here on OAE-2 at the Cenomanian-Turonian boundary (93.9 Ma using the timescale of Meyers et al. 2012), for which Fe isotope data are available from several proto-North Atlantic sites (Jenkyns et al. 2007; Clayton et al. 2007; Owens et al. 2012), including shallow shelf (Eastbourne, South Ferriby, UK; Tarfaya, Morocco), rifted margin (Demerara Rise, ODP sites 1258 and 1260), and deep- to mid-ocean depth settings (DSDP sites 105, 144, 367, and 603B); in addition, a shelf setting of the Western Tethys ocean has been studied (Furlo, Italy). In many cases these sites have extensive geochemical and isotopic datasets available that allow inferences to be made regarding Fe cycling and its relations to redox and euxinic conditions. A detailed chronostratigraphy has been developed for OAE-2, although estimates for the length of the event vary from ~400 kyr to ~1 my (e.g., Eldrett et al. 2015; Lanci et al. 2010; Laurin et al. 2016; Li et al. 2017b; Sageman et al. 2006).

Eastbourne, UK, contains the European Cenomanian-Turonian reference section, and the relatively low C_{org} contents of the shales suggests it, as well as a nearby locality at South Ferriby, represents an oxic shelf setting during OAE-2 (e.g., Tsikos et al. 2004). Sedimentation rates at Eastbourne were exceptionally high, capturing OAE-2 over ~14 m of stratigraphy. It is now recognized that two cooling events (*Plenus* events) are documented for OAE-2 at Eastbourne via C and O isotope and faunal changes (Fig. 6.6), confirmed by TEX_{86} data, as well as inferred drops in pCO_2 (Barclay et al. 2010; Jarvis et al. 2011; Jenkyns et al. 2017). Strong increases in Cr/Ca and V/Ca ratios in carbonates during the main (youngest) Plenus cooling event is taken to indicate a major oxidizing interval within OAE-2, which also correlates with a major increase in Fe_{tot}/Al ratios (Fig. 6.6; Owens et al. 2012; Jenkyns et al. 2017). Evidence for oxidizing conditions lies in the slightly positive δ^{56}Fe values for shales from Eastbourne and South Ferriby, almost all of which lie above the crustal average (Fig. 6.6; Jenkyns et al. 2007; Owens et al. 2012). The mildly elevated δ^{56}Fe values overlap with those observed in shelf sediments beneath the Peru margin OMZ that Scholz et al. (2014a, b) interpreted to reflect accumulation of Fe^{3+}-hydroxides that settled from a benthic Fe shuttle plume via partial Fe^{2+}_{aq} oxidation (see Sect. 5.4.4 and Fig. 5.19 in Chap. 5). Such a model, including the paleogeographic setting, fits quite well for the data from Eastbourne and South Ferriby, and is distinct from the negative δ^{56}Fe values produced by a euxinic sink for a benthic Fe shuttle (e.g., Severmann et al. 2008). Somewhat surprisingly, there is no correlation between δ^{56}Fe values and Fe_{tot}/Al ratios for shales at Eastbourne, perhaps indicating that sedimentation rates and oxidation rates were not correlated. It is also true, however, that the density of sampling for Fe isotope studies at Eastbourne and South Ferriby is much smaller than for other geochemical indicators, as is true for the other OAE-2 sections discussed below.

The majority of δ^{34}S values for seawater sulfate just prior to OAE-2 lie at ~+20‰ at Eastbourne (Fig. 6.6), overlapping the range seen in other pre-OEA-2 data, followed by an overall decrease in δ^{34}S values but increase in scatter at the early stages of OAE-2, after which there is an increase through the interval. The changes in δ^{34}S values across the interval has been modeled by Owens et al. (2013) to record an overall increase in pyrite burial that accompanied an expansion of euxinic conditions during OAE-2, perhaps an increase of 30 times in extent, a scenario supported by other models (e.g., Gomes et al. 2016). The negative excursion in δ^{34}S values within OAE-2 at Eastbourne is, however, unusual, and it is possible that the very high sedimentation rates at Eastbourne captured S isotope heterogeneity not seen in other sections, particularly during periods of low seawater sulfate contents. Jenkyns et al. (2017) suggests two explanations for the decrease in δ^{34}S values within OAE-2: an increase in magmatic input and/or oxidation of sedimentary pyrite during the Plenus cooling events, which are inferred to occur during decreases in pCO_2 and increases in pO_2. If the large swings in δ^{34}S values within OAE-2 record variations in pyrite burial, this is not apparently reflected in the Fe isotope record, although the scale sampling is quite coarse for Fe.

Changes in oceanic circulation across OAE-2 at Eastbourne are indicated in Nd isotope data, where a brief decrease in $\varepsilon_{Nd}(t)$ values immediately prior to OAE-2 is followed by a positive

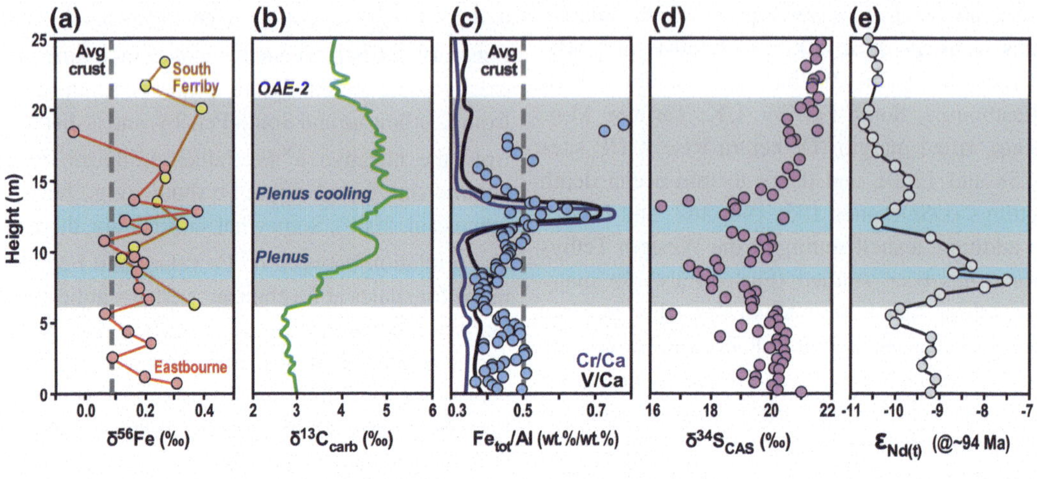

Fig. 6.6 Detailed geochemical changes across OAE-2 at the Eastbourne, UK, locality, which represents a proto-North Atlantic oxic shelf setting and is considered as the type European section for OAE-2. OAE-2 interval (grey bands) defined by positive $\delta^{13}C_{carb}$ excursion, and Plenus cooling events (light blue bands) defined by negative $\delta^{13}C_{carb}$ inflections within OAE-2, as well as faunal and O isotope changes (Jarvis et al. 2011; Jenkyns et al. 2017). **a** δ^{56}Fe values for shales from Eastbourne (pink/red), as well as the similar South Ferriby section (yellow/brown) are generally higher than those of average crust (Jenkyns et al. 2007; Owens et al. 2012). **b** $\delta^{13}C_{carb}$ values (PDB scale) for Eastbourne section (Tsikos et al. 2004) define the Cenomanian-Turonian Carbon Isotope Excursion (CIE), which is taken here, for simplicity, to be equal to OAE-2, although some workers define a slightly smaller extent for OAE-2 (e.g., Jenkyns et al. 2017). **c** Ratio to total Fe to Al (Fe_{tot}/Al) for shales from Eastbourne section (Owens et al. 2012). For reference, average crustal ratio of 0.5 is shown, although there is some uncertainty about local lithologic input. Generally, Fe_{tot}/Al ratios are lower than the crustal average throughout OAE-2, except for a positive excursion coincident with the major (younger) Plenus cooling event. Also shown are curves for Cr/Ca (purple) and V/Ca (black) ratios for carbonates from Eastbourne, where horizontal scale is equal to 0–25 μmol/mol for Cr/Ca and 0–5 μmol/mol for V/Ca (Jenkyns et al. 2017). The peak in Fe_{tot}/Al is coincident with the peak in Cr/Ca and V/Ca, indicating a significant oxidation event during the main (younger) Plenus cooling event. D: δ^{34}S values (CDT scale) for Carbonate Associated Sulfate (CAS) from the Eastbourne section (Owens et al. 2013), showing variable but decreasing values through the start of OAE-2, followed by increase in δ^{34}S and decrease in scatter after the second (main) Plenus cooling event. E: ε_{Nd}(t) values (relative to CHUR, where ^{143}Nd/^{144}Nd calculated at inferred depositional ages of 94 Ma) using fish teeth and bones from bulk chalk samples (Zheng et al. 2013), showing inflections at the start of OAE-2 and at the two Plenus cooling events

excursion at the start of the Plenus cooling events, followed by a decrease at the second (younger) Plenus event (Fig. 6.6e). Zheng et al. (2013) interpret these changes to indicate an increase in radiogenic Nd from volcanic sources leading up to the first Plenus event, and the following decrease in ε_{Nd}(t) might record a decrease in volcanic input and/or change in ocean circulation. These changes in inferred seawater Nd sources and circulation are not reflected in any systematic way in the Fe isotope record. Accepting the positive δ^{56}Fe values as indicating a paleoenvironment similar to that of the oxic zones of the Peru margin, as noted above, we might expect Fe isotope signatures to be correlated with different water masses, but this is not apparent in comparing the Fe and Nd isotope records at Eastbourne. It is possible, however, that the sampling density for Fe isotopes, relative to other geochemical data, is simply too sparse to attempt detailed comparisons.

Western Tethys exposures of the Cenomanian-Turonian boundary are well documented in Italy and Spain, and the locality at Furlo, Italy, and nearby correlatives, has been studied in detail, including Fe isotopes (e.g., Tsikos et al. 2004; Mort et al. 2007a; Jenkyns et al. 2007; Westermann et al. 2014; Du Vivier et al. 2015b; Owens

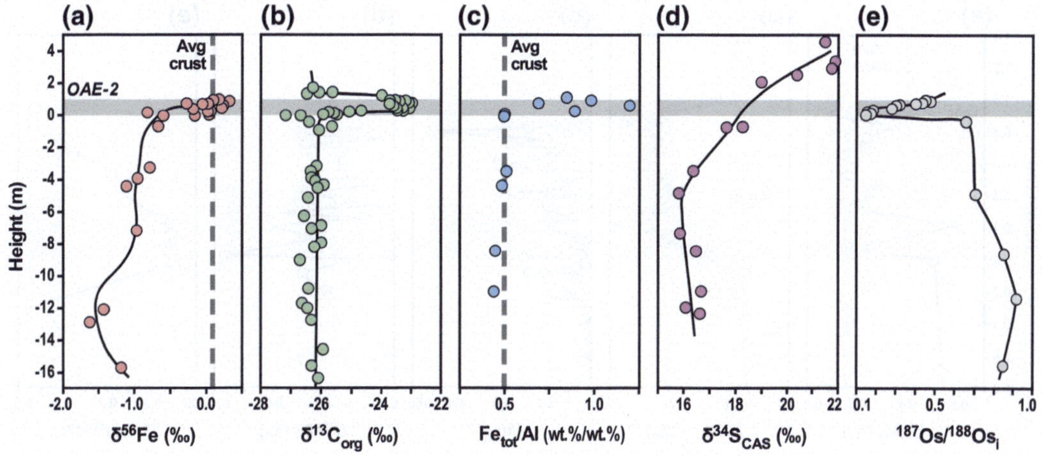

Fig. 6.7 Large-scale changes in geochemical trends across OAE-2 (horizontal grey band) at the Furlo, Italy locality, which represents a shelf setting in the Western Tethys Ocean. Using the timescale of Lanci et al. (2010), the base of section is ~95.5 Ma, and the top of section is ~93.0 Ma. **a** δ^{56}Fe values for shales, which are uniformly quite low (δ^{56}Fe < −1‰) in the carbonate-rich pre-OAE-2 section, and trend to values near average crust in the shale-rich OAE-2 section (Jenkyns et al. 2007). **b** $\delta^{13}C_{org}$ values (PDB scale) for C_{org} in shales, which are relatively homogeneous beneath OAE-2, and show a marked positive excursion within OAE-2, followed by a return to low values after OAE-2 (Jenkyns et al. 2007; Mort et al. 2007a; Du Vivier et al. 2015b; Owens et al. 2017b). **c** Ratio of total Fe to Al (Fe_{tot}/Al) for shales, which scatter about the average crustal value beneath OAE-2, but increase markedly within OAE-2 (Owens et al. 2017b). **d** δ^{34}S values (CDT scale) for CAS for carbonate units below and above OAE-2, showing initially low values, followed by strong increase through OAE-2; not shown are data for pyrite in shales, which show the same trend of increasing δ^{34}S values through OAE-2, including samples within the OAE-2 interval (Owens et al. 2012). **e** Changes in initial $^{187}Os/^{188}Os$ ratios for shales through OAE-2, which show a strong decrease immediately before OAE-2, followed by recovery within the OAE-2 interval, features that are seen on a global scale (Turgeon and Creaser 2008)

et al. 2017a). In contrast to the Eastbourne locality, sedimentation across OAE-2 occurred much more slowly and produced C_{org}-rich black shales. In Fig. 6.7 we plot the large-scale changes across OAE-2, which represents ~2.5 m.y. of time (Lanci et al. 2010). Black shales beneath OAE-2 record the lowest δ^{56}Fe values yet measured for Cretaceous marine sedimentary rocks (δ^{56}Fe < −1‰). Within OAE-2, however, δ^{56}Fe values increase markedly, and are well correlated with the increase in δ^{13}C values for C_{org} that marks the onset of OAE-2. In contrast with the δ^{56}Fe values that lie far below the crustal average in the pre-OAE-2 section, Fe_{tot}/Al ratios plot near the crustal average, but increase strongly within the OAE-2 level. δ^{34}S values for CAS start at very low values before OAE-2, some of the lowest measured for rocks of this age, and then begin to climb with the onset of OAE-2, reaching levels at the end of OAE-2 that are common in other sections (Fig. 6.7d). The trend of increasing δ^{34}S values through OAE-2 for CAS is also seen in data for pyrite (Owens et al. 2017a), although not plotted in Fig. 6.7 for simplicity. The C_{org}-rich nature of shales at Furlo has allowed isotopic analysis of a number of elements that are enriched in black shales, including Os, and the well-documented changes in initial $^{187}Os/^{188}Os$ ratios through OAE-2, with a strong decrease immediately prior to the event, are also seen at Furlo (Turgeon and Creaser 2008), interpreted to record a sudden increase in volcanic input, followed by recovery to more radiogenic compositions. Note that these changes are quite different from those shown for Nd isotopes at Eastbourne, likely reflecting the isotopic provinciality of Nd due to its short residence time relative to Os.

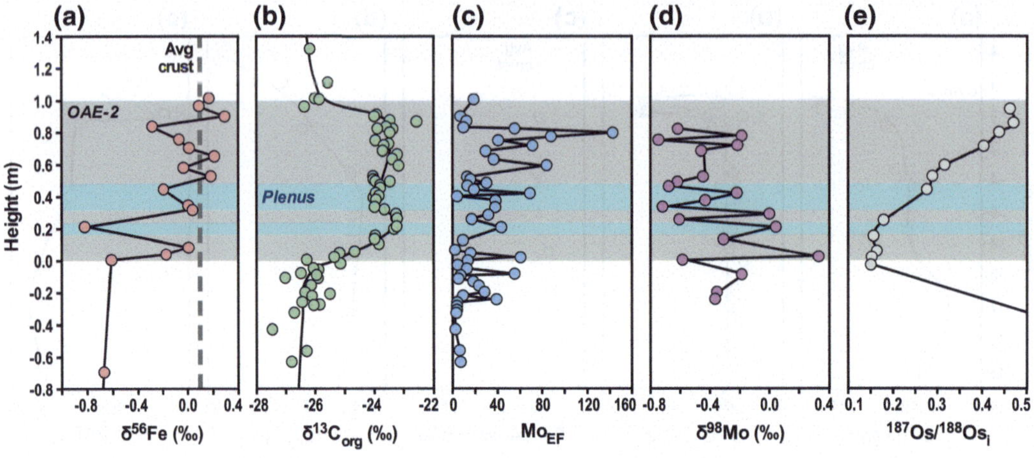

Fig. 6.8 Detailed changes in geochemical trends across OAE-2 at Furlo, Italy locality (see also Fig. 6.7). OAE-2 interval shown as grey band, with the first and second (major) Plenus cooling events as scaled from the Eastbourne section (Fig. 6.6). **a** δ^{56}Fe values for shales. **b** $\delta^{13}C_{org}$ values (PDB scale) for shales. **c** Enrichment factor for Mo (Mo_{EF}, defined as measured Mo/Al divided by Mo/Al of average upper crust). **d** Mo isotope compositions (reported as δ^{98}Mo using ^{98}Mo/^{95}Mo, relative to NIST 3134; Willbold et al. 2015). **e** Initial ^{187}Os/^{188}Os ratios for shales. The main (second) Plenus cooling event is shown at Furlo as a decrease in $\delta^{13}C_{org}$ values, similar to the C isotope changes for carbonate seen at Eastbourne (Fig. 6.6b), although the first Plenus interval is not seen in the $\delta^{13}C_{org}$ values at Furlo. The additional Mo concentration and isotopic data shown here highlight the cyclicity of Mo behavior, indicating transient euxinic conditions. The temporal changes in δ^{56}Fe values might also indicate cyclicity, although not closely correlated with Mo. Data sources as in Fig. 6.7, with additional data for "c" and "d" from Westermann et al. (2014)

In Fig. 6.8 we explore a detailed section from Furlo, adding Mo concentration and isotopic measurements made on the C_{org}-rich shales in the OAE-2 interval. Within OAE-2, δ^{56}Fe values oscillate but overall move toward those of average crust. At Furlo, the main Plenus cooling event is shown in δ^{13}C values for C_{org}, although the inflection seen in $\delta^{13}C_{carb}$ for the first Plenus interval at Eastbourne is not apparent in the C_{org} data. Iron speciation data from Furlo, in addition to redox-sensitive trace elements, indicate ferruginous but non-euxinic conditions immediately prior to OAE-2, followed by a shift to intermittently euxinic conditions within OAE-2, as marked by periodic spikes in enrichment factors for elements such as Mo (Fig. 6.8c; Westermann et al. 2014; Ostrander et al. 2017). Such a conclusion is supported by Mo isotope data, which are quite variable through OAE-2, and whose peak δ^{98}Mo values are significantly lower than those of modern seawater (δ^{98}Mo = +2.1‰ on the NIST 3134 scale; Nakagawa et al. 2012), indicating a significantly higher proportion of euxinic sinks relative to the modern ocean (Westermann et al. 2014). Considering both large-scale (Fig. 6.7) and fine-scale (Fig. 6.8) variations at Furlo suggests that a benthic Fe shuttle was active prior to OAE-2, as would be expected for the low δ^{56}Fe values at times of low seawater sulfate contents, which was then replaced by a sulfide-dominated system under euxinic conditions during OAE-2. This would produce higher δ^{56}Fe values in the water column during the euxinic conditions during OAE-2, either through FeS–Fe_{aq} exchange or formation of sedimentary pyrite (see Sect. 5.4.2 and Figs. 5.15 and 5.16 in Chap. 5). One difficulty in invoking a benthic Fe shuttle for the pre-OAE-2 section at Furlo is the lack of apparent Fe enrichment based on Fe_{tot}/Al ratios. It is possible that a lower than average crust Fe_{tot}/Al ratio is characteristic of the clastic input, which, if true, would indicate that Fe_{tot}/Al ~ 0.5 actually reflects Fe enrichment, but this remains unknown.

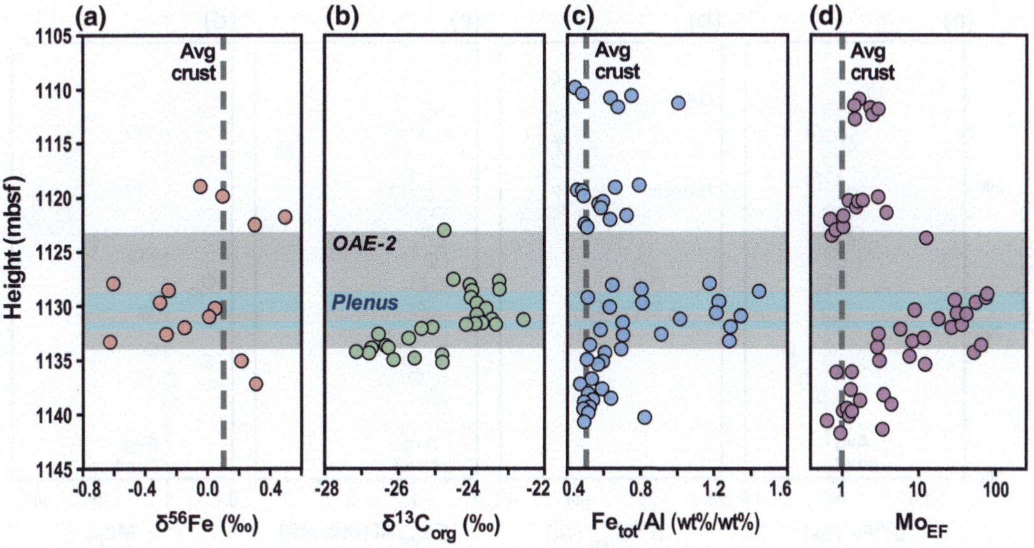

Fig. 6.9 Geochemical data across OAE-2 from DSDP core from site 603B, located east of the continental rise terrace off the east coast of the US. This site represents a deep-ocean (~4 km water depth) proto-North Atlantic setting during OAE-2 (Kuypers et al. 2004). Relative to an adjacent site (DSDP 105, 125 km SE), site 603B reflects hemipelagic sedimentation during OAE-2, with a larger continental influence. OAE-2 (grey band) identified based on C_{org} contents and C isotope compositions (Owens et al. 2012); Plenus cooling events are scaled from Eastbourne locality (Fig. 6.6) but are not identified at site 603B by C isotope data. **a** δ^{56}Fe values for shales (Owens et al. 2012). **b** $\delta^{13}C_{org}$ values (PDB scale) for shales (Kuypers et al. 2004; Owens et al. 2012). **c** Ratio of total Fe to Al (Fe_{tot}/Al) for shales (Owens et al. 2012). **d** Enrichment factor for Mo (Mo_{EF}) for shales (note log scale; Owens et al. 2012), as defined in Fig. 6.8. δ^{56}Fe values above and below OAE-2 are similar to those of average crust, but are distinctly negative within OAE-2. Fe_{tot}/Al ratios beneath OAE-2 are similar to, or slightly higher than, average crust, but extend to very high ratios within OAE-2, and then are low to moderately high after OAE-2. Mo_{EF} trends from normal crustal values (Mo_{EF} = 1) toward strong Mo enrichment through the early part of OAE-2, followed by a decline after OAE-2

Two DSDP cores off the east coast of the US, sites 603B and 105, transect the Cenomanian-Turonian boundary and have been analyzed for Fe isotopes (Figs. 6.9 and 6.10). Kuypers et al. (2004) interpreted the alternating C_{org}-poor and -rich shales to record fluctuating redox conditions in the water column at these sites, and, during euxinic periods, suggested that the redoxcline ascended to the photic zone. Both sites record the decrease in δ^{13}C values for C_{org} that defines OAE-2, although the data are more scattered than what is seen at Furlo. In both cores, δ^{56}Fe values are quite low within OAE-2, which Owens et al. (2012) interpreted to reflect a benthic Fe shuttle. This is quite different than the data from Eastbourne and Furlo, where δ^{56}Fe values within OAE-2 lie near crustal values or slightly higher. Importantly, the low δ^{56}Fe values at 603B and 105 are accompanied by both strong Fe enrichment (high Fe_{tot}/Al) and Mo enrichment (high Mo_{EF}; see caption for Fig. 6.8), indicating that at these sites, OAE-2 was a euxinic trap for a benthic Fe shuttle. This model is reasonable given the proximity of the locations to the continental margin at the time of deposition. The geochemical relations at sites 603B and 105 are analogous to those of the modern Black Sea, Gotland Deep, and Eastern Mediterranean, all of which are interpreted to record a euxinic trap of low-δ^{56}Fe Fe from a benthic Fe shuttle (see Sect. 5.4.4 and Fig. 5.19 from Chap. 5).

The last OAE-2 section we discuss is DSDP site 367 from the Cape Verde Basin (Fig. 6.11), which is interpreted to record deposition under strongly euxinic conditions, where, for example, biomarkers identify the occurrence of green

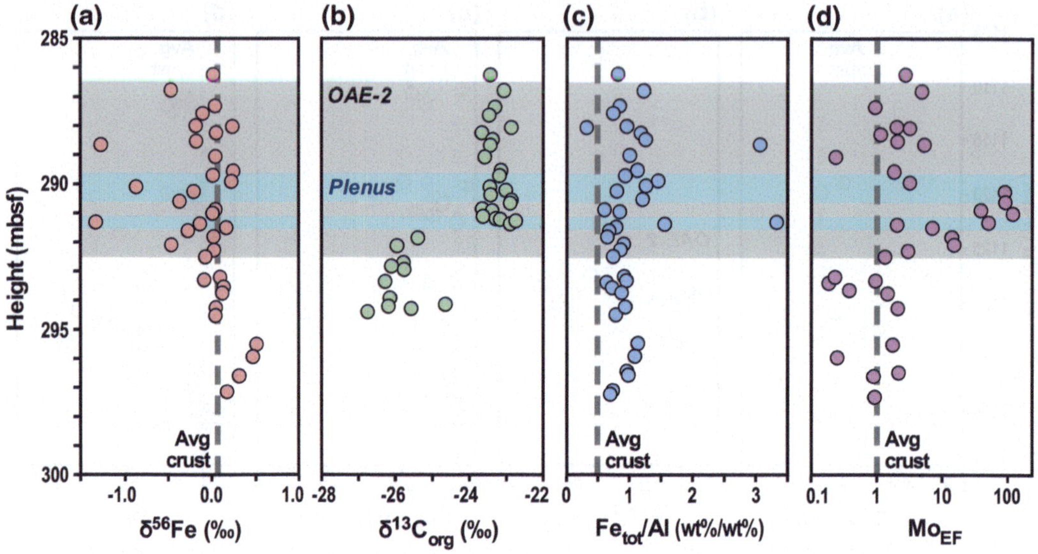

Fig. 6.10 Geochemical data across OAE-2 from DSDP core from site 105, located in the lower continental rise hills, ~125 km SE of site 603B (Fig. 6.9). This site represents pelagic deposition in a deep-ocean (~4 km) proto-North Atlantic setting during OAE-2 (Kuypers et al. 2004). Data sources same as in Fig. 6.9. **a** δ^{56}Fe values for shales. **b** $\delta^{13}C_{org}$ values (PDB scale) for shales. **c** Ratio of total Fe to Al (Fe_{tot}/Al) for shales. D: Enrichment factor for Mo (Mo_{EF}) for shales (note log scale), as defined in Fig. 6.8. Geochemical relations are similar to those of site 603B, although δ^{56}Fe values within OAE-2 extend to lower values. Both sites 603B and 105 show large amounts of Fe_{tot}/Al and Mo/Al enrichment within OAE-2, consistent with a benthic Fe shuttle sourced from the continental margin, as well as euxinic conditions

sulfur bacteria at shallow water depths (~15 m), indicating euxinic conditions virtually in the surface oceans (Kuypers et al. 2002). Extensive C isotope data document the onset of OAE-2, although note the subtleties that are associated with the Plenus cooling events. Owens et al. (2012) report Fe isotope data for this section, as well as other localities from the proto-North Atlantic. Pre-OAE-2 samples are characterized by slightly positive δ^{56}Fe values and Fe_{tot}/Al ratios near those of average crust, followed by a decrease in δ^{56}Fe values and strong increase in Fe_{tot}/Al ratios within OAE-2; these changes are similar to what was seen in DSDP sites 603B and 105, although the decreases in δ^{56}Fe values at site 367 are more modest. Very high resolution records for redox-sensitive trace elements and Mo isotopes are available for site 367, and these have been interpreted to record fluctuating redox and degree of euxinia (Westermann et al. 2014; Dickson et al. 2016). In particular, Dickson et al. (2016) point to fine-scale changes in enrichment factors for Mo and U (expressed as Mo_{EF}/U_{EF}), where high Mo_{EF}/U_{EF} records euxinic conditions relative to low-S anoxic conditions, which broadly correlate with peaks in δ^{98}Mo values (Fig. 6.11). Enrichment in U, but not corresponding enrichments in Mo (low Mo_{EF}/U_{EF}), are inferred to indicate anoxic but non-euxinic conditions. The maximum δ^{98}Mo values attained at site 367 (δ^{98}Mo ~ + 1‰) are significantly higher than those measured at Furlo (Fig. 6.8), consistent with the greater extent of euxinia at site 367.

The slightly negative δ^{56}Fe values for the C_{org}- and S-rich shales of the OAE-2 section at site 367 likely reflect a strong contribution from pyrite. Such compositions are similar to those measured for sedimentary pyrite in the Black Sea, and, as discussed in Sect. 5.3.3 (Chap. 5), where the negative δ^{56}Fe values may record kinetic Fe isotope fractionations upon precipitation from euxinic water columns (see also Fig. 5.13 in Chap. 5). Given the strong evidence

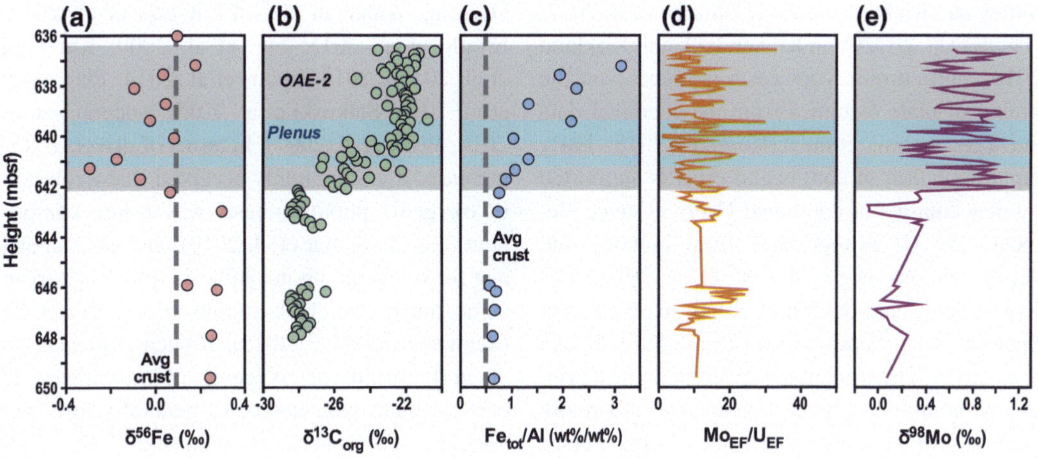

Fig. 6.11 Geochemical data across OAE-2 from DSDP core 367, located in the Cape Verde Basin off the coast of western Africa. This site represents a deep-ocean (~3.7 km water depth) proto-North Atlantic setting during OAE-2 under strongly euxinic conditions throughout the water column (Kuypers et al. 2002). **a** $\delta^{56}Fe$ values for shales (Owens et al. 2012), showing high values before OAE-2, followed by a decrease during OAE-2. Note sample plotted at 636 mbsf is next data point above a break in section. **b** $\delta^{13}C_{org}$ values (PDB scale) for shales (Dickson et al. 2016). **c** Ratio of total Fe to Al (Fe_{tot}/Al) for shales (Owens et al. 2012). **c** High-resolution data for ratio of Mo enrichment (Mo_{EF}) to U enrichment (U_{EF}) showing cyclicity of euxinic (high Mo_{EF}/U_{EF}) to anoxic (low Mo_{EF}/U_{EF}) conditions during OAE-2 (Westermann et al. 2014; Dickson et al. 2016); enrichment factors (EF) defined relative to Al, per Fig. 6.8 caption. **d** High-resolution record of Mo isotope variations (reported as $\delta^{98}Mo$ using $^{98}Mo/^{95}Mo$, relative to NIST 3134; Willbold et al. 2015), showing cyclical changes in extent of euxinia and/or seawater Mo isotope compositions during OAE-2 (Westermann et al. 2014; Dickson et al. 2016)

for periodic euxinia in virtually the entire water column at site 367 during OAE-2, it is possible that the Fe isotope compositions at this site record water column processes. Using the relations observed in the Black Sea, however, would suggest that Fe^{2+} in seawater at this site could have attained very high $\delta^{56}Fe$ values (Fig. 5.13 in Chap. 5), which is very difficult to test from the rock record. Fruitful studies for the future would be separation of different Fe phases in OAE-2 sediment samples for isotopic analysis, as well as close correlation with detailed isotopic records such as Mo isotopes.

6.3 Precambrian Earth: An Introduction

A vast literature exists on the chemical and biological evolution of the surface Earth over geologic time, and a comprehensive review of these subjects is far beyond the scope of this book. Our focus in this introduction will be on examples that ultimately bear on the Fe biogeochemical cycle in the ocean-atmosphere system, with an emphasis on the Precambrian marine system. Geologic data and modeling document the existence of a significant ocean since the earliest Archean (see, for example, review by Korenaga et al. 2017), with evidence for liquid water on the surface Earth extending back as far as ~4.2 to 4.3 Ga (e.g., Cavosie et al. 2005). The relative chemical fluxes to the oceans from mantle and continental sources have varied greatly over the last 4 b.y., the former scaling with mantle temperature and heat flow, and the later scaling with continental crustal volume and exposure. Mantle potential temperature (T_P) has been modeled as continuously decreasing or with inflections early in Earth history, depending upon the timing and vigor of plate tectonics (e.g., Labrosse and Jaupart 2007; Korenaga 2013). There is a substantial body of evidence that indicates the start of modern-style plate tectonics occurred about 3.2 to 3.0 Ga (e.g.,

Shirey and Richardson 2011; Dhuime et al. 2012; Naeraa et al. 2012; Van Kranendonk and Kirkland 2016), and current work casts inferences on the timing of plate tectonics from geochemical data into a geodynamic context (Korenaga 2018). Early rapid evolution of continental crust is supported by new datasets of combined U-Pb, O, trace element, and Hf isotope data from igneous and detrital zircons (e.g., Dhuime et al. 2012; Van Kranendonk and Kirkland 2016; Roberts and Spencer 2015; Hawkesworth et al. 2017; Iizuka et al. 2017). There are major differences, however, in interpretations of these data in terms of smooth increases in continental crustal volume versus periodicities that reflect episodic changes in mantle dynamics and continental amalgamation and dispersal. Ultimately, the details of the evolution of the mantle and continents have determined the geochemical changes in the ocean-atmosphere system, including the Fe cycle.

6.3.1 Broad Changes in the Surface Earth in the Precambrian

There is general consensus that Earth's atmosphere, and by extension, the oceans, transitioned from an anoxic to oxic state during the Precambrian (e.g., Holland 1984, 2006; Canfield 2005; Lyons et al. 2014, to name but a very small fraction of publications), although there continues to be much discussion over the timing, rates, and extent of changes in the oxidation state of the oceans and atmosphere. A major rise in atmospheric oxygen in the Paleoproterozoic at ~2.3 Ga, classically termed the "Great Oxidation Event" (GOE; Fig. 6.12) (e.g., Bekker et al. 2004; Holland 2006), was originally envisioned to record a step-wise increase in atmospheric O_2 from essentially anoxic conditions to perhaps ~0.01–0.1% Present Atmosphere Level (PAL). More recent work suggests that O_2 contents during the GOE may have reached very high levels, perhaps approaching 100% PAL (e.g., Karhu and Holland 1996; Bekker and Holland 2012). It is now recognized that there were likely periods of punctuated oxygen increases before the GOE as far back as ~3.2 Ga (e.g., Anbar et al. 2007; Czaja et al. 2012; Eroglu et al. 2015; Frei et al. 2009; Kurzweil et al. 2015a, 2016; Lyons et al. 2014; Planavsky et al. 2014; Satkoski et al. 2015; Voegelin et al. 2010; Wille et al. 2007). In terms of drivers of O_2 increase, it is commonly accepted that evolution of oxygenic photosynthesis was a key component (e.g., Farquhar et al. 2010), and an Archean age for oxygenic photosynthesis finds support in some molecular clock studies of cyanobacteria (Schirrmeister et al. 2015). If such evidence for an early origin of oxygenic photosynthesis is correct, then maintenance of generally low (but variable) atmospheric O_2 until the GOE might have reflected, for example, large oxygen sinks or other changes in overall electron balance (e.g., Kasting et al. 1993; Kump and Barley 2007; Gaillard et al. 2011), or nutrient limitation (e.g., Laakso and Schrag 2014; Ozaki et al. 2019). A major component of redox control is the overall balance of inorganic and organic C in the atmosphere-ocean-sediment system (e.g., Hayes and Waldbauer 2006; Husson and Peters 2017).

A wide variety of estimates have been made for post-GOE O_2 levels in the Proterozoic. Post-GOE decreases in redox-sensitive trace elements in shales, including Mo and U, suggest a decrease in O_2 after the GOE (e.g., Scott et al. 2008; Canfield et al. 2013; Partin et al. 2013a). Similar post-GOE decreases in O_2 have been estimated from temporal trends in the trace element contents of marine carbonates (e.g., Liu et al. 2016; Hardisty et al. 2017; Bellefroid et al. 2018), as well as isotopic variations (e.g., Planavsky et al. 2012b). Many of these records are interpreted to reflect atmospheric O_2 levels on the order of ~0.1 to ~1% PAL, which, if extended to the oceans, might suggest shallow-water O_2 levels as low as ~0.1% Present Ocean Level (POL), or about 0.3 μM (Tang et al. 2016). Not all proxy records, however, have been interpreted to record low post-GOE O_2 contents, with some estimates in the range of a few % PAL (Zhang et al. 2016; Yang et al. 2017). A recent review of proxy data from the mid-Proterozoic suggests oxygen levels <1% PAL (Planavsky et al. 2018a). Direct estimates of Proterozoic atmospheric O_2 contents via halite inclusions have produced controversial results from <10% PAL

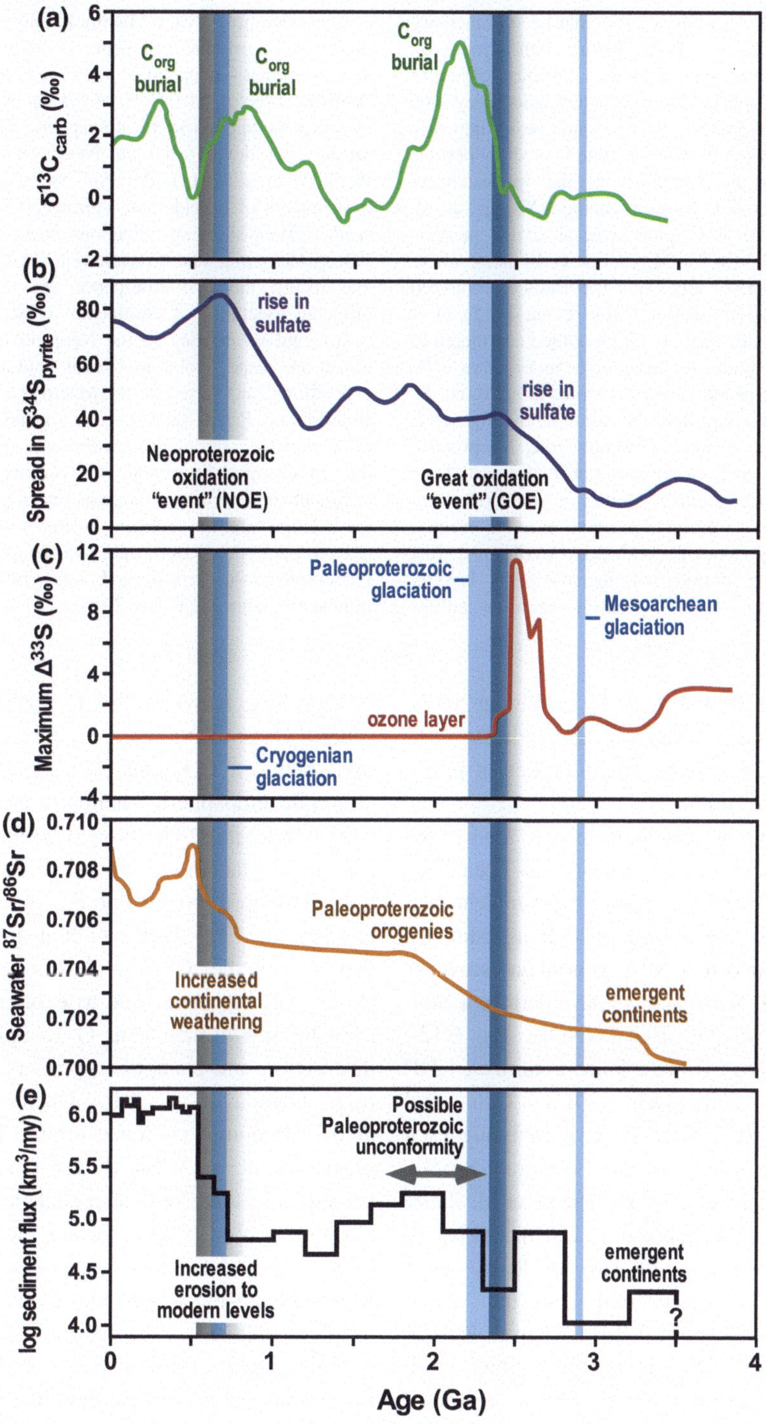
(a)
$\delta^{13}C_{carb}$ (‰)
6
4
2
0
-2
C_{org} burial
C_{org} burial
C_{org} burial
(b)
Spread in $\delta^{34}S_{pyrite}$ (‰)
80
60
40
20
0
rise in sulfate
rise in sulfate
Neoproterozoic oxidation "event" (NOE)
Great oxidation "event" (GOE)
(c)
Maximum $\Delta^{33}S$ (‰)
12
8
4
0
-4
Paleoproterozoic glaciation
Mesoarchean glaciation
ozone layer
Cryogenian glaciation
(d)
Seawater $^{87}Sr/^{86}Sr$
0.710
0.708
0.706
0.704
0.702
0.700
Paleoproterozoic orogenies
Increased continental weathering
emergent continents
(e)
log sediment flux (km^3/my)
6.0
5.5
5.0
4.5
4.0
Possible Paleoproterozoic unconformity
Increased erosion to modern levels
emergent continents
?
0
1
2
3
4
Age (Ga)

◄ **Fig. 6.12** Isotopic variations for marine sedimentary rocks and minerals over Earth history with a focus on Precambrian events and features. Periods of rapid increases in atmospheric O_2 shown in shaded grey bars for the Paleoproterozoic (GEO) and Neoproterozoic (NOE) from Lyons et al. (2014). Blue bars show periods of global or major glaciations in the Mesoarchean, Paleoproterozoic, and Neoproterozoic (Young et al. 1998; Hoffman 2013; Caquineau et al. 2018). **a** $\delta^{13}C$ values (PDB scale) for Ca–Mg marine carbonates, where excursions to positive values reflect periods of enhanced organic C burial (Krissansen-Totton et al. 2015). It is important to note that the curve is smoothed and therefore does not reflect significant excursions to negative $\delta^{13}C$ values that occur in the Neoproterozoic. In addition, the smoothing does not highlight the extent of the high-$\delta^{13}C$ maxima that occur in the Paleoproterozoic, Neoproterozoic, and Phanerozoic. Curve does not include data from Fe- or Mn-rich rich carbonates found in iron formations, many of which have highly negative $\delta^{13}C$ values. **b** Spread in $\delta^{34}S$ values for authigenic pyrite in marine sediments, noting periods of increase that broadly correspond to periods of increasing seawater sulfate contents (Kump 2012). **c** Maximum range in $\Delta^{33}S$ values for pyrite in marine sediments as an indicator of the degree of mass-independent fractionation of S isotopes (MIF-S); loss of MIF-S by ~2.3 Ga is interpreted to record a sufficient rise in atmospheric O_2 such that an ozone layer developed (data based on compilation by Williford et al. 2011). **d**: $^{87}Sr/^{86}Sr$ ratios for seawater from curves of Shields and Veizer (2002) and Satkoski et al. (2017), showing inflections in the early Archean, Paleoproterozoic, and Precambrian-Cambrian transition that record the first emergence of continents, major orogenic assembly of continental crust, and increased continental weathering in the Neoproterozoic-Cambrian transition, respectively. **e**: Global continental sediment depositional flux based on the Macrostrat database, from Husson and Peters (2017). Curve roughly scales with calculated C_{org} burial flux, with some additional peaks due to changes in organic C contents. Modeling of sediment survival curves suggests that significant parts of the Paleoproterozoic sediment inventory may be missing, reflecting a large unconformity, which would increase the Paleoproterozoic depositional rates relative to those plotted (Husson and Peters 2018)

to ~50% PAL (Blamey et al. 2016; Yeung 2017). Collectively, these findings might indicate variable O_2 levels over time or difficulties in proxy interpretation, or both.

Oxygenation of the atmosphere across the Neoproterozoic-Cambrian transition has been studied using redox-sensitive trace-element contents and their isotopic compositions, as well as estimates derived from the extent of C_{org} burial and seawater sulfate contents. Some authors have defined a Neoproterozoic Oxidation Event (NOE; Fig. 6.12), drawing an analogy with the Paleoproterozoic GOE (Shields-Zhou and Och 2011; Och and Shields-Zhou 2012). Many proxy records suggest a significant rise in O_2 in the Ediacaran, ~550–560 Ma (Canfield et al. 2007; Scott et al. 2008; Kendall et al. 2015b; Sahoo et al. 2016). The rise of atmospheric O_2 to near-modern levels, however, appears to have occurred in the Cambrian, ~520 Ma (Dahl et al. 2014; Chen et al. 2015; Wang et al. 2015; Wen et al. 2015; Xiang et al. 2017). Redox proxy records suggest significant spatial and temporal heterogeneity in O_2 during this transition (e.g., Li et al. 2010; Wen et al. 2015; Lau et al. 2017), which, perhaps, is not surprising given the wide range in marine oxygen contents that can be found in modern marine systems (see Chap. 5).

Coupling between the C cycle and atmospheric O_2 contents often produces inverse relations between CO_2 and O_2 (e.g., Berner 2006), although atmospheric chemistry and weathering add complexities. The broad correlation between the global glaciation ("Snowball Earth") periods in the Paleoproterozoic and Neoproterozoic (e.g., Evans et al. 1997; Hoffman et al. 1998; Hoffman 2013) and the GOE and NOE, respectively (Fig. 6.12) has been used to explicitly link CO_2 drawdown with increasing O_2, coupled with CH_4 destruction and changes in primary productivity (e.g., Harada et al. 2015; Daines and Lenton 2016). Although evidence for the Mesoarchean glaciation at ~2.9 Ga (Fig. 6.12) that is preserved in the Witwatersrand-Pongola basin (Young et al. 1998) is scarcer than that of the GOE- and NOE-associated glaciations, the Mesoarchean glaciation was followed by a significant marine transgression, suggesting a globally significant glaciation, as well as iron formation deposition (Smith et al. 2013), sharing similarities with the younger glacial episodes. It remains unclear if the Mesoarchean glaciation was accompanied by extensive CO_2 drawdown and increases in O_2, although such relations are plausible given the isotopic evidence for transient oxygenation during this time (e.g., Planavsky et al. 2014).

The C and S isotope records (Fig. 6.12) are broadly consistent with the environmental changes discussed above, although there are important differences between observed trends and what simple geochemical models might predict. The peaks in $\delta^{13}C_{carb}$ values in the Paleoproterozoic and Neoproterozoic (Fig. 6.12a) are traditionally interpreted to record increases in the proportion of C_{org} burial relative to inorganic C (e.g., Krissansen-Totton et al. 2015, and references within). It is important to note that the smoothed curves in Fig. 6.12 for C and S isotopes do not capture rapid and large changes in the marine record. The broad positive $\delta^{13}C_{carb}$ peaks seen in the Precambrian are not precisely aligned with the GOE and NOE; the Paleoproterozoic $\delta^{13}C_{carb}$ peak occurs after the GOE, just after the end of the Paleoproterozoic glaciation, and the Neoproterozoic $\delta^{13}C_{carb}$ peak occurs before the Neoproterozoic glaciation and the NOE. These "offsets" in part reflect smoothing, but such observations may also indicate that the $\delta^{13}C_{carb}$ record does not solely reflect burial of C_{org}, as highlighted in recent modeling (Daines et al. 2017; Miyazaki et al. 2018). The spread in $\delta^{34}S_{pyrite}$ values (Fig. 6.12b) is quite low in the early Archean, reflecting small S isotope fractionations between sulfate and pyrite that indicate very low seawater sulfate contents of perhaps a few μM (Crowe et al. 2014). The steady increase in the spread of $\delta^{34}S_{pyrite}$ values from ~3 Ga to the GOE has been commonly taken to reflect increasing seawater sulfate contents (e.g., Canfield 2005; Canfield and Farquhar 2009), perhaps reflecting the transient increases in oxygen levels inferred for this time period. Estimates for sulfate contents across the GOE suggest mM levels (Planavsky et al. 2012b). Between the GOE and ~1 Ga, the consistent spread in $\delta^{34}S_{pyrite}$ values may indicate low and broadly constant sulfate contents at the mM level (Kah et al. 2004), although on a fine-scale, S isotope compositions can be quite variable in this time period (e.g., Gomes et al. 2018), suggesting heterogeneity. The strong increase in the spread of $\delta^{34}S_{pyrite}$ values from ~1 Ga to the end of the Neoproterozoic is taken by many workers to indicate an increase in sulfate contents, consistent with the findings from fluid inclusion studies that the late Neoproterozoic oceans may have contained up to 15 mM sulfate prior to a decline in the Cambrian (Horita et al. 2002; Petrychenko et al. 2005). Recent diagenetic transport modeling of S cycling during the Proterozoic, however, points to sulfate concentrations remaining below 400 μm for much of the Precambrian (Fakhraee et al. 2019), demonstrating the difficulty of quantifying seawater sulfate contents.

Turning to mass-independent fractionation of S isotopes (MIF-S), expressed as $\Delta^{33}S$ values (Fig. 6.12c), non-zero $\Delta^{33}S$ values are common in pre-GOE marine sedimentary rocks (e.g., Farquhar et al. 2000, 2011), which are accepted by most workers to indicate low atmospheric O_2 contents, perhaps $<10^{-5}$ PAL (e.g., Pavlov and Kasting 2002), although such low levels in the Neoarchean may be pose a problem for the proxy results discussed above. The MIF-S record shows an excursion to highly positive $\Delta^{33}S$ values in the Neoarchean, before the GOE, and this end member is thought to be S^0 from atmospheric sources, whose isotopic composition was transferred to H_2S during pyrite formation (e.g., Farquhar et al. 2013). The loss of MIF-S is commonly taken as the defining feature of the GOE (e.g., Bekker et al. 2004), and indicative of growth of an ozone shield after the GOE.

As discussed in Sect. 6.1, the balance between continental and mantle inputs to the oceans affect the modern marine Fe cycle, and in the context of Earth evolution over the last 4 b.y., the chemistry of the oceans informs us about changes in continental exposure, erosion, C_{org} burial, and climate. The $^{87}Sr/^{86}Sr$ curve for the oceans has been challenging to reconstruct, particularly for Precambrian rocks, due to issues of clastic contamination that are exacerbated in old rocks, as well as post-depositional alteration (e.g., Shields and Veizer 2002). It is commonly assumed, for example, that $^{87}Sr/^{86}Sr$ ratios for seawater in most of the Archean lay close to that of the mantle, consistent with modeling that suggested the continents were largely submerged (Flament et al. 2008, 2013). More recent studies, however, show that early Archean seawater was much more radiogenic than previously thought, with a significant rise at ~3.2 Ga (Fig. 6.12d) that has been interpreted to record emergent and evolved continental crust that was produced in a convergent-margin setting and therefore indicates plate tectonics (Satkoski et al. 2016, 2017). Input of radiogenic Sr to the oceans is consistent with the high chemical alteration indices for early Archean rocks (e.g., Hessler and Lowe 2006), as well as evidence for high-$\delta^{18}O$ crust of 3.6–3.8 Ga age that indicates weathering (e.g., Bolhar et al. 2017). Post-Archean seawater $^{87}Sr/^{86}Sr$ ratios smoothly increase through the GOE, essentially ignoring proposed correlations between the emergence of evolved continental crust and the GOE (Lee et al. 2016; Smit and Mezger 2017), or evidence for a sudden increase in the $\delta^{18}O$ values for detrital zircons across the GOE (Valley et al. 2005), and instead peaks during the episode of major Paleoproterozoic orogenic activity (Fig. 6.12d). Seawater $^{87}Sr/^{86}Sr$ ratios remain relatively constant through the mid-Proterozoic but quickly ascend to their highest levels (higher than modern) across the Neoproterozoic-Cambrian transition (Fig. 6.12 d), which has been taken to indicate extensive continental weathering and an emergent terrestrial biosphere (e.g., Derry et al. 1992; Squire et al. 2006; Shields 2007).

The discussion above highlights the use of isotopic mass-balance models to infer burial of C_{org}, seawater sulfate, uplift and erosion, and the possible roles these factors played in determining climate and evolution of the biosphere. The advantage of such models is that they negate the need for determining absolute masses of element fluxes, instead relying on ratios of fluxes, such as the ratio of inorganic and organic C (e.g., Berner et al. 2003). Recently, however, a major effort has been underway to quantify global sedimentation rates over Earth history (Peters 2006). Reconstructed global sedimentation rates using the Macrostrat database offers an opportunity to directly test isotopic-based models (Husson and Peters 2017), and the results offer some surprising findings (Fig. 6.12e). Several aspects of the seawater $^{87}Sr/^{86}Sr$ curve correlate with pulses of increased sedimentation at ~3.2 Ga, ~1.8 Ga, and ~0.6 Ga. Using measured C_{org} contents and sedimentation rates, Husson and Peters (2017) show peaks of increased C_{org} burial at ~2.7 Ga and ~2.3 Ga, with the former corresponding to evidence noted above for transient increases in atmospheric O_2 in the Neoarchean and an increase in the spread of $\delta^{34}S$ values, and the latter corresponding to the GOE and global glaciation (but distinctly before the positive $\delta^{13}C$ Lomagundi excursion; Karhu and Holland 1996). An important consideration in the reconstructed sediment history is the possibility of a large missing inventory of Paleoproterozoic sediments, perhaps indicating a large Paleoproterozoic unconformity (Husson and Peters 2018); if true, the peak in sedimentation rates at ~1.8 Ga would have been even higher than that shown in Fig. 6.12e, and by extension, possibly much higher C_{org} burial rates. Through the Meso- and Neoproterozoic, sedimentation rates declined, consistent with the seawater $^{87}Sr/^{86}Sr$ curve, although Husson and Peters (2017) calculate an increase in C_{org} burial rates at ~1.0 Ga when additionally considering C_{org} contents, which corresponds with the Neoproterozoic increase in $\delta^{13}C_{carb}$ values (but does not explain the entire Proterozoic C isotope curve). The very rapid increase in sediment burial rates through the Neoproterozoic-Cambrian transition is well shown in Fig. 6.12e, corresponding closely to the increase in seawater $^{87}Sr/^{86}Sr$ ratios. As will be discussed below, the variations in sedimentation rates and C_{org} burial that may be inferred directly from the rock record provides us with additional guidance beyond that obtainable from isotopic records alone in terms of changes in climate, atmosphere, and microbial metabolisms that may be recorded in the C-S-Fe biogeochemical cycle.

6.3.2 Temporal Changes in Fe Abundance and Speciation

We next turn to the inventories of Fe measured in marine sedimentary rocks over time, with an eye toward changes in mass balance and redox state. As discussed in Chap. 5, it has been common to assess relative enrichments or depletions of Fe in marine sediments through normalization to Fe/Al or Fe/Ti ratios, and this also discussed above in Sect. 6.2. Although the Fe/Al (or Fe/Ti) ratios for local lithologic input to marine basins may be variable (see for example, Fig. 5.19 in Chap. 5), we assume, at least initially, a crustal average when considering broad compositional variations in Precambrian rocks. Based on the compilations and models of Condie (1993), the average Fe_{tot}/Al for Precambrian crust is taken to be 0.46 (Fe_{tot} ignores Fe^{2+} – Fe^{3+} speciation), and average Fe_{tot}/Ti is taken to be 11.23. These compositions would reflect the average detrital input to the ocean, ignoring biogeochemical diagenetic effects. For comparison, Raiswell et al. (2008) determined an average Fe_{tot}/Al ratio of 0.53 for Paleozoic shales deposited in oxic settings. In their compilation of the chemical compositions of sandstones and shales, Kump and Holland (1992) noted that virtually all Archean-age samples had Fe_{tot}/Ti ratios greater than those of igneous rocks or Phanerozoic sedimentary rocks, and the majority of samples had Fe_{tot}/Ti ratios over twice that of the igneous and Phanerozoic averages. They interpreted this to reflect net addition of hydrothermal Fe in Archean marine sedimentary basins, given the expected higher heat flow and seawater aqueous Fe contents in the Archean.

In Fig. 6.13 we compare the distribution of Fe_{tot}/Al ratios in fine-grained marine clastic rocks (shales/mudstones) of Paleoarchean to Neoproterozoic age using a large database that has its origins in the work of Bekker and Holland (2012), augmented with additional data (see caption). Relative to the crustal average, it is clear that rocks deposited before the GOE have significantly higher Fe_{tot}/Al ratios. The database is sparse for Paleoarchean rocks, but the data available suggests Fe enrichment relative to the crustal average Fe_{tot}/Al ratio (Fig. 6.13a). A large dataset is available for Mesoarchean rocks, which clearly show very high levels of Fe enrichment (Fig. 6.13b), extending down to Fe_{tot}/Al ratios that match those of average crust. The majority of the high-Fe_{tot}/Al lithologies belong to the very thick sequences of magnetic mudstones that are found in marine sections of this age, including the ~3.2 Ga Moodies Group of the Barberton greenstone belt (Heubeck et al. 2013; Heubeck and Lowe 1994) and the ~2.9 Ga Witwatersrand Supergroup of the Kaapvaal craton (Guy et al. 2012; Smith et al. 2013). The very high abundance of magnetite-rich mudstones and shales of this age range are unknown in younger sections of Precambrian marine sedimentary rocks. Neoarchean and pre-GOE Paleoproterozoic shales and mudstones have more modest Fe enrichments relative to the crustal average (Fig. 6.13b). In contrast, all of the samples of post-GOE age have Fe_{tot}/Al ratios that scatter about the crustal average, with the exception of a slight enrichment in some Neoproterozoic samples (Fig. 6.13d–e). Overall, however, it is clear that Fe enrichment has sporadically occurred in marine sedimentary environments prior to the GOE, most spectacularly in the Mesoarchean. We will address the origins of such enrichments using Fe isotopes in later sections.

The redox state of Fe in Precambrian marine shales is shown in Fig. 6.14, where we define F_{Fe3+} as the fraction of Fe^{3+} relative to total Fe (Fe^{2+} + Fe^{3+}), as determined by wet chemical analyses. For comparison, we assume an igneous rock F_{Fe3+} of 0.15, corresponding to moderately evolved magmas that formed within one log unit of the Fayalite-Magnetite-Quartz (FMQ) buffer (e.g., Bézos and Humler 2005), appropriate for oceanic or arc rocks that formed prior to the GOE (Nicklas et al. 2018; see also Sect. 1.1 in Chap. 1). There is a slight increase in F_{Fe3+} in pre-GOE shales from the Paleoarchean to the Neoarchean and early Paleoproterozoic (Fig. 6.14a–c), with the most significant increase seen in the youngest pre-GOE rocks. When compared to the Fe_{tot}/Al trends for these age groups (Fig. 6.13), it is clear that much of the Fe enrichment in the Mesoarchean lies in Fe^{2+} components. The Fe mineralogy of the Mesoarchean magnetic shales in the Witwatersrand Supergroup contains three groups of authigenic minerals, including magnetite (1/3 Fe^{2+}), chamosite, an Fe^{2+}-bearing chlorite (Hashman 2017), and siderite/ankerite (Fe^{2+} carbonate). As highlighted by Bekker and Holland (2012), F_{Fe3+} increases strongly after the GOE, particularly for Paleoproterozoic rocks (Fig. 6.14 d) as compared to Mesoproterozoic samples (Fig. 6.14e). This led them to suggest that

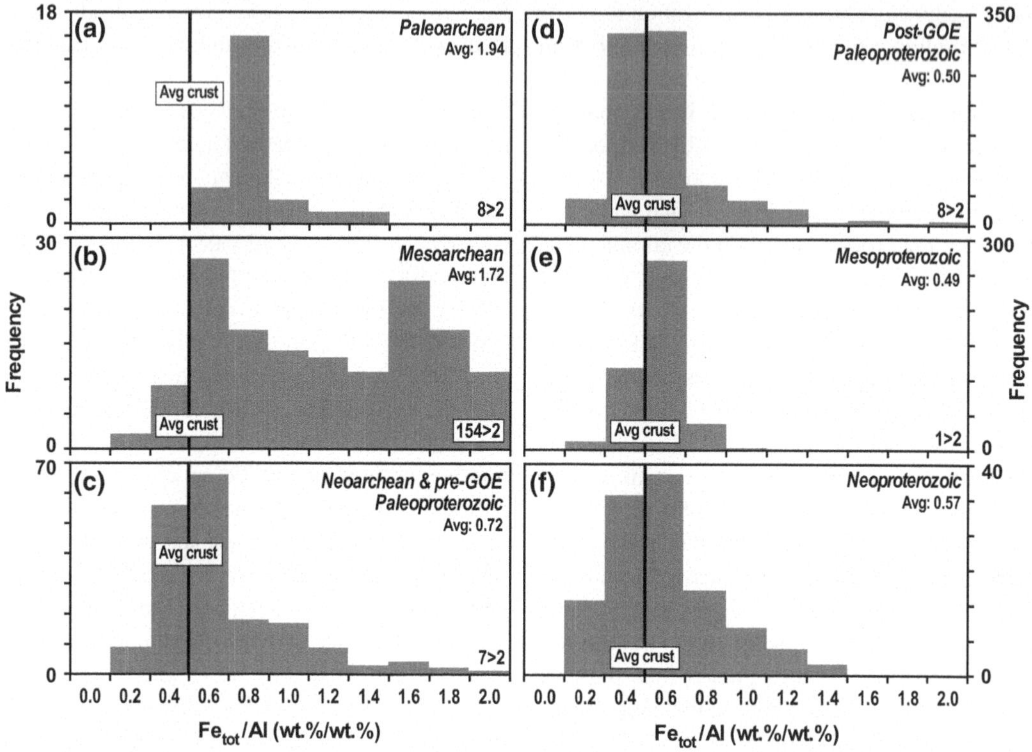

Fig. 6.13 Histograms showing range in total Fe to Al ratios (Fe_{tot}/Al) for Precambrian marine shales based on compilations and data from Yamaguchi (2002), Guy et al. (2012), Bekker and Holland (2012), and Kunzmann et al. (2017). For comparison, the average Fe_{tot}/Al ratio is shown for Precambrian crust ($Fe_{tot}/Al \sim 0.5$). Maximum scale is $Fe_{tot}/Al = 2$, and where the number of analyses that plot at $Fe_{tot}/Al > 2$ is noted in lower right corners (note very high number of samples, 154, for the Mesoarchean suite). To avoid excessive biases from shales in iron formations or Si-rich cherts, data only plotted for samples where $Al_2O_3 > 5$ wt%. Paleoarchean (**a**) and Mesoarchean (**b**) shales have Fe_{tot}/Al ratios significantly above the crustal average. Neoarchean shales, as well as Paleoproterozoic shales deposited before the Great Oxidation Event (GOE) (**c**) have Fe_{tot}/Al ratios slightly higher than the crustal average. Post-GOE Paleoproterozoic (**d**), Mesoproterozoic (**e**), and Neoproterozoic (**f**) shales have Fe_{tot}/Al ratios close to the crustal average. These relations highlight the Fe-enriched nature of pre-GOE marine sedimentary environments relative to those of post-GOE time

atmospheric O_2 contents in the GOE were extremely high, perhaps approaching modern levels, followed by a decline in the Mesoproterozoic. In the Neoproterozoic, F_{Fe3+} again increases, particularly for samples close to the NOE (Fig. 6.14f).

The trends in Fe redox for marine shales in Fig. 6.14 are broadly (but not completely) mirrored by changes in F_{Fe3+} measured for ophiolites and continental volcanic rocks (Stolper and Keller 2018). Ophiolites, which Stolper and Keller (2018) interpret to record oxygen contents of the deep oceans, have average F_{Fe3+} equal to the igneous average before the GOE ($\sim$0.15), but significantly increase after the GOE to $\sim$0.25. Interestingly, Neoproterozoic ophiolites have similar F_{Fe3+} as post-GOE Paleo- and Mesoproterozoic ophiolites, which Stolper and Keller (2018) take to indicate that final oxygenation of the deep oceans did not occur until the early Paleozoic. The redox evolution of continental volcanic rocks, however, is distinctly different, where F_{Fe3+} steadily increases after the GOE through the end of the Mesoproterozoic, suggesting a rise in atmospheric O_2 that was not mirrored in the deep ocean.

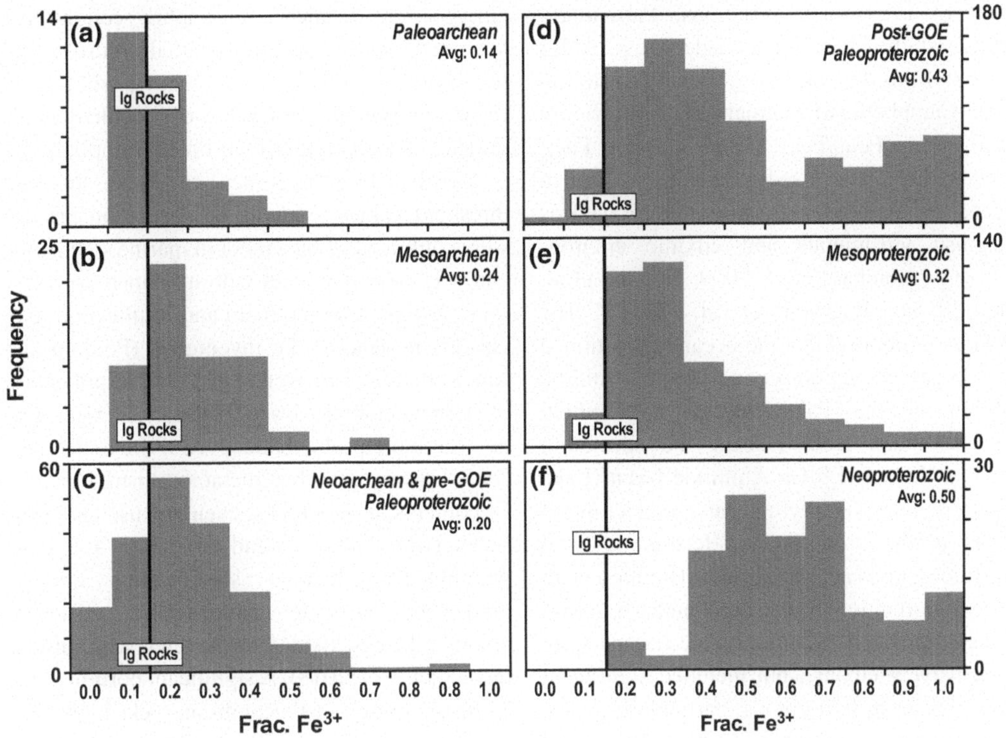

Fig. 6.14 Histograms showing fractions of Fe in Precambrian marine shales that are comprised of Fe^{3+} (Frac. Fe^{3+}, or "F_{Fe3+}"). For comparison, the fraction of Fe^{3+} in fresh igneous rocks (~0.15) is shown. Data and compilations from Yamaguchi (2002), Bekker and Holland (2012), and Kunzmann et al. (2017). This dataset is a subset of that plotted in Fig. 6.13 because relatively few samples have been analyzed for Fe^{3+} – Fe^{2+} speciation. For panels **a–e**, the fraction of Fe^{3+} is based on measured Fe^{3+} – Fe^{2+} abundances by wet chemistry. For panel **f**, the fraction of Fe^{3+} was calculated from the data of Kunzmann et al. (2017) using their reactive Fe extractions for oxide and magnetite fractions, assuming the oxide component is all Fe^{3+}, and the magnetite component is 2/3 Fe^{3+}, added to a lithologic component ($Fe_{tot}/Al = 0.5$) that is assumed to have an Fe^{3+} fraction of 0.15. The relations in **a–f** show that pre-GOE shales (panels **a–c**) have Fe^{3+} fractions similar to those of igneous rocks (perhaps increasing some with decreasing age), whereas post-GOE shales are significantly more oxidized (panels **d–f**)

Sequential extraction methods have been extensively employed in studies of modern and ancient fine-grained sedimentary rocks to distinguish highly reactive Fe pools, including those associated with sulfide, from non-reactive Fe pools such as those associated with clastic silicate minerals (e.g., Raiswell and Canfield 1998; Poulton and Canfield 2005; Raiswell et al. 2018). Application of these methods to modern marine sediments was discussed in Sect. 5.4.1 in Chap. 5. Threshold values for the ratio of highly reactive Fe (Fe_{HR}) to total Fe (Fe_{tot}) have been defined to distinguish oxic conditions ($Fe_{HR}/Fe_{tot} \leq 0.22$) from anoxic conditions ($Fe_{HR}/Fe_{tot} \geq 0.38$), based on samples from modern environments. In addition, samples that formed under anoxic and euxinic conditions are defined by high levels of pyrite Fe (Fe_{PY}) in the highly reactive pool ($Fe_{PY}/Fe_{HR} > 0.7$), whereas lower ratios of Fe_{PY}/Fe_{HR} are interpreted to record anoxic and ferruginous conditions. These parameters are often combined with assessments of total Fe enrichment in terms of Fe_{tot}/Al ratios, where ratios significantly in excess of the crustal average are generally assumed to reflect anoxic conditions (either euxinic or ferruginous).

The Fe speciation proxies have been applied to several well-preserved stratigraphic sequences of latest Archean and Proterozoic age, and we will discuss these in detail in later sections in the context of Fe isotope variations. Particularly extensive studies have been made of ~2.6–2.5 Ga rocks from the Hamersley Basin (Australia) and the Campbellrand-Malmani platform (South Africa), where Fe_{HR}/Fe_{tot}, Fe_{PY}/Fe_{HR}, and Fe_{tot}/Al ratios have been used to identify transient oxygenation in large anoxic oceans, as well as distinguish ferruginous and euxinic environments (e.g., Reinhard et al. 2009; Kendall et al. 2010, 2015a; Raiswell et al. 2011). The well-known proposal that the oceans transitioned from a largely ferruginous condition to a sulfidic condition in the Paleoproterozoic (Canfield 1998) has been studied using Fe speciation proxies in the 1.8–1.9 Ga Animikie basin (Lake Superior region), identifying spatial and temporal changes in the proportion of ferruginous and euxinic conditions across the basin (Poulton et al. 2010). Importantly, a growing database shows that the Meso- and Neoproterozoic oceans were not exclusively euxinic, but included significant regions that were ferruginous, particularly in the Neoproterozoic (Canfield et al. 2008; Planavsky et al. 2011; Guilbaud et al. 2015). Several detailed studies of Cryogenian and Ediacaran age sequences have spatially reconstructed continental margin basins during rifting of Rodinia, identifying shallow oxic regions that were underlain by near-shore euxinic zones, which in turn were underlain by deep-water ferruginous zones (Canfield et al. 2007; Johnston et al. 2010; Li et al. 2010; Huang et al. 2017). Such models are compelling because they are tied to sedimentary facies, and they paint a remarkably similar picture to what has been inferred for the Paleoproterozoic, where proximal euxinic waters were underlain by deeper ferruginous waters.

As the database for Fe speciation proxies grows, compilations have looked at statistical analysis of temporal trends. Sperling et al. (2015) suggest that there is no statistically significant trend in the proportion of oxic and anoxic environments as inferred by Fe speciation through the end of the Neoproterozoic into the Cambrian. Broadly this may be taken as support for largely ferruginous and not euxinic conditions in the Neoproterozoic. It is important to note, however, that large compilations may miss the details that emerge from detailed stratigraphic sections that have sedimentological and spatial controls, and in our discussion in later sections, we will tend to focus on specific examples of detailed stratigraphic sections rather than broad compilations.

Finally, it is important to stress that the threshold values used for Fe speciation proxies were calibrated against modern marine sediments, and it is clear that even within Phanerozoic sections, burial metamorphism and lithification may change the reactive Fe inventories (Poulton and Raiswell 2002; Raiswell et al. 2008). Application of Fe speciation proxies to Precambrian rocks may be even more problematic, even if the rocks have "only" seen greenschist metamorphism. Reactive Fe contents in such rocks, even greenschist facies rocks (which may extend to ~400 °C), have undoubtedly changed significantly through numerous prograde metamorphic reactions among silicates, oxides, carbonates, and sulfides (Slotznick et al. 2018). A significant portion of the Fe proxy studies of Precambrian rocks have been on greenschist-facies rocks, which may be cause for concern. Inferences of marine basin redox based on measured reactive Fe inventories will get progressively murkier the farther back in the rock record we go, as the possibilities of post-depositional changes via metamorphism and fluid-rock interaction increase, and we will highlight several cases of this in the sections below.

6.3.3 Differences in Marine Fe Pathways Between Modern and Ancient Earth

The large changes in the Precambrian marine C-S-Fe system discussed in Sects. 6.3.1 and 6.3.2 above undoubtedly changed the Fe pathways, and their isotopic fingerprints, in dramatic ways over Earth history. In this section we revisit two major marine Fe pathways discussed for the modern oceans in Sect. 5.5 of Chap. 5—evolution of mid-ocean-ridge hydrothermal plumes and benthic Fe fluxes from the continental margins, with an eye to how these would have been different in the Precambrian Earth. We ignore riverine and aeolian sources of Fe to the Precambrian oceans-although these are important in the modern oceans (see Sect. 5.5.2 in Chap. 5), these probably had minimal effect on the Fe isotope compositions of an Fe^{2+}_{aq}-rich Precambrian ocean, although they might have contributed to a low-Fe^{2+}_{aq}, oxygen-rich shallow ocean. For hydrothermal and benthic Fe fluxes, the lowered O_2 levels and smaller sulfate-sulfide inventories in the Precambrian marine system, especially in the Archean, would have produced quite different effects for these processes relative to those observed in the modern oceans. These changes would have affected the extent of Fe^{2+}_{aq} oxidation, the relative control of sulfides and oxides on isotopic exchange/fractionation with Fe^{2+}_{aq}, and the Fe mass balances in the oceans, all of which could be recorded in the Fe isotope compositions of Precambrian marine sedimentary rocks.

As discussed in Sect. 5.5.4 in Chap. 5, most modern marine hydrothermal systems are basalt-hosted, producing sulfide-rich particulates that have negative δ^{56}Fe values (see for example, Fig. 5.25) that push the remaining dissolved Fe (dFe) of the plume to higher δ^{56}Fe values. Modern ultramafic-hosted hydrothermal systems, such as the Rainbow system of the mid-Atlantic Ridge, are a better analog for Precambrian systems because they have higher Fe/S ratios, producing less sulfide precipitation. In Fig. 6.15 we show the dFe-δ^{56}Fe relations for the hydrothermal system of the East Pacific Rise that was sampled in the GEOTRACES GP16 transect, which illustrates the trend of increasing δ^{56}Fe and decreasing dFe during plume evolution that reflects the strong role that precipitating low-δ^{56}Fe sulfides exerts on the Fe isotope evolution of the plume. In the modern oceans, near-vent sulfide and oxide precipitation is extensive, producing rapid declines in dFe of the plume. For illustrative purposes, a Rayleigh fractionation curve is shown that broadly explains the modern plume data for a $\Delta^{56}Fe_{Fe\text{-}aq\ -\ solid}$ fractionation factor of +0.2‰ (grey arrow), as appropriate for a sulfide-oxide mixture where sulfide is the major precipitating phase.

In contrast, in the Precambrian oceans, oxidation of Fe^{2+}_{aq} in hydrothermal plumes would have been less extensive and sulfide contents much lower, particularly in the Archean when seawater sulfate contents were extremely low (e.g., Canfield and Farquhar 2009). Based on the modeling of Kump and Seyfried (2005), Precambrian hydrothermal fluids would have had much higher Fe/S ratios, lower sulfide contents, and much higher Fe^{2+}_{aq} contents. In the absence of significant seawater sulfate, redox processes that occur in modern MOR systems would have been absent, such as reduction of sulfate-bearing fluids as they descend into the oceanic crust (e.g., Alt 1995). This in turn changes temperature-pH relations, dramatically increasing Fe^{2+}_{aq} contents of emanating hydrothermal fluids. Kump and Seyfried (2005) estimate that Fe^{2+}_{aq} contents were likely >50 times higher in Precambrian MOR hydrothermal fluids as compared to modern systems. This is illustrated in Fig. 6.15 in the purple arrow, where an increase in the proportions of oxide to sulfide precipitation changes the sign of the Fe isotope fractionation factor, producing decreases in δ^{56}Fe values with decreasing dFe as the plume evolves, but maintains orders-of-magnitude higher dFe levels as compared to modern plumes. Importantly, the model illustrated in Fig. 6.15 for the Precambrian case ends at 80% oxidation/precipitation, but under low-O_2 conditions, it would not be expected that Fe^{2+}_{aq} would oxidize to such an extent. To the degree that oxidation of Fe^{2+}_{aq} was limited, δ^{56}Fe values and dFe contents of distal Precambrian hydrothermal plumes might not change significantly relative to their near-vent compositions. The expected δ^{56}Fe values and dFe contents for a hydrothermal end member in the Precambrian oceans will then entirely depend on estimates for the oxygen and sulfate contents of the oceans.

Benthic Fe fluxes are a significant source of Fe to the modern oceans (see Sect. 5.5.3 in Chap. 5), and under the low-oxygen conditions of the Precambrian, benthic fluxes should have been far more important. As discussed in Chap. 5, the benthic Fe flux is controlled by escape of pore

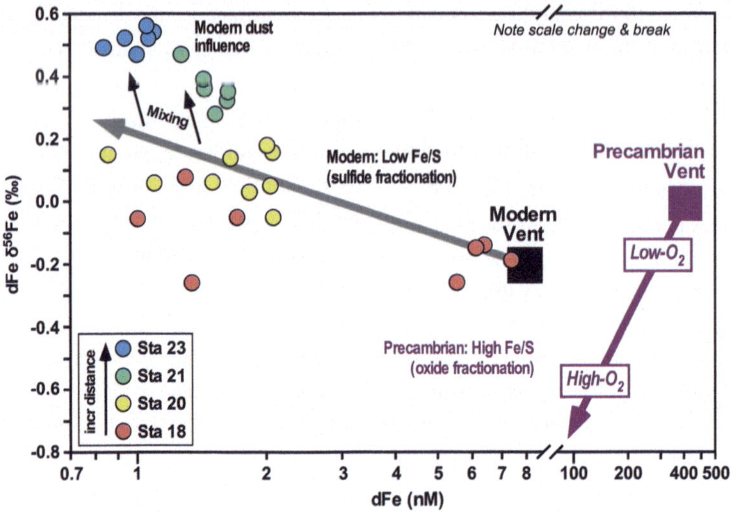

Fig. 6.15 Hydrothermal plume evolution from the GEOTRACES GP16 transect across the East Pacific Rise for samples of dFe that have $\delta^3He \geq 32$. See Fig. 5.27 in Chap. 5 for a variation on this plot (same data sources as here). Station 18 is at the vent, whereas Stations 20, 21, and 23 reflect increasing distance from the vent. Grey arrow shows a Rayleigh fractionation model for a sulfide-dominated fractionation assemblage ($\Delta^{56}Fe_{Fe\text{-}aq\ -\ solid}$ = + 0.2‰), appropriate for modern basalt- hosted MOR hydrothermal systems (low Fe/S). Vent fluid endmember is 8 nM dFe, $\delta^{56}Fe = -0.2$‰ (black box; see Fig. 5.24); this end member is fixed to where the Fe isotope fractionation is zero, and ignores higher Fe contents that could exist in the absence of Fe isotope fractionation. Arrow ends at 90% oxidation/precipitation. This model explains the broad increase in $\delta^{56}Fe$ values for dFe with decreasing dFe contents for Stations 18 and 20. For Stations 21 and 23, the shift to high $\delta^{56}Fe$ values reflects the influence of pFe-dFe exchange at low dFe contents from modern dust/particles (see Chap. 5). The purple arrow shows a Rayleigh fractionation model for a mixed oxide-sulfide fractionation assemblage ($\Delta^{56}Fe_{Fe\text{-}aq\ -\ solid}$ = −0.5‰), appropriate for a high-Fe and high-Fe/S MOR system that would be dominant in the Precambrian oceans due to low seawater sulfate contents. Vent fluid end member is 427 nM dFe (purple box), based on Fe concentration scaling from the model of Kump and Seyfried (2005). The initial $\delta^{56}Fe$ value is assumed to be zero, based on the expectation of higher temperature fluid-rock interaction and decrease in Fe isotope fractionation factors (e.g., Johnson et al. 2008b). Arrow ends at 80% oxidation/precipitation. This model indicates that Precambrian MOR hydrothermal plumes would tend to evolve to lower $\delta^{56}Fe$ values but retain higher dFe contents in the plume when oxidation (and sulfide precipitation) extent was less than in modern systems; importantly, however, under very low O_2 conditions, it is not expected that Precambrian plume evolution would extend to that shown, due to low extents of Fe^{2+}_{aq} oxidation. Note the log scale for dFe, and the very large scale change and break between dFe for the modern and Precambrian models

fluid Fe^{2+}_{aq} from marine sediments, which is a function of reactive Fe^{3+} contents, C_{org} contents, and bottom-water O_2 contents (see Fig. 5.21 in Chap. 5). In the modern oceans, bottom-water O_2 contents exert a major control on escape of pore water Fe^{2+}_{aq}, as shown by vertical pore water profiles (see Fig. 5.14 in Chap. 5). In modern marine sediments, however, much of the generated Fe^{2+} is retained in sediments as reactive S-bearing and non-S-bearing solid components (Fig. 5.14). In contrast, in the Precambrian oceans, decreased Fe^{2+}_{aq} oxidation and low sulfide contents will increase transport of pore water Fe^{2+} to the oceans, provided there is sufficient delivery of reactive Fe^{3+} and C_{org} to support DIR. Identification of benthic Fe fluxes in the Precambrian is important, because this helps establish the time DIR evolved, which is a deeply rooted metabolism that should have arisen early in Earth history (e.g., Vargas et al. 1998).

In Fig. 6.16 we illustrate possible changes in benthic Fe fluxes in the Precambrian oceans, accounting for the effects of expected differences in reactive Fe inventories relative to modern

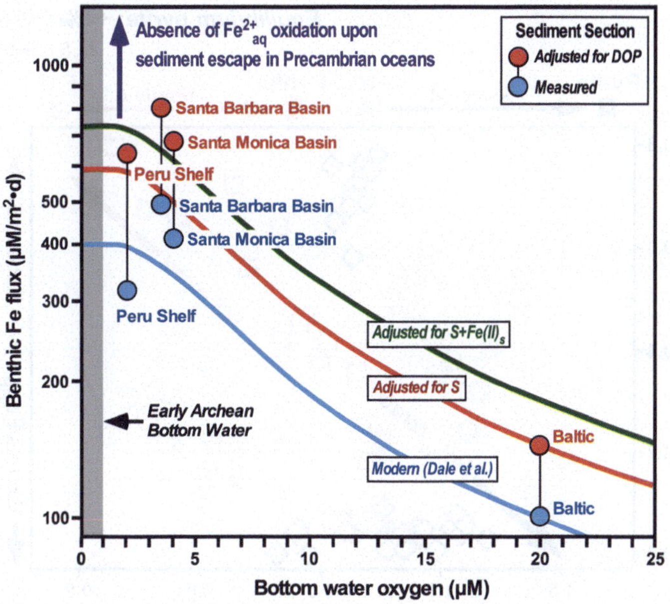

Fig. 6.16 Possible changes in benthic Fe flux-oxygen relations in the Precambrian oceans relative to the modern oceans (see Fig. 5.21 in Chap 5 for comparison). Blue curve same as that in Fig. 5.21, which approximately represents relations for the main marine trend (blue symbols; see Fig. 5.21 for data sources) using the model of Dale et al. (2015), where carbon oxidation rate (C_{ox}) set to 5 mmol m^{-2} d^{-1}, and , the maximum Fe flux (proportion of reactive Fe abundance) set to 400 μmol m^{-2} d^{-1}. Red symbols and curve reflect adjustment to remove the effects of the 47% of reactive Fe in modern marine sediments that is sequestered as sulfide using the data from Thamdrup et al. (1994), which assumes very low microbial sulfate reduction rates in low-sulfate Precambrian marine environments (particularly before the GOE). The green curve is a second adjustment of the Dale et al. curve, where the fraction of non-S-bound solid Fe(II) of Thamdrup et al. (1994) is considered mobile (see also Fig. 5.14 in Chap. 5); this is based on the assumption that in the high-Fe^{2+}_{aq} environments of Precambrian marine sediments, reactive Fe^{3+} will be converted to crystalline Fe^{3+}-oxide/hydroxides, minimizing the abundance of poorly crystalline Fe^{2+} – Fe^{3+} phases, releasing sorbed Fe^{2+}. Grey field for oxygen contents of early Archean bottom water from Satkoski et al. (2015). Finally, in the Precambrian, oxidation of Fe^{2+}_{aq} expelled from marine sediments will be minimal, which will further increase benthic Fe fluxes relative to modern environments (vertical purple arrow). These relations indicate that a benthic Fe flux in the Precambrian oceans, especially before the GOE, was likely to have been much more vigorous than modern fluxes, reflecting the lack of sulfide mineralization and higher proportion of mobilized Fe^{2+} in pre-GOE marine environments

marine sediments based on the relations in Fig. 5.14 in Chap. 5. In modern marine sediments where sulfate reduction rates are high, approximately half of the reactive Fe inventory is contained in sulfide species (e.g., Thamdrup et al. 1994), including FeS and pyrite (e.g., Fig. 5.14 c). The red curve in Fig. 6.16 reflects an adjustment to the maximum Fe flux of +47%, based on the Fe speciation data of Thamdrup et al. (1994), which assumes the moles of Fe contained in reactive Fe-S species are instead assigned to the Fe^{2+}_{aq} flux. In addition, in Fig. 6.16 we illustrate the effect of subtracting the Fe^{2+}-sulfide component at specific modern marine localities based on measured Fe-S speciation and degree of pyritization (DOP), shown in the red symbols (blue symbols are as measured).

We can further modify the modern benthic Fe flux curve to account for expected changes in the non-S $Fe(II)_{solid}$ pool of reactive Fe. It is well established that in the presence of Fe^{2+}_{aq}, unstable Fe^{3+} phases such as ferrihydrite will convert to more stable forms such as goethite (e.g., Boland et al. 2014), which will be accompanied by release of sorbed Fe^{2+}. This effect is significant, as conversion of ferrihydrite to more crystalline

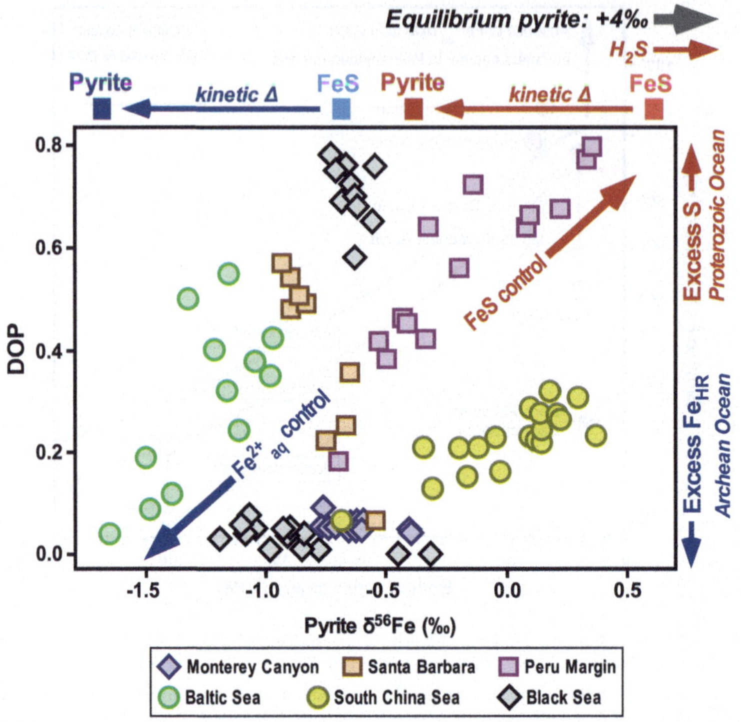

Fig. 6.17 Variations between $\delta^{56}Fe$ values for pyrite and Degree of Pyritization (DOP) for modern marine sediments as a predictor for compositions that might be measured in Proterozoic and Archean rocks. Many modern suites have a positive relation between DOP and $\delta^{56}Fe_{pyrite}$, reflecting competing controls by FeS (processes shown in red, producing high $\delta^{56}Fe_{pyrite}$) and Fe^{2+}_{aq} (processes shown in blue, producing low $\delta^{56}Fe_{pyrite}$). These are conditions that would be expected, for example, in a sulfide-rich Proterozoic ocean (high DOP) versus a sulfide-poor but Fe^{2+}_{aq}-rich Archean ocean (low DOP). The range in possible Fe isotope compositions are shown at the top of the figure in red and blue, where it is assumed that FeS is first formed, followed by conversion to pyrite. For high-DOP systems (red processes), the $\delta^{56}Fe$ value for FeS (light red box) assumes an Fe^{2+}_{aq} $\delta^{56}Fe$ value of zero and uses the equilibrium Fe isotope fractionation factor for aqueous Fe that consists of a mixture of Fe^{2+}_{aq} and FeS_{aq} from Wu et al. (2012a); a system where all aqueous Fe was FeS_{aq} due to high H_2S contents would produce even higher $\delta^{56}Fe$ values for FeS (right-facing "*H_2S*" red arrow). At equilibrium, the $\delta^{56}Fe$ values for pyrite at low temperatures are predicted to be very high, ~+4‰, assuming an Fe^{2+}_{aq} $\delta^{56}Fe$ value of zero, as shown by right-facing grey arrow at top (Rustad et al. 2010; Polyakov and Soultanov 2011; Blanchard et al. 2012; Syverson et al. 2013); see Chap. 3 for details. Conversion of FeS to pyrite is assumed to decrease $\delta^{56}Fe$ values by 1‰ via a kinetic isotope fractionation ("*kinetic Δ*"), based on the "Marine Trend" in Fig. 5.17 in Chap. 5. It is possible, however, that complete conversion of FeS to pyrite could produce no change in $\delta^{56}Fe$ values. For processes that would occur at low sulfide contents and high Fe^{2+}_{aq} contents (blue processes), it is assumed that the starting $\delta^{56}Fe$ value for FeS formed in equilibrium with Fe^{2+}_{aq} had a $\delta^{56}Fe$ value of −1‰, using the equilibrium Fe isotope fractionation factor for Fe^{2+}_{aq} and FeS from Wu et al. (2012a). When added to the kinetic isotope fractionation upon conversion of FeS to pyrite, this may produce very low $\delta^{56}Fe_{pyrite}$ values. Low $\delta^{56}Fe$ values for Fe^{2+}_{aq} may be produced by (1) oxidation of Fe^{2+}_{aq} at the sediment-seawater interface, (2) generation of Fe^{2+}_{aq} by DIR, or (3) exchange with Fe^{3+}-hydroxides (see Figs. 5.14, 5.15, and 5.17). Data from Severmann et al. (2006, 2008), Fehr et al. (2008, 2010), Scholz et al. (2014b), Henkel et al. (2016), and Lin et al. (2017)

phases generally reduces surface area an order of magnitude or more, which scales with the decrease in sorbed Fe^{2+}. Under conditions appropriate for an Fe^{2+}_{aq}-rich system in the Precambrian, significant phase conversion of reactive Fe^{3+} is expected. The green curve in

Fig. 6.16 assumes that all non-S $Fe(II)_{solid}$ is released into the Fe^{2+}_{aq} pool as a result of phase conversion of solid Fe^{2+} and Fe^{3+} to stable Fe^{3+} minerals, releasing sorbed Fe^{2+}. Finally, we schematically illustrate the trend of increased Fe^{2+}_{aq} flux in the absence of oxidation in the sediment near the sediment-seawater interface by the vertical purple arrow. It is important to note that even for the modern low-O_2 localities shown, significant oxidation of pore water Fe^{2+}_{aq} by O_2 still occurs, but in an anoxic ocean, O_2-induced oxidation would not be present. This would further increase the benthic Fe^{2+}_{aq} flux, although it is difficult to predict the magnitude of this increase.

The predictions in Fig. 6.16 are most appropriate for the Archean oceans, where seawater sulfate and O_2 was very low. In terms of Fe isotope mass balance, for a given Fe^{3+} and C_{org} flux to marine sediments, an increase in the proportion of Fe^{2+}_{aq} would moderate the lowering of $\delta^{56}Fe$ values. As discussed in Sect. 5.4.4 in Chap. 5, the modern flux-weighted $\delta^{56}Fe$ value for the benthic flux is −2.4‰ (see also Fig. 5.18). If, in the Archean, a higher proportion of Fe^{2+}_{aq} was released from marine sediments, the flux-weighted $\delta^{56}Fe$ value should have been less negative, perhaps in the range of ~−1 to −1.5‰. Alternatively, if Fe^{3+} and C_{org} supply was extremely high at certain times in the past, such as might occur during periods of transient oxidation in the presence of high primary productivity, it is possible that, despite a high Fe^{2+}_{aq} flux, $\delta^{56}Fe$ values remained low. For Precambrian periods where euxinic conditions were prevalent (e.g., Poulton and Canfield 2011), the presence of Fe-S species in marine sediments would likely have decreased the overall benthic Fe^{2+}_{aq} fluxes. Isotopic equilibration with Fe-S species, rather than Fe^{3+}-oxide/hydroxide, would be expected to increase the $\delta^{56}Fe$ values of the benthic Fe flux, possibly to near-zero or slightly positive values. Such compositions would be most likely when O_2 was low but seawater sulfate contents were high because of the absence of the influence of oxidation of Fe^{2+}_{aq} that lowers $\delta^{56}Fe$ values of the benthic flux. This might, for example, be characteristic of the mid-Proterozoic, as discussed above, although there are few Fe isotope studies from marine sedimentary rocks from this time period to test such proposals.

We finish this section with a discussion of the Fe isotope compositions of pyrite that might be expected in Proterozoic and Archean marine environments, because a significant portion of the database for Fe isotopes on Precambrian marine sedimentary rocks lies in pyrite from shales. The most important control on the Fe isotope compositions of sedimentary pyrite is the relative proportions of sulfide-bound and non-S reactive Fe, as discussed in detail in Sect. 5.4 of Chap. 5. Critically, as discussed in Chap. 3, Fe isotope fractionation during pyrite formation at marine temperatures is generally accompanied by significant kinetic effects. In Fig. 6.17 the relations between $\delta^{56}Fe$ values for pyrite ($\delta^{56}Fe_{pyrite}$) and Degree of Pyritization (DOP) for modern marine sediments are cast in terms of sulfide-dominated and non-sulfide Fe pathways and their resultant isotopic compositions. Under conditions of excess sulfide (euxinic conditions), as proposed for example by Canfield (1998) for the Proterozoic oceans, $\delta^{56}Fe$ values for Fe-S minerals will tend to be high, as will DOP (red labels, boxes, and arrows in Fig. 6.17). Sulfide-dominated conditions will produce moderately positive $\delta^{56}Fe$ values for FeS (mackinawite) under circum-neutral conditions where Fe isotope equilibrium is expected and aqueous Fe speciation has a large proportion of FeS_{aq} (Wu et al. 2012a). Based on inferred Fe isotope fractionations between FeS and pyrite in a variety of marine sediments, we assume that pyrite has a $\delta^{56}Fe$ that is ~1‰ lower than that of FeS (see "FeS Marine Trend" in Fig. 5.17 in Chap. 5), reflecting a kinetic isotope effect, and this would produce slightly negative $\delta^{56}Fe_{pyrite}$ values (Fig. 6.17). It is possible, however, that at very high degrees of pyritization, there is no Fe isotope fractionation between pyrite and FeS, as shown for data from the Peru OMZ in Fig. 5.17 in Chap. 5. In this case, $\delta^{56}Fe_{pyrite}$ values could become quite positive, and this is shown in Fig. 6.17 with the red arrow noting "FeS control".

In contrast to a euxinic Proterozoic ocean, a sulfide-poor, Fe^{2+}_{aq}-rich Archean ocean would tend to produce lower $\delta^{56}Fe_{pyrite}$ values (blue labels, boxes, and arrows in Fig. 6.17) because of the tendency for Fe^{2+}_{aq} to have low $\delta^{56}Fe$ values in the absence of sulfide, through isotopic exchange with Fe^{3+}-hydroxide/oxide, oxidation of Fe^{2+}_{aq}, or DIR. The high Fe/S ratio of hydrothermal fluids in the Archean would tend to have produced low $\delta^{56}Fe$ values in evolved hydrothermal plumes (Fig. 6.15), as would a vigorous benthic Fe shuttle (Fig. 6.16). Pore water Fe^{2+}_{aq} in low-sulfide systems is much higher than in sulfide systems, and may attain highly negative $\delta^{56}Fe$ values via DIR and/or oxidation near the sediment/seawater interface if the oceans contained O_2. In Fig. 6.17, we illustrate the $\delta^{56}Fe$ values for FeS that would be in equilibrium with Fe^{2+}_{aq} that had a $\delta^{56}Fe$ value of -1‰ in the light blue "FeS" box, as well as the $\delta^{56}Fe_{pyrite}$ value that might form assuming the 1‰ kinetic fractionation between FeS and pyrite. This illustrates that $\delta^{56}Fe_{pyrite}$ may attain very negative values, and such compositions do not demand an origin by DIR because of the effects of Fe^{2+}_{aq}-Fe^{3+}-hydroxide exchange (e.g., Fig. 5.14 in Chap. 5).

Of course, cases between high- and low-DOP are found in modern marine sediments, and in some cases (Baltic Sea, South China Sea, and Peru margin), the positive correlation between $\delta^{56}Fe_{pyrite}$ and DOP illustrates mixtures between sulfide-rich and sulfide-poor pyrite formation pathways. In other cases (Black Sea, Santa Barbara basin), there does not seem to be a relation between DOP and $\delta^{56}Fe_{pyrite}$. Scatter in the DOP-$\delta^{56}Fe_{pyrite}$ trends is to be expected because (1) the extent of kinetic Fe isotope fractionation between FeS and pyrite is likely variable (see Chap. 3), and (2) pore water Fe^{2+}_{aq} may have a wide range of Fe isotope compositions. In addition, pyrite formation pathways are quite varied (see Eqs. 5.10 through 5.16 in Chap. 5), providing opportunity for a diverse range of isotopic effects. Importantly, however, in no case does pyrite in modern marine sediments, regardless of DOP, show the highly positive $\delta^{56}Fe$ values expected for equilibrium fractionation between Fe^{2+}_{aq} and pyrite, which should produce $\delta^{56}Fe_{pyrite}$ values of +4‰ or higher (see Chap. 3). The most likely conditions for equilibrium pyrite formation would be at very slow rates of precipitation where Fe^{2+}_{aq} contents were very high but sulfide was very low, such as what might characterize the early Archean. Nevertheless, the relations in Fig. 6.17 provide strong evidence that pyrite rarely forms in Fe isotope equilibrium in modern natural systems, which adds to the complexity of interpreting Fe isotope data from pyrite in marine sedimentary rocks, modern or ancient (see, for example, recent discussion in Mansor and Fantle 2019).

6.3.4 Authigenic Fe Isotope and Reactive Fe Trends

The very large quantities of Fe that are expected to be cycled in Precambrian marine environments discussed in the previous section are best understood by combining Fe isotope data with reactive Fe speciation measurements, which, to date, have only been done on a limited number of marine shale sections (e.g., Kunzmann et al. 2015, 2017). As discussed in Sect. 6.3.2, we will use the commonly defined ratios of Fe_{tot}/Al, Fe_{HR}/Fe_{tot}, and Fe_{PY}/Fe_{HR} (e.g., Raiswell and Canfield 1998; Poulton and Canfield 2005; Raiswell et al. 2018), casting these into trends relative to $\delta^{56}Fe$ values (Fig. 6.18), focusing on those that record (1) net addition of authigenic Fe, as might be the case where hydrothermal Fe^{2+}_{aq} impacts a basin, and (2) differential Fe loss and gain across a basin as part of a benthic Fe shuttle (i.e., loss from shelf, gain in basin). We assume a "reference state" that can be taken to be equal to the local clastic input in terms of Fe_{tot}/Al, Fe_{HR}/Fe_{tot}, and Fe_{PY}/Fe_{HR}, which Fe isotope data can help constrain because this should equal the $\delta^{56}Fe$ value of average upper crust (~+0.1‰). The reader is referred to Chap. 5 for background on Fe isotope and mass-balance trends produced by these two processes.

Addition of an authigenic oxide component, derived by oxidation of Fe^{2+}_{aq}, will increase $\delta^{56}Fe$ values, where the increase is inversely proportional to the % oxidation of Fe^{2+}_{aq}. Composition 1 in Fig. 6.18 (panels a–c) records the $\delta^{56}Fe$, Fe_{tot}/Al, Fe_{HR}/Fe_{tot}, and Fe_{PY}/Fe_{HR} of the pure oxide component produced at low O_2 (low % oxidation of Fe^{2+}_{aq}). Mixture with an authigenic pyrite component will lower the $\delta^{56}Fe$ values (composition 1' in panels a–c), assuming mixture with pyrite that has Fe isotope compositions similar to those of modern euxinic environments ($\delta^{56}Fe$ ~ −0.7‰). Under conditions of a higher % oxidation of Fe^{2+}_{aq}, which would indicate high-O_2 conditions, greater enrichments in Fe_{tot}/Al and Fe_{HR}/Fe_{tot} are expected due to extensive conversion of Fe^{2+}_{aq} to oxides, as well as smaller increases in $\delta^{56}Fe$ values (composition 2 in panels a–c); mixing with an authigenic pyrite component moves the sample to composition 2'.

Several important conclusions may be seen in panels a–c for Fig. 6.18 in terms of Fe proxy interpretations. First, low-O_2 conditions will be identified by samples which have high-$\delta^{56}Fe$ values in the oxide component, but in samples that have both authigenic oxide and pyrite, the $\delta^{56}Fe$ values will be lowered due to the different Fe isotope fractionations produced by the two minerals. Second, accepting that high $\delta^{56}Fe$ values for oxides indicate low O_2 and hence highly ferruginous conditions, such samples may have relatively low Fe_{HR}/Fe_{tot} ratios due to the limited addition of authigenic oxide despite forming in highly ferruginous conditions. In contrast, low-$\delta^{56}Fe$ samples, reflecting greater extents of oxidation, will have higher Fe_{HR}/Fe_{tot} ratios due to high proportions of oxide addition. This highlights that interpretations based on Fe proxy data alone could be inconsistent with the constraints imposed by Fe isotopes, where high Fe_{HR}/Fe_{tot} ratios would be interpreted to record lower O_2 ("more ferruginous") yet their low-$\delta^{56}Fe$ values indicate a greater extent of oxidation and hence, in reality, high O_2 conditions. The interplay between Fe isotopes and Fe proxies are important consider where Fe proxy-based interpretations suggest redox-stratified water columns and possible changes in O_2 contents, although to date, only a few studies have taken this integrated approach

Turning to the relations produced by a benthic Fe shuttle, loss of low-$\delta^{56}Fe$ Fe^{2+}_{aq} on the shelf (red arrows in Fig. 6.18) will decrease Fe_{tot}/Al ratios and increase $\delta^{56}Fe$ values of the residual Fe. This will be accompanied by a decrease in Fe_{HR}/Fe_{tot} due to removal of oxide components but may increase Fe_{PY}/Fe_{HR} ratios due to lowering of Fe_{HR} and not an increase in Fe_{PY} (panels D-F in Fig. 6.18). In the absence of Fe isotope data, these changes in Fe proxy data could be mis-interpreted: A decrease in Fe_{HR}/Fe_{tot} might be inferred to indicate a transition from ferruginous to oxic conditions, whereas an increase in Fe_{PY}/Fe_{HR} ratios might be interpreted as an increase in euxinic conditions. These two interpretations are, of course, inconsistent, and the changes in Fe isotope compositions illustrate the reason for this: Fe did not behave as a closed system, and this is a good example of the usefulness of isotopic compositions for tracing net movement in or out of a system. As discussed in Chap. 5, the benthic Fe shuttle enriches Fe in the deep basin "trap", and this enrichment will have negative $\delta^{56}Fe$ values (blue arrows in Fig. 6.18). A ferruginous, non-euxinic trap would be characterized by high Fe_{HR}/Fe_{tot} and the presence of minerals such as magnetite and chamosite. A euxinic trap would be characterized by high Fe_{PY}/Fe_{HR} and sulfide minerals.

6.4 Precambrian Earth: The Neoproterozoic

The Neoproterozoic saw dramatic changes in the surface environments of the Earth. The list of biogeochemical and environmental changes through this interval is extensive: a rise in atmospheric and ocean oxygen (e.g., Canfield et al. 2007; Planavsky et al. 2014), a change from

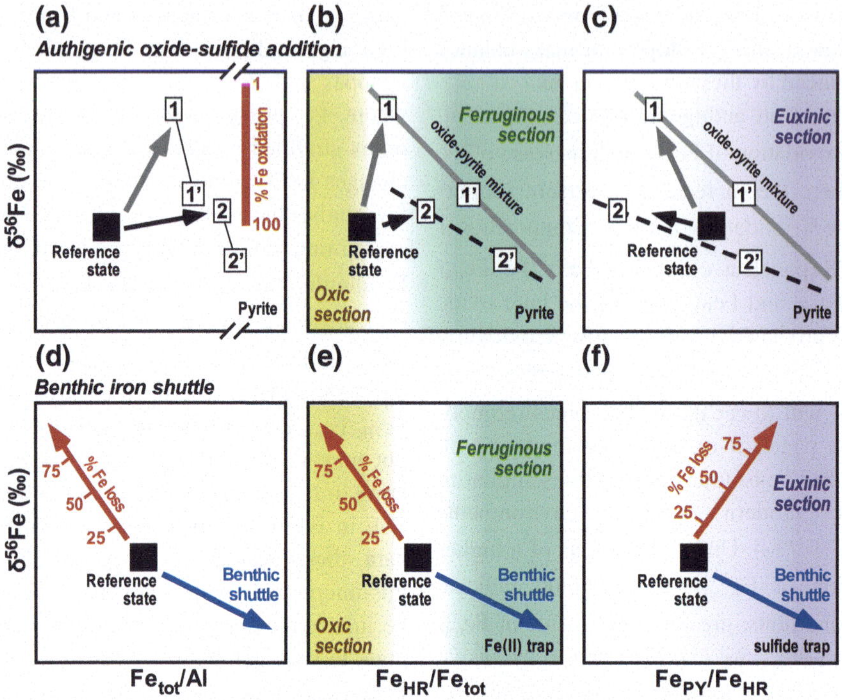

Fig. 6.18 Schematic trends in δ^{56}Fe values for bulk shales and reactive Fe species upon addition or subtraction of authigenic Fe, illustrated relative to a reference state that may be defined as the pure clastic input to a basin. Trends are intended to illustrate what an array of samples might record that were deposited in a variety of oxic, ferruginous, and euxinic conditions (note that mixing lines are technically not straight in some of the diagrams). Panels A-C show trends for net addition of authigenic Fe oxides and pyrite. Assuming the authigenic oxide component reflects oxidation of Fe^{2+}_{aq}, the δ^{56}Fe value of this component will be a function of the % oxidation (e.g., highest δ^{56}Fe for lowest % oxidation). For simplicity, a single moderately negative δ^{56}Fe value of −0.7‰ is assumed for pyrite. Partial oxidation of Fe^{2+}_{aq} will produce high-δ^{56}Fe and moderate Fe_{tot}/Al enrichment (thick grey arrow trending to composition 1 in panel **a**), but in basins that include a mixture of authigenic oxide and pyrite (recording a mixture between ferruginous and euxinic conditions), pyrite-rich samples may be displaced to lower δ^{56}Fe values, reflecting the distinct Fe isotope fractionation factors of oxides and sulfides (composition 1' in panel **a**). Larger extents of oxidation of Fe^{2+}_{aq} will produce lower δ^{56}Fe values for the oxide component but relatively high extents of Fe enrichment due to the greater proportion of conversion of Fe^{2+}_{aq} to oxide (thinner black arrow trending to composition 2). Addition of a low-δ^{56}Fe pyrite component would move to composition 2'. Panels B and C show equivalent mixing relations and compositional references for plots involving Fe_{HR}/Fe_{tot} and Fe_{PY}/Fe_{HR}. Trends involving Fe_{PY}/Fe_{HR} are complex to predict, but we assume that addition of only authigenic oxide will decrease Fe_{PY}/Fe_{HR}, followed by an increase upon pyrite addition. Panels D-F show trends for a benthic Fe shuttle, where loss of low-δ^{56}Fe Fe^{2+}_{aq} on the shelf produces an increase in δ^{56}Fe of the residual Fe as a function of % Fe lost (red arrow). In panel F, Fe loss is assumed to decrease Fe_{HR} at constant Fe_{PY}. The low-δ^{56}Fe benthic Fe shuttle Fe^{2+}_{aq} component is shown by the blue arrow, reflecting increasing addition of Fe relative to the reference state. The trapping mechanism for the benthic Fe shuttle could be Fe^{2+}-bearing oxides or silicates (magnetite, chamosite) or sulfides, depending on the nature of the deep basin (ferruginous versus euxinic)

sulfidic to ferruginous ocean chemistry (e.g., Anbar and Knoll 2002; Canfield et al. 2008), and eukaryote diversification and expansion (e.g., Erwin et al. 2011; Brocks et al. 2017). Superimposed on these changes was a series of extensive and long-lived global glaciations ("Snowball Earth") that define the Cryogenian period (e.g., Hoffman et al. 1998, 2017), as well as breakup of the equatorial-centered supercontinent Rodinia (e.g., Li et al. 2008). Below we look at the Fe isotope record of two lithologic groups in the Neoproterozoic, (1) shales/cherts, and (2) iron formations, with an eye to how such data may bear on reconstructing the surface environments in the Neoproterozoic.

6.4.1 Neoproterozoic Clastic Marine Sedimentary Rocks

The large environmental changes that occurred in the Neoproterozoic were accompanied by a large range in δ^{56}Fe values for marine sedimentary rocks and minerals, far beyond what is observed in Phanerozoic samples, including the Cretaceous OAEs. Whole-rock and mineral Fe isotope compositions for Neoproterozoic rocks virtually span the entire range measured on Earth from δ^{56}Fe = −2 to +4‰ (Fig. 6.19). For context, we first consider the record of paleoenvironmental conditions in the Neoproterozoic.

Weathering fluxes, sedimentation rates, and burial of C_{org} varied greatly during the Neoproterozoic. Although there is a rise in $^{87}Sr/^{86}Sr$ ratios for seawater from ~1000 to ~800 Ma (Fig. 6.19), relative to Sr isotope compositions across the Ediacaran-Cambrian transition, seawater Sr was relatively non radiogenic, indicating low continental fluxes to the oceans in the Tonian (e.g., Shields 2007). This is supported by the low clay contents in clastic sedimentary rocks of this age (Kennedy et al. 2006), and is consistent with the relative lack of orogenic activity based on detrital zircons (McKenzie et al. 2014), as well as non-radiogenic Os isotope compositions in sedimentary rocks prior to the Sturtian glaciation (Rooney et al. 2014). Macrostrat data indicate low burial rates of C_{org} in the Tonian (Husson and Peters 2017), which, given the moderately positive $\delta^{13}C_{carb}$ values, indicates that the total C reservoir was relatively small (low primary productivity) and that the water column was largely anoxic, retarding oxidation of settling organic matter (e.g., Derry et al. 1992). In contrast, the rapid increase in seawater $^{87}Sr/^{86}Sr$ ratios after the Cryogenian (Fig. 6.19) correlates with a rapid increase in sedimentation based on the Macrostrat database, as well as a rapid increase in C_{org} burial (Husson and Peters 2017). This also correlates with an increase in clay production through high post-Cryogenian continental weathering (Kennedy et al. 2006) and development of mountainous orogenic belts (Squire et al. 2006; McKenzie et al. 2014). To maintain the similar maximum $\delta^{13}C_{carb}$ values for the early and late Neoproterozoic (Fig. 6.19), in combination with a marked increase in C_{org} burial in the late Neoproterozoic, there must have been a concomitant increase in the size of the inorganic C reservoir, assuming a simple isotope mass-balance model. Broadly, these temporal trends point to increased erosion, nutrient delivery, and primary productivity toward the end of the Neoproterozoic, increasing the potential for periods of relatively high oxygenation (e.g., Canfield et al. 2007; Och and Shields-Zhou 2012). This is consistent with some interpretations of the large negative excursions in $\delta^{13}C_{carb}$ at the Marinoan and Gaskiers glaciations (Fig. 6.18) that invoke oxidation of dissolved organic C (e.g., Grotzinger et al. 2011). Oxygenation in the late Neoproterozoic is also indicated by multiple proxies, including Mo and U enrichments in Ediacaran and Cambrian shales (Fig. 6.19), as summarized in Sect. 6.3.1.

Many Neoproterozoic marine shales have positive δ^{56}Fe values for highly reactive Fe (Fe_{HR}), as do Ediacaran low-Fe cherts that have low clastic contents (Fig. 6.19a). To the degree that the positive δ^{56}Fe values may reflect authigenic Fe oxides, and the oceans were ferruginous, this has often been taken to indicate limited oxidation of Fe^{2+}_{aq} and hence limited oxygenation of the oceans (Fan et al. 2014; Kunzmann et al. 2017). A decrease in the average δ^{56}Fe values for Fe_{HR} from pre- to post-Cryogenian time led Kunzmann et al. (2017) to propose either a decrease in Fe^{2+}_{aq} in the oceans (leading to a greater extent of oxidation in a relative sense) or an increase in O_2 contents, where they favored the latter explanation. Fan et al. (2014) interpreted the slightly positive δ^{56}Fe values for

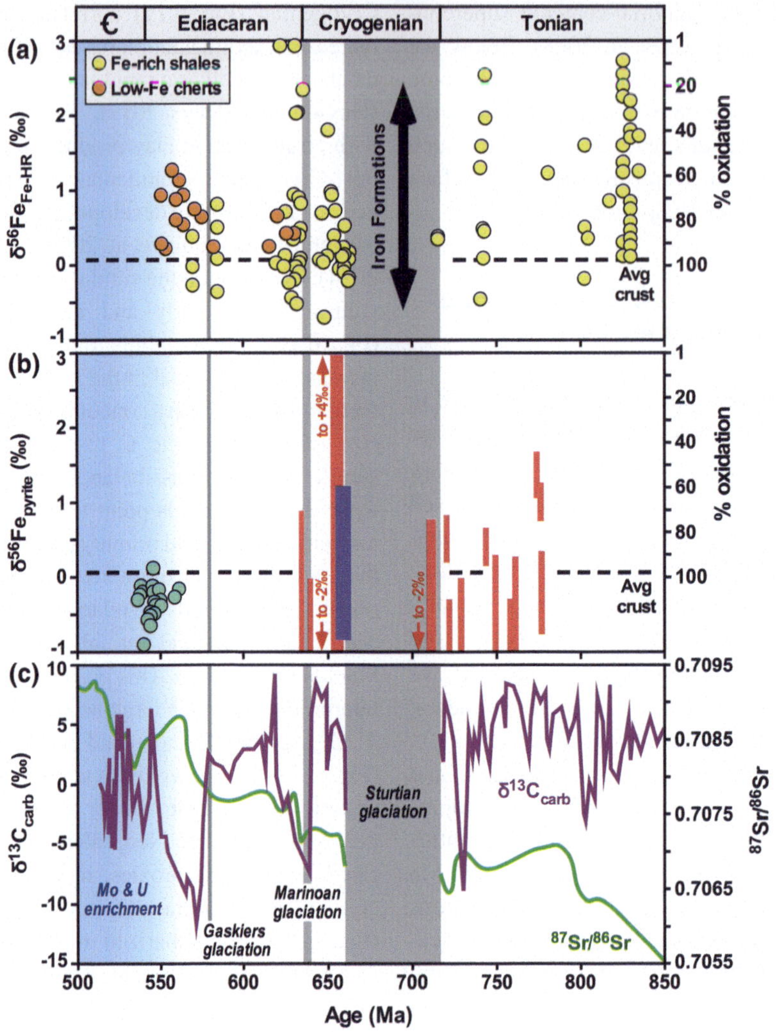

Fig. 6.19 Iron, C, and Sr isotope variations for marine sedimentary rocks for the later Neoproterozoic and earliest Cambrian. Vertical grey bands mark periods of major glaciation, including the "Snowball Earth" events (Sturtian and Marinoan of the Cryogenian), as well as the Gaskiers glaciation in the Ediacaran (timescale from Rooney et al. 2015; Prave et al. 2016; Pu et al. 2016; Cox et al. 2018; Macdonald et al. 2018). Shaded blue band in latest Ediacaran and early Cambrian notes period of strong enrichments in authigenic Mo and U contents in marine shales, interpreted to record a strong increase in seawater O_2 contents (see data and compilations by Partin et al. 2013b; Cheng et al. 2018). Horizontal dashed lines show δ^{56}Fe value for average crust (panels **a** and **b**). For panels **a** and **b**, % oxidation shown on right Y axis for a simple closed-system Fe isotope mass balance model, assuming an initial δ^{56}Fe value of zero for Fe^{2+}_{aq} and a +3‰ fractionation between Fe^{3+}-hydroxides and Fe^{2+}_{aq}. **a** δ^{56}Fe values for highly reactive Fe (Fe_{HR}) components of Fe-rich shales (yellow circles) and low-Fe cherts (brown circles). δ^{56}Fe values for Fe_{HR} of shales (Kunzmann et al. 2017) calculated based on Fe speciation determinations, which increases measured δ^{56}Fe values by an average of 0.4‰. δ^{56}Fe values for low-Fe cherts taken as measured in samples that contain no carbonate or phosphate, where Fe mineralogy is primarily hematite (Fan et al. 2014). Heavy double arrow shows range in the δ^{56}Fe values for the Fe oxide component for Sturtian iron formations (see Fig. 6.23). **b** δ^{56}Fe values for pyrite from marine shales. Data in red bands record range of in situ Fe isotope measurements by fs-LA (Tahata et al. 2015), which extend to δ^{56}Fe values as low as −2‰ and as high as +4‰, as noted. Data in purple band record δ^{56}Fe values measured for pyrite mineral separates (Zhang et al. 2015). δ^{56}Fe values in green circles are pyrite compositions estimated based on excess Fe above the crustal Fe_{tot}/Al ratio for sulfide-rich black shales (Kurzweil et al. 2015b). **c** δ^{13}C values (PDB scale; purple curve) and $^{87}Sr/^{86}Sr$ ratios (green curve) for Ca–Mg carbonates (Montañez et al. 1996; Sawaki et al. 2010; Macdonald et al. 2010; Cox et al. 2016b). The commonly high δ^{13}C values broadly indicate relatively high burial of C_{org}, punctuated by periods of very low δ^{13}C values that are broadly associated with glaciations and possible increases in O_2. The increase in $^{87}Sr/^{86}Sr$ ratios reflects increases in continental weathering after the Cryogenian

low-Fe cherts of Ediacaran age to reflect oxidation of Fe^{2+}_{aq} in a redox-stratified setting that correlates with proposed oxygenation during the Shuram negative $\delta^{13}C_{carb}$ excursion (Fig. 6.19). Fan et al. (2014) attempted to calculate authigenic components for Fe_{HR}, which produced $\delta^{56}Fe$ values as high as +3‰, which might indicate very low oxygen contents, but such calculations are subject to very large errors, given the very low Fe contents in their suite of low-Fe cherts.

There are, however, several complications with the broad interpretations of the Neoproterozoic Fe isotope data from shales and cherts. First, the large changes in the C cycle, as well as the occurrence of periods of major glaciation, suggests significant changes in oxygenation during the Neoproterozoic, and yet there is no systematic change in $\delta^{56}Fe$ values for Fe_{HR} in shales, as would be expected due to fundamental changes in the extent of Fe^{2+}_{aq} oxidation. There are temporal changes in average $\delta^{56}Fe$ values, as noted by Kunzmann et al. (2017), but there remains significant scatter in $\delta^{56}Fe$ values for any given time. In addition, the proposal for low O_2 in the Tonian, based on the positive $\delta^{56}Fe$ values, contrasts with inferences from, for example, recent Cr isotope data that suggest periods of oxygenation in the Tonian (Planavsky et al. 2014; Wei et al. 2018). Second, an oxidation model cannot explain the negative $\delta^{56}Fe$ values for Fe_{HR} in shales in the Neoproterozoic. It seems likely, therefore, given the generally negative $\delta^{56}Fe$ values measured for pyrite in modern marine sediments (see Sect. 5.4.3 in Chap. 5), that some of the scatter in $\delta^{56}Fe$ values for Fe_{HR} may reflect a mixture of authigenic components in the bulk samples, including oxide and sulfides; we will explore this in detail in the next section (Sect. 6.4.2).

Further complexity in the Fe isotope record for the Neoproterozoic is seen in data for pyrite (Fig. 6.19b), which tends to have, on average, significantly lower $\delta^{56}Fe$ values than those of Fe_{HR} in shales, although in some cases, pyrite has quite positive $\delta^{56}Fe$ values (Tahata et al. 2015; Zhang et al. 2015). In a detailed study of pyrite using in situ femtosecond laser ablation (fs-LA) of the Hecla Hoek Supergroup (Svalbard), Tahata et al. (2015) document large ranges in $\delta^{56}Fe$ values within individual samples, indicating a variety of Fe pathways during pyrite formation. Such findings are consistent with the pathway-dependence of Fe isotope fractionations during pyrite formation, as discussed above in Sect. 6.3.3 (see also Fig. 6.17). With a few exceptions, $\delta^{56}Fe_{pyrite}$ values for Tonian-age samples vary between zero and −1‰, overlapping the range observed in modern marine sediments. Post-Sturtian sediments have a very large range in $\delta^{56}Fe_{pyrite}$ values, from +4 to −2‰. When considered on a timescale much finer than that shown in Fig. 6.19, there is a broad trend of decreasing $\delta^{56}Fe_{pyrite}$ values with decreasing age across an 82 m section of the MacDonaldryggen Formation, which Tahata et al. (2015) interpret to record a rapid rise in O_2 upon emergence from the Sturtian glaciation. This follows their interpretation that pyrite in their samples commonly records the $\delta^{56}Fe$ values of precursor Fe^{3+}-oxides/hydroxides, and hence can be used, indirectly, to infer the % oxidation of Fe^{2+}_{aq}. This requires a very specific pathway for pyrite formation, where sulfide reacts with poorly crystalline Fe^{3+}-hydroxides (e.g., Eq. 5.13 in Sect. 5.4.3 in Chap. 5), and assumes no loss of Fe (no aqueous Fe^{2+}_{aq}). It is important to recall, however, that if pyrite forms in Fe isotope equilibrium with Fe^{2+}_{aq}, it should have very positive $\delta^{56}Fe$ values (see Chap. 3), and so distinguishing sulfidation of Fe^{3+}-hydroxide from equilibrium pyrite formation using Fe isotopes alone may be difficult. The $\delta^{56}Fe_{pyrite}$ values for the post-Marinoan Dracoisen Formation scatter about zero, which Tahata et al. (2015) take to indicate high O_2 levels through complete oxidation of Fe^{2+}_{aq}, assuming, again, that pyrite inherits the Fe isotope compositions of a Fe^{3+}-hydroxide precursor via a sulfidation mechanism. These interpretations are similar to those of Zhang et al. (2015), who analyzed bulk pyrite mineral separates from shales deposited between the Sturtian and Marinoan glaciations (purple band in

Fig. 6.19b), where high-$\delta^{56}Fe_{pyrite}$ values are inferred to record partial oxidation of Fe^{2+}_{aq} to Fe^{3+}-hydroxides, followed by conversion to pyrite through reaction with sulfide.

Ediacaran shales that have a large component of authigenic pyrite have negative $\delta^{56}Fe$ values (green circles, Fig. 6.19b), which Kurzweil et al. (2015b) interpret quite differently than the studies of Tahata et al. (2015) and Zhang et al. (2016). Co-variations in Fe_{tot}/Al, S, and whole-rock $\delta^{56}Fe$ values clearly indicate a major authigenic pyrite component that has negative $\delta^{56}Fe$ values of ~ −0.6‰. Increasing enrichment in authigenic pyrite is correlated with decreasing $\delta^{34}S_{pyrite}$ values, suggesting sulfate in excess and euxinic conditions. Kurzweil et al. (2015b) interpret the authigenic end member ($\delta^{56}Fe$ ~ −0.6‰) to reflect a slightly fractionated hydrothermal source, with minimal kinetic isotope fractionation upon pyrite formation. This would be consistent with the Precambrian hydrothermal evolution trend illustrated in Fig. 6.15, where high Fe/S ratios would suppress early sulfide formation, producing a decrease in the $\delta^{56}Fe$ values of hydrothermal Fe. Kurzweil et al. (2015b) suggest that the samples reflect relatively anoxic Ediacaran deep waters, although a role for seawater oxygenation at shallow levels is suggested by the moderately positive $\delta^{98}Mo$ values measured on the same samples, as well as negative $\delta^{34}S_{pyrite}$ values in samples that contain the greatest authigenic S and Fe enrichments.

6.4.2 Revisiting Reactive Fe Speciation and δ^{56}Fe

The complexities of the Fe isotope record in Fig. 6.19 in terms of authigenic mineral paragenesis in marine shales demonstrate that limited insights are gained when only looking at Fe isotope data in broad time plots. We therefore turn to a discussion of three detailed sections of Neoproterozoic marine shales of Tonian, Cryogenian, and Ediacaran age where reactive Fe contents are also available, using the interpretive context provided by Fig. 6.18 (Sect. 6.3.4).

The wide range in $\delta^{56}Fe$ values that have been calculated for Fe_{HR} in Tonian-age shales, extending from values equal to the crustal average to +3‰ (Fig. 6.19) poses several problems, as discussed in the previous section. In Fig. 6.20 we plot the calculated $\delta^{56}Fe$ values for Fe_{HR} for the Tonian Reefal Assemblage Formation based on sequential extraction methods, which in this suite, is dominated by Fe^{3+}-oxides/hydroxides (Kunzmann et al. 2017). The inverse correlation between Fe_{tot}/Al and $\delta^{56}Fe$ (panel A) is interpreted by Kunzmann et al. (2017) as Fe-loss from the shelf as a result of export of low-$\delta^{56}Fe$ Fe^{2+}_{aq} via a benthic Fe shuttle. This fits with the paleoenvironmental interpretation of the Reefal Assemblage Formation, which represents a proximal platform to fore slope sequence (Macdonald et al. 2012). Deposition of shales occurred during maximum flooding of the carbonate reef complex and was syndepositional with normal faulting. The shale sequence analyzed has virtually no pyrite nor carbonate (Kunzmann et al. 2017), indicating that it neither represents deep basin nor shallow-water sediments, but outer shelf to fore slope sediments.

The highest Fe_{HR}/Fe_{tot} samples suggest deposition in ferruginous conditions (Fig. 6.20b). Fe_{HR}/Fe_{tot} ratios, however, extend to very low ratios, which, by themselves, might be interpreted to record oxic conditions and hence a variety of redox states in the water column. The increase in $\delta^{56}Fe$ with decreasing Fe_{HR}/Fe_{tot} and Fe_{tot}/Al, however, can only be explained by loss of low-$\delta^{56}Fe$ Fe^{2+}_{aq}, and hence does not indicate a change from ferruginous to oxic conditions, but net Fe loss. Assuming an equilibrium Fe isotope fractionation between Fe^{3+}-oxides/hydroxides and Fe^{2+}_{aq} of ~+3‰ (see Chap. 3), the highest $\delta^{56}Fe$ samples would have experienced loss of >95–99% of their initial reactive Fe inventory using an equilibrium fractionation model. Importantly, the high $\delta^{56}Fe$ values for Fe_{HR} cannot be interpreted to reflect a small percent oxidation of Fe^{2+}_{aq}, the alternative way to produce high-$\delta^{56}Fe$ values. This deeper consideration of the data explains both the scatter in $\delta^{56}Fe$ values for Tonian-age

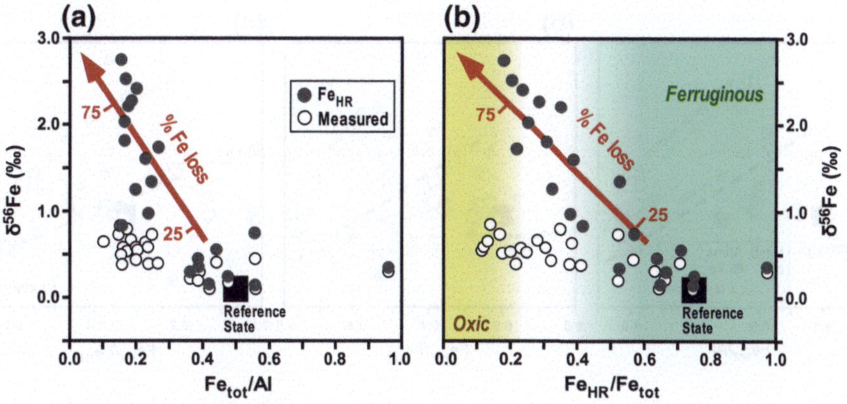

Fig. 6.20 δ^{56}Fe-reactive Fe relations for pre-Cryogenian (Tonian) shales of the Reefal Assemblage Formation, Canada (Kunzmann et al. 2017). The shale sequence was deposited in a shelf setting during syndepositional normal faulting, below storm wave base, during flooding of the underlying carbonate platform. δ^{56}Fe values are shown as measured (open circles) and for the calculated high-reactive Fe component (Fe_{HR}) based on extractions and methods in Kunzmann et al. (2017). Reactive Fe in this suite consists entirely of an oxide component, with negligible carbonate or sulfide components, simplifying interpretations. The negative relation between δ^{56}Fe for Fe_{HR} and decreasing Fe_{tot}/Al (panel A) is interpreted by Kunzmann et al. (2017) to reflect Fe loss as part of a benthic Fe shuttle. The "reference state" is chosen for δ^{56}Fe = 0.1‰ and Fe_{tot}/Al = 0.5 (average crust), and the Fe_{HR}/Fe_{tot} value (0.75) that corresponds to δ^{56}Fe = 0.1‰, which would be interpreted to reflect initially ferruginous conditions. Red arrows show Fe-loss trajectories for δ^{56}Fe for Fe_{HR} from 0.1 to 3‰ (see panels **d** and **e** in Fig. 6.18). Note that the loss of the oxide component (decrease in Fe_{HR}/Fe_{tot}; panel **b**) that accompanies the increase in δ^{56}Fe for Fe_{HR} and decrease in Fe_{tot}/Al would shift the interpretation of the depositional environment from ferruginous to oxic, whereas there was likely no actual change in the basin. This illustrates the danger of making paleoenvironmental interpretations of reactive Fe in systems that were not closed with respect to Fe

samples in Fig. 6.19, but also makes clear that the Fe isotope compositions cannot be interpreted, in this case, in terms of percent oxidation of Fe^{2+}_{aq} and hence a function of O_2 contents. Moreover, the δ^{56}Fe–Fe_{HR}/Fe_{tot} relations show that low Fe_{HR}/Fe_{tot} ratios, by themselves, cannot be taken as an unambiguous indicator of oxic conditions.

Data for Cryogenian-age shales deposited between the Sturtian and Marinoan glaciations are plotted in Fig. 6.21, and these represent a case of net Fe addition (Kunzmann et al. 2015, 2017), rather than the Fe-loss example shown for Tonian-age shales. Two sequences are illustrated, the MacDonaldryggen Formation (Svalbard) and Arena Formation (Greenland). Both sequences reflect deposition in middle and outer shelf environments. In terms of reactive Fe budgets, both suites are similar, although the Arena Formation suite contains, on average, higher magnetite and pyrite components. The choice for a reference state (clastic input) for Fe_{tot}/Al is a bit more difficult than for the Tonian example, as there are a large number of shales in the Cryogenian suite that have δ^{56}Fe values near the crustal average. To remain consistent with a net Fe addition model, we pick a Fe_{tot}/Al reference state that is slightly below that of the crustal average. Assuming a common clastic reference state, this suggests that authigenic Fe enrichment, based on Fe_{tot}/Al, is highest in the Arena Formation relative to the MacDonaldryggen Formation. It is not clear that both suites can be described by a common reference state for Fe_{HR}/Fe_{tot}, but staying consistent with the theme of net Fe addition, a relatively low Fe_{HR}/Fe_{tot} of ~0.2 is chosen; this would reflect the Fe_{HR}/Fe_{tot} of the clastic input and should not be taken to indicate an authigenic signal that records deposition in oxic conditions.

The δ^{56}Fe-reactive Fe relations for the Cryogenian suite are interpreted in terms of the multi-component mixing relations in Fig. 6.18

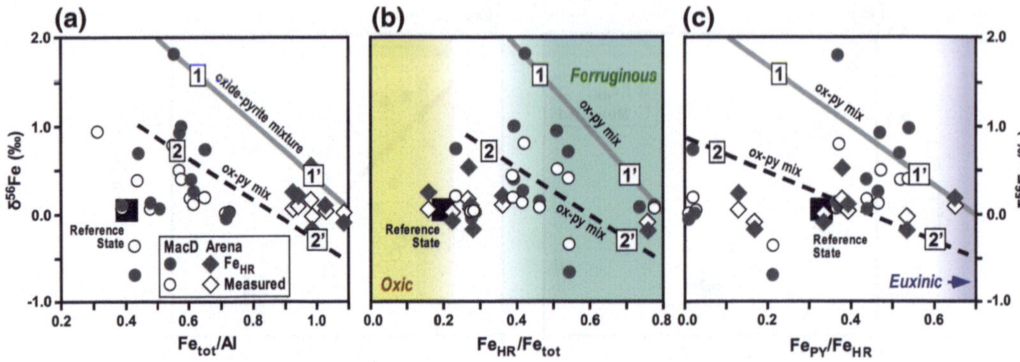

Fig. 6.21 δ^{56}Fe-reactive Fe relations for Cryogenian shales that were deposited between the Sturtian and Marinoan glaciations (Kunzmann et al. 2015, 2017). Data for measured δ^{56}Fe values (open symbols) and calculated δ^{56}Fe for Fe_{HR} (filled symbols) for two suites: MacDonaldryggen Formation, Svalbard (circles) and Arena Formation, Greenland (diamonds). Both suites were deposited in middle and outer shelf environments, at or below wave base, and consist of mixed clastic to carbonate lithologies. Kunzmann et al. (2015, 2017) interpret the sequences to have been largely deposited in ferruginous conditions, and they interpret the increases in δ^{56}Fe values to reflect addition of authigenic Fe; the choice of a moderately low Fe_{tot}/Al ratio (0.4) for the reference state was done to reflect this interpretation. Kunzmann et al. (2015), however, suggest that the Fe_{tot}/Al ratio of the detrital flux for the MacDonaldryggen Formation (but not the Arena Formation) may have been higher than the crustal average. The Arena Formation suite extends to higher magnetite and pyrite components compared to most of the MacDonaldryggen Formation samples, indicating different mixtures of authigenic minerals between the suites. See panels **a–c** in Fig. 6.18 for interpretations. The diagonal lines reflect mixing between authigenic oxide and pyrite for two cases: (1) a high-δ^{56}Fe array (1-1'; solid grey line) that records a small % oxidation of Fe^{2+}_{aq}, and (2) a low-δ^{56}Fe array (2-2'; dashed black line) that records a large % oxidation of Fe^{2+}_{aq}. Both arrays assume mixing with an authigenic pyrite component that has a δ^{56}Fe value of −0.7‰, the average for modern euxinic environments (see Chap. 5). Note that in the case of authigenic Fe enrichment, the initial reference state (δ^{56}Fe = 0.1‰) would be interpreted to lie in "oxic conditions" based on low Fe_{HR}/Fe_{tot}, whereas addition of Fe will move the interpretation to ferruginous conditions, the extent of change in δ^{56}Fe being a complex function of % oxidation of Fe^{2+}_{aq} and mixing between authigenic minerals

(panels a–c). In this context, the scatter in the data simultaneously reflect mixing between authigenic oxide and pyrite, where the oxide component has variable δ^{56}Fe values that reflect the % oxidation of Fe^{2+}_{aq}. As noted in the discussion of Fig. 6.18, a high-δ^{56}Fe oxide-pyrite mixing array (1-1' grey line in Fig. 6.21) would be appropriate for oxides that reflect only a small % oxidation of Fe^{2+}_{aq}, and hence relatively anoxic conditions. A low-δ^{56}Fe oxide-pyrite mixing array (2-2' grey line in Fig. 6.21) would reflect oxides that were produced by a high % oxidation of Fe^{2+}_{aq}, recording relatively oxic conditions. There is every reason to expect that the data reflect a range of oxide components that record a range of % oxidation of Fe^{2+}_{aq}, the result of which would produce scatter across the diagrams. The only hard constraint on possible oxide-pyrite mixing arrays lies in the maximum possible δ^{56}Fe value for the oxide component, which is δ^{56}Fe ~ +3‰ based on Fe isotope fractionation factors, which is approximated by the 1-1' line in Fig. 6.21.

The likelihood that the data for the Cryogenian shales record a wide range in δ^{56}Fe values for the oxide component (Fig. 6.21), based on the scatter in the diagrams, suggests a range of O_2 contents that controlled the % oxidation of Fe^{2+}_{aq}. Such a conclusion reflects the added dimension of interpretation that is possible when combining Fe isotope relations with reactive Fe proxy results. For samples that have high Fe_{PY}/Fe_{HR} ratios, their relatively low δ^{56}Fe values records the tendency of pyrite in marine sediments to express kinetic Fe isotope fractionations, as discussed in Sect. 5.4.3 in Chap. 5, as well as Chap. 3. The pyrite-rich endmember may indeed reflect euxinic conditions, but there are many examples of pyrite-rich modern marine sediments that were not deposited in euxinic conditions. Importantly, samples that might be described as mixtures along a low-δ^{56}Fe oxide-pyrite array (e.g., dashed black line 2-2' in Fig. 6.21) contain a component that records relatively oxic conditions (slightly positive δ^{56}Fe values for the oxides). Such information is not available if reactive Fe proxies are only considered in the absence of Fe isotope compositions. For the Cryogenian samples, it is possible to interpret the data as recording a continuum of oxide-pyrite mixing arrays, from those that indicate only a few % Fe^{2+}_{aq} oxidation (grey line 1-1') to those of near-complete Fe^{2+}_{aq} oxidation, the latter of which would be an array that passes close to a δ^{56}Fe value equal to the crustal average for the pure oxide component. Such an interpretation would suggest a large range of O_2 contents in Cryogenian seawater and/or variations in the supply of Fe^{2+}_{aq}.

The MacDonaldryggen Formation was studied by both Kunzmann et al. (2017) and Tahata et al. (2015), although using different sample suites and analytical approaches, and some clear inconsistencies emerge when the studies are compared. As discussed above, Tahata et al. (2015) interpret their in situ δ^{56}Fe values on pyrite to record sulfidation of a Fe^{3+}-hydroxide precursor, some of which have highly positive δ^{56}Fe values that would indicate small extents of oxidation of Fe^{2+}_{aq} (low O_2). For bulk samples, the relations between δ^{56}Fe and Fe_{PY}/Fe_{HR} from Kunzmann et al. (2017) fail to identify a high-δ^{56}Fe pyrite component, which should plot at high Fe_{PY}/Fe_{HR} and high δ^{56}Fe in the upper right corner of Fig. 6.21c. As discussed above, Tahata et al. (2015) identify a trend of decreasing δ^{56}Fe for pyrite with decreasing age in the MacDonaldryggen Formation, and it is possible that only the low-δ^{56}Fe pyrite component is present in the samples plotted in Fig. 6.21. It is also possible, however, that the very large size of pyrite grains analyzed by Tahata et al. (2015), sometimes approaching 400 μm, is sampling a pyrite component that is not representative of the range that occurs in bulk samples.

Variations between δ^{56}Fe values and reactive Fe components for the Ediacaran Dracoisen Formation (Fig. 6.22; Kunzmann et al. 2017) have broad similarities to those seen in Cryogenian shales (Fig. 6.21). Many of the δ^{56}Fe values for Fe_{HR} scatter between crustal values and $\sim$+1‰, which may be explained through oxide-pyrite mixing, where the oxide component that has δ^{56}Fe $\geq$+1‰ (array 2-2'). In a few cases, the oxide component appears to have quite positive δ^{56}Fe values of $\sim$+3‰ (array 1-1'); this might indicate very low O_2 contents. Surprisingly, many of these high-δ^{56}Fe samples have very low Fe_{tot}/Al ratios, below average crust. This could be interpreted to indicate that the clastic reference state had an unusually low Fe_{tot}/Al, as plotted in Fig. 6.22, and, if true, might suggest low % oxidation of Fe^{2+}_{aq}. Alternatively, the low Fe_{HR}/Fe_{tot} ratios could indicate Fe loss via a benthic Fe shuttle as the explanation for the high δ^{56}Fe values; in this case, the high-δ^{56}Fe values would not be interpreted to record limited % oxidation of Fe^{2+}_{aq} (low O_2 contents). These ambiguities highlight the difficulty of interpreting bulk samples of lithologies that contain multiple authigenic components.

A large portion of the data lies along mixing relations between a moderate δ^{56}Fe oxide component and a low-δ^{56}Fe pyrite component (array 2-2' in Fig. 6.22). This may indicate relatively high extents of oxidation of Fe^{2+}_{aq} for the oxide component. For the pyrite component, this lies within the range of δ^{56}Fe values measured in situ for pyrite in other samples of the Dracoisen Formation (Tahata et al. 2015), although in situ analyses also identified pyrite that had positive δ^{56}Fe values. In this regard, it is noteworthy that one sample has a high δ^{56}Fe value of $\sim$+2‰ for Fe_{HR} and a high Fe_{PY}/Fe_{HR} value ($\sim$0.5) in

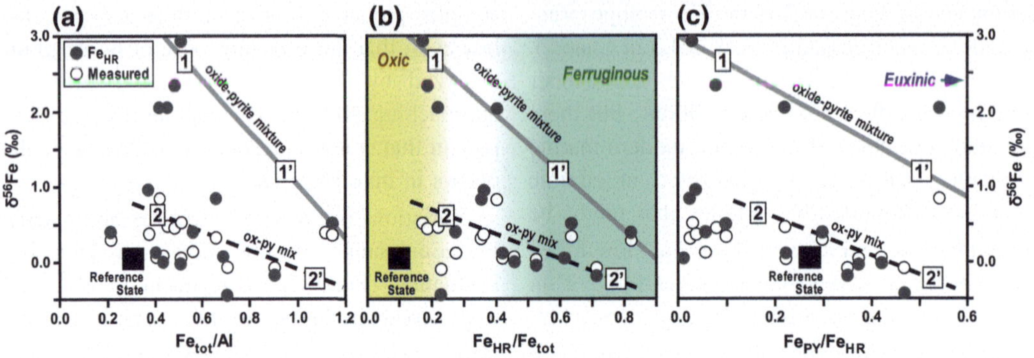

Fig. 6.22 δ^{56}Fe-reactive Fe relations for Ediacaran shales that were deposited after the Marinoan glaciation (Kunzmann et al. 2015, 2017). Data for measured δ^{56}Fe values (open circles) and calculated δ^{56}Fe for Fe_{HR} (filled circles) for the Dracoisen Formation, Svalbard. The shales were deposited on the terminal Marinoan cap carbonates during the post-glacial marine transgression on a mud-dominated platform. The highest δ^{56}Fe values for Fe_{HR} are restricted to section G436, which tends to have higher Al contents than the other sections of the Dracoisen Formation. Interpretive trends are the same as those shown in Fig. 6.21, where enrichment in authigenic Fe is thought to have occurred through accumulation of Fe oxides that reflect partial oxidation of Fe^{2+}_{aq} (Kunzmann et al. 2017), although it is important to note that this model requires a reference state of relatively low Fe_{tot}/Al ratios for this sample suite, below the crustal average; as discussed in the text, it is difficult to pick reference states for this suite. Mixing between authigenic oxide and pyrite is shown in panels **b** and **c**, and the majority of data scatter about a mixing relation appropriate for a large % oxidation of Fe^{2+}_{aq} (high O_2) for the oxide component (2-2'; black dashed lines)

Fig. 6.22, which is consistent with mixing with a high-δ^{56}Fe pyrite component. This also highlights, however, the difficulty in equating mineral Fe isotope compositions measured in situ with those inferred from bulk samples when the in situ measurements may be biased toward a certain crystal size because of analytical constraints. Because of the uncertainty in choosing appropriate reference states for Fe_{tot}/Al and Fe_{HR}/Fe_{tot} for the bulk shales of the Dracoisen Formation, it is difficult to understand the origin of the high δ^{56}Fe samples and hence how the relations in Fig. 6.22 may be used to constrain O_2 contents.

It seems clear from the discussion in this section that interpretation of Fe isotope data for fine-grained marine sediments such as shales can be difficult, although the combination with Fe speciation/proxy data is an important step forward. There is clear evidence for significant oxidation of Fe^{2+}_{aq} via the moderately elevated (just above crustal compositions) δ^{56}Fe values inferred for oxide components. This is consistent with the relatively high Fe^{3+} fractions of many Neoproterozoic shales (Fig. 6.14), which seem to be well explained by extensive oxidation under relatively high O_2 conditions. It is important, however, that authigenic components be estimated not only through consideration of Fe_{tot}/Al ratios (where it may be challenging to accurately estimate local clastic input), but also by evaluating the various Fe_{HR} components. Iron isotope data in turn can help constrain redox interpretations of Fe_{HR} (e.g., oxic, ferruginous, euxinic) by identifying samples that were open with respect to reactive Fe (loss, gain) and identifying the mineral-dependent Fe isotope fractionation factors that govern bulk δ^{56}Fe values.

That said, the key data that are often missing in reactive Fe or Fe isotope studies of shales/mudstones is determining the bulk sample Fe^{3+} abundances through classical wet chemistry, which would help greatly in constraining the oxide abundance beyond reliance on extraction methods. In addition, bulk-sample XRD analysis can also provide excellent constraints on mineralogy in fine-grained rocks, although, with a few exceptions, this is rarely done. In situ isotopic analysis has clear importance in constraining the Fe isotope budget of samples, although in the case of fine-grained rocks, where much of the inventory of specific Fe minerals may be too small to access by in situ methods, there will likely be mis-match in interpretations using various approaches.

Finally, it is important to note that in the transition that occurred in the late Neoproterozoic to the Cambrian from a ferruginous to an oxic ocean, as well as the possibility of locally euxinic conditions, the $\delta^{56}Fe$ values of seawater may have been quite variable, particularly as seawater Fe contents became low. It is probably safe to assume that unfractionated hydrothermal vent fluids had $\delta^{56}Fe$ values near the crustal average, which, under high Fe/S conditions, would evolve toward moderately negative $\delta^{56}Fe$ values (Fig. 6.15), where the most negative compositions would be predicted to be found in the most oxygen-rich environments. A decrease in overall seawater dFe contents across the Neoproterozoic-Cambrian boundary, would, however, produce conditions where extensive formation of authigenic minerals could influence the $\delta^{56}Fe$ values for dFe. The large changes in $\delta^{34}S$ values for pyrite in the Ediacaran and Cambrian (e.g., Fig. 6.12b) have been taken to record periods of extensive pyrite burial (e.g., Och and Shields-Zhou 2012; Sansjofre et al. 2016), which could have produced a large outflux of Fe from seawater that had negative $\delta^{56}Fe$ values (Fig. 6.17). Under low-oxygen conditions, pyrite burial efficiency would have been high (e.g., Laakso and Schrag 2014), as would burial of C_{org} (Daines et al. 2017). This suggests that in the transition from a high-Fe_{aq}, low-O_2 state to a low-Fe_{aq}, high-O_2 state, extensive formation of authigenic pyrite might significantly raise the $\delta^{56}Fe$ values of Fe_{aq} in seawater above the slightly negative $\delta^{56}Fe$ values expected for a ferruginous ocean. This, of course, greatly increases the complexity in interpreting the data discussed above. The answer to the question of what were the $\delta^{56}Fe$ values for Neoproterozoic seawater must be answered through studies of authigenic minerals that can be demonstrated to have formed in isotope equilibrium with seawater, which probably rules out minerals that experience large kinetic isotope fractionations such as pyrite, but might be possible for oxides, Fe carbonates, and possibly Fe phosphates.

6.4.3 Neoproterozoic Iron Formations (IFs)

Global glaciation ("Snowball Earth"), iron formation (IF) deposition, and a rise in oxygen have long been interwoven in models for the late Neoproterozoic. Kirschvink (1992) proposed that Neoproterozoic Iron Formations (NIFs) formed by a combination of buildup of Fe^{2+}_{aq} in the oceans during maximum glaciation, followed by rapid oxidation upon emergence from the global "ice house" and exposure of an Fe^{2+}_{aq}-rich ocean to an oxygenated atmosphere. This model offered an explanation for why significant volumes of IFs reappear in the Precambrian rock record after a ~1 b.y. gap in time following deposition of the late Paleoproterozoic IFs. Subsequent studies, however, have shown that virtually all NIFs were syndepositional with the Sturtian glaciation, rather than post-glaciation, as in the Kirschvink model (e.g., Hoffman et al. 1998, 2017; Cox et al. 2016a, b). Numerous proposals have been made for the trigger of Cryogenian glaciations, involving CO_2 drawdown, weathering changes, and large igneous activity (e.g., Hoffman et al. 1998, 2017; Schrag et al. 2002; Swanson-Hysell et al. 2010; Rooney et al. 2014; Cox et al. 2016a, b; Macdonald and Wordsworth 2017; Macdonald et al. 2018). Models for emergence from global glaciations, on the other hand, generally invoke an increase in CO_2, perhaps to levels as high as 100 times PAL (e.g., Hoffman et al. 2017).

The majority of NIFs were deposited in fault-bounded basins that developed during breakup of Rodinia and are closely associated with Sturtian-age glaciogenic deposits (e.g., Cox et al. 2013). Many depositional models for NIFs call upon accumulation in isolated, tectonically- bound basins, trapping Fe^{3+}-oxides generated by oxidation of Fe^{2+}_{aq} through interaction with glacial brines that transported O_2 from the surface during periods of major glaciation (Cox et al. 2013;

Lechte and Wallace 2015, 2016). Some recent models, however, propose that oxidation only occurred after deglaciation (e.g., Baldwin et al. 2016), similar to the original model of Kirschvink (1992). In all cases, NIFs contain significant siliciclastic components when compared to Paleoproterozoic and older IFs, generally several wt% Al_2O_3 or higher. Mineral banding is rare, in contrast to the "Banded Iron Formations" (BIFs) of the Archean and Paleoproterozoic. The Fe mineralogy of NIFs is generally simple compared to Neoproterozoic shales. Authigenic Fe is largely hematite, and authigenic Fe^{2+}-bearing phases are very rare, where magnetite, chamosite, and pyrite are generally absent, although siderite does occur in some sections (e.g., Klein 2005); this contrasts with the very high abundance of magnetite and siderite in older IFs. Therefore, despite sometimes relatively high siliciclastic levels, the restricted range in authigenic minerals in most NIFs greatly simplifies interpretations relative to the Neoproterozoic shales discussed in the previous section. The very low levels of C_{org} in NIFs has been taken to indicate oxidation by O_2, rather than biological processes such as photoferrotrophy (e.g., Cox et al. 2013).

Detailed geochronology for the Cryogenian (Rooney et al. 2015; Prave et al. 2016; Cox et al. 2018) documents a long period of ~57 m.y. for the Sturtian glaciation (717–660 Ma), followed by an interglacial period of ~21 m.y., terminated by the relatively short-lived Marinoan glaciation (639–635 Ma). Although it is clear that most NIFs were deposited during the Sturtian glaciation, NIF deposits are not uniformly distributed throughout Sturtian sections, nor do NIFs occupy the major portion of Sturtian stratigraphy. In some sections, NIF deposition is terminated by an increase in glacially-associated clastic deposition (e.g., Baldwin et al. 2016), suggesting a change from deep- to shallow-water deposition, and in other cases, NIF sections are interpreted to have been deposited during marine transgressions (e.g., Halverson et al. 2011; Cox et al. 2016a).

We divide our discussion of the Fe isotope compositions of NIFs into those that show no trend in $\delta^{56}Fe$ with stratigraphic height (Fig. 6.23a) and those that have a stratigraphic trend of increasing $\delta^{56}Fe$ with increasing stratigraphic height (Fig. 6.23b); no NIF appears to have a stratigraphic trend of decreasing $\delta^{56}Fe$ with increasing height. The three sections of the Fulu IF (China) studied by Busigny et al. (2018) have a relatively homogenous Fe isotope composition over the entire stratigraphic height of each section, where calculated $\delta^{56}Fe$ values for the hematite component (obtained by mass-balance calculations using Fe_{tot}/Al ratios) largely range between +1 and +2‰. The exception are samples from the base of the Mengzhai section, which is dominated by siderite and has slightly negative $\delta^{56}Fe$ values, consistent with equilibrium with Fe^{2+}_{aq} that had a near-zero $\delta^{56}Fe$ value (Wiesli et al. 2004). Busigny et al. (2018) apply the dispersion/reaction model of Czaja et al. (2012) to calculate seawater O_2 contents that might produce the moderately positive $\delta^{56}Fe$ values and conclude seawater O_2 was <0.4 nM, which is extremely low, lower than what is detectable in modern marine Oxygen Minimum Zones (OMZs). It is not appropriate, however, to apply the model of Czaja et al. (2012) to the depositional setting of the Fulu IF, as the original model was developed for the photic zone in contact with an anoxic atmosphere in an open (ice-free) continental margin of Archean age and assumes water depths, photic zone depths, Fe^{2+}_{aq} transport, and pH conditions that are unlikely to be directly applicable to NIF deposition in a subglacial environment. Nevertheless, the moderately positive $\delta^{56}Fe$ values are reasonably interpreted as reflecting partial oxidation of Fe^{2+}_{aq} and hence relatively low O_2 contents, but probably not as low as they infer using the model of Czaja et al. (2012). Busigny et al. (2018) propose O_2 was sourced to trapped gas bubbles in glacial ice that was released upon melting of the base of the ice sheets. Given the relatively constant $\delta^{56}Fe$ values across the three sections, the extent of Fe^{2+}_{aq} oxidation must have been very constant, suggesting relatively stable marine and glacial conditions.

The Sperry Wash section of the Kingston Peak IF (USA) has relatively constant $\delta^{56}Fe$

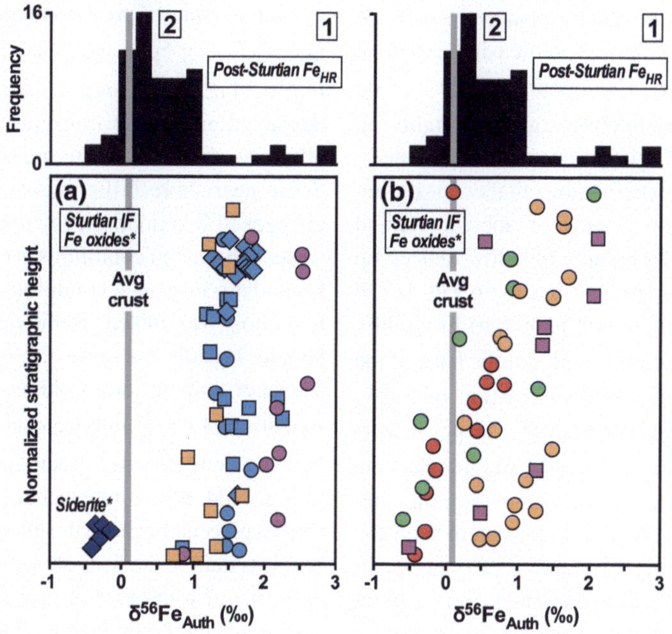

Fig. 6.23 δ^{56}Fe values for authigenic components of Sturtian Iron Formations (IFs) in lower panels, placed in relative stratigraphic order (normalized to total height of analyzed sections). In upper panels, histograms show δ^{56}Fe values for oxides in post-Sturtian Fe-rich shales and Fe-poor cherts (same histogram on both sides). Numbers "1" and "2" in boxes on histograms indicate Ediacaran authigenic oxide components plotted in Fig. 6.22. For IFs data points, except where noted (*), δ^{56}Fe values plotted are for the oxide component of the IFs. A: IFs where there is no systematic change in δ^{56}Fe values with stratigraphic height ($R^2 < 0.3$). Symbols and localities as follows: Fulu IF (blue; circles: Jiangkou; squares: Fengmu; diamonds: Mengzhai—note dark blue indicates siderite-rich samples); Kingston Peak Formation IF (tan; squares: Sperry Wash). B: IFs where there are general trends of increasing δ^{56}Fe values with increasing stratigraphic height ($R^2 > 0.3$). Symbols and localities as follows: Holowilena IF (purple/pink; circles: Oraparinna, squares: Homestead—note data for Homestead samples are bulk compositions); Rapitan IF (red); Tatonduk IF (green); Kingston Peak Formation IF (tan; circles: Kingston Range). Relative to late-Sturtian IFs, the δ^{56}Fe values for the oxide component of post-Sturtian shales and cherts are significantly lower than the IFs, particularly the δ^{56}Fe values at the top of the IF sections. *Data sources* Fulu (Busigny et al. 2018), Holowilena (Cox et al. 2016a), Tatonduk (Cox et al. 2016a), Rapitan (Halverson et al. 2011), Kingston Peak (Lechte et al. 2018). Data for post-Sturtian Fe oxides from sources given in Fig. 6.19

values with stratigraphic height, overlapping the Fe isotope compositions of the Fulu IF (Lechte et al. 2018; Fig. 6.23a). The Kingston Range IF is different from most NIFs in that the Fe-rich sections are thin (few meters) and occur throughout a thick clastic section (hundreds of meters), whereas most NIFs consist of thick sections of IF that are bounded by clastic sections. Lechte et al. (2018) interpret the moderately positive δ^{56}Fe values to indicate limited oxidation of Fe^{2+}_{aq} that occurred when oxygenated glaciogenic fluids interacted with ferruginous seawater. They suggest that IF deposition occurred during short periods when O_2-bearing brines were released into restricted marine basins that were Fe^{2+}_{aq} rich. Supply of oxidants was therefore directly tied to glacial processes.

Several NIF sections show broad correlations of increasing δ^{56}Fe values with increasing stratigraphic height ($R^2 > 0.3$), including the Rapitan (Canada), Tatonduk (USA), Holowilena (Australia), and the Kingston Range section of the Kingston Peak Formation (USA) (Fig. 6.23b). Such relations have been generally taken to record a transgressive marine sequence, where IF deposition records deeper, high-Fe^{2+}_{aq} seawater that had low O_2 contents, expressed by an increase in $\delta^{56}Fe$ values up section, reflecting decreasing extent of Fe^{2+}_{aq} oxidation (e.g., Halverson et al. 2011; Cox et al. 2016a). This interpretation highlights an unusual aspect of glacial-deglacial sequences, where the marine transgressions that occur during deglaciation provide a view of ancient vertical gradients in the water column (shallow to deep up section). If the stratigraphic trends in Fig. 6.23b are taken at face value, the near-zero to slightly negative $\delta^{56}Fe$ values for oxides at the base of the sections would imply 100% oxidation of Fe^{2+}_{aq}, and hence very high levels of O_2. In the marine transgression model, this would suggest that the shallow Cryogenian oceans were significantly oxygenated. Such a model is consistent with the appearance of highly elevated $\delta^{53}Cr$ values in NIFs (Frei et al. 2009). Deeper waters, on the other hand, increasingly sampled during the transgression, would have been relatively anoxic, consistent with the proxy data discussed above (e.g., Canfield et al. 2008).

There has been much discussion about the paleogeography that might be inferred for NIFs and how this may affect inferences of global versus local marine conditions. It is important to stress that the presence of extensive glaciation in marine settings, accompanied by the local controls imposed by preservation in extensional basins during rifting of Rodinia, presents challenges in interpreting geochemical data in terms of global processes (e.g., Li et al. 2010, 2012a). Figure 6.24 brings together NIF models discussed in Halverson et al. (2011), Cox et al. (2016a), and Busigny et al. (2018). During periods of glaciation, penetration of O_2 from the atmosphere is limited beneath grounded marine ice sheets, and NIF deposition in these settings would be associated with limited oxidation of Fe^{2+}_{aq}, producing high $\delta^{56}Fe$ values for oxides. NIF sections that might record more open-ocean deposition, with greater exchange between seawater and atmospheric O_2, would be expected to have been associated with greater extents of oxidation of Fe^{2+}_{aq} and hence less positive $\delta^{56}Fe$ values for oxides (right side of Fig. 6.24a). In contrast, during periods of deglaciation, marine transgression would generally raise the oxic/anoxic transition in the ocean. Some near-shore NIFs might not see significant changes in % oxidation; essentially the limited O_2 under sub-glacial conditions is replaced by limited O_2 under non-glacial conditions if the oxic/anoxic transition was raised. Removal of glacial cover should broadly increase O_2 exchange between seawater and the atmosphere, increasing the % oxidation of Fe^{2+}_{aq} and decreasing $\delta^{56}Fe$ values of Fe^{3+}-oxide/hydroxide precipitates. As illustrated in Fig. 6.24, this results in a measurable change in the deep ocean because the effects of glacial cover was less than in the near-shore example, but this is only one of a number of paleogeographic reconstructions that could be envisioned.

Finally, we return to considering changes in Fe isotope compositions between Sturtian NIFs and post-Sturtian units in Fig. 6.23. The majority of $\delta^{56}Fe$ values for Fe_{HR} in post-Sturtian shales and Fe-poor cherts are significantly less than those of NIFs, especially those emplaced at the end of the Sturtian that recorded stratigraphic increases in $\delta^{56}Fe$. The histograms in Fig. 6.23 additionally show the two inferred oxide components in Ediacaran shales (Fig. 6.22), where component "1" has high-$\delta^{56}Fe$ values that indicate limited oxidation, but component "2", where the majority of the data plot, records a lower-$\delta^{56}Fe$ value for oxides and hence greater extent of oxidation. These relations provide qualitative evidence for an increase in O_2 after the end of Sturtian glaciation and continuing into the Ediacaran (e.g., Scott et al. 2008; Kendall et al. 2015b; Sahoo et al. 2016; Xiang et al. 2017; Lau et al. 2017). Final rise of marine O_2 then occurred in the Cambrian (e.g., Dahl et al. 2014; Chen et al. 2015; Wang et al. 2015; Wen et al. 2015; Cheng et al. 2017), based on other proxies, although relatively few Fe isotope data are available for this time period.

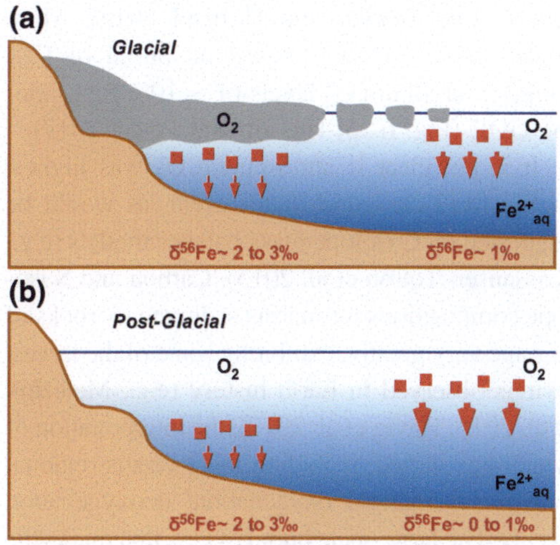

Fig. 6.24 Illustration of depositional environments for iron oxide components in Neoproterozoic iron formations and Fe-bearing shales and cherts, drawing upon models from Halverson et al. (2011), Cox et al. (2016a), and Busigny et al. (2018). Marine gradients show deep-water Fe^{2+}_{aq}-rich zones in blue and O_2-rich shallow zones in white. Glacial ice shown in grey. Horizontal blue lines note ocean surface. Ferric oxide/hydroxides formed in the water column via oxidation of upwelling Fe^{2+}_{aq} shown as red squares; settling of these oxide/hydroxides shown by vertical red arrows, where the size of the arrows is proportional to the extent of oxidation of Fe^{2+}_{aq}. During glacial times in the Cryogenian (**a**), the extent of Fe^{2+}_{aq} oxidation was limited, particularly beneath marine ice sheets where O_2 penetration through glacial ice was limited; limited Fe^{2+}_{aq} oxidation produced ferric oxide/hydroxides that had high δ^{56}Fe values of ~ +2 to +3‰. In relatively ice-free zones, atmospheric O_2 exchange with the shallow oceans was higher, producing larger extents of Fe^{2+}_{aq} oxidation and lower δ^{56}Fe values of ~ +1‰. In inter- or post-glacial periods (Cryogenian or Ediacaran), oxygen availability was increased due to higher rates of atmosphere-ocean O_2 exchange in the absence of glacial ice, as well as generally increased atmospheric O_2 contents (**b**). Marine transgressions upon de-glaciation would have shallowed the redoxcline, moving Fe^{2+}_{aq} deep waters onto the continental shelf; oxidation of these waters may have been limited, producing high δ^{56}Fe values of ~ +2 to +3‰ in some cases. Generally higher O_2 contents, however, are expected to broadly produce higher extents of Fe^{2+}_{aq} oxidation, producing δ^{56}Fe values of ~0 to +1‰ in post-Cryogenian time, particularly during periods of deposition of high-Mo and -U shales in the later Ediacaran and early Cambrian (see also Fig. 6.19)

6.5 Precambrian Earth: The Paleoproterozoic and Neoarchean Transition Through the GOE

The "Great Oxidation Event" (GOE) of the Paleoproterozoic is classically defined as the transition from an anoxic to slightly oxic world based on numerous lines of evidence, including the loss of mass-independent fractionation of S isotopes (MIF-S), loss of mobility of Fe in paleosols, loss of detrital uraninite, siderite, and pyrite, and appearance of continental red beds (see, for example, Holland 2006). The loss of MIF-S is a common marker for the time of the GOE, ~2.3–2.4 Ga, and this was initially based on studies from the Eastern Transvaal Basin (e.g., Bekker et al. 2004, and references within), but it has since become clear that atmospheric O_2 levels fluctuated significantly before and after the GOE (see, for example, Lyons et al. 2014), and, in fact, new geochronological constraints across the GOE show that the canonical MIF-S signal for the GOE was not globally coincident, but instead varied dependent on S pathways on a regional basis (e.g., Reinhard et al. 2013; Gumsley et al. 2017; Philippot et al. 2018). Many workers, therefore, consider the GOE not so much as an "event" but a transition that spanned up to several hundred m.y. and involved complex changes in marine redox, O_2 levels in the oceans and atmosphere, CO_2 levels, weathering, C_{org} burial, and global glaciations, all of which profoundly impacted the Fe biogeochemical cycle.

Although here we take the position that the GOE was not a single event, much of the geochemical literature on the Paleoproterozoic and Neoarchean has been cast in this framework, and we will loosely retain this in our organization here for familiarity. Our discussion in this section first focuses on post-GOE sedimentary rocks of ~2.4 to 1.8 Ga age (Paleoproterozoic). Although the Mesoproterozoic is a time period where there is currently great research focus, documenting relatively low levels of O_2 and high seawater Fe^{2+}_{aq} (see, for example, recent discussions by Diamond and Lyons 2018; Planavsky et al. 2018a), there is relatively little published Fe isotope data for sedimentary rocks of Mesoproterozoic age. Following our discussion of post-GOE Paleoproterozoic rocks, we then explore possible changes in the Fe cycle during surface weathering through a comparison of pre- and post-GOE paleosols. As we step deeper into an anoxic world, it becomes clear we need to revisit the issue of Fe^{2+}_{aq} oxidation, as application of models developed for the modern world can no longer adequately describe oxidation through the transition from the Archean to the Proterozoic. Next, we will focus on the very large iron formation (IF) deposits that span the Archean-Proterozoic transition, a lithology that has figured prominently in discussions of the ancient Fe cycle. Finally, we integrate the results obtained from diverse lithologies (shales and Ca–Mg carbonates) that were deposited on the continental margins at the end of the Archean, a record that is complementary to that provided by contemporaneous IFs.

6.5.1 The Post-GOE Sedimentary Record

Loss of MIF-S occurred between ~2.43 Ga and ~2.32 Ga, but possibly as late as ~2.25 Ga (Philippot et al. 2018). The canonical 2.32 Ga age based on data from the Eastern Transvaal Basin (e.g., Bekker et al. 2004, and references within) correlates with a marked increase in the burial flux of C_{org} using the Macrostrat database (Fig. 6.25), and Husson and Peters (2017) correlate this increase to an increase in atmospheric O_2. C_{org} burial rates varied between ~2.3 Ga and the time of the positive $\delta^{13}C$ excursion for carbonates of the Lomagundi-Jatuli event, as well as the negative $\delta^{13}C$ excursion for C_{org} of the Shunga event, although it is important to stress that current geochronology has significant uncertainties for specific geologic units that prevent precise temporal alignment of all data sets. There have been a wide variety of estimates for atmospheric O_2 levels after the GOE, with some proposals that O_2 approached modern levels (e.g., Bekker and Holland 2012). More conservative estimates based on burial of C_{org} suggest maximum O_2 levels of ~10% PAL after the GOE (Fig. 6.25; Husson and Peters 2017).

It is not clear if atmospheric O_2 was highest during the Lomagundi-Jatuli event, as would be suggested by C isotope mass-balance models (e.g., Krissansen-Totton et al. 2015). Carbon and S isotope compositions for marine sedimentary rocks of this age vary greatly, exhibiting some of the largest swings observed in Earth history (e.g., Melezhik et al. 2015; Havig et al. 2017), the interpretation of which has varied, including local versus regional signals or multiple oxygenation/ deoxygenation events. The large-scale burial of C_{org} that marks the Shunga event is associated with a strong decrease in $\delta^{13}C$ values for C_{org}, which has been interpreted in a variety of ways, including extensive oxidation of C_{org} (Kump et al. 2011), similar to proposals for the negative $\delta^{13}C$ excursions of the late Neoproterozoic and early Cambrian (see Sect. 6.4.1), as well as anaerobic oxidation of methane coupled to sulfate reduction (Ossa Ossa et al. 2018b). Several studies argue for substantial seawater sulfate levels during the Lomagundi-Jatuli event, perhaps up to ~10 mM (one-third of modern levels), which would argue for very high atmospheric O_2 contents (e.g., Planavsky et al. 2012b; Reuschel et al. 2012; Scott et al. 2014; Blättler et al. 2016). Following the Lomagundi-Jatuli event, seawater sulfate levels are inferred to have markedly decreased during deoxygenation beginning ~2.05 Ga forward (e.g., Planavsky et al. 2012b; Scott et al. 2014; Ossa Ossa et al. 2018a), dropping to sub-mM levels in the Mesoproterozoic (e.g., Luo et al. 2015), although such decreases may have been variable on a regional scale (Paiste et al. 2018). These changes in inferred O_2 levels, as well as inferred changes in euxinic conditions, are mirrored in Mo isotope changes recorded in black shales from this time interval, where low $\delta^{98}Mo$ values are measured in 2.32 Ga rocks, followed by a peak during the Lomagundi-Jatuli event, followed by a decrease in $\delta^{98}Mo$ during the Shunga event (Asael et al. 2018).

Early studies of Fe isotope compositions of marine sedimentary rocks in the period 2.4 to

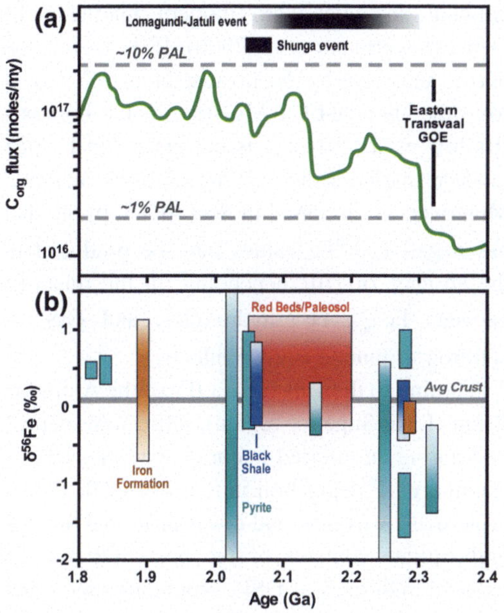

Fig. 6.25 Temporal variations in organic carbon (C_{org}) burial fluxes (**a**) and δ^{56}Fe values (**b**) for marine and terrestrial sedimentary rocks after the Great Oxidation Event (GOE). C_{org} fluxes (green curve) from Husson and Peters (2017), which roughly scales between ~1 and ~10% Present Atmospheric Level (PAL) for atmospheric O_2 using their oxygen model; other models for atmospheric O_2 during this time interval suggest more extreme changes (e.g., Lyons et al. 2014). The GOE, as defined by the loss of mass-independent isotope fractionation of S (MIF-S) from the Eastern Transvaal basin, is from Bekker et al. (2004), although new studies show that the timing of the GOE varied among different continents (e.g., Philippot et al. 2018). The timing for the positive carbonate δ^{13}C excursion for the Lomagundi-Jatuli event from Martin et al. (2013), and the time of the negative C_{org} δ^{13}C excursion of the Shunga event from Kump et al. (2011); both events are interpreted to record strong increases in atmospheric O_2. For δ^{56}Fe values, 413 data points are represented as gradient boxes for clarity, where the darkest colors correspond to the highest density of points, and light shading corresponds to a low density of points. Data for IFs include analyses of WR and oxides, which broadly overlap. Data for pyrite, all from C_{org}-rich black shales, include mineral separates and in situ analyses. Samples of continental red beds and paleosols are WR analyses. Isotopic data from Rouxel et al. (2005), Yamaguchi et al. (2005, 2007), Frost et al. (2007), Hyslop et al. (2008), Planavsky et al. (2009, 2012a), Nishizawa et al. (2010), Asael et al. (2013), Virtasalo et al. (2015), and Bankole et al. (2016), with depositional ages updated using new geochronology for this time period, as summarized in Gumsley et al. (2017) and Philippot et al. (2018)

1.8 Ga suggested δ^{56}Fe values tended to be positive, although ranging from −0.5 to +1.0‰ (Rouxel et al. 2005). These data include those obtained on sedimentary pyrite, which Rouxel et al. (2005) interpreted to directly record the Fe isotope compositions of coeval seawater, as well as a few analyses of oxides. Rouxel et al. (2005) suggested that the positive δ^{56}Fe values reflected an absence of the extensive Fe oxide precipitation that marked the pre-GOE period. Subsequent studies document a much wider range in δ^{56}Fe values for pyrite from this period, extending to values less than −2‰, as well as positive values that exceed +1.5‰ (Fig. 6.25). There is no evidence for the highly positive δ^{56}Fe values measured for Neoproterozoic shales (Fig. 6.19) in the time interval of 2.4−1.8 Ga, which might be broadly taken to indicate higher overall levels of O_2 than in the Neoproterozoic. In part, however, this contrast may reflect the paucity of shale data from 2.4−1.8 Ga that has full chemistry available, including reactive Fe inventories, as compared to the studies of Neoproterozoic rocks; this biases the discussion of data from Neoproterozoic units toward oxide components in a manner that is not possible to match when considering shales of 2.4−1.8 Ga age. We must therefore keep the mineralogical-isotopic distinctions between oxides and sulfides in mind when looking at the Paleoproterozoic Fe isotope record in Fig. 6.25. An exception is the study of the Shunga-age black shales of the Zaonega Formation from the Onega Basin (Asael et al. 2013), which applied detailed chemical, mineralogical, and Fe speciation analysis to the samples measured for isotopic compositions, allowing more confident inference of various authigenic components. Asael et al. (2013) noted that δ^{56}Fe values broadly change from moderately positive values to slightly negative values up-section, correlating with an increase in Fe_{PY} and shift from anoxic (partial oxidation of Fe^{2+}_{aq}) to euxinic conditions (complete uptake of seawater Fe).

For pyrite from black shales of 2.4−1.8 Ga age, these appear to be just as variable in their δ^{56}Fe values as pyrite from Neoproterozoic shales (compare Figs. 6.19 and 6.25). An in situ Fe isotope study of pyrite from the Shunga-age (or slightly post-Shunga) Talvivaara Formation (Finland) by Virtasalo et al. (2015) document an exceptionally wide range of $\delta^{56}Fe$ values for pyrite, from −3.6 to +1.7‰. As is the case for pyrite from Neoproterozoic shales, the very large range in $\delta^{56}Fe$ values in the Talvivaara Formation underscores the large pathway-dependency of Fe isotope compositions of sedimentary pyrite. The depositional environment of the Talvivaara Formation is interpreted to reflect a tidal to near-shore marine setting, recording the interaction of Fe and S fluxes from riverine and seawater sources. Pyrite that has highly negative $\delta^{56}Fe$ values are thought to reflect kinetic isotope effects upon rapid pyrite formation, where abundant sulfide was supplied by DSR. In contrast, pyrite that has positive $\delta^{56}Fe$ values are interpreted to record sulfidation of Fe^{3+}-hydroxides that formed by partial oxidation of Fe^{2+}_{aq} by low levels of atmospheric O_2, echoing the theme of the discussion of Neoproterozoic pyrite above. As discussed above, although it is possible high-$\delta^{56}Fe$ pyrite could record sulfidation of high-$\delta^{56}Fe$ Fe^{3+}-hydroxides, this requires a major interpretive step, and it is also possible that highly positive $\delta^{56}Fe$ values for pyrite reflects formation under equilibrium conditions (see Chap. 3). It would, of course, be much better to directly analyze the Fe isotope compositions of Fe^{3+}-oxide/hydroxide, although this approach would likely rule out study of C_{org}-rich shales.

Nishizawa et al. (2010) studied pyrite from the ~2.25 Ga Kazput Formation, Turee Creek Group (Australia) via in situ methods and found a wide range in $\delta^{56}Fe$ values from −2 to +0.5‰. Philippot et al. (2018) document the persistence of small MIF-S signals up through the Kazput Formation, as part of their documentation of a long transition for the GOE. The pyrites analyzed by Nishizawa et al. (2010) are from stromatolitic carbonates, and they interpret the generally negative $\delta^{56}Fe$ values to record a low-$\delta^{56}Fe$ Fe^{2+}_{aq} component that was produced by DIR, drawing analogies with modern marine sediments (Severmann et al. 2006). If, however, the pyrite analyzed by Nishizawa et al. (2010) is from the abundant Ca–Mg stromatolite horizons (Barlow et al. 2016), it is not clear that Fe-rich modern marine sediments are a strong analogy. Moreover, as discussed in Sect. 6.3.3, pyrite that has negative $\delta^{56}Fe$ values may be produced in the absence of DIR depending on the relations between Fe^{2+}_{aq}, Fe^{3+}-hydroxides, and FeS, in addition to kinetic isotope effects.

Turning to the 1.8–1.9 Ga IFs of the Animikie Basin (Lake Superior region), these were deposited during an inferred major change in seawater chemistry. IF deposition in the post-GOE Paleoproterozoic was rare, and it was proposed that the high sulfate contents of the post-GOE oceans induced high rates of DSR, producing increasing sulfide in the oceans, which in turn caused the cessation of major IF deposition until the late Neoproterozoic (Canfield 1998; Poulton et al. 2004). Subsequent work on the Animikie Basin has focused on spatial variability within the basin, recognizing that the tectonics of the Penokean Orogeny to the south may have limited water exchange within the basins, as well as with the open oceans (Pufahl et al. 2010; Poulton et al. 2010). These studies document spatial and temporal changes in ferruginous and euxinic zones in the basin based on Fe speciation, with periodic incursions of these reduced zones into shallow oxygenated waters. This interpretation is supported by Mo isotope studies of the Rove Formation, which overlies the IFs, where the highest $\delta^{98}Mo$ values are recorded in the most euxinic shales, but shales with lower Fe_{PY}/Fe_{HR} ratios had more variable $\delta^{98}Mo$ values (Kendall et al. 2011). Recent work on the deepest portion of the basin suggests periodic oxygenation of deep-water ferruginous and euxinic zones, which has produced a very wide range in Mo isotope compositions (Planavsky et al. 2018b). Transport of oxygenated, shallow-water-derived metals was accomplished by a Fe–Mn oxyhydroxide shuttle from the basin margin. There are no published Fe isotope data for the shales of the Animikie Basin.

Planavsky et al. (2009, 2012a) report Fe isotope data for the Gunflint and Biwabik IFs, with a focus on correlating $\delta^{56}Fe$ values with the presence or absence of Ce anomalies as a constraint on redox. For stromatolitic units, hematite that is texturally associated with microbial filaments generally has slightly positive $\delta^{56}Fe$ values but lack significant Ce anomalies, which they took to indicate oxidation of Fe^{2+}_{aq} without oxygen. They therefore suggested that the filament-associate hematite was formed via partial oxidation of Fe^{2+}_{aq} by photoferrotrophs, drawing upon the experimental work of Croal et al. (2004). Such an interpretation would indicate very low O_2 contents during deposition of the stromatolitic units. Units that do not contain microfossils, including siderite-dominated IF, had lower $\delta^{56}Fe$ values, including Mn-bearing units that had strongly negative Ce anomalies. These compositions were taken to indicate more complete oxidation, as modeled using a Rayleigh fractionation model. This interpretation, however, contrasts with that of Lepot et al. (2017), who argued that the intra-fossil Fe minerals in the Gunflint IF were more consistent with oxygenic photosynthesizing microbes.

A more complex view of O_2 levels during Gunflint IF time comes from the study of Fralick et al. (2017), who compared redox-sensitive trace-element contents and Cr isotope compositions of the upper Gunflint IF that experienced meteoric ground water alteration shortly after deposition, with deeper sections that were below the meteoric alteration zone. They concluded that atmospheric O_2 levels were significant at the time of meteoric alteration, consistent with the conclusion of Lepot et al. (2017) that Fe biomineralization was produced by oxygenic photosynthesis. The Cr isotope results from the meteoric alteration zone of the Gunflint IF, however, stand in marked contrast to the restricted range in Cr isotope compositions measured in Paleoproterozoic through early Neoproterozoic shales that have been interpreted to record low atmospheric O_2 contents (Cole et al. 2016). The study of Fralick et al. (2017) highlights the important differences in inferred redox conditions that may be produced when comparing deep-water lithologies to surface conditions. Finally, it is noteworthy that the inferred higher O_2 levels during deposition of the Gunflint IF by Fralick et al. (2017) aligns with the Cr isotope study of Babechuk et al. (2017) of the broadly coeval (1.85 Ga) Flin Flon paleosol.

Two studies have traced the effects of contact metamorphism of the ~1.8 Ga Biwabik IF by the 1.1 Ga Duluth gabbro complex (Frost et al. 2007; Hyslop et al. 2008). The lack of a correlation between spread in $\delta^{56}Fe$ values and proximity to the igneous contact led Frost et al. (2007) to conclude that, for bulk samples, the original sedimentary $\delta^{56}Fe$ values remained intact. Combining O and Fe isotope fractionations led Hyslop et al. (2008) to conclude that at high grades of metamorphism, Fe isotope fractionations between magnetite and Fe-silicates lowered the original sedimentary $\delta^{56}Fe$ values for magnetite by ~0.2–0.3‰, a small, but significant amount. Much greater decreases in $\delta^{18}O$ values were observed for magnetite, reflecting re-equilibration with quartz at high temperatures, demonstrating distinct isotopic responses for O and Fe isotopes during metamorphism that reflect the differences in mass balances and isotopic exchange rates. Broadly, the data from Frost et al. (2007) and Hyslop et al. (2008) indicate $\delta^{56}Fe$ values for magnetite from zero to slightly positive, overlapping the results obtained by Planavsky et al. (2009, 2012a) outside the contact metamorphic zone. These results are generally consistent with partial oxidation of Fe^{2+}_{aq}, suggestive of low O_2 levels, although the slightly negative $\delta^{56}Fe$ values are suggestive of more extensive oxidation. We will return to the data from the contact metamorphism zone of the Biwabik IF in our discussion of the high-grade metamorphic terranes of the Eoarchean in Sect. 6.6.3.

We now turn to red beds, which represent fluvial, lacustrine, or aeolian environments and therefore may, potentially, represent a more direct assessment of atmosphere redox than marine sediments. It has long been recognized that red beds permanently appear in the rock record by about 2.3 Ga (e.g., Chandler 1980), and these have been among multiple lines of evidence used to constrain the timing of the GOE. As shown in

Fig. 6.14, post-GOE Paleoproterozoic sedimentary rocks have very high fractions of Fe^{3+} in their Fe budgets, which has been taken to indicate high levels of atmospheric oxygen, at least immediately after the GOE (Bekker and Holland 2012). High levels of Fe^{3+}-oxide/hydroxides in the red beds might reflect oxidation during early diagenesis, or weathering of oxidized source rocks. Bankole et al. (2016) studied the ~2.1 Ga FA Formation from the Franceville Basin (Gabon), which is largely comprised of sandstones, and host well-known U mineralization. There is some uncertainty in the depositional age of the FA Formation. The overlying Franceville units are <2.1 Ga in age and record negative $\delta^{13}C$ excursions for C_{org} that are correlated with the Shunga event (Kump et al. 2011; Ossa Ossa et al. 2018a), placing a minimum age constraint on deposition of the FA Formation, but its maximum depositional age is more difficult to determine.

Bankole et al. (2016) interpreted the range in $\delta^{56}Fe$ values measured for WR samples of the FA Formation to record mixing between a high-$\delta^{56}Fe$ authigenic Fe oxide component and detrital silicate minerals that had a near-zero $\delta^{56}Fe$ value (Fig. 6.26). Importantly, the study by Bankole et al. (2016) included detailed mineralogical characterization via XRD and bulk rock chemistry, including Fe speciation. They observed a correlation of increasing $\delta^{56}Fe$ with decreasing total Fe contents, as well as decreasing Fe_{tot}/Al ratios. Bankole et al. (2016), however, do not interpret these relations to indicate loss of a low-$\delta^{56}Fe$ component, but instead point to a correlation of increasing $\delta^{56}Fe$ with increasing Fe_{tot}/Mg ratio to argue that the high $\delta^{56}Fe$ (but low-Fe_{tot}) samples record net hematite addition through partial oxidation of Fe^{2+}_{aq}-bearing pore water. In this interpretation, the samples that have the lowest Fe_{tot}/Al ratios would be most sensitive to addition of an authigenic Fe^{3+}-oxide/hydroxide component. The model put forth by Bankole et al. (2016) uses the inferred oxidized component as an indirect proxy for atmospheric O_2, and they argue that the oxide minerals were not detrital. In addition, a correlation of increasing $\delta^{56}Fe$ and

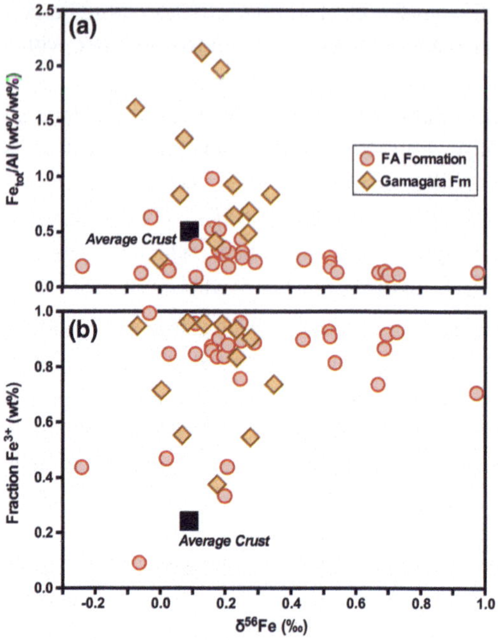

Fig. 6.26 Variations in $\delta^{56}Fe$ values for ~2.1–2.2 Ga red beds relative to fraction of Fe^{3+} and Fe_{tot}/Al ratios. Data for the ~2.1 Ga FA Formation from the Franceville Basin (Gabon) from Bankole et al. (2016), shown in red symbols. Data for the 2.15 Ga Gamagara Formation (South Africa) from Yamaguchi et al. (2005) and Yamaguchi and Ohmoto (2006), shown in brown symbols, with age assignment based on the geochronology of Gumsley et al. (2017) and Philippot et al. (2018), which places it in the middle of the Lomagund-Jatuli event. Compositions for average crust taken to be the same as igneous rocks for $\delta^{56}Fe$ and fraction of Fe^{3+}, and Proterozoic shales from Condie (1993)

increasing Cr/Fe_{tot} ratios is taken by Bankole et al. (2016) to indicate mixing with an oxidized fluid component. Although oxidation could have occurred during early diagenesis, it is also possible that oxidation occurred during later U mineralization at ~2.0 Ga.

A very different scenario was offered for the 2.15 Ga Gamagara Formation, Griqualand West region (South Africa) by Yamaguchi et al. (2005) and Yamaguchi and Ohmoto (2006). The Gamagara Formation sits on a major erosional unconformity of regional extent that includes deeply weathered early Paleoproterozoic IFs and lavas (Holland and Beukes 1990; Beukes et al. 2002), reflecting an extensive weathering profile that was originally estimated to be ~2.2 Ga based on the age of lavas in the Ongeluk Formation. New geochronology for the Ongeluk lavas show they are 2.43 Ga in age (Gumsley et al. 2017), and this suggests the oxidized weathering surface may be older than 2.2 Ga, although the depositional age of the Gamagara is still interpreted to be less than 2.2 Ga. Based on relations between Fe_{tot}/Ti and Fe_{tot}/Al, as well as trace-element ratios, Yamaguchi and Ohmoto (2006) suggest that the very high Fe_{tot}/Al ratios reflect Fe enrichment during weathering of the source, rather than post-depositional addition of authigenic Fe^{3+}, as is interpreted for the Franceville Basin study by Bankole et al. (2016). The moderately positive $\delta^{56}Fe$ values for the Gamagara Formation (Fig. 6.26) are therefore taken to record those of the underlying weathered laterites under a moderately oxidizing atmosphere, in addition to possible loss of a low-$\delta^{56}Fe$ component during soil formation on the source rocks. Overall, the relatively restricted range in $\delta^{56}Fe$ values for the Gamagara Formation, relative to the FA Formation (Fig. 6.26), is consistent with the contrasting origins proposed for these units. A limited range in $\delta^{56}Fe$ is expected for in situ oxidative weathering of the Gamagara source rocks relative to that produced by the variable oxidation of Fe^{2+}_{aq} model proposed for the FA Formation. Importantly, a detrital source for oxides in the Gamagara Formation focuses attention on the age of oxidative weathering of the source terrane, which, using the new age for the Ongeluk lavas, might have occurred at the earliest stages of the GOE; this possibility, however, remains untested.

6.5.2 Changes in Weathering Across the GOE

The geochemical effects of modern weathering regimes have been reviewed in Sect. 5.1 of Chap. 5, which shows a large range of conditions in which Fe may be removed or added to soils, some of which are associated with large Fe isotope effects, and others where large changes in Fe contents are not associated with significant changes in Fe isotope compositions. A key component to Fe mobility in a modern oxic world is complexation of Fe^{3+} via organic ligands, as well as redoximorphic processes that produce reduced regions where Fe^{2+}_{aq} is mobile. Under anoxic surface conditions, however, thermodynamic modeling shows that weathering produces Fe^{2+}-smectites, followed by Fe^{2+}-carbonates at more advanced stages of weathering (Hao et al. 2017), and this has been confirmed in experimental studies (Murakami et al. 2004). Therefore, weathering under anoxic conditions would be expected to produce Fe isotope variations that are distinct from those in modern soils.

The use of paleosols in constraining changes in atmospheric O_2 focuses on the relative retentivity of Fe (normalized to an immobile element such as Al or Ti), as well as Mn, where paleosols that lost Fe are interpreted to record anoxic conditions where Fe^{2+}_{aq} was stable, allowing removal via pedogenic processes, whereas paleosols that retained Fe reflect "rusting in place" under oxic conditions (e.g., Rye and Holland 1998). Extending such data to atmospheric constraints requires modeling weathering reactions and estimating CO_2 fluxes. Recent models account for complex Fe oxidation kinetic relations, in addition to water flow, temperature, and pH, and support early proposals of an increase in atmospheric O_2 from the Neoarchean to the Paleoproterozoic (e.g., Kanzaki and Murakami 2016). In addition, redox models based on Fe have been tested using redox-sensitive trace elements (Murakami et al. 2016). Anoxic weathering produces distinct products dependent on source rock compositions, where it is estimated that weathering of tonalite-trondhjemite-granodiorite produces a greater proportion of Fe^{2+}_{aq} than weathering of basalts (Hao et al. 2017), suggesting that Fe retention or gain in paleosols may be compositionally dependent. Central to modeling Fe redox in paleosols is calculation of the moles of oxygen required to oxidize the observed Fe inventory, but little focus has been paid to how open (or closed) a paleosol system may have been to Fe gain/loss external to the paleosol; it is in this last aspect that Fe isotopes may contribute greatly and potentially change estimates of atmospheric O_2.

The 2.15 Ga Hekpoort paleosol (age revised based on Gumsley et al. 2017; Philippot et al. 2018) was developed on the regionally extensive Hekpoort basalt (Fe_2O_3* ~ 11 wt%, MgO ~ 7 wt%) in the Transvaal and Griqualand West regions of Botswana and South Africa. Initial work by Rye and Holland (1998) suggested that the paleosol had lost Fe, which might indicate formation under anoxic conditions, but later work on a complete section of the paleosol at Gaborone showed both Fe loss and gain, where Yang and Holland (2003) suggested that, over the complete section, Fe losses and gains balanced and hence the Hekpoort paleosol was an example of one of the oldest post-GOE paleosols that showed no net Fe transport. They therefore interpreted the paleosol to have formed under oxic conditions, and they estimated atmospheric O_2 to have been up to ~1% PAL. In contrast, Beukes et al. (2002) argued that Fe mobilization within the paleosol suggests the presence of organic acids derived from land biota (see also Ohmoto 1996), raising the possibility that the Hekpoort paleosol was an open system for Fe transport via groundwater flow coupled with Fe-organic complexes, and hence the low atmospheric O_2 levels calculated by Yang and Holland (2003) were incorrect.

Yamaguchi et al. (2007) determined Fe isotope compositions on the same Gaborone section studied by Yang and Holland (2003) and found that in all but one sample, δ^{56}Fe values were significantly higher than that of the protolith (Fig. 6.27), indicating that Fe must have behaved as an open system, in contrast to the calculated zero net gain/loss by Yang and Holland (2003) based on elemental abundances. This demonstrates the utility of Fe isotopes in identifying open or closed-system behavior, similar to that highlighted in the discussion above on Fe speciation proxies. Surprisingly, sections of the Hekpoort paleosol that had undergone, on an elemental basis, both Fe loss (decreased Fe_{tot}/Al),

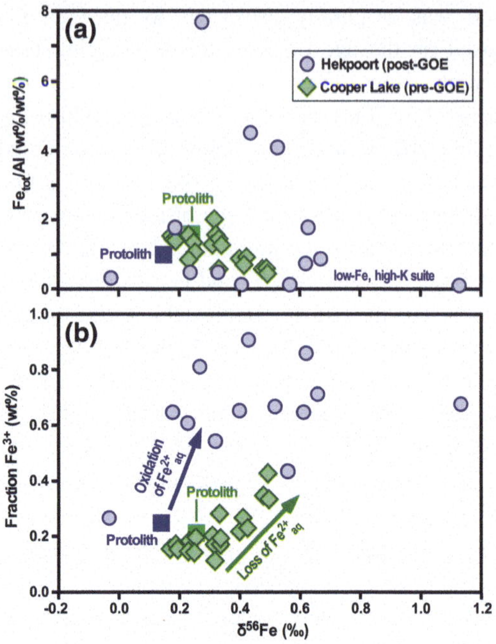

Fig. 6.27 Variations in δ^{56}Fe values for two Paleoproterozoic paleosols relative to fraction of Fe^{3+} and Fe_{tot}/Al ratios. The 2.15 Ga Hekpoort paleosol (Gaborone, Botswana) formed after the GOE (purple symbols; Yamaguchi et al. 2007), and the 2.45 Ga Cooper Lake Lake paleosol (Ontario, Canada) formed prior to the GOE (green symbols; Babechuk et al. 2019). For the Hekpoort paleosol, the protolith is basaltic lava (Fe_2O_3* ~11 wt%, MgO ~ 7 wt%), and Yamaguchi et al. (2007) noted two suites–an Fe-depleted suite of low Fe_{tot}/Al ratios (which is enriched in K_2O) and an Fe-enriched suite that had high Fe_{tot}/Al ratios (normal K_2O contents). A K-metasomatic event for the Hekpoort paleosol is dated at ~1.9 Ga (MacFarlane and Holland 1991). The high fraction of Fe^{3+} and positive δ^{56}Fe values are thought to record oxidation of Fe^{2+}_{aq} from O_2 bearing groundwater (post GOE). For the Cooper Lake paleosol, the protolith is an Fe-rich mafic dike (Fe_2O_3* ~20 wt%, MgO ~6 wt%), and, relative to the protolith (green box), δ^{56}Fe values increase with increasing fraction of Fe^{3+} and decreasing Fe_{tot}/Al, interpreted to reflect loss of low-δ^{56}Fe Fe^{2+}_{aq} under low-O_2 conditions prior to the GOE

as well as gain (increased Fe_{tot}/Al), had positive δ^{56}Fe values (Fig. 6.27a). The broadly positive correlation between fraction of Fe^{3+} and δ^{56}Fe (Fig. 6.27b) led Yamaguchi et al. (2007) to infer that extensive groundwater flow occurred where partial oxidation of Fe^{2+}_{aq} produced high-δ^{56}Fe Fe^{3+}-oxides in the measured paleosol section, a model that is similar to that proposed for the FA Formation red beds by Bankole et al. (2016). Such a model would imply much greater O_2 contents than what were calculated by Yang and Holland (2003), dependent upon the percent oxidation and total Fe flux through the paleosol. Yamaguchi et al. (2007) estimated that O_2 contents may have been as high as 10% PAL, about an order-of-magnitude higher than that estimated by Yang and Holland (2003), who assumed a closed system and hence smaller Fe^{2+} inventory.

Using the large dataset of Yang and Holland (2003) clearly shows that the Fe-loss sections of the Hekpoort paleosol corresponds to a sericite alteration zone, whereas the Fe-gain sections show no evidence for K addition. Studying a separate location of the same unit, MacFarlane and Holland (1991) showed that sericite alteration had a ^{87}Rb-^{87}Sr age of 1925 ± 32 Ma and a very radiogenic initial $^{87}Sr/^{86}Sr$ ratio of 0.718, far higher than that measured for the Hekpoort basalt protolith. These findings suggest extensive fluid flow at ~1.9 Ga that circulated though radiogenic rocks, possibly the overlying clastic sediments of the Olifantshoek Group. For the low-Fe group, it remains unclear if the K-metasomatic event correlates with Fe^{2+} oxidation (and removal), or if the groundwater flow model of Fe^{2+}_{aq} oxidation invoked by Yamaguchi et al. (2007) occurred at 2.15 Ga, followed by a sericite alteration event at 1.9 Ga. The lack of K-addition in the high-Fe group suggests that it is this suite of data that is most likely to contain information about redox conditions at the time of paleosol formation; in this case, the mass-balance calculations of Yamaguchi et al. (2007) provide a reasonably confident estimate of O_2, which is much higher than that of Yang and Holland (2003). Nevertheless, the complexity of the Hekpoort paleosol highlights the issue that the age of a redox proxy may have significant uncertainties for rocks that saw complex histories, including circulation of post-formation groundwater.

A detailed Cr and Fe isotope study of the 2.45 Ga Cooper Lake paleosol (Ontario, Canada) by Babechuk et al. (2019) offers an opportunity to investigate the isotopic effects of weathering immediately before the GOE, under presumably low-O_2 conditions. The Cooper Lake paleosol was developed on a sediment-hosted mafic dike that has high total Fe contents (~20 wt% Fe_2O_3*), and Babechuk et al. (2019) argue that Fe was only mobilized as Fe^{2+} based on Fe–Mg–Mn relations. A small increase in $\delta^{56}Fe$ values and accompanying decrease in Fe_{tot}/Al ratios (Fig. 6.27a) suggests loss of a low-$\delta^{56}Fe$ Fe^{2+}_{aq} component. In contrast, Cr isotope compositions are homogenous throughout the paleosol, which Babechuk et al. (2019) took to indicate little redox cycling. Based on Fe^{2+}/Al and Fe^{3+}/Al relations, Babechuk et al. (2019) take the apparent increase in the fraction of Fe^{3+} in Fig. 6.27b to indicate loss of Fe^{2+} under low-O_2 conditions. This contrasts with a model where oxidation of Fe^{2+}_{aq} produces positive $\delta^{56}Fe$ values for oxides, as is the case for the younger Hekpoort paleosol and the red beds of the FA Formation (above).

Based on the observed increase in $\delta^{56}Fe$ values with decreasing Fe contents, Babechuk et al. (2019) suggest that riverine Fe^{2+}_{aq} had low $\delta^{56}Fe$ values prior to the GOE. This contrasts with the commonly positive $\delta^{56}Fe$ values of the dissolved component in modern rivers, which in part reflects organic Fe complexation (see Sect. 5.2.1 in Chap. 5). Babechuk et al. (2019) further posit that low-$\delta^{56}Fe$ riverine Fe^{2+}_{aq} may be an Fe source for IFs that have negative $\delta^{56}Fe$ values. This is an important proposal, and we will turn to the question of Fe sources for pre-GOE IFs in Sect. 6.5.4. It is not yet clear that weathering of Fe^{2+}-silicates under anoxic conditions will supply large quantities of low-$\delta^{56}Fe$ Fe^{2+}_{aq} delivery to the oceans. This is not consistent with predicted Fe isotope fractionations among Fe^{2+}_{aq}, Fe^{2+}-silicates, and Fe^{2+}-carbonates. Considering the $\beta^{56/54}$ factors (see discussion in Chap. 3) for olivine, Fe^{2+}-phyllosilicates, and siderite indicates that at room temperature, Fe^{2+}_{aq} should have $\delta^{56}Fe$ values that are ~0.5‰ higher than ferrous Fe weathering products (Polyakov and Mineev 2000; Rustad et al. 2010). This is opposite that suggested by

Babechuk et al. (2019), and would predict that riverine Fe^{2+}_{aq} input to the oceans, under anoxic weathering conditions, would instead have positive $\delta^{56}Fe$ values. The predicted fractionations between Fe^{2+}_{aq} and siderite based on $\beta^{56/54}$ factors has been confirmed experimentally (Wiesli et al. 2004), lending some support to the use of $\beta^{56/54}$ factors in predicting the isotopic effects of anoxic weathering, although experimental determination of the Fe isotope fractionation between Fe^{2+}-smectites and igneous Fe silicates under anoxic conditions is clearly needed.

6.5.3 Moving to a Low-Oxygen World: Key Issues of Fe Mass Balance, Fe Isotope Fractionation Factors, Fe^{2+}_{aq} Oxidation, and the Age of Redox Proxies

Part of the challenge in interpreting Fe isotope variations in the ancient Earth, particularly as we move to pre-GOE surface environments that contained little or no oxygen, is the profoundly different Fe biogeochemical cycle relative to that of today, in addition to the fact that we are interpreting marine processes from the rock record, where the fluid component is gone. We have already discussed above some of the differences between modern and ancient Fe cycles in Sect. 6.3.3, and here we add several other important distinctions that bear on how Fe isotopes may be modeled for the Earth before the GOE. The concepts discussed in this section are particularly import for understanding the very large IFs that were deposited in the Neoarchean and Paleoproterozoic, a sequence of rocks we have not yet encountered as we progressively step back in the rock record.

The principles of isotopic mass balance equally apply in the modern and ancient Earth, although the reservoir scales must change dramatically. For example, a 2‰ range in $\delta^{56}Fe$ values in modern marine environments might only require isotopic mass balance among relatively small Fe reservoirs due to the very low total dissolved Fe inventory in the modern oceans. In the Archean, however, a 2‰ range in $\delta^{56}Fe$ values in Fe-rich lithologies such as IFs requires isotopic exchange/ fractionation that must have involved very large reservoir(s). It is important, therefore, to consider the lithologies and Fe mass balance that contain various Fe isotope compositions, avoiding the tendency to only focus on the $\delta^{56}Fe$ values.

We illustrate the dynamic range of Fe contents and $\delta^{56}Fe$ values that must be considered as we move into the pre-GOE Earth in Fig. 6.28, which plots WR compositions for pre-GOE IFs, jaspers, and Ca–Mg (low Fe) carbonates. These lithologies span a ~5‰ range in $\delta^{56}Fe$ values and four orders-of-magnitude in Fe contents. Some clear groupings emerge: Paleoarchean jaspers tend to have high $\delta^{56}Fe$ values, extending to relatively low Fe contents, whereas pre-GOE Paleoproterozoic to Mesoarchean IFs largely extend from near-zero $\delta^{56}Fe$ values to highly negative $\delta^{56}Fe$ values at very high Fe contents. Within the IF group, many data scatter about a $\delta^{56}Fe$ value of zero (red circles with pink fill), and these are the very large Marra Mamba, Brockman, and Kuruman IFs of the Hamersley-Transvaal basin (e.g., Beukes and Gutzmer 2008). In contrast, the slightly younger Mn IFs from these basins, including the Koegas Group and Hotazel Formation (red circles with grey fill), have distinctly lower $\delta^{56}Fe$ values, generally between −1 and −2‰. The Mesoarchean IFs define a group that overlaps the previous two, although on average, clearly have negative $\delta^{56}Fe$ values. At the other end of the Fe concentration spectrum, although completely overlapping the $\delta^{56}Fe$ values with IFs, are the large Neoarchean Ca–Mg carbonate platforms of the Hamersley-Transvaal basin, whose $\delta^{56}Fe$ values generally range from near-zero to −2‰ but at several orders-of-magnitude lower Fe contents than the IFs. The relations in Fig. 6.28 make it clear that we need to consider processes that might make negative $\delta^{56}Fe$ values in samples of very low-Fe contents, as well as processes that would make negative $\delta^{56}Fe$ values in very Fe-rich samples; looking to a common process to explain

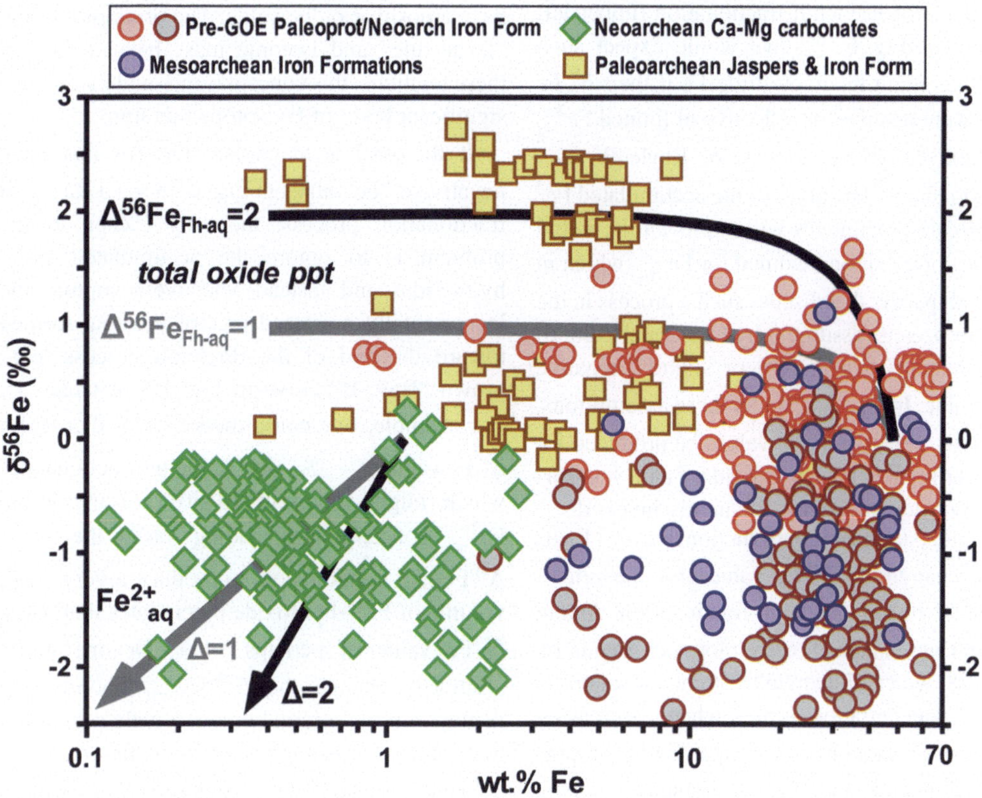

Fig. 6.28 Relations between $\delta^{56}Fe$ values and total Fe contents (log scale) for WR samples of Pre-GOE Paleoproterozoic and Neoarchean iron formations (IFs) that have low Mn contents (red circles with pink fill), Paleoproterozoic Mn IFs (red circles with grey fill), Neoarchean Ca–Mg carbonates (green diamonds), Mesoarchean IFs (purple circles), and Paleoarchean jaspers and IFs (brown/yellow squares). Also shown are Rayleigh fractionation lines for oxidation of Fe^{2+}_{aq} to produce Fe^{3+}-hydroxide (ferrihydrite, of "Fh") precipitates that have been commonly invoked in the literature on these rocks. The Rayleigh fractionation model for total oxide/hydroxide precipitates is shown on a relative scale in the upper two curves, illustrated for two different Fh-Fe^{2+}_{aq} fractionation factors ($\Delta^{56}Fe_{Fh\text{-}aq}$), and fractionation progresses from left to right, where 100% oxidation/precipitation produces $\delta^{56}Fe$ values equal to the assumed initial Fe^{2+}_{aq} (here taken to be $\delta^{56}Fe = 0$, fixed on the diagram to the Fe content of hematite). The two arrows in the lower left quadrant show the Rayleigh trends for Fe^{2+}_{aq}, scaled correctly to relative changes in Fe contents, with an arbitrary starting Fe concentration; fractionation progresses in the direction of the arrows, from right to left. A key point of the diagram is the very wide range in Fe contents for various lithologies that have negative $\delta^{56}Fe$ values, which cannot be explained by a single process. Data from Rouxel et al. (2005), von Blanckenburg et al. (2008), Planavsky et al. (2009, 2012a, 2014), Tsikos et al. (2010), Czaja et al. (2010, 2012, 2018), Li et al. (2013b, 2015), Satkoski et al. (2015), Haugaard et al. (2016), Kurzweil et al. (2016), Hashman (2017), Smith et al. (2017), Eroglu et al. (2018), Lantink et al. (2018), and Thibon et al. (2019)

all of the negative $\delta^{56}Fe$ values in Fig. 6.28 is unlikely to be correct.

Also shown in Fig. 6.28 are the relative trends of Rayleigh fractionation models, using a range of $^{56}Fe/^{54}Fe$ fractionations between Fe^{3+}-hydroxide (ferrihydrite, of "Fh") and Fe^{2+}_{aq} ($\Delta^{56}Fe_{Fh\text{-}aq}$) that have been commonly used in the Fe isotope literature on rocks of this age. We cannot strictly plot such models on Fig. 6.28, but we can scale the relative isotopic-concentration effects to glean some insights into what is, and what is not, possible. Based on these relations, the positive $\delta^{56}Fe$ values of low-Fe jaspers could indeed reflect the initial Fe^{3+}-hydroxide precipitates produced at low extents of Rayleigh fractionation (upper left quadrant of Fig. 6.28)—we would expect such rocks to contain relatively little Fe if their δ^{56}Fe values were positive, as reflective of limited Fe^{2+}_{aq} oxidation. At greater extents of Rayleigh fractionation, the δ^{56}Fe values of the accumulated Fe^{3+}-hydroxide precipitates will approach the initial isotopic composition assumed for Fe^{2+}_{aq}, which in almost all papers that discuss such a process in the Archean oceans, assume a hydrothermal composition that has a near-zero or slightly negative δ^{56}Fe value. In no cases, under these assumptions, can highly negative δ^{56}Fe values be produced for the accumulated Fe^{3+}-hydroxides, and yet, surprisingly, such a model is frequently invoked.

To understand this contradiction, we next focus on the changes in δ^{56}Fe values for the remaining Fe^{2+}_{aq} during Rayleigh fractionation. Again, although we cannot directly plot such trends on Fig. 6.28, we can illustrate the trends in a correct relative concentration scale, where decreasing δ^{56}Fe values should be accompanied by proportionally decreasing Fe^{2+}_{aq} contents. Using $\Delta^{56}Fe_{Fh\text{-}aq}$ fractionations that have been commonly used in the literature suggests that, for example, δ^{56}Fe values of −2‰ for Fe^{2+}_{aq} could be produced by Rayleigh fractionation, accompanied by a roughly order-of-magnitude decrease in Fe^{2+}_{aq} contents. If, for example, such low levels of aqueous Fe were directly incorporated into shallow-water Ca–Mg carbonates, Rayleigh fractionation involving extensive precipitation Fe^{3+}-hydroxides could explain the negative δ^{56}Fe values of these low-Fe lithologies. Critically, however, the complementary reservoir is a thick pile of Fe^{3+}-hydroxide precipitates that have a near-zero δ^{56}Fe value that represents an order-of-magnitude greater Fe inventory than that contained in the low-δ^{56}Fe Ca–Mg carbonates. Such relations in Fe abundance and isotopic composition are required by the isotopic mass balance imposed by the Rayleigh fractionation model (see discussion Sect. 1.2.3 in Chap. 1). This example shows that mass balance does not require that the complement to a negative δ^{56}Fe value component must have a positive δ^{56}Fe value. The principles of the "lever rule" and isotopic mass balance govern these relations; this concept is mis-understood in a significant body of Fe isotope literature.

If the goal is to explain the IFs that have negative δ^{56}Fe values in Fig. 6.28 via a Rayleigh fractionation process, the only escape to the problem is to ignore the accumulated Fe^{3+}-hydroxides and instead selectively capture the incrementally produced precipitates that formed toward the end of the Rayleigh process (e.g., brown "Ray. B" curve in Fig. 1.5 of Chap. 1). For example, one could model Fe^{2+}_{aq} that had a δ^{56}Fe value of −2‰ after Rayleigh fractionation, which might represent 10–30% of the initial Fe^{2+}_{aq} concentration, adding back the same $\Delta^{56}Fe_{Fh\text{-}aq}$ fractionation, to produce a very small volume of Fe^{3+}-hydroxide precipitates that had a δ^{56}Fe value of perhaps −1‰. Or one might arbitrarily change the $\Delta^{56}Fe_{Fh\text{-}aq}$ fractionation factor from the original non-zero value (required to produce any isotopic shift in the first place) to $\Delta^{56}Fe_{Fh\text{-}aq} = 0$ (100% oxidation) to capture a Fe^{3+}-hydroxide precipitate that had a δ^{56}Fe value of −2‰ (identical to the Fe^{2+}_{aq}). The difficulty with either of these scenarios is that the low-δ^{56}Fe Fe^{3+}-hydroxides represent only a very small portion of the total Fe^{3+}-hydroxide precipitates that are produced, requiring them to be selectively isolated from the majority of the precipitated mass, assuming we stay true to the mass-balance imposed by the Rayleigh model. The problem is most severe if the $\Delta^{56}Fe_{Fh\text{-}aq}$ fractionation is kept consistently positive in the model, meaning that only a very small amount of only moderately low-δ^{56}Fe Fe^{3+} hydroxides are produced. We conclude that it is simply not possible to make large volumes of Fe-rich rocks that have negative δ^{56}Fe values using a Rayleigh model without special pleading for changing fractionation factors or selective isolation of precipitates of the desired Fe isotope composition. This important conclusion directly follows from the isotopic mass balances imposed by the model.

Although we have focused on a Rayleigh model because of its common use in the literature, such a model is actually inappropriate for an open marine system. The photic zone of a pre-GOE ocean, for example, received input Fe fluxes (e.g., hydrothermally-sourced Fe^{2+}_{aq}) and had output fluxes (e.g., Fe precipitates). Rayleigh fractionation, by definition, is a model of reaction in a "closed box", which might proceed via equilibrium or kinetic isotope fractionation, where the end products do not maintain isotopic equilibrium with the reactants. A Rayleigh model does not allow for a continued input of Fe^{2+}_{aq}, for example, from "outside the box". Instead, it is more appropriate to model Fe^{2+}_{aq} oxidation in a pre-GOE ocean using an open-system flux model, and, given the slow rates of oxidation of Fe^{2+}_{aq} and precipitation in a low-O_2 world, it should be expected that Fe isotope equilibrium occurs between Fe^{2+}_{aq} and ferrihydrite during precipitation (see Chap. 3).

A dispersion/reaction model for Fe^{2+}_{aq} oxidation in the Archean oceans was developed by Czaja et al. (2012) to address the shortcomings discussed above that make a Rayleigh model inappropriate for open-marine systems, particularly for the pre-GOE Earth. The model of Czaja et al. (2012) allows quantifying photic zone O_2 contents and production rates for a given measured $\delta^{56}Fe$ value for Fe^{3+}-hydroxide precipitates or an inferred Fe isotope composition for Fe^{2+}_{aq}. This is not to say that oxidation of Fe^{2+}_{aq} could not also occur via photoferrotrophy (e.g., Kappler et al. 2005), which is also modeled by Czaja et al. (2012), but here we focus on oxidation by O_2 for simplicity. In the case of oxidation by photosynthetically produced O_2 in the photic zone, downward oxygen transport was modeled to have occurred via eddy diffusion from the photic zone, which reacts with upwardly transported Fe^{2+}_{aq} from assumed hydrothermal sources below a redoxcline. Importantly, an upper boundary condition for O_2 levels in the photic zone was set to be in equilibrium with an atmosphere that contained 10^{-5} PAL O_2 to stay consistent with models for MIF-S (Pavlov and Kasting 2002); this means that the results as presented in Czaja et al. (2012) cannot be applied without modification to conditions where atmospheric O_2 was significantly higher. Isotopic equilibrium between Fe^{2+}_{aq} and Fe^{3+}-hydroxides was assumed to occur during precipitation, using the range of $\Delta^{56}Fe_{Fh\text{-}aq}$ fractionation factors measured for Fe–Si gels by Wu et al. (2012b). Using the input parameters discussed in Czaja et al. (2012), the model reaches a dynamic steady-state condition after $\sim 10^3$ yrs.

In Fig. 6.29 we illustrate an example of the dispersion/reaction model of Czaja et al. (2012) where Fe^{2+}_{aq} oxidation rates are such that not all upwelling Fe^{2+}_{aq} from beneath the photic zone is oxidized, which produces a gradient in Fe^{2+}_{aq} concentrations and $\delta^{56}Fe$ values for Fe^{2+}_{aq} and Fe^{3+}-hydroxide precipitates. Equilibrium Fe isotope fractionation between Fe^{2+}_{aq} and Fe^{3+}-hydroxide precipitates produces the similar topology of the depth-$\delta^{56}Fe$ curves for the two phases, and the case shown uses $\Delta^{56}Fe_{Fh\text{-}aq} = +4‰$, the highest fractionation measured for Fe–Si gels by Wu et al. (2012b). Settling Fe^{3+}-hydroxides, under steady-state conditions, have an integrated $\delta^{56}Fe$ value of +1.2‰ in the example shown, which reflects an O_2 production rate of 0.2% of modern values in the photic zone. Use of a smaller $\Delta^{56}Fe_{Fh\text{-}aq}$ fractionation factor would produce a lower integrated $\delta^{56}Fe$ value for Fe^{3+}-hydroxides, and this remains a significant uncertainty in applying the model to rocks, although if an estimate for the Fe/Si ratio of the precipitates can be made, this uncertainty will decrease based on the Fe/Si-$\Delta^{56}Fe_{Fh\text{-}aq}$ relations from Wu et al. (2012b).

For the case shown in Fig. 6.29, all O_2 produced is consumed through oxidation but the oxidation rate of Fe^{2+}_{aq}, relative to the influx of Fe^{2+}_{aq} from below the photic zone is such that Fe^{2+}_{aq} exists up to the top of the photic zone. As the oxidation rate increases, however, relative to a fixed influx of Fe^{2+}_{aq}, the Fe^{2+}_{aq} contents in the

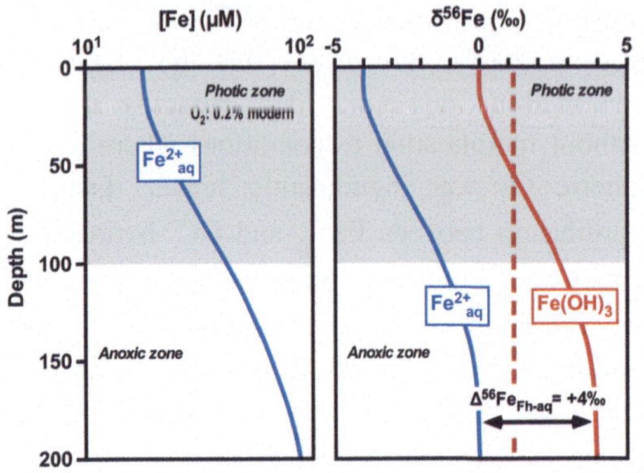

Fig. 6.29 Illustration of the dispersion/reaction model of Czaja et al. (2012), which describes the Fe isotope compositions of Fe^{2+}_{aq} and Fe^{3+}-hydroxide precipitates during oxidation of upwelling Fe^{2+}_{aq} into the photic zone. Example shown represents conditions where oxidation occurs via reaction of upwelling Fe^{2+}_{aq} from beneath the photic zone (e.g., hydrothermal source with $\delta^{56}Fe = 0$) with O_2 produced by oxygenic photosynthesis in the photic zone. Atmosphere O_2 is set to 10^{-5} PAL. The conditions illustrated are for a case where Fe^{2+}_{aq} oxidation rates are less than upwelling Fe^{2+}_{aq} rates, allowing Fe^{2+}_{aq} to coexist with Fe^{3+}-hydroxides in the photic zone. Maintenance of Fe isotope equilibrium between Fe^{2+}_{aq} and Fe^{3+}-hydroxides is assumed, using a $\Delta^{56}Fe_{Fh\text{-}aq}$ fractionation of +4‰ (see Wu et al. 2012b). The integrated $\delta^{56}Fe$ value of Fe^{3+}-hydroxides that settle from the photic zone (producing Fe^{3+}-oxide deposits on the seafloor) is shown by the vertical dashed red line, which, in this case, has a $\delta^{56}Fe$ value of +1.2‰, and corresponds to O_2 production rates of 0.2% of modern. Higher O_2 contents will shift the integrated $\delta^{56}Fe$ value of Fe^{3+}-hydroxides toward that of the source Fe^{2+}_{aq} ($\delta^{56}Fe = 0$), making the $\delta^{56}Fe$ values of Fe^{3+}-hydroxides an insensitive indicator of O_2 contents. The opposite occurs for the $\delta^{56}Fe$ values of Fe^{2+}_{aq} in the photic zone, where these become a very sensitive indicator of O_2 levels at high O_2 conditions. In this case, the best rocks to be studied would not be Fe_{3+}-hydroxides but a lithology that would capture Fe^{2+}_{aq}, such as shallow water Ca–Mg carbonates

photic zone will decrease to negligible levels. When the flux of Fe^{3+}-hydroxide out of the photic zone approaches the rate of Fe^{2+}_{aq} influx to the photic zone, the photic zone is effectively devoid of Fe^{2+}_{aq} and free O_2 will begin to accumulate in the upper-most part of water column. At this point, the integrated $\delta^{56}Fe$ values for Fe^{3+}-hydroxides will essentially match that of the influx (e.g., hydrothermal Fe^{2+}_{aq}), meaning the model loses sensitivity in the relation between $\delta^{56}Fe$ of Fe^{3+}-hydroxide precipitates and O_2 contents (see discussion in Li et al. 2013a). Note that this may occur at very low O_2 contents, dependent on the assumed Fe fluxes; it is therefore not possible to conclude that a $\delta^{56}Fe$ value of zero (=hydrothermal source) means that there were very high levels of O_2. This important conclusion follows the principles of isotopic mass-balance—as the $\delta^{56}Fe$ values of the Fe^{3+} precipitates approach that of the starting material, changes in the driving force (O_2) cannot be sensitively detected. These relations mean that the dispersion/reaction model, as originally developed, is most useful for inferring O_2 contents using Fe isotope compositions of the Fe^{3+}-hydroxide precipitates under conditions of very low O_2; the model is not applicable to periods of relatively high inferred O_2, such as the Paleoproterozoic Lomagundi-Jatuli event or the Neoproterozoic and Cambrian rise in O_2. We will therefore only apply this model to detecting very low levels of surface-ocean O_2 as might have occurred prior to the GOE.

This limitation to low-O_2 conditions does not apply to the Fe^{2+}_{aq} component of the model, and this is clear by inspection of the vertical gradients produced by the model in Fig. 6.29. At the highest extent of oxidation, which occurs at the top of the photic zone, the $\delta^{56}Fe$ values of the Fe^{3+}-hydroxide precipitates match that of the input Fe. But if our focus is shifted to the Fe^{2+}_{aq} component, this becomes highly negative at high extents of oxidation, and hence this component becomes most sensitive to oxidation as the Fe^{3+}-hydroxide component becomes insensitive. These relations follow the basic principles of isotopic mass balance where the largest shifts in isotopic compositions will occur in the minor component of the system. We will revisit using the $\delta^{56}Fe$ values of photic-zone Fe^{2+}_{aq} under high-O_2 conditions when discussing data from the carbonate platform of the Neoarchean Ghaap Group below.

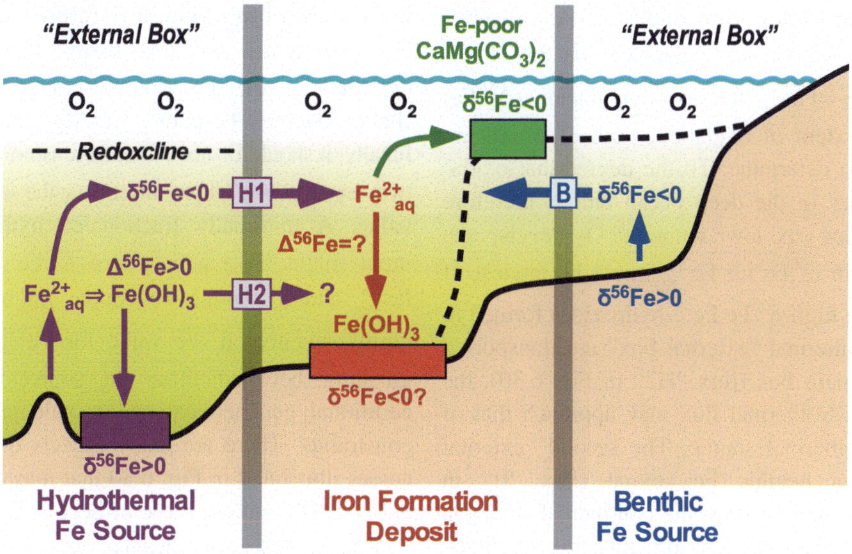

Fig. 6.30 Illustration of a marine basin cross section showing two mechanisms that may produce Iron Formations (IFs) that have negative $\delta^{56}Fe$ values, both of which require an external source of Fe^{2+}_{aq} that had negative $\delta^{56}Fe$ values. In the illustration, O_2 exists in the photic zone above a redoxcline, below which Fe^{2+}_{aq} concentrations increase downward (yellow shading). The dispersion/reaction model from Fig. 6.29 can be envisioned to operate in the central box, but instead of a $\delta^{56}Fe = 0$ input for Fe^{2+}_{aq}, two "external boxes" are shown that contain processes that can generate Fe^{2+}_{aq} that has negative $\delta^{56}Fe$ values: (1) an evolved hydrothermal fluid from the deep ocean (mid-ocean ridge) that previously precipitated Fe^{3+}-hydroxides (purple coloring) in the left "external box" (flux H1), or (2) a benthic Fe source on the continental shelf that generated low-$\delta^{56}Fe$ Fe^{2+}_{aq} by DIR (blue coloring) in the right "external box" (flux B). For the hydrothermal box, oxidation of Fe^{2+}_{aq} must be associated with a positive Fe^{3+}-hydroxide—Fe^{2+}_{aq} fractionation factor ($\Delta^{56}Fe > 0$) to make low-$\delta^{56}Fe$ Fe^{2+}_{aq} that can exit the box. By mass-balance, Fe^{3+}-hydroxides in the hydrothermal box likely have positive $\delta^{56}Fe$ values. It is also possible that some of the high-$\delta^{56}Fe$ Fe^{3+}-hydroxides are transported into the central box (flux H2). For the benthic Fe source box, liberation of low-$\delta^{56}Fe$ Fe^{2+}_{aq} by DIR should make a positive $\delta^{56}Fe$ residuum on the continental shelf. Formation of IF deposits (red coloring) occurs in the central box by oxidation of low-$\delta^{56}Fe$ Fe^{2+}_{aq}, followed by precipitation of Fe^{3+}-hydroxides; it is not clear what the Fe^{3+}-hydroxide—Fe^{2+}_{aq} fractionation factor would be ($\Delta^{56}Fe$ = ?) upon oxidation and precipitation of $Fe(OH)_3$ in the central box during formation of the IF deposit. If $\Delta^{56}Fe = 0$, the negative $\delta^{56}Fe$ value of Fe^{2+}_{aq} will be directly transferred to the IF deposit. If, however, $\Delta^{56}Fe > 0$ upon oxidation and precipitation, as would occur where oxidation was limited, the negative $\delta^{56}Fe$ signal from Fe^{2+}_{aq} may be erased or reversed. Finally, in the upper water column, where Fe^{2+}_{aq} contents are very low, $\delta^{56}Fe$ values for aqueous Fe will become highly negative (see Fig. 6.29), which may explain the negative $\delta^{56}Fe$ values of Ca–Mg carbonates (green coloring) on a shallow-water carbonate platform (dashed line profile)

The dispersion/reaction model shows that the only way to produce large volumes of Fe-rich rocks (Fe^{3+}-hydroxide precipitates) that have negative $\delta^{56}Fe$ values (e.g., data that plot in the lower-right quadrant in Fig. 6.28) is to change the Fe isotope composition of the input Fe^{2+}_{aq}. In Fig. 6.30 we illustrate two processes that may do this. The dispersion/reaction model can be envisioned to operate in the center box, but two "external boxes" are shown that may produce a Fe^{2+}_{aq} input that has negative $\delta^{56}Fe$ values. One "external box" is an oceanic hydrothermal Fe source, presumably a mid-ocean ridge, where partial oxidation produces high-$\delta^{56}Fe$ $Fe(OH)_3$ precipitates (which are not preserved), producing Fe^{2+}_{aq} that has negative $\delta^{56}Fe$ values (flux "H1" in Fig. 6.30). This is essentially the "Precambrian Vent" trajectory illustrated in Fig. 6.15.

The extent of lowering of the $\delta^{56}Fe$ value for Fe^{2+}_{aq} by this process depends on the Fe^{3+}-hydroxide—Fe^{2+}_{aq} fractionation factor ($\Delta^{56}Fe$) and the extent of oxidation, both of which are difficult to determine. To the degree that oxidation occurs in the deep ocean and is therefore limited due to low ambient O_2 levels, the decrease in $\delta^{56}Fe$ for Fe^{2+}_{aq} might be modest. If any of the high-$\delta^{56}Fe$ Fe^{3+}-hydroxides formed in the hydrothermal "external box" are transported into the main box (flux "H2" in Fig. 6.30), the overall hydrothermal flux may approach that of the hydrothermal source. The second "external box" is a benthic Fe source (flux "B" in Fig. 6.30) generated on the continental shelf via DIR. In the pre-GOE Earth, this flux might be quite large, as shown in Fig. 6.16. Loss of low-$\delta^{56}Fe$ Fe^{2+}_{aq} from continental shelf sediments should produce a residual Fe pool in shelf sediments that had positive $\delta^{56}Fe$ values (see, for example, Fig. 6.20).

The dispersion/reaction model in Fig. 6.29 can be directly scaled to accommodate a negative $\delta^{56}Fe$ Fe^{2+}_{aq} input from the "external boxes" of Fig. 6.30. If, for example, the input Fe^{2+}_{aq} had $\delta^{56}Fe = -1‰$, the integrated $\delta^{56}Fe$ of Fe^{3+}-hydroxide precipitates from the photic zone would decrease in the example shown from +1.2 to +0.2‰. To produce Fe^{3+}-hydroxide precipitates that had a $\delta^{56}Fe$ value of −2‰ for the case shown in Fig. 6.29 would require an input Fe^{2+}_{aq} $\delta^{56}Fe$ value of −3.2‰, which is quite negative. If complete oxidation occurs, such that there is no free Fe^{2+}_{aq} in the photic zone, the dispersion/reaction model produces an integrated Fe^{3+}-hydroxide flux that is equal to the input Fe^{2+}_{aq} for $\delta^{56}Fe$, as discussed above. The effective $\Delta^{56}Fe$ fractionation upon formation of the IF deposit, therefore, is difficult to constrain and requires an estimate for the O_2 content during oxidation. Bringing our Fe-$\delta^{56}Fe$ relations from Fig. 6.28 back into the discussion, $\delta^{56}Fe$ values for Fe^{2+}_{aq} in the shallow photic zone will become very low in the central box of Fig. 6.30 (green), as will the Fe^{2+}_{aq} contents, and this seems likely to be recorded in the Fe isotope compositions of shallow-water Fe-poor Ca–Mg carbonates. Finally, it should be noted that there is no reason the two "external box" fluxes must have the same $\delta^{56}Fe$ value. A minimally fractionated hydrothermal input might have a near-zero $\delta^{56}Fe$ value for Fe^{2+}_{aq} and a benthic Fe source might have a highly negative $\delta^{56}Fe$ value for Fe^{2+}_{aq}. Distinguishing between these Fe sources requires additional geochemical and/or paleogeographic constraints. There are many aspects of the processes illustrated in Fig. 6.30 that may be tested, and we will address these in Sect. 6.5.4.

Finally, before we discuss the pre-GOE rock record, we conclude this section by highlighting the issue that the age of the redox signals measured in rocks may not be coincident with the depositional age of a rock. We have already noted above the issue of K-metasomatism of paleosols, although it is not yet clear that such events changed Fe oxidation states and/or Fe isotope compositions. Sericite alteration is not restricted to paleosols, and, in fact, has been proposed to record post-depositional element mobilization in red beds (Land et al. 2018). There are only two commonly used geochronologic systems that may directly tie age to redox state, the ^{187}Re-^{187}Os system and the U-Pb system (^{238}U-^{206}Pb, ^{235}U-^{207}Pb), and there are a few examples (though not many) where such constraints have confirmed the fidelity of ancient Re and U enrichments (e.g., Anbar et al. 2007; Satkoski et al. 2015), but in others, have shown post-depositional remobilization (e.g., Yang et al. 2017). The application of U-Pb geochronology to U isotope studies of redox changes over Earth history (e.g., Wang et al. 2018) is a logical venue for future work. Although a much more difficult isotope system, the ^{138}La-^{138}Ce system has been used to determine the age of Ce anomalies, which in some cases have been shown to reflect hydrothermal processes up to 2 b.y. after sediment deposition (Hayashi et al. 2004).

For Fe oxidation, it has been shown in some cases that high proportions of Fe^{3+} interpreted to reflect an oxic early Archean ocean are in fact a Phanerozoic feature, reflecting recent circulation of oxygenated groundwaters (Li et al. 2012b). For Fe-rich rocks, such effects might not produce significant changes in WR $\delta^{56}Fe$ values, given the low solubility of aqueous Fe in the presence of oxygen (an exception being complexation via organic ligands, or conditions of low pH). The same may not be true, however, for redox signals comprised of trace elements, given solubility and abundance differences as compared to Fe. There is great focus on the use of redox-sensitive trace elements to follow changes over Earth history (e.g., Robbins et al. 2016). To date, however, there has been little discussion about the possibility that the age of the proxy in some cases may be younger than the depositional age of the rock. Recent work on stable Cr isotopes, for example, suggests that some early Archean "oxidation signals" instead reflect recent weathering (Albut et al. 2018). It seems likely that the greatest potential impact by post-depositional redox changes will be on rocks that are interpreted to contain only a small signal for ancient O_2 because they were deposited before the GOE. As we step into the pre-GOE Earth it is worth remembering that rocks of the Archean spent at least 2 b.y. of their "lifetime" in an oxygenated world.

6.5.4 Early Paleoproterozoic Iron Formations (IFs)

As chemical sediments that have very low to insignificant clastic components, the very large IFs of the early Paleoproterozoic and Neoarchean have been intensively studied as potential recorders of seawater chemistry. Compared to the Neoproterozoic IFs, IFs of late Neoarchean to early Paleoproterozoic age are much larger in volume, have much lower clastic components, and are mineralogically much more diverse compared to the hematite-dominated Neoproterozoic IFs. In the classification of Gross (1980), these large IFs have been described as "Superior-type" to reflect their depositional environment on continental margin basins. Interpretations of the genesis of these IFs have varied greatly, where earlier work tended to view IFs as formed by abiologic processes, with no role for biology (e.g., Klein 2005), but these have given way to an increasing recognition that biology was likely involved during oxidation of Fe^{2+}_{aq} in the shallow oceans, as well as diagenetic processes in the IF sediment on the seafloor (e.g., Konhauser et al. 2017).

It has been generally accepted for two decades that the source of Fe to IFs are mid-ocean ridge (MOR) hydrothermal fluids, based on REE + Y characteristics (e.g., Derry and Jacobsen 1990; Bau and Möller 1993). The very high sorption coefficient for the REEs on Fe^{3+}-hydroxides that precipitate from MOR vent fluids, up to 10^6 (e.g., Quinn et al. 2006, and references within), forms the basis for using REE + Y characteristics of IFs to infer a MOR hydrothermal source for Fe. This interpretation was confirmed by early Nd isotope measurements that showed positive $\varepsilon_{Nd}(t)$ values which overlapped those of contemporaneous mantle (e.g., Bau et al. 1997), although it has since been recognized that Nd isotope compositions of IFs often reflect a mixture between MOR and continental sources (Alexander et al. 2009; Viehmann et al. 2015).

Neodymium isotopes potentially allow us to distinguish between the two "external box" fluxes of Fe^{2+}_{aq} shown in Fig. 6.30, which in turn should allow distinction between Fe derived from MOR hydrothermal sources (fluxes "H1" and "H2" in Fig. 6.30) or benthic Fe sources (flux "B" in Fig. 6.30), based on the inferred paleogeography. In Fig. 6.31 we illustrate Nd isotope data for the Brockman and Marra Mamba IFs from the Hamersley Basin (Australia), along with Nd–Fe isotope data for a subset of samples from the Dales Gorge member of the Brockman IF. The majority of $\varepsilon_{Nd}(t)$ values for the older Marra Mamba IF lie near zero, which could reflect a mixture of primitive mantle (positive $\varepsilon_{Nd}(t)$) and old continental crust (negative $\varepsilon_{Nd}(t)$). The spread in $\varepsilon_{Nd}(t)$ values for the Brockman IF, however, is much larger, suggesting different mixing proportions of continental and MOR

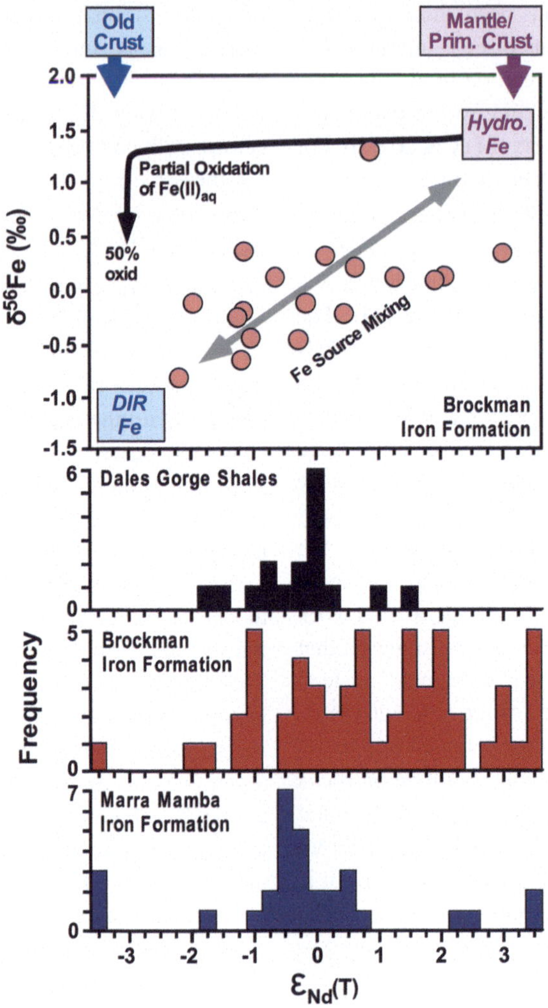

Fig. 6.31 Neodymium and Fe isotope data from the Marra Mamba and Brockman IFs and shales of the Hamersley Basin, Western Australia, adapted from Konhauser et al. (2017). $\varepsilon_{Nd}(t)$ values for the IFs scatter between a mantle/primitive crust composition ($\varepsilon_{Nd}(t) \sim 3$) and those of old crust ($\varepsilon_{Nd}(t) \sim -3$), indicating a mixture of REE sources. The intermediate $\varepsilon_{Nd}(t)$ values for the Dales Gorge member shales indicate local continental crust may have had near-zero $\varepsilon_{Nd}(t)$ values, which would increase the continental component of REEs relative to an assumed continental $\varepsilon_{Nd}(t)$ value of −3. The positive correlation between $\varepsilon_{Nd}(t)$ and $\delta^{56}Fe$ seen in the 2.48 Ga Dales Gorge member of the Brockman IF is interpreted to reflect water-mass mixing (grey double arrow) between a hydrothermal (high-$\varepsilon_{Nd}(t)$, high-$\delta^{56}Fe$) Fe source and Fe sourced to an Fe benthic shuttle generated by DIR on the continental shelf (low-$\varepsilon_{Nd}(t)$, low-$\delta^{56}Fe$). These relations indicate that relative to the two "external boxes" in Fig. 6.30, the hydrothermal Fe source can be identified through positive $\varepsilon_{Nd}(t)$ values, whereas the benthic Fe source should have negative $\varepsilon_{Nd}(t)$ values. Note, however, that a range in $\varepsilon_{Nd}(t)$ values for the continental source is indicated by the range in Nd isotope compositions of the shales, which would add scatter to the $\varepsilon_{Nd}(t)$-$\delta^{56}Fe$ relations. The solid black arrow indicates the trend for a Rayleigh fractionation model of oxidation of Fe^{2+}_{aq} from a hydrothermal source, which has a strong "L-shaped" pattern due to the very high K_d for REEs in Fe^{3+}-hydroxides, which is very sensitive to a continental seawater component relative to changes in Fe^{3+}-hydroxide mass and Fe isotope compositions. Data from Jacobsen and Pimentel-Klose (1988), Alibert and McCulloch (1993), and Li et al. (2015)

hydrothermal sources. The restricted range in $\varepsilon_{Nd}(t)$ values for shales in the Dales Gorge Member of the Brockman IF could either reflect consistent mixing of high- and low-$\varepsilon_{Nd}(t)$ sources, or the existence of a younger continental crust component of intermediate $\varepsilon_{Nd}(t)$ values.

The very high distribution coefficient (K_d) for Nd (and other REEs) in Fe^{3+}-hydroxides that precipitate from MOR fluids, relative to the mass of Fe^{3+}-hydroxide precipitated during Fe^{2+}_{aq} oxidation, produces highly non-linear trends on a $\varepsilon_{Nd}(t)$-$\delta^{56}Fe$ diagram for oxidation of hydrothermal fluids, relative to mixing between water masses (or populations of Fe^{3+}-hydroxides), which Li et al. (2015) use to distinguish between the two "external box" cases in Fig. 6.30. Oxidation of hydrothermal Fe^{2+}_{aq} produces Fe^{3+}-hydroxide precipitates that will rapidly become dominated by low-$\varepsilon_{Nd}(t)$ continentally-derived REEs, due to highly efficient scavenging, before any significant decrease occurs in $\delta^{56}Fe$ values, producing a distinctive "L-shaped" trend on a diagram of $\varepsilon_{Nd}(t)$-$\delta^{56}Fe$ (upper diagram in Fig. 6.31). That the samples from the Brockman IF that have been analyzed for both $\varepsilon_{Nd}(t)$ and $\delta^{56}Fe$ values do not lie along such an "L-trend" argues instead for mixing of water-mass particulates, one end member of which reflects Fe^{3+}-hydroxide precipitates from a MOR hydrothermal source (high-$\varepsilon_{Nd}(t)$, high-$\delta^{56}Fe$; flux "H2" in Fig. 6.30), and another end member that reflects Fe^{3+}-precipitates from a benthic Fe source (low-$\varepsilon_{Nd}(t)$, low-$\delta^{56}Fe$; flux "B" in Fig. 6.30). The scatter in the data likely records some heterogeneity in $\varepsilon_{Nd}(t)$ values for the crustal endmember, as suggested by the range in $\varepsilon_{Nd}(t)$ compositions for the larger dataset of shales and IFs, in addition to possible variations in the Fe^{2+}_{aq}—Fe^{3+}-hydroxide fractionation factor (dependent on O_2 contents; see Fig. 6.29). These mixing relations provide strong support for contributions from both a benthic Fe shuttle and MOR hydrothermal fluids to the Brockman IF. Based on extensive micro-sampling at various scales, Li et al. (2015) concluded that Fe sources to the Brockman IF were stable over periods of 10^0 to 10^3 yrs, indicating that the inferred Fe mixing relations within the basin reflect circulation changes on periods of $\sim 10^4$ years.

We next turn to the issue of Fe isotope equilibrium between IF deposits and ancient seawater, which bears on the use of IFs to infer ancient seawater chemistry or paleo-environmental conditions. Magnetite and siderite in IFs, for example, have been used to constrain pCO_2 in the atmosphere (Ohmoto et al. 2006; Rosing et al. 2010), which requires formation in equilibrium with seawater. Siderite has been used to estimate Fe^{2+}_{aq} contents of the ancient oceans, based on assumed equilibrium (Sumner 1997). Experimental studies of magnetite and siderite formation in IF precursors suggests that these minerals may form in equilibrium with each other during early burial diagenesis through reaction of Fe^{3+}-hydroxides and C_{org} (Posth et al. 2013), and it has been proposed that magnetite may directly precipitate from seawater through interactions between Fe^{3+}-hydroxides and Fe^{2+}_{aq} (Li et al. 2017a). In contrast, some workers argue that magnetite in IFs is exclusively of metamorphic origin, forming at high temperatures (Rasmussen and Muhling 2018). The isotopic compositions of IF minerals inform us about likely paragenetic sequences and allow us to test common models that suggest the primary IF precipitate was Fe^{3+}- and Fe^{2+}-Si-bearing gels, followed by a sequence of diagenetic conversion to magnetite and siderite (e.g., Klein 2005; Beukes and Gutzmer 2008). The very large database that is now available for Fe isotope fractionation factors, derived from theory and experiment (see Chap. 3), provides us with the means to test these proposals, focusing on the likelihood that IF minerals such as hematite, magnetite, and siderite formed in isotopic equilibrium with the ancient oceans, and, by extension of equilibrium thermodynamics, with each other. We first focus below on Fe isotopes, but then add C, O, and Sr isotopes to provide additional constraints on IF mineral paragenesis.

The likelihood that IF minerals formed in Fe isotope equilibrium with seawater may be evaluated by calculating the $\delta^{56}Fe$ values for Fe^{2+}_{aq} in seawater, using the measured $\delta^{56}Fe$ values and

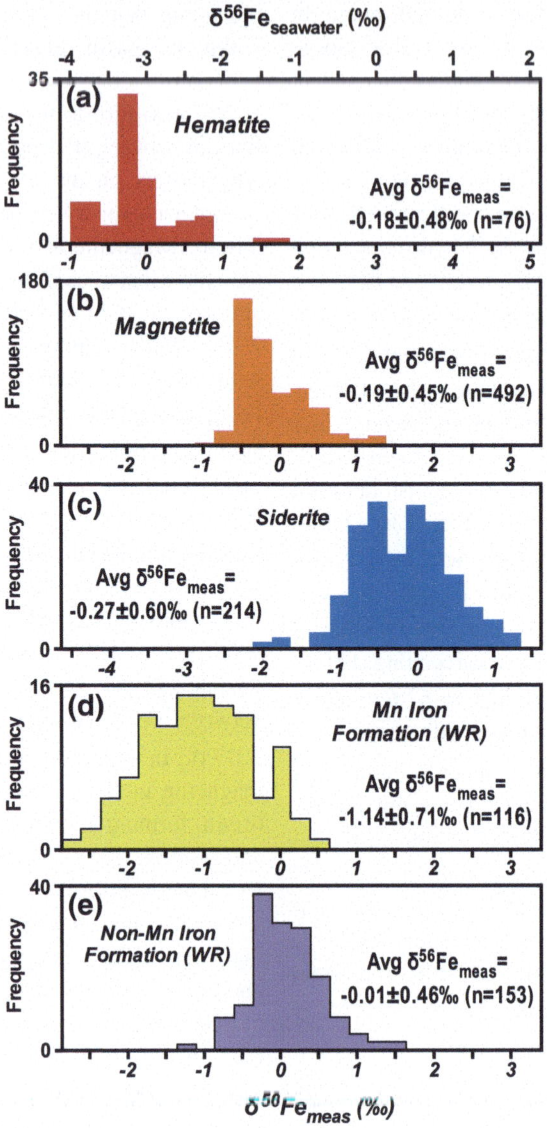

Fig. 6.32 Histograms of δ^{56}Fe values for the 2.4–2.6 Ga iron formations (IFs) of the Hamersley-Transvaal basin, referenced to the Fe isotope composition of seawater Fe^{2+}_{aq} that would be required for Fe isotope equilibrium (scale for Fe^{2+}_{aq} in seawater at top of figure). Diagram intended to evaluate if minerals in IFs formed in Fe isotope equilibrium with seawater, as well as in equilibrium with each other (readily seen when referenced to a common fluid). Scale for measured δ^{56}Fe values shown at the bottom of each histogram. Calculated δ^{56}Fe values for seawater Fe^{2+}_{aq} made using the following mineral-Fe^{2+}_{aq} ^{56}Fe/^{54}Fe fractionation factors: hematite-Fe^{2+}_{aq} = +3‰, magnetite-Fe^{2+}_{aq} = +1.3‰, and siderite-Fe^{2+}_{aq} = −0.5‰ (see compilations in Chap. 3). Panels **a**, **b**, and **c** show data for pure hematite, magnetite, and siderite (measured in mineral separates, estimated end member, or in situ), respectively. Panel D shows WR data for Mn IFs and panel E shows WR data for non-Mn IFs; in both cases reference to seawater compositions made on the assumption that magnetite is the dominant Fe phase. Data from Johnson et al. (2003, 2008a), Rouxel et al. (2005), Steinhoefel et al. (2010), Heimann et al. (2010), Tsikos et al. (2010), Craddock and Dauphas (2011), Planavsky et al. (2012a), Li et al. (2013b, 2015), Haugaard et al. (2016), Kurzweil et al. (2016), Lantink et al. (2018), and Thibon et al. (2019)

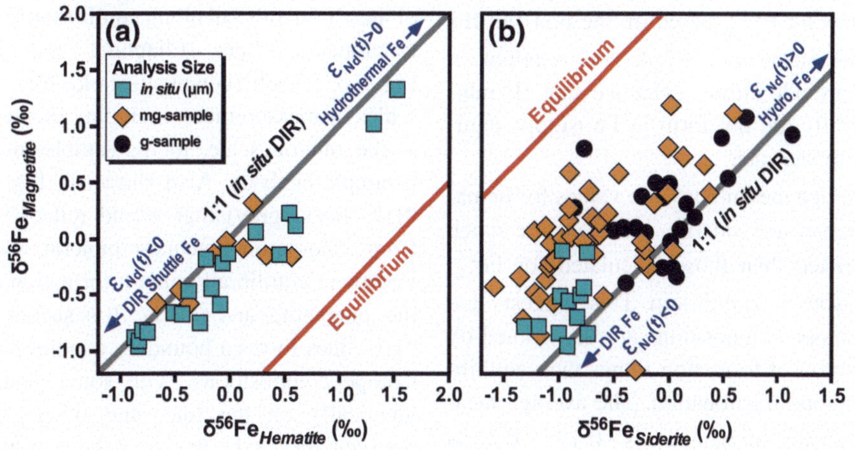

Fig. 6.33 Iron isotope variations among coexisting hematite, magnetite, and siderite (same sample) from the 2.5 Ga Dales Gorge Member of the Brockman IF Australia) and Kuruman IF (South Africa), as measured on various scales, including the μm scale via in situ analysis by laser ablation, mg-size bulk analysis, and g-size bulk analysis. The equilibrium hematite-magnetite and siderite-magnetite Fe isotope fractionations at low temperature (~25 °C) shown in red lines (Wiesli et al. 2004; Wu et al. 2010; Frierdich et al. 2014b), and none of the minerals measured plot along the Fe isotope equilibrium lines. The grey lines (marked "1:1") reflect an origin through reduction of a common precursor Fe^{3+}-oxide/hydroxide via "in situ" DIR in the sediment prior to lithification. The broad variations in absolute $\delta^{56}Fe$ values, from positive to negative values, are interpreted to reflect mixing between a hydrothermal (positive $\delta^{56}Fe$, positive $\varepsilon_{Nd}(t)$), and a benthic Fe source via a DIR shuttle from continental shelves (negative $\delta^{56}Fe$, negative $\varepsilon_{Nd}(t)$); see Figs. 6.30 and 6.31. Figure adapted from Konhauser et al. (2017). *Data sources* Johnson et al. (2003, 2008a), Steinhoefel et al. (2010), Heimann et al. (2010), Craddock and Dauphas (2011)

applying the appropriate mineral-fluid fractionation factors. Converting measured isotopic compositions to a common fluid also provides a ready test for equilibrium among multiple minerals because the conversion also accounts for mineral-mineral isotopic fractionation. Figure 6.32 shows the distribution of $\delta^{56}Fe$ values for the 2.4–2.6 Ga IFs of the Hamersley-Transvaal basin for hematite, magnetite, and siderite (Panels a–c), as well as WR samples from Mn IF and non-Mn IF. Extensive in situ studies of minerals in these IFs, as well as analyses of mineral separates or concentrates, additionally provides an opportunity to evaluate inter-mineral Fe isotope equilibrium on the mineral scale. As discussed in Beukes and Gutzmer (2008), the Hamersley IFs (e.g., Marra Mamba, Brockman, and Boolgeeda) are correlated with the Transvaal IFs (e.g., Kuruman and Griquatown), so we consider the dataset for these units collectively. The slightly younger Mn IFs (Koegas and Hotazal) are considered separately. The histograms for the WR samples (Fig. 6.32d and e) are calculated assuming all of the Fe is magnetite, the dominant Fe mineral in the rocks.

It is immediately apparent in Fig. 6.32 that hematite, magnetite, and siderite are far from Fe isotope equilibrium with each other on a coarse level. If these minerals formed in rough equilibrium with each other, the histograms would be aligned when referenced to Fe^{2+}_{aq}. The sampling of hematite, magnetite, and siderite shown in Fig. 6.32 panels a–c is random relative to stratigraphic height and therefore cannot be explained by systematic changes in $\delta^{56}Fe$ values for seawater Fe^{2+}_{aq}. This broad disequilibrium is expressed by a wide range of inferred $\delta^{56}Fe$ values for seawater Fe^{2+}_{aq}, exceeding the range measured in natural fluids, from $\delta^{56}Fe = -4$ to +2‰. For the WR samples, the Mn IF samples, assuming magnetite was in equilibrium with seawater, suggests Fe^{2+}_{aq} $\delta^{56}Fe$ values between −1 and −3‰, whereas the inferred seawater

$\delta^{56}Fe$ values for Fe^{2+}_{aq} based on the non-Mn IFs is ~ −1‰. Taken as a whole, the relations in Fig. 6.32 provide strong evidence that IF minerals generally did not form in Fe isotope equilibrium with seawater.

The average measured $\delta^{56}Fe$ values for hematite, magnetite, and siderite, however, are much more restricted than those calculated for Fe^{2+}_{aq} assuming isotope equilibrium. This suggests that their Fe isotope compositions are determined by the pathway(s) of formation rather than equilibrium Fe isotope fractionation. The average measured $\delta^{56}Fe$ value for hematite is −0.18 ± 0.48‰ (n = 76), which overlaps that measured for magnetite (−0.19 ± 0.45‰, n = 492) and siderite (−0.27 ± 0.60‰, n = 214). Although there is a 1 −2‰ range in measured $\delta^{56}Fe$ values, identifying a range of Fe sources (e.g., hydrothermal, benthic), it is striking that the average measured $\delta^{56}Fe$ values are similar. This suggests that perhaps the Fe^{2+}_{aq}-bearing minerals (magnetite and siderite) inherited the Fe isotope composition(s) of early precipitated hematite or its Fe^{3+}-hydroxide precursors. This is consistent with the commonly invoked paragenetic sequence for these minerals where Fe^{3+}-hydroxides and hematite are thought to be the earliest and most primary IF precipitate (Beukes and Gutzmer 2008). Turning to the WR samples, which have greater uncertainty in terms of mineralogical control, the range of measured $\delta^{56}Fe$ values for WR samples of non-Mn IF overlaps that determined on pure magnetite (panels E and B in Fig. 6.32). In contrast, the range in measured $\delta^{56}Fe$ values for WR samples of Mn IFs is clearly more variable, with a significantly more negative average $\delta^{56}Fe$ value (−1.14 ± 0.71‰, n = 116), suggesting a very different origin as compared to the non-Mn IFs. This will be explored in Sect. 6.5.5.

A test of the broad conclusions regarding Fe isotope equilibrium among hematite, magnetite, and siderite in IFs is provided through comparison of co-existing minerals in individual samples (Fig. 6.33). As discussed in Chap. 3, the Fe isotope fractionations among these minerals are well constrained. On an individual sample basis, none of these minerals fall along equilibrium Fe isotope fractionation lines (diagonal red lines in Fig. 6.33). Such a conclusion holds for a variety of scales of measurement, including in situ analysis at the micron scale, to mg-sample analysis, to g-sample analysis. Also shown in Fig. 6.33 are "1:1" lines (grey) that would reflect formation from a common precursor mineral with no Fe isotope re-equilibration upon formation of hematite, magnetite, and siderite. It is striking that the "1:1" lines forms a boundary condition to the Fe isotope compositions, with some data plotting identically on the line, and others scattering between the "1:1" line and the equilibrium Fe isotope fractionation line. Importantly, in essentially no case do data plot significantly in a space that is outside the area bounded by the "1:1" and equilibrium lines, demonstrating that the "1:1" line is a fundamental constraint on Fe formation pathways in the rocks.

Integrating the concepts from Figs. 6.30 and 6.31 with the diagenetic reactions among Fe inventories in marine sediments in Sect. 5.4.1 of Chap. 5, data that plot along the "1:1" lines in Fig. 6.33 are interpreted to record phase conversion of initial Fe^{3+}-hydroxide precipitates and reduction by DIR, where the DIR-produced magnetite and siderite inherited the Fe isotope compositions of the precursor Fe^{3+}-hydroxides (see also discussion in Konhauser et al. 2017). Samples that plot with positive $\delta^{56}Fe$ values are inferred to reflect initial Fe^{3+}-hydroxides that formed by precipitation from a hydrothermal source, and for the magnetite and hematite Fe isotope compositions that were determined by in situ analysis, these have positive $\varepsilon_{Nd}(t)$ values, confirming their hydrothermal origin (Li et al. 2013b, 2015). These would record flux "H2" in Fig. 6.30. For samples that plot at low $\delta^{56}Fe$ values, these record Fe^{3+}-hydroxides that precipitated from a low-$\delta^{56}Fe$ benthic Fe shuttle (flux "B" in Fig. 6.30). Similarly, the same samples analyzed by in situ methods for hematite and magnetite that have low $\delta^{56}Fe$ values also have negative $\varepsilon_{Nd}(t)$ values, confirming the benthic flux origin from the continental shelf (Li et al. 2013b, 2015). This highlights the importance of bringing additional constraints to inferring Fe sources such as Nd isotopes.

The diagenetic reactions that may explain Fe isotope compositions that plot on the "1:1" lines are: Conversion of Fe^{3+}-hydroxides to hematite:

$$2Fe(OH)_3 \rightarrow Fe_2O_3 + 3H_2O \qquad (6.5)$$

Reduction of Fe^{3+}-hydroxides by DIR to produce magnetite:

$$12Fe(OH)_3 + CH_2O \rightarrow 4Fe_3O_4 + CO_2 + 19H_2O \qquad (6.6)$$

Reduction of Fe^{3+}-hydroxides by DIR to produce siderite:

$$4Fe(OH)_3 + CH_2O + 3HCO_3^- \rightarrow 4FeCO_3 + 3OH^- + 7H_2O \qquad (6.7)$$

Note that in Eqs. 6.5–6.7, no free Fe^{2+}_{aq} is present, and hence there would be no change in $\delta^{56}Fe$ values expected from the precursor Fe^{3+}-hydroxides. Isotopic equilibrium between minerals would also be limited in the absence of Fe^{2+}_{aq}, given the importance of coupled electron and atom exchange in systems that contain Fe^{3+}-oxide/hydroxide and aqueous Fe^{2+} (see Chap. 3). Solid-state conversion of Fe^{3+}-hydroxides to magnetite by DIR, accompanied by minimal change in $\delta^{56}Fe$ values, has been observed in modern lake sediments (Percak-Dennett et al. 2013), as expected where minimal Fe^{2+}_{aq} is present. Note that in the case of siderite formation (Eq. 6.7), solid-state conversion of Fe requires additional carbonate from ambient seawater to produce the 1:1 Fe:C stoichiometry of siderite.

We can, however, write reactions that produce free Fe^{2+}_{aq}, which, if present, would move co-existing minerals toward Fe isotope equilibrium due to the presence of a catalytic exchange medium (Fe^{2+}_{aq}). For Eq. 6.6, magnetite is formed with no free Fe^{2+}_{aq} when the C:Fe ratio of reactants is 1:12 as written, but at higher organic C proportions, free Fe^{2+}_{aq} could exist if balanced by decreased magnetite production. In the case of siderite production, a change in the proportion of inorganic C to organic C will be directly correlated with the abundance of free Fe^{2+}_{aq}:

$$4Fe(OH)_3 + CH_2O + 2HCO_3^- \rightarrow 3FeCO_3 + Fe^{2+}_{aq} + 4OH^- + 6H_2O \qquad (6.8)$$

$$4Fe(OH)_3 + CH_2O + HCO_3^- \rightarrow 2FeCO_3 + 2Fe^{2+}_{aq} + 5OH^- + 5H_2O \qquad (6.9)$$

A large number of siderite-magnetite pairs in Fig. 6.33 scatter between the "1:1" and equilibrium lines, and it seems likely those data reflect reactions that included Fe^{2+}_{aq} to allow partial equilibration between the minerals. In addition, the different stoichiometries of Eqs. 6.7–6.9 predict distinct changes in $\delta^{13}C$ values, reflecting the different proportions of inorganic and organic C (discussed below). A prediction would be that the closer a siderite-magnetite pair plots toward the equilibrium fractionation line, the greater amount of Fe^{2+}_{aq} was present to facilitate re-equilibration, which would correspond to a decreasing $\delta^{13}C$ value due to decreasing seawater carbonate. Unfortunately, no C isotope data are available for the samples plotted in Fig. 6.33b, but this would be a fruitful avenue of future work, particularly the combination of in situ C and Fe isotope analysis on Fe-carbonates.

Although we have written the reactions above without Si present, the primary IF precipitate was likely a Fe^{3+}-Si gel, a conclusion supported by experimental studies as well as Si isotope data for cherts in IFs relative to Fe-free chert (Percak-Dennett et al. 2011; Zheng et al. 2016; Konhauser et al. 2017). In fact, a key observation is that abiologic conversion of Fe^{3+}-Si gels to magnetite is strongly inhibited relative to pure Fe^{3+}-hydroxides, and only DIR of Fe^{3+}-Si gels has been shown to be able to produce sufficient Fe^{2+} to attain magnetite stoichiometry (Reddy et al. 2016). Such experimental results, as well as the Fe isotope observations among coexisting minerals shown in Fig. 6.33, provide very strong evidence for in situ DIR in IF sediment during lithification, early in the seafloor diagenetic sequence. This is a second stage of DIR in BIF

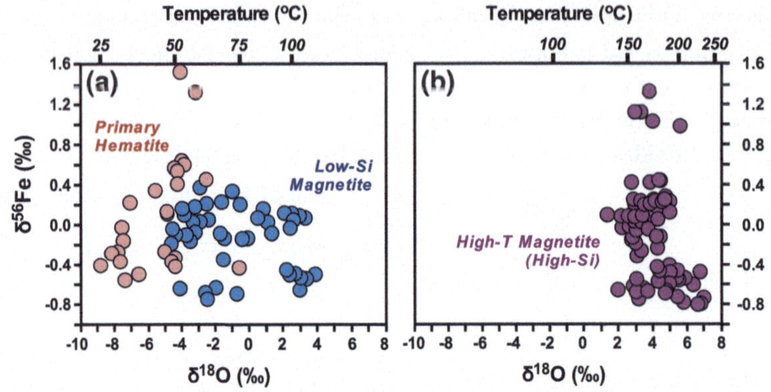

Fig. 6.34 Iron-O isotope variations in coexisting hematite and magnetite from the Dales Gorge Member of the Brockman IF (Australia). $\delta^{18}O$ values reported relative to SMOW (data from Li et al. 2013b). Hematite is the most primary precipitate based on petrographic relations (Huberty et al. 2012; Li et al. 2013b). The earliest-formed magnetite is fine-grained and has low Si contents, whereas high-temperature magnetite is course-grained and/or contains high levels of Si (Huberty et al. 2012). Oxygen isotope temperatures were obtained using hematite-quartz and magnetite-quartz fractionations; those for high-temperature magnetite use the average measured $\delta^{18}O$ value for quartz ($\delta^{18}O \sim +21‰$), reflecting the last equilibration of quartz during metamorphism (panel **b**), whereas low-temperature magnetite and hematite assumes a $\delta^{18}O$ value for quartz of +23‰ (panel **a**; see discussion in Li et al. 2013b). The relative O isotope temperatures confirm hematite is the most primary phase, followed by early diagenetic formation of low-Si magnetite, followed by formation of high-Si magnetite under hydrothermal/metamorphic conditions. Importantly, because Fe solubility is low for hematite and magnetite in hydrothermal fluids, there are no significant changes expected in $\delta^{56}Fe$ values for these minerals, although O isotopes are highly sensitive to temperature changes and fluid-mineral exchange. Figure adapted from Konhauser et al. (2017). Oxygen isotope fractionation factors from Kita et al. (1985), Mandernack et al. (1999), and Bao and Koch (1999)

genesis, reflecting formation of the final diagenetic mineral assemblage in the deep basin. The initial role of DIR, however, lies in producing a low-$\delta^{56}Fe$ Fe^{2+}_{aq} flux via a benthic Fe shuttle that is transported from the shelf to the deep basin, as identified by the low-$\delta^{56}Fe$ and -$\varepsilon_{Nd}(t)$ end member in Fig. 6.31. The importance of DIR in driving the Fe cycle in the early Paleoproterozoic was stressed by Johnson et al. (2008b), who argued that DIR would have been most vigorous prior to the rise in seawater sulfate contents that occurred before the GOE, at which point high sulfide contents would react with Fe^{3+}-hydroxides, removing their availability to DIR.

We next explore the relations between Fe and O isotope compositions for hematite and magnetite as determined by in situ isotopic analysis, which address the issue of oxide paragenesis in IFs. The petrographic evidence for hematite as the earliest oxide phase in the Dales Gorge member of the Brockman IF (Li et al. 2013b) is confirmed by the low measured $\delta^{18}O$ values (Fig. 6.34). These relations also argue against a post-depositional origin for hematite via oxidation by hydrothermal fluids (Rasmussen et al. 2014). Magnetite is a replacement to hematite, an observation that is common in IFs of this age (Klein 2005; Beukes and Gutzmer 2008), and this is confirmed by the slightly higher $\delta^{18}O$ values for low-Si magnetite (Fig. 6.34). It is now recognized that IF magnetite may occur as an essentially pure Fe_3O_4 phase, but also as a higher temperature form that has 1–2 mol% Si substitution (Huberty et al. 2012). Based on a study magnetite chemistry from a wide range of IFs, Huberty et al. (2012) documented that low-Si magnetite is always earlier relative to high-Si magnetite, the latter of which commonly occurs as a hydrothermal replacement. This paragenetic sequence is well reflected in the $\delta^{18}O$ values, where high-Si magnetite equilibrated at temperatures of ~150 to ~250 °C (Fig. 6.34).

Importantly, there appears to be no change in the range of $\delta^{56}Fe$ values with increasing paragenetic stage (hematite to low-Si magnetite to high-Si magnetite), suggesting that despite extensive O isotope exchange, Fe isotopes are relatively unaffected. This is consistent with the studies of the Biwabik IF discussed in Sect. 6.5.1, where Hyslop et al. (2008) found that only at very high grades of metamorphism (>400 °C) did Fe isotope exchange occur between magnetite and Fe silicates. The lack of change in $\delta^{56}Fe$ values for high-Si magnetite, despite its hydrothermal origin, is consistent with the low solubility for Fe in magnetite-fluid systems (Chou and Eugster 1977). Making the connection back to Nd isotopes, which played a prominent role in distinguishing hydrothermal from benthic sources, as discussed above, it is important to note that Li et al. (2015) documented via in situ analysis that the high-Si magnetite contains very little Nd. This observation shows that, at least in this sample suite, Nd isotopes can be used to infer Fe sources in the marine basin because they were not compromised by later high-temperature hydrothermal alteration. Collectively, these observations confirm that Fe isotope compositions of IFs, even those that have seen extensive mineral reactions at elevated temperatures of 250–300 °C, can be confidently used to infer marine processes. This is a remarkable, though perhaps under-appreciated, conclusion.

The reactions in Eqs. 6.7–6.9 describe the interplay between C sources and Fe^{2+}_{aq} during DIR, which produce a clear set of predicted changes in $\delta^{13}C$ and $\delta^{18}O$ values for siderite in IFs (Fig. 6.35). As discussed in Heimann et al. (2010), changes in the proportion of C sources will change the $\delta^{13}C$ values for siderite, but also $\delta^{18}O$ values, given the fact that any O inherited from DIR of Fe^{3+}-hydroxides will have much lower $\delta^{18}O$ values than those that reflect O isotope equilibrium between carbonate and seawater. This reflects the dramatically different carbonate—water and Fe-oxide/hydroxide—water $^{18}O/^{16}O$ fractionations, but has not been generally recognized in the IF literature. The $\delta^{13}C$ values for siderite produced by DIR where the ratio of organic C (C_{org}) to seawater C (C_{SW}) is unity (Eq. 6.9) will be ~−14‰, reflecting equal contributions from organic and inorganic C, whereas the $\delta^{13}C$ values for a C_{org}:C_{SW} ratio of 1:2 (Eq. 6.8) and C_{org}:C_{SW} ratio of 1:3 (Eq. 6.7) will be progressively higher, as will the $\delta^{18}O$ values (Heimann et al. 2010).

Superimposed on these mixing relations are additional effects such as re-equilibration during burial, which will tend to decrease $\delta^{18}O$ values relative to those predicted by Eqs. 6.7–6.9, as shown by the left-facing arrow at the top of Fig. 6.35a, which illustrates the shift in $\delta^{18}O$ that would occur by a 25 °C temperature increase during burial diagenesis. Despite these complications, DIR offers an explanation for the anomalously low $\delta^{13}C$ and $\delta^{18}O$ values for IF siderites that have been previously difficult to explain, extending earlier interpretations based on C isotopes alone (e.g., Baur et al. 1985). Water-column stratification in $\delta^{13}C$ or $\delta^{18}O$ values cannot explain the results (see discussion in Heimann et al. 2010), and therefore the relations for Fe-carbonates in Fig. 6.35 must reflect those that formed in the IF sediment prior to lithification. This conclusion is reinforced by Sr isotope studies on the same samples by Johnson et al. (2013a), who showed that Fe-carbonates from the Kuruman IF could not have formed in equilibrium with seawater, in contrast to the underlying Ca–Mg carbonates that record seawater compositions. Strontium isotopes provide a vigorous test of equilibrium with seawater because $^{87}Sr/^{86}Sr$ ratios are not fractionated upon precipitation (using standard normalization during analysis), nor cycled by biology, and the $^{87}Sr/^{86}Sr$ seawater curve is well established for the Neoarchean-Paleoproterozoic.

The relations between $\delta^{13}C$ and wt% C_{org} are well explained by DIR, where the lowest $\delta^{13}C$ and C_{org} contents tend to occur in the oxide-facies IF relative to the siderite-facies IF (Fig. 6.35b). At low ratios of C_{org}:C_{SW} (e.g., Eq. 6.9), abundant free Fe^{2+}_{aq} is generated, which, if reacted with Fe^{3+}-hydroxides, will produce the high magnetite abundances found in oxide-facies IF relative to siderite-facies IF. Such relations, in addition to those shown in Figs. 6.33 and 6.34, document the electron equivalents that are contained in both magnetite

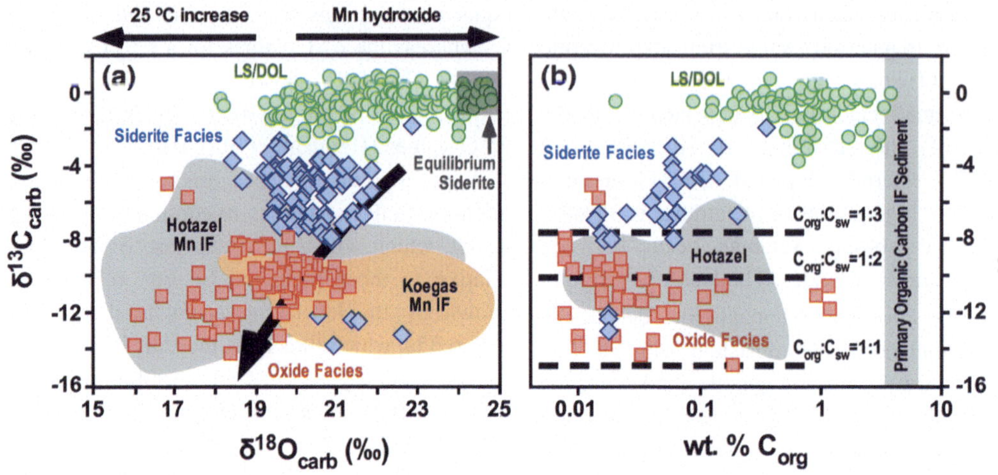

Fig. 6.35 Variations among $\delta^{18}O$ (SMOW scale) and $\delta^{13}C$ (PDB scale) for carbonates (**a**), and $\delta^{13}C_{carb}$–WR wt % C_{org} (**b**) relations for the Kuruman IF (siderite facies: blue diamonds, oxide facies: red squares) and Ca–Mg carbonates of the Campbellrand-Malmami platform (green circles), South Africa. Also shown are fields for Mn–Fe carbonates from the Hotazel Mn IF (grey field) and Koegas Group Mn IFs (brown field), South Africa. Transparent grey box in upper right corner of panel **a** shows $\delta^{13}C$ and $\delta^{18}O$ values for siderite that would be in equilibrium with seawater, indicating that none of the Fe carbonates from the IFs formed in C or O isotope equilibrium with seawater (seawater $\delta^{18}O = -2$‰; Zakharov and Bindeman 2019). Heavy black curved arrow in **a** shows mixing line between seawater carbonate and carbonate produced by DIR at various ratios of organic C to seawater carbon (see Eqs. 6.7–6.9 in text; horizontal dashed arrows indicate associated C_{org}:C_{SW} ratios), where the correlated shifts in $\delta^{13}C$ and $\delta^{18}O$ reflect inheritance of isotopic compositions from C_{org} and Fe^{3+}-hydroxides, respectively. Estimate for primary C_{org} content in the Kuruman IF, prior to DIR, is ~5 wt% (Konhauser et al. 2017). Left-facing horizontal arrow at top shows $\delta^{18}O$ shift that would occur with a 25 °C temperature increase from ~25 to ~50 °C, illustrating the effect of early burial diagenesis. See Heimann et al. (2010) for details of C and O isotope fractionation factors for siderite and Ca–Mg carbonates. Right-facing arrow at top shows $\delta^{18}O$ shift that would occur if the $\delta^{18}O$ values of IF carbonates were inherited from Mn hydroxides as opposed to Fe^{3+}-hydroxides (see Frierdich et al. 2016 for O isotope fractionation factors). Figure adapted from Konhauser et al. (2017). *Data sources* cited in Heimann et al. (2010), with Mn IF data from Tsikos et al. (2003), Schneiderhan et al. (2006), and Nel (2013)

and siderite in IFs. As discussed in Konhauser et al. (2017), the low measured C_{org} contents for IFs cannot be used to reject a role for biology in their formation because the majority of the original C_{org} that accumulated in the IF sediment has been respired via DIR. Respiration of C_{org} can be thought of as a transfer electrons to Fe^{3+}-hydroxides, converting them to magnetite and siderite. Reconstruction of the average original C_{org} contents in IF sediments based on $\delta^{13}C$ values of siderite is equivalent to 1.3 wt% C for oxide-facies IF and 3.2 wt% C for siderite-facies IF. An additional 3.4 wt% C_{org} is calculated from electron balance via conversion of Fe^{3+}-hydroxides to magnetite for oxide-facies IF, which produces a total of 4.7 wt% C_{org} in the original IF sediment when combine with the C_{org} equivalents contained in siderite. Higher levels of initial C_{org} are calculated for siderite-facies IF based on their higher Fe^{2+} abundance as compared to oxide-facies IF. Overall, these calculations suggest that the IF sediment of the large Hamersley- Transvaal units originally contained ~5 wt% C_{org}, similar to the levels measured for Neoarchean and Paleoproterozoic black shales. These reconstructed C_{org} levels provide strong support for the DIR reactions proposed by Konhauser et al. (2005), and demonstrate that the measured C_{org} contents in IFs are likely several orders of magnitude lower than what was deposited with the original sediment.

Iron carbonates from the Mn IFs of the Koegas Group and Hotazel Formation, which overlie the Kuruman IF, have overlap in their $\delta^{13}C$ and $\delta^{18}O$ relations with the non-Mn IFs (Kuruman), but also extend to higher $\delta^{18}O$ values yet maintain their low-$\delta^{13}C$ values (Fig. 6.35). Carbonates in Mn IFs are dominated by Mn-siderite, rhodochrosite, and kutnohorite (e.g., Johnson et al. 2016), and the negative $\delta^{13}C$ values of these minerals have been commonly interpreted to reflect microbial Mn reduction (e.g., Tsikos et al. 2003). The extension to higher $\delta^{18}O$ values, at negative $\delta^{13}C$ values, of some samples from the Koegas and Hotazel IFs may be explained by a shift in the electron acceptor from Fe^{3+}-hydroxides to Mn^{4+}/Mn^{3+} oxides/hydroxides, which should correspond to an increase in $\delta^{18}O$ values, given the positive $^{18}O/^{16}O$ fractionation between Mn^{4+}/Mn^{3+}-oxide/hydroxide and Fe^{3+}-hydroxide (Frierdich et al. 2016). If this interpretation is correct, there should be a correlation between $\delta^{13}C$ and Mn/Fe ratio of the carbonates in Mn IFs, although such detailed studies have not yet been done. We will investigate the unique geochemical pathways for Fe and their isotopic effects in Mn IFs in the next section.

6.5.5 Early Paleoproterozoic Rise of Mn Redox

Special attention has been paid to the occurrence of Mn-rich marine sedimentary rocks, including Mn IFs, because very high Eh conditions are required to oxidize Mn^{2+} relative to Fe^{2+}, and thus the appearance of Mn-rich deposits is often taken to indicate highly oxidizing conditions, likely involving high O_2 (e.g., Roy 2006). The late Neoarchean to early Paleoproterozoic stratigraphy in the Hamersley and Transvaal basins show a clear trend toward increasing Mn contents in IFs with decreasing age, as well as an overall decrease in IF abundance (e.g., Beukes and Gutzmer 2008), suggesting increasingly oxidized marine conditions leading up to the GOE, culminating in the largest Mn IF deposits in the world (Beukes et al. 2016). Deposition of Mn-rich sequences continues past the GOE, including units we have already discussed such as the Francevillian sequence.

Oxidation of hydrothermal Mn^{2+}_{aq} occurs through several reactions, including oxidation of Mn^{2+} to Mn^{3+}, and oxidation of Mn^{3+} to Mn^{4+} (e.g., Morgan 2005):

$$4Mn^{2+} + O_2 + 6H_2O \rightarrow 4MnOOH + 8H^+ \quad (6.10)$$

$$4MnOOH + O_2 \rightarrow 4MnO_2 + 2H_2O \quad (6.11)$$

Under conditions of 100% PAL for O_2, the oxidation half-lives are relatively long for Mn oxidation, on the order of 10^2 days, several orders of magnitude longer than for Fe^{2+}_{aq}, reflecting a strong kinetic inhibition. In contrast, microbial oxidation of Mn^{2+} is much more rapid (e.g., Nealson et al. 1988; Tebo 1991), and this is commonly taken as evidence that formation of Mn oxides likely requires a biological role. In addition to microbial oxidation in the presence of O_2, an alternative proposal is that the large Paleoproterozoic Mn deposits were produced by photosynthetic oxidation of Mn^{2+} under anaerobic conditions, where Mn^{2+} served as the electron donor (Johnson et al. 2013b).

Transport of Mn oxides and C_{org} to the seafloor will support microbial reduction of Mn, which has long been recognized as an early pathway to C_{org} degradation in marine sediments (e.g., Lovley and Phillips 1988; Lovley 1991; Thamdrup et al. 2000), given the high energy yield that is available:

$$2MnO_2 + CH_2O + H_2O \rightarrow 2Mn^{2+} + HCO_3^- + 3OH^- \quad (6.12)$$

A major end product of microbial reduction of Mn oxide is Mn-bearing carbonates, which, in addition to Mn-silicates, dominates the mineralogy of Mn IFs (e.g., Johnson et al. 2016), and the presence of Mn carbonates in the ancient rock record has been taken to indicate high-O_2 conditions in the water column (e.g., Calvert and Pedersen 1993). In modern marine sediments that contain high levels of reactive Mn and Fe oxides, DIR also occurs in the same sediment sections, although lower in the sediment pile, reflecting the lower energy yield of Fe reduction as compared to Mn reduction. Writing the reaction for DIR in the same format as above (formation of carbonate) produces:

$$4Fe(OH)_3 + CH_2O \rightarrow 4Fe^{2+} + HCO_3^- + 7OH^- + 3H_2O \quad (6.13)$$

There are numerous examples of the same bacterial species reducing either Mn or Fe (e.g., Lovley and Phillips 1988; Myers and Nealson 1988), confirming the expectation that both processes will occur in sediments that contain Mn- and Fe-oxide/hydroxides. The two reactions above suggest that the product of simultaneous Mn and Fe reduction in the same sediment section will be mixed Fe–Mn carbonates, and these are common in Mn IFs. Importantly, however, is recognition that Mn reduction consumes C_{org} at twice the rate as Fe reduction, as shown by comparison of Eqs. 6.12 and 6.13, as well as the fact that Mn reduction occurs first in down-depth trends in marine sediments (e.g., Sørensen and Jørgensen 1987; Hyun et al. 2017). This in turn indicates that where Mn reduction occurs, C_{org} may become limited for DIR, and we may infer that the percent Fe^{3+} reduction by DIR may decrease where Mn reduction occurs. This would produce a significant shift toward more negative $\delta^{56}Fe$ values for Fe^{2+}_{aq} released by DIR in the presence of Mn reduction. In addition, to the degree that the presence of Mn oxides indicates higher Eh conditions, greater extents of oxidation of Fe^{2+}_{aq} that escapes the marine sediment is expected, which will further lower $\delta^{56}Fe$ values of pore water that escapes the sediment section. Importantly, all of these factors suggest that the $\delta^{56}Fe$ values of a benthic Fe shuttle will be significantly lower in the presence of a Mn cycle than in the absence of Mn redox chemistry.

Coexistence of Mn^{2+}_{aq} and Fe^{3+}-oxide/ hydroxides, or Fe^{2+}_{aq} and Mn^{4+}-oxide/ hydroxides, produces extensive redox cycling (e.g., Dellwig et al. 2010, and references within) through reactions such as:

$$MnO_2 + 2Fe^{2+} + 2H_2O \rightarrow Mn^{2+} + 2FeOOH + 2H^+ \quad (6.14)$$

The intimate redox connection between Fe and Mn, therefore, has led to proposals that Mn and Fe are shuttled together on basin-wide scales in the redoxcline of anoxic basins (Tebo 1991; Dellwig et al. 2010). Isotopic tracer studies confirm extensive Fe isotope exchange in reactions such as that of Eq. 6.14, as well as conversion of Fe^{3+}-hydroxides to secondary minerals such as magnetite (Schaefer et al. 2017). Such findings have clear implications for Fe isotope studies of Mn–Fe sediments. In addition to potential redox cycling in the water column, studies of modern marine sediments suggest that Fe and Mn may be redox cycled hundreds of times in sediment sections before final burial in the rock record (Canfield et al. 1993).

Isotopic-composition trends for early Paleoproterozoic IFs of the Transvaal Basin show clear systematic changes from early (Kuruman IF) to late (Hotazel IF), reflecting increasing Mn contents and decreasing average $\delta^{56}Fe$ values with decreasing age (Fig. 6.36). Deposition of these IFs occurred from ~2.52 to ~2.39 Ga, overlapping the earliest estimates of the GOE based on MIF-S using current timescales (Gumsley et al. 2017). Detailed geologic and geochemical studies of this sequence can be found in Tsikos and Moore (1997), Tsikos et al. (2001, 2003, 2010), Schneiderhan et al. (2006), Beukes and Gutzmer (2008), Schröder et al. (2011), Kunzmann et al. (2014), Kurzweil et al. (2016), Lantink et al. (2018), and Thibon et al. (2019). The oldest unit shown in Fig. 6.36, the Kuruman IF, is also the largest IF, and was

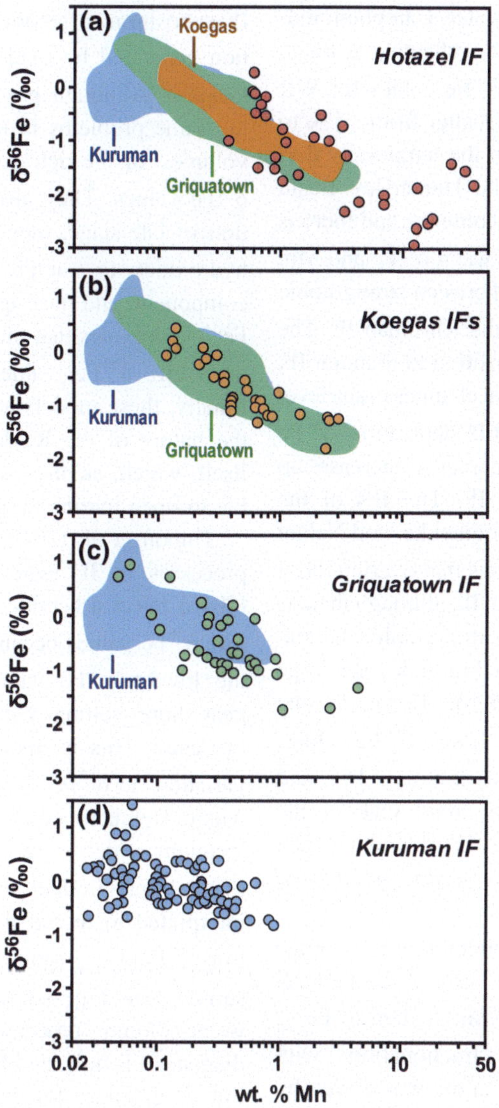

Fig. 6.36 Changes in WR δ^{56}Fe-Mn trends for the early Paleoproterozoic IF sequences of the Transvaal Basin that demonstrate the temporal transition from non-Mn IF deposition to Mn IF deposition (units shown in stratigraphic sequence from oldest to youngest in panels D to A). Note log scale for Mn, in wt%. Correlation coefficients for logarithmic relations between δ^{56}Fe and Mn are: Kuruman, $R^2 = 0.21$; Griquatown, $R^2 = 0.54$; Koegas, $R^2 = 0.75$; Hotazel, $R^2 = 0.50$. The increase in Mn contents upward in the stratigraphy is correlated with changes in mineralogy, where Mn is largely hosted in Mn-siderite, in addition to subordinate Mn-oxide/hydroxide and silicates. Data from Kurzweil et al. (2016) and Thibon et al. (2019), and in both studies the decreasing δ^{56}Fe values with decreasing age is interpreted to record an increased extent of Fe^{3+}-oxide/hydroxide precipitation via Rayleigh fractionation. Similar relations are seen if δ^{56}Fe is plotted relative to Fe/Mn ratios, although such relations are defined almost entirely by Mn contents. There is little correlation with Fe contents: correlation coefficients for δ^{56}Fe–Fe (wt%) are: Kuruman, $R^2 = 0.04$ (linear), $R^2 = 0.03$ (log); Griquatown, $R^2 = 0.04$ (linear), $R^2 = 0.05$ (log); Koegas, $R^2 = 0.10$ (linear), $R^2 = 0.10$ (log); Hotazel, $R^2 = 0.24$ (linear), $R^2 = 0.22$ (log). With the exception of the Hotazel IF, there is no correlation between Fe and Mn contents. For the Hotazel IF, Fe–Mn concentration relations have a correlation coefficient of $R^2 = 0.55$ (linear), $R^2 = 0.67$ (log)

deposited on the shallow-water Campbellrand-Malmani carbonate platform, reflecting a transgressive marine sequence. $\delta^{56}Fe$ values for WR samples of the Kuruman IF scatter from −1‰ to 1‰, similar to that found in the temporally correlative Brockman IF of the Hamersley Basin (Australia), at very low Mn contents, and there is no significant correlation between $\delta^{56}Fe$ and Mn. There is also no correlation between stratigraphic height and $\delta^{56}Fe$ values in the Kuruman IF. The Griquatown IF records a transition to granular IF, interpreted to indicate cyclical changes in sea level and an overall shallowing of IF deposition. $\delta^{56}Fe$ values decrease and Mn contents increase up section in the Griquatown IF. The IFs of the Koegas Group (Doradale, Rooinekke, and Nelani units) have $\delta^{56}Fe$-Mn relations that overlap those of the underlying Griquatown IF, although there is no systematic change with stratigraphic height. The youngest IF plotted in Fig. 6.36, the large Hotazel IF of the Kalahari Mn Field, has the highest Mn contents and the lowest $\delta^{56}Fe$ values, extending to $\delta^{56}Fe$ values as low as −3‰, and there is no clear stratigraphic trend. Overall, the Griquatown through Hotazel IFs define a strong logarithmic correlation between Mn contents and $\delta^{56}Fe$ values.

Thibon et al. (2019) model the Fe isotope variations for the Kuruman-Griquatown-Hotazel sequence to record an increasing extent of Fe^{2+}_{aq} oxidation via Rayleigh fractionation with decreasing age, which they argue was driven by an increase in O_2 contents and accompanying decrease in residence time for Fe^{2+}_{aq} in seawater. They reject a role for biological processes based on low measured C_{org} contents, although we addressed this issue in the previous section, and is not considered a valid argument against microbial diagenesis. A similar model was argued for the Koegas Group IFs by Kurzweil et al. (2016), who also interpreted the negative $\delta^{56}Fe$ values to indicate extensive Fe^{3+}-hydroxide precipitation. These models are shown in Fig. 6.30, illustrated as "flux H1", which, in this case, would require 100% oxidation of the low-$\delta^{56}Fe$ Fe^{2+}_{aq} from "flux H1" at the site of IF deposition to record the low-$\delta^{56}Fe$ composition generated by oxidation of Fe^{2+}_{aq} in a distal location. Although possible, these models have the same problems in terms of generating large volumes of Fe-rich rocks that have negative $\delta^{56}Fe$ values. They also require selective isolation of late-stage, incrementally-produced Fe^{3+}-hydroxides to capture a small low-$\delta^{56}Fe$ IF component, and must ignore the large volume of Fe^{3+}-hydroxides that must have formed and have near-zero $\delta^{56}Fe$ values, as discussed above. Finally, these models are likely to be limited by the nature of the Rayleigh fractionation model itself, which, as discussed above, is not applicable to open marine systems.

Thibon et al. (2019) only consider abiologic processes in IF genesis, and Kurzweil et al. (2016) reject a benthic Fe source to explain the low $\delta^{56}Fe$ values because the samples that have the lowest $\delta^{56}Fe$ values were deposited in a near-shore setting based on sedimentological evidence. This is distinct from the geographic variations in $\delta^{56}Fe$ values from the Kuruman IF, where Heimann et al. (2010) showed that the locations which were most proximal to the continental margin had the highest $\delta^{56}Fe$ values, interpreted to record the residue after loss of a low-$\delta^{56}Fe$ Fe^{2+}_{aq} component via a benthic Fe shuttle (see Fig. 6.30). In addition to inferred water-column processes, extensive microbial diagenesis is indicated for the Mn IFs by the very low $\delta^{13}C$ values for the Mn IFs (see Fig. 6.35), as discussed in Tsikos et al. (2001, 2003), Schneiderhan et al. (2006), and Nel (2013).

The preceding discussion highlights the difficulty in interpreting the Fe isotope data for Mn IFs. On the one hand, the higher Eh conditions and greater extent of oxidation of seawater Fe^{2+}_{aq} that is expected offers a ready explanation for the very negative $\delta^{56}Fe$ values of Mn IFs. On the other hand, we face the same limitations discussed above in terms of choosing the appropriate model, the fact that only a small portion of incrementally produced Fe^{3+}-hydroxides must be captured in the IF, and what to do with the large

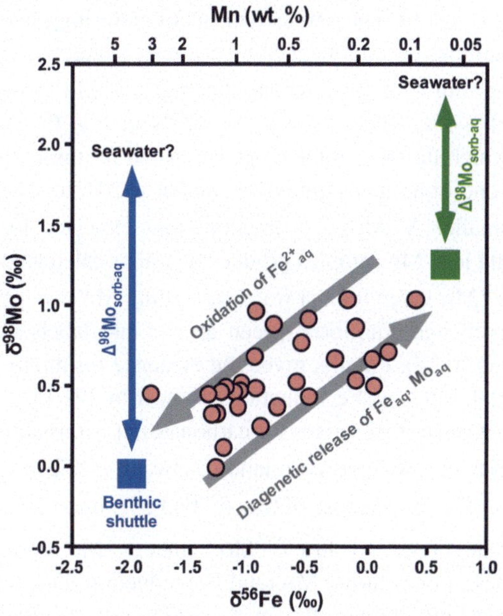

Fig. 6.37 Iron and Mo isotope data from the Koegas Group Mn IF, Transvaal Basin, South Africa. $\delta^{56}Fe$ and $\delta^{98}Mo$ ("NIST + 0.25" reference frame) are positively correlated, and negatively correlated with the log of Mn contents (see Fig. 6.36). Scale for Mn contents at top based on regression of $\delta^{56}Fe$-Mn relations ($R^2 = 0.75$; Fig. 6.36). Kurzweil et al. (2016) interpret the high-$\delta^{56}Fe$, -$\delta^{98}Mo$ samples to record precipitation in the deep anoxic part of the basin, where Mn oxide/hydroxides were not stable, whereas the low-$\delta^{56}Fe$, -$\delta^{98}Mo$ samples are thought to record formation above a redoxcline where Fe^{3+}-hydroxides and Mn-oxide/hydroxides precipitated. They invoke extensive oxidation of Fe^{2+}_{aq} to produce negative $\delta^{56}Fe$ values, and high levels of O_2 to produce Mn oxides that sorbed Mo from seawater that had low-$\delta^{98}Mo$ values. Other interpretations are shown, including Mo isotope equilibrium with seawater for a Fe^{3+}-hydroxide end member (green box and arrow), assuming $\Delta^{98}Mo_{sorb-aq} = -1.1‰$ (Goldberg et al. 2009). For Mn-rich samples, which have a molar Fe:Mn ratio near unity, we estimate an overall $\Delta^{98}Mo_{sorb-aq}$ fractionation of -1.9‰, obtained by scaling the Mo isotope fractionation for ferrihydrite relative to that of Mn oxides (blue box and arrow; Barling and Anbar 2004; Wasylenki et al. 2008; Goldberg et al. 2009). Assumption of equilibrium Mo isotope exchange produces a seawater $\delta^{98}Mo$ value of ~ +2‰, which lies in the range of other estimates for seawater at this time and is consistent with an oxygenated ocean (Kurzweil et al. 2015). However, the low-$\delta^{56}Fe$, -$\delta^{98}Mo$ component is also consistent with a Fe–Mn benthic shuttle, shown in blue box (see Fig. 6.30; Goldberg et al. 2012). Additional interpretations for a positive correlation between $\delta^{56}Fe$ and $\delta^{98}Mo$ include loss of low-$\delta^{56}Fe$, -$\delta^{98}Mo$ aqueous Fe and Mo during Fe and Mn reduction via microbial diagenesis. Data from Kurzweil et al. (2016)

volume of near-zero Fe^{3+}-hydroxide deposits that are not seen in the rock record; this last issue is quite serious based on the well-preserved stratigraphic records for these sequences. We next turn to another isotope system that is well suited to understanding Mn redox cycles, stable Mo isotopes. The large, and distinct, Mo isotope fractionations that occur during sorption to Mn- and Fe^{3+}-oxides/hydroxides (see review by Kendall et al. 2017) make the combination of Mo and Fe isotope studies of IFs, including Mn IFs, particularly attractive.

Kurzweil et al. (2016) noted that $\delta^{56}Fe$, $\delta^{98}Mo$, and Mn contents were correlated in the Koegas Group Mn IFs, and we summarize these relations, with possible interpretations, in Fig. 6.37. They interpret the low-$\delta^{56}Fe$, -$\delta^{98}Mo$, high-Mn compositions to reflect a prior history of extensive Fe^{2+}_{aq} oxidation at a distal location, which produced residual Fe^{2+}_{aq} that had negative $\delta^{56}Fe$ values ("flux H1" in Fig. 6.30), where Eh conditions reached sufficient levels to induce co-precipitation of Mn and Fe oxides. These oxides would have had very low $\delta^{98}Mo$ values due to the large negative $\Delta^{98}Mo_{sorb-aq}$ fractionation upon sorption of Mo (Barling and Anbar 2004; Wasylenki et al. 2008). Samples that have lower Mn contents are inferred to have formed at lower Eh conditions, perhaps below a redoxcline, where Fe–Mn redox cycling (see Eq. 6.14) produced high-$\delta^{56}Fe$ Fe^{3+}-hydroxides. Importantly, Kurzweil et al. (2016) infer that the absence of Mn oxides allowed incorporation of high-$\delta^{98}Mo$ seawater Mo into the IF precipitates, although the mechanism of this incorporation is not clear, nor is it clear that this would occur without isotopic fractionation, as required in their model. For the high-$\delta^{56}Fe$, -$\delta^{98}Mo$, low-Mn component, redox cycling via Eq. 6.14 may only produce limited Fe isotope changes, and in the absence of external sources of Mn or Fe, is a zero-sum issue. Experimental studies show that the Fe^{3+}-hydroxides produced by Fe–Mn redox cycling are lepidocrocite or goethite (e.g., Dellwig et al. 2010; Schaefer et al. 2017), reflecting phase conversion to crystalline forms. Although the

Fe^{2+}_{aq}-lepidocrocite Fe isotope fractionation factor has not been measured, the Fe^{2+}_{aq}-goethite $^{56}Fe/^{54}Fe$ fractionation is well established at −1.1‰ (Beard et al. 2010; Frierdich et al. 2014a). Based on a maximum molar Mn:Fe ratio of 1:1, the stoichiometry of Eq. 6.14 indicates that only half of the Fe^{2+}_{aq} is converted to FeOOH, which, based on isotopic mass balance using the Fe^{2+}_{aq}-goethite fractionation factor, only results in an increase in $\delta^{56}Fe$ for the FeOOH precipitate of ∼ 0.5 to 0.6‰.

Several additional interpretations are illustrated in Fig. 6.37. Although Kurzweil et al. (2016) do not favor a benthic source for Fe or Mo, such a source would provide Fe^{2+}_{aq} that has negative $\delta^{56}Fe$ values ("flux B" in Fig. 6.30), as discussed extensively in Sect. 5.4.4 of Chap. 5 (see also Fig. 5.18). Studies of modern marine sediments that have undergone microbial Mn reduction show that porewater Mo has very low $\delta^{98}Mo$ values due to release of sorbed Mo upon reduction, as much as 2‰ lower than seawater Mo (Goldberg et al. 2012). Respiration of C_{org} is also a source for low-$\delta^{98}Mo$ Mo, given the large Mo isotope fractionation between aqueous Mo and Mo sorbed organic C (King et al. 2018). In addition, as noted above, the presence of Mn reduction will tend to shift the $\delta^{56}Fe$ values of the benthic Fe flux to lower $\delta^{56}Fe$ values if C_{org} becomes limited, as well as the effects of oxidation of pore water Fe^{2+}_{aq} (see discussion in Sect. 5.4.4 in Chap. 5). Transport of a low-$\delta^{56}Fe$ Fe–Mn shuttle is one explanation, therefore, for the correlation between decreasing $\delta^{56}Fe$ and increasing Mn contents. Importantly, pore-water profiles in modern marine sediments show exponential relations in Mn^{2+} and Fe^{2+} contents (Thamdrup et al. 2000), consistent with the logarithmic relations between $\delta^{56}Fe$, $\delta^{98}Mo$, and Mn contents (Figs. 6.36 and 6.37).

Alternatively, the low $\delta^{98}Mo$ values may not reflect coupled Fe-Mo behavior but instead could in part be driven by mineralogical changes, where high-Mn samples will have a larger $\Delta^{98}Mo_{sorb-aq}$ fractionation than low-Mn samples (Barling and Anbar 2004; Wasylenki et al. 2008; Goldberg et al. 2009). The low-$\delta^{98}Mo$ values of the high-Mn samples could reflect Mo isotope exchange equilibrium with seawater Mo that had a $\delta^{98}Mo$ value of ∼2‰ (blue arrows and labels in Fig. 6.37), which lies within the range estimated for high-O_2 conditions at this time (Kurzweil et al. 2015a). The smaller $\Delta^{98}Mo_{sorb-aq}$ fractionations expected for the low-Mn samples produce a similar calculated $\delta^{98}Mo$ value for sea water, assuming Mo isotope exchange equilibria (green arrows and labels in Fig. 6.37). Finally, given the evidence for microbial Mn and Fe reduction in the Mn IFs (e.g., negative $\delta^{13}C$ values for carbonate), it is possible that the positive correlation between $\delta^{56}Fe$ and $\delta^{98}Mo$ reflects loss of low-$\delta^{56}Fe$ Fe^{2+}_{aq} upon Fe^{3+} reduction, and low-$\delta^{98}Mo$ aqueous Mo upon desorption during Mn (and Fe) reduction. Distinguishing among this wide variety of isotopic pathways will require additional data, including micro-scale C, O, and Fe isotope analysis, in addition to other tracers of water-column pathways such as Nd isotopes, or Sr isotope analysis of carbonates to assess equilibrium with seawater. Regardless of remaining uncertainties it seems clear that the Mn IFs record a substantial change in the Fe (and Mn, of course) geochemical cycle that is entirely consistent with increasing Eh conditions leading up to the GOE, likely reflecting increasing O_2 in the marine system. These concepts will arise again in our discussions of Mesoarchean IFs below.

6.5.6 Paleoproterozoic and Neoarchean Continental Margins: Relations Between Shales, Carbonate Platforms, and IFs

The 2.50 Ga Mt. McRae Shale of the Hamersley Basin (Australia) underlies the Brockman IF, and has figured prominently in studies of transient oxygenation events ("whiffs") before the GOE (e.g., Anbar et al. 2007; Kaufman et al. 2007). Anomalous enrichments in redox-sensitive trace elements such as Re and Mo in the Upper Shale

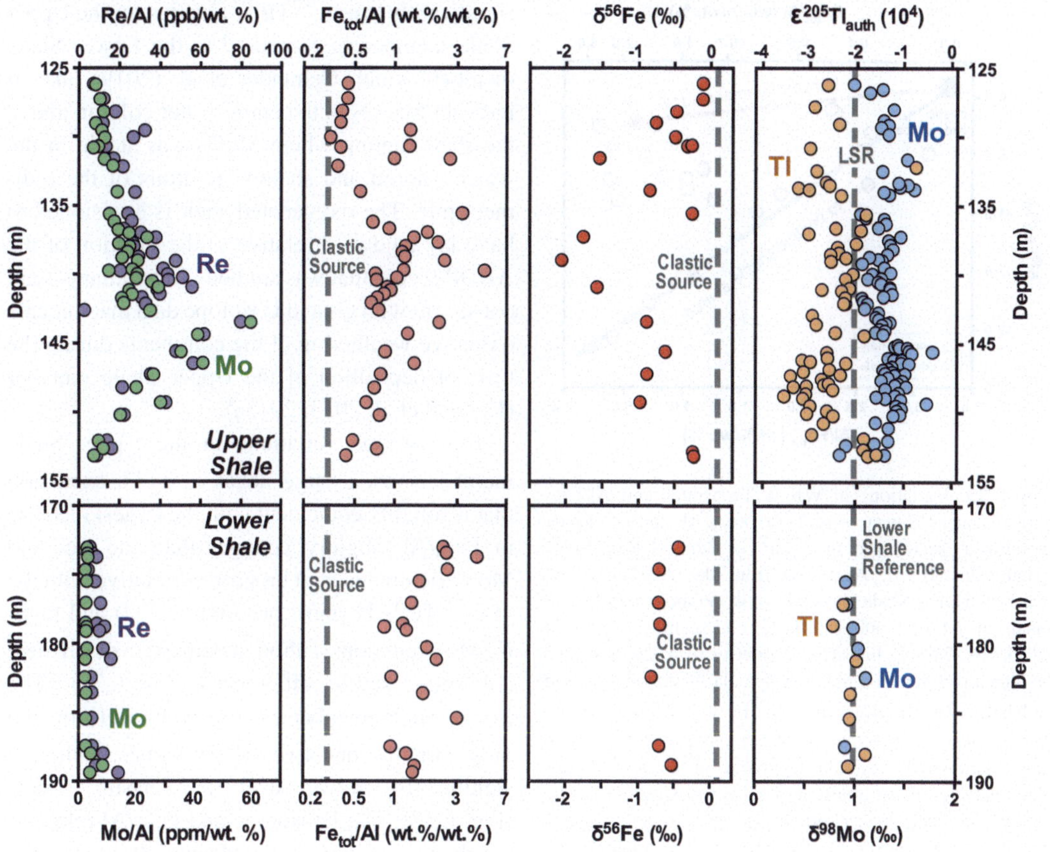

Fig. 6.38 Stratigraphic variations in redox-sensitive trace elements, Fe contents, and Fe, Mo, and Tl isotope compositions for the 2.50 Ga Mt. McRae Shale (Hamersley Basin, Australia) from the ABDP-9 core for the Upper Shale member (top panels) and Lower Shale member (lower panels). Reactive Fe species indicate the Upper Shale member was deposited in euxinic conditions (high Fe_{HR}/Fe_{tot}, high Fe_{PY}/Fe_{HR}), whereas the Lower Shale member was deposited under ferruginous conditions (high Fe_{HR}/Fe_{tot}, low Fe_{PY}/Fe_{HR}). For panels showing multiple datasets, upper scales are for Re/Al and $\varepsilon^{205}Tl_{auth}$, and lower scales are for Mo/Al and δ^{98}Mo. For reference, the average upper crustal ratios for Re/Al (ppb/wt%) is 0.12 and Mo/Al (ppm/wt%) is 0.19, which are indistinguishable from the zero point of the vertical axes at this scale. Several reference compositions are shown by vertical dashed lines: the estimated Fe_{tot}/Al ratio for the clastic input for this sequence is 0.3 (see Fig. 6.39), and the δ^{56}Fe value of the clastic source is taken to be 0.09‰. δ^{98}Mo values reported relative to the "NIST + 0.25‰" reference frame, and $\varepsilon^{205}Tl_{auth}$ values are the $^{205}Tl/^{203}Tl$ ratio calculated for the authigenic component, reported relative to NIST 997, in parts per 10^4. Reference frames for Tl and Mo isotope data are shown based on the Lower Shale member ("LSR"), following the approach taken by Ostrander et al. (2019). Note that Fe_{tot}/Al is plotted on log scale. Data from Anbar et al. (2007), Reinhard et al. (2009), Duan (2010), Duan et al. (2010), Kendall et al. (2015a), and Ostrander et al. (2019)

member of the Mt. McRae Shale (Fig. 6.38) correlate with a decrease in δ^{34}S values for pyrite and a change in Δ^{33}S-Δ^{36}S relations, interpreted to record fundamental changes in the S cycle that accompanied enrichment in Re and Mo during transient O_2 increases (Anbar et al. 2007; Kaufman et al. 2007). Studies of reactive Fe inventories of the Mt. McRae Shale suggest that the Lower Shale member was deposited in relatively anoxic and ferruginous conditions, whereas deposition of the Upper Shale member occurred under euxinic conditions, implying a mid-depth euxinic zone that was overlain by oxygenated waters and underlain by ferruginous

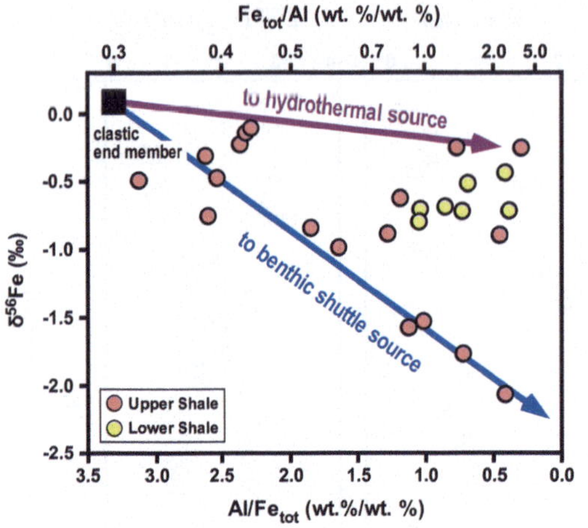

Fig. 6.39 Variations in WR $\delta^{56}Fe$ values and Al/Fe_{tot} ratios (linear lower scale) or Fe_{tot}/Al ratios (upper non-linear scale) for the 2.50 Ga Mt. McRae Shale (Hamersley Basin, Australia) from the ABDP-9 core. Data for Upper Shale shown in pink symbols, and Lower Shale in yellow symbols. Clastic end member shown in black box. Mixing lines (linear relative to Al/Fe_{tot}) shown in blue arrow for mixing between a clastic component and a benthic Fe shuttle source ($\delta^{56}Fe \sim -2.3‰$), and in purple arrow between a clastic component and a non-fractionated hydrothermal source ($\delta^{56}Fe \sim -0.3‰$). Data between these trends may reflect mixing between a non-fractionated hydrothermal and benthic Fe sources, variably fractionated hydrothermal sources, or benthic Fe sources that had less negative $\delta^{56}Fe$ values. Notably, the very low-$\delta^{56}Fe$ benthic Fe source corresponds with the maximum enrichment in Re and Mo (see Fig. 6.38). Data and interpretations from Duan (2010)

(low-sulfide) waters (Reinhard et al. 2009; Raiswell et al. 2011). This chemically and redox-stratified model is very similar to those proposed for the Neoproterozoic oceans (see Sect. 6.4.2).

Isotopic studies of the Mt. McRae shale are consistent with transient oxygenation. $\delta^{98}Mo$ values for the Lower Shale are moderately positive, continuing a trend of increasing $\delta^{98}Mo$ with decreasing age seen in the underlying units of the Hamersley Group (Kurzweil et al. 2015a), and markedly increase in the Upper Shale member to $\delta^{98}Mo$ values of $\sim +1.7‰$ (Fig. 6.38). This is interpreted to record an increase in MnO_2 deposition elsewhere in the basin, indicating higher Eh conditions (Duan et al. 2010; Ostrander et al. 2019). New data from Tl isotopes indicate significantly lower $^{205}Tl/^{203}Tl$ ratios for the Upper Shale member as compared to the Lower Shale member, which Ostrander et al. (2019) take to indicate free O_2 in the entire water column above the shelf region, where MnO_2 was stable in the water column and shallow portions of the sediment pile. The oxygenated shelf is envisioned to have lay landward relative to the location of the ABDP-9 core that was studied. These changes are consistent with Os and U isotope data that indicate oxidative weathering of the continents during the time of deposition of the Upper Shale member (Kendall et al. 2013, 2015a).

Iron isotope variations in the Upper Shale member show a large negative $\delta^{56}Fe$ excursion that is roughly coincident with the largest increase in Fe_{tot}/Al, slightly younger than the Re- and Mo-enrichments, and broadly correlative with the low-$^{205}Tl/^{203}Tl$ (most negative $\varepsilon^{205}Tl$) and high-$\delta^{98}Mo$ excursions, although perhaps in detail there are some subtle differences (Fig. 6.38). The Lower Shale member has relatively uniform, and only slightly negative $\delta^{56}Fe$ values, although significantly lower than the clastic source (Fig. 6.38). The Fe isotope and Fe_{tot}/Al relations are shown in Fig. 6.39, and Duan (2010) interpret these to record mixing between a clastic source and two authigenic components: (1) a benthic Fe source that had a very low $\delta^{56}Fe$ value of $\sim -2.3‰$ (blue arrow in Fig. 6.39) and (2) a hydrothermal source that had a $\delta^{56}Fe$ value between 0.1 and $-0.3‰$ (purple arrow). The data for the Lower Shale member is most consistent with a hydrothermally-dominated authigenic component, and some high-Fe_{tot}/Al and slightly negative $\delta^{56}Fe$ samples from the Upper Shale member were also interpreted to record a hydrothermal authigenic component by Duan (2010). Data that plot between these mixing lines might record an evolved hydrothermal fluid that had undergone oxidation of Fe^{2+}_{aq} or a mixture of benthic and hydrothermal sources, or a benthic source that was only modestly negative in its $\delta^{56}Fe$ value.

The contrasting interpretations for the Mt. McRae Shale and the slightly younger Koegas and

Hotazel Mn IFs highlight how inferences of Fe sources may in part be dependent on lithology, where identification of authigenic Fe enrichment in shales makes use of clastic reference frames (e.g., Fe_{tot}/Al) and reactive Fe inventories. A correlation between decreasing $\delta^{56}Fe$ values and increasing Fe_{tot}/Al and Fe_{HR}/Fe_{tot} in shales has almost universally been taken to indicate a benthic Fe shuttle driven by DIR (see, for example, Sect. 6.4.2). Although in principle extensive oxidation of hydrothermal Fe^{2+}_{aq} might be invoked to produce highly negative $\delta^{56}Fe$ values for the Mt. McRae Shale, this is not a logical explanation for high authigenic Fe enrichment (high Fe_{tot}/Al) that has highly negative $\delta^{56}Fe$ values, as extensive oxidation of Fe^{2+}_{aq} would have rapidly depleted the Fe contents of the hydrothermal component. Such extensive oxidation and decrease in Fe^{2+}_{aq} contents would be hard to detect relative to the clastic Fe_{tot}/Al ratios. In addition, the hydrothermal component inferred for the Mt. McRae Shale was not evolved, as it had only slightly negative $\delta^{56}Fe$ values at high Fe_{tot}/Al ratios, indicting limited oxidation of Fe^{2+}_{aq}. This is an important observation, as the Tl and Mo isotope data suggest that deposition of the Mt. McRae Shale occurred under the same highly oxidizing conditions as deposition of the Mn IFs. Ironically, therefore, the origin of authigenic enrichment of Fe (benthic Fe shuttle, hydrothermal) may in some cases be better determined in clastic rocks than the chemical sedimentary rocks (IFs) due to the lack of an Fe enrichment reference frame in the IFs.

A benthic Fe source for the Mt. McRae Shale that has highly negative $\delta^{56}Fe$ values during the inferred time of MnO_2 deposition (based on Mo and Tl isotopes) directly follows the relations implied by Eqs. 6.12 and 6.13 above. If MnO_2 was stable in both the water column and shallow sediment sections on continental shelves during Mt. McRae Shale deposition, as proposed by Ostrander et al. (2019), microbial Mn reduction would have consumed large quantities of C_{org} before it was available to DIR. As discussed above, this suggests that the Fe isotope composition of a benthic Fe shuttle would shift to more negative $\delta^{56}Fe$ values, reflecting both the limitation of extent of DIR reduction in the presence of Mn reduction, and the greater likelihood that higher Eh conditions would oxidize escaping porewater Fe^{2+}_{aq}. The Mt. McRae Shale data, therefore, provide key insights into the inter-relations of the Mn and Fe cycles in the time leading up to the GOE and the changes in Fe isotope compositions that result.

The euxinic conditions inferred for deposition of the Upper Shale member of the Mt. McRae Shale indicate a sulfide trapping mechanism for the benthic Fe shuttle, similar to that invoked for the modern Black Sea (Severmann et al. 2008). A euxinic condition also favors complete sequestration of the Mo isotope composition of seawater Mo (e.g., Kendall et al. 2017), producing an authigenic component that has low $\delta^{56}Fe$ and high $\delta^{98}Mo$ values (Fig. 6.38). In contrast, the negative $\delta^{56}Fe$ component for the Mn IFs of the Koegas Group has a near-zero $\delta^{98}Mo$ value (Fig. 6.37), which Kurzweil et al. (2016) interpret to reflect incorporation of a low-$\delta^{98}Mo$ (relative to sea water) sorbed component in settling MnO_2 particles, a conclusion supported by the fact that the low-$\delta^{98}Mo$ samples are Mn rich. Assuming equilibrium between seawater Mo and Mo sorbed to MnO_2, the measured $\delta^{98}Mo$ values of the Mn-rich samples of the Koegas Group IFs suggest a $\delta^{98}Mo$ for seawater of ~+1.8 to +2.0‰ (Fig. 6.37), as noted above, which is similar to that inferred from the euxinic sections of the Mt. McRae Shale. These relations highlight the well-established differences in Mo isotope compositions recorded by euxinic (sulfide) systems relative to oxides (e.g., Kendall et al. 2017), and show the potential decoupling, and independent tests, that exist for Fe-Mo isotope variations as a function of lithology.

We continue our discussion of Neoarchean continental margins by turning to the Ghaap Group (and correlative Chuniespoort Group) of South Africa, which records marine sedimentation for the entire Neoarchean in the Transvaal and Griqualand West basins. There are a number

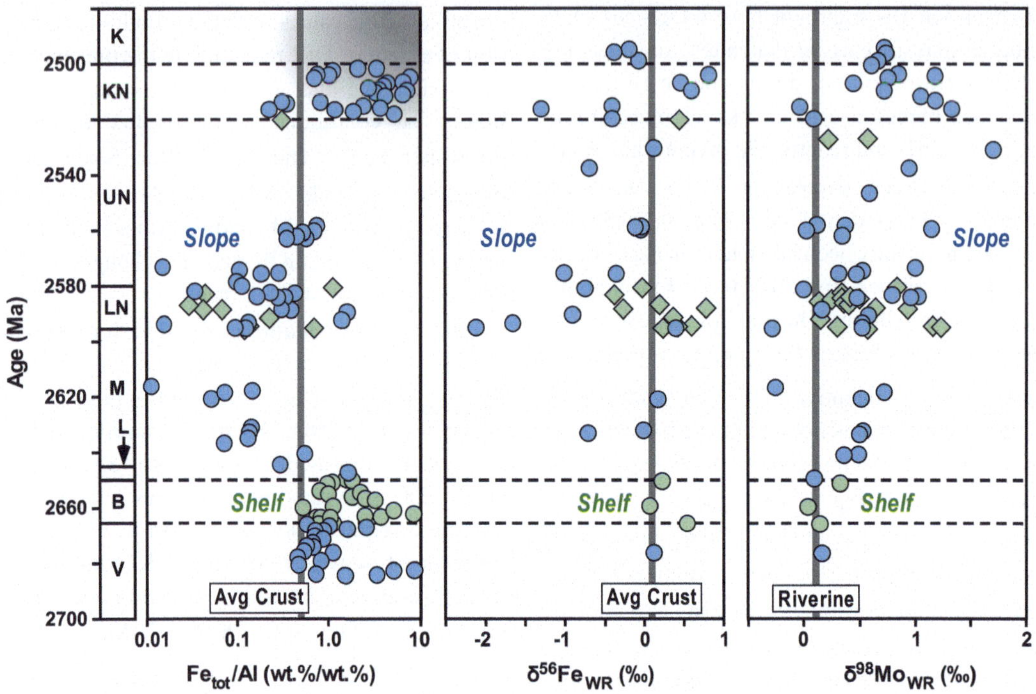

Fig. 6.40 Stratigraphic variations for Fe_{tot}/Al ratios and $\delta^{56}Fe$ and $\delta^{98}Mo$ ("NIST + 0.25‰" scale) values for shales of the Ghaap Group, South Africa. Data colored blue for samples from the continental slope and green from the continental shelf. Circles from the GKP01 and GKF01 cores (Schröder et al. 2006), which generally represent slope paleoenvironments, and diamonds from the KMF-5 and BH-1 cores and Kuruman Kop outcrop (Eroglu et al. 2015), which generally represent shelf paleoenvironments. Data brought to a common timescale using the correlations of Czaja et al. (2012) and Eroglu et al. (2018). Stratigraphic units based on the GKP01 core: V, Vryburg; B Boomplaas; L, Lokamonna; M, Monteville; LN, Lower Nauga; UP, Upper Nauga; KN, Klein Naute; K, Kuruman. The upper part of the Klein Naute is temporally correlative with the Mt. McRae Shale of the Hamersley Basin (Fig. 6.38). Major transitions shown in horizontal dashed lines, including the shallow shelf setting of the Boomplaas Formation, followed by marine transgression during Lokamona deposition, and the major transgression recorded in the Klein Naute and Kuruman formations. Fe_{tot}/Al ratios shown, with average crust reference (0.5; vertical grey line). Fe_{tot}/Al ratios for parts of the Klein Naute and all Kuruman IF samples plot off scale at $Fe_{tot}/Al > 10$, schematically shown by shaded grey region. Average crustal $\delta^{56}Fe$ shown and estimated riverine input shown for $\delta^{98}Mo$ (Wille et al. 2007) in vertical grey lines. High-Fe_{tot}/Al and near-zero or slightly positive $\delta^{56}Fe$ and $\delta^{98}Mo$ values in the lower units suggest deposition under anoxic conditions, whereas low-Fe_{tot}/Al, -$\delta^{56}Fe$ but positive-$\delta^{98}Mo$ values in the middle and upper sections record deposition in increasingly oxygenated conditions. Marine transgression recorded in the Klein Naute Formation indicates ferruginous deep-water conditions, but elevated $\delta^{98}Mo$ values indicate overlying waters were oxic. Data from Schröder et al. (2006), Wille et al. (2007), Czaja et al. (2012), and Eroglu et al. (2015, 2018)

of lines of evidence that the Hamersley and Transvaal basins were contiguous or at least formed on the same continental margin (Cheney 1996; Beukes and Gutzmer 2008). The Transvaal (and Griqualand West) Basin, however, records excellent basin-slope-shelf transects and experienced a lower metamorphic grade, compared to the Hamersley Basin, although recent drilling of the Hamersley Group has been aimed at better understanding basin-slope-shelf transitions (e.g., Koehler et al. 2018). Relative to the preceding discussion, the Mt. McRae Shale is temporally correlative with the Klein Naute Formation in the upper part of the Ghaap Group.

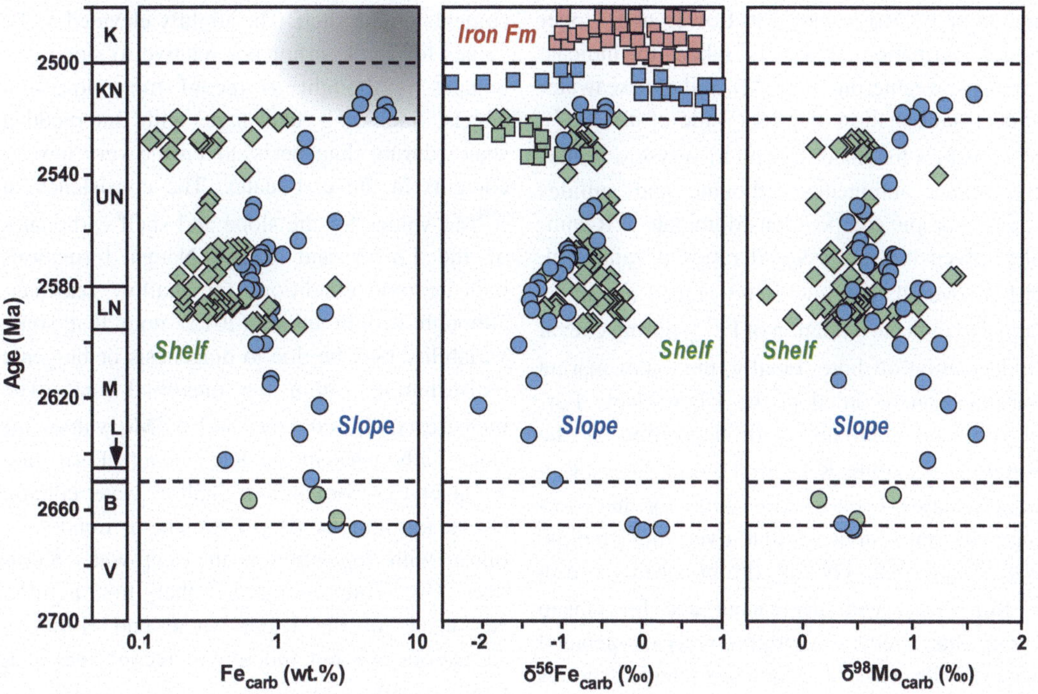

Fig. 6.41 Stratigraphic variations for carbonates of the Ghaap Group, South Africa, showing carbonate Fe contents (log scale) and Fe and Mo isotope compositions (Fe_{carb}, $\delta^{56}Fe_{carb}$, and $\delta^{98}Mo_{carb}$; "NIST + 0.25‰" scale). Data symbols and colors, as well as stratigraphic section and timescale, same as in Fig. 6.40. Additional Fe isotope data shown in squares, with pink showing data from the Kuruman IF. Shaded grey area shows off-scale Fe contents for some Klein Naute and all Kuruman IF carbonates. Fe_{carb} contents are markedly lower for shelf carbonates relative to slope carbonates, indicating greater exchange of Fe with the open ocean, as well as deeper water, for the slope carbonates. The $\delta^{56}Fe_{carb}$ and $\delta^{98}Mo_{carb}$ values vary antithetically for the slope carbonates, indicating the influence of IF deposition off shore (Czaja et al. 2012), whereas these relations are more scattered for the shelf carbonates, indicating less exchange with the open oceans and local continental influences (Eroglu et al. 2018). The scatter in $\delta^{56}Fe_{carb}$ values for the Kuruman IF reflect variable roles for microbial diagenesis (Heimann et al. 2010). Data from Johnson et al. (2003), von Blanckenburg et al. (2008), Heimann et al. (2010), Voegelin et al. (2010), Czaja et al. (2012), and Eroglu et al. (2015, 2018)

Figure 6.40 illustrates Fe_{tot}/Al, δ^{56}Fe, and δ^{98}Mo variations for shales from the Ghaap Group, divided by samples from the paleo slope and shelf. The oldest unit, the Vryburg Formation, represents a transgressive clastic sequence that records initial flooding of the Kaapvaal Craton, followed by deposition of the shallow-water, stromatolite-rich Boomplaas Formation (Schröder et al. 2006; Sumner and Beukes 2006). The relatively high Fe_{tot}/Al ratios for shales from these units suggests deposition under anoxic conditions in deep- and shallow-water environments, and the near-zero δ^{56}Fe and δ^{98}Mo values indicate little redox cycling, consistent with the interpretation of relatively anoxic conditions (Wille et al. 2007; Czaja et al. 2012). Deep-water sedimentation is represented by the overlying Lokammona and Monteville formations, reflecting a slope/turbidite environment, and strongly decreasing Fe_{tot}/Al and increasing δ^{98}Mo is thought to record progressively increasing oxic conditions (Wille et al. 2007). Large changes in Fe and Mo isotope compositions for shales occur in the Lower and Upper Nauga formations. The low Fe_{tot}/Al ratios and high $\delta^{98}Mo$ values for Lower and Upper Nauga formation shales is taken to indicate extensive weathering (high Al), and oxic seawater at deep and shallow depths (Wille et al. 2007). $\delta^{56}Fe$ values for shales extend to very low values and reflect authigenic carbonate and sulfides whose isotopic compositions dominate WR samples, given the low Fe_{tot}/Al ratios (Czaja et al. 2012); overall, the low-$\delta^{56}Fe$ values are interpreted to indicate oxidation of Fe^{2+}_{aq} during coeval IF deposition offshore. Finally, the major marine transgression recorded in the Klein Naute Formation, and continued with deposition of the Kuruman IF, coincides with a strong increase in Fe_{tot}/Al ratios. High-$\delta^{98}Mo$ values for these ferruginous units indicate that overlying shallow waters were oxic. Overall, the broad increase in $\delta^{98}Mo$ values with decreasing age for Ghaap Group shales reflects a progressively oxygenated ocean (Wille et al. 2007), and this interpretation is supported by changes in reactive Fe inventories in shales from the upper part of the section (Kendall et al. 2010).

There is an exceptionally large database for Fe and Mo isotope compositions of Ghaap Group carbonates (Fig. 6.41), which provides a more direct record of water column compositions than shales. Assuming the Fe and Mo isotope compositions of the low-Fe, Ca–Mg carbonates largely record seawater compositions (e.g., Voegelin et al. 2010; Czaja et al. 2012), the decrease in $\delta^{56}Fe$ values through the Monteville Formation that is accompanied by an increase in $\delta^{98}Mo$ values indicates progressively more oxic conditions in the water column, consistent with the data from coeval shales previously noted. Comparison of slope and shelf carbonates in the Lower and Upper Nauga formations show overlapping Fe and Mo isotope compositions, but much lower Fe contents for the shelf samples (Fig. 6.41). Eroglu et al. (2018) modeled this concentration contrast, coupled with Fe isotope variations, to indicate seawater Fe^{2+}_{aq} contents of 10^2–10^3 μM on the slope but three-fold lower contents on the shelf. The slightly elevated $\delta^{56}Fe$ values for shelf carbonates relative to slope carbonates is thought to record more localized effects, including exchange with interbedded shales during diagenesis, given the very low Fe contents in the carbonates. The consistency in $\delta^{98}Mo$ values for the slope and shelf carbonates of the Lower and Upper Nauga formations indicate oxic conditions for shallow seawater, although Eroglu et al. (2015) note that some variability may be due to diagenesis or biogenic redistribution within the microbial mats. The increases in Fe contents and $\delta^{98}Mo$ values for slope carbonates in the Klein Naute Formation are taken to record an oxic shallow water column above ferruginous deeper waters, a trend continued with deposition of the deep-water Kuruman IF. Note, however, that the high-Fe carbonates of the Klein Naute and Kuruman formations are not thought to record seawater, but instead the wide range in $\delta^{56}Fe$ values in part reflects production of siderite/ankerite by DIR (see Fig. 6.33). This interpretation is supported by their highly radiogenic $^{87}Sr/^{86}Sr$ ratios, which indicate that they did not form in equilibrium with seawater (Heimann et al. 2010; Johnson et al. 2013a). No Mo isotope data have been published for the Kuruman IF carbonates.

Czaja et al. (2012) recognized that covariation in $\delta^{56}Fe$ and $\delta^{98}Mo$ values for the low-Fe Ghaap Group carbonates provides strong constraints on Mo and Fe pathways and the extent of oxygenation of the photic zone. Covariation in Fe and Mo isotopes in carbonates are particularly useful because there are a limited range of values for key parameters that may produce such trends, including seawater Fe and Mo contents, sorption coefficients, and fractionation factors for oxidation of Fe^{2+}_{aq} and sorption of Mo to Fe^{3+}-hydroxides. Czaja et al. (2012) modeled the coupled Fe-Mo isotope variations through oxidation of Fe^{2+}_{aq} to form Fe^{3+}-hydroxide precipitates that accumulated as IFs offshore. They pointed to the coeval Marra Mamba IF in the Hamersley Basin as an

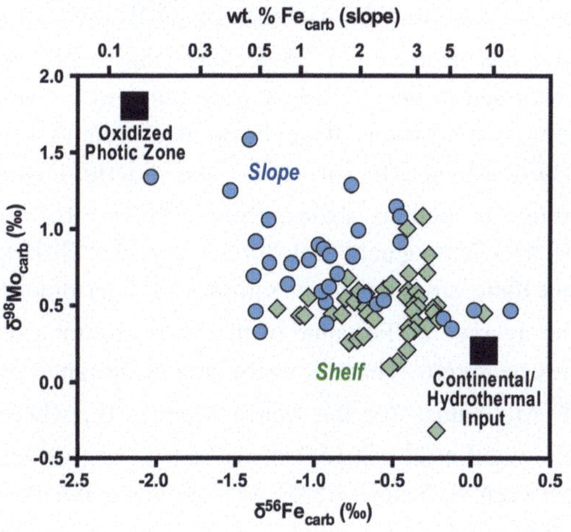

Fig. 6.42 Variations in $\delta^{56}Fe_{carb}$ and $\delta^{98}Mo_{carb}$ ("NIST + 0.25‰" scale) values for Ghaap Group carbonates, South Africa. Symbols and data sources from Fig. 6.41. Scale for carbonate Fe contents at top from log regression of slope carbonates ($R^2 = 0.5$; Czaja et al. 2012). The $\delta^{56}Fe_{carb}$-$\delta^{98}Mo_{carb}$ array for slope carbonates is interpreted by Czaja et al. (2012) to indicate variable contributions from continental/hydrothermal sources and a low-Fe, low-δ^{56}Fe, high-δ^{98}Mo component that reflects an oxidized photic zone. Using the dispersion/reaction model shown in Fig. 6.29 to simultaneously account for changes in $\delta^{56}Fe_{carb}$ and $\delta^{98}Mo_{carb}$ values allowed Czaja et al. (2012) to estimate photic zone O_2 contents of ~35 μM, which is ~12% Present Ocean Level (POL). Such high O_2 levels in the photic zone are consistent with the inferences of water column oxygenation during deposition of the Mt. McRae Shale (time correlative with the upper Klein Naute Formation) based on Mo-Tl isotope variations (see Fig. 6.38; Ostrander et al. 2019), as well as the broad inferences of oxygenation based on redox-sensitive trace elements measured for the upper Ghaap Group (Kendall et al. 2010). The persistence of MIF-S through the Ghaap Group, however, indicates that atmospheric O_2 contents remained low ($<10^{-5}$ PAL; Ono et al. 2009), highlighting decoupling of photic zone and atmospheric O_2 levels. The $\delta^{56}Fe_{carb}$-$\delta^{98}Mo_{carb}$ variations for the shelf carbonates are additionally interpreted to reflect local continental influences and possibly diagenetic effects (Eroglu et al. 2015, 2018)

example repository of the Fe-rich deposits that would provide mass-balance to the Fe-poor seawater reservoir on the Ca–Mg carbonate platform. We have previously discussed the dispersion/reaction model of Czaja et al. (2012) primarily in terms of the Fe isotope compositions of the accumulated Fe^{3+}-oxide/hydroxide precipitates during formation of IFs (Sect. 6.5.3 and Fig. 6.29), noting that application of the model using Fe-rich precipitates becomes insensitive to O_2 as oxygen levels become high. If, however, our focus changes to a lithology that reflects direct sampling of Fe^{2+}_{aq} in the low-Fe photic zone under conditions of high O_2, the model may be well applied to understanding processes in an oxic photic zone (see Sect. 6.5.3).

The covariation in δ^{56}Fe and δ^{98}Mo for Ca–Mg (Fe-poor) platform carbonates (Fig. 6.42) is best explained by relatively high photic zone O_2 contents of ~35 μM, which is ~12% of Present Ocean Levels (POL). The highest O_2 contents are inferred for the largest negative δ^{56}Fe and largest positive δ^{98}Mo excursions, which occur twice in the Ghaap Group, first during deposition of the Monteville Formation, and second with the deposition of the Klein Naute Formation (Fig. 6.41). The second excursion in the Klein Naute Formation is time-correlative with deposition of the Mt. McRae Shale of the Hamersley Basin, which, as discussed above, is inferred to record a shelf environment where the entire water column contained significant O_2 (see Fig. 6.38; Ostrander et al. 2019). If the stratigraphic cycles in Fe and Mo isotope compositions of Fe-poor carbonates reflect seawater compositions, these results suggest periods of high- and low-O_2 conditions in the photic zone in the Neoarchean. In contrast, the consistent MIF-S throughout the Ghaap Group indicates that atmospheric O_2 contents remained low, less than 10^{-5} PAL (Ono et al. 2009), highlighting the decoupling that may occur between oxygenation of the photic zone and the atmosphere prior to the GOE (e.g., Olson et al. 2013). Based on δ^{13}C values for slope and shelf carbonates in the Ghaap Group, the slope samples are more likely to record open-ocean isotopic compositions than the shelf samples, which have some added complexity due to restricted water exchange (Fischer et al. 2009; Eroglu et al. 2017).

The large changes in δ^{56}Fe and δ^{98}Mo values that are measured in the Ghaap Group of South Africa also occur in similar-age rocks of the Hamersley Basin (Australia), although the records show distinct differences. The Jeerinah Formation of the Hamersley Basin spans approximately 100 m.y. of deposition and is temporally correlative with the Vryburg through lower Monteville formations of the Ghaap Group. Czaja et al. (2010) noted consistently negative δ^{56}Fe values and high Fe contents for deep-water samples of the Jeerinah Formation, but positive δ^{56}Fe values and low Fe contents for shallow-water samples, which they interpreted to indicate a DIR-driven benthic shuttle, analogous to the Fe isotope mass balances suggested for the Black Sea (Severmann et al. 2008; see also Fig. 6.30). Based on very low δ^{13}C values for C_{org} in the Jeerinah Formation, which Eigenbrode and Freeman (2006) interpreted to have been produced by anaerobic oxidation of methane or aerobic methanotrophy (see also Hayes 1994), Czaja et al. (2010) suggested that DIR may have been coupled to these other metabolisms, which would indicate a highly complex microbial ecology at that time. Such conclusions align with recent evidence for oxygenation during deposition of the Jeerinah Formation based on Se concentrations and N and Se isotopes (Koehler et al. 2018). In detail, however, the oxygenation recorded in the Jeerinah Formation is earlier than the first negative-δ^{56}Fe, positive-δ^{98}Mo excursion in the Ghaap Group, possibly indicating heterogeneity in oxygenation environments, ambiguity in proxy interpretation, errors in timescales, or a combination of such issues. These issues notwithstanding, the broad increase in δ^{98}Mo values for Ghaap Group shales (Fig. 6.40) is broadly correlative with a similar increase in time-correlative units in the Hamersley Basin (Kurzweil et al. 2015a).

Czaja et al. (2012) point to the mid-Neoarchean Marra Mamba IF of the Hamersley Basin as an example of the Fe^{3+}-oxide/hydroxide inventory that is required to provide Fe and Mo isotope mass balance for the Ghaap Group carbonates plotted in Fig. 6.42. Isotopic mass balance for Fe shows that highly negative δ^{56}Fe values for a low-Fe^{2+}_{aq} photic zone may be balanced by an IF inventory that has near-zero δ^{56}Fe values (Fig. 6.29), as discussed in Sect. 6.5.3. A wide range of Fe isotope compositions have been measured in the Marra Mamba IF for oxide- and siderite-facies units, as well as shales, from $\delta^{56}Fe = -0.7$ to +0.4‰ (Yamaguchi et al. 2005; Czaja et al. 2010), but there are insufficient samples to determine if the average δ^{56}Fe value of the Marra Mamba IF lies near zero. There is a similarly wide range of δ^{98}Mo values for the Marra Mamba IF, where Kurzweil et al. (2015a) measured δ^{98}Mo values between +0.5 and +1.5‰. At first glance, positive δ^{98}Mo values for the Marra Mamba IF would seem to be inconsistent with the Mo isotope mass balance required for the high-δ^{98}Mo carbonates of the Ghaap Group (Fig. 6.42). Kurzweil et al. (2015a) therefore suggested that a low-δ^{98}Mo sink in the Hamersley Basin was more likely to be C_{org} rather than Fe^{3+}-hydroxides because the Mo contents of the oxide-facies IFs are so low.

Looking for the IF record that mass-balances the Ghaap Group carbonates is complicated by the issue of changes in Mo isotope compositions of IFs during diagenesis. Microbial respiration in the Marra Mamba IF was extensive, as indicated strong correlations between $\delta^{13}C_{carb}$ and $\delta^{18}O_{carb}$ (Baur et al. 1985; Kurzweil et al. 2015a), similar to those seen in the Kuruman IF that are interpreted to record DIR (see Fig. 6.35). As discussed in Sect. 6.5.4, reaction between initial Fe^{3+}-hydroxides and C_{org} in the primary IF sediment will produce magnetite and siderite, and this is likely to be accompanied by release of low-δ^{98}Mo sorbed Mo, producing high-δ^{98}Mo values in the residual IF that is not representative of the original sediment. Of course, this then raises the question of where did the low-δ^{98}Mo Mo go in such a scenario. Further complexity arises if we consider the likely spatial distribution of microbial diagenesis in the IF sediment. It seems probable that this will be greatest in near-coastal regions, assuming C_{org} delivery is highest in shelf and slope settings due to high nutrient abundances, relative to the deep ocean. It is possible, therefore, that in the case of sorption of Mo to Fe^{3+}-hydroxides, the low-

$\delta^{98}Mo$ Fe^{3+}-hydroxide inventory that mass-balances high-$\delta^{98}Mo$, low-Fe carbonates are Fe^{3+}-hydroxides deposited on oceanic crust further from the slope, where C_{org} delivery was low; this record, unfortunately, would probably be lost via subduction. This is a difficult (if not impossible) hypothesis to test, and hence is not satisfactory. But these issues raise an important flag: development of complete isotope mass balance models for marine basins is difficult where uncertainties remain in how representative are the exposed sections and the possibility of missing important offshore sedimentary reservoirs.

We conclude our discussion in this section by highlighting several studies of early Neoarchean IFs. The large early Paleoproterozoic IFs discussed above (Sect. 6.5.4) formed in large passive-margin basins, whereas early Neoarchean IFs tend to be smaller and more directly associated with marine hydrothermal and volcanic systems, sometimes referred to as "Algoma-type" IFs (e.g., Gross 1980). A number of Neoarchean IFs have positive $\delta^{56}Fe$ values that have been taken to indicate partial oxidation of hydrothermal Fe^{2+}_{aq} under limited O_2 contents (Fabre et al. 2011; Planavsky et al. 2012a; Czaja et al. 2018). Pyrite is abundant in many Algoma-type Neoarchean IFs, again in contrast to the large early Paleoproterozoic IFs where pyrite is rare, and detailed studies show that hydrothermal pyrite may have positive $\delta^{56}Fe$ values that record equilibrium Fe isotope fractionation (Czaja et al. 2018). This is a much different interpretation than that proposed for the positive $\delta^{56}Fe$ values of pyrite in Neoarchean shales, for example, which were inferred to reflect sulfidation of Fe^{3+}-hydroxides that formed by partial oxidation of Fe^{2+}_{aq} under low-O_2 conditions (see discussion in Sect. 6.4.1). Other Neoarchean IFs, however, have near-zero average $\delta^{56}Fe$ values that are thought to record complete oxidation of Fe^{2+}_{aq} under higher-O_2 conditions (Mendes et al. 2017). And finally, highly negative $\delta^{56}Fe$ values of −1 to −2‰ have been reported for Mn-rich IFs from exhalative deposits of the Abitibi greenstone belt (Canada); these are taken to reflect extensive oxidation of hydrothermal Fe^{2+}_{aq} (Hiebert et al. 2018), and hence high-O_2 contents via a "low $\delta^{56}Fe$" signal (e.g., Fig. 6.30). The data from the Abitibi greenstone belt show remarkable similarities to the Mn-IF deposits of early Paleoproterozoic age (Koegas Group, Hotazel IF; see Sect. 6.5.5). An oxidation model for the highly negative $\delta^{56}Fe$ values faces the same mass-balance issues discussed above in Sect. 6.5.3 in terms of producing Fe-rich inventories that have low-$\delta^{56}Fe$ values, although the smaller size of some of Algoma-type IFs may mitigate this concern compared to the very large early Paleoproterozoic IFs. The Fe isotope data from the early Neoarchean IFs therefore paints an inconsistent picture: some IFs are thought to have formed under low-O_2 conditions, others under very high-O_2 conditions, particularly those that formed Mn deposits.

6.6 Precambrian Earth: The Early Archean Record

Our review of Fe redox thus far has shown multiple lines of evidence for oxygenation of the shallow oceans in the Neoarchean and early Paleoproterozoic, prior to the GOE. The extensive stratigraphic record for this time period, coupled with relatively low grades of regional metamorphism for many sections (e.g., <150 °C for the Ghaap Group), has allowed us to combine multiple redox proxies, which greatly strengthens interpretations. As we move deeper into geologic time, however, the rock record becomes sparser and the metamorphic grade higher, clouding our view. Below we highlight key marine sedimentary sequences of Mesoarchean and Paleoarchean age that demonstrate very ancient Fe redox cycling. We conclude this section with discussion of the Eoarchean, which is represented by high-grade metamorphic terranes that add significant additional complexity to our interpretations but are important to the question of when Fe redox cycling began on Earth.

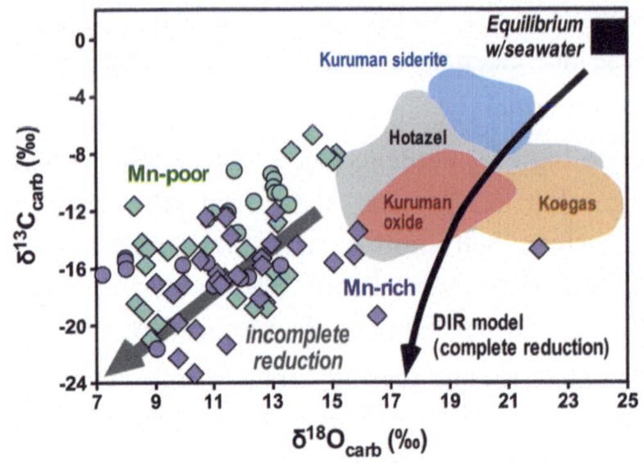

Fig. 6.43 Variations in $\delta^{18}O$ (SMOW scale) and $\delta^{13}C$ (PDB scale) values for Mesoarchean carbonates in Fe-rich shales and IFs from the Witwatersrand (circles) and Pongola (diamonds) supergroups, southern Africa. Samples that are Mn-rich (>1 wt% Mn) shown in purple colors, and those that are Mn-poor are shown in green colors. For comparison, fields shown for early Paleoproterozoic IFs (siderite- and oxide-facies of the Kuruman IF, Koegas Group Mn IFs, and Hotazel Mn IF) from Fig. 6.35. Also shown from Fig. 6.35 is box for equilibrium precipitation of siderite from seawater, as well as model for DIR from Heimann et al. (2010) based on Eqs. 6.7 through 6.9 in Sect. 6.5.4. The DIR model assumes complete reduction of $Fe(OH)_3$ and O isotope equilibration between porewater (produced by $Fe(OH)_3$ dehydration) and siderite. Incomplete reduction and/or decreased equilibration with pore fluid will shift $\delta^{18}O$ values, at a given $\delta^{13}C$ value to the left (lower $\delta^{18}O$), reflecting a greater inheritance from the $\delta^{18}O$ value of $Fe(OH)_3$. Although the very low $\delta^{18}O$ values for the Mesoarchean carbonates could in part be explained by increased temperatures during diagenesis, the magnitude of the lowering in $\delta^{18}O$ values exceeds that expected for temperature alone (see Fig. 6.35). Relative to the early Paleoproterozoic IFs, the Mesoarchean units have much greater C_{org} utilization (identified by the very low $\delta^{13}C$ values) and more limited extent of Fe^{3+} reduction and pore-fluid equilibration (identified by the very low $\delta^{18}O$ values). These changes in mineral reactions are reflected in the very high ratio of magnetite to Fe-carbonate in the Mesoarchean IFs as compared to the early Paleoproterozoic IFs. Data from Smith (2007), Cochrane (2009), Smith et al. (2013), and Ossa Ossa et al. (2018b)

6.6.1 The Mesoarchean Witwatersrand and Pongola Basins

The oldest well-preserved marine basin of Mesoarchean age is represented in the ~3 Ga Witwatersrand and Pongola supergroups of southern Africa, reflecting a sequence of ~10 km of marine clastic and chemical sedimentary rocks where excellent proximal to distal context is preserved. These rocks have played a prominent role in discussions of the early Earth, where detrital pyrite in the Witwatersrand Au-bearing gravels has been long cited as evidence for an anoxic Archean atmosphere (e.g., Frimmel 2005), and the oldest-known significant glacial deposits (Young et al. 1998) suggest major changes in atmospheric CO_2 contents and associated climate changes in the Mesoarchean. The Witwatersrand and Pongola supergroups are correlated and have been interpreted to broadly record the same continental margin or basin setting (Beukes and Cairncross 1991; Nhleko 2003), although the Pongola sequence was deposited in a more outboard location relative to the Witwatersrand (e.g., Nhleko 2003; Smith 2007). In contrast to the Neoarchean and early Paleoproterozoic sequences discussed above, the Witwatersrand and Pongola supergroups contain IFs that are relatively thin, but, notably, they contain thick sequences of Fe-rich magnetic shales that are absent in younger sequences.

The degree of Fe enrichment in shales of the Witwatersrand and Pongola supergroups is remarkably high, with an average Fe_{tot}/Al ratio of 1.72, far higher than the crustal average (~0.5), and many shales contain $Fe_{tot}/Al > 2$ (see Fig. 6.13). Such exceptionally high Fe_{tot}/Al ratios would be taken to indicate anoxic ferruginous marine conditions (e.g., Raiswell and Canfield 1998; Poulton and Canfield 2005, 2011; Raiswell et al. 2018), suggesting that seawater Fe contents were much higher than in the Neoarchean (compare panels B and C in Fig. 6.13). Although shales plotted in Fig. 6.14 are dominated by Fe^{2+}-bearing minerals, the average fraction of Fe^{3+} is higher than the average in the Paleoarchean and Neoarchean, reflecting the high abundance of magnetite in the Mesoarchean shales. There is a significant correlation between wt% Al_2O_3 and fraction of Fe^{3+} in the data sets used for Figs. 6.13 and 6.14, where lithologies that contain <5 wt% Al_2O_3 have Fe^{3+} fractions exceeding 0.45. This corresponds with high magnetite abundance in the IFs of the Witwatersrand and Pongola supergroups (Smith 2007; Smith et al. 2013). These observations pose a significant interpretive challenge: how do we reconcile the inferred strongly ferruginous marine conditions based on very high Fe_{tot}/Al ratios, with the elevated authigenic Fe^{3+} contents contained in magnetite that would indicate oxidation? In addition, Mn-rich shales and IFs are found in both the Witwatersrand and Pongola supergroups (Smith 2007; Cochrane 2009), suggesting that a Mn redox cycle operated at this time, perhaps indicating high-Eh conditions. Collectively, these observations indicate that significant Fe redox contrasts must have existed in the Witwatersrand and Pongola basins and that very large quantities of Fe were involved in redox cycling.

We first focus on C and O isotope relations among Fe- and Mn-bearing carbonates in the Witwatersrand and Pongola Fe-rich shales and IFs (Fig. 6.43), as this bears on the primary fluxes of Fe, Mn, and C to the seafloor, as well as potential microbial diagenesis. As discussed above, diagenetic reactions, sometimes ignored in proxy studies, are key to understanding C_{org} delivery to the seafloor, and evidence for primary C_{org} contents in marine sediments will lie in the electron equivalents contained in Fe^{2+}- and Mn^{2+}-bearing minerals. Compared to the early Paleoproterozoic IFs (Fig. 6.35), the Witwatersrand and Pongola supergroup Fe–Mn carbonates have much lower $\delta^{13}C$ and $\delta^{18}O$ values, indicating that their origin is dominated by microbial respiration and are therefore authigenic in origin (Fig. 6.43). This is a very important conclusion, as it potentially bears on redox signals such as Mo isotopes, where Mo may be desorbed from Fe–Mn oxides during reduction. The C and O isotope compositions contrast with those of Ca–Mg stromatolitic carbonates from the Chobeni Formation (Pongola Supergroup), which have much higher $\delta^{18}O$ and $\delta^{13}C$ values, closer to those expected for sea water, although Sr isotope data from the Chobeni Formation samples indicate significant post-depositional alteration (Siahi et al. 2018).

The strong decrease in $\delta^{18}O$ and $\delta^{13}C$ values of the Witwatersrand-Pongola Fe–Mn carbonates relative to what would be expected in equilibrium with seawater does not fit the DIR model of Heimann et al. (2010) that was proposed to explain the C and O isotope compositions of the early Paleoproterozoic Kuruman IF carbonates. For $\delta^{13}C$ values approaching −24‰, nearly complete C_{org} utilization is required, but in the DIR model of Heimann et al. (2010), $\delta^{18}O$ values only decrease to ~ +17‰ at complete C_{org} utilization. This is based on the assumption of Heimann et al. (2010) that under conditions of complete Fe^{3+} reduction, isotopic exchange was restricted to pore fluids that were liberated from reduction of Fe^{3+}-hydroxides. In contrast, the lower $\delta^{18}O$ values of the Witwatersrand-Pongola Fe–Mn carbonates could possibly be produced under conditions of partial Fe^{3+} reduction, where Fe^{3+}-hydroxides that were not reduced reacted with excess Fe^{2+}_{aq} to form magnetite, likely limiting O isotope equilibrium between siderite and pore fluid. This would shift the $\delta^{18}O$ values of siderite toward a greater inheritance from Fe^{3+}-hydroxides, toward lower-than-expected $\delta^{18}O$ values based on the Heimann et al. (2010) model. This scenario can be expressed by an extension of Eqs. 6.7 through 6.9 to:

$$4Fe(OH)_3 + CH_2O \rightarrow 4FeCO_3 + 3Fe^{2+}_{aq} + 6OH^- + 4H_2O \quad (6.15)$$

followed by a magnetite-forming reaction:

$$3Fe^{2+}_{aq} + 6OH^- + 6Fe(OH)_3 \rightarrow 3Fe_3O_4 + 12H_2O \quad (6.16)$$

In this case, we would expect magnetite to be in O isotope equilibrium with pore fluids, but perhaps not siderite, dependent upon the rate of siderite and magnetite formation. In addition, the sequence of reactions for Eqs. 6.15 and 6.16 suggests magnetite is later than siderite, which may limit O isotope equilibration between siderite and magnetite. This paragenetic sequence is sometimes observed in the Witwatersrand and Pongola supergroups, but many examples exist where magnetite precedes siderite, adding complexity (Smith 2007). Alternatively, the very low $\delta^{18}O$ values may reflect diagenetic mineral formation at high temperature, as favored by Ossa Ossa et al. (2018b). Equilibration under metamorphic conditions, however, is an unlikely explanation for all of the changes in $\delta^{18}O$ values based on petrographic relations for the carbonates

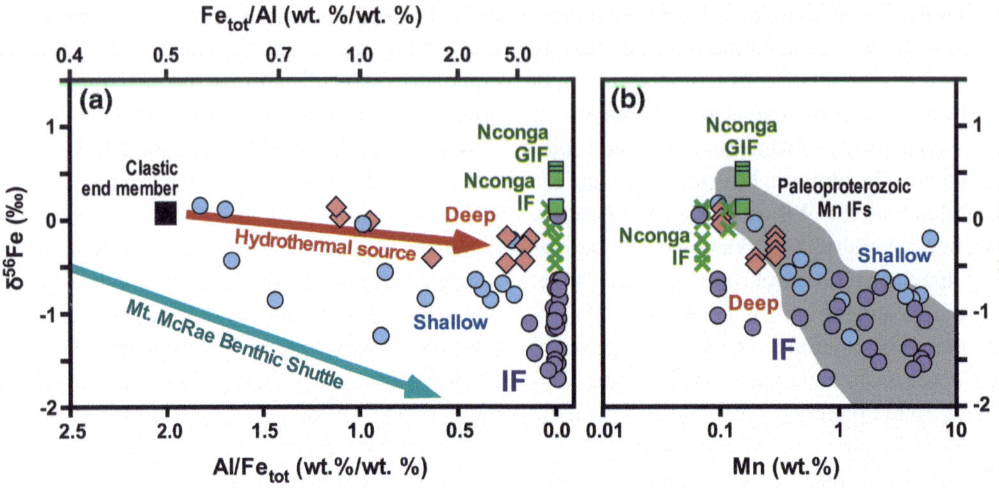

Fig. 6.44 Relations among WR δ^{56}Fe, Fe_{tot}/Al (Al/Fe_{tot}), and Mn contents for shale and IF samples from the Sinqeni, Ntombe, Nconga, and Thalu formations, Mozaan Group, Pongola Supergroup (southern Africa). Shallow-water shales shown in blue circles, whereas deep-water shales shown in red diamonds. Samples from the Sinqeni IF shown in purple circles. Samples from the Nconga Formation shown in green, divided into granular iron formation (GIF) in squares and micritic iron formation (IF) in "X" symbols. Scale in panel A is linear with respect to Al/Fe_{tot} to illustrate linear mixing trends, and equivalent Fe_{tot}/Al is shown on upper scale. Clastic end member set to Fe_{tot}/Al = 0.5 (average crust), but this may be too low given generally high Fe_{tot}/Al ratios for Mesoarchean shales (see Fig. 6.13). A possible hydrothermal source trend is shown by red arrow, with only slight decrease in δ^{56}Fe values with increasing Fe enrichment. Trend for DIR-driven benthic Fe shuttle inferred for the 2.5 Ga Mt. McRae Shale shown in teal color from Fig. 6.38. The relatively high δ^{56}Fe values for the Nconga Formation require a different set of formation pathways than the Sinqeni, Ntombe, and Thalu formations. Mn contents (panel B) plotted on log scale, and grey field outlines data for early Paleoproterozoic Mn IFs (Koegas Group and Hotazel formations) from Fig. 6.36; it is important to note that the data for the early Paleoproterozoic Mn IFs extend to higher Mn contents and lower δ^{56}Fe values than the scale shown here. Data from Planavsky et al. (2014), Smith et al. (2017), and Ossa Ossa et al. (2018b)

(Smith 2007; Cochrane 2009). These issues can be addressed by detailed in situ C and O isotope analysis that is tied to study of paragenetic sequences, but this has yet to be done. Broadly, therefore, we will take Eqs. 6.15 and 6.16 to show that very low δ^{13}C values for Fe–Mn carbonates in the Witwatersrand and Pongola supergroup Fe–Mn-rich shales and IFs should be generally associated with extensive authigenic magnetite formation, which is a clear characteristic of the sequences.

The C-O isotope relations and required authigenic mineral formation reactions provide important paleoenvironmental constraints on deposition of the Fe- and Mn-rich sediments of the Witwatersrand and Pongola supergroups. A relatively large flux to the seafloor of Fe^{3+}- and Mn^{4+}/Mn^{3+}-hydroxides is required during initial sediment deposition, presumably reflecting processes that occurred in the shallow ocean. As discussed above in Sect. 6.5.5, the inference of Mn^{4+}/Mn^{3+}-hydroxide deposition bears importantly on inferring high Eh conditions, and the Mesoarchean must have shared similarities with the early Paleoproterozoic in this regard. Production of Mn carbonates through reduction of Mn^{4+}/Mn^{3+}-hydroxides, however, under the conditions of near-complete C_{org} utilization that are required to explain the very low δ^{13}C values, produces excess aqueous Mn. This is illustrated by extending the reaction of Eq. 6.12 to illustrate an excess of Mn^{2+} during carbonate formation:

$$2Mn^{2+} + HCO_3^{-} + 3OH^{-} \rightarrow MnCO_3 + Mn^{2+} + 2OH^{-} + H_2O \quad (6.17)$$

Unless sequestered in other authigenic Mn^{2+} minerals, possible loss of pore water Mn^{2+} would indicate that the initial Mn^{4+}/Mn^{3+}-hydroxide contents may have been significantly higher than suggested by the measured Mn contents. This will bear on Fe-Mo isotope relations discussed below. Additional complexity is introduced by evidence for post-formation mobilization of Mn in sections of the Pongola Supergroup (Ossa Ossa et al. 2016).

The Sinqeni, Ntombe, and Thalu formations of the Mozaan Group, Pongola Supergroup, have been studied for Fe and Mo isotope variations (Planavsky et al. 2014; Ossa Ossa et al. 2018b), and we first focus on the relations between $\delta^{56}Fe$ and Fe_{tot}/Al (Al/Fe_{tot}) and Mn contents (Fig. 6.44). Shales deposited in deep- and shallow-water settings all have Fe_{tot}/Al ratios greater than those of average crust, consistent with the high-Fe enrichments that are characteristic for the Mesoarchean (Fig. 6.13), but samples from a shallow-water setting generally have lower $\delta^{56}Fe$ values at a given Fe_{tot}/Al (Al/Fe_{tot}) ratio than those from a deep-water setting. Ossa Ossa et al. (2018b) interpret this to indicate oxidation of Fe^{2+}_{aq} in an O_2-rich shallow photic zone that was accompanied by increasing Mn hydroxide formation. This scenario is similar to that proposed for the early Paleoproterozoic Koegas and Hotazel Mn IFs discussed above in Sect. 6.5.5 (Kurzweil et al. 2016; Thibon et al. 2019), and is the same model proposed by Planavsky et al. (2014) for the Sinqeni IF. Importantly, the $\delta^{56}Fe$ values and Mn contents for the Mozaan Group samples do not extend to as extreme compositions as the early Paleoproterozoic Mn IFs, which is qualitatively consistent with less oxidation of Fe^{2+}_{aq} for the Mesoarchean. Nevertheless, the inferred presence of free O_2, and attainment of Eh conditions that produced Mn^{4+}/Mn^{3+}-hydroxides in the photic zone, indicate remarkably oxic conditions for the Mesoarchean. This conclusion matches that of Eickmann et al. (2018) based on the negative

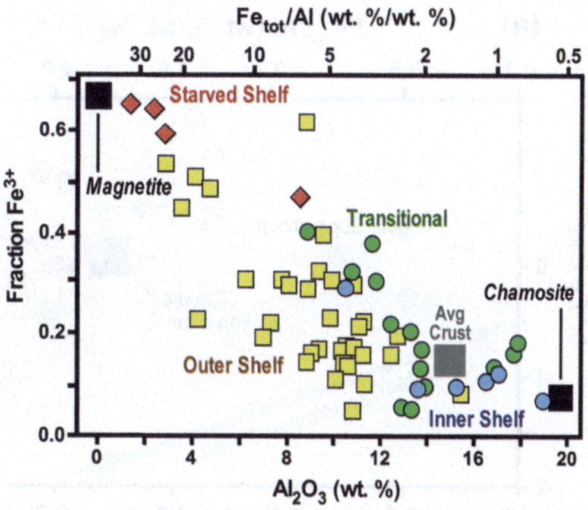

Fig. 6.45 Variations in Al_2O_3 contents and the fraction of Fe^{3+} (Fe^{3+}/Fe_{tot}) for shales and IFs from the Rietkuil, Coronation, Promise, Partktown, Palmietfontein, Roodepoort, and Water Tower formations of the Witwatersrand Supergroup (southern Africa), divided by depofacies, progressing from proximal to distal locations in the basin (Inner Shelf, Transitional (shelf), Outer Shelf, Starved Shelf; the latter includes IFs). Classification simplified from that of Guy (2012). The spread in the data reflect mixing between an Fe^{2+}-rich component and an Fe^{3+}-rich component. The Fe^{2+}-rich component dominates the proximal samples and includes detrital Fe-silicates, as well as authigenic Fe^{2+}-rich silicates (chamosite; lower right black box). The distal component is dominated by magnetite (upper left black box). The inferred mineralogical relations based on Fe^{3+}–Fe^{2+} speciation are confirmed from XRD and SEM-EDS data. The detrital component is assumed to be Mesoarchean average crust for Al_2O_3 and fraction of Fe^{3+} equal to igneous rocks (grey box; Condie 1993). Upper scale is Fe_{tot}/Al based on regression of data relative to Al_2O_3 ($R^2 = 0.79$), which shows that all samples have high Fe_{tot}/Al ratios relative to average crust, as is typical of Mesoarchean shales. The Fe redox-depofacies relations are opposite of that expected for a redox-stratified water column, where proximal shallow-water sediments should have high fractions of Fe^{3+} and distal deep-water sediments should have low fractions of Fe^{3+}. This is interpreted to indicate DIR on the inner shelf that released Fe^{2+}_{aq} to the water column that was oxidized to Fe^{3+}-hydroxides, followed by transport via a benthic Fe shuttle to the deep basin where authigenic conversion occurred to magnetite. Data and interpretations from Yamaguchi (2002), Smith (2007), Guy (2012), and Hashman (2017)

$\delta^{56}Fe$ values for pyrite from the Pongola Supergroup, which they use to infer extensive Fe^{2+}_{aq} oxidation, although, as discussed in Sect. 6.3.3, the large kinetic Fe isotope fractionations that can

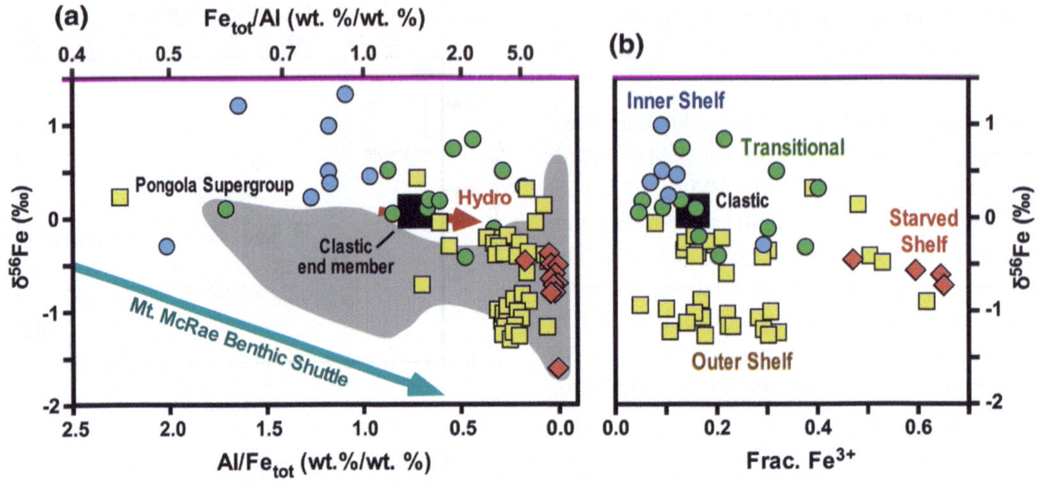

Fig. 6.46 Relations among δ^{56}Fe, Fe_{tot}/Al (Al/Fe_{tot}), and fraction of Fe^{3+} contents for shale and IF samples from nine cores from the Witwatersrand Supergroup (see Fig. 6.45), Southern Africa. Isotopic data largely on WR samples, with some additional data on magnetite and pyrite. Depofacies identify proximal to distal changes, and are the same as used in Fig. 6.45. Because multiple formations are grouped for simplicity, some formation-specific relations are obscured, such as the Rietkuil Formation, which forms a separate low-δ^{56}Fe Outer Shelf group (data cluster near label). Overall, there is a trend of decreasing δ^{56}Fe with increasing fraction of Fe^{3+} within depofacies groups, and a general trend of decreasing δ^{56}Fe and increasing Fe_{tot}/Al. The clastic endmember is based on average crustal Al_2O_3 contents from Fig. 6.45, equivalent to Fe_{tot}/Al = 1.5, which is much higher than that of average crust but consistent with the Fe-enriched nature of Mesoarchean shales (see Fig. 6.13). In this context, all Inner Shelf samples are Fe-depleted and generally have positive δ^{56}Fe values, reflecting loss of low-δ^{56}Fe Fe^{2+}_{aq} by a DIR-driven benthic Fe shuttle. Trapping of the low-δ^{56}Fe shuttled Fe is recorded in samples from the Outer Shelf and Starved Shelf depofacies. Also shown is the trend for the Mt. McRae Shale benthic Fe shuttle, as well as grey field for the Pongola Supergroup sample suite (see Figs. 6.38 and 6.44). Although the δ^{56}Fe values for the benthic Fe shuttle recorded in the Witwatersrand Supergroup are not as low as those of the Mt. McRae Shale, the mass of Fe transported in the Witwatersrand basin was likely much larger, but the Fe isotope shift was muted due to the higher seawater Fe^{2+}_{aq} contents in the Mesoarchean. Data from Yamaguchi et al. (2005), Smith (2007), Guy et al. (2012), Smith et al. (2013), and Hashman (2017)

occur during pyrite formation make connections to the Fe isotope composition of seawater difficult using pyrite.

Quite different Fe pathways were inferred by Smith et al. (2017) for the Nconga Formation, where relatively high δ^{56}Fe values were measured for granular iron formation (GIF) and micritic IF. The Nconga Formation contains the oldest known GIF, where textural relations suggest microbial oxidation of Fe^{2+}_{aq} and precipitation of Fe oxides on shallow-water oncoids. Chemolithoautotrophic Fe^{2+}_{aq} oxidation under microaerobic conditions (Konhauser et al. 2002) may be described by:

$$6Fe^{2+}_{aq} + 0.5O_2 + CO_2 + 16H_2O \rightarrow CH_2O + 6Fe(OH)_3 + 12H^+ \quad (6.18)$$

Alternatively, anoxygenic photoferrotrophy has often been invoked as an early Earth pathway for oxidation of Fe^{2+}_{aq}, given the deeply rooted nature of this form of photosynthesis (e.g., Widdel et al. 1993; Xiong et al. 2000), which may be described by (Kappler et al. 2005):

$$4Fe^{2+}_{aq} + CO_2 + 11H_2O + h\nu \rightarrow CH_2O + 4Fe(OH)_3 + 8H^+ \quad (6.19)$$

The lack of Fe-carbonates in the GIF units of the Nconga Formation led Smith et al. (2017) to suggest that the photic zone had low Fe^{2+}_{aq} contents, but the positive $\delta^{56}Fe$ values would imply limited O_2, likely less than is suggested by the redox conditions required for Mn redox. The $\delta^{56}Fe$ values for the bounding micritic IFs are only slightly negative, which Smith et al. (2017) interpret to record a hydrothermal input that is only slightly fractionated.

Extensive drill core are available from the Witwatersrand Supergroup that provide spatial and temporal sampling within the basin, and detailed studies of nine cores by Yamaguchi (2002), Smith (2007), Guy (2012), and Hashman (2017) set geochemical variations into a depofacies context. Figure 6.45 shows variations in the fraction of Fe^{3+} based on Fe^{3+}–Fe^{2+} speciation data relative to clastic input, represented by Al_2O_3 contents, using a simplified version of the depofacies model of Guy (2012). From proximal to distal depofacies, clastic components decrease (decreasing Al_2O_3), Fe_{tot}/Al increases, and the fraction of Fe^{3+} (Fe^{3+}/Fe_{tot}) increases. All of the samples have Fe_{tot}/Al ratios greater than those of average crust, consistent with the generally Fe-rich character of Mesoarchean shales (see Fig. 6.13). The geochemical changes with depofacies are accompanied by changes in the major authigenic Fe minerals, from primarily Fe^{2+}-silicates (chamosite) to magnetite for proximal to distal samples, respectively. These changes are opposite those expected for deposition in a redox-stratified water column. Authigenic Fe minerals that were deposited in a proximal, shallow-water setting would be expected to contain a large Fe^{3+}-hydroxide component, yet proximal sediments have very little Fe^{3+}. The very high Fe^{3+} contents of the distal depofacies, both in an absolute and relative sense, is also not expected for authigenic mineral formation in a deep-water anoxic (ferruginous) setting. These relations highlight the importance of determining Fe^{3+}–Fe^{2+} speciation, which is a missing component of many redox proxy studies.

The high Fe^{3+} contents of the deep-water samples requires a mechanism for juxtaposition of oxidized and reduced components, and this may be accomplished via a benthic Fe shuttle, where DIR in the proximal settings reduced Fe^{3+}-hydroxides, producing Fe^{2+}_{aq}, some of which was lost to the water column, and some retained in the proximal sediments. Shelf-generated Fe^{2+}_{aq} underwent oxidation to Fe^{3+}-hydroxides during transport to the deep basin (Hashman 2017), similar to what has been argued for benthic Fe transport in modern basins (e.g., Canfield et al. 1993; Anderson and Raiswell 2004; Lyons and Severmann 2006). This is consistent with the evidence for oxygenation of the photic zone in the Pongola Supergroup samples discussed previously. Settling of Fe^{3+}-hydroxides in the deep basin was later converted to magnetite through a second stage of DIR (Smith et al. 2013; Hashman 2017). The relations in Fig. 6.45 are therefore interpreted to indicate DIR in both a proximal shelf setting that supplies a benthic Fe flux to the deep basin, followed by DIR in the deep basin sediment, a model that is the same as that proposed for the early Paleoproterozoic IFs discussed above.

The Fe_{tot}/Al-$\delta^{56}Fe$ relations for the Witwatersrand Supergroup are distinct from those of the Pongola Supergroup (Fig. 6.46) and are consistent with a benthic Fe shuttle model (Hashman 2017). Given the relatively high clastic Fe_{tot}/Al ratio of the Witwatersrand suite, the generally positive $\delta^{56}Fe$ values for the Inner Shelf samples are consistent with loss of Fe^{2+}_{aq} via DIR, and this is a likely explanation for the positive $\delta^{56}Fe$ values for some of the Transitional samples. The low fraction of Fe^{3+} for most samples that have positive $\delta^{56}Fe$ values (Fig. 6.46b) indicates that these isotopic compositions cannot be explained by accumulation of Fe^{3+}-hydroxides. The negative $\delta^{56}Fe$ values of the high Fe_{tot}/Al samples, including some Outer Shelf samples (especially the Rietkuil Formation; Fig. 6.46), as well as the Starved Shelf samples, are interpreted to reflect trapping of the low-$\delta^{56}Fe$ benthic Fe shuttle in the deeper portions of the basin, but this model requires that the shuttled Fe was in the form of Fe^{3+}-hydroxides to explain the high fraction of Fe^{3+}. To maintain relatively low $\delta^{56}Fe$ values would require near complete oxidation during the conversion of shuttled Fe^{2+}_{aq} to Fe^{3+}-hydroxides,

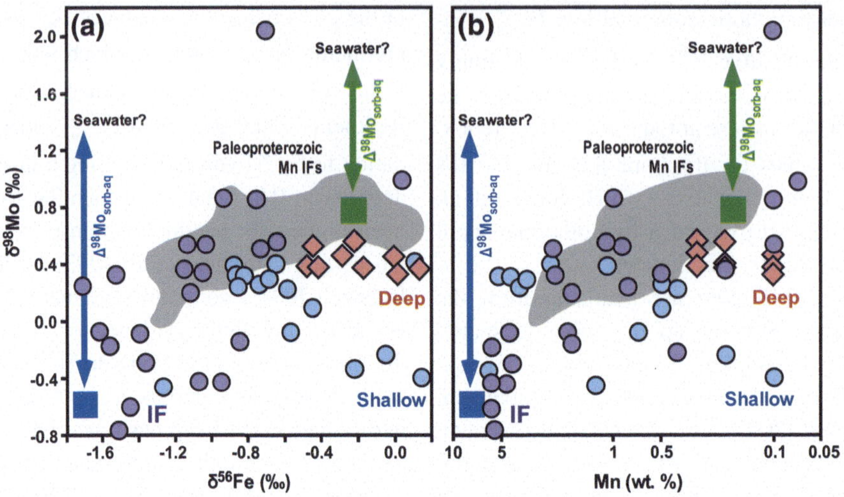

Fig. 6.47 δ^{56}Fe-δ^{98}Mo–Mn relations for shales and IFs of the Pongola Supergroup (symbols same as in Fig. 6.44), Southern Africa. Also shown are fields for the early Paleoproterozoic Mn IFs of the Koegas Group (see Fig. 6.37; two extreme values omitted to better represent the majority of the data). Also shown are the $\Delta^{98}Mo_{sorb\text{-}aqueous}$ fractionation factors for Mn-rich (blue arrow) and Fe-rich (green arrow) samples from Fig. 6.37, adjusted in placement for δ^{98}Mo to better align with the data shown here. In part the scatter for the shale data reflects variable clastic and authigenic components. Measured data from the IFs most closely represent the authigenic Fe and Mo isotope compositions. Relative to the early Paleoproterozoic Mn IFs, the δ^{98}Mo values for the Pongola Supergroup IFs extend to significantly more negative values, suggesting lower seawater δ^{98}Mo values in the Mesoarchean, assuming equilibrium exchange between Mn oxide and seawater (vertical blue arrow). Slightly higher inferred seawater δ^{98}Mo values might be indicated by the Fe-rich IFs, assuming equilibrium Mo isotope fractionation. Although the δ^{56}Fe values shown for the Pongola Supergroup IFs extend to low values, significantly lower δ^{56}Fe values are measured for the early Paleoproterozoic Hotazel Mn IF but there are no Mo isotope data for this suite (see Fig. 6.36). Planavsky et al. (2014) and Ossa Ossa et al. (2018b) interpret the negative δ^{56}Fe values to record extensive oxidation of Fe^{2+}_{aq} under relatively high O_2 conditions. The negative correlation between log Mn versus δ^{98}Mo for the Pongola Supergroup IFs has an R^2 of 0.59; a slightly better correlation is seen for log Fe/Mn ($R^2 = 0.63$), but there is no correlation between δ^{98}Mo or δ^{56}Fe values and Fe contents. Data from Planavsky et al. (2014) and Ossa Ossa et al. (2018b). δ^{98}Mo values relative to the "NIST + 0.25" reference frame

which qualitatively suggests relatively oxidizing conditions. Because, however, the authigenic Fe minerals in the deep basin are dominated by magnetite, some mechanism is required to convert Fe^{3+}-hydroxides to magnetite, suggesting a second stage of DIR in the deep basin, consistent with studies of magnetite paragenesis (Smith et al. 2013).

The shift to negative δ^{56}Fe values for the distal Witwatersrand Supergroup samples is less than that seen in the Mt. McRae Shale (Fig. 6.46), but this does not indicate that a benthic shuttle in the Witwatersrand basin was less important. As noted above, the very high overall Fe_{tot}/Al ratios of Mesoarchean shales suggests much higher seawater Fe^{2+}_{aq} contents than in the Neoarchean or pre-GOE Paleoproterozoic, and this would mute any isotopic signatures of a benthic Fe shuttle in the Mesoarchean. It could be, in fact, that the moles of Fe transported via a benthic Fe shuttle in the Witwatersrand basin was much higher than in the Neoarchean-Paleoproterozoic Hamersley-Transvaal basins. Independent support for a benthic Fe shuttle in the Witwatersrand-Pongola basin lies in the highly non-radiogenic Nd isotope compositions of the IFs, indicating a continental source for authigenic Fe (Alexander et al. 2008).

We conclude this section by considering Mo-Fe isotope variations for the Pongola Supergroup, which provide similarities, and contrasts, with those discussed above for the early Paleoproterozoic Mn IFs in Sect. 6.5.5. Broadly decreasing $\delta^{98}Mo$ and $\delta^{56}Fe$ values with increasing Mn contents for shales and IFs of the Pongola Supergroup (Fig. 6.47) have been taken to indicate progressive oxidation of hydrothermally sourced Fe^{2+}_{aq}, accompanied by increasing Mn^{4+}/Mn^{3+}-hydroxide formation under increasing Eh conditions (Planavsky et al. 2014; Ossa Ossa et al. 2018b). Although these studies focused on changes in Fe/Mn ratios, such changes are entirely driven by changes in Mn contents, as there is no correlation between Mo and Fe isotope compositions and Fe contents. For the shales, although there is a broad correlation of decreasing $\delta^{56}Fe$ values with increasing Fe_{tot}/Al (Fig. 6.44), there is no correlation between $\delta^{98}Mo$ values and Al contents or Mn/Al or Mo/Al ratios. Planavsky et al. (2014) presented laser ablation data which showed that the bulk-sample Mo inventory could be accounted for in magnetite and siderite, both of which would be authigenic minerals that record extensive microbial diagenesis based on C–O isotope relations (Fig. 6.43). Under the conditions of high C_{org} utilization that is indicated by the C and O isotope data, escape of aqueous Mn and Mo during diagenesis could have altered Mn contents and Mo isotope compositions (e.g., Eq. 6.17), leading to scatter in the $\delta^{98}Mo$ Mn relations (Fig. 6.47). There is insufficient data to consider correlations between $\delta^{13}C$ and $\delta^{98}Mo$ values that might test the effects of diagenesis. Loss of sorbed Mo during reduction of either Fe^{3+}-hydroxides or Mn^{4+}/Mn^{3+}-hydroxides would tend to increase $\delta^{98}Mo$ in the residual sediment (Goldberg et al. 2012), which would suggest lower seawater $\delta^{98}Mo$ values than might be inferred from the relations in Fig. 6.47; this issue would be most likely to apply to the samples that have high-$\delta^{98}Mo$, -$\delta^{56}Fe$, and low-Mn contents, which would imply that the inferred $\delta^{98}Mo$ for seawater for the green arrow in Fig. 6.47 may be too high due to diagenetic effects.

The minimal clastic component of the Pongola Supergroup IFs suggests that interpretation of the authigenic Mo components may be more straightforward, although there remains considerable scatter in the $\delta^{56}Fe$-$\delta^{98}Mo$ relations (Fig. 6.47) that may reflect loss of aqueous Mn, Mo, and Fe during microbial diagenesis. The very low $\delta^{13}C$ and $\delta^{18}O$ values in Fig. 6.43 therefore predict a large scatter in the $\delta^{56}Fe$-$\delta^{98}Mo$ relations in Fig. 6.47. In addition, the possibility of loss of aqueous Mn, Mo, and Fe may indicate that the inferred $\delta^{98}Mo$ for seawater shown for the Mn-rich end member (blue arrow in Fig. 6.47) is a maximum value. An alternative interpretation of the relations in Fig. 6.47 is that the high-Mn, low-$\delta^{98}Mo$, -$\delta^{56}Fe$ component reflects that of a benthic shuttle, where Fe^{3+}-hydroxides and Mn^{4+}/Mn^{3+}-hydroxides were shuttled from a proximal setting to the deep basin. With the available data, this cannot be independently tested, but it again highlights the multiple pathways by which Fe-rich rocks may attain negative $\delta^{56}Fe$ (and $\delta^{98}Mo$) values in a marine system. Importantly, either model (oxidation of Fe^{2+}_{aq} or a benthic shuttle) require large redox contrasts in the basin, with the likely presence of free O_2 in shallow seawater. The greater scatter in the Fe–Mo isotope signals for the Pongola Supergroup relative to the early Paleoproterozoic Mn IFs might indicate redox heterogeneity in the Mesoarchean basin, or more likely, the effects of microbial diagenesis under conditions of high C_{org} utilization.

The discussion above paints a picture of strong Fe and Mn redox cycling that involved large inventories of Fe. Oxidative Fe pathways may have included oxidation by O_2 generated by oxygenic photosynthesis, photoferrotrophy, or chemolithoautotrophy. Independent evidence for free O_2 comes from temporal changes in U isotope compositions at $\sim$3 Ga, as recorded by data from the same units discussed above from the Pongola Supergroup (Wang et al. 2018). Microbially-catalyzed Mn^{2+} oxidation is also indicated based on the abundance of Mn in the Fe-rich sequences, as well as Mo isotope data. Reductive pathways include heterotrophic Fe^{3+} (DIR) and Mn^{4+}/Mn^{3+} respiration coupled to oxidation of C_{org}. In addition, the oldest highly negative $\delta^{13}C$ excursion for C_{org} is recorded in the Witwatersrand Supergroup strata, indicating an active CH_4 cycle, possibly including methanogenesis and aerobic methanotrophy, and a wide range in $\delta^{34}S$ values suggests microbial sulfate reduction occurred under sulfate-limited conditions (Guy et al. 2012). It is impressive that such an apparently diverse ecosystem existed >500 m.y. before the GOE.

6.6.2 The Paleoarchean Barberton Greenstone Belt and Pilbara Craton

The ~3.6 to ~3.2 Ga Barberton Greenstone Belt (BGB) is a structurally complex volcano-sedimentary sequence exposed in South Africa and Swaziland that includes, from oldest to youngest, the Onverwacht, Fig Tree, and Moodies groups (e.g., Lowe and Byerly 1999, 2007; Byerly et al. 2019). These groups record broad transitions from mafic volcanic sequences to quartz-poor clastic deposition to quartzose terrigenous sequences. Several Fe isotope studies have focused on the Mapepe Formation of the Fig Tree Group in the southern part of the BGB, including jasper-rich sections in the Manzimnyama syncline and barite-rich sections in Barite Valley that record deposition in a variety of marine environments (Satkoski et al. 2015; Galić et al. 2017; Busigny et al. 2017). In addition, siderite-rich clastic units of the Sheba Formation in the northern BGB that are time correlative with the Mapepe Formation have been studied for Fe isotopes (Yamaguchi et al. 2005). Regression of $\delta^{56}Fe$-Al/Fe_{tot} and $\delta^{56}Fe$-Fe^{2+}/Fe_{tot} relations in the Sheba Formation (Yamaguchi 2002; Yamaguchi et al. 2005) to a pure siderite component suggests that 3.2 Ga anoxic seawater had a $\delta^{56}Fe$ value of $0.20 \pm 0.15‰$ using the Fe^{2+}_{aq}-siderite fractionation factor of Wiesli et al. (2004), which overlaps that of average crust and mantle.

The BARB-4 core through the Manzimnyama jaspilite and associated clastic units in the Mapepe Formation has been the subject of S, Fe, Nd, Hf, and U-Th–Pb isotope studies, as well as detailed fluid inclusion, WR geochemistry, and magnetic studies (Zentner 2014; Satkoski et al. 2015; Farber et al. 2016; Galić et al. 2017; Garçon et al. 2017; Wabo et al. 2018). Clastic sections include pyrite but little Fe oxide, whereas jasper-rich ("IF") sections include pure hematite + chert layers of variable Fe content, siderite-rich layers, and magnetite + siderite layers. For the jasper layers, low-Fe sections consist of granules of hematite + chert encased in chert-rich groundmass, which are interpreted to record deposition in a shallow-water setting, whereas Fe-rich layers are finely laminated and contain relatively few chert layers, interpreted to record deep-water deposition (Satkoski et al. 2015). $\delta^{34}S$ and $\Delta^{33}S$ values for pyrite from the clastic sections are almost all positive, which has been taken to indicate limited seawater sulfate levels and deep-water sources of S, where the $\delta^{34}S$-$\Delta^{33}S$ relations were produced by microbial reduction of S^0 (Galić et al. 2017). Samples from the clastic layers have initial ε_{Nd} values of −2 to +1, and initial ε_{Hf} values of −1 to +12, which Garçon et al. (2017) interpret to record derivation from an intermediate-composition crustal source that was ~300–400 m.y. older than the depositional age. Geochemical, fluid inclusion, and textural data provide evidence for post-depositional fluid-flow and metasomatism, and ^{238}U–^{206}Pb geochronology identifies a ~2 Ga fluid-alteration event that was likely sourced to Bushveld hydrothermal fluids (Zentner 2014; Satkoski et al. 2015; Farber et al. 2016).

The origin of hematite in the Manzimnyama jaspilite suggests oxidation of Fe^{2+}_{aq}, but, as discussed in the previous section, this might have occurred by O_2 or anaerobic pathways, and Fe isotopes alone cannot distinguish these (e.g., Croal et al. 2004; Kappler et al. 2010). Satkoski et al. (2015) addressed this ambiguity by combining Fe isotope analysis with U-Th–Pb geochronology to identify changes in primary U contents that could indicate the presence or absence of O_2 (Fig. 6.48). $\delta^{56}Fe$ values for the shallow-water, low-Fe jasper are, on average, lower than those of the deep-water, high-Fe jasper. Assuming a seawater Fe^{2+}_{aq} source that had a $\delta^{56}Fe$ value equal to that of average crust/mantle, this suggests that the shallow water, low-Fe jaspers reflect a greater extent

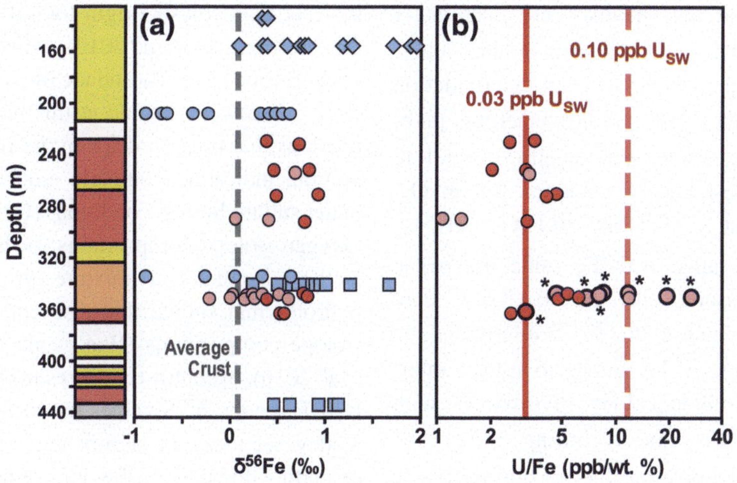

Fig. 6.48 Stratigraphic variations in δ^{56}Fe values and U/Fe ratios (log scale) for jaspers (pink and red symbols) and pyrite (blue symbols) from the BARB-4 core through the 3.23 Ga Mapepe Formation of the Fig Tree Group, Barberton Greenstone belt, southern Africa. Simplified stratigraphic section on left: pink: jaspers/IFs, tan: mixed clastic + jasper section, yellow: clastics, white: damaged core section, grey: underlying Mendon Formation (from Zentner 2014; Satkoski et al. 2015; Galić et al. 2017). Jasper samples are pure hematite + chert from the IF layers, divided by low-Fe (light pink symbols) and high-Fe (red symbols). The low-Fe jasper corresponds to shallow-water facies, and the high-Fe jasper reflects deep-water facies. For U/Fe ratios, large symbols with heavy lines and adjacent asterisks are those that were isotopically closed, or nearly closed, in terms of ^{238}U-^{206}Pb geochronology, and these provide the most robust estimates for primary seawater U contents (U_{SW}, shown in solid red line for high-Fe jasper, and dashed pink line for low-Fe jasper). The shallow-water low-Fe jasper has slightly positive δ^{56}Fe values and high inferred U_{SW}, indicating higher O_2 contents, relative to deep-water high-Fe jasper, whose higher δ^{56}Fe values and lower inferred U_{SW} contents indicate lower O_2 contents. U_{SW} for the high-Fe jasper based on single sample that was isotopically closed for U-Th–Pb, whereas U_{SW} for the low-Fe jasper based on the average of the range measured for samples that were isotopically closed for U-Th–Pb. Pyrite δ^{56}Fe values obtained via in situ analysis of clastic or mixed IF + clastic section. Layered pyrite shown in squares, disseminated pyrite in circles, and aggregate pyrite in diamonds. Data from Satkoski et al. (2015) and Galić et al. (2017)

of Fe^{2+}_{aq} oxidation than the deep-water jaspers. The very high sorption coefficients of U^{6+}_{aq} to Fe^{3+}-oxide/hydroxide (e.g., Wazne et al. 2003; Davis et al. 2004) makes U contents in IFs and jaspers a potentially sensitive indicator of seawater U contents and hence O_2 (e.g., Partin et al. 2013b). The possibility of post-depositional fluid flow, however, may mobilize U, particularly if fluid-flow events occurred after the GOE and hence were oxidizing, increasing aqueous U solubility. Jaspers in the upper section of the BARB-4 core (depths between ~230 and ~260 m) are cut by Bushveld-age secondary veins (Satkoski et al. 2015). In contrast, many of the samples in the lower jasper section (320 m and deeper), which is not cut by extensive veins, have remained isotopically closed with respect to U-Th–Pb geochronology, particularly the Fe-poor shallow-water jasper facies (Satkoski et al. 2015). When considering only the samples that are isotopically closed (or nearly closed) for U-Th–Pb geochronology, primary seawater U contents (U_{SW}) may be estimated at 0.10 ppb for the shallow-water jaspers and 0.03 ppb for the deep-water jaspers, significantly lower than that of the modern oceans (~3 ppb U; Barnes and Cochran 1990). The high-inferred U_{SW} contents for the shallow-water jasper is robust given the high proportion of samples that were closed with respect to U-Th–Pb geochronology, and the estimated U_{SW} is a conservative one based on the average U/Fe ratio (Fig. 6.48). The contrast in inferred U_{SW} for the shallow- and deep-water jaspers agrees with the contrast in $\delta^{56}Fe$ values, which provides an independent test that oxidation of Fe^{2+}_{aq} occurred by O_2 and not anaerobic pathways. The Manzimnyama jaspilite therefore records the oldest known evidence for an O_2-stratified ocean at 3.2 Ga (Satkoski et al. 2015).

A very wide range of $\delta^{56}Fe$ values for pyrite has been measured in the BARB-4 core (Fig. 6.48), and Galić et al. (2017) interpreted the generally positive $\delta^{56}Fe$ values to reflect interaction between sulfide and Fe^{3+}-hydroxides, with the scatter in $\delta^{56}Fe$ values recording shifts in the Fe^{2+}_{aq}-pyrite fractionation factor during pyrite formation, including potential kinetic isotope effects. Although the positive $\delta^{34}S$ values indicate limited sulfide, the positive $\delta^{56}Fe$ values for the jaspers indicates a limited Fe^{3+}-hydroxide inventory relative to seawater Fe sources, and hence it is possible that there was sufficient sulfide to consume reactive Fe^{3+}. Pyrite, which is restricted to the clastic sections, attains highly positive $\delta^{56}Fe$ values, higher than those measured in the jaspers, but it is not possible to confidently assume this reflects inheritance from Fe^{3+}-hydroxides, given the complexity of pyrite formation pathways discussed above. The pathway-specific nature of Fe isotope compositions of pyrite is underscored by the wide range in $\delta^{56}Fe$ values measured via in situ analysis within single samples (Galić et al. 2017). Nevertheless, the generally positive $\delta^{56}Fe$ values for pyrite in the BARB-4 core is consistent with a Fe^{3+}-hydroxide source for pyrite Fe, although the positive $\delta^{56}Fe$ values for pyrite are also consistent with formation under equilibrium conditions (Chap. 3).

The Mapepe Formation has been studied for S and Fe isotope variations in the Barite Valley region (Bao et al. 2007; Roerdink et al. 2012; Satkoski et al. 2016; Busigny et al. 2017), a locality that is broadly correlative with the section represented by the BARB-4 core, but which has some distinct differences (Zentner 2014). Multiple S isotopes on barite in the Mapepe Formation are taken to indicate a photolysis origin for sulfate (Bao et al. 2007; Roerdink et al. 2012), where $\delta^{34}S$ values range from +3 to +7‰ and are interpreted to reflect those of seawater sulfate at this time. This range overlaps that measured for pyrite in the BARB-4 section, indicating complete reduction of sulfate under sulfate-limited conditions (Galić et al. 2017). Oxygen isotope compositions for barite in Barite Valley document a mixture of seawater and hydrothermal sulfate that also correlates with Sr isotope compositions (Bao et al. 2007; Satkoski et al. 2016). In contrast to the results for pyrite in the BARB-4 core, $\delta^{34}S$ values for pyrite in the Barite Valley sections are significantly more negative, extending down to −13‰, indicating much higher levels of seawater sulfate and local redox variability not seen in the BARB-4 core (Roerdink et al. 2013; Busigny et al. 2017). A wide range in $\delta^{56}Fe$ values were measured for pyrite in the bedded barites at the Barite Valley locality (BBDP2 core), from −2 to 0‰, and Busigny et al. (2017) ascribe these compositions to post-depositional fluid alteration at ~300 °C, based on the assumption that the Fe isotope contrast between pyrite and Fe carbonate in the samples records an equilibrium fractionation. They reject a control by kinetic isotope fractionation upon pyrite formation, but, given the discussion in Sect. 6.3.3, disequilibrium effects seem likely to have played a role in determining the Fe isotope compositions measured. The origin of the large contrast in $\delta^{56}Fe$ values measured for pyrite from the BARB-4 core and the Barite Valley location remains unclear.

In addition to pyrite, Busigny et al. (2017) measured three ferruginous cherts that contain hematite and had measured $\delta^{56}Fe$ values of +0.2, +1.2, and +2.1‰, and they extrapolated these data to a pure oxide endmember $\delta^{56}Fe$ value of ~+2‰. Busigny et al. (2017) interpret the high inferred $\delta^{56}Fe$ value for oxides to indicate very low O_2 contents during oxidation of seawater Fe^{2+}_{aq}. They further suggest that the lower $\delta^{56}Fe$ values for oxides measured by Satkoski et al. (2015) in the BARB-4 core may be due to accidental analysis of siderite, a suggestion that is unwarranted because Satkoski et al. analyzed

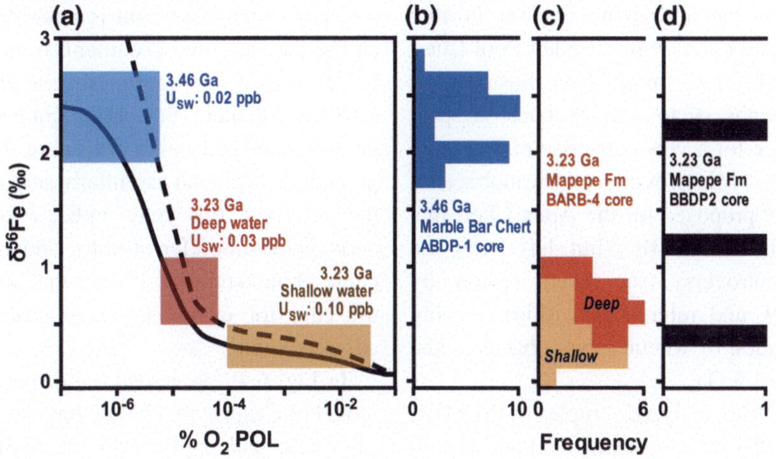

Fig. 6.49 Comparison of the dispersion/reaction model of Czaja et al. (2012), as calculated for Fe^{3+}-hydroxide precipitates (panel **a**), with the range in measured $\delta^{56}Fe$ values for hematite + chert jaspers from the 3.46 Ga Marble Bar Chert from the ABDP-1 core (Pilbara Craton, Australia, panel **b**), high-Fe (deep water) and low-Fe (shallow water) jaspers from the 3.23 Ga Mapepe Formation (Barberton Greenstone Belt, southern Africa) from the BARB-4 core (panel **c**), and hematite-bearing ferruginous cherts from the Mapepe Formation from the BBDP2 core (panel **d**). The largest datasets for measured $\delta^{56}Fe$ values ("B" and "C") are used to broadly outline possible ranges in % O_2 in the photic zone relative to Present Ocean Levels (POL). Two fractionation factors are illustrated in "A", $\Delta^{56}Fe_{oxide\text{-}fluid}$ = + 4‰ (dashed upper curve) and $\Delta^{56}Fe_{oxide\text{-}fluid}$ = +2.5‰ (solid lower curve), which encompass the range measured for Fe^{3+}-Si hydroxides (Wu et al. 2012b). The inflection in the relation between $\delta^{56}Fe$ for Fe^{3+}-hydroxide and O_2 occurs when all Fe^{2+}_{aq} is consumed in the photic zone (here assumed to be 200 m deep), which makes the model insensitive to O_2 for the Fe^{3+}-hydroxide component as the $\delta^{56}Fe$ values approach that of the input Fe^{2+}_{aq} (here assumed to be $\delta^{56}Fe$ = 0‰; see also Fig. 6.29). The relative O_2 contents inferred from the $\delta^{56}Fe$ values, where increasing O_2 is calculated for the Marble Bar Chert and the high-Fe and low-Fe jasper from the BARB-4 core corresponds well with differences in inferred U_{SW} contents based on U-Th–Pb geochronology, confirming the presence of O_2 in seawater by 3.2 Ga. Data from Li et al. (2013a), Satkoski et al. (2015), and Busigny et al. (2017)

jaspers that were pure hematite + chert. A more likely explanation for the contrasting results for jaspers between the two localities is heterogeneity in redox conditions, given the dramatically different S (and Fe) isotope results for pyrite in the two sections that suggests different Fe–S geochemistry.

The Pilbara Craton, Australia, has played a prominent role in studies of early Earth surface environments (e.g., Van Kranendonk 2006, 2014; Van Kranendonk et al. 2019). A key focus has been on the 3.5 to 3.4 Ga units, including, from oldest to youngest, the Dresser Formation, Marble Bar Chert (MBC), Apex Chert, and Strelley Pool Chert (Van Kranendonk et al. 2007b; Smithies et al. 2007). The 3.48 Ga Dresser Formation contains some of the earliest evidence for life in non-high-grade terranes (Van Kranendonk et al. 2008), including the oldest terrestrial hot springs that likely supported life (Djokic et al. 2017). Surface exposures of the overlying MBC of 3.46 Ga age (Glikson et al. 2016) contains a thick section of hematite + chert jasper, but it was unclear if hematite was a primary mineral or if it formed by later oxidation of precursor minerals such as siderite (Van Kranendonk 2006). Hoashi et al. (2009) studied hematite in the MBC from the ABDP-1 core that was drilled to obtain samples below the Phanerozoic weathering level, and they argued that the occurrence of hematite in the core showed the MBC hematite was primary in nature and indicated deposition in an oxygenated ocean. By extension, this might indicate that the stromatolites in the underlying Dresser Formation, as well as the overlying Strelley Pool Chert (Allwood et al. 2006), might have hosted oxygenic phototrophs such as cyanobacteria, providing evidence for a very early rise of oxygenic photosynthesis. Microfossils of cyanobacteria were famously proposed for the Apex Chert that directly overlies the MBC, but has been the subject of controversy (see recent review by Sugitani 2019, and references within), leaving open the question of whether cyanobacteria had evolved by ~3.5 Ga.

Li et al. (2013a) analyzed samples of the MBC from the ABDP-1 core from two sections, as well as outcrop samples, for Fe isotope compositions and U-Th–Pb geochronology, where they measured very high $\delta^{56}Fe$ values of +1.8 to +2.7‰. Such compositions indicate very limited extents of oxidation of Fe^{2+}_{aq}, arguing against an oxygen-rich ocean as advocated by Hoashi et al. (2009). No systematic differences in $\delta^{56}Fe$ values were found between the primary hematite + chert jasper and adjacent secondary chert veins, and no correlation was observed between $\delta^{56}Fe$ values and Fe contents. Moreover, the same range in $\delta^{56}Fe$ values was measured for outcrop and core samples. These relations were interpreted to indicate no significant Fe changes during either early chert replacement or Phanerozoic weathering, as expected given the low Fe solubility in most fluids. Li et al. (2013a) did observe small differences in $\delta^{56}Fe$ values on the m-scale within the ABDP-1 core, but the broadly similar $\delta^{56}Fe$ values were consistent with the uniform moderate- to deep-water depositional environment of the MBC. U-Th–Pb geochronology showed open-system behavior with respect to Pb addition, more extensive that what Satkoski et al. (2015) observed in the BARB-4 core of the Mapepe Formation, but ancient (> 1 Ga) U mobility could be ruled out using $^{208}Pb/^{204}Pb$-$^{206}Pb/^{204}Pb$ relations. It is possible, however, that some U contents were elevated in the Phanerozoic through circulation of oxygenated groundwaters, a process seen in the overlying Apex Basalt (Li et al. 2012a) that was well constrained based on the restricted range in Th/U ratios of igneous rocks; no such constraint is possible for jaspers, and so the measured U contents of the MBC must be taken as a maximum. Nevertheless, Li et al. (2013a) calculate that U_{SW} contents were no greater than 0.02 ppb, confirming that the MBC precipitated from an essentially anoxic ocean, and they suggest this may indicate oxidation by anoxygenic photoferrotrophy and not via oxygenic photosynthesis. If true, this would weaken the case for cyanobacteria microfossils in the adjacent strata.

In Fig. 6.49 we revisit the dispersion/reaction model of Czaja et al. (2012) that was illustrated in Fig. 6.29, and apply it to the hematite + chert jaspers from the 3.46 Ga Marble Bar Chert and 3.23 Ga Mapepe Formation. As discussed in Sect. 6.5.3, under the conditions of low photic zone O_2 in the Paleoarchean, the dispersion/reaction model is best applied using the Fe isotope compositions of the Fe^{3+}-hydroxide precipitates, where the model is most sensitive when incomplete oxidation of Fe^{2+}_{aq} occurs in the photic zone. The inflection in the $\delta^{56}Fe$-O_2 curves occurs at the point where photic zone Fe^{2+}_{aq} is essentially consumed, where the $\delta^{56}Fe$ values of Fe^{3+}-hydroxide precipitates approach that of the input Fe^{2+}_{aq}, and the model becomes insensitive with regard to using the Fe isotope compositions of oxides.

The dispersion/reaction model suggests very low seawater O_2 contents for deposition of the Marble Bar Chert, $<10^{-5}$% O_2 relative to Present Ocean Levels (POL). Such O_2 levels are essentially anoxic, and this is confirmed by the very low estimated U_{SW} values. The lower $\delta^{56}Fe$ values for the deep-water, high-Fe jasper from the BARB-4 core of the Mapepe Formation indicate higher O_2 levels, ~10^{-5} to 10^{-4}% POL, and the still lower $\delta^{56}Fe$ values for the shallow-water, low-Fe jaspers of the BARB-4 core suggest $>10^{-4}$% POL, perhaps as high as 0.1% POL. The results for the three measured $\delta^{56}Fe$ values for the Mapepe Formation from the BBDP2 core are varied, although Busigny et al. (2017) suggest an extrapolation to a single high-$\delta^{56}Fe$ values of ~ + 2‰,

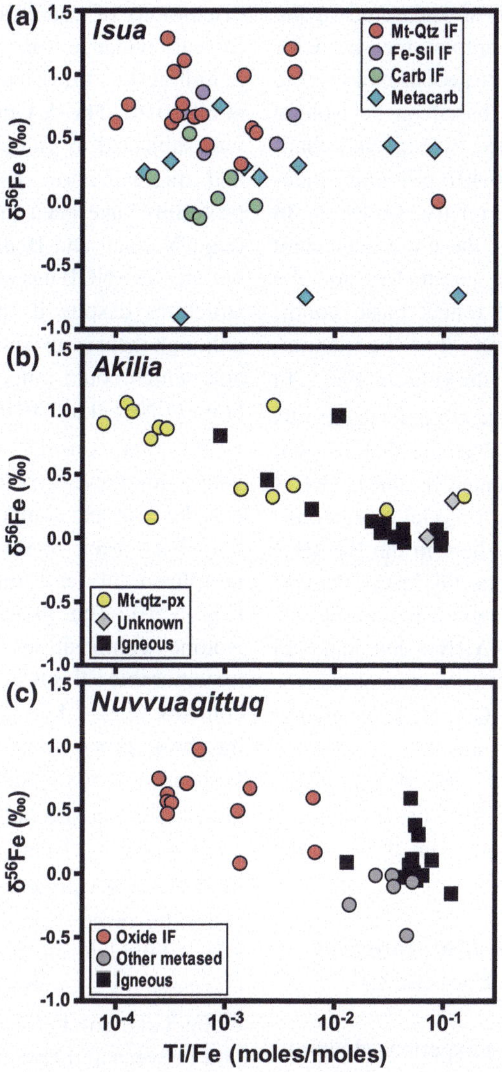

Fig. 6.50 Comparison of bulk-sample δ^{56}Fe values for supracrustal and igneous rocks from the Isua Supracrustal Belt (Greenland, **a**), Akilia Association (Greenland, **b**), and Nuvvuagittuq Greenstone Belt (Canada, **c**), relative to molar Ti/Fe ratios (note log scale). Supracrustal samples with low Ti/Fe ratios most closely represent clastic-free chemical sediments that have REE + Y contents similar to seawater (e.g., Dauphas et al. 2007a, b). For the Isua samples, all are from the low-strain region, and there are broad groupings for different IF units, with decreasing δ^{56}Fe values in the order Mt-Qtz IF > Fe-Sil IF > Carb IF. This order is consistent with that expected for Fe isotope fractionation factors for magnetite, Fe-silicate, and Fe-carbonate (see Chap. 3). Metacarbonates not clearly associated with IFs have highly scattered δ^{56}Fe values. For the Akilia Association, most samples come from Akilia Island, and identification of supracrustal units has been intensely debated. Samples likely to reflect igneous protoliths have a significantly wider range in δ^{56}Fe values than those measured for pristine igneous rocks, suggesting Fe isotope changes during metamorphism/metasomatism. Magnetite-quartz lithologies that are proposed to reflect an IF protolith have δ^{56}Fe values that vary greatly and are correlated with Ti/Fe ratios, possibly identifying a low-Ti/Fe, high-δ^{56}Fe authigenic end member. δ^{56}Fe values for samples from the Nuvvuagittuq Greenstone Belt also vary with Ti/Fe, where a low-Ti/Fe end member is characterized by high δ^{56}Fe values, similar to that seen in the datasets from Isua and Akilia. Igneous samples from Nuvvuagittuq have a significant spread in δ^{56}Fe values, although less than that of Akilia. Data from Dauphas et al. (2004, 2007a, b), Herrick (2007), O'Neil et al. (2007), and Whitehouse et al. (2015)

significantly higher than those measured in the BARB-4 core. For those samples measured for both Fe isotopes and U Th Pb geochronology, the relative order of O_2 levels based on Fe isotope compositions is accompanied by expected changes in U_{SW} contents (Fig. 6.49), providing independent confirmation of the relative O_2 levels. In terms of absolute O_2 levels, these are dependent upon how well the input parameters for the dispersion/ reaction model reflect basin conditions, as well as the choice of $^{56}Fe/^{54}Fe$ fractionation factor between Fe^{3+}-hydroxide and Fe^{2+}_{aq}. In this sense, it is most realistic to compare the results for the same local basin, such as the shallow- and deep-water portions represented by the BARB-4 core. The dispersion/reaction model more confidently quantifies the O_2 gradient in the BARB-4 core, rather than the absolute O_2 levels. Nevertheless, the up to three-orders-of-magnitude O_2 gradient recorded in the BARB-4 led Satkoski et al. (2015) to suggest that oxygenic photosynthesis had evolved by 3.2 Ga. There is greater uncertainty in comparing absolute O_2 levels inferred from different units (e.g., MBC), or different sections of the Mapepe Formation (e.g., results from the BBDP2 core with those of the BARB-4 core).

6.6.3 The High-Grade Metamorphic Terranes of the Eoarchean

All Eoarchean terranes have experienced significant metamorphism of at least amphibolite facies, presenting new challenges for interpreting Fe isotopes in (meta) sedimentary rocks that we have not yet discussed in detail. A major focus of research on Eoarchean rocks has been on the Isua Supracrustal Belt (ISB) and Akilia Association (AA) of SW Greenland, as well as the Nuvvuagittuq Greenstone Belt (NGB) of the NE Superior Province of Canada. The low-strain regions of the ISB are particularly attractive because of enhanced preservation of sedimentary structures, although multiple metamorphic events involved heating in excess of 500 °C (see reviews by Nutman and Bennett 2019; Nutman et al. 2019). The ISB has figured prominently in discussions of chemical, isotopic, and morphological evidence for an early biosphere (e.g., Schidlowski 2001; Rosing and Frei 2004; Nutman et al. 2016). The AA predates the main ISB and was subjected to granulite-facies metamorphism and migmatization, and proposed sedimentary protoliths have been the subject of great debate (e.g., Nutman and Bennett 2019, and references within). The NGB has well-preserved sedimentary structures despite its high metamorphic grade, although there is no consensus on the depositional age, which could range from ~3.8 to ~4.3 Ga (e.g., O'Neil et al. 2019).

We first consider variations in WR $\delta^{56}Fe$ values for these terranes, based on the premise that Fe isotope compositions of bulk samples may have been relatively unaffected by metamorphism, at least on the hand-sample scale (Fig. 6.50). The greatest changes in bulk Fe isotope compositions would be expected if metamorphic fluids contained high chloride contents (e.g., Chou and Eugster 1977), and it has been proposed that exsolution of Cl^--rich Fe-bearing fluids at sub-solidus igneous temperatures will increase bulk-sample $\delta^{56}Fe$ values by a few tenths ‰ (Heimann et al. 2008). The range in WR $\delta^{56}Fe$ values for the ISB, AA, and NGB suites, however, exceeds that expected for loss of a Fe–Cl fluid during prograde metamorphism. As discussed by Dauphas et al. (2004, 2007a, b), $\delta^{56}Fe$-Ti/Fe (or Fe/Ti) variations allow distinction between authigenic sedimentary components (low Ti/Fe) and igneous/clastic components (high Ti/Fe), as well as some assessment of Fe mobility. For the metasedimentary units of the ISB that are generally accepted to reflect oxide-silicate IFs (e.g., Dymek and Klein 1998), the data suggest positive $\delta^{56}Fe$ values between +0.5 and +1.3‰ for the authigenic (low Ti/Fe) component (Fig. 6.50a). The $\delta^{56}Fe$ values, at low Ti/Fe, for carbonate IF samples from Isua, however, are significantly lower, suggesting a mineralogical control that is consistent with expected Fe isotope fractionation factors for oxides versus carbonates (see Chap. 3). Isua metacarbonates have a wide range in $\delta^{56}Fe$

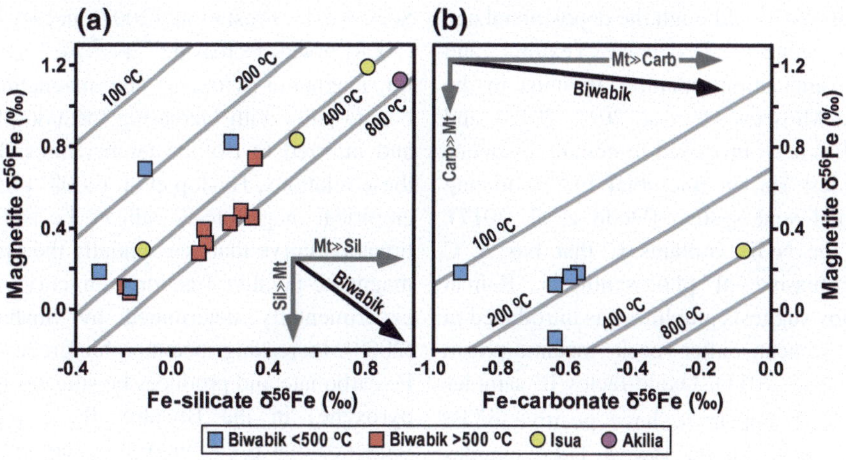

Fig. 6.51 Isotopic fractionations among magnetite, Fe-silicate (pyroxene), and Fe-carbonate (ankerite-siderite) for IFs that experienced high-grade metamorphism, including the ~1.9 Ga Biwabik IF (Lake Superior region) and ~3.8 Ga IFs from Isua and Akilia (Greenland). The contact-metamorphic zone of the Biwabik IF shows systematic decreases in magnetite—Fe silicate and magnetite—Fe carbonate $^{56}Fe/^{54}Fe$ fractionations with increasing metamorphic grade and decreasing quartz-magnetite $^{18}O/^{16}O$ fractionations (Frost et al. 2007; Hyslop et al. 2008). Equilibrium $^{56}Fe/^{54}Fe$ fractionation relations for magnetite-Fe silicate (diagonal solid grey lines in "a") from Hyslop et al. (2008), and are identical within error of the experimental determination for magnetite-fayalite from Shahar et al. (2008). Equilibrium $^{56}Fe/^{54}Fe$ fractionation between magnetite and siderite (diagonal solid grey lines in "b") combines the results of Wiesli et al. (2004) and Frierdich et al. (2014b), linearly extrapolated to zero fractionation at infinite temperature as $10^6/T^2$ (T as K). Grey arrows show isotopic equilibration trajectories for different Fe molar balances among magnetite, Fe-silicate, and Fe-carbonate. Black arrows show equilibration trajectories for the Biwabik IF, where the magnetite-Fe silicate trend is based on the average molar proportion for samples heated >500 °C, and the magnetite-Fe carbonate trend is based on average molar proportions for samples heated <500 °C. These results show that the $\delta^{56}Fe$ values for magnetite will slightly decrease with increasing temperature, as noted by Hyslop et al. (2008). Because, however, Fe-carbonates are reacted out with increasing metamorphic grade, the low molar proportion of Fe-carbonate in IFs at high temperatures will produce very large increases in their $\delta^{56}Fe$ values relative to those of magnetite, indicating that primary $\delta^{56}Fe$ values for Fe-carbonates cannot be inferred from measurements of high-grade IFs. Data from Dauphas et al. (2007b), Herrick (2007), Frost et al. (2007), and Hyslop et al. (2008)

values, with some suggestion of a bi-modal grouping (Fig. 5.20a).

The AA is most famous for the proposal that it records C isotope compositions which indicate photosynthesis >3.8 Ga (Mojzsis et al. 1996). There is extensive debate over such conclusions, where criticism has focused on the age, occurrence of C, and the nature of the protolith (e.g., Fedo and Whitehouse 2002; Lepland et al. 2005; Whitehouse et al. 2009). Dauphas et al. (2007b) interpreted the positive $\delta^{56}Fe$ values for WR samples from magnetite-quartz-pyroxene rocks from Aklia Island to support the sedimentary (IF) protolith proposed by Mojzsis et al. (1996), and $\delta^{56}Fe$-Ti/Fe relations suggest a high-$\delta^{56}Fe$ authigenic component (Fig. 6.50b). Additional Fe isotope studies by Whitehouse et al. (2015), however, show that units interpreted to reflect an ultramafic igneous protolith have anomalously high $\delta^{56}Fe$ values relative to that expected for igneous rocks, and they argue that Fe isotopes in such high-grade and deformed terranes cannot be used to unambiguously infer a sedimentary protolith. They suggest that the positive $\delta^{56}Fe$ values measured in bulk samples instead records metasomatic effects.

Identification of sedimentary protoliths in the NGB is less controversial than for the AA (e.g.,

O'Neil et al. 2019), although the depositional age remains a subject of debate. Textural and chemical compositions identify IF units in the NGB (e.g., Mloszewska et al. 2012, 2013), and the NGB has been proposed to contain evidence for the oldest known microbial Fe^{2+}-oxidizing hydrothermal vent system (Dodd et al. 2017). Although the NGB contains C that has $\delta^{13}C$ values permissive of photosynthesis, Raman spectroscopy suggests graphite was introduced in the later Archean, after peak metamorphism (Papineau et al. 2011). Oxide-facies IF samples from the NGB appear to have positive $\delta^{56}Fe$ values of $\sim+1$‰ for the lowest Ti/Fe samples (Fig. 6.50c), which has been taken to indicate partial oxidation of hydrothermal Fe^{2+} under low-O_2 conditions (Dauphas et al. 2007b; O'Neil et al. 2007). Such a conclusion is consistent with inferences from the magnetite-quartz IF from the ISB, as well as the magnetite-quartz-pyroxene units from the AA that were proposed to reflect an IF protolith (although this is controversial). It is somewhat concerning, however, that some of the NGB samples identified as igneous units have $\delta^{56}Fe$ values that are significantly higher than those of average crust (Fig. 6.50c), raising the possibility of Fe isotope changes during high-grade metamorphism.

Numerous studies of the Fe isotope compositions of individual minerals from sedimentary protoliths from the ISB, AA, and NGB have been motivated by the goal of isolating authigenic components (Dauphas et al. 2004, 2007a, b; Herrick 2007; Whitehouse and Fedo 2007; Craddock and Dauphas 2011; Czaja et al. 2013; Yoshiya et al. 2015a). In only a few cases, however, have co-existing mineral pairs been analyzed to assess the degree of Fe isotope exchange during the high grades of metamorphism that these terranes experienced. In Fig. 6.51 we consider inter-mineral Fe isotope variations among magnetite, Fe-silicate (pyroxene), and Fe-carbonate for samples from the ISB and AA, within the context of a larger dataset that is available for the contact metamorphic zone of the $\sim$1.9 Ga Biwabik IF (see Sect. 6.5.1). Frost et al. (2007) and Hyslop et al. (2008) noted decreasing $^{56}Fe/^{54}Fe$ fractionations for magnetite-pyroxene and magnetite-Fe carbonate pairs with increasing metamorphic grade and inferred O isotope temperatures. Based on these relations, Hyslop et al. (2008) proposed an empirical magnetite-Fe silicate Fe isotope fractionation curve that is essentially the same as the magnetite-fayalite fractionation curve that was experimentally determined by Shahar et al. (2008). Increasing metamorphic grade consumes Fe carbonate and produces Fe silicates (primarily pyroxene) in the Biwabik IF, a paragenetic sequence that is common in metamorphosed IFs (e.g., Klein 2005). This is accompanied by a slight decrease in $\delta^{56}Fe$ values for magnetite through isotopic exchange with pyroxene. A very different response is seen for Fe carbonates, however, where very large increases in $\delta^{56}Fe$ values occur through exchange with magnetite, driven by the relatively low Fe fraction that is contained in carbonate. These relations suggest that measured $\delta^{56}Fe$ values for magnetite in the ISB, AA, and NGB suites would be minimums relative to those of magnetite that formed authigenically, prior to significant metamorphism. Importantly, however, the magnetite-carbonate relations from the Biwabik IF indicate that measured $\delta^{56}Fe$ values for Fe carbonates cannot be taken as indicative of those that reflect pre-metamorphic carbonate formation.

Craddock and Dauphas (2011) interpreted $\delta^{56}Fe$-$\delta^{13}C$ relations among ISB carbonates to indicate DIR at 3.8 Ga, an important finding as this would place development of this metabolism much earlier than what is inferred from the discussions above. They focused on two aspects of the carbonate data from the ISB, first the moderately negative $\delta^{13}C$ values between −4 and −6‰ for Fe-rich carbonates, which might indicate respiration of C_{org}, and the positive $\delta^{56}Fe$ values for Fe-carbonates that indicated disequilibrium with expected seawater compositions and hence a possible microbial origin. The positive $\delta^{56}Fe$ values were interpreted in the framework

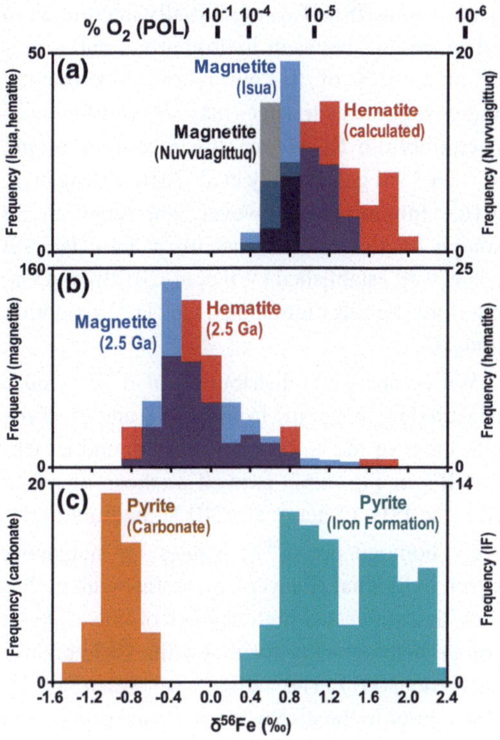

Fig. 6.52 Distribution of δ^{56}Fe values for (**a**) oxide minerals from ~3.8 Ga IFs (Isua and Nuvvuagittuq), (**b**) oxide minerals from ~2.5 Ga IFs (Kuruman and Brockman IFs), and (**c**) pyrite from different lithologies from Isua. Overlapping distributions shown by combined shading (e.g., blue and red overlap shown as purple). Top scale shows % O_2 relative to Present Ocean Level (POL) using dispersion/reaction model of Czaja et al. (2012) and a ^{56}Fe/^{54}Fe fractionation factor between Fe^{3+}-hydroxide and Fe^{2+}_{aq} of +3‰. For panel A, distribution for magnetite as measured, whereas distribution for hematite is calculated assuming the Fe^{2+} component of magnetite had a δ^{56}Fe value of zero ($\delta^{56}Fe_{hematite} = \delta^{56}Fe_{magnetite}/\frac{2}{3}$). For panels B and C, data shown as measured. The positive δ^{56}Fe values for ~3.8 Ga IFs (measured magnetite or inferred hematite values) indicate low O_2 contents for seawater, between ~10^{-5} and $10^{-6\%}$ POL, in contrast to the significantly lower δ^{56}Fe values for the large ~2.5 Ga IFs that indicate much high oxygen contents and different Fe biogeochemical cycling. The bi-modal distribution of δ^{56}Fe values for pyrite from Isua between carbonate- and IF-lithologies indicate different Fe pathways for pyrite formation at 3.8 Ga. Data from Johnson et al. (2003, 2008a), Rouxel et al. (2005), Dauphas et al. (2007a), Herrick (2007), Whitehouse and Fedo (2007), Steinhoefel et al. (2010), Craddock and Dauphas (2011), Czaja et al. (2013), Li et al. (2013b, 2015), and Yoshiya et al. (2015a)

set by Heimann et al. (2010) for the 2.5 Ga Kuruman IF, where positive δ^{56}Fe values for siderite/ankerite were inferred to record DIR of Fe^{3+}-hydroxides that had previously lost a low-δ^{56}Fe Fe^{2+}_{aq} component. An alternative interpretation is that the positive δ^{56}Fe values for Fe-carbonate from the ISB reflected complete reduction by DIR of Fe^{3+}-hydroxides that had positive δ^{56}Fe values; this finds support in the inferred positive δ^{56}Fe values of the authigenic component of Isua IFs (Fig. 6.50). The relations for the Biwabik IF in Fig. 6.51, however, indicate that positive δ^{56}Fe values for Fe carbonate in high-grade metamorphic rocks are likely to reflect equilibration with magnetite and Fe-silicate, which negates the Fe isotope evidence for DIR invoked by Craddock and Dauphas (2011). The Fe isotope compositions of Fe carbonates in these high-grade rocks cannot, therefore be used to infer disequilibrium with seawater.

A wide variety of models have been invoked for carbonate paragenesis in the ISB. During prograde metamorphism, siderite breakdown occurs between ~350 and ~450 °C (French 1971) via the reaction:

$$6FeCO_3 \rightarrow 2Fe_3O_4 + C + 5CO_2(\pm CO) \quad (6.20)$$

During prograde metamorphism, the molar proportion of Fe contained in siderite will decrease with increasing temperature, producing large changes in the δ^{56}Fe values of the remaining siderite upon exchange with an increasing magnetite (± Fe silicate) reservoir (see Fig. 6.51b). In terms of C isotope compositions, concomitant loss of CO_2 will decrease the δ^{13}C values of the remaining C (siderite and graphite), an effect recognized since the early work of Schidlowski et al. (1979). Much of the debate on δ^{13}C values for C in the ISB-AA suites has focused on graphite and the potential preservation of photosynthetic C isotope compositions, and a number of studies have concluded that loss of CO_2 during prograde metamorphism may explain some of the negative δ^{13}C values measured for graphite (e.g., Eiler et al.

1997; van Zuilen et al. 2003). A decrease in $\delta^{13}C$ values for graphite with progressive CO_2 loss could be accompanied by changes in the $\delta^{13}C$ values for siderite, given graphite-carbonate $^{13}C/^{12}C$ fractionations at elevated temperatures and C isotope mass-balance constraints (Kitchen and Valley 1995; Baumgartner and Valley 2001), and this is supported by evidence for graphite-carbonate C isotope equilibrium in the ISB (Ueno et al. 2002). In addition to metamorphic reactions for protoliths that involved authigenic siderite, other workers have argued that many (if not all) of the Isua carbonates were produced by metasomatic reactions (Rose et al. 1996), although Nutman et al. (2010) counter that seawater-like REE + Y contents can be taken to indicate that some ISB carbonates are sedimentary in origin. Collectively, however, these considerations suggest that moderately negative $\delta^{13}C$ values for Fe-carbonates cannot be taken at face value to indicate a role for DIR, given the high metamorphic grade and complex metamorphic/ metasomatic history of even the low-strain regions of the ISB. This is not to say that the C and Fe isotope compositions of Fe-carbonates in high-grade metamorphic terranes cannot be understood, but this seems likely to require detailed petrologic studies as well as assessment of isotopic exchange among all C and Fe components.

Magnetite paragenesis in IFs has been classically assumed to reflect reaction between poorly crystallized Fe^{2+} and Fe^{3+} gels and/or reaction between Fe^{3+}-hydroxides and Fe^{2+}_{aq} in non-lithified sediments (e.g., Klein 2005). Metamorphic breakdown of Fe-carbonates (Eq. 6.20), however, raises the possibility that magnetite in the ISB IFs is at least in part metamorphic in origin. In addition, Nutman et al. (2017) question the assumption that magnetite in ISB IFs ultimately had its origin in Fe^{3+}-hydroxide precipitates, instead calling upon breakdown of authigenic Fe^{2+}–Fe^{3+} clays that formed via partial oxidation of Fe^{2+}_{aq}. The likely presence of aqueous Si supports the concept of authigenic Fe^{2+}–Fe^{3+}-Si gels as a precursor sediment, and this finds support in the range of Si isotope compositions measured for ISB IFs (Heck et al. 2011); the large range in $\delta^{30}Si$ values for the ISB IFs were initially interpreted to reflect mixing between hydrothermal and continental sources of Si, but recent experimental studies suggest such ranges may reflect differences in equilibrium Si isotope fractionations among Fe^{2+}–Fe^{3+}-Si gels (Reddy et al. 2016; Zheng et al. 2016). Importantly, however, the range in Fe isotope fractionation factors for at least Fe^{3+}-Si gels is well established (Wu et al. 2012b) and can therefore be accommodated in Fe^{2+}-oxidation models.

We compare the distribution of $\delta^{56}Fe$ values measured for magnetite from the ISB and NGB IFs with those of the ~2.5 Ga Kuruman and Brockman IFs in Fig. 6.52, as well as those of pyrite from the ISB. Czaja et al. (2013) measured relatively homogenous $\delta^{56}Fe$ values for magnetite within individual IF layers, a contrast with earlier work that suggested high degrees of fine-scale Fe isotope heterogeneity for magnetite (Whitehouse and Fedo 2007). The earlier work was done by SIMS, prior to the discovery of crystal orientation effects for certain high-symmetry minerals, including magnetite (Kita et al. 2011), whereas the work of Czaja et al. (2013) involved fs-LA, which is not affected by crystal orientation. In addition, Czaja et al. (2013) considered the relations between the $\delta^{56}Fe$ value measured for magnetite and possible precursor Fe^{3+}-hydroxides, and the calculated $\delta^{56}Fe$ values for hematite in Fig. 6.52a reflect their model of simple addition of 1/3 Fe^{2+} with a $\delta^{56}Fe$ value of zero to 2/3 for hematite (see their Fig. 4). As discussed above, the effects of metamorphic equilibration and/or production of metamorphic magnetite will tend to lower the $\delta^{56}Fe$ values and hence the calculated $\delta^{56}Fe$ values for hematite in Fig. 6.52a should be considered minimums. Variations in photic-zone O_2 relative to the $\delta^{56}Fe$ values calculated for hematite are shown along the top scale, using the dispersion/reaction model discussed in Sect. 6.6.2 (see also Fig. 6.49). Calculated O_2 levels lie between 10^{-6} and 10^{-5}% POL, which is essentially anoxic. It is important to note, however, that the confidence with which O_2 estimates may be made based on the ISB is much lower than that which was possible for the Marble Bar Chert and

Mapepe Formation because of complexity in magnetite paragenesis and metamorphic equilibration for the ISB samples. Nevertheless, the inferred very low photic-zone O_2 contents at ~3.8 Ga are consistent with estimates for 3.5–3.2 Ga as discussed in Sect. 6.6.2. These estimates are in line with recent work based on U abundances in the ISB that suggest very low (although not zero) O_2 contents during deposition of the IFs (Frei et al. 2016).

At this point, it is useful to bring back the data from the large late Neoarchean and early Paleoproterozoic IFs for context. In contrast to magnetite from the ISB, magnetite from the ~2.5 Ga IFs of the Kuruman and Brockman formations largely have $\delta^{56}Fe$ values that are lower than that of average crust (Fig. 6.52b). As discussed in Sect. 6.5.4, hematite and magnetite from the ~2.5 Ga IFs are generally not in Fe isotope equilibrium (Fig. 6.32), nor are they in Fe isotope equilibrium with siderite (Fig. 6.33). The contrast in Fe isotope compositions for oxides in the ISB relative to the ~2.5 Ga IFs suggests a major change in the Fe redox cycle, where, as discussed above, DIR played a major role in IF genesis at ~2.5 Ga. Given evidence for relatively high photic-zone O_2 contents at the end of the Neoarchean and early Paleoproterozoic (see Sects. 6.5.4, 6.5.5, and 6.5.6), complete or near-complete oxidation of Fe^{2+}_{aq}, generated by a benthic Fe shuttle and/or sourced to hydrothermal fluids, would be expected, relative to the low extents of oxidation inferred for the Paleoarchean and Eoarchean. Importantly, these contrasts correlate with the size of the IFs compared in Fig. 6.52a and b, where the Isua IFs have an "Fe footprint" of $\sim 4 \times 10^{11}$ mol km^{-2}, approximately two-orders-of-magnitude smaller than that of the ~2.5 Ha Hamersley and Transvaal basins, whose "Fe footprints" are $\sim 2 \times 10^{13}$ mol km^{-2} (Czaja et al. 2013). These relations suggest a far more limited Fe redox cycle in the Eoarchean, including a limited role for biologically-driven Fe redox cycling, as compared to younger sequences.

In situ analysis of pyrite from a wide variety of lithologies from the ISB demonstrate a surprisingly large range in $\delta^{56}Fe$ values, comparable to that observed in younger Precambrian sediments or modern marine sediments, compared to the relatively restricted range measured for oxides (Fig. 6.52c). Focusing on two major lithologies, pyrite from carbonate units have consistently negative $\delta^{56}Fe$ values, whereas those from IF units have positive (and variable) $\delta^{56}Fe$ values (Whitehouse and Fedo 2007; Yoshiya et al. 2015a). Yoshiya et al. (2015a) ascribe the positive $\delta^{56}Fe$ values for pyrite in the IFs to reflect inheritance from a high-$\delta^{56}Fe$ Fe^{3+}-oxide precursor that was produced by small extents of oxidation of Fe^{2+}_{aq}. This interpretation is possible, as the $\delta^{56}Fe$ values for pyrite in the IFs broadly overlap those for the calculated hematite component, although a large fraction of the $\delta^{56}Fe$ values for pyrite are higher than those measured for magnetite (Fig. 6.52). The negative $\delta^{56}Fe$ values for pyrite in the carbonate units is ascribed by Yoshiya et al. (2015a) to a low-$\delta^{56}Fe$ Fe^{2+}_{aq} source generated by DIR. These interpretations mirror those put forth by Yoshiya et al. (2015b) to explain similarly large ranges in $\delta^{56}Fe$ values for pyrite from the Barberton Greenstone Belt.

As discussed above, however, it remains unclear if the C and Fe isotope compositions for ISB carbonates can be confidently used to support a DIR origin. Moreover, we are again faced with the quandary that a wide variety of Fe isotope fractionations are associated with pyrite formation, making it difficult to uniquely ascribe Fe sources and pathways based on $\delta^{56}Fe$ values alone. In the case of the ISB, these uncertainties are compounded by the fact that at least some pyrite post-dates peak metamorphism in the ISB, as suggested by U-Th–Pb geochronology and Pb isotope compositions of sulfides (Frei et al. 1999; Frei and Rosing 2001; Whitehouse et al. 2005). The origin(s) of the wide range in $\delta^{56}Fe$ values for ISB pyrite therefore remains unclear, and combined S-Fe–Pb isotope studies may be required to fully understand pyrite paragenesis and Fe pathways in these high-grade metamorphic rocks before they can be used to infer ancient marine processes.

6.7 Precambrian Earth: Synthesis of the Eoarchean Through Paleoproterozoic

We synthesize our discussion of Precambrian Earth in this last section, with special attention to the Eoarchean through Paleoproterozoic, given the abundant evidence for very large changes in Fe redox in this time period. Although our focus in the later sections of this chapter has been on the ancient surface Earth, it is important to stress that mantle processes have been perhaps the fundamental driver of changes in the Precambrian biogeochemical cycles of Fe through delivery of Fe^{2+}_{aq} from marine hydrothermal systems, as well as C from magmatic sources. The size of the biosphere in the past was subject to limitation of key nutrients, of which P is one of the most important over geologic timescales (Tyrell 1999), and P delivery to (and removal from) the oceans has been the subject of much research (e.g., Planavsky et al. 2010; Jones et al. 2015; Laakso and Schrag 2017; Poulton 2017; Reinhard et al. 2017). Such work shows that changes in the history of nutrient delivery over time was tied to the evolution of continental crust, which in turn was a function of when modern style plate tectonics arose to allow production of significant volumes of evolved magmas (e.g., Campbell and Taylor 1983).

It is difficult to constrain Fe^{2+}_{aq} contents and fluxes in Precambrian marine environments, although it seems clear that MOR hydrothermal fluids had much higher Fe^{2+}_{aq} contents in the Precambrian, particularly prior to development of significant seawater sulfate contents (see Sect. 6.3.3). We now recognize that estimates of seawater Fe^{2+}_{aq} based on siderite solubility (e.g., Holland 1984; Sumner 1997) are not valid to the degree that siderite reflects DIR of Fe^{3+}-hydroxide and did not form in equilibrium with seawater (Johnson et al. 2013a). Estimates for Precambrian seawater Fe^{2+}_{aq} contents and fluxes have been made based on IF deposition rates (e.g., Isley 1995), as well as scaling to ancient hydrothermal fluxes (e.g., Lowell and Keller 2003), where the latter produce exponentially decreasing Fe^{2+}_{aq} contents with decreasing age, reflecting the exponential decrease in radioactive heat production (e.g., Labrosse and Jaupart 2007). If, however, development of plate tectonics is considered, the transition from a "stagnant lid" tectonic regime (e.g., O'Neill and Debaille 2014) to modern-style plate tectonics where crust production at MORs is balanced by density-driven subduction (e.g., Korenaga 2013), there will be an inflection in the mantle potential temperature (T_P) curve, where peak T_P occurs during the onset of plate tectonics (Herzberg et al. 2010; Korenaga 2013). This suggests that marine Fe^{2+}_{aq} fluxes might have been lower prior to the start of plate tectonics, peaking at the time where T_P was a maximum, then slowly declining with decreasing age prior to the GOE.

A large part of our discussion of the Precambrian Earth has been on the Fe isotope effects of oxidation of Fe^{2+}_{aq}. Prior to appearance of free O_2, oxidation of Fe^{2+}_{aq} could have occurred by anoxygenic photosynthesis (photo-ferrotrophy; Widdel et al. 1993), and during this time the size of the biosphere would have been strongly limited by the availability of electron donors (e.g., Ward et al. 2019). The remarkable biological innovation of oxygenic photosynthesis, where the electron donor shifted to H_2O, removed electron donor limitations, although nutrient availability may have initially placed constraints on the size of the biosphere (e.g., Hayes and Waldbauer 2006). As discussed in the sections above, multiple lines of evidence point to development of oxygenic photosynthesis significantly earlier than the GOE, which finds support in some molecular clock studies of cyanobacteria (e.g., Schirrmeister et al. 2015), although questioned by some interpretations of the molecular phylogenetic record that place the origin of oxygenic photosynthesis much

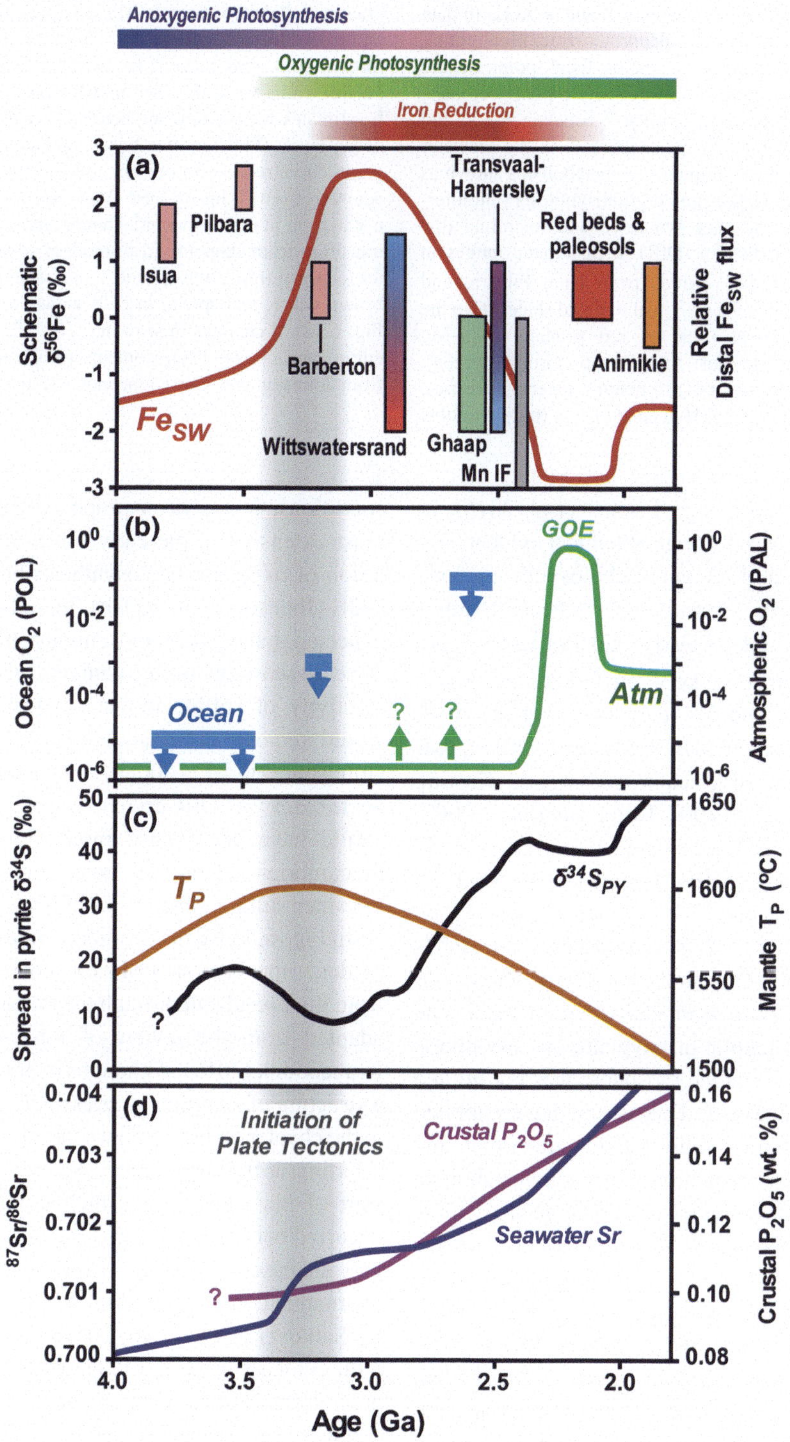
Anoxygenic Photosynthesis
Oxygenic Photosynthesis
Iron Reduction
(a)
Schematic δ56Fe (‰)
Relative Distal FeSW flux
Isua
Pilbara
Barberton
Wittswatersrand
Transvaal-Hamersley
Ghaap
Mn IF
Red beds & paleosols
Animikie
FeSW
(b)
Ocean O2 (POL)
Atmospheric O2 (PAL)
GOE
Ocean
Atm
(c)
Spread in pyrite δ34S (‰)
Mantle TP (°C)
TP
δ34SPY
(d)
87Sr/86Sr
Crustal P2O5 (wt. %)
Initiation of Plate Tectonics
Crustal P2O5
Seawater Sr
Age (Ga)

◀ **Fig. 6.53** Synthesis of geochemical and solid Earth data for the Eoarchean through Paleoproterozoic, highlighting biogeochemical changes. A major focal point of the diagram is the initiation of modern-style plate tectonics, which is taken to have occurred between ~3.4 and 3.2 Ga based on the first major increase in the seawater $^{87}Sr/^{86}Sr$ curve. Inferred microbial metabolisms shown as bars along top for anoxygenic photosynthesis (photoferrotrophy), oxygenic photosynthesis, and microbial dissimilatory iron reduction (DIR). **a** Schematic ranges in $\delta^{56}Fe$ values for Fe^{3+}-oxides from Isua, Pilbara, and Barberton (pink boxes), proximal to distal shales from the Witwatersrand Basin (blue to red gradient), Ca–Mg (low-Fe) carbonates of the Ghaap Group (green), hydrothermal (positive ε_{Nd}) to benthic Fe (negative ε_{Nd}) mixing for the Transvaal-Hamersley basin (purple to blue gradient), Paleoproterozoic Mn IFs (grey box), red beds and paleosols (red box), and IFs from the Animikie basin (brown box). Red curve is an interpretive distal seawater Fe flux (relative scale). See text for basis of the figure. **b** estimates for photic-zone ocean O_2 relative to present ocean levels (POL) in blue boxes, and atmospheric O_2 in green curve relative to present atmosphere levels (PAL), adapted from Johnson and Van Kranendonk (2019). **c** Variation in mantle potential temperature (T_P) through a transition from stagnant-lid to modern-style plate tectonics (adapted from Johnson and Van Kranendonk 2019) in brown curve, and spread in $\delta^{34}S$ values for pyrite (from Fig. 6.12). **d** changes in seawater $^{87}Sr/^{86}Sr$ ratios (purple curve) and crustal P_2O_5 contents (pink curve), adapted from Johnson and Van Kranendonk (2019)

closer to the GOE (e.g., Fischer et al. 2016). If, however, we accept the geochemical evidence for an early origin of oxygenic photosynthesis, long before the GOE, keeping O_2 "in check" requires large reduced sinks. Clearly oxidation of Fe^{2+}_{aq} is a sink, but others likely included interaction between CH_4 and O_2 (e.g., Olson et al. 2013; Daines and Lenton 2016; Ozaki et al. 2017), oxidation of reduced volcanic gases (e.g., Kasting et al. 1993; Holland 2002; Kump and Barley 2007; Gaillard et al. 2011), and oxidation of early mafic crust (e.g., Kamber 2010; Lee et al. 2016; Smit and Mezger 2017).

The products of photosynthesis are Fe^{3+}-hydroxides and C_{org}, and very large quantities of these would have been delivered to the ocean floor under a regime of oxygenic photosynthesis during times when the biosphere was not limited by nutrient delivery. This would have supplied the reactants needed to support microbial dissimilatory iron reduction (DIR), a metabolism that we have discussed in great detail. Unlike photosynthesis, which is restricted to Bacteria, DIR occurs over a wide range of prokaryotes in the Bacteria and Archaea, under a wide range of pH and temperature and has likely been a metabolism that was active in the early Earth (e.g., Vargas et al. 1998; Weber et al. 2006; Amenabar et al. 2017). DIR is not as commonly discussed in the geochemical literature, however, compared to other respiration pathways such as sulfate reduction. Because DIR occurs under anaerobic conditions, it has been proposed that DIR was most extensive in the early Earth after the evolution of oxygenic photosynthesis but before the GOE (Johnson et al. 2008b). In addition, DIR is expected to have been most important prior to the increase seawater sulfate contents, given the high reactivity of sulfide to Fe^{3+}-hydroxides, which would remove the availability of Fe^{3+} to DIR (Johnson et al. 2008b). As discussed in Sect. 6.3.3, a DIR-driven benthic Fe shuttle would have been much more vigorous in the Precambrian relative to today, especially when seawater sulfate contents were low.

In Fig. 6.53 we bring together multiple lines of evidence that bear on the Fe biogeochemical cycle from the Eoarchean through the Paleoproterozoic, adapted from the review of Johnson and Van Kranendonk (2019). Although so far we have used a structure of stepping progressively back in time in the chapter, in Fig. 6.53 we take the approach of "starting at the beginning". A key context is the start of plate tectonics, which is taken to have occurred between ~3.3 and 3.1 Ga, based on the first increase in seawater $^{87}Sr/^{86}Sr$ ratios beyond the mantle composition (Fig. 6.53d), as proposed by Satkoski et al. (2016, 2017). This age aligns with evidence from the Pilbara Craton (Smithies et al. 2005; Van Kranendonk et al. 2007a), and is equal to, or slightly earlier than, that proposed from a variety of datasets (e.g., Shirey and Richardson 2011; Dhuime et al. 2012; Naeraa et al. 2012; Van Kranendonk and Kirkland 2016).

The 3.3-3.1 Ga increase in seawater $^{87}Sr/^{86}Sr$ ratios correlates with a marked increase in land surface area for the "fast plate" model of Korenaga et al. (2017), and immediately precedes the inferred increase in crustal P_2O_5 contents from Greber et al. (2017), suggesting a marked increase in continental weathering, evolution of evolved magmas, and increased nutrient delivery to the oceans. This is independently confirmed by high chemical alteration indices for 3.2 Ga rocks from the Barberton Greenstone Belt (Hessler and Lowe 2006). Although Flament et al. (2008) suggested that the continents were largely submerged in the early Archean, their conclusion is a direct result of an assumed high crustal geotherm, but this is not supported by petrologic or thermobarometric studies of Archean terrains (Burke and Kidd 1978; Galer and Mezger 1998; Brown 2007; Satkoski et al. 2017). We therefore take the increase in seawater $^{87}Sr/^{86}Sr$ as overriding support for significant continental weathering in the late Paleoarchean-early Mesoarchean.

In depicting seawater Fe^{2+}_{aq}, we schematically illustrate a distal seawater Fe^{2+}_{aq} flux (Fe_{SW}), specifically envisioning a flux that is far from a MOR plume source to avoid complications of near-vent precipitation. Initially Fe_{SW} fluxes are proposed to have increased as T_P increased prior to development of plate tectonics. Small extents of oxidation of a relatively small reservoir of Fe^{2+}_{aq} in the Eoarchean can explain both the positive $\delta^{56}Fe$ values for the Isua and Pilbara sequences and their relatively small volume, when the oceans and atmosphere were essentially anoxic (Fig. 6.53a, b). Initiation of plate tectonics and evolution of oxygenic photosynthesis is proposed at ~3.2 Ga based on the Fe isotope and U content evidence for a marine O_2 gradient at 3.2 Ga from the Barberton Greenstone Belt, and we suggest a rapid increase in Fe_{SW} as an extensive MOR system was developed at the peak in T_P. Importantly, the limited spread in the $\delta^{34}S$ values for pyrite indicates low seawater sulfate contents, allowing high Fe_{SW}. The common occurrence of MIF-S during this time indicates low atmospheric O_2 despite increased ocean O_2 (Fig. 6.53b), demonstrating decoupling between the atmosphere and the shallow oceans.

The early Mesoarchean was a period when we would expect a benthic Fe shuttle to have been most vigorous relative to any other time in Earth history, given the evidence for both high Fe_{SW} and oxygenation of the photic zone. Oxygenic photosynthesis was producing Fe^{3+}-hydroxides (via oxidation of Fe^{2+}_{aq}) and C_{org} at this time, but seawater sulfate remained low, and hence Fe^{3+} was abundantly available to DIR. In Fig. 6.53 we illustrate the proximal to distal Fe isotope variations seen in the Wittwatersrand basin that are interpreted to reflect an extensive benthic Fe shuttle. As discussed in Sect. 6.6.1, not all workers, however, interpret the negative $\delta^{56}Fe$ values in the Fe-rich lithologies of the Wittwatersrand-Pongola basins to reflect a benthic shuttle, instead arguing for abiologic oxidation of Fe^{2+}_{aq}. Toward the end of the Mesoarchean, seawater sulfate contents began to increase, as marked by the increasing spread in $\delta^{34}S$ values for pyrite (Fig. 6.53c), and declining T_P suggests decreasing Fe_{SW}.

Multiple lines of evidence indicate a significant oxygenation of the oceans in the Neoarchean through early Paleoproterozoic, as discussed in Sect. 6.5.6, and Fe-Mo isotope modeling of the low-Fe Ca–Mg carbonates of the Ghaap Group showed photic-zone O_2 contents on the order of ~10% relative to today, yet the atmosphere remained essentially anoxic (Fig. 6.53b). A benthic Fe shuttle was invoked to explain the negative $\delta^{56}Fe$ values found in shales and IFs of the Transvaal-Hamersley basin, the latter of which was supported by Nd isotope data that allowed distinction between Fe shuttled from the shelf and distal MOR hydrothermal sources. Carbon, O, and Fe isotope studies of IF minerals document extensive DIR in IF sediment prior to lithification. Although seawater sulfate contents were likely higher in the late Neoarchean to early Paleoproterozoic as compared to earlier time, based on the spread in $\delta^{34}S$ values for pyrite, sulfides are quite rare in the large IFs of this time period, indicating that Fe^{3+} would have still been extensively available in at least some marine environments to support DIR. Nevertheless, Fe speciation studies of shales demonstrate that euxinic zones began to become common, interposed with oxic and ferruginous zones. Slightly younger Mn IFs extend to some of the lowest $\delta^{56}Fe$ values yet measured for Precambrian rocks, perhaps indicating very high Eh conditions.

The GOE should have been accompanied by a strong decrease in Fe_{SW}, changing the nature of the marine Fe isotope record, but its effects are well preserved in red beds and paleosols, which tend to have slightly positive $\delta^{56}Fe$ values (Fig. 6.53). This could be taken to indicate limited atmospheric O_2 relative to that of today, although a wide range of $\delta^{56}Fe$ values are measured for modern soils, highlighting the importance of organic ligands in mobilizing Fe during pedogenesis under high-O_2 conditions. Evidence for limited oxygenation of marine basins in the Paleoproterozoic comes from studies of the Animikie Basin, where moderately positive $\delta^{56}Fe$ values for IFs suggest limited oxidation of Fe^{2+}_{aq}. Such conclusions are consistent with the idea that the very high O_2 levels during the GOE declined in post-GOE time, returning the oceans to a largely ferruginous condition. Atmospheric O_2 levels have been suggested to have decreased after the GOE, with possibly low levels throughout the Mesoproterozoic.

6.8 Chapter Summary

- The Fe isotope compositions of Cenozoic to Late Cretaceous Fe–Mn crusts document a very wide range in inferred seawater $\delta^{56}Fe$ values. In some cases, crusts far from continental and hydrothermal sources are homogenous in their Fe isotope compositions, reflecting a consistent dust input on timescales of 10^1 m.y. that ignored large climactic and tectonic changes. In other cases, proximity to hydrothermal sources record large swings in $\delta^{56}Fe$ values over short time periods, reflecting variable capture of hydrothermal Fe on a regional scale. One crust that provides a nearly 80 m.y. record of seawater includes highly positive $\delta^{56}Fe$ excursions not seen in modern seawater, which might record the effects of Fe fertilization during large mid-Cenozoic ignimbrite flare-ups that would correlate with global cooling, increased weathering, and decreased pCO_2. Alternatively, highly positive $\delta^{56}Fe$ values for Cenozoic seawater could reflect extensive pyrite burial during periods of expanded euxinia and high seawater sulfate contents.
- Marine sediments deposited during Cretaceous oceanic anoxic event OAE-2 record bulk-sample $\delta^{56}Fe$ values from moderately negative to moderately positive, deviating from the crustal average composition of clastic input. In several sections, negative $\delta^{56}Fe$ excursions correlate with Fe enrichment (high Fe_{tot}/Al) and Mo enrichment, suggesting a euxinic trap of a benthic Fe shuttle during OAE-2. One section shows highly negative $\delta^{56}Fe$ values in carbonate-rich rocks prior to OAE-2 but no enrichment in Fe_{tot}/Al ratios, which is inconsistent with a benthic Fe shuttle source. In other sections, including those from oxic shelf settings, $\delta^{56}Fe$ values are slightly positive but there is no clear correlation with Fe_{tot}/Al, suggesting partial oxidation of Fe^{2+}_{aq}. In cases where high Fe_{tot}/Al ratios are measured with little deviation in $\delta^{56}Fe$ values from the crustal average, a hydrothermal source may be inferred. Overall, the Fe isotope data for OAE-2 indicate a larger extent of Fe cycling than in the modern oceans, consistent with anoxic and euxinic conditions and regional heterogeneity.
- Precambrian marine shales/mudstones deposited before the GOE may be highly enriched in Fe_{tot}/Al relative to the crustal average, indicating high Fe contents in marine systems, particularly in the Mesoarchean. The fraction of Fe^{3+} (relative to total Fe) for marine shales/mudstones clusters about (or is slightly higher than) the average for igneous rocks prior to the GOE, but post-GOE marine

shales/mudstones are significantly more oxidized.

- Hydrothermal and benthic Fe fluxes to the Precambrian oceans should have been much higher than that of today, reflecting fundamental differences in oxygen and sulfide levels. The high Fe/S ratios of Precambrian hydrothermal fluids would have increased Fe^{2+}_{aq} contents and suppressed Fe-S precipitation. Benthic Fe fluxes should have been very high when seawater sulfate and oxygen contents were low, assuming a supply of reactive Fe^{3+} and C_{org}. This would reflect the decreased retention of Fe^{2+} in marine sediments when sulfide diagenesis was low, low extent of sorbed Fe^{2+}, and minimal oxidation of pore water Fe^{2+}_{aq} under low-O_2 conditions.
- Using modern marine sediments for insight, the $\delta^{56}Fe$ values for pyrite in Proterozoic and Archean marine sedimentary rocks are likely to record a wide variety of Fe pathways during formation. In general, where sulfide diagenesis was high, pyrite should have higher $\delta^{56}Fe$ values than where sulfide diagenesis was minimal. Very low $\delta^{56}Fe$ values for pyrite may reflect exchange between Fe^{2+}_{aq} and Fe^{3+}-hydroxides under low-sulfide conditions, and may not be diagnostic of DIR. In contrast, pyrite that formed under equilibrium conditions should have very high $\delta^{56}Fe$ values, and such conditions would have been most likely when Fe^{2+}_{aq} was high and sulfide low, such as in the early Archean. The strong pathway dependence of pyrite formation, and common expression of kinetic isotope effects, indicates that Fe isotope compositions of pyrite are a poor proxy for water column redox processes.
- Studies of highly reactive Fe inventories that are used to infer oxic, ferruginous, and euxinic conditions in ancient marine shales will benefit from addition of Fe isotope data that can help avoid mis-interpretations of Fe proxy data, due to the constraints provided by isotopic data in terms of net Fe loss or gain.
- A very wide range in $\delta^{56}Fe$ values are measured for Neoproterozoic shales, and these are best interpreted in the context of reactive Fe measurements that identify authigenic oxide and sulfide components. For example, the wide range of Fe isotope compositions for some Tonian-age shales do not reflect partial oxidation of seawater Fe^{2+}_{aq} but instead record loss of a low-$\delta^{56}Fe$ Fe^{2+} component as part of a benthic Fe shuttle. A similarly wide range in $\delta^{56}Fe$ values for Cryogenian-age shales reflects mixtures between authigenic oxide and sulfide components during net Fe addition, where the wide range of Fe isotope compositions of the inferred oxide components records heterogeneous O_2 levels in the Cryogenian oceans. The large changes in Fe isotope compositions for Ediacaran-age shales seem likely to reflect decreased seawater Fe^{2+} contents during oxygenation of the oceans, where Fe isotope compositions of seawater became sensitive to not just the extent of Fe^{2+}_{aq} oxidation, but other factors such as pyrite burial and changes in Fe/S ratios of hydrothermal fluids.
- Authigenic components of Cryogenian-age iron formations (IFs) have $\delta^{56}Fe$ values that range from the crustal average to ~2.5‰, and in several cases $\delta^{56}Fe$ values increase with increasing stratigraphic height, which is interpreted to record shallowing of the redoxcline during the marine transgressions that occurred during deglaciation. Overall, the $\delta^{56}Fe$ values of the IFs are higher than those of post-Sturtian shales, consistent with an increase in oxygenation of the shallow oceans after the Cryogenian.
- IFs and reds of post-GOE Paleoproterozoic age have slightly positive average $\delta^{56}Fe$ values, qualitatively consistent with moderately oxidizing conditions, supported by Cr and Mo isotope studies. Some red beds that contain detrital oxides have $\delta^{56}Fe$ values that are only slightly higher than that of average crust, suggesting weathering of source terranes under oxidizing conditions. The highly positive $\delta^{56}Fe$ values that occur in some Neoproterozoic shales and IFs are absent in the post-GOE

Paleoproterozoic, possibly indicating higher oxidation state in some environments after the GOE. Pyrite has a wide range in δ^{56}Fe values in post-GOE Paleoproterozoic shales, similar to what is found in the Neoproterozoic, reflecting a variety of Fe pathways during pyrite formation.

- Paleosols formed before and after the GOE have similarly positive δ^{56}Fe values, despite forming under very different redox conditions. Weathering after the GOE may produce Fe enrichment and high-δ^{56}Fe values for oxides through partial oxidation of groundwater Fe^{2+}_{aq}. Prior to the GOE, weathering under anoxic conditions is proposed to produce Fe^{2+}_{aq} that has negative δ^{56}Fe values, which might characterize the Fe isotope compositions of riverine Fe prior to the rise of atmospheric O_2, although it is not yet clear this is consistent with expected Fe isotope fractionation factors between Fe^{2+}_{aq} and Fe^{2+}-bearing weathering products.
- Despite the common application of Rayleigh models to Fe isotope data, open-system marine systems require use of a flux-based dispersion/reaction model, which, at low levels of O_2, produces sensitive relations between O_2 contents and Fe isotope compositions of Fe^{3+} precipitates. At high O_2 levels, the most sensitive relation between Fe isotopes and O_2 is in lithologies that capture photic zone Fe^{2+}_{aq}, such as shallow-water Ca–Mg carbonates.
- Based on Fe–Nd isotope relations, some of the very large IFs of early Paleoproterozoic and late Neoarchean age record a mixture of two Fe sources: (1) MOR hydrothermal (high-δ^{56}Fe, -ε_{Nd}(t)), and (2) DIR-driven benthic Fe shuttle (low-δ^{56}Fe, -ε_{Nd}(t)). Hematite, magnetite, and siderite in these IFs did not form in Fe isotope equilibrium with seawater, nor with each other, but instead record authigenic conversion of Fe^{3+}-hydroxides by DIR, where Fe isotope compositions are largely inherited from the precursor Fe^{3+}-hydroxides; these may reflect mixtures between hydrothermal and benthic Fe sources. The authigenic origin of these minerals is confirmed by C, O, and Sr isotope compositions.
- Early Paleoproterozoic Mn IFs deposited before the GOE have some of the lowest δ^{56}Fe values measured in the rock record, and these reflect deposition under very high Eh conditions. Coexistence of microbial Mn and Fe reduction will tend to lower δ^{56}Fe values for a DIR-driven benthic Fe shuttle, and high Eh conditions will favor higher degrees of Fe^{2+}_{aq} oxidation, further lowering δ^{56}Fe values for a benthic Fe shuttle, as well as providing a means for lowering δ^{56}Fe values for a hydrothermal component. Correlations between Fe and Mo isotopes for Mn IFs support an origin via an Fe–Mn benthic shuttle at high-Eh marine conditions, but this is only one explanation for the Fe-Mo isotope variations.
- Iron and Mo isotope data on Neoarchean shales show broadly increasing oxygenation of the photic zone of the oceans from 2.7 to 2.5 Ga. Coeval shallow-water Ca–Mg carbonates have Fe–Mo isotope variations that suggest photic zone O_2 contents as high as $\sim$12% Present Ocean Levels (POL) in the later Neoarchean. Correlations between Fe_{tot}/Al and δ^{56}Fe values in shales deposited at the end of the Neoarchean are well explained by a benthic Fe shuttle that had a very low δ^{56}Fe value, consistent with Mo and Tl isotope data that indicate high O_2 in the oceans, where MnO_2 was stable on the continental shelf.
- Mesoarchean shales have the highest Fe_{tot}/Al ratios measured in the Precambrian rock record, suggesting a highly ferruginous ocean that had Fe^{2+}_{aq} higher than at any other period, yet their high abundance of magnetite indicates significant Fe oxidation. Evidence for high-Eh conditions in the shallow oceans includes the abundance of Mn-rich shales and IFs, and correlations between Fe and Mo isotope compositions. Microbial Fe and Mn reduction is documented in the C and O isotope compositions of carbonates, indicating

that Fe and Mn were subject to active redox cycling. Proximal to distal transects additionally support the presence of vigorous benthic Fe shuttling. Collectively, a variety of isotopic data indicate a highly diverse microbial ecosystem at ~3 Ga.

- Paleoarchean jaspers of 3.2 and 3.5 Ga age record temporal changes in Fe isotope compositions and seawater U contents (calculated using U-Th–Pb geochronology) that indicate essentially anoxic seawater at 3.5 Ga, but a significant O_2 gradient in seawater by 3.2 Ga, where shallow seawater attained O_2 contents up to ~0.1% Present Ocean Levels (POL). This may indicate that oxygenic photosynthesis evolved prior to 3.2 Ga.
- Eoarchean metasedimentary rocks, including IFs, were subjected to high grades of metamorphism that induced small decreases in the $\delta^{56}Fe$ values for magnetite, but large increases in the $\delta^{56}Fe$ values for Fe-carbonates. Based on analyses of magnetite, low-temperature Fe^{3+}-hydroxide precursors likely had positive $\delta^{56}Fe$ values that were produced by anaerobic Fe^{2+}_{aq} oxidation pathways, although there are greater uncertainties relative to younger units due to metamorphism. Breakdown of Fe-carbonates during prograde metamorphism, however, produced large changes in Fe isotope compositions, preventing their use in inferring marine Fe pathways.
- Integrating inferences from solid Earth evolution and the biogeochemical record prior to the GOE suggests that the start of plate tectonics ~3.2 Ga would have been associated with high nutrient fluxes, supporting an increasing biosphere size, as well as high seawater Fe^{2+}_{aq} fluxes. Anoxygenic photosynthesis in the Eoarchean and Paleoarchean gave way to oxygenic photosynthesis by the Mesoarchean, as well as microbial Fe^{3+} reduction, where very large masses of Fe were cycled in redox-stratified oceans. Increasing seawater sulfate contents leading up to the GOE, as well as increasing Eh conditions, led to a decline in seawater Fe contents and a decrease in biological cycling of Fe before the GOE, relative to earlier periods.

References

Adams DD, Hurtgen MT, Sageman BB (2010) Volcanic triggering of a biogeochemical cascade during Oceanic Anoxic Event 2. Nat Geosci 3(3):201–204. https://doi.org/10.1038/ngeo743

Albut G, Babechuk MG, Kleinhanns IC, Benger M, Beukes NJ, Steinhilber B, Smith AJB, Kruger SJ, Schoenberg R (2018) Modern rather than Mesoarchaean oxidative weathering responsible for the heavy stable Cr isotopic signatures of the 2.95 Ga old Ijzermijn iron formation (South Africa). Geochim Cosmochim Acta 228:157–189. https://doi.org/10.1016/j.gca.2018.02.034

Alexander BW, Bau M, Andersson P (2009) Neodymium isotopes in Archean seawater and implications for the marine Nd cycle in Earth's early oceans. Earth Planet Sci Lett 283(1–4):144–155. https://doi.org/10.1016/j.epsl.2009.04.004

Alexander BW, Bau M, Andersson P, Dulski P (2008) Continentally-derived solutes in shallow Archean seawater: rare earth element and Nd isotope evidence in iron formation from the 2.9 Ga Pongola Supergroup, South Africa. Geochim Cosmochim Acta 72(2):378–394. https://doi.org/10.1016/j.gca.2007.10.028

Algeo TJ, Luo GM, Song HY, Lyons TW, Canfield DE (2015) Reconstruction of secular variation in seawater sulfate concentrations. Biogeosciences 12(7):2131–2151. https://doi.org/10.5194/bg-12-2131-2015

Alibert C, McCulloch MT (1993) Rare earth element and neodymium isotopic compositions of the banded iron-formations and associated shales from Hamersley, Western Australia. Geochim Cosmochim Acta 57:187–204

Allwood AC, Walter MR, Kamber BS, Marshall CP, Burch IW (2006) Stromatolite reef from the Early Archaean era of Australia. Nature 441(7094):714–718. https://doi.org/10.1038/nature04764

Alt JC (1995) Sulfur isotopic profile through the oceanic crust: sulfur mobility and seawater-crustal sulfur exchange during hydrothermal alteration. Geology 23(7):585–588

Amenabar MJ, Shock EL, Roden EE, Peters JW, Boyd ES (2017) Microbial substrate preference dictated by energy demand rather than supply. Nat Geosci 10:577–581

Anbar AD, Duan Y, Lyons TW, Arnold GL, Kendall B, Creaser RA, Kaufman AJ, Gordon GW, Scott C, Garvin J, Buick R (2007) A whiff of oxygen before the great oxidation event? Science 317:1903–1906. https://doi.org/10.1126/science.1140325

Anbar AD, Knoll AH (2002) Proterozoic Ocean chemistry and evolution: a bioinorganic bridge? Science 297:1137–1142

Anderson TF, Raiswell R (2004) Sources and mechanisms for the enrichment of highly reactive iron in euxinic black sea sediments. Am J Sci 304:203–233

Asael D, Rouxel O, Poulton SW, Lyons TW, Bekker A (2018) Molybdenum record from black shales indicates oscillating atmospheric oxygen levels in the early Paleoproterozoic. Am J Sci 318 (3):275–299. https://doi.org/10.2475/03.2018.01

Asael D, Tissot FLH, Reinhard CT, Rouxel O, Dauphas N, Lyons TW, Ponzevera E, Liorzou C, Chéron S (2013) Coupled molybdenum, iron and uranium stable isotopes as oceanic paleoredox proxies during the Paleoproterozoic Shunga Event. Chem Geol 362:193–210. https://doi.org/10.1016/j.chemgeo.2013.08.003

Babechuk MG, Kleinhanns IC, Schoenberg R (2017) Chromium geochemistry of the ca. 1.85 Ga Flin Flon paleosol. Geobiology 15(1):30–50. https://doi.org/10.1111/gbi.12203

Babechuk MG, Weimar NE, Kleinhanns IC, Eroglu S, Swanner ED, Kenny GG, Kamber BS, Schoenberg R (2019) Pervasively anoxic surface conditions at the onset of the Great Oxidation Event: new multi-proxy constraints from the Cooper Lake paleosol. Precambrian Res 323:126–163. https://doi.org/10.1016/j.precamres.2018.12.029

Baldwin GJ, Turner EC, Kamber BS (2016) Tectonic controls on distribution and stratigraphy of the Cryogenian Rapitan iron formation, northwestern Canada. Precambrian Res 278:303–322. https://doi.org/10.1016/j.precamres.2016.03.014

Bankole OM, El Albani A, Meunier A, Rouxel OJ, Gauthier-Lafaye F, Bekker A (2016) Origin of red beds in the Paleoproterozoic Franceville Basin, Gabon, and implications for sandstone-hosted uranium mineralization. Am J Sci 316 (9):839–872. https://doi.org/10.2475/09.2016.02

Bao H, Koch PL (1999) Oxygen isotope fractionation in ferric oxide-water systems: low temperature synthesis. Geochim Cosmochim Acta 63(5):559–613

Bao H, Rumble D, Lowe DR (2007) The five stable isotope compositions of Fig Tree barites: Implications on sulfur cycle in ca. 3.2 Ga oceans. Geochim Cosmochim Acta 71(20):4868–4879. https://doi.org/10.1016/j.gca.2007.05.032

Barclay RS, McElwain JC, Sageman BB (2010) Carbon sequestration activated by a volcanic CO2 pulse during Ocean Anoxic Event 2. Nat Geosci 3(3):205–208. https://doi.org/10.1038/ngeo757

Barling J, Anbar AD (2004) Molybdenum isotope fractionation during adsorption by manganese oxides. Earth Planet Sci Lett 217(3–4):315–329. https://doi.org/10.1016/s0012-821x(03)00608-3

Barlow E, Van Kranendonk MJ, Yamaguchi KE, Ikehara M, Lepland A (2016) Lithostratigraphic analysis of a new stromatolite-thrombolite reef from across the rise of atmospheric oxygen in the Paleoproterozoic Turee Creek Group, Western Australia. Geobiology 14(4):317–343. https://doi.org/10.1111/gbi.12175

Barnes CE, Cochran JK (1990) Uranium removal in oceanic sediments and the oceanic U balance. Earth Planet Sci Lett 97:94–101

Bau M, Höhndorf A, Dulski P, Beukes NJ (1997) Sources of rare-earth elements and iron in paleoproterozoic iron-formations from the transvaal supergroup, South Africa: evidence from Neodymium Isotopes. J Geol 105:121–129

Bau M, Möller P (1993) Rare earth element systematics of the chemically precipitated component in Early Precambrian iron formations and the evolution of the terrestrial atmosphere-hydrosphere-lithosphere system. Geochim Cosmochim Acta 57:2239–2249

Baumgartner LP, Valley JW (2001) Stable isotope transport and contact metamorphic fluid flow. Rev Mineral Geochem 43(1):415–467

Baur ME, Hayes JM, Studley SA, Walter MR (1985) Millimeter-scale variations of stable isotope abundances in carbonates from banded iron-formations in the Hamersley Group of Western Australia. Econ Geol 80(2):270–282. https://doi.org/10.2113/gsecongeo.80.2.270

Beard BL, Handler RM, Scherer MM, Wu L, Czaja AD, Heimann A, Johnson CM (2010) Iron isotope fractionation between aqueous ferrous iron and goethite. Earth Planet Sci Lett 295(1–2):241–250. https://doi.org/10.1016/j.epsl.2010.04.006

Beard BL, Johnson CM, Von Damm KL, Poulson RL (2003) Iron isotope constraints on Fe cycling and mass balance in oxygenated Earth oceans. Geology 31 (7):629–632

Becker TW, Conrad CP, Buffett B, Müller RD (2009) Past and present seafloor age distributions and the temporal evolution of plate tectonic heat transport. Earth Planet Sci Lett 278(3–4):233–242. https://doi.org/10.1016/j.epsl.2008.12.007

Bekker A, Holland HD (2012) Oxygen overshoot and recovery during the early Paleoproterozoic. Earth Planet Sci Lett 317–318:295–304. https://doi.org/10.1016/j.epsl.2011.12.012

Bekker A, Holland HD, Wang P-L, Rumble D III, Stein HJ, Hannah JL, Coetzee LL, Beukes NJ (2004) Dating the rise of atmospheric oxygen. Nature 427:117–120

Bellefroid EJ, Hood AVS, Hoffman PF, Thomas MD, Reinhard CT, Planavsky NJ (2018) Constraints on Paleoproterozoic atmospheric oxygen levels. Proc Natl Acad Sci USA 115 (32):8104–8109. https://doi.org/10.1073/pnas.1806216115

Berner RA (2006) GEOCARBSULF: A combined model for Phanerozoic atmospheric O2 and CO2. Geochim Cosmochim Acta 70 (23):5653–5664. https://doi.org/10.1016/j.gca.2005.11.032

Berner RA, Beerling DJ, Dudley R, Robinson JM, Wildman RA (2003) Phanerozoicatmosphericoxygen. Annu Rev Earth Planet Sci 31(1):105–134. https://doi.org/10.1146/annurev.earth.31.100901.141329

Beukes NJ, Cairncross B (1991) A lithostratigraphic-sedimentological reference profile for the Late Archaean Mozaan Group, Pongola Sequence:

application to sequence stratigraphy and correlation with the Witwatersrand Supergroup. S Afr J Geol 94 (1):44–69

Beukes NJ, Dorland H, Gutzmer J, Nedachi M, Ohmoto H (2002) Tropical laterites, life on land, and the history of atmospheric oxygen in the Paleoproterozoic. Geology 30(6):491–494

Beukes NJ, Gutzmer J (2008) Origin and paleoenvironmental significance of major iron formations at the Archean-Paleoproterozoic boundary. SEG Rev 15:5–17

Beukes NJ, Swindell EPW, Wabo H (2016) Manganese deposits of Africa. Episodes 39(2). https://doi.org/10.18814/epiiugs/2016/v39i2/95779

Bézos A, Humler E (2005) The Fe3+/ΣFe ratios of MORB glasses and their implications for mantle melting. Geochim Cosmochim Acta 69 (3):711–725. https://doi.org/10.1016/j.gca.2004.07.026

Blamey NJF, Brand U, Parnell J, Spear N, Lécuyer C, Benison K, Meng F, Ni P (2016) Paradigm shift in determining Neoproterozoic atmospheric oxygen. Geology 44(8):651–654. https://doi.org/10.1130/g37937.1

Blanchard M, Poitrasson F, Méheut M, Lazzeri M, Mauri F, Balan E (2012) Comment on "New data on equilibrium iron isotope fractionation among sulfides: Constraints on mechanisms of sulfide formation in hydrothermal and igneous systems" by V.B. Polyakov and D.M. Soultanov. Geochim Cosmochim Acta 87:356–359. https://doi.org/10.1016/j.gca.2012.01.048

Blättler CL, Jenkyns HC, Reynard LM, Henderson GM (2011) Significant increases in global weathering during Oceanic Anoxic Events 1a and 2 indicated by calcium isotopes. Earth Planet Sci Lett 309(1–2):77–88. https://doi.org/10.1016/j.epsl.2011.06.029

Blättler CL, Kump LR, Fischer WW, Paris G, Kasbohm JJ, Higgins JA (2016) Constraints on ocean carbonate chemistry and pCO2 in the Archaean and Palaeoproterozoic. Nat Geosci 10(1):41–45. https://doi.org/10.1038/ngeo2844

Boland DD, Collins RN, Miller CJ, Glover CJ, Waite TD (2014) Effect of solution and solid-phase conditions on the Fe(II)-accelerated transformation of ferrihydrite to lepidocrocite and goethite. Environ Sci Technol 48 (10):5477–5485. https://doi.org/10.1021/es4043275

Bolhar R, Hofmann A, Kemp AIS, Whitehouse MJ, Wind S, Kamber BS (2017) Juvenile crust formation in the Zimbabwe Craton deduced from the O-Hf isotopic record of 3.8–3.1 Ga detrital zircons. Geochim Cosmochim Acta 215:432–446. https://doi.org/10.1016/j.gca.2017.07.008

Brocks JJ, Jarrett AJM, Sirantoine E, Hallmann C, Hoshino Y, Liyanage T (2017) The rise of algae in Cryogenian oceans and the emergence of animals. Nature 548(7669):578–581. https://doi.org/10.1038/nature23457

Brown M (2007) Metamorphic conditions in orogenic belts: a record of secular change. Int Geol Rev 49:193–234

Burke K, Kidd WS (1978) Were Archean continental geothermal gradients much steeper than those of today? Nature 272(5650):240–241

Burton KW, Ling H-F, O'Nions RK (1997) Closure of the Central American Isthmus and its effect on deep-water formation in the North Atlantic. Nature 386:382–385

Busigny V, Marin-Carbonne J, Muller E, Cartigny P, Rollion-Bard C, Assayag N, Philippot P (2017) Iron and sulfur isotope constraints on redox conditions associated with the 3.2 Ga barite deposits of the Mapepe Formation (Barberton Greenstone Belt, South Africa). Geochim Cosmochim Acta 210:247–266. https://doi.org/10.1016/j.gca.2017.05.002

Busigny V, Planavsky NJ, Goldbaum E, Lechte MA, Feng L, Lyons TW (2018) Origin of the Neoproterozoic Fulu iron formation, South China: Insights from iron isotopes and rare earth element patterns. Geochim Cosmochim Acta 242:123–142. https://doi.org/10.1016/j.gca.2018.09.006

Byerly GR, Lowe DR, Heubeck C (2019) Geologic evolution of the Barberton Greenstone Belt—a unique record of crustal development, surface processes, and early life 3.55–3.20 Ga. In: Earth's oldest rocks, pp 569–613. https://doi.org/10.1016/b978-0-444-63901-1.00024-1

Calvert SE, Pedersen TF (1993) Geochemistry of recent oxic and anoxic marine sediments: implications for the geological record. Mar Geol 113:67–88

Campbell IH, Taylor SR (1983) No water, No granites-No oceans, No continents. Geophys Res Lett 10 (11):1061–1064

Canfield DE (1998) A new model for Proterozoic ocean chemistry. Nature 396:450–453

Canfield DE (2005) The early history of atmospheric oxygen: homage to Robert M. Garrels. Annu Rev Earth Planet Sci 33(1):1–36. https://doi.org/10.1146/annurev.earth.33.092203.122711

Canfield DE, Farquhar J (2009) Animal evolution, bioturbation, and the sulfate concentration of the oceans. Proc Natl Acad Sci USA 106(20):8123–8127. https://doi.org/10.1073/pnas.0902037106

Canfield DE, Ngombi-Pemba L, Hammarlund EU, Bengtson S, Chaussidon M, Gauthier-Lafaye F, Meunier A, Riboulleau A, Rollion-Bard C, Rouxel O, Asael D, Pierson-Wickmann AC, El Albani A (2013) Oxygen dynamics in the aftermath of the Great Oxidation of Earth's atmosphere. Proc Natl Acad Sci USA 110 (42):16736–16741. https://doi.org/10.1073/pnas.1315570110

Canfield DE, Poulton SW, Knoll AH, Narbonne GM, Ross G, Goldberg T, Strauss H (2008) Ferruginous conditions dominated later neoproterozoic deep-water chemistry. Science 321(5891):949–952

Canfield DE, Poulton SW, Narbonne GM (2007) Late-neoproterozoic deep-ocean oxygenation and the rise of animal life. Science 315(5808):92–95. https://doi.org/10.1126/science.1135013

Canfield DE, Thamdrup B, Hansen JW (1993) The anaerobic degradation of organic matter in Danish coastal sediments: iron reduction, manganese reduction, and sulfate reduction. Geochim Cosmochim Acta 57:3867–3883

Caquineau T, Paquette J-L, Philippot P (2018) U-Pb detrital zircon geochronology of the Turee Creek Group, Hamersley Basin, Western Australia: timing and correlation of the Paleoproterozoic glaciations. Precambrian Res 307:34–50. https://doi.org/10.1016/j.precamres.2018.01.003

Cather SM, Dunbar NW, McDowell FW, McIntosh WC, Scholle PA (2009) Climate forcing by iron fertilization from repeated ignimbrite eruptions: the icehouse–silicic large igneous province (SLIP) hypothesis. Geosphere 5(3):315–324. https://doi.org/10.1130/ges00188.1

Caves JK, Jost AB, Lau KV, Maher K (2016) Cenozoic carbon cycle imbalances and a variable weathering feedback. Earth Planet Sci Lett 450:152–163. https://doi.org/10.1016/j.epsl.2016.06.035

Cavosie AJ, Wilde SA, Valley JW, F. EIM (2005) Magmatic δ18O in 4400–3900 Ma detrital zircons: a record of the alteration and recycling of crust in the Early Archean. Earth Planet Sci Lett 235:663–681

Chandler FW (1980) Proterozoic redbed sequences of Canada. Energy, Mines and Resources Canada, Geological Survey of Canada Bulletin 311:1–53

Chen T-Y, Ling H-F, Hu R, Frank M, Jiang S-Y (2013) Lead isotope provinciality of central North Pacific Deep Water over the Cenozoic. Geochem Geophys Geosyst 14(5):1523–1537. https://doi.org/10.1002/ggge.20114

Chen X, Ling HF, Vance D, Shields-Zhou GA, Zhu M, Poulton SW, Och LM, Jiang SY, Li D, Cremonese L, Archer C (2015) Rise to modern levels of ocean oxygenation coincided with the Cambrian radiation of animals. Nat Commun 6:7142. https://doi.org/10.1038/ncomms8142

Cheney ES (1996) Sequence stratigraphy and plate tectonic significance of the Transvaal succession of southern Africa and its equivalent in Western Australia. Precambrian Res 79:3–24

Cheng M, Li C, Chen X, Zhou L, Algeo TJ, Ling H-F, Feng L-J, Jin C-S (2018) Delayed Neoproterozoic oceanic oxygenation: evidence from Mo isotopes of the Cryogenian Datangpo formation. Precambrian Res 319:187–197. https://doi.org/10.1016/j.precamres.2017.12.007

Cheng M, Li C, Zhou L, Feng L, Algeo TJ, Zhang F, Romaniello S, Jin C, Ling H, Jiang S (2017) Transient deep-water oxygenation in the early Cambrian Nanhua Basin, South China. Geochim Cosmochim Acta 210:42–58. https://doi.org/10.1016/j.gca.2017.04.032

Chou I-M, Eugster HP (1977) Solubility of magnetite in supercritical chloride solutions. Am J Sci 277:1296–1314

Christensen JN, Halliday AN, Godfrey LV, Hein JR (1997) Climate and Ocean dynamics and the lead isotopic records in Pacific Ferromanganese Crusts. Science 277:913–918

Chu N, Johnson C, Beard B, German C, Nesbitt R, Frank M, Bohn M, Kubik P, Usui A, Graham I (2006) Evidence for hydrothermal venting in Fe isotope compositions of the deep Pacific Ocean through time. Earth Planet Sci Lett 245(1–2):202–217. https://doi.org/10.1016/j.epsl.2006.02.043

Clarkson MO, Stirling CH, Jenkyns HC, Dickson AJ, Porcelli D, Moy CM, Pogge von Strandmann PAE, Cooke IR, Lenton TM (2018) Uranium isotope evidence for two episodes of deoxygenation during Oceanic Anoxic Event 2. Proc Natl Acad Sci USA 115(12):2918–2923. https://doi.org/10.1073/pnas.1715278115

Clayton RE, Nederbrag AJ, Malinovsky D, Andersson P, Thurow J (2007) 10. Data report: iron isotope geochemistry of mid-cretaceous organic-rich sediments at demerara rise (ODP Leg 207). Proc ODP Sci Results 207:1–14. https://doi.org/10.2973/odp.proc.sr.207.109.2006

Cochrane JM (2009) Diagenetic carbonates and biogeochemical cycling of organic matter in selected Archean-Paleoterozoic sedimentary successions of the Kaapvaal Craton. University of Johannesburg, South Africa

Cole DB, Reinhard CT, Wang X, Gueguen B, Halverson GP, Gibson T, Hodgskiss MSW, McKenzie NR, Lyons TW, Planavsky NJ (2016) A shale-hosted Cr isotope record of low atmospheric oxygen during the Proterozoic. Geology 44(7):555–558. https://doi.org/10.1130/g37787.1

Condie KC (1993) Chemical composition and evolution of the upper continental crust: contrasting results from surface samples and shales. Chem Geol 104:1–37

Conrad CP, Lithgow-Bertelloni C (2007) Faster seafloor spreading and lithosphere production during the mid-Cenozoic. Geology 35(1). https://doi.org/10.1130/g22759a.1

Conway TM, John SG (2014) Quantification of dissolved iron sources to the North Atlantic Ocean. Nature 511(7508):212–215. https://doi.org/10.1038/nature13482

Cox GM, Halverson GP, Minarik WG, Le Heron DP, Macdonald FA, Bellefroid EJ, Strauss JV (2013) Neoproterozoic iron formation: an evaluation of its temporal, environmental and tectonic significance. Chem Geol 362:232–249. https://doi.org/10.1016/j.chemgeo.2013.08.002

Cox GM, Halverson GP, Poirier A, Le Heron D, Strauss JV, Stevenson R (2016a) A model for Cryogenian iron formation. Earth Planet Sci Lett 433:280–292. https://doi.org/10.1016/j.epsl.2015.11.003

Cox GM, Halverson GP, Stevenson RK, Vokaty M, Poirier A, Kunzmann M, Li Z-X, Denyszyn SW, Strauss JV, Macdonald FA (2016b) Continental flood basalt weathering as a trigger for Neoproterozoic Snowball Earth. Earth Planet Sci Lett 446:89–99. https://doi.org/10.1016/j.epsl.2016.04.016

Cox GM, Isakson V, Hoffman PF, Gernon TM, Schmitz MD, Shahin S, Collins AS, Preiss W, Blades ML, Mitchell RN, Nordsvan A (2018) South Australian U-Pb zircon (CA-ID-TIMS) age supports globally synchronous Sturtian deglaciation. Precambrian Res 315:257–263. https://doi.org/10.1016/j.precamres.2018.07.007

Craddock PR, Dauphas N (2011) Iron and carbon isotope evidence for microbial iron respiration throughout the Archean. Earth Planet Sci Lett 303(1–2):121–132. https://doi.org/10.1016/j.epsl.2010.12.045

Cramer BS, Miller KG, Barrett PJ, Wright JD (2011) Late Cretaceous–Neogene trends in deep ocean temperature and continental ice volume: reconciling records of benthic foraminiferal geochemistry (δ18O and Mg/Ca) with sea level history. J Geophys Res 116(C12). https://doi.org/10.1029/2011jc007255

Cramer BS, Toggweiler JR, Wright JD, Katz ME, Miller KG (2009) Ocean overturning since the Late Cretaceous: inferences from a new benthic foraminiferal isotope compilation. Paleoceanography 24(4). https://doi.org/10.1029/2008pa001683

Croal LR, Johnson CM, Beard BL, Newman DK (2004) Iron isotope fractionation by Fe(II)-oxidizing photoautotrophic bacteria 1 1Associate editor: D. E. Canfield. Geochim Cosmochim Acta 68(6):1227–1242. https://doi.org/10.1016/j.gca.2003.09.011

Crowe SA, Paris G, Katsev S, Jones C, Kim S, Zerkle AL, Nomosatryo S, Fowie DA, Adkins JF, Sessions AL, Farquhar J, Canfield DE (2014) Sulfate was a trace constituent of Archean seawater. Science 346 (6210):735–739

Czaja AD, Johnson CM, Beard BL, Eigenbrode JL, Freeman KH, Yamaguchi KE (2010) Iron and carbon isotope evidence for ecosystem and environmental diversity in the ~2.7 to 2.5 Ga Hamersley Province, Western Australia. Earth Planet Sci Lett 292(1–2):170–180. https://doi.org/10.1016/j.epsl.2010.01.032

Czaja AD, Johnson CM, Beard BL, Roden EE, Li W, Moorbath S (2013) Biological Fe oxidation controlled deposition of banded iron formation in the ca. 3770 Ma Isua Supracrustal Belt (West Greenland). Earth Planet Sci Lett 363:192–203. https://doi.org/10.1016/j.epsl.2012.12.025

Czaja AD, Johnson CM, Roden EE, Beard BL, Voegelin AR, Nägler TF, Beukes NJ, Wille M (2012) Evidence for free oxygen in the Neoarchean ocean based on coupled iron–molybdenum isotope fractionation. Geochim Cosmochim Acta 86:118–137. https://doi.org/10.1016/j.gca.2012.03.007

Czaja AD, Van Kranendonk MJ, Beard BL, Johnson CM (2018) A multistage origin for Neoarchean layered hematite-magnetite iron formation from the Weld Range, Yilgarn Craton, Western Australia. Chem Geol 488:125–137. https://doi.org/10.1016/j.chemgeo.2018.04.019

Dahl TW, Boyle RA, Canfield DE, Connelly JN, Gill BC, Lenton TM, Bizzarro M (2014) Uranium isotopes distinguish two geochemically distinct stages during the later Cambrian SPICE event. Earth Planet Sci Lett 401:313–326. https://doi.org/10.1016/j.epsl.2014.05.043

Daines SJ, Lenton TM (2016) The effect of widespread early aerobic marine ecosystems on methane cycling and the Great Oxidation. Earth Planet Sci Lett 434:42–51. https://doi.org/10.1016/j.epsl.2015.11.021

Daines SJ, Mills BJ, Lenton TM (2017) Atmospheric oxygen regulation at low Proterozoic levels by incomplete oxidative weathering of sedimentary organic carbon. Nat Commun 8:14379. https://doi.org/10.1038/ncomms14379

Dale AW, Nickelsen L, Scholz F, Hensen C, Oschlies A, Wallmann K (2015) A revised global estimate of dissolved iron fluxes from marine sediments. Global Biogeochem Cycles 29:691–707. https://doi.org/10.1002/2014GB005017

Dauphas N, Cates NL, Mojzsis SJ, Busigny V (2007a) Identification of chemical sedimentary protoliths using iron isotopes in the >3750 Ma Nuvvuagittuq supracrustal belt, Canada. Earth Planet Sci Lett 254(3–4):358–376. https://doi.org/10.1016/j.epsl.2006.11.042

Dauphas N, van Zuilen M, Busigny V, Lepland A, Wadhwa M, Janney PE (2007b) Iron isotope, major and trace element characterization of early Archean supracrustal rocks from SW Greenland: protolith identification and metamorphic overprint. Geochim Cosmochim Acta 71(19):4745–4770. https://doi.org/10.1016/j.gca.2007.07.019

Dauphas N, van Zuilen M, Wadhwa M, Davis AM, Marty B, Janney PE (2004) Clues from Fe isotope variations on the origin of early Archean BIFs from Greenland. Science 306:2077–2080

Davis AS, Gray LB, Clague DA, Hein JR (2002) The Line Islands revisited: New 40Ar/39Ar geochronologic evidence for episodes of volcanism due to lithospheric extension. Geochem Geophys Geosyst 3 (3):1–28. https://doi.org/10.1029/2001GC000190

Davis JA, Meece DE, Kohler M, Curtis GP (2004) Approaches to surface complexation modeling of Uranium(VI) adsorption on aquifer sediments. Geochim Cosmochim Acta 68(18):3621–3641. https://doi.org/10.1016/j.gca.2004.03.003

De Vleeschouwer D, Vahlenkamp M, Crucifix M, Pälike H (2017) Alternating Southern and Northern Hemisphere climate response to astronomical forcing during the past 35 m.y. Geology 45(4):375–378. https://doi.org/10.1130/g38663.1

DeConto RM, Pollard D (2003) Rapid Cenozoic glaciation of Antarctica induced by declining atmospheric CO. Nature 421:245–249

DeConto RM, Pollard D, Wilson PA, Palike H, Lear CH, Pagani M (2008) Thresholds for Cenozoic bipolar glaciation. Nature 455(7213):652–656. https://doi.org/10.1038/nature07337

Delaney ML, Boyle RA (1986) Lithium in foraminiferal shells: implications for high-temperature hydrothermal circulation fluxes and oceanic crustal generation rates. Earth Planet Sci Lett 80:91–105

Dellwig O, Leipe T, März C, Glockzin M, Pollehne F, Schnetger B, Yakushev EV, Böttcher ME, Brumsack H-J (2010) A new particulate Mn–Fe–P-shuttle at the redoxcline of anoxic basins. Geochim Cosmochim Acta 74(24):7100–7115. https://doi.org/10.1016/j.gca.2010.09.017

Derry LA, France-Lanord C (1996) Neogene growth of the sedimentary organic carbon reservoir.

Paleoceanography 11(3):267–275. https://doi.org/10.1029/95pa03839

Derry LA, Jacobsen SB (1990) The chemical evolution of Precambrian seawater: evidence from REEs in banded iron formations. Geochim Cosmochim Acta 54:2965–2977

Derry LA, Kaufman AJ, Jacobsen SB (1992) Sedimentary cycling and environmental change in the Late Proterozoic: evidence from stable and radiogenic isotopes. Geochim Cosmochim Acta 56:131–1329

Dessert C, Dupré B, Gaillardet J, François LM, Allègre CJ (2003) Basalt weathering laws and the impact of basalt weathering on the global carbon cycle. Chem Geol 202(3–4):257–273. https://doi.org/10.1016/j.chemgeo.2002.10.001

Dhuime B, Hawkesworth CJ, Cawood PA, Storey CD (2012) A change in the geodynamics of continental growth 3 Billion years ago. Science 335(6074):1334–1336

Diamond CW, Lyons TW (2018) Mid-Proterozoic redox evolution and the possibility of transient oxygenation events. Emerg Top Life Sci 2(2):235–245. https://doi.org/10.1042/etls20170146

Dickson AJ, Cohen AS (2012) A molybdenum isotope record of Eocene Thermal Maximum 2: implications for global ocean redox during the early Eocene. Paleoceanography 27(3):n/a-n/a. https://doi.org/10.1029/2012pa002346

Dickson AJ, Jenkyns HC, Porcelli D, van den Boorn S, Idiz E (2016) Basin-scale controls on the molybdenum-isotope composition of seawater during Oceanic Anoxic Event 2 (Late Cretaceous). Geochim Cosmochim Acta 178:291–306. https://doi.org/10.1016/j.gca.2015.12.036

Dickson AJ, Saker-Clark M, Jenkyns HC, Bottini C, Erba E, Russo F, Gorbanenko O, Naafs BDA, Pancost RD, Robinson SA, van den Boorn SHJM, Idiz E, Heimhofer U (2017) A Southern Hemisphere record of global trace-metal drawdown and orbital modulation of organic-matter burial across the Cenomanian-Turonian boundary (Ocean Drilling Program Site 1138, Kerguelen Plateau). Sedimentology 64(1):186–203. https://doi.org/10.1111/sed.12303

Djokic T, Van Kranendonk MJ, Campbell KA, Walter MR, Ward CR (2017) Earliest signs of life on land preserved in ca. 3.5 Ga hot spring deposits. Nat Commun 8:15263. https://doi.org/10.1038/ncomms15263

Dodd MS, Papineau D, Grenne T, Slack JF, Rittner M, Pirajno F, O'Neil J, Little CT (2017) Evidence for early life in Earth's oldest hydrothermal vent precipitates. Nature 543(7643):60–64. https://doi.org/10.1038/nature21377

Du Vivier ADC, Jacobson AD, Lehn GO, Selby D, Hurtgen MT, Sageman BB (2015a) Ca isotope stratigraphy across the Cenomanian-Turonian OAE 2: Links between volcanism, seawater geochemistry, and the carbonate fractionation factor. Earth Planet Sci Lett 416:121–131. https://doi.org/10.1016/j.epsl.2015.02.001

Du Vivier ADC, Selby D, Condon DJ, Takashima R, Nishi H (2015b) Pacific 187 Os/188 Os isotope chemistry and U-Pb geochronology: synchroneity of global Os isotope change across OAE 2. Earth Planet Sci Lett 428:204–216. https://doi.org/10.1016/j.epsl.2015.07.020

Du Vivier ADC, Selby D, Sageman BB, Jarvis I, Gröcke DR, Voigt S (2014) Marine 187Os/188Os isotope stratigraphy reveals the interaction of volcanism and ocean circulation during Oceanic Anoxic Event 2. Earth Planet Sci Lett 389:23–33. https://doi.org/10.1016/j.epsl.2013.12.024

Duan Y (2010) Isotope systematics of iron and molybdenum in black shales: inferring redox evolution in ancient sedimentary environments. Arizona State University

Duan Y, Anbar AD, Arnold GL, Lyons TW, Gordon GW, Kendall B (2010) Molybdenum isotope evidence for mild environmental oxygenation before the Great Oxidation Event. Geochim Cosmochim Acta 74(23):6655–6668. https://doi.org/10.1016/j.gca.2010.08.035

Duggen S, Olgun N, Croot P, Hoffmann L, Dietze H, Delmelle P, Teschner C (2010) The role of airborne volcanic ash for the surface ocean biogeochemical iron-cycle: a review. Biogeosciences 7:827–844

Dymek RF, Klein C (1998) Chemistry, petrology and origin of banded iron-formation lithologies from the 3800 Ma Isua Supracrustal Belt, West Greenland. Precambrian Res 39:247–302

Eickmann B, Hofmann A, Wille M, Bui TH, Wing BA, Schoenberg R (2018) Isotopic evidence for oxygenated Mesoarchaean shallow oceans. Nat Geosci 11(2):133–138. https://doi.org/10.1038/s41561-017-0036-x

Eigenbrode JL, Freeman KH (2006) Late Archean rise of aerobic microbial ecosystems. Proc Natl Acad Sci USA 103(43):15759–15764. https://doi.org/10.1073/pnas.0607540103

Eiler JM, Mojzsis SJ, Arrhenius G (1997) Carbon isotope evidence for early life. Nature 386:665

Eldrett JS, Ma C, Bergman SC, Lutz B, Gregory FJ, Dodsworth P, Phipps M, Hardas P, Minisini D, Ozkan A, Ramezani J, Bowring SA, Kamo SL, Ferguson K, Macaulay C, Kelly AE (2015) An astronomically calibrated stratigraphy of the Cenomanian, Turonian and earliest Coniacian from the Cretaceous Western Interior Seaway, USA: implications for global chronostratigraphy. Cretac Res 56:316–344. https://doi.org/10.1016/j.cretres.2015.04.010

Ellwood MJ, Hutchins DA, Lohan MC, Milne A, Nasemann P, Nodder SD, Sander SG, Strzepek R, Wilhelm SW, Boyd PW (2015) Iron stable isotopes track pelagic iron cycling during a subtropical phytoplankton bloom. Proc Natl Acad Sci USA 112(1):E15–20. https://doi.org/10.1073/pnas.1421576112

Eroglu S, Schoenberg R, Pascarelli S, Beukes NJ, Kleinhanns IC, Swanner ED (2018) Open ocean vs. continentally-derived iron cycles along the Neoarchean Campbellrand-Malmani Carbonate platform,

South Africa. Am J Sci 318(4):367–408. https://doi.org/10.2475/04.2018.01
Eroglu S, Schoenberg R, Wille M, Beukes N, Taubald H (2015) Geochemical stratigraphy, sedimentology, and Mo isotope systematics of the ca. 2.58–2.50 Ga-old Transvaal Supergroup carbonate platform, South Africa. Precambrian Res 266:27–46. https://doi.org/10.1016/j.precamres.2015.04.014
Eroglu S, van Zuilen MA, Taubald H, Drost K, Wille M, Swanner ED, Beukes NJ, Schoenberg R (2017) Depth-dependent δ13C trends in platform and slope settings of the Campbellrand-Malmani carbonate platform and possible implications for Early Earth oxygenation. Precambrian Res 302:122–139. https://doi.org/10.1016/j.precamres.2017.09.018
Erwin DH, Laflamme M, Tweedt SM, Sperling EA, Pisani D, Peterson KJ (2011) The Cambrian conundrum: early divergence and later ecological success in the early history of animals. Science 334:1091–1097
Evans DA, Beukes NJ, Kirschvink JL (1997) Low-latitude glaciation in the Palaeoproterozoic era. Nature 386:262–266
Fabre S, Nédélec A, Poitrasson F, Strauss H, Thomazo C, Nogueira A (2011) Iron and sulphur isotopes from the Carajás mining province (Pará, Brazil): implications for the oxidation of the ocean and the atmosphere across the Archaean-Proterozoic transition. Chem Geol 289(1–2):124–139. https://doi.org/10.1016/j.chemgeo.2011.07.019
Fakhraee M, Hancisse O, Canfield DE, Crowe SA, Katsev S (2019) Proterozoic seawater sulfate scarcity and the evolution of ocean–atmosphere chemistry. Nat Geosci 12(5):375–380. https://doi.org/10.1038/s41561-019-0351-5
Fan H, Zhu X, Wen H, Yan B, Li J, Feng L (2014) Oxygenation of Ediacaran Ocean recorded by iron isotopes. Geochim Cosmochim Acta 140:80–94. https://doi.org/10.1016/j.gca.2014.05.029
Farber K, Dziggel A, Meyer FM, Harris C (2016) Petrology, geochemistry and fluid inclusion analysis of altered komatiites of the Mendon Formation in the BARB4 drill core, Barberton Greenstone Belt, South Africa. S Afr J Geol 119(4):639–654. https://doi.org/10.2113/gssajg.119.4.639
Farquhar J, Bao H, Thiemens M (2000) Atmospheric Influence of Earth's Earliest Sulfur Cycle. Science 289:756–758
Farquhar J, Cliff J, Zerkle AL, Kamyshny A, Poulton SW, Claire M, Adams D, Harms B (2013) Pathways for Neoarchean pyrite formation constrained by mass-independent sulfur isotopes. Proc Natl Acad Sci USA 110(44):17638–17643. https://doi.org/10.1073/pnas.1218851110
Farquhar J, Zerkle AL, Bekker A (2010) Geological constraints on the origin of oxygenic photosynthesis. J Geodyn 107:11–36
Farquhar J, Zerkle AL, Bekker A (2011) Geological constraints on the origin of oxygenic photosynthesis. Photosynth Res 107(1):11–36. https://doi.org/10.1007/s11120-010-9594-0
Fedo CM, Whitehouse MJ (2002) Mesasomatic origin of quartz-pyroxene rock, Akilia, Greenland, and implications for Earth's earliest life. Science 296:1448–1452
Fehr MA, Andersson PS, Hålenius U, Gustafsson Ö, Mörth C-M (2010) Iron enrichments and Fe isotopic compositions of surface sediments from the Gotland Deep, Baltic Sea. Chemical Geology 277(3–4):310–322. https://doi.org/10.1016/j.chemgeo.2010.08.014
Fehr MA, Andersson PS, Hålenius U, Mörth C-M (2008) Iron isotope variations in Holocene sediments of the Gotland Deep, Baltic Sea. Geochim Cosmochim Acta 72(3):807–826. https://doi.org/10.1016/j.gca.2007.11.033
Ferrari L, Valencia-Moreno M, Bryan S (2007) Magmatism and tectonics of the Sierra Madre Occidental and its relation with the evolution of the western margin of North America. In: Special Paper 422: Geology of México: Celebrating the Centenary of the Geological Society of México, pp 1–39. https://doi.org/10.1130/2007.2422(01)
Fischer WW, Hemp J, Johnson JE (2016) Evolution of oxygenic photosynthesis. Annu Rev Earth Planet Sci 44(1):647–683. https://doi.org/10.1146/annurev-earth-060313-054810
Fischer WW, Schroeder S, Lacassie JP, Beukes NJ, Goldberg T, Strauss H, Horstmann UE, Schrag DP, Knoll AH (2009) Isotopic constraints on the Late Archean carbon cycle from the Transvaal Supergroup along the western margin of the Kaapvaal Craton, South Africa. Precambrian Res 169(1–4):15–27. https://doi.org/10.1016/j.precamres.2008.10.010
Flament N, Coltice N, Rey PF (2008) A case for late-Archaean continental emergence from thermal evolution models and hypsometry. Earth Planet Sci Lett 275(3–4):326–336. https://doi.org/10.1016/j.epsl.2008.08.029
Flament N, Coltice N, Rey PF (2013) The evolution of the 87Sr/86Sr of marine carbonates does not constrain continental growth. Precambrian Res 229:177–188. https://doi.org/10.1016/j.precamres.2011.10.009
Fralick P, Planavsky N, Burton J, Jarvis I, Addison WD, Barrett TJ, Brumpton GR (2017) Geochemistry of Paleoproterozoic Gunflint Formation carbonate: implications for hydrosphere-atmosphere evolution. Precambrian Res 290:126–146. https://doi.org/10.1016/j.precamres.2016.12.014
France-Lanord C, Derry LA (1997) Organic carbon burial forcing of the carbon cycle from Himalayan erosion. Nature 390:65–67
Frank M (2002) Radiogenic isotopes: traces of past ocean circulation and erosional input. Rev Geophys 40(1):1–38
Frank M, O'Nions RK, Hein JR, Banakar VK (1999) 60 Myr records of major elements and Pb–Nd isotopes from hydrogenous ferromanganese crusts: reconstruction of seawater paleochemistry. Geochim Cosmochim Acta 63(11/12):1689–1708
Frei R, Bridgwater D, Rosing M, Stecher O (1999) Controversial Pb-Pb and Sm-Nd isotope results in the early Archean Isua (West Greenland) oxide iron

formation: preservation of primary signatures versus secondary disturbances. Geochim Cosmochim Acta 63 (3/4):473–488

Frei R, Crowe SA, Bau M, Polat A, Fowle DA, Dossing LN (2016) Oxidative elemental cycling under the low O_2 Eoarchean atmosphere. Sci Rep 6:21058. https://doi.org/10.1038/srep21058

Frei R, Gaucher C, Poulton SW, Canfield DE (2009) Fluctuations in Precambrian atmospheric oxygenation recorded by chromium isotopes. Nature 461 (7261):250–253. https://doi.org/10.1038/nature08266

Frei R, Rosing MT (2001) The least radiogenic terrestrial leads; implications for the early Archean crustal evolution and hydrothermal–metasomatic processes in the Isua Supracrustal Belt (West Greenland). Chem Geol 181:47–66

French BM (1971) Stability relations of siderite ($FeCO_3$) in the system Fe-C-O. Am J Sci 271:37–78

Friedrich O, Norris RD, Erbacher J (2012) Evolution of middle to Late Cretaceous oceans—A 55 m.y. record of Earth's temperature and carbon cycle. Geology 40 (2):107–110. https://doi.org/10.1130/g32701.1

Frierdich AJ, Beard BL, Reddy TR, Scherer MM, Johnson CM (2014a) Iron isotope fractionation between aqueous Fe(II) and goethite revisited: new insights based on a multi-direction approach to equilibrium and isotopic exchange rate modification. Geochim Cosmochim Acta 139:383–398. https://doi.org/10.1016/j.gca.2014.05.001

Frierdich AJ, Beard BL, Scherer MM, Johnson CM (2014b) Determination of the Fe(II)aq–magnetite equilibrium iron isotope fractionation factor using the three-isotope method and a multi-direction approach to equilibrium. Earth Planet Sci Lett 391:77–86. https://doi.org/10.1016/j.epsl.2014.01.032

Frierdich AJ, Spicuzza MJ, Scherer MM (2016) Oxygen isotope evidence for Mn(II)-catalyzed recrystallization of manganite (gamma-MnOOH). Environ Sci Technol 50(12):6374–6380. https://doi.org/10.1021/acs.est.6b01463

Frimmel HE (2005) Archaean atmospheric evolution: evidence from the Witwatersrand gold fields, South Africa. Earth Sci Rev 70(1–2):1–46. https://doi.org/10.1016/j.earscirev.2004.10.003

Frogner P, Gislason SR, Oskarsson N (2001) Fertilizing potential of volcanic ash in ocean surface water. Geology 29(6):487–490

Frost CD, von Blanckenburg F, Schoenberg R, Frost BR, Swapp SM (2007) Preservation of Fe isotope heterogeneities during diagenesis and metamorphism of banded iron formation. Contrib Miner Petrol 153 (2):211–235. https://doi.org/10.1007/s00410-006-0141-0

Gaillard F, Scaillet B, Arndt NT (2011) Atmospheric oxygenation caused by a change in volcanic degassing pressure. Nature 478(7368):229–232. https://doi.org/10.1038/nature10460

Galer SJG, Mezger K (1998) Metamorphism, denudation and sea level in the Archean and cooling of the Earth. Precambrian Res 92(1998):389–412

Galić A, Mason PRD, Mogollón JM, Wolthers M, Vroon PZ, Whitehouse MJ (2017) Pyrite in a sulfate-poor Paleoarchean basin was derived predominantly from elemental sulfur: evidence from 3.2 Ga sediments in the Barberton Greenstone Belt, Kaapvaal Craton. Chem Geol 449:135–146. https://doi.org/10.1016/j.chemgeo.2016.12.006

Gall L, Williams HM, Siebert C, Halliday AN, Herrington RJ, Hein JR (2013) Nickel isotopic compositions of ferromanganese crusts and the constancy of deep ocean inputs and continental weathering effects over the Cenozoic. Earth Planet Sci Lett 375:148–155. https://doi.org/10.1016/j.epsl.2013.05.019

Garçon M, Carlson RW, Shirey SB, Arndt NT, Horan MF, Mock TD (2017) Erosion of Archean continents: the Sm-Nd and Lu-Hf isotopic record of Barberton sedimentary rocks. Geochim Cosmochim Acta 206:216–235. https://doi.org/10.1016/j.gca.2017.03.006

Garrels RM, Lerman A (1981) Phanerozoic cycles of sedimentary carbon and sulfur. Proc Natl Acad Sci USA 78(8):4652–4656

Glikson A, Hickman A, Evans NJ, Kirkland CL, Park J-W, Rapp R, Romano S (2016) A new ~3.46 Ga asteroid impact ejecta unit at Marble Bar, Pilbara Craton, Western Australia: a petrological, microprobe and laser ablation ICPMS study. Precambrian Res 279:103–122. https://doi.org/10.1016/j.precamres.2016.04.003

Goldberg T, Archer C, Vance D, Poulton SW (2009) Mo isotope fractionation during adsorption to Fe (oxyhydr)oxides. Geochim Cosmochim Acta 73(21):6502–6516. https://doi.org/10.1016/j.gca.2009.08.004

Goldberg T, Archer C, Vance D, Thamdrup B, McAnena A, Poulton SW (2012) Controls on Mo isotope fractionations in a Mn-rich anoxic marine sediment, Gullmar Fjord, Sweden. Chem Geol 296–297:73–82. https://doi.org/10.1016/j.chemgeo.2011.12.020

Goldberg T, Poulton SW, Wagner T, Kolonic SF, Rehkämper M (2016) Molybdenum drawdown during Cretaceous Oceanic Anoxic Event 2. Earth Planet Sci Lett 440:81–91. https://doi.org/10.1016/j.epsl.2016.02.006

Gomes ML, Fike DA, Bergmann KD, Jones C, Knoll AH (2018) Environmental insights from high-resolution (SIMS) sulfur isotope analyses of sulfides in Proterozoic microbialites with diverse mat textures. Geobiology 16(1):17–34. https://doi.org/10.1111/gbi.12265

Gomes ML, Hurtgen MT, Sageman BB (2016) Biogeochemical sulfur cycling during Cretaceous oceanic anoxic events: a comparison of OAE1a and OAE2. Paleoceanography 31:233–251. https://doi.org/10.1002/2015PA002869

Goto KT, Anbar AD, Gordon GW, Romaniello SJ, Shimoda G, Takaya Y, Tokumaru A, Nozaki T, Suzuki K, Machida S, Hanyu T, Usui A (2014) Uranium isotope systematics of ferromanganese crusts in the Pacific Ocean: implications for the marine 238 U/ 235 U isotope system. Geochim Cosmochim Acta 146:43–58. https://doi.org/10.1016/j.gca.2014.10.003

Greber ND, Dauphas N, Bekker A, Ptacek MP, Bindeman IN, Hofmann A (2017) Titanium isotopic evidence for felsic crust and plate tectonics 3.5 billion years ago. Science 357(6357):1271–1274. https://doi.org/10.1126/science.aan8086

Gross GA (1980) A classification of iron formations based on depositional environments. Can Mineral 18:215–222

Grotzinger JP, Fike DA, Fischer WW (2011) Enigmatic origin of the largest-known carbon isotope excursion in Earth's history. Nat Geosci 4(5):285–292. https://doi.org/10.1038/ngeo1138

Guilbaud R, Poulton SW, Butterfield NJ, Zhu M, Shields-Zhou GA (2015) A global transition to ferruginous conditions in the early Neoproterozoic oceans. Nat Geosci 8(6):466–470. https://doi.org/10.1038/ngeo2434

Gumsley AP, Chamberlain KR, Bleeker W, Soderlund U, de Kock MO, Larsson ER, Bekker A (2017) Timing and tempo of the Great Oxidation Event. Proc Natl Acad Sci USA 114(8):1811–1816. https://doi.org/10.1073/pnas.1608824114

Guy BM (2012) Pyrite in the Mesoarchean Witwatersrand Supergroup. University of Johannesburg, South Africa

Guy BM, Ono S, Gutzmer J, Kaufman AJ, Lin Y, Fogel ML, Beukes NJ (2012) A multiple sulfur and organic carbon isotope record from non-conglomeratic sedimentary rocks of the Mesoarchean Witwatersrand Supergroup, South Africa. Precambrian Res 216–219:208–231. https://doi.org/10.1016/j.precamres.2012.06.018

Halverson GP, Poitrasson F, Hoffman PF, Nédélec A, Montel J-M, Kirby J (2011) Fe isotope and trace element geochemistry of the Neoproterozoic syn-glacial Rapitan iron formation. Earth Planet Sci Lett 309(1–2):100–112. https://doi.org/10.1016/j.epsl.2011.06.021

Handoh IC, Lenton TM (2003) Periodic mid-Cretaceous oceanic anoxic events linked by oscillations of the phosphorus and oxygen biogeochemical cycles. Global Biogeochem Cycles 17(4):n/a-n/a. https://doi.org/10.1029/2003gb002039

Hao J, Sverjensky DA, Hazen RM (2017) A model for late Archean chemical weathering and world average river water. Earth Planet Sci Lett 457:191–203. https://doi.org/10.1016/j.epsl.2016.10.021

Harada M, Tajika E, Sekine Y (2015) Transition to an oxygen-rich atmosphere with an extensive overshoot triggered by the Paleoproterozoic Snowball Earth. Earth Planet Sci Lett 419:178–186. https://doi.org/10.1016/j.epsl.2015.03.005

Hardisty DS, Lu Z, Bekker A, Diamond CW, Gill BC, Jiang G, Kah LC, Knoll AH, Loyd SJ, Osburn MR, Planavsky NJ, Wang C, Zhou X, Lyons TW (2017) Perspectives on Proterozoic surface ocean redox from iodine contents in ancient and recent carbonate. Earth Planet Sci Lett 463:159–170. https://doi.org/10.1016/j.epsl.2017.01.032

Hashman BM (2017) A mesoarchean microbial iron shuttle, Witwatersrand basin, South Africa. M.S., University of Wisconsin-Madison

Haugaard R, Pecoits E, Lalonde S, Rouxel O, Konhauser K (2016) The Joffre banded iron formation, Hamersley Group, Western Australia: assessing the palaeoenvironment through detailed petrology and chemostratigraphy. Precambrian Res 273:12–37. https://doi.org/10.1016/j.precamres.2015.10.024

Havig JR, Hamilton TL, Bachan A, Kump LR (2017) Sulfur and carbon isotopic evidence for metabolic pathway evolution and a four-stepped Earth system progression across the Archean and Paleoproterozoic. Earth Sci Rev 174:1–21. https://doi.org/10.1016/j.earscirev.2017.06.014

Hawkesworth CJ, Cawood PA, Dhuime B, Kemp AIS (2017) Earth's continental lithosphere through time. Annu Rev Earth Planet Sci 45:169–198. https://doi.org/10.1146/annurev-earth-063016-020525

Hayashi T, Tanimizu M, Tanaka T (2004) Origin of negative Ce anomalies in Barberton sedimentary rocks, deduced from La–Ce and Sm–Nd isotope systematics. Precambrian Res 135(4):345–357. https://doi.org/10.1016/j.precamres.2004.09.004

Hayes JM (1994) Global methanotrophy at the Archean–Proterozoic transition. In: Bengtson S (ed) Early life on earth. Columbia University Press, Nobel Symposium, pp 220–236

Hayes JM, Waldbauer JR (2006) The carbon cycle and associated redox processes through time. Philos Trans R Soc Lond B Biol Sci 361(1470):931–950. https://doi.org/10.1098/rstb.2006.1840

Heck PR, Huberty JM, Kita NT, Ushikubo T, Kozdon R, Valley JW (2011) SIMS analyses of silicon and oxygen isotope ratios for quartz from Archean and Paleoproterozoic banded iron formations. Geochim Cosmochim Acta 75(20):5879–5891. https://doi.org/10.1016/j.gca.2011.07.023

Heimann A, Beard BL, Johnson CM (2008) The role of volatile exsolution and sub-solidus fluid/rock interactions in producing high 56Fe/54Fe ratios in siliceous igneous rocks. Geochim Cosmochim Acta 72(17):4379–4396. https://doi.org/10.1016/j.gca.2008.06.009

Heimann A, Johnson CM, Beard BL, Valley JW, Roden EE, Spicuzza MJ, Beukes NJ (2010) Fe, C, and O isotope compositions of banded iron formation carbonates demonstrate a major role for dissimilatory iron reduction in ~2.5 Ga marine environments. Earth Planet Sci Lett 294(1–2):8–18. https://doi.org/10.1016/j.epsl.2010.02.015

Henkel S, Kasten S, Poulton SW, Staubwasser M (2016) Determination of the stable iron isotopic composition of sequentially leached iron phases in marine sediments. Chem Geol 421:93–102. https://doi.org/10.1016/j.chemgeo.2015.12.003

Herrick MJ (2007) Isotopic studies of the 3.7-3.8 Ga ISUA banded iron formation provide insight into early archean geochemical cycles. M.S., University of Wisconsin-Madison,

Herzberg C, Condie K, Korenaga J (2010) Thermal history of the Earth and its petrological expression. Earth Planet Sci Lett 292(1–2):79–88. https://doi.org/10.1016/j.epsl.2010.01.022

Hessler AM, Lowe DR (2006) Weathering and sediment generation in the Archean: An integrated study of the evolution of siliciclastic sedimentary rocks of the 3.2 Ga Moodies Group, Barberton Greenstone Belt, South Africa. Precambrian Research 151 (3–4):185–210. https://doi.org/10.1016/j.precamres.2006.08.008

Heubeck C, Engelhardt J, Byerly GR, Zeh A, Sell B, Luber T, Lowe DR (2013) Timing of deposition and deformation of the Moodies Group (Barberton Greenstone Belt, South Africa): very-high-resolution of Archaean surface processes. Precambrian Res 231:236–262. https://doi.org/10.1016/j.precamres.2013.03.021

Heubeck C, Lowe DR (1994) Depositional and tectonic setting of the Archean Moodies Group, Barberton Greenstone Belt, South Africa. Precambrian Res 68:257–290

Hiebert RS, Bekker A, Houlé MG, Rouxel OJ (2018) Depositional setting of the Late Archean Fe oxide- and sulfide-bearing chert and graphitic argillite in the Shaw Dome, Abitibi Greenstone Belt, Canada. Precambrian Res 311:98–116. https://doi.org/10.1016/j.precamres.2018.04.004

Hoashi M, Bevacqua DC, Otake T, Watanabe Y, Hickman AH, Utsunomiya S, Ohmoto H (2009) Primary haematite formation in an oxygenated sea 3.46 billion years ago. Nat Geosci 2(4):301–306. https://doi.org/10.1038/ngeo465

Hoffman PF (2013) The great oxidation and a siderian Snowball Earth: MIF-S based correlation of Paleoproterozoic glacial epochs. Chem Geol 362:143–156. https://doi.org/10.1016/j.chemgeo.2013.04.018

Hoffman PF, Abbot DS, Ashkenazy Y, Benn DI, Brocks JJ, Cohen PA, Cox GM, Cheveling JR, Donnadieu Y, Erwin DH, Fairchild IJ, Ferreira D, Goodman JC, Halverson GP, Jansen MF, Le Hir G, Love GD, Macdonald FA, Maloof AC, Partin CA, Ramstein G, Rose BEJ, Rose CV, Sadler PM, Tziperman E, Voigt A, Warren SG (2017) Snowball Earth climate dynamics and Cryogenian geology-geobiology. Science Advances 3:1–43

Hoffman PF, Kaufman AJ, Halverson GP, Schrag DP (1998) A neoproterozoic Snowball Earth. Science 281 (5381):1342–1346. https://doi.org/10.1126/science.281.5381.1342

Holland HD (1984) The chemical evolution of the atmosphere and oceans. Princeton University Press, Princeton

Holland HD (2002) Volcanic gases, black smokers, and the great oxidation event. Geochim Cosmochim Acta 66(21):3811–3826

Holland HD (2006) The oxygenation of the atmosphere and oceans. Philos Trans R Soc Lond B Biol Sci 361:903–915

Holland HD, Beukes NJ (1990) A paleoweathering profile from Griqualand West, South Africa: evidence for a dramatic rise in atmospheric oxygen between 2.2 and 1.9 bybp. Am J Sci 290-A:1–34

Holmden C, Jacobson AD, Sageman BB, Hurtgen MT (2016) Response of the Cr isotope proxy to Cretaceous Ocean Anoxic Event 2 in a pelagic carbonate succession from the Western Interior Seaway. Geochim Cosmochim Acta 186:277–295. https://doi.org/10.1016/j.gca.2016.04.039

Hong SK, Lee YI (2012) Evaluation of atmospheric carbon dioxide concentrations during the Cretaceous. Earth Planet Sci Lett 327–328:23–28. https://doi.org/10.1016/j.epsl.2012.01.014

Horita J, Zimmerman A, Holland HD (2002) Chemical evolution of seawater during the Phanerozoic: implications from the record of marine evaporites. Geochim Cosmochim Acta 66(21):3733–3756. https://doi.org/10.1016/S0016-7037(01)00884-5

Horner TJ, Williams HM, Hein JR, Saito MA, Burton KW, Halliday AN, Nielsen SG (2015) Persistence of deeply sourced iron in the Pacific Ocean. Proc Natl Acad Sci USA 112(5):1292–1297. https://doi.org/10.1073/pnas.1420188112

Huang J, Feng L, Chu X, Sun T, Wen H, Qin L, Shen Y (2017) A predominantly ferruginous condition in the Ediacaran deep ocean: geochemistry of black shales in the Ediacaran Doushantuo Formation, South China. Precambrian Res 295:12–23. https://doi.org/10.1016/j.precamres.2017.04.019

Huberty JM, Konishi H, Heck PR, Fournelle JH, Valley JW, Xu H (2012) Silician magnetite from the Dales Gorge Member of the Brockman Iron Formation, Hamersley Group, Western Australia. Am Mineral 97(1):26–37. https://doi.org/10.2138/am.2012.3864

Husson JM, Peters SE (2017) Atmospheric oxygenation driven by unsteady growth of the continental sedimentary reservoir. Earth Planet Sci Lett 460:68–75. https://doi.org/10.1016/j.epsl.2016.12.012

Husson JM, Peters SE (2018) Nature of the sedimentary rock record and its implications for Earth system evolution. Emerg Top Life Sci 2(2):125–136. https://doi.org/10.1042/etls20170152

Hyslop EV, Valley JW, Johnson CM, Beard BL (2008) The effects of metamorphism on O and Fe isotope compositions in the Biwabik Iron Formation, northern Minnesota. Contrib Miner Petrol 155(3):313–328. https://doi.org/10.1007/s00410-007-0244-2

Hyun J-H, Kim S-H, Mok J-S, Cho H, Lee T, Vandieken V, Thamdrup B (2017) Manganese and iron reduction dominate organic carbon oxidation in surface sediments of the deep Ulleung Basin, East Sea. Biogeosciences 14(4):941–958. https://doi.org/10.5194/bg-14-941-2017

Iizuka T, Yamaguchi T, Itano K, Hibiya Y, Suzuki K (2017) What Hf isotopes in zircon tell us about crust–mantle evolution. Lithos 274–275:304–327. https://doi.org/10.1016/j.lithos.2017.01.006

Isley AE (1995) Hydrothermal plumes and the delivery of iron to banded iron formation. J Geol 103:169–185

Jacobsen SB, Pimentel-Klose MR (1988) Nd isotopic variations in preCambrian banded iron formations. Geophys Res Lett 15(4):393–396

Jacobson AD, Blum JD (2000) Ca/Sr and 87Sr/86Sr geochemistry of disseminated calcite in Himalayan silicate rocks from Nanga Parbat: Influence on river-water chemistry. Geology 28:463–466

Jacobson AD, Blum JD, Chamberlain CP, Poage MA, Sloan VF (2002) Ca/Sr and Sr isotope systematics of a Himalayan glacial chronosequence: carbonate versus silicate weathering rates as a function of landscape surface age. Geochim Cosmochim Acta 66:13–27

Jarvis I, Lignum JS, Gröcke DR, Jenkyns HC, Pearce MA (2011) Black shale deposition, atmospheric CO2 drawdown, and cooling during the Cenomanian-Turonian Oceanic Anoxic Event. Paleoceanography 26(3):n/a-n/a. https://doi.org/10.1029/2010pa002081

Jenkyns HC (2010) Geochemistry of oceanic anoxic events. Geochem Geophys Geosyst 11(3):n/a-n/a. https://doi.org/10.1029/2009gc002788

Jenkyns HC (2018) Transient cooling episodes during Cretaceous Oceanic Anoxic Events with special reference to OAE 1a (Early Aptian). Philos Trans A Math Phys Eng Sci 376(2130). https://doi.org/10.1098/rsta.2017.0073

Jenkyns HC, Dickson AJ, Ruhl M, van den Boorn SHJM, Heimhofer U (2017) Basalt-seawater interaction, the Plenus Cold Event, enhanced weathering and geochemical change: deconstructing Oceanic Anoxic Event 2 (Cenomanian-Turonian, Late Cretaceous). Sedimentology 64(1):16–43. https://doi.org/10.1111/sed.12305

Jenkyns HC, Matthews A, Tsikos H, Erel Y (2007) Nitrate reduction, sulfate reduction, and sedimentary iron isotope evolution during the Cenomanian-Turonian oceanic anoxic event. Paleoceanography 22(3):n/a-n/a. https://doi.org/10.1029/2006pa001355

Jicha BR, Scholl DW, Rea DK (2009) Circum-Pacific arc flare-ups and global cooling near the Eocene-Oligocene boundary. Geology 37(4):303–306. https://doi.org/10.1130/g25392a.1

Johnson CM, Beard BL, Beukes NJ, Klein C, O'Leary JM (2003) Ancient geochemical cycling in the Earth as inferred from Fe isotope studies of banded iron formations from the Transvaal Craton. Contrib Miner Petrol 144(5):523–547. https://doi.org/10.1007/s00410-002-0418-x

Johnson CM, Beard BL, Klein C, Beukes NJ, Roden EE (2008a) Iron isotopes constrain biologic and abiologic processes in banded iron formation genesis. Geochim Cosmochim Acta 72(1):151–169. https://doi.org/10.1016/j.gca.2007.10.013

Johnson CM, Beard BL, Roden EE (2008b) The iron isotope fingerprints of redox and biogeochemical cycling in modern and ancient earth. Annu Rev Earth Planet Sci 36(1):457–493. https://doi.org/10.1146/annurev.earth.36.031207.124139

Johnson CM, Ludois JM, Beard BL, Beukes NJ, Heimann A (2013a) Iron formation carbonates: paleoceanographic proxy or recorder of microbial diagenesis? Geology 41(11):1147–1150. https://doi.org/10.1130/g34698.1

Johnson CM, Van Kranendonk MJ (2019) Ancient life and plate tectonics. In: Kolb VM (ed) Hanbook of astrobiology. CRC Press

Johnson JE, Webb SM, Ma C, Fischer WW (2016) Manganese mineralogy and diagenesis in the sedimentary rock record. Geochim Cosmochim Acta 173:210–231. https://doi.org/10.1016/j.gca.2015.10.027

Johnson JE, Webb SM, Thomas K, Ono S, Kirschvink JL, Fischer WW (2013b) Manganese-oxidizing photosynthesis before the rise of cyanobacteria. Proc Natl Acad Sci USA 110(28):11238–11243. https://doi.org/10.1073/pnas.1305530110

Johnston DT, Poulton SW, Dehler C, Porter S, Husson J, Canfield DE, Knoll AH (2010) An emerging picture of Neoproterozoic ocean chemistry: insights from the Chuar Group, Grand Canyon, USA. Earth Planet Sci Lett 290(1–2):64–73. https://doi.org/10.1016/j.epsl.2009.11.059

Jones C, Nomosatryo S, Crowe SA, Bjerrum CJ, Canfield DE (2015) Iron oxides, divalent cations, silica, and the early earth phosphorus crisis. Geology 43(2):135–138. https://doi.org/10.1130/g36044.1

Jones CE, Jenkyns HC (2001) Seawater strontium isotopes, oceanic anoxic events, and seafloor hydrothermal activity in the Jurassic and Cretaceous. Am J Sci 301:112–149

Kah LC, Lyons TW, Frank TD (2004) Low marine sulphate and protracted oxygenation of the Proterozoic biosphere. Nature 431:834–838

Kamber BS (2010) Archean mafic–ultramafic volcanic landmasses and their effect on ocean–atmosphere chemistry. Chem Geol 274(1–2):19–28. https://doi.org/10.1016/j.chemgeo.2010.03.009

Kanzaki Y, Murakami T (2016) Estimates of atmospheric O 2 in the Paleoproterozoic from paleosols. Geochim Cosmochim Acta 174:263–290. https://doi.org/10.1016/j.gca.2015.11.022

Kappler A, Johnson CM, Crosby HA, Beard BL, Newman DK (2010) Evidence for equilibrium iron isotope fractionation by nitrate-reducing iron(II)-oxidizing bacteria. Geochim Cosmochim Acta 74(10):2826–2842. https://doi.org/10.1016/j.gca.2010.02.017

Kappler A, Pasquero C, Konhauser KO, Newman DK (2005) Deposition of banded iron formations by anoxygenic phototrophic Fe(II)-oxidizing bacteria. Geology 33(11). https://doi.org/10.1130/g21658.1

Karhu JA, Holland HD (1996) Carbon isotopes and the rise of atmospheric oxygen. Geology 24(10):867–870

Kasting JF, Eggler DH, Raeburn SP (1993) Mantle redox evolution and the oxidation state of the Archean atmosphere. J Geol 101:245–257

Kaufman AJ, Johnston DT, Farquhar J, Masterson AL, Lyons TW, Bates S, Anbar AD, Arnold GL, Garvin J,

Buick R (2007) Late Archean biospheric oxygenation and atmospheric evolution. Science 317:1900–1903
Kendall B, Brennecka GA, Weyer S, Anbar AD (2013) Uranium isotope fractionation suggests oxidative uranium mobilization at 2.50 Ga. Chem Geol 362:105–114. https://doi.org/10.1016/j.chemgeo.2013.08.010
Kendall B, Creaser RA, Reinhard CT, Lyons TW, Anbar AD (2015a) Transient episodes of mild environmental oxygenation and oxidative continental weathering during the late Archean. Sci Adv 1(10). https://doi.org/10.1126/sciadv.1500777
Kendall B, Dahl TW, Anbar AD (2017) The stable isotope geochemistry of molybdenum. Rev Miner Geochem 82(1):683–732. https://doi.org/10.2138/rmg.2017.82.16
Kendall B, Gordon GW, Poulton SW, Anbar AD (2011) Molybdenum isotope constraints on the extent of late Paleoproterozoic ocean euxinia. Earth Planet Sci Lett 307(3–4):450–460. https://doi.org/10.1016/j.epsl.2011.05.019
Kendall B, Komiya T, Lyons TW, Bates SM, Gordon GW, Romaniello SJ, Jiang G, Creaser RA, Xiao S, McFadden K, Sawaki Y, Tahata M, Shu D, Han J, Li Y, Chu X, Anbar AD (2015b) Uranium and molybdenum isotope evidence for an episode of widespread ocean oxygenation during the late Ediacaran Period. Geochim Cosmochim Acta 156:173–193. https://doi.org/10.1016/j.gca.2015.02.025
Kendall B, Reinhard CT, Lyons TW, Kaufman AJ, Poulton SW, Anbar AD (2010) Pervasive oxygenation along late Archaean ocean margins. Nat Geosci 3 (9):647–652. https://doi.org/10.1038/ngeo942
Kennedy M, Droser M, Mayer LM, Pevear D, Mrofka D (2006) Late Precambrian oxygenation; inception of the clay mineral factory. Science 311(5766):1446–1449
Kent DV, Muttoni G (2013) Modulation of Late Cretaceous and Cenozoic climate by variable drawdown of atmospheric pCO_2 from weathering of basaltic provinces on continents drifting through the equatorial humid belt. Clim Past 9(2):525–546. https://doi.org/10.5194/cp-9-525-2013
King EK, Perakis SS, Pett-Ridge JC (2018) Molybdenum isotope fractionation during adsorption to organic matter. Geochim Cosmochim Acta 222:584–598. https://doi.org/10.1016/j.gca.2017.11.014
Kirschivink JL (1992) Late Proterozoic low-latitude global glaciation: the snow-ball earth. The Proterozoic biosphere: a multi-disciplinary study. Cambridge University Press, New York, pp 51–52
Kita I, Taguchi S, Matsubaya O (1985) Oxygen isotope fractionation between amorphous silica and water at 34–93 °C. Nature 314:83–84
Kita NT, Huberty JM, Kozdon R, Beard BL, Valley JW (2011) High-precision SIMS oxygen, sulfur and iron stable isotope analyses of geological materials: accuracy, surface topography and crystal orientation. Surf Interface Anal 43(1–2):427–431. https://doi.org/10.1002/sia.3424
Kitchen NE, Valley JW (1995) Carbon isotope thermometry in marbles of the Adirondack Mountains, New York. J Metamorph Geol 13:577–594
Klein C (2005) Some Precambrian banded iron-formations (BIFs) from around the world: their age, geologic setting, mineralogy, metamorphism, geochemistry, and origins. Am Miner 90(10):1473–1499. https://doi.org/10.2138/am.2005.1871
Klemm V, Levasseur S, Frank M, Hein J, Halliday A (2005) Osmium isotope stratigraphy of a marine ferromanganese crust. Earth Planet Sci Lett 238(1–2):42–48. https://doi.org/10.1016/j.epsl.2005.07.016
Klemm V, Reynolds B, Frank M, Pettke T, Halliday AN (2007) Cenozoic changes in atmospheric lead recorded in central Pacific ferromanganese crusts. Earth Planet Sci Lett 253(1–2):57–66. https://doi.org/10.1016/j.epsl.2006.10.018
Koehler MC, Buick R, Kipp MA, Stueken EE, Zaloumis J (2018) Transient surface ocean oxygenation recorded in the approximately 2.66-Ga Jeerinah Formation, Australia. Proc Natl Acad Sci USA 115(30):7711–7716. https://doi.org/10.1073/pnas.1720820115
Kohn MJ, Strömberg CAE, Madden RH, Dunn RE, Evans S, Palacios A, Carlini AA (2015) Quasi-static Eocene-Oligocene climate in Patagonia promotes slow faunal evolution and mid-Cenozoic global cooling. Palaeogeogr Palaeoclimatol Palaeoecol 435:24–37. https://doi.org/10.1016/j.palaeo.2015.05.028
Konhauser KO, Hamade T, Raiswell R, Morris RC, Ferris FG, Southam G, Canfield DE (2002) Could bacteria have formed the Precambrian banded iron formations? Geology 30(12):1079–1082
Konhauser KO, Newman DK, Kappler A (2005) The potential significance of microbial Fe(III) reduction during deposition of Precambrian banded iron formations. Geobiology 3:167–177
Konhauser KO, Planavsky NJ, Hardisty DS, Robbins LJ, Warchola TJ, Haugaard R, Lalonde SV, Partin CA, Oonk PBH, Tsikos H, Lyons TW, Bekker A, Johnson CM (2017) Iron formations: a global record of Neoarchaean to Palaeoproterozoic environmental history. Earth Sci Rev 172:140–177. https://doi.org/10.1016/j.earscirev.2017.06.012
Korenaga J (2013) Initiation and evolution of plate tectonics on earth: theories and observations. Annu Rev Earth Planet Sci 41(1):117–151. https://doi.org/10.1146/annurev-earth-050212-124208
Korenaga J (2018) Crustal evolution and mantle dynamics through Earth history. Philos Trans A Math Phys Eng Sci 376(2132). https://doi.org/10.1098/rsta.2017.0408
Korenaga J, Planavsky NJ, Evans DAD (2017) Global water cycle and the coevolution of the Earth's interior and surface environment. Philos Trans A Math Phys Eng Sci 375(2094). https://doi.org/10.1098/rsta.2015.0393
Kraal P, Slomp CP, Forster A, Kuypers MMM (2010) Phosphorus cycling from the margin to abyssal depths in the proto-Atlantic during oceanic anoxic event 2. Palaeogeogr Palaeoclimatol Palaeoecol 295(1–2):42–54. https://doi.org/10.1016/j.palaeo.2010.05.014

Krissansen-Totton J, Buick R, Catling DC (2015) A statistical analysis of the carbon isotope record from the Archean to Phanerozoic and implications for the rise of oxygen. Am J Sci 315(4):275–316. https://doi.org/10.2475/04.2015.01

Kump LR (2012) Sulfur isotopes and the stepwise oxygenation of the biosphere. Elements 8(6):410–411. https://doi.org/10.2113/gselements.8.6.410

Kump LR, Arthur MA (1999) Interpreting carbon-isotope excursions: carbonates and organic matter. Chem Geol 161:181–198

Kump LR, Barley ME (2007) Increased subaerial volcanism and the rise of atmospheric oxygen 2.5 billion years ago. Nature 448(7157):1033–1036. https://doi.org/10.1038/nature06058

Kump LR, Holland HD (1992) Iron in Precambrian rocks: implications for the global oxygen budget of the ancient Earth. Geochim Cosmochim Acta 56:3217–3223

Kump LR, Junium C, Arthur MA, Brasier A, Fallick A, Melezhik V, Lepland A, Crne AE, Luo GM (2011) Isotopic evidence for massive oxidation of organic matter following the Great Oxidation Event. Science 334(6063):1694–1696. https://doi.org/10.1126/science.1213999

Kump LR, Seyfried WE (2005) Hydrothermal Fe fluxes during the Precambrian: effect of low oceanic sulfate concentrations and low hydrostatic pressure on the composition of black smokers. Earth Planet Sci Lett 235(3–4):654–662. https://doi.org/10.1016/j.epsl.2005.04.040

Kunzmann M, Gibson TM, Halverson GP, Hodgskiss MSW, Bui TH, Carozza DA, Sperling EA, Poirier A, Cox GM, Wing BA (2017) Iron isotope biogeochemistry of Neoproterozoic marine shales. Geochim Cosmochim Acta 209:85–105. https://doi.org/10.1016/j.gca.2017.04.003

Kunzmann M, Gutzmer J, Beukes NJ, Halverson GP (2014) Depositional environment and lithostratigraphy of the paleoproterozoic Mooidraai Formation, Kalahari Manganese Field, South Africa. S Afr J Geol 117 (2):173–192. https://doi.org/10.2113/gssajg

Kunzmann M, Halverson GP, Scott C, Minarik WG, Wing BA (2015) Geochemistry of Neoproterozoic black shales from Svalbard: implications for oceanic redox conditions spanning Cryogenian glaciations. Chem Geol 417:383–393. https://doi.org/10.1016/j.chemgeo.2015.10.022

Kurtz AC, Kump LR, Arthur MA, Zachos JC, Paytan A (2003) Early Cenozoic decoupling of the global carbon and sulfur cycles. Paleoceanography 18(4): n/a-n/a. https://doi.org/10.1029/2003pa000908

Kurzweil F, Wille M, Gantert N, Beukes NJ, Schoenberg R (2016) Manganese oxide shuttling in pre-GOE oceans–evidence from molybdenum and iron isotopes. Earth Planet Sci Lett 452:69–78. https://doi.org/10.1016/j.epsl.2016.07.013

Kurzweil F, Wille M, Schoenberg R, Taubald H, Van Kranendonk MJ (2015a) Continuously increasing δ98Mo values in Neoarchean black shales and iron formations from the Hamersley Basin. Geochim Cosmochim Acta 164:523–542. https://doi.org/10.1016/j.gca.2015.05.009

Kurzweil F, Drost K, Pašava J, Wille M, Taubald H, Schoeckle D, Schoenberg R (2015b) Coupled sulfur, iron and molybdenum isotope data from black shales of the Teplá- Barrandian unit argue against deep ocean oxygenation during the Ediacaran. Geochim Cosmochim Acta 171:121–142. http://dx.doi.org/10.1016/j.gca.2015.08.022

Kuypers MMM, Lourens LJ, Rijpstra WIC, Pancost RD, Nijenhuis IA, Sinninghe Damsté JS (2004) Orbital forcing of organic carbon burial in the proto-North Atlantic during oceanic anoxic event 2. Earth Planet Sci Lett 228(3–4):465–482. https://doi.org/10.1016/j.epsl.2004.09.037

Kuypers MMM, Pancost RD, Nijenhuis IA, Sinninghe Damsté JS (2002) Enhanced productivity led to increased organic carbon burial in the euxinic North Atlantic basin during the late Cenomanian oceanic anoxic event. Paleoceanography 17(4):3-1–3-13. https://doi.org/10.1029/2000pa000569

Laakso TA, Schrag DP (2014) Regulation of atmospheric oxygen during the Proterozoic. Earth Planet Sci Lett 388:81–91. https://doi.org/10.1016/j.epsl.2013.11.049

Laakso TA, Schrag DP (2017) A theory of atmospheric oxygen. Geobiology 15(3):366–384. https://doi.org/10.1111/gbi.12230

Labrosse S, Jaupart C (2007) Thermal evolution of the Earth: Secular changes and fluctuations of plate characteristics. Earth Planet Sci Lett 260(3–4):465–481. https://doi.org/10.1016/j.epsl.2007.05.046

Ladant J-B, Donnadieu Y, Bopp L, Lear CH, Wilson PA (2018) Meridional contrasts in productivity changes driven by the opening of drake passage. Paleoceanogr Paleoclimatology 33(3):302–317. https://doi.org/10.1002/2017pa003211

Lanci L, Muttoni G, Erba E (2010) Astronomical tuning of the cenomanian scaglia bianca formation at Furlo, Italy. Earth Planet Sci Lett 292(1–2):231–237. https://doi.org/10.1016/j.epsl.2010.01.041

Land JS, Tsikos H, Cousins D, Luvizotto G, Zack T (2018) Origin of red beds and paleosols in the Palaeoproterozoic Transvaal and Olifansthoek Supergroups of South Africa: provenance versus metasomatic controls. Geol J 53(1):191–202. https://doi.org/10.1002/gj.2885

Lantink ML, Oonk PBH, Floor GH, Tsikos H, Mason PRD (2018) Fe isotopes of a 2.4 Ga hematite-rich IF constrain marine redox conditions around the GOE. Precambrian Res 305:218–235. https://doi.org/10.1016/j.precamres.2017.12.025

Lau KV, Macdonald FA, Maher K, Payne JL (2017) Uranium isotope evidence for temporary ocean oxygenation in the aftermath of the Sturtian Snowball Earth. Earth Planet Sci Lett 458:282–292. https://doi.org/10.1016/j.epsl.2016.10.043

Laurin J, Meyers SR, Galeotti S, Lanci L (2016) Frequency modulation reveals the phasing of orbital eccentricity during Cretaceous Oceanic Anoxic

Event II and the Eocene hyperthermals. Earth Planet Sci Lett 442:143–156. https://doi.org/10.1016/j.epsl.2016.02.047
Leavitt WD, Halevy I, Bradley AS, Johnston DT (2013) Influence of sulfate reduction rates on the Phanerozoic sulfur isotope record. Proc Natl Acad Sci USA 110 (28):11244–11249. https://doi.org/10.1073/pnas.1218874110
Lechler M, Pogge von Strandmann PAE, Jenkyns HC, Prosser G, Parente M (2015) Lithium-isotope evidence for enhanced silicate weathering during OAE 1a (Early Aptian Selli event). Earth Planet Sci Lett 432:210–222. https://doi.org/10.1016/j.epsl.2015.09.052
Lechte M, Wallace M (2016) Sub–ice shelf ironstone deposition during the Neoproterozoic Sturtian glaciation. Geology 44(11):891–894. https://doi.org/10.1130/g38495.1
Lechte MA, Wallace MW (2015) Sedimentary and tectonic history of the Holowilena Ironstone, a Neoproterozoic iron formation in South Australia. Sed Geol 329:211–224. https://doi.org/10.1016/j.sedgeo.2015.09.014
Lechte MA, Wallace MW, AvS Hood, Planavsky N (2018) Cryogenian iron formations in the glaciogenic Kingston Peak Formation, California. Precambrian Res 310:443–462. https://doi.org/10.1016/j.precamres.2018.04.003
Leckie RM, Bralower TJ, Cashman R (2002) Oceanic anoxic events and plankton evolution: biotic response to tectonic forcing during the mid-Cretaceous. Paleoceanography 17(3):13-11–13-29. https://doi.org/10.1029/2001pa000623
Lee C-D, Halliday AN, Hein JR, Burton KW, Christensen JN, Gunther D (1999) Hafnium isotope stratigraphy of ferromanganese crusts. Science 285:1052–1054
Lee C-TA, Yeung LY, McKenzie NR, Yokoyama Y, Ozaki K, Lenardic A (2016) Two-step rise of atmospheric oxygen linked to the growth of continents. Nat Geosci 9(6):417–424. https://doi.org/10.1038/ngeo2707
Lepland A, van Zuilen MA, Arrhenius G, Whitehouse MJ, Fedo CM (2005) Questioning the evidence for Earth's earliest life—Akilia revisited. Geology 33(1). https://doi.org/10.1130/g20890.1
Lepot K, Addad A, Knoll AH, Wang J, Troadec D, Beche A, Javaux EJ (2017) Iron minerals within specific microfossil morphospecies of the 1.88 Ga Gunflint Formation. Nat Commun 8:14890. https://doi.org/10.1038/ncomms14890
Levasseur S, Frank M, Hein JR, Halliday AN (2004) The global variation in the iron isotope composition of marine hydrogenetic ferromanganese deposits: implications for seawater chemistry? Earth Planet Sci Lett 224(1–2):91–105. https://doi.org/10.1016/j.epsl.2004.05.010
Li C, Love GD, Lyons TW, Fike DA, Sessions AL, Chu X (2010) A stratified redox model for the Ediacaran Ocean. Science 328:80–83
Li C, Love GD, Lyons TW, Scott CT, Feng L, Huang J, Chang H, Zhang Q, Chu X (2012a) Evidence for a redox stratified Cryogenian marine basin, Datangpo Formation, South China. Earth Planet Sci Lett 331–332:246–256. https://doi.org/10.1016/j.epsl.2012.03.018
Li G, Elderfield H (2013) Evolution of carbon cycle over the past 100 million years. Geochim Cosmochim Acta 103:11–25. https://doi.org/10.1016/j.gca.2012.10.014
Li G, Hartmann J, Derry LA, West AJ, You C-F, Long X, Zhan T, Li L, Li G, Qiu W, Li T, Liu L, Chen Y, Ji J, Zhao L, Chen J (2016) Temperature dependence of basalt weathering. Earth Planet Sci Lett 443:59–69. https://doi.org/10.1016/j.epsl.2016.03.015
Li W, Beard BL, Johnson CM (2015) Biologically recycled continental iron is a major component in banded iron formations. Proc Natl Acad Sci USA 112 (27):8193–8198. https://doi.org/10.1073/pnas.1505515112
Li W, Czaja AD, Van Kranendonk MJ, Beard BL, Roden EE, Johnson CM (2013a) An anoxic, Fe(II)-rich, U-poor ocean 3.46 billion years ago. Geochim Cosmochim Acta 120:65–79. https://doi.org/10.1016/j.gca.2013.06.033
Li W, Huberty JM, Beard BL, Kita NT, Valley JW, Johnson CM (2013b) Contrasting behavior of oxygen and iron isotopes in banded iron formations revealed by in situ isotopic analysis. Earth Planet Sci Lett 384:132–143. https://doi.org/10.1016/j.epsl.2013.10.014
Li W, Johnson CM, Beard BL (2012b) U-Th–Pb isotope data indicate phanerozoic age for oxidation of the 3.4 Ga Apex Basalt. Earth Planet Sci Lett 319–320:197–206
Li Y-L, Konhauser KO, Zhai M (2017a) The formation of magnetite in the early Archean oceans. Earth Planet Sci Lett 466:103–114. https://doi.org/10.1016/j.epsl.2017.03.013
Li Y-X, Montañez IP, Liu Z, Ma L (2017b) Astronomical constraints on global carbon-cycle perturbation during Oceanic Anoxic Event 2 (OAE2). Earth Planet Sci Lett 462:35–46. https://doi.org/10.1016/j.epsl.2017.01.007
Li ZX, Bogdanova SV, Collins AS, Davidson A, De Waele B, Ernst RE, Fitzsimons ICW, Fuck RA, Gladkochub DP, Jacobs J, Karlstrom KE, Lu S, Natapov LM, Pease V, Pisarevsky SA, Thrane K, Vernikovsky V (2008) Assembly, configuration, and break-up history of Rodinia: a synthesis. Precambrian Res 160(1–2):179–210. https://doi.org/10.1016/j.precamres.2007.04.021
Lin Z, Sun X, Lu Y, Strauss H, Xu L, Gong J, Teichert BMA, Lu R, Lu H, Sun W, Peckmann J (2017) The enrichment of heavy iron isotopes in authigenic pyrite as a possible indicator of sulfate-driven anaerobic oxidation of methane: insights from the South China Sea. Chem Geol 449:15–29. https://doi.org/10.1016/j.chemgeo.2016.11.032
Ling H-F, Burton KW, O'Nions RK, Kamber BS, von Blanckenburg F, Gibb AJ, Hein JR (1997) Evolution

of Nd and Pb isotopes in Central Pacific seawater from ferromanganese crusts. Earth Planet Sci Lett 146:1–12

Liu XM, Kah LC, Knoll AH, Cui H, Kaufman AJ, Shahar A, Hazen RM (2016) Tracing Earth's O2 evolution using Zn/Fe ratios in marine carbonates. Geochem Perspect Lett 2(1):24–34. https://doi.org/10.7185/geochemlet.1603

Lovley DR (1991) Dissimilatory Fe(III) and Mn(IV) Reduction. Microbiol Rev 55(2):259–287

Lovley DR, Phillips EJP (1988) Novel mode of microbial energy metabolism: organic carbon oxidation coupled to dissimilatory reduction of iron or manganese. Appl Environ Microbiol 54(6):1472–1480

Lowe DR, Byerly GR (1999) Stratigraphy of the west-central part of the Baberton Greenstone Belt, South Africa. Special Papers-Geological Society of America:1–39

Lowe DR, Byerly GR (2007) Chapter 5.3 An overview of the geology of the Barberton Greenstone Belt and vicinity: implications for early crustal development. In: Earth's oldest rocks. Developments in Precambrian geology, pp 481–526. https://doi.org/10.1016/s0166-2635(07)15053-2

Lowell RP, Keller SM (2003) High-temperature seafloor hydrothermal circulation over geologic time and archean banded iron formations. Geophys Res Lett 30(7). https://doi.org/10.1029/2002gl016536

Luo G, Ono S, Huang J, Algeo TJ, Li C, Zhou L, Robinson A, Lyons TW, Xie S (2015) Decline in oceanic sulfate levels during the early Mesoproterozoic. Precambrian Res 258:36–47. https://doi.org/10.1016/j.precamres.2014.12.014

Lyle M, Barron J, Bralower TJ, Huber M, Olivarez Lyle A, Ravelo AC, Rea DK, Wilson PA (2008) Pacific Ocean and Cenozoic evolution of climate. Rev Geophys 46(2). https://doi.org/10.1029/2005rg000190

Lyons TW, Reinhard CT, Planavsky NJ (2014) The rise of oxygen in Earth's early ocean and atmosphere. Nature 506(7488):307–315. https://doi.org/10.1038/nature13068

Lyons TW, Severmann S (2006) A critical look at iron paleoredox proxies: new insights from modern euxinic marine basins. Geochim Cosmochim Acta 70(23):5698–5722. https://doi.org/10.1016/j.gca.2006.08.021

Macdonald FA, Halverson GP, Strauss JV, Smith EF, Cox G, Sperling EA, Roots CF (2012) Early neoproterozoic basin formation in Yukon, Canada: implications for the make-up and break-up of Rodinia. Geosci Can 39:77–99

Macdonald FA, Schmitz MD, Crowley JL, Roots CF, Jones DS, Maloof AC, Strauss JV, Cohen PA, Johnston DT, Schrag DP (2010) Calibrating the cryogenian. Science 327(5970):1241–1243. https://doi.org/10.1126/science.1183325

Macdonald FA, Schmitz MD, Strauss JV, Halverson GP, Gibson TM, Eyster A, Cox G, Mamrol P, Crowley JL (2018) Cryogenian of Yukon. Precambrian Res 319:114–143. https://doi.org/10.1016/j.precamres.2017.08.015

Macdonald FA, Wordsworth R (2017) Initiation of Snowball Earth with volcanic sulfur aerosol emissions. Geophys Res Lett 44:1–9. https://doi.org/10.1002/2016GL072335

MacFarlane AW, Holland HD (1991) The timing of alkali metasomatism in paleosols. Can Miner 29:1043–1050

Mandernack KW, Bazylinski DA, Shanks WC III, Bullen TD (1999) Oxygen and Iron Isotope Studies of Magnetite Produced by Magnetotactic Bacteria. Science 285(5435):1892–1896. https://doi.org/10.1126/science.285.5435.1892

Mansor M, Fantle MS (2019) A novel framework for interpreting pyrite-based Fe isotope records of the past. Geochim Cosmochim Acta 253:39–62. https://doi.org/10.1016/j.gca.2019.03.017

Marcus MA, Edwards KJ, Gueguen B, Fakra SC, Horn G, Jelinski NA, Rouxel O, Sorensen J, Toner BM (2015) Iron mineral structure, reactivity, and isotopic composition in a South Pacific Gyre ferromanganese nodule over 4 Ma. Geochim Cosmochim Acta 171:61–79. https://doi.org/10.1016/j.gca.2015.08.021

Martin AP, Condon DJ, Prave AR, Lepland A (2013) A review of temporal constraints for the Palaeoproterozoic large, positive carbonate carbon isotope excursion (the Lomagundi-Jatuli Event). Earth Sci Rev 127:242–261. https://doi.org/10.1016/j.earscirev.2013.10.006

McArthur JM, Howarth RJ, Bailey TR (2001) Strontium isotope stratigraphy: lowess version 3: Best fit to the marine Sr-isotope curve for 0–509 Ma and accompanying look-up table deriving numerical age. J Geol 109(155–170)

McDowell FW, Housh TB, Wark DA (1999) Nature of the crust beneath west-central Chihuahua, Mexico, based upon Sr, Nd, and Pb isotopic compositions at the Tomóchic volcanic center. GSA Bull 111(6):823–830

McKenzie NR, Hughes NC, Gill BC, Myrow PM (2014) Plate tectonic influences on Neoproterozoic–early Paleozoic climate and animal evolution. Geology 42(2):127–130. https://doi.org/10.1130/g34962.1

Melezhik VA, Fallick AE, Brasier AT, Lepland A (2015) Carbonate deposition in the Palaeoproterozoic Onega basin from Fennoscandia: a spotlight on the transition from the Lomagundi-Jatuli to Shunga events. Earth Sci Rev 147:65–98. https://doi.org/10.1016/j.earscirev.2015.05.005

Mendes M, Lobato LM, Kunzmann M, Halverson GP, Rosière CA (2017) Iron isotope and REE+Y composition of the Cauê banded iron formation and related iron ores of the Quadrilátero Ferrífero, Brazil. Miner Depos 52(2):159–180. https://doi.org/10.1007/s00126-016-0649-9

Meyers SR, Siewert SE, Singer BS, Sageman BB, Condon DJ, Obradovich JD, Jicha BR, Sawyer DA (2012) Intercalibration of radioisotopic and astrochronologic time scales for the Cenomanian-Turonian boundary interval, Western Interior Basin, USA. Geology 40(1):7–10. https://doi.org/10.1130/g32261.1

Meynadier L, Allègre C, O'Nions RK (2008) Plate tectonics, radiogenic isotopic tracers and paleoceanography. Earth Planet Sci Lett 272(3–4):513–522. https://doi.org/10.1016/j.epsl.2008.04.029

Mills JV, Gomes ML, Kristall B, Sageman BB, Jacobson AD, Hurtgen MT (2017) Massive volcanism, evaporite deposition, and the chemical evolution of the Early Cretaceous ocean. Geology 45(5):475–478. https://doi.org/10.1130/g38667.1

Misra S, Froelich PN (2012) Lithium isotope history of cenozoic seawater: changes in silicate weathering and reverse weathering. Science 335(6070):818–823. https://doi.org/10.1126/science.1214697

Miyazaki Y, Planavsky NJ, Bolton EW, Reinhard CT (2018) Making sense of massive carbon isotope excursions with an inverse carbon cycle model. J Geophys Res Biogeosci 123(8):2485–2496. https://doi.org/10.1029/2018jg004416

Mloszewska AM, Mojzsis SJ, Pecoits E, Papineau D, Dauphas N, Konhauser KO (2013) Chemical sedimentary protoliths in the >3.75 Ga Nuvvuagittuq Supracrustal Belt (Québec, Canada). Gondwana Res 23(2):574–594. https://doi.org/10.1016/j.gr.2012.11.005

Mloszewska AM, Pecoits E, Cates NL, Mojzsis SJ, O'Neil J, Robbins LJ, Konhauser KO (2012) The composition of Earth's oldest iron formations: the Nuvvuagittuq Supracrustal Belt (Québec, Canada). Earth Planet Sci Lett 317–318:331–342. https://doi.org/10.1016/j.epsl.2011.11.020

Mojzsis SJ, Arrhenius G, McKeegan KD, Harrison TM, Nutman AP, Friend CRL (1996) Evidence for life on Earth before 3800 million years ago. Nature 384:55–59

Montañez IP, Banner JL, Osleger DA, Borg LE, Bosserman PJ (1996) Integrated Sr isotope variations and sea-level history of Middle to Upper Cambrian platform carbonates: implications for the evolution of Cambrian seawater 87Sr/86Sr. Geology 24 (10):917–920

Monteiro FM, Pancost RD, Ridgwell A, Donnadieu Y (2012) Nutrients as the dominant control on the spread of anoxia and euxinia across the Cenomanian-Turonian oceanic anoxic event (OAE2): Model-data comparison. Paleoceanography 27(4). https://doi.org/10.1029/2012pa002351

Moore TC Jr, Wade BS, Westerhold T, Erhardt AM, Coxall HK, Baldauf J, Wagner M (2014) Equatorial Pacific productivity changes near the Eocene-Oligocene boundary. Paleoceanography 29:825–844. https://doi.org/10.1002/2014PA002656

Morgan JJ (2005) Kinetics of reaction between O2 and Mn(II) species in aqueous solutions. Geochim Cosmochim Acta 69(1):35–48. https://doi.org/10.1016/j.gca.2004.06.013

Mort H, Jacquat O, Adatte T, Steinmann P, Föllmi K, Matera V, Berner Z, Stüben D (2007a) The Cenomanian/Turonian anoxic event at the Bonarelli Level in Italy and Spain: enhanced productivity and/or better preservation? Cretac Res 28(4):597–612. https://doi.org/10.1016/j.cretres.2006.09.003

Mort HP, Adatte T, Föllmi KB, Keller G, Steinmann P, Matera V, Berner Z, Stüben D (2007b) Phosphorus and the roles of productivity and nutrient recycling during oceanic anoxic event 2. Geology 35(6). https://doi.org/10.1130/g23475a.1

Müller RD, Sdrolias M, Gaina C, Steinberger B, Heine C (2008) Long-term sea-level fluctuations driven by ocean basin dynamics. Science 319:1357–1362. https://doi.org/10.1126/science.1151540

Murakami T, Ito J-I, Utsunomiya S, Kasama T, Kozai N, Ohnuki T (2004) Anoxic dissolution processes of biotite: implications for Fe behavior during Archean weathering. Earth Planet Sci Lett 224(1–2):117–129. https://doi.org/10.1016/j.epsl.2004.04.040

Murakami T, Matsuura K, Kanzaki Y (2016) Behaviors of trace elements in Neoarchean and Paleoproterozoic paleosols: implications for atmospheric oxygen evolution and continental oxidative weathering. Geochim Cosmochim Acta 192:203–219. https://doi.org/10.1016/j.gca.2016.07.008

Myers CR, Nealson KH (1988) Microbial reduction of manganese oxides: interactions with iron and sulfur. Geochim Cosmochim Acta 52:2727–2732

Naeraa T, Schersten A, Rosing MT, Kemp AI, Hoffmann JE, Kokfelt TF, Whitehouse MJ (2012) Hafnium isotope evidence for a transition in the dynamics of continental growth 3.2 Gyr ago. Nature 485 (7400):627–630. https://doi.org/10.1038/nature11140

Nakagawa Y, Takano S, Firdaus ML, Norisuye K, Hirata T, Vance D, Sohrin Y (2012) The molybdenum isotopic composition of the modern ocean. Geochem J 44:131–141

Nealson KH, Tebo BM, Rosson RA (1988) Occurence and mechanisms of microbial oxidation of manganese. In: Advances in applied microbiology, vol 33. Academic Press, Inc., pp 279–318

Nel BP (2013) Petrography and geochemistry of iron formations of the Paleoproterozoic Koegas Subgroup, Transvaal Supergroup, Griqualand West. University of Johannesburg, South Africa

Nhleko N (2003) The pongola supergroup in Swaziland. Rand Afrikaans University

Nicklas RW, Puchtel IS, Ash RD (2018) Redox state of the Archean mantle: evidence from V partitioning in 3.5–2.4 Ga komatiites. Geochim Cosmochim Acta 222:447–466. https://doi.org/10.1016/j.gca.2017.11.002

Nielsen SG, Mar-Gerrison S, Gannoun A, LaRowe D, Klemm V, Halliday AN, Burton KW, Hein JR (2009) Thallium isotope evidence for a permanent increase in marine organic carbon export in the early Eocene. Earth Planet Sci Lett 278(3–4):297–307. https://doi.org/10.1016/j.epsl.2008.12.010

Nishizawa M, Yamamoto H, Ueno Y, Tsuruoka S, Shibuya T, Sawaki Y, Yamamoto S, Kon Y, Kitajima K, Komiya T, Maruyama S, Hirata T (2010) Grain-scale iron isotopic distribution of pyrite from Precambrian shallow marine carbonate revealed by a femtosecond laser ablation multicollector ICP-MS technique: possible proxy for the redox state of ancient seawater. Geochim Cosmochim Acta 74 (9):2760–2778. https://doi.org/10.1016/j.gca.2010.02.014

Nutman AP, Bennett VC (2019) The 3.9–3.6 Ga itsaq gneiss complex of Greenland. In: Earth's oldest rocks, pp 375–399. https://doi.org/10.1016/b978-0-444-63901-1.00017-4

Nutman AP, Bennett VC, Friend CR, Van Kranendonk MJ, Chivas AR (2016) Rapid emergence of life shown by discovery of 3,700-million-year-old microbial structures. Nature 537(7621):535–538. https://doi.org/10.1038/nature19355

Nutman AP, Bennett VC, Friend CRL (2017) Seeing through the magnetite: reassessing Eoarchean atmosphere composition from Isua (Greenland) $\geq$ 3.7 Ga banded iron formations. Geosci Front 8(6):1233–1240. https://doi.org/10.1016/j.gsf.2017.02.008

Nutman AP, Bennett VC, Friend CRL (2019) Eoarchean life from the isua supracrustal belt (Greenland). In: Earth's oldest rocks, pp 965–983. https://doi.org/10.1016/b978-0-444-63901-1.00039-3

Nutman AP, Friend CRL, Bennett VC, Wright D, Norman MD (2010) $\geq$3700 Ma pre-metamorphic dolomite formed by microbial mediation in the Isua supracrustal belt (W. Greenland): simple evidence for early life? Precambrian Res 183(4):725–737. https://doi.org/10.1016/j.precamres.2010.08.006

O'Brien CL, Robinson SA, Pancost RD, Sinninghe Damsté JS, Schouten S, Lunt DJ, Alsenz H, Bornemann A, Bottini C, Brassell SC, Farnsworth A, Forster A, Huber BT, Inglis GN, Jenkyns HC, Linnert C, Littler K, Markwick P, McAnena A, Mutterlose J, Naafs BDA, Püttmann W, Sluijs A, van Helmond NAGM, Vellekoop J, Wagner T, Wrobel NE (2017) Cretaceous sea-surface temperature evolution: constraints from TEX 86 and planktonic foraminiferal oxygen isotopes. Earth Sci Rev 172:224–247. https://doi.org/10.1016/j.earscirev.2017.07.012

O'Neil J, Carlson RW, Papineau D, Levine EY, Francis D (2019) The Nuvvuagittuq Greenstone Belt. In: Earth's oldest rocks, pp 349–374. https://doi.org/10.1016/b978-0-444-63901-1.00016-2

O'Neil J, Maurice C, Stevenson RK, Larocque J, Cloquet C, David J, Francis D (2007) Chapter 3.4 The geology of the 3.8 Ga Nuvvuagittuq (Porpoise Cove) Greenstone Belt, Northeastern Superior Province, Canada. In: Earth's oldest rocks. Developments in Precambrian geology, pp 219–250. https://doi.org/10.1016/s0166-2635(07)15034-9

O'Neill C, Debaille V (2014) The evolution of Hadean-Eoarchaean geodynamics. Earth Planet Sci Lett 406:49–58

O'Nions RK, Frank M, von Blanckenburg F, Ling H-F (1998) Secular variation of Nd and Pb isotopes in ferromanganese crusts from the Atlantic, Indian and Pacific Oceans. Earth Planet Sci Lett 155:15–28

Och LM, Shields-Zhou GA (2012) The Neoproterozoic oxygenation event: environmental perturbations and biogeochemical cycling. Earth Sci Rev 110(1–4):26–57. https://doi.org/10.1016/j.earscirev.2011.09.004

Ohmoto H (1996) Evidence in pre-2.2 Ga paleosols for the early evolution of atmospheric oxygen and terrestrial biota Geology 24(12):1135–1138

Ohmoto H, Watanabe Y, Yamaguchi KE, Naraoka H, Haruna M, Kakegawa T, Hayashi K-i, Kato Y (2006) Chemical and biological evolution of early Earth: constraints from banded iron formations. In: Memoir 198: evolution of early earth's atmosphere, hydrosphere, and biosphere-constraints from ore deposits, pp 291–331. https://doi.org/10.1130/2006.1198(17)

Olson SL, Kump LR, Kasting JF (2013) Quantifying the areal extent and dissolved oxygen concentrations of Archean oxygen oases. Chem Geol 362:35–43. https://doi.org/10.1016/j.chemgeo.2013.08.012

Ono S, Kaufman AJ, Farquhar J, Sumner DY, Beukes NJ (2009) Lithofacies control on multiple-sulfur isotope records and Neoarchean sulfur cycles. Precambrian Res 169(1–4):58–67. https://doi.org/10.1016/j.precamres.2008.10.013

Ossa Ossa F, Eickmann B, Hofmann A, Planavsky NJ, Asael D, Pambo F, Bekker A (2018a) Two-step deoxygenation at the end of the Paleoproterozoic Lomagundi Event. Earth Planet Sci Lett 486:70–83. https://doi.org/10.1016/j.epsl.2018.01.009

Ossa Ossa F, Hofmann A, Vidal O, Kramers JD, Belyanin G, Cavalazzi B (2016) Unusual manganese enrichment in the Mesoarchean Mozaan Group, Pongola Supergroup, South Africa. Precambrian Res 281:414–433. https://doi.org/10.1016/j.precamres.2016.06.009

Ossa Ossa F, Hofmann A, Wille M, Spangenberg JE, Bekker A, Poulton SW, Eickmann B, Schoenberg R (2018b) Aerobic iron and manganese cycling in a redox-stratified Mesoarchean epicontinental sea. Earth Planet Sci Lett 500:28–40. https://doi.org/10.1016/j.epsl.2018.07.044

Ostrander CM, Nielsen SG, Owens JD, Kendall B, Gordon GW, Romaniello SJ, Anbar AD (2019) Fully oxygenated water columns over continental shelves before the Great Oxidation Event. Nat Geosci 12 (3):186–191. https://doi.org/10.1038/s41561-019-0309-7

Ostrander CM, Owens JD, Nielsen SG (2017) Constraining the rate of oceanic deoxygenation leading up to a Cretaceous Oceanic Anoxic Event (OAE-2: ~ 94 Ma). Sci Adv 3:1–5

Owens JD, Gill BC, Jenkyns HC, Bates SM, Severmann S, Kuypers MM, Woodfine RG, Lyons TW (2013) Sulfur isotopes track the global extent and dynamics of euxinia during Cretaceous Oceanic Anoxic Event 2. Proc Natl Acad Sci USA 110(46):18407–18412. https://doi.org/10.1073/pnas.1305304110

Owens JD, Lyons TW, Hardisty DS, Lowery CM, Lu Z, Lee B, Jenkyns HC, Heimhofer U (2017a) Patterns of local and global redox variability during the Cenomanian-Turonian Boundary Event (Oceanic Anoxic Event 2) recorded in carbonates and shales from central Italy. Sedimentology 64(1):168–185. https://doi.org/10.1111/sed.12352

Owens JD, Lyons TW, Li X, Macleod KG, Gordon G, Kuypers MMM, Anbar A, Kuhnt W, Severmann S (2012) Iron isotope and trace metal records of iron cycling in the proto-North Atlantic during the Cenomanian-Turonian oceanic anoxic event (OAE-2). Paleoceanography 27(3):n/a-n/a. https://doi.org/10.1029/2012pa002328

Owens JD, Lyons TW, Lowery CM (2018) Quantifying the missing sink for global organic carbon burial during a Cretaceous oceanic anoxic event. Earth Planet Sci Lett 499:83–94. https://doi.org/10.1016/j.epsl.2018.07.021

Owens JD, Nielsen SG, Horner TJ, Ostrander CM, Peterson LC (2017b) Thallium-isotopic compositions of euxinic sediments as a proxy for global manganese-oxide burial. Geochim Cosmochim Acta 213:291–307. https://doi.org/10.1016/j.gca.2017.06.041

Owens JD, Reinhard CT, Rohrssen M, Love GD, Lyons TW (2016) Empirical links between trace metal cycling and marine microbial ecology during a large perturbation to Earth's carbon cycle. Earth Planet Sci Lett 449:407–417. https://doi.org/10.1016/j.epsl.2016.05.046

Ozaki K, Reinhard CT, Tajika E (2019) A sluggish mid-Proterozoic biosphere and its effect on Earth's redox balance. Geobiology 17(1):3–11. https://doi.org/10.1111/gbi.12317

Ozaki K, Tajika E, Hong PK, Nakagawa Y, Reinhard CT (2017) Effects of primitive photosynthesis on Earth's early climate system. Nat Geosci 11(1):55–59. https://doi.org/10.1038/s41561-017-0031-2

Pagani M, Huber M, Zhonghui L, Bohaty SM, Henderiks J, Sijp W, Krishnan S, DeConto RM (2011) The role of carbon dioxide during the onset of Antarctic glaciation. Science 334(6060):1261–1264

Pagani M, Zachos JC, Freeman KH, Tipple B, Bohaty S (2005) Marked decline in atmospheric carbon dioxide concentrations during the Paleogene. Science 309:600–603

Paiste K, Lepland A, Zerkle AL, Kirsimäe K, Izon G, Patel NK, McLean F, Kreitsmann T, Mänd K, Bui TH, Romashkin AE, Rychanchik DV, Prave AR (2018) Multiple sulphur isotope records tracking basinal and global processes in the 1.98 Ga Zaonega formation, NW Russia. Chem Geol 499:151–164. https://doi.org/10.1016/j.chemgeo.2018.09.025

Papineau D, De Gregorio BT, Cody GD, O'Neil J, Steele A, Stroud RM, Fogel ML (2011) Young poorly crystalline graphite in the >3.8-Gyr-old Nuvvuagittuq banded iron formation. Nat Geosci 4(6):376–379. https://doi.org/10.1038/ngeo1155

Partin CA, Bekker A, Planavsky NJ, Scott CT, Gill BC, Li C, Podkovyrov V, Maslov A, Konhauser KO, Lalonde SV, Love GD, Poulton SW, Lyons TW (2013a) Large-scale fluctuations in Precambrian atmospheric and oceanic oxygen levels from the record of U in shales. Earth Planet Sci Lett 369–370:284–293. https://doi.org/10.1016/j.epsl.2013.03.031

Partin CA, Lalonde SV, Planavsky NJ, Bekker A, Rouxel OJ, Lyons TW, Konhauser KO (2013b) Uranium in iron formations and the rise of atmospheric oxygen. Chem Geol 362:82–90. https://doi.org/10.1016/j.chemgeo.2013.09.005

Pavlov AA, Kasting JF (2002) Mass-Independent fractionation of sulfur isotopes in Archean sediments: strong evidence for an anoxic Archean atmosphere. Astrobiology 2(1):27–41

Paytan A, Kastner M, Campbell D, Thiemens MH (1998) Sulfur isotopic composition of Cenozoic seawater sulfate. Science 282:1459–1462. https://doi.org/10.1126/science.282.5393.1459

Paytan A, Kastner M, Camphell D, Thiemens MH (2004) Seawater sulfur isotope fluctuations in the cretaceous. Science 304:1663–1665

Pearson PN (2012) Oxygen isotopes in foramenifera: overview and historical review. Paleontol Soc Pap 18:1–38

Percak-Dennett EM, Beard BL, Xu H, Konishi H, Johnson CM, Roden EE (2011) Iron isotope fractionation during microbial dissimilatory iron oxide reduction in simulated Archaean seawater. Geobiology 9(3):205–220. https://doi.org/10.1111/j.1472-4669.2011.00277.x

Percak-Dennett EM, Loizeau J-L, Beard BL, Johnson CM, Roden EE (2013) Iron isotope geochemistry of biogenic magnetite-bearing sediments from the Bay of Vidy, Lake Geneva. Chem Geol 360–361:32–40. https://doi.org/10.1016/j.chemgeo.2013.10.008

Peters SE (2006) Macrostratigraphy of North America. J Geol 114(4):391–412

Petrychenko OY, Peryt TM, Chechel EI (2005) Early Cambrian seawater chemistry from fluid inclusions in halite from Siberian evaporites. Chem Geol 219(1–4):149–161. https://doi.org/10.1016/j.chemgeo.2005.02.003

Philippot P, Avila JN, Killingsworth BA, Tessalina S, Baton F, Caquineau T, Muller E, Pecoits E, Cartigny P, Lalonde SV, Ireland TR, Thomazo C, van Kranendonk MJ, Busigny V (2018) Globally asynchronous sulphur isotope signals require re-definition of the Great Oxidation Event. Nat Commun 9(1):2245. https://doi.org/10.1038/s41467-018-04621-x

Planavsky N, Rouxel O, Bekker A, Shapiro R, Fralick P, Knudsen A (2009) Iron-oxidizing microbial ecosystems thrived in late Paleoproterozoic redox-stratified oceans. Earth Planet Sci Lett 286(1–2):230–242. https://doi.org/10.1016/j.epsl.2009.06.033

Planavsky N, Rouxel OJ, Bekker A, Hofmann A, Little CTS, Lyons TW (2012a) Iron isotope composition of some Archean and Proterozoic iron formations. Geochim Cosmochim Acta 80:158–169. https://doi.org/10.1016/j.gca.2011.12.001

Planavsky NJ, Asael D, Hofmann A, Reinhard CT, Lalonde SV, Knudsen A, Wang X, Ossa Ossa F, Pecoits E, Smith AJB, Beukes NJ, Bekker A, Johnson TM, Konhauser KO, Lyons TW, Rouxel OJ (2014) Evidence for oxygenic photosynthesis half a billion years before the Great Oxidation Event. Nat Geosci 7(4):283–286. https://doi.org/10.1038/ngeo2122

Planavsky NJ, Bekker A, Hofmann A, Owens JD, Lyons TW (2012b) Sulfur record of rising and falling marine oxygen and sulfate levels during the Lomagundi event. Proc Natl Acad Sci USA 109(45):18300–18305. https://doi.org/10.1073/pnas.1120387109

Planavsky NJ, Cole DB, Isson TT, Reinhard CT, Crockford PW, Sheldon ND, Lyons TW (2018a) A case for low atmospheric oxygen levels during Earth's middle history. Emerg Top Life Sci 2(2):149–159. https://doi.org/10.1042/etls20170161

Planavsky NJ, McGoldrick P, Scott CT, Li C, Reinhard CT, Kelly AE, Chu X, Bekker A, Love GD, Lyons TW (2011) Widespread iron-rich conditions in the mid-Proterozoic ocean. Nature 477(7365):448–451. https://doi.org/10.1038/nature10327

Planavsky NJ, Rouxel OJ, Bekker A, Lalonde SV, Konhauser KO, Reinhard CT, Lyons TW (2010) The evolution of the marine phosphate reservoir. Nature 467(7319):1088–1090. https://doi.org/10.1038/nature09485

Planavsky NJ, Slack JF, Cannon WF, O'Connell B, Isson TT, Asael D, Jackson JC, Hardisty DS, Lyons TW, Bekker A (2018b) Evidence for episodic oxygenation in a weakly redox-buffered deep mid-Proterozoic ocean. Chem Geol 483:581–594. https://doi.org/10.1016/j.chemgeo.2018.03.028

Pogge von Strandmann PAE, Jenkyns HC, Woodfine RG (2013) Lithium isotope evidence for enhanced weathering during Oceanic Anoxic Event 2. Nat Geosci 6 (8):668–672. https://doi.org/10.1038/ngeo1875

Polyakov VB, Mineev SD (2000) The use of Mossbauer spectroscopy in stable isotope geochemistry. Geochim Cosmochim Acta 64(5):849–865

Polyakov VB, Soultanov DM (2011) New data on equilibrium iron isotope fractionation among sulfides: constraints on mechanisms of sulfide formation in hydrothermal and igneous systems. Geochim Cosmochim Acta 75(7):1957–1974. https://doi.org/10.1016/j.gca.2011.01.019

Posth NR, Köhler I, D. Swanner E, Schröder C, Wellmann E, Binder B, Konhauser KO, Neumann U, Berthold C, Nowak M, Kappler A (2013) Simulating Precambrian banded iron formation diagenesis. Chem Geol 362:66–73. https://doi.org/10.1016/j.chemgeo.2013.05.031

Poulsen CJ, Tabor C, White JD (2015) Long-term climate forcing by atmospheric oxygen concentrations. Science 348(6240):1238–1241

Poulton SW (2017) Early phosphorus redigested. Nat Geosci 10(2):75–76. https://doi.org/10.1038/ngeo2884

Poulton SW, Canfield DE (2005) Development of a sequential extraction procedure for iron: implications for iron partitioning in continentally derived particulates. Chem Geol 214(3–4):209–221. https://doi.org/10.1016/j.chemgeo.2004.09.003

Poulton SW, Canfield DE (2011) Ferruginous conditions: a dominant feature of the ocean through earth's history. Elements 7(2):107–112. https://doi.org/10.2113/gselements.7.2.107

Poulton SW, Fralick PW, Canfield DE (2004) The transition to a sulphidic ocean approximately 1.84 billion years ago. Nature 431(7005):173–177. https://doi.org/10.1038/nature02912

Poulton SW, Fralick PW, Canfield DE (2010) Spatial variability in oceanic redox structure 1.8 billion years ago. Nat Geosci 3(7):486–490. https://doi.org/10.1038/ngeo889

Poulton SW, Raiswell R (2002) The low-temperature geochemical cycle of iron: from continental fluxes to marine sediment deposition. Am J Sci 302:774–805

Prave AR, Condon DJ, Hoffmann KH, Tapster S, Fallick AE (2016) Duration and nature of the end-Cryogenian (Marinoan) glaciation. Geology 44 (8):631–634. https://doi.org/10.1130/g38089.1

Pu JP, Bowring SA, Ramezani J, Myrow P, Raub TD, Landing E, Mills A, Hodgin E, Macdonald FA (2016) Dodging snowballs: geochronology of the Gaskiers glaciation and the first appearance of the Ediacaran biota. Geology 44(11):955–958. https://doi.org/10.1130/g38284.1

Pufahl PK, Hiatt EE, Kyser TK (2010) Does the Paleoproterozoic Animikie Basin record the sulfidic ocean transition? Geology 38(7):659–662. https://doi.org/10.1130/g30747.1

Quinn KA, Byrne RH, Schijf J (2006) Sorption of yttrium and rare earth elements by amorphous ferric hydroxide: influence of pH and ionic strength. Mar Chem 99 (1–4):128–150. https://doi.org/10.1016/j.marchem.2005.05.011

Raiswell R, Canfield DE (1998) Sources of iron for pyrite formation in marine sediments. Am J Sci 298:219–245

Raiswell R, Hardisty DS, Lyons TW, Canfield DE, Owens JD, Planavsky NJ, Poulton SW, Reinhard CT (2018) The iron paleoredox proxies: a guide to the pitfalls, problems and proper practice. Am J Sci 318 (5):491–526. https://doi.org/10.2475/05.2018.03

Raiswell R, Newton R, Bottrell SH, Coburn PM, Briggs DEG, Bond DPG, Poulton SW (2008) Turbidite depositional influences on the diagenesis of Beecher's Trilobite Bed and the Hunsruck Slate; sites of soft tissue pyritization. Am J Sci 308(2):105–129. https://doi.org/10.2475/02.2008.01

Raiswell R, Reinhard CT, Derkowski A, Owens J, Bottrell SH, Anbar AD, Lyons TW (2011) Formation of syngenetic and early diagenetic iron minerals in the late Archean Mt. McRae Shale, Hamersley Basin, Australia: New insights on the patterns, controls and paleoenvironmental implications of authigenic mineral formation. Geochim Cosmochim Acta 75(4):1072–1087. https://doi.org/10.1016/j.gca.2010.11.013

Raitzsch M, Hönisch B (2013) Cenozoic boron isotope variations in benthic foraminifers. Geology 41 (5):591–594. https://doi.org/10.1130/g34031.1

Rasmussen B, Krapez B, Meier DB (2014) Replacement origin for hematite in 2.5 Ga banded iron formation: evidence for postdepositional oxidation of iron-bearing minerals. Geol Soc Am Bull 126(3–4):438-446. https://doi.org/10.1130/b30944.1

Rasmussen B, Muhling JR (2018) Making magnetite late again: evidence for widespread magnetite growth by thermal decomposition of siderite in Hamersley banded iron formations. Precambrian Res 306:64–93. https://doi.org/10.1016/j.precamres.2017.12.017
Raymo ME, Ruddiman WF (1992) Tectonic forcing of late Cenozoic climate. Nature 359:117–122
Reddy TR, Zheng X-Y, Roden EE, Beard BL, Johnson CM (2016) Silicon isotope fractionation during microbial reduction of Fe(III)–Si gels under Archean seawater conditions and implications for iron formation genesis. Geochim Cosmochim Acta 190:85–99. https://doi.org/10.1016/j.gca.2016.06.035
Reinhard CT, Planavsky NJ, Gill BC, Ozaki K, Robbins LJ, Lyons TW, Fischer WW, Wang C, Cole DB, Konhauser KO (2017) Evolution of the global phosphorus cycle. Nature 541(7637):386–389. https://doi.org/10.1038/nature20772
Reinhard CT, Planavsky NJ, Lyons TW (2013) Long-term sedimentary recycling of rare sulphur isotope anomalies. Nature 497(7447):100–103. https://doi.org/10.1038/nature12021
Reinhard CT, Raiswell R, Scott C, Anbar AD, Lyons TW (2009) A late Archean sulfidic sea stimulated by early oxidative weathering of the continents. Science 326 (5953):713–716. https://doi.org/10.1126/science.1176711
Rennie VCF, Paris G, Sessions AL, Abramovich S, Turchyn AV, Adkins JF (2018) Cenozoic record of δ34S in foraminiferal calcite implies an early Eocene shift to deep-ocean sulfide burial. Nat Geosci 11 (10):761–765. https://doi.org/10.1038/s41561-018-0200-y
Reusch DN (2011) New Caledonian carbon sinks at the onset of Antarctic glaciation. Geology 39(9):807–810. https://doi.org/10.1130/g31981.1
Reuschel M, Melezhik VA, Whitehouse MJ, Lepland A, Fallick AE, Strauss H (2012) Isotopic evidence for a sizeable seawater sulfate reservoir at 2.1Ga. Precambrian Res 192–195:78–88. https://doi.org/10.1016/j.precamres.2011.10.013
Reynolds BC, Frank M, O'Nions RK (1999) Nd- and Pb-isotope time series from Atlantic ferromanganese crusts: implications for changes in provenance and paleocirculation. Earth Planet Sci Lett 173:381–396
Richter FM, Rowley DB, DePaolo DJ (1992) Sr isotope evolution of seawater: the role of tectonics. Earth Planet Sci Lett 109:11–23
Robbins LJ, Lalonde SV, Planavsky NJ, Partin CA, Reinhard CT, Kendall B, Scott C, Hardisty DS, Gill BC, Alessi DS, Dupont CL, Saito MA, Crowe SA, Poulton SW, Bekker A, Lyons TW, Konhauser KO (2016) Trace elements at the intersection of marine biological and geochemical evolution. Earth Sci Rev 163:323–348. https://doi.org/10.1016/j.earscirev.2016.10.013
Roberts NMW, Spencer CJ (2015) The zircon archive of continent formation through time. Geol Soc Lond Spec Publ 389(1):197–225. https://doi.org/10.1144/sp389.14
Roerdink DL, Mason PRD, Farquhar J, Reimer T (2012) Multiple sulfur isotopes in Paleoarchean barites identify an important role for microbial sulfate reduction in the early marine environment. Earth Planet Sci Lett 331–332:177–186. https://doi.org/10.1016/j.epsl.2012.03.020
Roerdink DL, Mason PRD, Whitehouse MJ, Reimer T (2013) High-resolution quadruple sulfur isotope analyses of 3.2 Ga pyrite from the Barberton Greenstone Belt in South Africa reveal distinct environmental controls on sulfide isotopic arrays. Geochim Cosmochim Acta 117:203–215. https://doi.org/10.1016/j.gca.2013.04.027
Rooney AD, Macdonald FA, Strauss JV, Dudas FO, Hallmann C, Selby D (2014) Re-Os geochronology and coupled Os-Sr isotope constraints on the Sturtian Snowball Earth. Proc Natl Acad Sci USA 111(1):51–56. https://doi.org/10.1073/pnas.1317266110
Rooney AD, Strauss JV, Brandon AD, Macdonald FA (2015) A Cryogenian chronology: two long-lasting synchronous Neoproterozoic glaciations. Geology 43 (5):459–462. https://doi.org/10.1130/g36511.1
Rose NM, Rosing MT, Bridgwater D (1996) The origin of metacarbonate rocks in the Archaean Isua Supracrustal Belt, West Greenland. Am J Sci 296:1004–1044
Rosing MT, Bird DK, Sleep NH, Bjerrum CJ (2010) No climate paradox under the faint early Sun. Nature 464 (7289):744–747. https://doi.org/10.1038/nature08955
Rosing MT, Frei R (2004) U-rich Archaean sea-floor sediments from Greenland–indications of >3700 Ma oxygenic photosynthesis. Earth Planet Sci Lett 217(3–4):237–244. https://doi.org/10.1016/s0012-821x(03)00609-5
Rouxel OJ, Bekker A, Edwards KJ (2005) Iron Isotope Constraints on the Archean and Paleoproterozoic ocean redox state. Science 307:1088–1091
Rowley DB (2002) Rate of plate creation and destruction: 180 Ma to present. GSA Bull 114(8):927–933
Roy S (2006) Sedimentary manganese metallogenesis in response to the evolution of the Earth system. Earth Sci Rev 77(4):273–305. https://doi.org/10.1016/j.earscirev.2006.03.004
Rustad JR, Casey WH, Yin Q-Z, Bylaska EJ, Felmy AR, Bogatko SA, Jackson VE, Dixon DA (2010) Isotopic fractionation of Mg2+ (aq), Ca2 + (aq), and Fe2 + (aq) with carbonate minerals. Geochim Cosmochim Acta 74(22):6301–6323. https://doi.org/10.1016/j.gca.2010.08.018
Rye R, Holland HD (1998) Paleosols and the evolution of the atmospheric oxygen: a critical review. Am J Sci 298:621–672
Sageman BB, Meyers SR, Arthur MA (2006) Orbital time scale and new C-isotope record for Cenomanian-Turonian boundary stratotype. Geology 34(2). https://doi.org/10.1130/g22074.1
Sahoo SK, Planavsky NJ, Jiang G, Kendall B, Owens JD, Wang X, Shi X, Anbar AD, Lyons TW (2016) Oceanic oxygenation events in the anoxic Ediacaran ocean. Geobiology 14(5):457–468. https://doi.org/10.1111/gbi.12182

Sansjofre P, Cartigny P, Trindade RI, Nogueira AC, Agrinier P, Ader M (2016) Multiple sulfur isotope evidence for massive oceanic sulfate depletion in the aftermath of Snowball Earth. Nat Commun 7:12192. https://doi.org/10.1038/ncomms12192

Satkoski AM, Beukes NJ, Li W, Beard BL, Johnson CM (2015) A redox-stratified ocean 3.2 billion years ago. Earth Planet Sci Lett 430:43–53. https://doi.org/10.1016/j.epsl.2015.08.007

Satkoski AM, Fralick P, Beard BL, Johnson CM (2017) Initiation of modern-style plate tectonics recorded in Mesoarchean marine chemical sediments. Geochim Cosmochim Acta 209:216–232. https://doi.org/10.1016/j.gca.2017.04.024

Satkoski AM, Lowe DR, Beard BL, Coleman ML, Johnson CM (2016) A high continental weathering flux into Paleoarchean seawater revealed by strontium isotope analysis of 3.26 Ga barite. Earth Planet Sci Lett 454:28–35. https://doi.org/10.1016/j.epsl.2016.08.032

Sawaki Y, Ohno T, Tahata M, Komiya T, Hirata T, Maruyama S, Windley BF, Han J, Shu D, Li Y (2010) The Ediacaran radiogenic Sr isotope excursion in the Doushantuo formation in the three Gorges area, South China. Precambrian Res 176(1–4):46–64. https://doi.org/10.1016/j.precamres.2009.10.006

Schaefer MV, Handler RM, Scherer MM (2017) Fe(II) reduction of pyrolusite (beta-MnO2) and secondary mineral evolution. Geochem Trans 18(1):7. https://doi.org/10.1186/s12932-017-0045-0

Schidlowski M (2001) Carbon isotopes as biogeochemical recorders of life over 3.8 Ga of earth history: evolution of a concept. Precambrian Res 106:117–134

Schidlowski M, Appel PWU, Eichmann R, Junge CE (1979) Carbon isotope geochemistry of the 3.7 x 109-yr-old Isua sediments, West Greenland: implications for the Archaean carbon and oxygen cycles. Geochim Cosmochim Acta 43:189–199

Schirrmeister BE, Gugger M, Donoghue PC (2015) Cyanobacteria and the great oxidation event: evidence from genes and fossils. Palaeontology 58(5):769–785. https://doi.org/10.1111/pala.12178

Schneiderhan EA, Gutzmer J, Strauss H, Mezger K, Beukes NJ (2006) The chemostratigraphy of a Paleoproterozoic MnF-BIF succession–the Voëlwater Subgroup of the Transvaal Supergroup in Griqualand West, South Africa. S Afr J Geol 109:31–48

Scholz F, Severmann S, McManus J, Hensen C (2014a) Beyond the Black Sea paradigm: the sedimentary fingerprint of an open-marine iron shuttle. Geochim Cosmochim Acta 127:368–380. https://doi.org/10.1016/j.gca.2013.11.041

Scholz F, Severmann S, McManus J, Noffke A, Lomnitz U, Hensen C (2014b) On the isotope composition of reactive iron in marine sediments: redox shuttle versus early diagenesis. Chem Geol 389:48–59. https://doi.org/10.1016/j.chemgeo.2014.09.009

Schrag DP, Berner RA, Hoffman PF, Halverson GP (2002) On the initiation of a Snowball Earth. Geochem Geophys Geosyst 3(6):1–21. https://doi.org/10.1029/2001GC000219

Schröder S, Bedorf D, Beukes NJ, Gutzmer J (2011) From BIF to red beds: Sedimentology and sequence stratigraphy of the Paleoproterozoic Koegas Subgroup (South Africa). Sed Geol 236(1–2):25–44. https://doi.org/10.1016/j.sedgeo.2010.11.007

Schröder S, Lacassie JP, Beukes NJ (2006) Stratigraphic and geochemical framework of the Agouron drill cores, Transvaal Supergroup (Neoarchean-Paleoproterozoic, South Africa). S Afr J Geol 109:23–54

Scott C, Lyons TW, Bekker A, Shen Y, Poulton SW, Chu X, Anbar AD (2008) Tracing the stepwise oxygenation of the Proterozoic ocean. Nature 452 (7186):456–459. https://doi.org/10.1038/nature06811

Scott C, Wing BA, Bekker A, Planavsky NJ, Medvedev P, Bates SM, Yun M, Lyons TW (2014) Pyrite multiple-sulfur isotope evidence for rapid expansion and contraction of the early Paleoproterozoic seawater sulfate reservoir. Earth Planet Sci Lett 389:95–104. https://doi.org/10.1016/j.epsl.2013.12.010

Seton M, Gaina C, Muller RD, Heine C (2009) Mid-Cretaceous seafloor spreading pulse: fact or fiction? Geology 37(8):687–690. https://doi.org/10.1130/g25624a.1

Severmann S, Johnson CM, Beard BL, McManus J (2006) The effect of early diagenesis on the Fe isotope compositions of porewaters and authigenic minerals in continental margin sediments. Geochim Cosmochim Acta 70(8):2006–2022. https://doi.org/10.1016/j.gca.2006.01.007

Severmann S, Lyons TW, Anbar A, McManus J, Gordon G (2008) Modern iron isotope perspective on the benthic iron shuttle and the redox evolution of ancient oceans. Geology 36(6). https://doi.org/10.1130/g24670a.1

Shahar A, Young ED, Manning CE (2008) Equilibrium high-temperature Fe isotope fractionation between fayalite and magnetite: an experimental calibration. Earth Planet Sci Lett 268(3–4):330–338. https://doi.org/10.1016/j.epsl.2008.01.026

Shields G, Veizer J (2002) Precambrian marine carbonate isotope database: Version 1.1. Geochem Geophys Geosyst 3(6):1–12. https://doi.org/10.1029/2001gc000266

Shields GA (2007) A normalised seawater strontium isotope curve: possible implications for Neoproterozoic-Cambrian weathering rates and the further oxygenation of the Earth. eEarth 2:35–42

Shields-Zhou G, Och L (2011) The case for a Neoproterozoic oxygenation event: geochemical evidence and biological consequences. GSA Today 21(3):4–11. https://doi.org/10.1130/gsatg102a.1

Shirey SB, Richardson SH (2011) Start of the Wilson cycle at 3 Ga shown by diamonds from subcontinental mantle. Science 333(6041):434–436. https://doi.org/10.1126/science.1206275

Siahi M, Hofmann A, Master S, Wilson A, Mayr C (2018) Trace element and stable (C, O) and radiogenic (Sr) isotope geochemistry of stromatolitic carbonate rocks of the Mesoarchaean Pongola Supergroup:

implications for seawater composition. Chem Geol 476:389–406. https://doi.org/10.1016/j.chemgeo.2017.11.036

Siebert C, Nägler TF, von Blanckenburg F, Kramers JD (2003) Molybdenum isotope records as a potential new proxy for paleoceanography. Earth Planet Sci Lett 211(1–2):159–171. https://doi.org/10.1016/s0012-821x(03)00189-4

Slotznick SP, Eiler JM, Fischer WW (2018) The effects of metamorphism on iron mineralogy and the iron speciation redox proxy. Geochim Cosmochim Acta 224:96–115. https://doi.org/10.1016/j.gca.2017.12.003

Smit MA, Mezger K (2017) Earth's early O2 cycle suppressed by primitive continents. Nat Geosci 10 (10):788–792. https://doi.org/10.1038/ngeo3030

Smith AJB (2007) The paleo-environmental significance of the iron-formations and iron-rich mudstones of the Mesoarchean Witwatersrand-Mozaan Basin. University of Johannesburg, South Africa

Smith AJB, Beukes NJ, Gutzmer J (2013) The composition and depositional environments of Mesoarchean iron formations of the West Rand Group of the Witwatersrand Supergroup, South Africa. Econ Geol 108:111–134

Smith AJB, Beukes NJ, Gutzmer J, Czaja AD, Johnson CM, Nhleko N (2017) Oncoidal granular iron formation in the Mesoarchaean Pongola Supergroup, Southern Africa: textural and geochemical evidence for biological activity during iron deposition. Geobiology 15(6):731–749. https://doi.org/10.1111/gbi.12248

Smithies RH, Champion DC, Van Kranendonk MJ (2007) Chapter 4.2 The oldest well-preserved felsic volcanic rocks on earth: geochemical clues to the early evolution of the pilbara supergroup and implications for the growth of a Paleoarchean protocontinent. In: Earth's oldest rocks. Developments in Precambrian geology, pp 339–367. https://doi.org/10.1016/s0166-2635(07)15042-8

Smithies RH, Champion DC, Van Kranendonk MJ, Howard HM, HIckman AH (2005) Modern-style subduction processes in the Mesoarchean: geochemical evidence from the 3.12 Ga Whundo intraoceanic arc. Earth Planet Sci Lett 231:221–237

Sørensen J, Jørgensen BB (1987) Early diagenesis in sediments from Danish coastal waters: microbial activity and Mn-Fe-S geochemistry. Geochim Cosmochim Acta 51:1583–1590

Sperling EA, Wolock CJ, Morgan AS, Gill BC, Kunzmann M, Halverson GP, Macdonald FA, Knoll AH, Johnston DT (2015) Statistical analysis of iron geochemical data suggests limited late Proterozoic oxygenation. Nature 523(7561):451–454. https://doi.org/10.1038/nature14589

Squire R, Campbell I, Allen C, Wilson C (2006) Did the Transgondwanan Supermountain trigger the explosive radiation of animals on Earth? Earth Planet Sci Lett 250(1–2):116–133. https://doi.org/10.1016/j.epsl.2006.07.032

Steinhoefel G, von Blanckenburg F, Horn I, Konhauser KO, Beukes NJ, Gutzmer J (2010) Deciphering formation processes of banded iron formations from the Transvaal and the Hamersley successions by combined Si and Fe isotope analysis using UV femtosecond laser ablation. Geochim Cosmochim Acta 74(9):2677–2696. https://doi.org/10.1016/j.gca.2010.01.028

Stolper DA, Keller CB (2018) A record of deep-ocean dissolved O2 from the oxidation state of iron in submarine basalts. Nature 553(7688):323–327. https://doi.org/10.1038/nature25009

Sugitani K (2019) Early Archean (Pre-3.0 Ga) cellularly preserved microfossils and microfossil-like structures from the Pilbara Craton, Western Australia—a review. In: Earth's oldest rocks, pp 1007–1028. https://doi.org/10.1016/b978-0-444-63901-1.00041-1

Sumner DY (1997) Carbonate precipitation and oxygen stratification in Late Archean seawater as deduced from facies and stratigraphy of the Gamohaan and Frisco formations, transvaal supergroup, South Africa. Am J Sci 297:455–487

Sumner DY, Beukes NJ (2006) Sequence Stratigraphic Development of the Neoarchean Transvaal carbonate platform, Kaapvaal Craton, South Africa. S Afr J Geol 109:11–22

Swanson-Hysell NL, Rose CV, Calmet CC, Halverson GP, Hurtgen MT, Maloof AC (2010) Cryogenian glaciation and the onset of carbon-isotope decoupling. Science 328(5978):608–611. https://doi.org/10.1126/science.1184508

Sweere TC, Dickson AJ, Jenkyns HC, Porcelli D, Elrick M, van den Boorn SHJM, Henderson GM (2018) Isotopic evidence for changes in the zinc cycle during Oceanic Anoxic Event 2 (Late Cretaceous). Geology 46(5):463–466. https://doi.org/10.1130/g40226.1

Syverson DD, Borrok DM, Seyfried WE (2013) Experimental determination of equilibrium Fe isotopic fractionation between pyrite and dissolved Fe under hydrothermal conditions. Geochim Cosmochim Acta 122:170–183. https://doi.org/10.1016/j.gca.2013.08.027

Tahata M, Sawaki Y, Yoshiya K, Nishizawa M, Komiya T, Hirata T, Yoshida N, Maruyama S, Windley BF (2015) The marine environments encompassing the Neoproterozoic glaciations: evidence from C, Sr and Fe isotope ratios in the Hecla Hoek Supergroup in Svalbard. Precambrian Res 263:19–42. https://doi.org/10.1016/j.precamres.2015.03.007

Takashima R, Nishi H, Huber BT, Leckie RM (2006) Greenhouse world and the mesozoic ocean. Oceanography 19:82–92

Tang D, Shi X, Wang X, Jiang G (2016) Extremely low oxygen concentration in mid-Proterozoic shallow seawaters. Precambrian Res 276:145–157. https://doi.org/10.1016/j.precamres.2016.02.005

Tebo BM (1991) Manganese(II) oxidation in the suboxic zone of the Black Sea. Deep Sea Res Part A Oceanogr Res Pap 38:S883–S905. https://doi.org/10.1016/s0198-0149(10)80015-9

Tessin A, Sheldon ND, Hendy I, Chappaz A (2016) Iron limitation in the Western Interior Seaway during the late cretaceous OAE 3 and its role in phosphorus recycling and enhanced organic matter preservation. Earth Planet Sci Lett 449:135–144. https://doi.org/10.1016/j.epsl.2016.05.043

Thamdrup B, Fossing H, Jorgensen BB (1994) Manganese, iron, and sulfur cycling in a coastal marine sediment, Aarhus Bay, Denmark. Geochim Cosmochim Acta 58(23):5115–5129

Thamdrup B, Rossello-Mora R, Amann R (2000) Microbial manganese and sulfate reduction in Black Sea shelf sediments. Appl Environ Microbiol 66:2888–2897

Them TR 2nd, Gill BC, Caruthers AH, Gerhardt AM, Grocke DR, Lyons TW, Marroquin SM, Nielsen SG, Trabucho Alexandre JP, Owens JD (2018) Thallium isotopes reveal protracted anoxia during the Toarcian (Early Jurassic) associated with volcanism, carbon burial, and mass extinction. Proc Natl Acad Sci USA 115(26):6596–6601. https://doi.org/10.1073/pnas.1803478115

Thibon F, Blichert-Toft J, Tsikos H, Foden J, Albalat E, Albarede F (2019) Dynamics of oceanic iron prior to the Great Oxygenation Event. Earth Planet Sci Lett 506:360–370. https://doi.org/10.1016/j.epsl.2018.11.016

Torres MA, West AJ, Li G (2014) Sulphide oxidation and carbonate dissolution as a source of CO2 over geological timescales. Nature 507(7492):346–349. https://doi.org/10.1038/nature13030

Trabucho Alexandre J, Tuenter E, Henstra GA, van der Zwan KJ, van de Wal RSW, Dijkstra HA, de Boer PL (2010) The mid-Cretaceous North Atlantic nutrient trap: black shales and OAEs. Paleoceanography 25(4): n/a-n/a. https://doi.org/10.1029/2010pa001925

Tsikos H, Beukes NJ, Moore JM, Harris C (2003) Deposition, diagenesis, and secondary enrichment of metals in the Paleoproterozoic Hotazel iron formation, Kalahari Manganese Field, South Africa. Econ Geol 98:1449–1462

Tsikos H, Jenkyns HC, Walsworth-Bell B, Petrizzo MR, Forster A, Kolonic S, Erba E, Premoli Silva I, Baas M, Wagner T, Sinninghe Damste JS (2004) Carbon-isotope stratigraphy recorded by the Cenomanian-Turonian Oceanic Anoxic Event: correlation and implications based on three key localities. J Geol Soc Lond 161:711–719

Tsikos H, Matthews A, Erel Y, Moore JM (2010) Iron isotopes constrain biogeochemical redox cycling of iron and manganese in a Palaeoproterozoic stratified basin. Earth Planet Sci Lett 298(1–2):125–134. https://doi.org/10.1016/j.epsl.2010.07.032

Tsikos H, Moore JM (1997) Petrography and Geochemistry of the Paleoproterozoic Hotazel iron-formation, Kalahari manganese field, South Africa: implications for precambrian manganese metallogenesis. Econ Geol 92:87–97

Tsikos H, Moore JM, Harris C (2001) Geochemistry of the Palaeoproterozoic mooidraai formation: Fe-rich limestone as end member of iron formation deposition, Kalahari manganese field, Transvaal Supergroup, South Africa. J Afr Earth Sci 32(1):19–27

Turgeon SC, Creaser RA (2008) Cretaceous oceanic anoxic event 2 triggered by a massive magmatic episode. Nature 454(7202):323–326. https://doi.org/10.1038/nature07076

Tyrell T (1999) The relative influences of nitrogen and phosphorus on oceanic primary production. Nature 400:525–531

Ueno Y, Yurimoto H, Yoshioka H, Komiya T, Maruyama S (2002) Ion microprobe analysis of graphite from ca. 3.8 Ga metasediments, Isua supracrustal belt, West Greenland: relationship between metamorphism and carbon isotopic composition. Geochim Cosmochim Acta 66(7):1257

Valley JW, Lackey JS, Cavosie AJ, Clechenko CC, Spicuzza MJ, Basei MAS, Bindeman IN, Ferreira VP, Sial AN, King EM, Peck WH, Sinha AK, Wei CS (2005) 4.4 billion years of crustal maturation: oxygen isotope ratios of magmatic zircon. Contrib Miner Petrol 150(6):561–580. https://doi.org/10.1007/s00410-005-

Van Cappellen P, Ingall EI (1994) Benthic phosphorus regeneration, net primary production, and ocean anoxia: a model of the coupled marine biogeochemical cycles of carbon and phosphorus. Paleoceanography 9:677–692

van de Flierdt T, Frank M, Halliday AN, Hein JR, Hattendorf B, Günther D, Kubik PW (2004a) Deep and bottom water export from the Southern Ocean to the Pacific over the past 38 million years. Paleoceanography 19(1):n/a-n/a. https://doi.org/10.1029/2003pa000923

van de Flierdt T, Frank M, Lee D-C, Halliday AN, Reynolds BC, Hein JR (2004b) New constraints on the sources and behavior of neodymium and hafnium in seawater from Pacific Ocean ferromanganese crusts. Geochim Cosmochim Acta 68(19):3827–3843. https://doi.org/10.1016/j.gca.2004.03.009

van Helmond NAGM, Ruvalcaba I, Baroni AS, Sinninghe Damste JS, Slomp CP (2014) Spatial extent and degree of oxygen depletion in the deep proto-North Atlantic basin during Oceanic Anoxic Event 2. Geochem Geophys Geosyst 15:4254–4266. https://doi.org/10.1002/2014GC005528

Van Kranendonk MJ (2006) Volcanic degassing, hydrothermal circulation and the flourishing of early life on Earth: a review of the evidence from c. 3490-3240 Ma rocks of the Pilbara Supergroup, Pilbara Craton, Western Australia. Earth Sci Rev 74 (3–4):197–240. https://doi.org/10.1016/j.earscirev.2005.09.005

Van Kranendonk MJ (2014) Earth's early atmosphere and surface environments: a review. In: Earth's early atmosphere and surface environment. Geol Soc Am Spec Pap 105–130. https://doi.org/10.1130/2014.2504(12)

Van Kranendonk MJ, Kirkland CL (2016) Conditioned duality of the Earth system: geochemical tracing of the supercontinent cycle through Earth history. Earth Sci Rev 160:171–187. https://doi.org/10.1016/j.earscirev.2016.05.009

Van Kranendonk MJ, Philippot P, Lepot K, Bodorkos S, Pirajno F (2008) Geological setting of Earth's oldest fossils in the ca. 3.5 Ga Dresser Formation, Pilbara Craton, Western Australia. Precambrian Res 167(1–2):93–124. https://doi.org/10.1016/j.precamres.2008.07.003

Van Kranendonk MJ, Smithies RH, Champion DC (2019) Paleoarchean development of a continental nucleus. In: Earth's oldest rocks, pp 437–462. https://doi.org/10.1016/b978-0-444-63901-1.00019-8

Van Kranendonk MJ, Smithies RH, Hickman AH, Champion DC (2007a) Chapter 4.1 Paleoarchean development of a continental nucleus: the East Pilbara Terrane of the Pilbara Craton, Western Australia. Developments in Precambrian geology 15:307–337

Van Kranendonk MJ, Smithies RH, Hickman AH, Champion DC (2007b) Chapter 4.1 Paleoarchean development of a continental nucleus: the East Pilbara Terrane of the Pilbara Craton, Western Australia. In: Earth's oldest rocks. Developments in Precambrian geology, pp 307–337. https://doi.org/10.1016/s0166-2635(07)15041-6

van Zuilen MA, Lepland A, Teranes J, Finarelli J, Wahlen M, Arrhenius G (2003) Graphite and carbonates in the 3.8 Ga old Isua Supracrustal Belt, Southern West Greenland. Precambrian Res 126(3–4):331–348. https://doi.org/10.1016/s0301-9268(03)00103-7

Vargas M, Kashefi K, Blunt-Harris EL, Lovley DR (1998) Microbiological evidence for Fe(III) reduction on early earth. Nature 395:65–67

Viehmann S, Bau M, Hoffmann JE, Münker C (2015) Geochemistry of the Krivoy Rog banded iron formation, Ukraine, and the impact of peak episodes of increased global magmatic activity on the trace element composition of Precambrian seawater. Precambrian Res 270:165–180. https://doi.org/10.1016/j.precamres.2015.09.015

Virtasalo JJ, Laitala JJ, Lahtinen R, Whitehouse MJ (2015) Pyritic event beds and sulfidized Fe (oxyhydr) oxide aggregates in metalliferous black mudstones of the Paleoproterozoic Talvivaara formation, Finland. Earth Planet Sci Lett 432:449–460. https://doi.org/10.1016/j.epsl.2015.09.010

Voegelin AR, Nägler TF, Beukes NJ, Lacassie JP (2010) Molybdenum isotopes in late Archean carbonate rocks: implications for early Earth oxygenation. Precambrian Res 182(1–2):70–82. https://doi.org/10.1016/j.precamres.2010.07.001

von Blanckenburg F, Mamberti M, Schoenberg R, Kamber BS, Webb GE (2008) The iron isotope composition of microbial carbonate. Chem Geol 249(1–2):113–128. https://doi.org/10.1016/j.chemgeo.2007.12.001

Wabo H, Maré LP, Beukes NJ, Kruger SJ, Humbert F, de Kock MO (2018) Mineral transformations during thermal demagnetization of sideritic jasper mesobands in jaspilites of the ~3.25 Ga Fig Tree Group in the Barberton Greenstone Belt, Kaapvaal craton (South Africa). S Afr J Geol 121(2):131–140. https://doi.org/10.25131/sajg.121.0010

Walker JCG, Hays PB, Kasting JF (1981) A negative feedback mechanism for the long-term stabilization of Earth's surface temperature. J Geophys Res 86(C10). https://doi.org/10.1029/jc086ic10p09776

Wang X, Planavsky NJ, Hofmann A, Saupe EE, De Corte BP, Philippot P, LaLonde SV, Jemison NE, Zou H, Ossa FO, Rybacki K, Alfimova N, Larson MJ, Tsikos H, Fralick PW, Johnson TM, Knudsen AC, Reinhard CT, Konhauser KO (2018) A Mesoarchean shift in uranium isotope systematics. Geochim Cosmochim Acta 238:438–452. https://doi.org/10.1016/j.gca.2018.07.024

Wang X, Planavsky NJ, Reinhard CT, Hein JR, Johnson TM (2016a) A Cenozoic seawater redox record derived from 238U/235U in ferromanganese crusts. Am J Sci 316(1):64–83. https://doi.org/10.2475/01.2016.02

Wang X, Reinhard CT, Planavsky NJ, Owens JD, Lyons TW, Johnson TM (2016b) Sedimentary chromium isotopic compositions across the Cretaceous OAE2 at Demerara Rise Site 1258. Chem Geol 429:85–92. https://doi.org/10.1016/j.chemgeo.2016.03.006

Wang X, Shi X, Zhao X, Tang D (2015) Increase of seawater Mo inventory and ocean oxygenation during the early Cambrian. Palaeogeogr Palaeoclimatol Palaeoecol 440:621–631. https://doi.org/10.1016/j.palaeo.2015.09.003

Wanner C, Sonnenthal EL, Liu X-M (2014) Seawater δ7Li: a direct proxy for global CO2 consumption by continental silicate weathering? Chem Geol 381:154–167. https://doi.org/10.1016/j.chemgeo.2014.05.005

Ward LM, Rasmussen B, Fischer WW (2019) Primary productivity was limited by electron donors prior to the advent of oxygenic photosynthesis. J Geophys Res Biogeosci 124(2):211–226. https://doi.org/10.1029/2018jg004679

Wasylenki LE, Rolfe BA, Weeks CL, Spiro TG, Anbar AD (2008) Experimental investigation of the effects of temperature and ionic strength on Mo isotope fractionation during adsorption to manganese oxides. Geochim Cosmochim Acta 72(24):5997–6005. https://doi.org/10.1016/j.gca.2008.08.027

Wazne M, Korfiatis GP, Meng X (2003) Carbonate effects on hexavalent Uranium adsorption by iron oxyhyroxide. Environ Sci Technol 37:3619–3624

Weber KA, Achenbach LA, Coates JD (2006) Microorganisms pumping iron: anaerobic microbial iron oxidation and reduction. Nat Rev Microbiol 4:752–764

Wei W, Frei R, Klaebe R, Li D, Wei G-Y, Ling H-F (2018) Redox condition in the Nanhua Basin during the waning of the Sturtian glaciation: a chromium-isotope perspective. Precambrian Res 319:198–210. https://doi.org/10.1016/j.precamres.2018.02.009

Wen H, Fan H, Zhang Y, Cloquet C, Carignan J (2015) Reconstruction of early Cambrian ocean chemistry

from Mo isotopes. Geochim Cosmochim Acta 164:1–16. https://doi.org/10.1016/j.gca.2015.05.008
Westerhold T, Röhl U, Donner B, Zachos JC (2018) Global extent of early eocene hyperthermal events: a new Pacific benthic foraminiferal isotope record from Shatsky Rise (ODP Site 1209). Paleoceanogr Paleoclimatology 33(6):626–642. https://doi.org/10.1029/2017pa003306
Westermann S, Vance D, Cameron V, Archer C, Robinson SA (2014) Heterogeneous oxygenation states in the Atlantic and Tethys oceans during Oceanic Anoxic Event 2. Earth Planet Sci Lett 404:178–189. https://doi.org/10.1016/j.epsl.2014.07.018
Whitehouse MJ, Fedo CM (2007) Microscale heterogeneity of Fe isotopes in >3.71 Ga banded iron formation from the Isua Greenstone Belt, Southwest Greenland. Geology 35(8). https://doi.org/10.1130/g23582a.1
Whitehouse MJ, Kamber BS, Fedo CM, Lepland A (2005) Integrated Pb- and S-isotope investigation of sulphide minerals from the early Archaean of southwest Greenland. Chem Geol 222(1–2):112–131. https://doi.org/10.1016/j.chemgeo.2005.06.004
Whitehouse MJ, Myers JS, Fedo CM (2009) The Akilia Controversy: field, structural and geochronological evidence questions interpretations of >3.8 Ga life in SW Greenland. J Geol Soc 166(2):335–348. https://doi.org/10.1144/0016-76492008-070
Whitehouse MJ, Schoenberg R, Fedo CM, Kamber BS (2015) Does a heavy Fe-isotope composition of Akilia quartz-amphibole-pyroxene rocks necessitate a BIF origin? Astrobiology 15(10):816–824. https://doi.org/10.1089/ast.2015.1362
Widdel F, Schnell S, Heising S, Ehrenreich A, Assmus B, Schink B (1993) Ferrous iron oxidation by anoxygenic phototrophic bacteria. Nature 362:834–836
Wiesli RA, Beard BL, Johnson CM (2004) Experimental determination of Fe isotope fractionation between aqueous Fe(II), siderite and "green rust" in abiotic systems. Chem Geol 211(3–4):343–362. https://doi.org/10.1016/j.chemgeo.2004.07.002
Willbold M, Hibbert K, Lai Y-J, Freymuth H, Hin RC, Coath C, Vils F, Elliott T (2015) High-precision mass-dependent molybdenum isotope variations in magmatic rocks determined by Double-Spike MC-ICP-MS. Geostand Geoanalytical Res n/a-n/a. https://doi.org/10.1111/ggr.12109
Wille M, Kramers JD, Nägler TF, Beukes NJ, Schröder S, Meisel T, Lacassie JP, Voegelin AR (2007) Evidence for a gradual rise of oxygen between 2.6 and 2.5 Ga from Mo isotopes and Re-PGE signatures in shales. Geochim Cosmochim Acta 71(10):2417–2435. https://doi.org/10.1016/j.gca.2007.02.019
Williford KH, Van Kranendonk MJ, Ushikubo T, Kozdon R, Valley JW (2011) Constraining atmospheric oxygen and seawater sulfate concentrations during Paleoproterozoic glaciation: in situ sulfur three-isotope microanalysis of pyrite from the Turee Creek Group, Western Australia. Geochim Cosmochim Acta 75(19):5686–5705. https://doi.org/10.1016/j.gca.2011.07.010
Wortmann UG, Paytan A (2012) Rapid variability of seawater chemistry over the past 130 million years. Science 337(6092):334–336. https://doi.org/10.1126/science.1220656
Wu L, Beard BL, Roden EE, Kennedy CB, Johnson CM (2010) Stable Fe isotope fractionations produced by aqueous Fe(II)-hematite surface interactions. Geochim Cosmochim Acta 74(15):4249–4265. https://doi.org/10.1016/j.gca.2010.04.060
Wu L, Druschel G, Findlay A, Beard BL, Johnson CM (2012a) Experimental determination of iron isotope fractionations among Fe2þ–FeSaq–Mackinawite at low temperatures: implications for the rock record. Geochim Cosmochim Acta 89:46–61. https://doi.org/10.1016/j.gca.2012.04.047
Wu L, Percak-Dennett EM, Beard BL, Roden EE, Johnson CM (2012b) Stable iron isotope fractionation between aqueous Fe(II) and model Archean ocean Fe–Si coprecipitates and implications for iron isotope variations in the ancient rock record. Geochim Cosmochim Acta 84:14–28. https://doi.org/10.1016/j.gca.2012.01.007
Xiang L, Schoepfer SD, Shen S-z, Cao C-q, Zhang H (2017) Evolution of oceanic molybdenum and uranium reservoir size around the Ediacaran-Cambrian transition: evidence from western Zhejiang, South China. Earth Planet Sci Lett 464:84–94. https://doi.org/10.1016/j.epsl.2017.02.012
Xiong J, Fischer WM, Inoue K, Nakahara M, Bauer CE (2000) Molecular evidence for the early evolution of photosynthesis. Science 289:1724–1730
Yamaguchi KE (2002) Geochemistry of Archean-Paleoproterozoic black shales: the early evolution of the atmosphere, oceans, and biosphere. The Pennsylvania State University, PhD
Yamaguchi KE, Johnson CM, Beard BL, Beukes NJ, Gutzmer J, Ohmoto H (2007) Isotopic evidence for iron mobilization during Paleoproterozoic lateritization of the Hekpoort paleosol profile from Gaborone, Botswana. Earth Planet Sci Lett 256(3–4):577–587. https://doi.org/10.1016/j.epsl.2007.02.010
Yamaguchi KE, Johnson CM, Beard BL, Ohmoto H (2005) Biogeochemical cycling of iron in the Archean-Paleoproterozoic earth: constraints from iron isotope variations in sedimentary rocks from the Kaapvaal and Pilbara Cratons. Chem Geol 218(1–2):135–169. https://doi.org/10.1016/j.chemgeo.2005.01.020
Yamaguchi KE, Ohmoto H (2006) Geochemical and isotopic constraints on the origin of Paleoproterozoic red shales of the Gamagara/Mapedi formation, Postmasburg Group, South Africa. S Afr J Geol 109:123–138
Yang S, Kendall B, Lu X, Zhang F, Zheng W (2017) Uranium isotope compositions of mid-Proterozoic black shales: Evidence for an episode of increased ocean oxygenation at 1.36 Ga and evaluation of the effect of post-depositional hydrothermal fluid flow. Precambrian Res 298:187–201. https://doi.org/10.1016/j.precamres.2017.06.016

Yang W, Holland HD (2003) The Hekpoort paleosol profile in Strata 1 at Gaborone, Botswana: soil formation during the Great Oxidation Event. Am J Sci 303:187–220

Yao W, Paytan A, Wortmann UG (2018) Large-scale ocean deoxygenation during the Paleocene-Eocene Thermal Maximum. Science 361:804–806

Yeung LY (2017) Low oxygen and argon in the Neoproterozoic atmosphere at 815 Ma. Earth Planet Sci Lett 480:66–74. https://doi.org/10.1016/j.epsl.2017.09.044

Yoshiya K, Sawaki Y, Hirata T, Maruyama S, Komiya T (2015a) In-situ iron isotope analysis of pyrites in ∼ 3.7 Ga sedimentary protoliths from the Isua supracrustal belt, southern West Greenland. Chem Geol 401:126–139. https://doi.org/10.1016/j.chemgeo.2015.02.022

Yoshiya K, Sawaki Y, Shibuya T, Yamamoto S, Komiya T, Hirata T, Maruyama S (2015b) In-situ iron isotope analyses of pyrites from 3.5 to 3.2 Ga sedimentary rocks of the Barberton Greenstone Belt, Kaapvaal Craton. Chem Geol 403:58–73. https://doi.org/10.1016/j.chemgeo.2015.03.007

Young GM, von Brumm V, Gold DJC, Minter WEL (1998) Earth's oldest reported glaciation: physical and chemical evidence from the Archean Mozaan Group (∼2.9 Ga) of South Africa. J Geol 106:523–538

Zachos JC, Pagani M, Sloan L, Thomas E, Billups K (2001) Trends, rhythms, and aberrations in global climate 65 Ma to present. Science 2001:686–693

Zakharov DO, Bindeman IN (2019) Triple oxygen and hydrogen isotopic study of hydrothermally altered rocks from the 2.43–2.41 Ga Vetreny belt, Russia: an insight into the early Paleoproterozoic seawater. Geochim Cosmochim Acta 248:185–209. https://doi.org/10.1016/j.gca.2019.01.014

Zeng Z, Pike M, Tice MM, Kelly C, Marcantonio F, Xu G, Maulana I (2018) Iron fertilization of primary productivity by volcanic ash in the Late Cretaceous (Cenomanian) Western Interior Seaway. Geology 46 (10):859–862. https://doi.org/10.1130/g45304.1

Zentner DB (2014) Stratigraphy, sedimentology and provenance of the ca. 3.26 Ga Mapepe formation in the Manzimnyama Syncline, Barberton Greenstone Belt, South Africa. M.S., Stanford University,

Zhang F, Zhu X, Yan B, Kendall B, Peng X, Li J, Algeo TJ, Romaniello S (2015) Oxygenation of a Cryogenian ocean (Nanhua Basin, South China) revealed by pyrite Fe isotope compositions. Earth Planet Sci Lett 429:11–19. https://doi.org/10.1016/j.epsl.2015.07.021

Zhang S, Wang X, Wang H, Bjerrum CJ, Hammarlund EU, Costa MM, Connelly JN, Zhang B, Su J, Canfield DE (2016) Sufficient oxygen for animal respiration 1,400 million years ago. Proc Natl Acad Sci USA 113(7):1731–1736. https://doi.org/10.1073/pnas.1523449113

Zhang YG, Pagani M, Liu Z, Bohaty SM, Deconto R (2013) A 40-million-year history of atmospheric CO(2). Philos Trans A Math Phys Eng Sci 371(2001):20130096. https://doi.org/10.1098/rsta.2013.0096

Zheng X-Y, Beard BL, Reddy TR, Roden EE, Johnson CM (2016) Abiologic silicon isotope fractionation between aqueous Si and Fe(III)–Si gel in simulated Archean seawater: implications for Si isotope records in Precambrian sedimentary rocks. Geochim Cosmochim Acta 187:102–122. https://doi.org/10.1016/j.gca.2016.05.012

Zheng X-Y, Jenkyns HC, Gale AS, Ward DJ, Henderson GM (2013) Changing ocean circulation and hydrothermal inputs during Ocean Anoxic Event 2 (Cenomanian–Turonian): Evidence from Nd-isotopes in the European shelf sea. Earth Planet Sci Lett 375:338–348. https://doi.org/10.1016/j.epsl.2013.05.053

Zhou X, Jenkyns HC, Lu W, Hardisty DS, Owens JD, Lyons TW, Lu Z (2017) Organically bound iodine as a bottom-water redox proxy: preliminary validation and application. Chem Geol 457:95–106. https://doi.org/10.1016/j.chemgeo.2017.03.016

Zhou X, Jenkyns HC, Owens JD, Junium CK, Zheng X-Y, Sageman BB, Hardisty DS, Lyons TW, Ridgwell A, Lu Z (2015) Upper ocean oxygenation dynamics from I/Ca ratios during the Cenomanian-Turonian OAE 2. Paleoceanography 30 (5):510–526. https://doi.org/10.1002/2014pa002741

Zhu X-K, O'Nions RK, Guo Y, Reynolds BC (2000) Secular variation of iron isotopes in North Atlantic deep water. Science 287:2000–2002

The manufacturer's authorised representative in the EU is Springer Nature Customer Service Centre GmbH, Europaplatz 3, 69115 Heidelberg, Germany. If you have any concerns regarding our products, please contact ProductSafety@springernature.com

Printed and bound by CPI Group (UK) Ltd, Croydon, CR0 4YY
30/09/2026
02225446-0020